T0327690

Next Generation Multiple Access

IEEE Press
445 Hoes Lane
Piscataway, NJ 08854
IEEE Press Editorial Board
Sarah Spurgeon, *Editor in Chief*

Jón Atli Benediktsson
Anjan Bose
James Duncan
Amin Moeness
Desineni Subbaram Naidu
Behzad Razavi
Jim Lyke
Hai Li
Brian Johnson
Jeffrey Reed
Diomidis Spinellis
Adam Drobot
Tom Robertazzi
Ahmet Murat Tekalp

Next Generation Multiple Access

Edited by

Yuanwei Liu
Queen Mary University of London
UK

Liang Liu
Hong Kong Polytechnic University
Hong Kong
China

Zhiguo Ding
University of Manchester
UK

Xuemin Shen
University of Waterloo
Ontario
Canada

IEEE PRESS
WILEY

Copyright © 2024 by The Institute of Electrical and Electronics Engineers, Inc.
All rights reserved.

Published by John Wiley & Sons, Inc., Hoboken, New Jersey.
Published simultaneously in Canada.

No part of this publication may be reproduced, stored in a retrieval system, or transmitted in any form or by any means, electronic, mechanical, photocopying, recording, scanning, or otherwise, except as permitted under Section 107 or 108 of the 1976 United States Copyright Act, without either the prior written permission of the Publisher, or authorization through payment of the appropriate per-copy fee to the Copyright Clearance Center, Inc., 222 Rosewood Drive, Danvers, MA 01923, (978) 750-8400, fax (978) 750-4470, or on the web at www.copyright.com. Requests to the Publisher for permission should be addressed to the Permissions Department, John Wiley & Sons, Inc., 111 River Street, Hoboken, NJ 07030, (201) 748-6011, fax (201) 748-6008, or online at http://www.wiley.com/go/permission.

Trademarks: Wiley and the Wiley logo are trademarks or registered trademarks of John Wiley & Sons, Inc. and/or its affiliates in the United States and other countries and may not be used without written permission. All other trademarks are the property of their respective owners. John Wiley & Sons, Inc. is not associated with any product or vendor mentioned in this book.

Limit of Liability/Disclaimer of Warranty: While the publisher and author have used their best efforts in preparing this book, they make no representations or warranties with respect to the accuracy or completeness of the contents of this book and specifically disclaim any implied warranties of merchantability or fitness for a particular purpose. No warranty may be created or extended by sales representatives or written sales materials. The advice and strategies contained herein may not be suitable for your situation. You should consult with a professional where appropriate. Further, readers should be aware that websites listed in this work may have changed or disappeared between when this work was written and when it is read. Neither the publisher nor authors shall be liable for any loss of profit or any other commercial damages, including but not limited to special, incidental, consequential, or other damages.

For general information on our other products and services or for technical support, please contact our Customer Care Department within the United States at (800) 762-2974, outside the United States at (317) 572-3993 or fax (317) 572-4002.

Wiley also publishes its books in a variety of electronic formats. Some content that appears in print may not be available in electronic formats. For more information about Wiley products, visit our web site at www.wiley.com.

Library of Congress Cataloging-in-Publication Data:

Names: Liu, Yuanwei, author. | Liu, Liang (Professor), author. | Ding, Zhiguo, author. | Shen, X. (Xuemin), 1958-author.
Title: Next generation multiple access / Yuanwei Liu, Liang Liu, Zhiguo Ding, Xuemin (Sherman) Shen.
Description: Hoboken, New Jersey : Wiley, [2024] | Includes index.
Identifiers: LCCN 2023045953 (print) | LCCN 2023045954 (ebook) | ISBN 9781394180493 (hardback) | ISBN 9781394180509 (adobe pdf) | ISBN 9781394180516 (epub)
Subjects: LCSH: Multiple access protocols (Computer network protocols)
Classification: LCC TK5105.5 .L569 2024 (print) | LCC TK5105.5 (ebook) | DDC 004.6/2–dc23/eng/20231130
LC record available at https://lccn.loc.gov/2023045953
LC ebook record available at https://lccn.loc.gov/2023045954

Cover Design: Wiley
Cover Image: © Ralf Hiemisch/Getty Images

Set in 9.5/12.5pt STIXTwoText by Straive, Chennai, India

Contents

About the Editors

Yuanwei Liu received the PhD degree in electrical engineering from the Queen Mary University of London, UK, in 2016. He was with the Department of Informatics, King's College London, from 2016 to 2017, where he was a Post-Doctoral Research Fellow. He has been a Senior Lecturer (Associate Professor) with the School of Electronic Engineering and Computer Science, Queen Mary University of London, since August 2021, where he was a Lecturer (Assistant Professor) from 2017 to 2021. His research interests include non-orthogonal multiple access, reconfigurable intelligent surfaces, integrated sensing and communications, and machine learning. Yuanwei Liu has been a Web of Science Highly Cited Researcher since 2021, an IEEE Communication Society Distinguished Lecturer, an IEEE Vehicular Technology Society Distinguished Lecturer, and the academic Chair for the Next-Generation Multiple Access Emerging Technology Initiative. He was listed as one of 35 Innovators Under 35 China in 2022 by MIT Technology Review. He received the IEEE ComSoc Outstanding Young Researcher Award for EMEA in 2020. He received the 2020 IEEE Signal Processing and Computing for Communications (SPCC) Technical Committee Early Achievement Award and IEEE Communication Theory Technical Committee (CTTC) 2021 Early Achievement Award. He received the IEEE ComSoc Outstanding Nominee for Best Young Professionals Award in 2021. He is the co-recipient of the Best Student Paper Award in IEEE VTC2022-Fall, the Best Paper Award in ISWCS 2022, and the 2022 IEEE SPCC-TC Best Paper Award. He serves as the Co-Editor-in-Chief of IEEE ComSoc TC Newsletter, an Area Editor of IEEE Communications Letters, an Editor of IEEE Communications Surveys & Tutorials, IEEE Transactions on Wireless Communications, IEEE Transactions on Network Science and Engineering, and IEEE Transactions on Communications. He serves as the Guest Editor for IEEE JSAC on Next-Generation Multiple Access, IEEE JSTSP on Intelligent Signal Processing and Learning for Next-Generation Multiple Access, and IEEE Network on Next Generation Multiple Access for 6G. He serves as the Publicity Co-Chair for IEEE VTC 2019-Fall, Symposium Co-Chair for Cognitive

Radio & AI-enabled networks for IEEE GLOBECOM 2022 and Communication Theory for IEEE GLOBECOM 2023. He serves as the chair of the Special Interest Group (SIG) in SPCC Technical Committee on Signal Processing Techniques for Next-Generation Multiple Access, the vice-chair of SIG in SPCC Technical Committee on Near Field Communications for Next Generation Mobile Networks, and the vice-chair of SIG WTC on Reconfigurable Intelligent Surfaces for Smart Radio Environments.

Liang Liu received the BEng degree from the School of Electronic and Information Engineering at Tianjin University in 2010 and the PhD degree from the Department of Electrical and Computer Engineering at National University of Singapore (NUS) in 2014. He was a Post-Doctoral Fellow at the University of Toronto from 2015 to 2017 and a research fellow at NUS from 2017 to 2019. Currently, he is an Assistant Professor in the Department of Electrical and Electronic Engineering at The Hong Kong Polytechnic University (PolyU). His research interests include wireless communications and networking, advanced signal processing and optimization techniques, and Internet-of-Things (IoT). Dr. Liang LIU is the recipient of the 2021 IEEE Signal Processing Society Best Paper Award, the 2017 IEEE Signal Processing Society Young Author Best Paper Award, the Best Student Paper Award for 2022 IEEE International Conference on Acoustics, Speech and Signal Processing (ICASSP), and the Best Paper Award for 2011 International Conference on Wireless Communications and Signal Processing (WCSP). He was listed in Highly Cited Researchers, also known as "World's Most Influential Scientific Minds", by Clarivate Analytics (Thomson Reuters) in 2018. He is an editor for IEEE Transactions on Wireless Communications and was a leading guest editor for IEEE Wireless Communications' special issue on "Massive Machine-Type Communications for IoT."

Zhiguo Ding received the BEng in electrical engineering from the Beijing University of Posts and Telecommunications in 2000 and his PhD degree in Electrical Engineering from Imperial College London in 2005. He is currently a Professor in Communications at the University of Manchester. Previously, he had been working at Queen's University Belfast, Imperial College, Newcastle University and Lancaster University. From October 2012 to September 2024, he has also been an academic visitor in Prof. Vincent Poor's group at Princeton University. Dr. Ding research interests are machine learning, B5G networks, cooperative and energy harvesting networks and statistical signal processing. His h-index is 100, and his work receives 44,000+ Google citations. He is serving as an Area Editor for the IEEE TWC and OJ-COMS, an Editor for IEEE TVT, COMST, and OJ-SP, and was an Editor for IEEE TCOM, IEEE WCL, IEEE CL, and WCMC. He received the best paper award of IET ICWMC-2009 and IEEE WCSP-2014, the EU Marie Curie Fellowship 2012–2014, the Top IEEE TVT Editor 2017, IEEE Heinrich Hertz Award 2018, IEEE Jack Neubauer Memorial Award 2018, IEEE

Best Signal Processing Letter Award 2018, Alexander von Humboldt Foundation Friedrich Wilhelm Bessel Research Award 2020, and IEEE SPCC Technical Recognition Award 2021. He is a member of the Global Research Advisory Board of Yonsei University, a Web of Science Highly Cited Researcher in two disciplines (2019–2023), an IEEE ComSoc Distinguished Lecturer, and a Fellow of the IEEE.

Xuemin Shen received the PhD degree in electrical engineering from Rutgers University, New Brunswick, NJ, USA, in 1990. He is a University Professor with the Department of Electrical and Computer Engineering, University of Waterloo, Canada. His research focuses on network resource management, wireless network security, the Internet of Things, 5G and beyond, and vehicular networks. Dr. Shen is a registered Professional Engineer of Ontario, Canada, an Engineering Institute of Canada Fellow, a Canadian Academy of Engineering Fellow, a Royal Society of Canada Fellow, a Chinese Academy of Engineering Foreign Member, and a Distinguished Lecturer of the IEEE Vehicular Technology Society and Communications Society. Dr. Shen received "West Lake Friendship Award" from "Zhejiang Province in 2023," President's Excellence in Research from the University of Waterloo in 2022, the Canadian Award for Telecommunications Research from the Canadian Society of Information Theory (CSIT) in 2021, the R.A. Fessenden Award in 2019 from IEEE, Canada, Award of Merit from the Federation of Chinese Canadian Professionals (Ontario) in 2019, James Evans Avant Garde Award in 2018 from the IEEE Vehicular Technology Society, Joseph LoCicero Award in 2015 and Education Award in 2017 from the IEEE Communications Society (ComSoc), and Technical Recognition Award from Wireless Communications Technical Committee (2019) and AHSN Technical Committee (2013). He has also received the Excellent Graduate Supervision Award in 2006 from the University of Waterloo and the Premier's Research Excellence Award (PREA) in 2003 from the Province of Ontario, Canada. He serves/served as the General Chair for the 6G Global Conference'23, and ACM "Mobihoc'15," Technical Program Committee Chair/Co-Chair for IEEE Globecom'24, 16 and 07, IEEE Infocom'14, IEEE VTC'10 Fall, and the Chair for the IEEE ComSoc Technical Committee on Wireless Communications. Dr. Shen is the President of the IEEE ComSoc. He was the Vice President for Technical and Educational Activities, Vice President for Publications, Member-at-Large on the Board of Governors, Chair of the Distinguished Lecturer Selection Committee, and Member of the IEEE Fellow Selection Committee of the ComSoc. Dr. Shen served as the Editor-in-Chief of the IEEE IoT Journal, IEEE Network, and IET Communications.

List of Contributors

Faouzi Bellili
Department of Electrical and Computer Engineering
University of Manitoba
Winnipeg
Manitoba
Canada

Xinyu Bian
Department of Electronic and Computer Engineering
The Hong Kong University of Science and Technology
Hong Kong
China

Xiaowen Cao
School of Science and Engineering (SSE) and Future Network of Intelligence Institute (FNii)
The Chinese University of Hong Kong (Shenzhen)
Shenzhen
China

Xuan Chen
Department of Electronics and Communication Engineering
Guangzhou University
Guangzhou
China

Yilong Chen
School of Science and Engineering (SSE) and Future Network of Intelligence Institute (FNii)
The Chinese University of Hong Kong (Shenzhen)
Shenzhen
China

Zhilin Chen
The Edward S. Rogers Sr. Department of Electrical and Computer Engineering
University of Toronto
Toronto
Ontario
Canada

Jinho Choi
School of Information Technology
Deakin University
Burwood
Victoria
Australia

Shuguang Cui
School of Science and Engineering (SSE) and Future Network of Intelligence Institute (FNii)
The Chinese University of Hong Kong (Shenzhen)
Shenzhen
China

and

Data-Driven Information System Lab
Shenzhen Research Institute of Big Data
Shenzhen
China

and

Peng Cheng Laboratory
Shenzhen
China

Ying Cui
IoT Thrust
Information Hub
Hong Kong University of Science and Technology (Guangzhou)
Guangzhou
China

Ruoqi Deng
School of Electronics
Peking University
Beijing
China

Boya Di
School of Electronics
Peking University
Beijing
China

Zhiguo Ding
School of Electrical and Electronic Engineering
University of Manchester
Manchester
UK

Yaru Fu
Department of Electronic Engineering and Computer Science
School of Science and Technology
Hong Kong Metropolitan University
Hong Kong
China

Dongning Guo
Department of Electrical and Computer Engineering
Robert R. McCormick School of Engineering and Applied Science
Northwestern University
Evanston
IL
USA

Wenfang Guo
School of Electronic Engineering
Beijing University of Posts and Telecommunications
Beijing
China

Wudan Han
Department of Electrical and Computer Engineering
Western University
London
Ontario
Canada

Ekram Hossain
Department of Electrical and Computer Engineering
University of Manitoba
Winnipeg
Manitoba
Canada

Kaibin Huang
Department of Electrical and Electronic Engineering
The University of Hong Kong
Hong Kong SAR
China

Shuchao Jiang
Key Laboratory for Information Science of Electromagnetic Waves (MoE)
Department of Communication Science and Engineering
Fudan University
Shanghai
China

Wuyang Jiang
Department of Communications and Signals
School of Urban Railway Transportation
Shanghai University of Engineering Science
Shanghai
China

Eduard Jorswieck
Institute for Communications Technology
Technical University of Braunschweig
Braunschweig
Germany

Bo Li
Communication Research Center
Harbin Institute of Technology
Harbin
China

Qiang Li
Department of Electronic Engineering
Jinan University
Guangzhou
China

Tianya Li
Department of Electronic Engineering
School of Electronic Information and Electrical Engineering
Shanghai Jiao Tong University
Shanghai
China

Yang Li
State Key Laboratory of Internet of Things for Smart City
University of Macau
Macao
China

Liang Liu
Department of Electrical and Electronic Engineering
The Hong Kong Polytechnic University
Hong Kong
China

Lina Liu
Department of Electrical and Computer Engineering
Robert R. McCormick School of Engineering and Applied Science
Northwestern University
Evanston
IL
USA

Wang Liu
IoT Thrust
Information Hub
Hong Kong University of Science and Technology (Guangzhou)
Guangzhou
China

Ya-Feng Liu
Institute of Computational Mathematics and Scientific/Engineering Computing
Academy of Mathematics and Systems Science
Chinese Academy of Sciences
Beijing
China

Yuanwei Liu
School of Electronic Engineering and Computer Science
Queen Mary University of London
London
UK

Yuyi Mao
Department of Electronic and Information Engineering
The Hong Kong Polytechnic University
Hong Kong
China

Jie Mei
Department of Electrical and Computer Engineering
Western University
London
Ontario
Canada

Amine Mezghani
Department of Electrical and Computer Engineering
University of Manitoba
Winnipeg
Manitoba
Canada

Xidong Mu
School of Electronic Engineering and Computer Science
Queen Mary University of London
London
UK

Dusit Niyato
School of Computer Science and Engineering
Nanyang Technological University
Singapore
Singapore

Liping Qian
College of Information Engineering
Zhejiang University of Technology
Hangzhou
China

Tony Q. S. Quek
Department of Information Systems Technology and Design
Information Systems Technology and Design
Singapore University of Technology and Design
Singapore
Singapore

Ju Ren
Department of Computer Science and Technology
Tsinghua University
Beijing
China

Sepehr Rezvani
Institute for Communications Technology
Technical University of Braunschweig
Braunschweig
Germany

and

Department of Telecommunication Systems
Technical University of Berlin
Berlin
Germany

and

Department of Wireless Communications and Networks
Fraunhofer Institute for Telecommunications
Heinrich-Hertz-Institute
Berlin
Germany

Zhichao Shao
National Key Laboratory of Wireless Communications
University of Electronic Science and Technology of China
Chengdu
China

Boxiao Shen
Department of Electronic Engineering
School of Electronic Information and Electrical Engineering
Shanghai Jiao Tong University
Shanghai
China

Xuemin Shen
Department of Electrical and Computer Engineering
University of Waterloo
Waterloo
Ontario
Canada

Yuanming Shi
School of Information Science and Technology
ShanghaiTech University
Shanghai
China

Zheng Shi
School of Intelligent Systems Science and Engineering
Jinan University
Zhuhai
China

Volodymyr Shyianov
Department of Electrical and Computer Engineering
University of Manitoba
Winnipeg
Manitoba
Canada

Foad Sohrabi
Nokia Bell Labs
Murray Hill
NJ
USA

Lingyang Song
School of Electronics
Peking University
Beijing
China

Tianzhu Song
School of Electronic Engineering
Beijing University of Posts and Telecommunications
Beijing
China

Bowen Tan
IoT Thrust, Information Hub
Hong Kong University of Science and Technology (Guangzhou)
Guangzhou
China

Jie Tang
School of Electronic and Information Engineering
South China University of Technology
Guangzhou
China

Jiaheng Wang
National Mobile Communications Research Laboratory
Southeast University
Nanjing
China

and

Pervasive Communication Research Center
Purple Mountain Laboratories
Nanjing
China

Xianbin Wang
Department of Electrical and Computer Engineering
Western University
London
Ontario
Canada

Xin Wang
Key Laboratory for Information Science of Electromagnetic Waves (MoE)
Department of Communication Science and Engineering
Fudan University
Shanghai
China

Yuan Wang
National Mobile Communications Research Laboratory
Southeast University
Nanjing
China

and

Pervasive Communication Research Center
Purple Mountain Laboratories
Nanjing
China

Zhaolin Wang
School of Electronic Engineering and Computer Science
Queen Mary University of London
London
UK

Ziyue Wang
Institute of Computational Mathematics and Scientific/Engineering Computing
Academy of Mathematics and Systems Science
Chinese Academy of Sciences
Beijing
China

and

School of Mathematical Sciences
University of Chinese Academy of Sciences
Beijing
China

Miaowen Wen
Department of Electronic and Information Engineering
South China University of Technology
Guangzhou
China

Kai-Kit Wong
Department of Electronic and Electrical Engineering
University College London
London
UK

Yongpeng Wu
Department of Electronic Engineering
School of Electronic Information and Electrical Engineering
Shanghai Jiao Tong University
Shanghai
China

Yuan Wu
State Key Laboratory of Internet of Things for Smart City
University of Macau
Macao
China

and

Zhuhai UM Science and Technology Research Institute
Zhuhai
China

Xinyu Xie
Department of Electronic Engineering
School of Electronic Information and Electrical Engineering
Shanghai Jiao Tong University
Shanghai
China

Chongbin Xu
Key Laboratory for Information Science of Electromagnetic Waves (MoE)
Department of Communication Science and Engineering
Fudan University
Shanghai
China

Jie Xu
School of Science and Engineering (SSE) and Future Network of Intelligence Institute (FNii)
The Chinese University of Hong Kong (Shenzhen)
Shenzhen
China

Yao Xu
School of Electronic and Information Engineering
Nanjing University of Information Science and Technology
Nanjing
China

Lu Yin
School of Electronic Engineering
Beijing University of Posts and Telecommunications
Beijing
China

Wei Yu
The Edward S. Rogers Sr. Department of Electrical and Computer Engineering
University of Toronto
Toronto
Ontario
Canada

Xiaojun Yuan
National Key Laboratory of Wireless Communications
University of Electronic Science and Technology of China
Chengdu
China

Sheng Yue
Department of Computer Science and Technology
Tsinghua University
Beijing
China

Jun Zhang
Department of Electronic and Computer Engineering
The Hong Kong University of Science and Technology
Hong Kong
China

Nan Zhao
School of Information and Communication Engineering
Dalian University of Technology
Dalian
China

Tianying Zhong
National Mobile Communications Research Laboratory
Southeast University
Nanjing
China

Yong Zhou
School of Information Science and Technology
ShanghaiTech University
Shanghai
China

Guangxu Zhu
Data-Driven Information System Lab
Shenzhen Research Institute of Big Data
Shenzhen
China

Yinan Zou
School of Information Science and Technology
ShanghaiTech University
Shanghai
China

Preface

Multiple access (MA) has long been the "pearls in the crown" for each generation of mobile communications networks. Compared to 1G–5G, the next generation of mobile communications network imposes much more stringent requirements, which calls for the development of advanced MA technologies, namely next-generation multiple access (NGMA). The key concept of NGMA is to intelligently accommodate multiple terminals and multiple services in the allotted resource blocks in the most efficient manner possible, considering metrics such as resource efficiency, connectivity, coverage, and intelligence. In this book, we explore the road to developing NGMA with a focus on non-orthogonal multiple access (NOMA), massive access, and other possible MA candidates. This book consists of three parts. In Part I, we discuss the evolution of NOMA toward NGMA with the aid of advanced modulation techniques, power allocation and resource management strategies, as well as NOMA-empowered new wireless applications. In Part II, we discuss about massive IoT connectivity from the perspective of capacity limits, random access schemes, device activity detection in massive IoT connectivity, and deep learning for massive access. In Part III, we focus on advanced emerging MA techniques, which can be used in the next-generation mobile networks. We believe that this book will provide readers with a clear picture of the development of NGMA toward next-generation mobile networks to support ubiquitous and massive connectivity.

Yuanwei Liu
Liang Liu
Zhiguo Ding
Xuemin Shen

Acknowledgments

We would like to express our sincere gratitude to all the colleagues who contributed to the work and projects that led to this work.

We would also like to particularly thank all the editorial staff from Wiley for producing this book.

Yuanwei Liu
Liang Liu
Zhiguo Ding
Xuemin Shen

1

Next Generation Multiple Access Toward 6G

Yuanwei Liu[1], Liang Liu[2], Zhiguo Ding[3], and Xuemin Shen[4]

[1] *School of Electronic Engineering and Computer Science, Queen Mary University of London, London, UK*
[2] *Department of Electrical and Electronic Engineering, The Hong Kong Polytechnic University, Hong Kong, China*
[3] *School of Electrical and Electronic Engineering, University of Manchester, Manchester, UK*
[4] *Department of Electrical and Computer Engineering, University of Waterloo, Waterloo, Ontario, Canada*

1.1 The Road to NGMA

Since the feasibility of wireless communications was first demonstrated at the end of the 19th century, wireless communication technologies have rapidly developed and significantly changed human life and society. Until today, there have been five generations of wireless networks, which enable diverse ways to communicate with each other (e.g., text, voice, and video). By the end of 2023, it is predicted that the number of mobile users and Internet-enabled devices will reach 13.1 billion and 29.3 billion, respectively. Given the explosively increasing number of connected devices and the emergence of revolutionary killer applications (e.g., autonomous driving, telemedicine, metaverse, etc.), there are stringent requirements to be satisfied by future wireless networks, such as Tb/s-order peak data rates, extremely low latency, ultra-high reliability, and massive connectivity.

Considering the aforementioned targets, growing research efforts are being devoted to achieving them in next-generation wireless networks (e.g., sixth generation (6G) and beyond). More importantly, in contrast to the current wireless networks, which have mainly focused on providing communication services within terrestrial coverage, next-generation wireless networks' vision goes beyond this and can be summarized as follows:

- *Human–machine–things connections*: Next-generation wireless networks have to shift from connecting humans only to connecting humans/machines/things, thus facilitating beneficial interactions between different types of devices and realizing promising applications, such as smart cities and smart factories.

Next Generation Multiple Access, First Edition.
Edited by Yuanwei Liu, Liang Liu, Zhiguo Ding, and Xuemin Shen.
© 2024 The Institute of Electrical and Electronics Engineers, Inc. Published 2024 by John Wiley & Sons, Inc.

- *Ubiquitous space-air-ground-sea coverage*: Instead of merely providing terrestrial coverage, next-generation wireless networks aim to achieve flawless information flow over space/air/ground/sea with the integration of heterogeneous infrastructure, such as satellite/drone/underwater vehicle-based non-terrestrial networks and cellular/WiFi-based terrestrial networks.
- *Multi-functionality integration*: Compared to current communication-oriented wireless networks, next-generation wireless networks are expected to integrate other diverse functionalities, including but not limited to radio frequency (RF) sensing, imaging, computing, and localization.
- *Native intelligent networks*: On the one hand, artificial intelligence (AI) will play an unprecedented important role in improving the performance of next-generation wireless networks. On the other hand, next-generation wireless networks also have to support seamless AI services. This necessitates the development of native AI for next-generation wireless networks.

However, considering the fact that the available radio resources are limited and the emerging requirements are quite stringent, the realization of the above exciting vision of next-generation wireless networks is nontrivial and requires advanced technologies to be developed. Among others, multiple access (MA) is one of the fundamental technologies in wireless networks, which enables a large number of user terminals to be simultaneously served given the available radio resources. Given the advantages of low complexity and interference avoidance, orthogonal multiple access (OMA) schemes have been extensively employed in practical wireless communication systems, such as frequency division multiple access (FDMA) in the first generation (1G), time division multiple access (TDMA) in the second generation (2G), code division multiple access (CDMA) with orthogonal codes in the third generation (3G), and orthogonal frequency division multiple access (OFDMA) in the fourth generation (4G) and fifth generation (5G), where users are allocated with orthogonal frequency/time/code resource blocks. As discussed above, next-generation wireless networks not only have to satisfy stringent communication requirements but also have to connect heterogeneous types of devices, provide ubiquitous coverage, integrate diverse functionalities, and support native intelligence. In line with this, communication-oriented MA schemes are expected to be replaced by advanced MA schemes, namely next generation multiple access (NGMA).

The key concept of NGMA is to intelligently accommodate multiple terminals and multiple services in the allotted resource blocks in the most efficient manner possible considering metrics such as resource efficiency, connectivity, coverage, and intelligence. In contrast to previous MA schemes, which are mainly employed in cellular systems, NGMA is expected to be applicable to a wide range of wireless systems, including but not limit to cellular systems, WiFi, satellite systems, unmanned systems, and radar/sensing/monitoring systems, thus realizing the attractive vision of next-generation wireless networks. Given these promising features, in the past few years, the development of NGMA has been pursued from different viewpoints by multiple disciplines, including information theory, communication theory, wireless networking, signal processing, machine learning, big data, and hardware design, comprising both theoretical and experimental perspectives. The main research route toward NGMA can be summarized as follows. On the one hand, a paradigm shift in MA design can be observed from grant-based OMA to non-orthogonal multiple access (NOMA)/massive assess and other promising MA candidates, thus significantly improving the resource efficiency and supporting massive connectivity. On the other hand, new techniques (e.g., smart antenna, random access, and advanced modulation and channel coding schemes) and advanced machine learning (ML) tools have been exploited for NGMA to satisfy the stringent requirements and intelligently support new services. In the following, we will give a brief overview of the two main promising NGMA candidates, namely NOMA and massive access.

1.2 Non-Orthogonal Multiple Access

Different from OMA, the key idea of NOMA is to allow different users to share the same resource blocks. To deal with the resulting interference caused by the non-orthogonal resource allocation, superposition coding (SC), and successive interference cancelation (SIC) techniques have to be employed at the transmitters and receivers, respectively. Although NOMA increases the transmitter and receiver complexity, significant benefits can be achieved, such as supporting massive connectivity, achieving high spectral efficiency, and guaranteeing user fairness. Given the above advantages, growing research efforts have been devoted into NOMA. Generally speaking, NOMA can be loosely classified into power-domain (PD)-NOMA and code-domain (CD)-NOMA.

- *PD-NOMA*: The key idea of PD-NOMA is to serve multiple users in the same time/frequency/code resources and distinguish them in the power domain. SC and SIC are the two key technologies in PD-NOMA, which have been proven to be capacity-achieving in the single-antenna broadcast (BC) and MAC. For broadband communications over frequency-selective fading channels, where the channel coherence bandwidth is smaller than the system bandwidth, PD-NOMA can be straightforwardly integrated with OFDMA by assigning multiple users to each OFDMA subcarrier and serving them with PD-NOMA. This approach was adopted in multiuser superposition transmission (MUST), which was incorporated into LTE-A for simultaneously supporting two users on the same OFDMA subcarrier. Another application of PD-NOMA is layered division multiplexing (LDM), which was included in the digital TV standard (ATSC 3.0) to deliver multiple superpositioned data streams for TV broadcasting.

- *CD-NOMA*: Inspired by CDMA, where multiple users are served via the same time/frequency resources and distinguished by the allocated dedicated user-specific spreading sequences, CD-NOMA was proposed, whose key idea is still to serve multiple users in the same time/frequency resources but employing user-specific spreading sequences, which are either sparse sequences or non-orthogonal cross-correlation sequences having low cross-correlation. At the receiver, multiuser detection (MUD) is usually carried out in an iterative manner using MP-based algorithms. The family of CD-NOMA schemes has many members, such as low-density signature (LDS)-CDMA, LDS-OFDM, sparse code multiple access (SCMA), and pattern division multiple access (PDMA).

The motivation for treating NOMA as the one of the most promising candidates for NGMA can be explained as follows. On the one hand, we expect the overloaded regime to be an important use case for next-generation wireless networks, for which NOMA is a promising technology for supporting massive connectivity. On the other hand, the existing research contributions have shown that NOMA provides a higher degree of compatibility and flexibility. This enables the synergistic integration of NOMA with other components of next generation networks, such as multi-antenna techniques, multi-functionality integration and other physical layer techniques. However, the evolution of NOMA toward NGMA also imposes many challenges. The recent advances in developing NOMA for NGMA will be provided in the Part I of this book.

1.3 Massive Access

The next wave of wireless technologies will proliferate in connecting sensors, machines, and robots for myriad new Internet-of-Things (IoT) applications. To achieve this goal, massive machine-type communications (mMTC) has been defined as a key use case for 5G networks. A generic scenario for mMTC involves a massive number of IoT devices, among which a small number of them become active at each time slot. The sporadic traffic pattern in IoT systems is due to the fact that often IoT devices are designed to sleep most of the time in order to save energy and are activated only when triggered by external events. In the above massive connectivity setup, the core problem is how to detect the active users from a large number of users as quickly and accurately as possible such that we can schedule them to transmit their critical data with the minimum delay. In the conventional cellular network designed for human-type communications, the contention-based random access schemes, e.g., ALOHA, are widely used, where the users have to compete for the grant from the base stations for data transmission. However, in IoT systems with a large number of devices, the collision probability for competing for the transmission grant will be very high, which leads to huge access delay. To tackle this issue, the grant-free random access scheme, where the users directly send their data to the base stations without waiting for their permissions, is now deemed as a low-latency solution for massive IoT connectivity. In general, there are three ways for accommodating the active IoT devices to transmit under the grant-free random access scheme.

- *Compressed sensing-based random access*: Because of the sporadic user activity, the joint problem for device activity detection and channel estimation can be cast as a compressed sensing problem. Algorithms such as approximate message passing (AMP) can be applied to detect the active devices from a large number of devices and estimate their channels in massive connectivity.
- *Covariance-based random access*: In some IoT application scenario, we merely aim to detect the active devices, without the need to know their channels. Recently, it has been shown that the covariance matrix of the received signals is sufficient for detecting the active devices. Because the task of channel estimation is not considered, the covariance-based method can detect the active devices within a shorter time period as compared to the compressed sensing-based method, which also aims to estimate the channels of the active devices.
- *Unsourced random access*: In some IoT application scenario, we are not interested in detecting which subset of users are active. Instead, we just would like to decode their messages. This belongs to the unsourced random access problem. In the literature, various codebooks have been designed for unsourced random access, such that user messages can be correctly decoded even if a large number of users use the same codebook.

1.4 Book Outline

This book provides a comprehensive overview of the novel technologies for developing NGMA, with a particular focus on the NOMA, massive access, and other new MA technologies. The rest of this book consists of three parts.

- *Part I – Evolution of NOMA toward NGMA*: This first part discusses the evolution of NOMA toward NGMA with the aid of advanced modulation techniques, power allocation, and resource management strategies, as well as

NOMA-empowered new wireless applications. In particular, Chapter 2 provides a comprehensive overview of index modulation techniques for NOMA/NGMA and outlines the recent research progress. Chapter 3 further studies the resource allocation issue for a multi-channel NOMA system and investigates the impact of practical modulations and imperfect SIC on the NOMA performance. Chapter 4 studies the optimal resource allocation for NGMA with a particular focus on the power allocation for different downlink NOMA scenarios. Chapter 5 explores the potential of employing cooperative NOMA in future device-to-device communications. As a further advance, Chapter 6 proposes a novel multi-scale NOMA technique for efficiently facilitating high-precision positioning services in cellular communication systems. Chapter 7 investigates NOMA-aware wireless content caching networks, where NOMA is exploited for enhancing the performance of wireless content delivery via cache placement, recommendation decisions, and resource management. Chapter 8 studies the role of NOMA in improving the performance of mobile edge computing, which constitutes a promising solution for supporting digital twin (DT) and metaverse in future wireless networks. Chapter 9 investigates the employment of NOMA for facilitating integrated sensing and communications in future wireless networks, where both NOMA-based downlink and uplink frameworks are proposed.

- *Part II – Massive Access for NGMA*: The second part aims to discuss about massive IoT connectivity. In particular, Chapter 10 characterizes the capacity of the many access channels, where the number of users scales with the channel coding block length. Chapter 11 overviews the existing random access schemes for mMTC. Chapter 12 presents several compressed sensing based device activity detection and channel estimation algorithms by utilizing the sporadic user activity in massive IoT connectivity. The compressed sensing problem arising from joint device activity detection and channel estimation is also solved by algorithm unrolling in Chapter 13. Chapter 14 further considers the joint activity detection, channel estimation, and data decoding problem in grant-free random access scheme, and shows that the above problem can be solved by a bilinear inference problem by exploiting the sparsity in user activity. Chapter 15 also considers this problem when the IoT users are not perfectly synchronized, and proposes a Turbo AMP algorithm to solve this problem. Besides the above compressed sensing-based methods, Chapter 16 studies the covariance-based user activity detection scheme in single-cell and multi-cell massive multiple-input multiple-output (MIMO) systems, where both theoretical analysis about the minimum pilot sequence length and practical algorithm design are provided. Moreover, Chapter 17 discusses the employment of deep learning for massive access. Chapter 18 presents both theoretical foundation and practical algorithms for the unsourced random access problem.

- *Part III – Other Advanced Emerging MA Techniques for NGMA*: This third part focuses on those advanced emerging multiple access techniques which can be used in the next-generation mobile networks. In particular, Chapter 19 is to illustrate an advanced and important variant of spatial division multiple access, termed holographic-pattern division multiple access, by utilizing the holographic patterns offered by reconfigurable holographic surfaces as bandwidth resources for serving users. Chapter 20 is to describe a multiple access technique which is tailored to support federated learning and is termed over-the-air computation, where the superimposing nature of the communication medium is used for the model aggregation for federated learning. Chapter 21 is to present a general framework of multiple access techniques, termed multidimensional multiple access, where the resources from multiple domains are efficiently and intelligently exploited based on the dynamic requirements of the networks and users. Chapter 22 is to focus on the application of multiple access techniques to the design of federated learning, where advanced resource allocation is carried out to improve the convergence rate of federated learning by using the feature of multi-access wireless networks.

Part I

Evolution of NOMA Towards NGMA

2

Modulation Techniques for NGMA/NOMA

Xuan Chen[1], *Qiang Li*[2], *and Miaowen Wen*[3]

[1] *Department of Electronics and Communication Engineering, Guangzhou University, Guangzhou, China*
[2] *Department of Electronic Engineering, Jinan University, Guangzhou, China*
[3] *Department of Electronic and Information Engineering, South China University of Technology, Guangzhou, China*

2.1 Introduction

Next generation multiple access (NGMA) is committed to enabling a tremendous number of users/devices to be efficiently and flexibly connected with the network over the given wireless radio resources. Here, non-orthogonal techniques such as non-orthogonal multiple access (NOMA) and sparse code multiple access (SCMA) can achieve massive access by allowing multiple users to share the same resources (Liu et al., 2022a,b). However, NOMA is susceptible to inter-user interference (IUI) arising from the multiplexing in the power domain, and therefore how to effectively separate the superposed signals appears as a very challenging task, especially with the high IUI. Recently, plenty of works have provided alternative ways to transmit information in contrast to traditional digital modulation schemes that rely on the amplitude/phase/frequency to map information bits, i.e., index modulation (IM) (Basar et al., 2017, Cheng et al., 2018). Specifically, IM refers to the modulation scheme employing the indices of transmit entities for information embedding, such as antenna indices in space, subcarrier indices in frequency, and codebook indices in code. Owing to this special mechanism for data transmission, IM has the following properties: (i) highly spectrum-efficient; (ii) highly energy-efficient; (iii) high compatibility (i.e., without increasing hardware complexity for digital systems). Recalling the dilemma faced by NOMA schemes, one can easily observe that IM can provide a good solution to alleviate the high IUI in NOMA, i.e., creating a completely new dimension for data transmission.

Next Generation Multiple Access, First Edition.
Edited by Yuanwei Liu, Liang Liu, Zhiguo Ding, and Xuemin Shen.
© 2024 The Institute of Electrical and Electronics Engineers, Inc. Published 2024 by John Wiley & Sons, Inc.

At present, multi-antenna techniques are envisioned to be an indispensable component of NGMA, thereby suggesting that the design of multiple-input multiple-output NOMA (MIMO–NOMA) is a promising direction toward NGMA. Spatial modulation (SM), a competitive multi-antenna IM technique, can be employed in NOMA systems without increasing the power consumption and implementation complexity. By presetting the system parameters, SM-aided NOMA scheme can mitigate or even completely avoid the high IUI in a controllable manner (Li et al., 2019, Yang et al., 2019). On the other hand, for broadband communications over frequency selective fading channels, NOMA can be straightforwardly integrated with orthogonal frequency-division multiple access (OFDMA) to serve multiple users, while applying IM to NOMA-OFDMA systems can achieve the advantages attained by SM-aided NOMA (Chen et al., 2020a,b). Moreover, in the literature, numerous studies have shown that application of IM in various domains enables an attractive trade-off among spectrum/energy efficiency, transceiver complexity, interference immunity, and transmission reliability (Basar et al., 2017, Cheng et al., 2018). In summary, we can conclude that the synergistic integration of NOMA/SCMA with IM could provide a higher degree of compatibility and flexibility for user access, and support a massive number of users in a more spectrum- and energy-efficient manner for NGMA.

The objective of this chapter is to provide a comprehensive overview of prominent members of the IM-NGMA scheme and to outline the recent research progress. Specifically, Section 2.2 focuses on space-domain IM for NGAM, where the antenna indices in space is explored to aid the information transmission. Section 2.3 deals with the use of frequency-domain IM techniques for NGMA, particularly considering the downlink transmission. Sections 2.4 and 2.5 present the application of code-domain and power-domain IM techniques in NGMA, respectively. Finally, Section 2.6 concludes this chapter.

2.2 Space-Domain IM for NGMA

SM is a representative IM technique in the space domain (Wen et al., 2019). It works with a single radio-frequency (RF) chain and conveys information via the antenna index. The SM signal constellation consists of:

- The conventional amplitude phase modulation (APM) constellation;
- The antenna index constellation.

Resorting to the SM philosophy, several researchers have developed numerous variants of the SM technology, which typically include generalized SM (GSM), quadrature SM (QSM), generalized QSM (GQSM), receive SM (RSM), receive

QSM (RQSM), and generalized RQSM (GRQSM). In this section, we present the system models of various space-domain IM-multiple access (IMMA) systems and the recent literature on space-domain IMMA.

2.2.1 SM-Based NOMA

SM-based NOMA can contribute to both downlink and uplink multiuser communications. For uplink transmission, each user employs SM to transmit its own information to a base station (BS) independently; for downlink transmission, the BS broadcasts the superposition of multiple SM symbols for all users. In this subsection, we will focus only on the downlink case. Depending on the number of RF chains at the BS, SM-based downlink NOMA can be categorized into multi-RF and single-RF schemes.

2.2.1.1 Multi-RF Schemes

In multi-RF schemes, the data for each user are first encoded into an SM symbol (vector) independently, then the SM symbols for all users are superposed according to the NOMA principle, and finally the combined symbol is transmitted as in classical multi-antenna transmission (Wang and Cao, 2017).

Figure 2.1 shows a transmitter structure of SM-based NOMA with multi-RF chains for downlink transmission, where a N_t-antenna BS is dedicated to serve K users.[1] Each user is equipped with N_r receive antennas, and the channel from the BS to the kth user is given by $\mathbf{H}^k \in \mathbb{C}^{N_r \times N_t}$, where $k = 1, \ldots, K$. For the kth user, the data to be sent is divided into blocks, each containing $\log_2(N_t) + \log_2(M)$ bits.

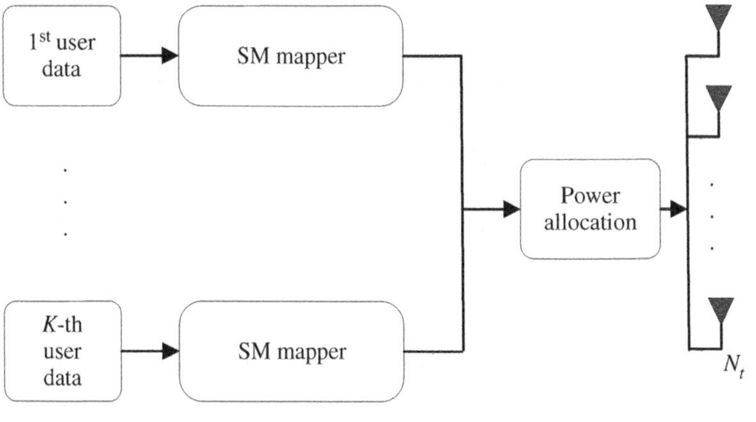

Figure 2.1 Transmitter structure of SM-based downlink NOMA with multi-RF chains.

1 N_t in this chapter is assumed to be an arbitrary integer power of two.

According to the principle of SM, the first $\log_2(N_t)$ bits are used to determine the index of a selected antenna, and the remaining $\log_2(M)$ bits are used for selecting an M-ary APM symbol. Thus, the signal vector transmitted over N_t antennas for the kth user can be expressed as

$$\mathbf{x}^k \triangleq [0, \ldots, 0, \underset{\substack{\uparrow \\ j_k\text{th position}}}{s_k}, 0, \ldots, 0]^T \in \mathbb{C}^{N_t \times 1}, \tag{2.1}$$

where $1 \leq j_k \leq N_t$ denotes the selected antenna index, and s_k denotes the M-ary APM symbol. Obviously, the total transmit rate of K users in multi-RF and SM based NOMA is $K(\log_2(N_t) + \log_2(M))$ bits per channel use (bpcu). Due to the NOMA principle, the BS combines the K signal vectors via

$$\mathbf{x} = \sum_{k=1}^{K} \sqrt{\rho_k}\mathbf{x}^k, \tag{2.2}$$

where ρ_k denotes the power allocation factor for the kth user with $\sum_{k=1}^{K} \rho_k = 1$. A higher power allocation factor is often assigned to a far user, while a lower factor a near user. Without loss of generality, we assume $\rho_1 > \cdots > \rho_K$, i.e., the first user enjoys the highest transmit power.

The received signal at the kth user is given by

$$\begin{aligned} \mathbf{y}^k &= \mathbf{H}^k\mathbf{x} + \mathbf{w}^k \\ &= \sqrt{\rho_k}\mathbf{h}_{j_k}^k s_k + \sum_{l \neq k} \sqrt{\rho_l}\mathbf{h}_{j_l}^k s_l + \mathbf{w}^k, \end{aligned} \tag{2.3}$$

where $\mathbf{w}^k \in \mathbb{C}^{N_r \times 1}$ denotes the complex additive white Gaussian noise (AWGN) vector at the kth user, $\mathbf{h}_{j_k}^k$ is the j_kth column of $\mathbf{H}^k$, and the second term on the right side of the equation denotes the IUI. For the first user, the successive interference cancelation (SIC) procedure is unnecessary and its data can be directly detected by the conventional SM detectors, e.g., the maximum likelihood (ML) detector formulated as

$$(\hat{j}_1, \hat{s}_1) = \arg\max_{j,s} \|\mathbf{y}^1 - \sqrt{\rho_1}\mathbf{h}_j^1 s\|^2, \tag{2.4}$$

where $\hat{j}_1$ and $\hat{s}_1$ are the estimates of j_1 and s_1, respectively. Obviously, the search complexity is of the order $\sim \mathcal{O}(MN_t)$. However, for the kth user ($k > 1$), data detection becomes more complicated compared with the first user since the SIC procedure is required to remove the interference from the previous $k-1$ users. After SIC, the received signal at the kth user can be updated as

$$\tilde{\mathbf{y}}^k = \mathbf{y}^k - \sum_{l=1}^{k-1} \sqrt{\rho_l}\mathbf{h}_{\hat{j}_{l,k}}^k \hat{s}_{l,k}, \tag{2.5}$$

where $(\hat{j}_{l,k}, \hat{s}_{l,k}), 1 \leq l \leq k-1$ denotes the detected data of the lth user at the kth user. From (2.5), the kth user is able to detect its own data as

$$(\hat{j}_k, \hat{s}_k) = \arg\max_{j,s} \|\tilde{\mathbf{y}}^k - \sqrt{\rho_k}\mathbf{h}_j^k s\|^2, \tag{2.6}$$

where $\hat{j}_k$ and $\hat{s}_k$ are the estimates of j_k and s_k, respectively.

As described above, the multi-RF SM-based NOMA is a straightforward combination of SM and NOMA, which requires multiple RF chains at the BS and needs to perform the SIC procedure at the users.

2.2.1.2 Single-RF Schemes

SM can be integrated with NOMA by using only a single RF chain at the BS (Chen et al., 2017). Figure 2.2 presents a transmitter structure of SM-based NOMA with a single RF chain for K-user downlink transmission. As seen from Fig. 2.2, the data for the kth user is divided into two parts, namely $\mathbf{b}_1^k$ and $\mathbf{b}_2^k$. The first part contributes partially to the active antenna selection, and the second part, consisting of $\log_2(M)$ bits, is mapped into an M-ary APM symbol. Note that per transmission, only a single transmit antenna is activated, and the index of the active antenna is determined jointly by $\mathbf{b}_1^1, \dots, \mathbf{b}_1^K$. The active antenna then transmits the superposed symbol $s = \sum_{k=1}^K \sqrt{\rho_k} s_k$, where s_k is an M-ary APM symbol determined by $\mathbf{b}_2^k$, and ρ_k denotes the power allocation factor for the kth user with $\sum_{k=1}^K \rho_k = 1$. By assuming that the tth antenna is active, the transmit vector is given by

$$\mathbf{x} \triangleq [0, \dots, 0, \underset{\substack{\uparrow \\ t\text{th position}}}{s}, 0, \dots, 0]^T \in \mathbb{C}^{N_t \times 1}, \tag{2.7}$$

where N_t denotes the number of transmit antennas at the BS. Obviously, the total transmit rate of single-RF SM-based NOMA is $\log_2(N_t) + K\log_2(M)$ bpcu. The received signal at the kth user can be expressed as

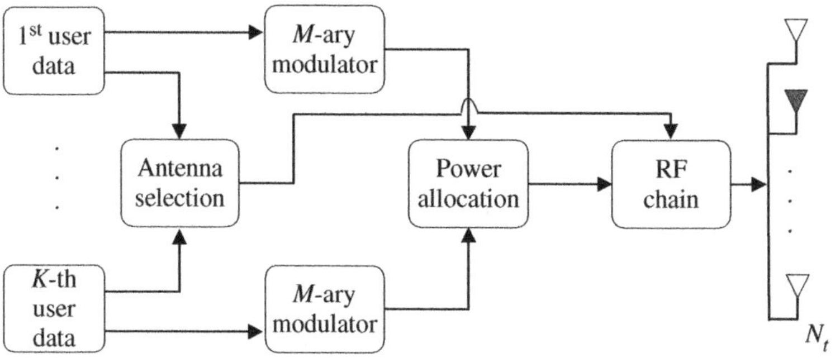

Figure 2.2 Transmitter structure of single-RF and SM-based downlink NOMA.

$$\begin{aligned}\mathbf{y}^k &= \mathbf{H}^k\mathbf{x} + \mathbf{w}^k \\ &= \mathbf{h}_t^k\sqrt{\rho_k}s_k + \mathbf{h}_t^k\sum_{l\neq k}\sqrt{\rho_l}s_l + \mathbf{w}^k,\end{aligned} \tag{2.8}$$

where $\mathbf{H}^k$ denotes the channel matrix from the BS to the kth user, $\mathbf{h}_t^k$ is the tth column of $\mathbf{H}^k$, and $\mathbf{w}^k$ denotes the complex AWGN vector at the kth user.

The detection of the single-RF SM-based NOMA is similar to that of the multi-RF case, except that the active antenna index is identical for all users. Hence, while performing SIC, each user employs the strongest user signal to detect the index of the active antenna. This index does not have to be re-detected again when canceling other users' interference. Therefore, the SIC complexity is considerably reduced.

There is a special case of single-RF SM-based NOMA, termed cooperative relaying system (CRS)-SM-NOMA, in which two information-bearing units of SM are utilized to convey the messages for two users, respectively, and the near use also acts as a relay to improve the performance of the far user (Li et al., 2019). A system model of CRS-SM-NOMA is given in Fig. 2.3, where a BS serves two users (A and B) simultaneously. All nodes operate in the half-duplex mode. Both the BS and user A have N_t transmit antennas, while users A and B are equipped with N_r^A and N_r^B receive antennas, respectively. There is no constraint on N_t and N_r^A. The channel matrices of BS-to-A, BS-to-B, and A-to-B links are denoted by $\mathbf{H} \in \mathbb{C}^{N_r^A \times N_t}$, $\mathbf{G} \in \mathbb{C}^{N_r^B \times N_t}$, and $\mathbf{F} \in \mathbb{C}^{N_r^B \times N_t}$, respectively.

In CRS-SM-NOMA, there are two phases involved in each transmission of $b_1 + b_2$ bits to users A and B, where b_1 (b_2) bits are dedicated for user A (B). During the first time slot, the BS generates an SM symbol

$$\mathbf{x} = \Bigg[\underbrace{0,\ldots,0}_{l-1},s,\underbrace{0,\ldots,0}_{N_t-l}\Bigg]^T, \tag{2.9}$$

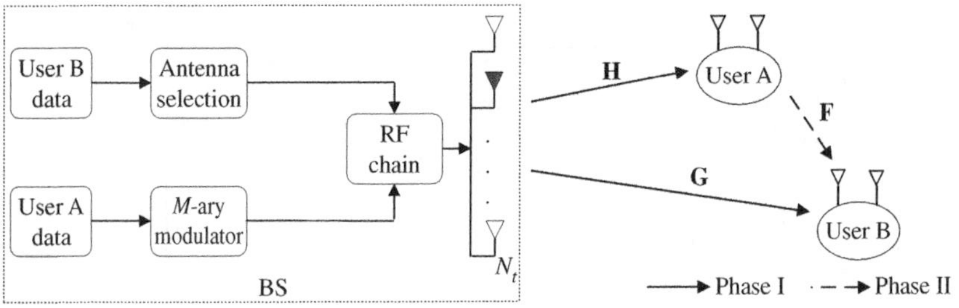

Figure 2.3 Block diagram of SM-NOMA with a single RF chain and without SIC.

and broadcasts it to users A and B, where s is an M-ary APM symbol with unit average power determined by the b_1 bits, and $l \in \{1, \ldots, N_t\}$ denotes the active antenna index determined by the b_2 bits. Obviously, we have $b_1 = \log_2(M)$ and $b_2 = \log_2(N_t)$. By normalizing the transmit power to unity, the received signals at users A and B can be expressed as

$$\mathbf{y}^{\mathrm{SA}} = \mathbf{H}\mathbf{x} + \mathbf{w}^{\mathrm{SA}} \tag{2.10}$$

and

$$\mathbf{y}^{\mathrm{SB}} = \mathbf{G}\mathbf{x} + \mathbf{w}^{\mathrm{SB}}, \tag{2.11}$$

respectively, where $\mathbf{w}^{SA} \in \mathbb{C}^{N_r^A \times 1}$ and $\mathbf{w}^{SB} \in \mathbb{C}^{N_r^B \times 1}$ are the AWGN vectors at users A and B, respectively. User B preserves the received signal $\mathbf{y}^{SB}$ for decoding purposes at the second phase, while user A adopts ML detection to jointly decode s and l, namely

$$\begin{aligned}(\hat{s}, \hat{l}) &= \arg\min_{s,l} \left\|\mathbf{y}^{\mathrm{SA}} - \mathbf{H}\mathbf{x}\right\|^2 \\ &= \arg\min_{s,l} \left\|\mathbf{y}^{\mathrm{SA}} - \mathbf{h}_l s\right\|^2,\end{aligned} \tag{2.12}$$

where $\hat{s}$ and $\hat{l}$ are the estimates of s and l at user A, respectively. User A then recovers its own information bits from $\hat{s}$ and preserves $\hat{l}$ for cooperation.

At the second phase, user A forwards $\hat{l}$ to B via a space-shift keying (SSK) signal as

$$\mathbf{y}^{\mathrm{AB}} = \mathbf{f}_{\hat{l}} + \mathbf{w}^{\mathrm{AB}}, \tag{2.13}$$

where $\mathbf{w}^{\mathrm{AB}} \in \mathbb{C}^{N_r^B \times 1}$ is the AWGN vector at user B. By combining $\mathbf{y}^{\mathrm{AB}}$ in (2.13) and $\mathbf{y}^{\mathrm{SB}}$ in (2.11), user B obtains its own message from

$$\tilde{l} = \arg\min_{l} \left\{ \left\|\mathbf{y}^{\mathrm{AB}} - \mathbf{f}_l\right\|^2 + \min_{s} \left\|\mathbf{y}^{\mathrm{SB}} - \mathbf{g}_l s\right\|^2 \right\}, \tag{2.14}$$

where $\tilde{l}$ is the estimate of l at user B.

With CRS-SM-NOMA, the information for the two users can be conveyed in the same time/frequency/code channel. Moreover, there is no interference between the two users; thus, it is possible to improve the performance. Further, the superposition coding at the BS and the complicated SIC signal processing at the receiver is avoided; hence, it decreases the computational complexity and reduces the decoding latency of multi-antenna NOMA systems (Zhong et al., 2018).

Figure 2.4 depicts the performance comparison among CRS-SM-NOMA, CRS-NOMA, and SM-OMA. As shown in the figure, user B of SM-OMA performs the worst among all schemes without cooperating with user A. Both users of CRS-SM-NOMA outperform those of CRS-NOMA.

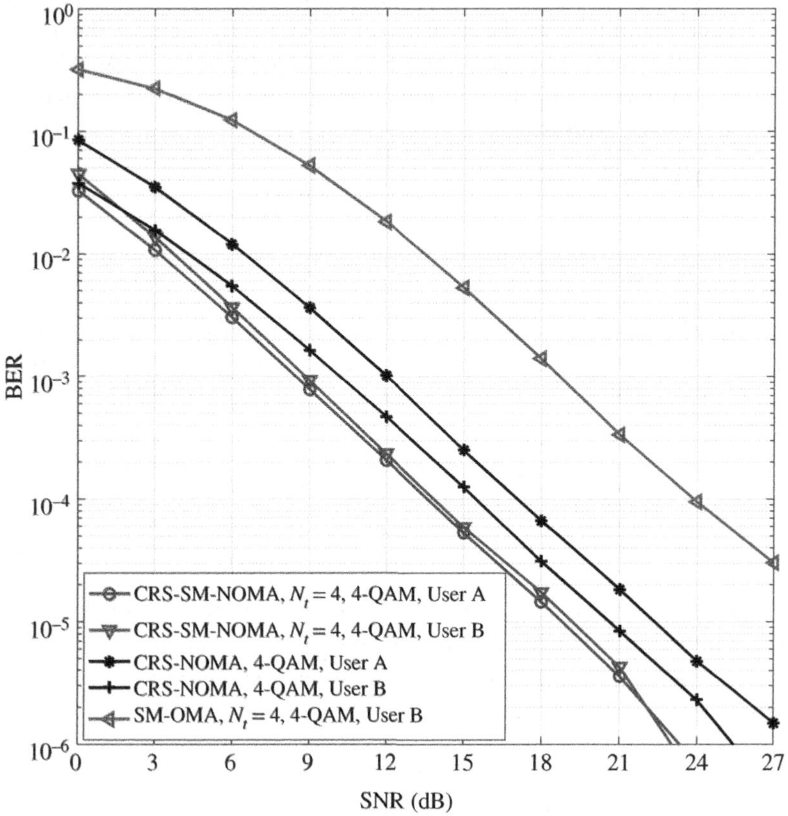

Figure 2.4 Performance comparison among CRS-SM-NOMA, CRS-NOMA, and SM-OMA, where the number of receive antennas is 2 for both users. In CRS-SM-NOMA, the BS uses 4 transmit antennas and 4-QAM. In CRS-NOMA, the BS uses one antenna to transmit the superposition coded symbol with the power allocation factor of 0.22. In SM-OMA, the BS transmits two SM signals to users A and B, which extract the 4-QAM symbol and the index bits as the desired information, respectively, in two consecutive time slots.

2.2.1.3 Recent Developments in SM-NOMA

In this part, we review other studies that combine SM and NOMA. In Al-Ansi et al. (2022), a generalized single-RF SM-NOMA was proposed, where each user can contribute zero or more bits to select the shared transmit antenna index. In Li et al. (2022), NOMA with GSM, where the data for each user is encoded into a GSM symbol, was presented to improve the system spectral efficiency (SE). In Hong et al. (2020), the authors discussed QSM-based NOMA, which doubles the number of index bits for each user by applying IM into both the in-phase (I) and quadrature (Q) components of the data symbol. Further, GQSM-NOMA was proposed in Li et al. (2020a) by integrating the concepts of GSM and QSM. References Si et al. (2020) and Yu et al. (2021) investigated the potential of SM-NOMA in relay networks for downlink and uplink transmission, respectively. Finally, in Liu et al. (2020), SM-NOMA was studied in multi-carrier grant-free uplink systems, where user-activity and data are detected jointly.

2.2.2 RSM-Based NOMA

In RSM-based NOMA, the IM concept is applied at the receiver side. Thanks to the simple form of the received RSM signals, RSM-based NOMA enjoys a low-complexity receiver architecture, which are highly desirable for downlink transmission. Figure 2.5 shows a block diagram of RSM-based downlink NOMA, where a N_t-antenna BS severs K N_r-antenna users and the data for each user is encoded into a RSM symbol. Specifically, for the kth user, the signal vector generated by the RSM mapper is

$$\mathbf{c}^k \triangleq [0, \ldots, 0, \underset{\substack{\uparrow \\ t\text{th position}}}{s}, 0, \ldots, 0]^T \in \mathbb{C}^{N_r \times 1}, \tag{2.15}$$

where $t \in \{1, \ldots, N_r\}$ and s denote the index of an active receive antenna and an M-APM symbol, carrying $\log_2(N_r)$ bits and $\log_2(M)$ bits, respectively. Then, the RSM symbol is precoded and normalized, i.e.,

$$\mathbf{x}^k = \mathbf{P}^k \mathbf{D}^k \mathbf{c}^k, \tag{2.16}$$

where $\mathbf{P}^k = [\mathbf{p}_1^k, \ldots, \mathbf{p}_{N_r}^k] \in \mathbb{C}^{N_t \times N_r}$ is the precoding matrix, and $\mathbf{D}^k = \text{diag}([d_1^k, \ldots, d_{N_r}^k]^T) \in \mathbb{C}^{N_r \times N_r}$ denotes the normalization diagonal matrix with

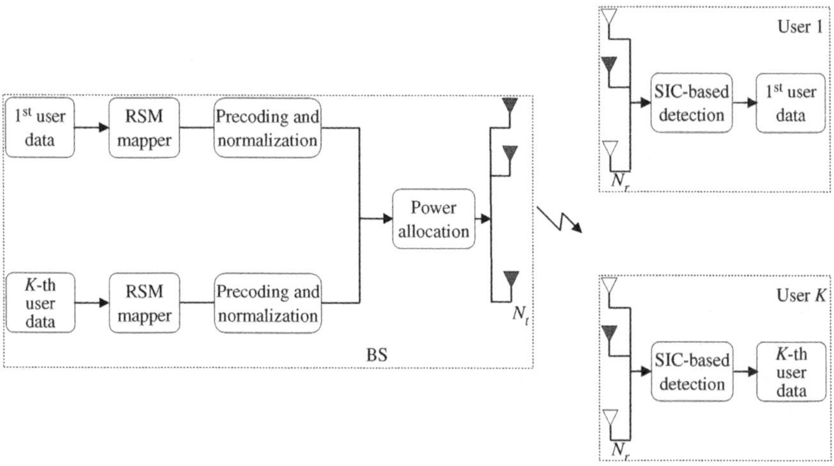

Figure 2.5 Block diagram of RSM-based downlink NOMA.

$d_\tau^k = 1/\sqrt{||\mathbf{p}_\tau^k||^2}, \tau = 1, \dots, N_r$. For simplicity, we apply the zero-forcing precoding and thereby have $\mathbf{P}^k = (\mathbf{H}^k)^H(\mathbf{H}^k(\mathbf{H}^k)^H)^{-1}$, where $\mathbf{H}^k \in \mathbb{C}^{N_r \times N_t}$ denotes the channel matrix from the BS to the kth user. With the NOMA principle, the BS transmits the superposed symbol

$$\mathbf{x} = \sum_{k=1}^{K} \sqrt{\rho_k}\mathbf{x}^k, \tag{2.17}$$

to all users, where ρ_k denotes the power allocation factor for the kth user with $\sum_{k=1}^{K} \rho_k = 1$.

The received signal at the kth user is given by

$$\begin{aligned} \mathbf{y}^k &= \mathbf{H}^k\mathbf{x} + \mathbf{w}^k \\ &= \sqrt{\rho_k}\mathbf{D}^k\mathbf{c}^k + \mathbf{H}^k\sum_{l \neq k}^{K} \sqrt{\rho_l}\mathbf{P}^{(l)}\mathbf{D}^{(l)}\mathbf{c}^k + \mathbf{w}^k, \end{aligned} \tag{2.18}$$

where $\mathbf{w}^k \in \mathbb{C}^{N_r \times 1}$ is the AWGN vector at the kth user. From (2.18), SIC-based detection can be employed to recover the transmitted data. In particular, since $\mathbf{D}^k$ is a diagonal matrix, after the SIC process, we can first estimate the index of the active receive antenna via an energy detector and then detect the APM symbol received at that antenna, reducing the detection complexity (Yang et al., 2019).

RSM-NOMA is further extended by combining NOMA with GRQSM in Li et al. (2020b) to improve the SE, where the data for each user is encoded into GRQSM symbols. In Kumar et al. (2021), a three-user cooperative RSM-NOMA system over two time slots was investigated for multiple-input multiple-output downlink scenarios. The BS transmits information for the two near users in the odd and even time slots, respectively, while for the third (far) user in both odd and even time slots. In each odd (even) time slot, the first part of the information bits of the first (second) user is used to select the index of an active receive antenna, and the remaining bits are mapped to a modulated symbol that is superposed with the symbol of the third user according to the NOMA principle.

2.2.3 SM-Aided SCMA

SCMA is one of the promising techniques supporting massive connectivity and high SE. Figure 2.6 shows the system model of uplink SM-SCMA with K users sharing N orthogonal resource elements (OREs), where each user is equipped with N_t transmit antennas and the BS N_r receive antennas. The kth user has a multidimensional codebook, $\mathbf{C}^k = [\mathbf{c}_1^k, \dots, \mathbf{c}_M^k] \in \mathbb{C}^{N \times M}$, where $\mathbf{c}_m^k \in \mathbb{C}^{N \times 1}$ is the mth codeword of $\mathbf{C}^k$ with $k = 1, \dots, K$ and $m = 1, \dots, M$. It should be noted that each codeword is sparse with d_v non-zero elements. The positions of zero and nonzero

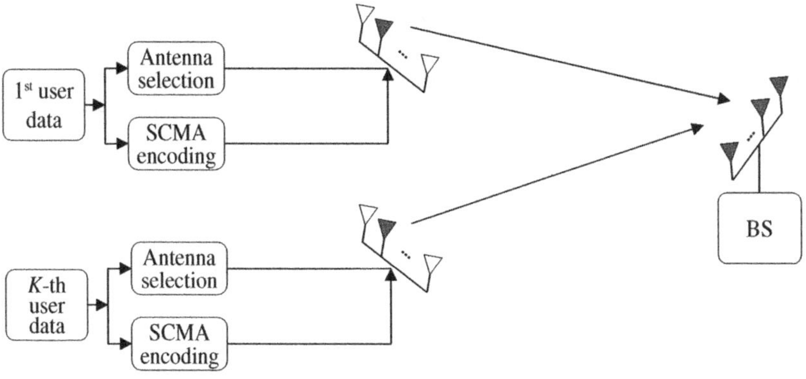

Figure 2.6 Block diagram of uplink SM-SCMA.

elements are fixed for a codebook (i.e., for a user), and vary from codebook to another to provide a fixed number of overlapped users denoted by d_f per ORE.

For the kth user in SM-SCMA, the first $\log_2(N_t)$ bits determine the index of the transmit antenna to be activated, $j_k \in \{1, \dots, N_t\}$, while the remaining $\log_2(M)$ bits are mapped to choose a corresponding codebook, $\mathbf{c}^k \in \mathbf{C}^k$, to be transmitted from that active antenna. The received signal at the rth receive antenna of the BS is given by

$$\mathbf{y}_r = \sum_{k=1}^{K} \text{diag}\left(\mathbf{h}_{rj_k}\right)\mathbf{c}^k + \mathbf{w}_r, \tag{2.19}$$

where $\mathbf{y}_r = [y_{r,1}, \dots, y_{r,N}]^T \in \mathbb{C}^{N\times1}$, $\mathbf{w}_r \in \mathbb{C}^{N\times1}$ is the AWGN vector, and $\mathbf{h}_{rj_k} = [h_{rj_k,1}, \dots, h_{rj_k,N}]^T$ with $h_{rj_k,n}, n = 1, \dots, N$ being the channel coefficient from the j_kth antenna of the kth user to the rth antenna of the BS for the nth ORE. The SE of each user is $\log_2(N_t) + \log_2(M)$ bpcu.

At the BS, the index of the active antenna and the transmitted codeword should be estimated for all users. Obviously, the optimal ML detector jointly performs an exhaustive search for all possible combinations of transmit antennas and codewords for all users, namely

$$\left(\hat{\mathbf{C}}, \hat{\mathbf{j}}\right) = \underset{j_k, \mathbf{c}^k, k=1,\dots,K}{\arg\min} \left\{ \sum_{r=1}^{N_r} \left\| \mathbf{y}_r - \sum_{k=1}^{K} \text{diag}\left(\mathbf{h}_{rj_k}\right)\mathbf{c}^k \right\|^2 \right\}, \tag{2.20}$$

where $\hat{\mathbf{j}} = [\hat{j}_1, \dots, \hat{j}_K]^T$ and $\hat{\mathbf{C}} = [\hat{\mathbf{c}}^1, \dots, \hat{\mathbf{c}}^K]$ with $\hat{j}_k$ and $\hat{\mathbf{c}}^k$ being the estimates of j_k and $\mathbf{c}^k$, respectively. Although the ML detector provides the optimum error performance, it has an impractically high decoding complexity. Thanks to the sparsity of the codebooks, message passing algorithm (MPA), which is able to achieve

Table 2.1 Advantages and disadvantages of spatial-domain IM for NGMA.

Schemes	Advantages	Disadvantages
SM-based NOMA (Multi-RF)	High SE	High detection complexity
SM-based NOMA (Single-RF)	Single-RF chain at Tx	Low SE
RSM-based NOMA	Low detection complexity	Multiple receive antennas required
SM-aided SCMA	High SE	High detection complexity

near-ML performance with much lower complexity, can be applied at the BS to decompose the superimposed codewords of the multiusers (Pan et al., 2018).

To further lower the detection complexity, three low-complexity algorithms that achieve different trade-offs between error performance and computational complexity were developed in Al-Nahhal et al. (2020). SCMA was combined with GSM in Al-Nahhal et al. (2019) and with low density signature (LDS) in Liu et al. (2017). In Liu et al. (2019), SM-LDS was investigated with multi-carrier (MC) signaling to combat frequency-selective fading. Reference Xiang et al. (2021) investigated low-complexity detection for the scheme in Liu et al. (2019). Space-time coded GSM with LDS was proposed in Liu et al. (2021) to exploit the transmit diversity in the spatial and frequency domains.

In Table 2.1, we summarize the key advantages and disadvantages of spatial-domain IM for NGMA.

2.3 Frequency-Domain IM for NGMA

OFDM with IM (OFDM-IM) is a representative IM technique in the frequency domain, which extends the SM principle to OFDM subcarriers. Specifically, partial subcarriers are activated to carry information symbols, and the indices of inactive subcarriers convey information via IM (Basar et al., 2013). However, notice that not all subcarriers in OFDM-IM are activated to carry constellation symbols, which greatly limits the SE. For the sake of enabling IM while activating all subcarriers, dual-mode aided OFDM-IM (DM-OFDM-IM) intentionally activates the idle subcarriers of OFDM-IM to transmit symbols drawn from a constellation different from that employed by the "active" subcarriers (Mao et al., 2017). In summary, OFDM/IM and DM-OFDM-IM are typical and promising IM techniques in the frequency domain. This section focuses on the capability of NOMA for information transfer via the frequency-domain IM. Therefore, in the following, we review some advanced and representative frequency-domain NGMA techniques.

2.3.1 NOMA with Frequency-Domain IM

2.3.1.1 OFDM-IM NOMA

The OFDM-IM principle was applied to a NOMA system, forming a scheme called OFDM-IM NOMA in Almohamad et al. (2021) and Arslan et al. (2020). OFDM-IM NOMA is an energy-efficient, spectral-efficient, and flexible scheme. This flexibility is provided by both NOMA and OFDM-IM with an adjustable power allocation coefficient and an adjustable subcarrier activation ratio, respectively.

A possible transmitter structure for OFDM-IM NOMA is illustrated in Fig. 2.7, consisting of one BS and $\mathcal{U}$ users. For brevity, it is assumed that all nodes are configured with a single transmit/receive antenna, and the channel coefficients of links BS $\rightarrow$ user k is defined as $\mathbf{h}_k$ with $k \in \{1, 2, \dots, \mathcal{U}\}$. The BS generates OFDM-IM signals to serve $\mathcal{U}$ users independently through the same frequency and time slots by SC. For the kth user, a total of m_k bits are transmitted, where $k \in \{1, 2, \dots, \mathcal{U}\}$. These m_k bits are split into g subblocks, each containing $p_k = m_k/g$ bits. Considering that the IM process for each subblock is independent and identical, let us take the βth subblock as an example to conduct the follow-up analysis, where $\beta \in \{1, \dots, g\}$. For clarity, in the βth subblock, we define $p_k = p_{k,1} + p_{k,2}$, where $p_{k,1}$ bits are used to determine L_k active subcarriers out of N available ones, i.e., $p_{k,1} = \lfloor \log_2 (N, L_k) \rfloor$; while $p_{k,2}$ bits are mapped onto conventional signal constellations (M_k-phase-shift keying (PSK)/quadrature amplitude modulation (QAM)), which are carried by L_k selected subcarriers, i.e., $p_{k,2} = L_k \log_2(M_k)$.

Assuming that the interleaved subcarrier grouping is used to form g subblocks, the subcarriers within each subcarrier subblock are independently faded, and then

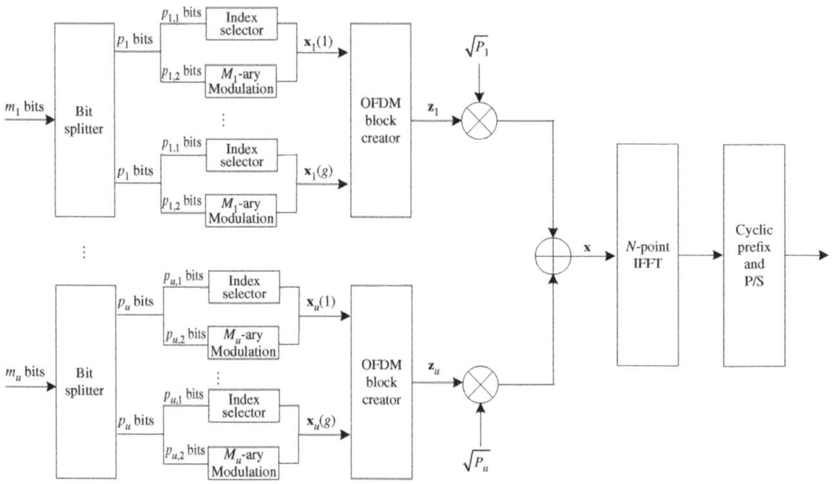

Figure 2.7 Block diagrams of the transmitter in the OFDM-IM NOMA scheme.

the indices of the selected subcarriers can be written as

$$I_k(\beta) = \{i_{k,1}(\beta), \dots, i_{k,L_k}(\beta)\}, \tag{2.21}$$

where $i_{k,\gamma}(\beta) \in \{\beta, \dots, \beta + (\gamma - 1)g, \dots, \beta + (N-1)g\}$ for $\gamma = 1, \dots, L_k$. The PSK/QAM symbol vector carrying $p_{k,2}$ bits is represented by

$$\mathbf{s}_k(\beta) = [s_{k,1}(\beta), s_{k,2}(\beta), \dots, s_{k,L_k}(\beta)]^T, \tag{2.22}$$

where $s_{k,\gamma}(\beta) \in \mathcal{X}_k$ denotes the modulated symbol of user k, and $\mathcal{X}_k$ represents the complex PSK/QAM constellation set of cardinality M_k. Afterwards, the transmitted vector for the βth subblock can be written as

$$\mathbf{x}_k(\beta) = [x_k(\beta), \dots, x_k(\beta + (\lambda - 1)g), \dots, x_k(\beta + (N-1)g)]^T, \tag{2.23}$$

where $x_k(\beta + (\lambda - 1)g) \in \{0, \mathcal{X}_k\}$ with $\lambda = \{1, \dots, N\}$ is created by $I_k(\beta)$ and $\mathbf{s}_k(\beta)$. All OFDM subblocks are concatenated to generate a new OFDM block, thus generating $\mathbf{z}_k = \left[\mathbf{x}_k^T(1)\mathbf{x}_k^T(2) \cdots \mathbf{x}_k^T(g)\right]^T$. Furthermore, SC is applied to obtain the overall transmission vector as

$$\mathbf{x} = \sum_{k=1}^{\mathcal{U}} \sqrt{P_k}\mathbf{z}_k, \tag{2.24}$$

The remaining procedures are the same as those of the classical OFDM, including the inverse fast Fourier transform (IFFT), cyclic prefix (CP) appending, and parallel-to-series conversion.

At the receiving side, after removing the CP and performing the fast Fourier transform (FFT), the frequency-domain received signals at the kth user can be expressed as

$$\mathbf{y}_k = \mathbf{X}\mathbf{h}_k + \mathbf{w}_k, \tag{2.25}$$

where $\mathbf{X} = \text{diag}(\mathbf{x})$, $\mathbf{h}_k$ is the $N \times 1$ frequency-domain channel vector, and $\mathbf{w}_k$ is the $N \times 1$ frequency-domain noise vector. From (2.25), the optimal ML detector for the $\mathcal{U}$th user (the farthest user) can be expressed as

$$(\hat{\mathcal{I}}_{\mathcal{U}}, \hat{\mathcal{S}}_{\mathcal{U}}) = \arg\min_{\mathcal{I}_{\mathcal{U}}, \mathcal{S}_{\mathcal{U}}} \|\mathbf{y} - \mathbf{X}\mathbf{h}_{\mathcal{U}}\|^2, \tag{2.26}$$

where $\mathcal{I}_{\mathcal{U}} = [I_{\mathcal{U}}(1)I_{\mathcal{U}}(2) \dots I_{\mathcal{U}}(g)]$ and $\mathcal{S}_{\mathcal{U}} = [\mathbf{s}_{\mathcal{U}}(1)\mathbf{s}_{\mathcal{U}}(2) \dots \mathbf{s}_{\mathcal{U}}(g)]$. When $k < \mathcal{U}$, the SIC detection will be performed, i.e., each user should decode and cancel the signals of the other users that are farther than him. Specifically, user k will first detect the signal vector of user $\mathcal{U}$, $(\mathcal{I}_{\mathcal{U}}, \mathcal{S}_{\mathcal{U}})$, assuming all other signals as noise. The detected signal vector $(\hat{\mathcal{I}}_{\mathcal{U}}, \hat{\mathcal{S}}_{\mathcal{U}})$ will then be subtracted from the received signal vector. This same procedure is repeated till user k extracts its own signal. The detailed detection is illustrated in Fig. 2.8.

Note that the OFDM-IM NOMA described above is a general case. Currently, some special schemes are proposed by adjusting the system parameters, such as L_k and M_k. For clarity, we summarize these cases as follows,

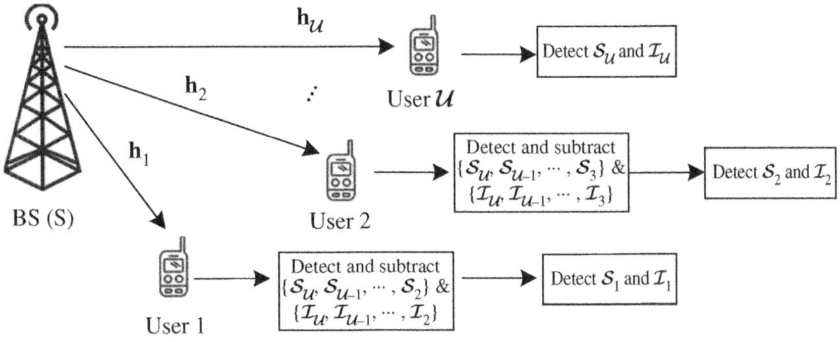

Figure 2.8 Block diagrams of the receivers in the OFDM-IM NOMA scheme.

- **Case I:** $\mathcal{I}_1 = \mathcal{I}_2 = \cdots = \mathcal{I}_K$, the index bit is determined by a certain user (Chen et al., 2020b).
- **Case II:** $\exists k, L_k = N$, the kth user's information can be carried by all subcarriers (Tusha et al., 2020).
- **Case III:** $\mathcal{I}_1 = \mathcal{I}_2 = \cdots = \mathcal{I}_K$ and $M_k = 0$, the kth user's information can be only carried by the indices of subcarriers (Chen et al., 2020a).

2.3.1.2 DM-OFDM NOMA

DM-OFDM NOMA is proposed as an extension of OFDM-IM NOMA, based on the following two reasons: on one hand, the high IUI cannot be avoided for the OFDM-IM NOMA with fully activated subcarriers, thus requiring the SIC process during the transmission (Tusha et al., 2020); while as for the OFDM-IM NOMA with idle subcarriers, the transmission rate exists a bottleneck for OFDM-IM NOMA (Chen et al., 2020a,b). Therefore, a DM-OFDM NOMA scheme as illustrated in Fig. 2.9 is considered, in which the concept of dual-mode constellation modulation proposed in Mao et al. (2017) is applied to increase the data rate by using all subcarriers. Constrained by the dual mode, $\mathcal{U} = 2$ users are served in the DM-OFDM NOMA; while the multi-mode OFDM scheme designed in Wen et al. (2017) is considered, $\mathcal{U} > 2$ users can be introduced.

In Fig. 2.9, g subblocks are created to perform the data mapping, where each subblock contains $p_k = m_k/g$ bits for $k = A, B$. Different from OFDM-IM NOMA, the indices are only determined by one user, i.e., U_A (user A) or U_B (user B). In the follow-up analysis, we assume that U_B is far user. Here, for U_A, the input bits p_A is only carried by the constellation symbols. For U_B, we have $p_B = p_{B,1} + p_{B,2}$, where $p_{B,1}$ bits represent the additional information bits conveyed by the indices of active subcarriers, whereas $p_{B,2}$ bits are mapped onto the conventional constellation symbols. Notice that the $p_{B,1}$ input bits are assigned to select L_B active subcarriers out of N available ones to realize the corresponding IM procedures,

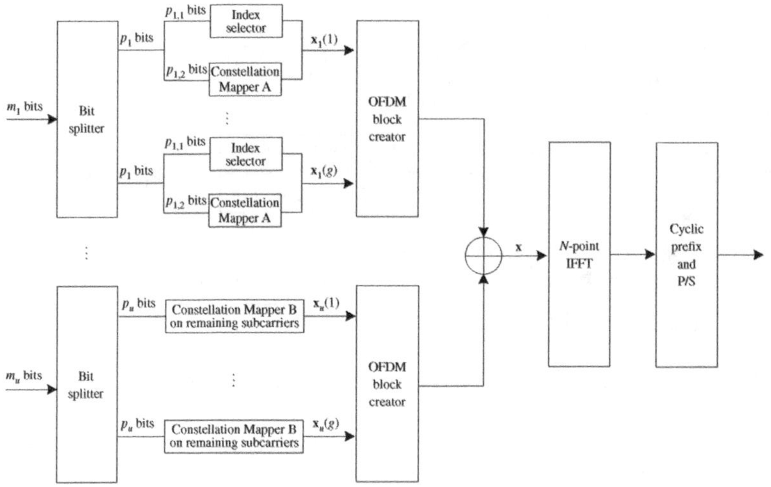

Figure 2.9 Block diagram of the DM-OFDM NOMA transmitter.

i.e., $p_{B,1} = \lfloor \log_2(C(N, L_B)) \rfloor$. The indices of the activated subcarriers for the βth subblock are expressed as

$$I(\beta) = \{i_1(\beta), \dots, i_\gamma(\beta) \dots, i_{L_k}(\beta)\}, \tag{2.27}$$

where $i_\gamma(\beta) \in \{1, 2, \dots, N\}$ for $\gamma = 1, 2, \dots, L_B$ and $\beta = 1, 2, \dots, g$.

According to (2.27), it is clear that $(N - L_B)$ subcarriers are assigned to U_A and the remaining subcarriers are allocated to U_B. To distinguish the subcarriers of users A and B clearly, different modulation modes (Modes I and II) are emplyed to encode the incoming bit stream for users A and B, respectively. For DM-OFDM NOMA with PSK modulation, Modes I and II can be generated by the rotation of a basic M-ary PSK constellation (Wen et al., 2018), i.e., $M_A = M_B = M$; while for DM-OFDM NOMA with QAM modulation, linear transformation should be performed on the basic M-ary QAM constellation to form multiple modes as described in Li et al. (2017). Therefore, we have $p_A = (N - L_B)\log_2(M_A)$ and $p_{B,2} = L_B \log_2(M_B)$. Furthermore, the PSK/QAM constellation vector carrying p_A and $p_{B,2}$ bits can be expressed as

$$\begin{aligned} \mathbf{s}_A(\beta) &= \left[s_{A,1}(\beta), \dots, s_{A,\gamma}(\beta), \dots, s_{A,N-L_B}(\beta)\right]^T \\ \mathbf{s}_B(\beta) &= \left[s_{B,1}(\beta), \dots, s_{B,\gamma}(\beta), \dots, s_{B,L_B}(\beta)\right]^T, \end{aligned} \tag{2.28}$$

where $s_{A,\gamma}(\beta) \in \mathcal{X}_A$ and $s_{B,\gamma}(\beta) \in \mathcal{X}_B$ denote the modulated symbol of U_A and U_B, while $\mathcal{X}_A$ and $\mathcal{X}_B$ represent the Modes I and II constellation sets of cardinality M, respectively. When $I(\beta)$, $\mathbf{s}_A(\beta)$, and $\mathbf{s}_B(\beta)$ are determined, the transmitted vector

Table 2.2 Advantages and disadvantages of frequency-domain IM for NGMA.

Schemes	Advantages	Disadvantages
OFDM-IM NOMA with fully activated subcarriers	High SE	High IUI
OFDM-IM NOMA with idle subcarriers	Low IUI	Low SE
DM-OFDM NOMA	No IUI	Low SE

can be written as

$$\mathbf{x}(\beta) = \left[x_1(\beta), x_2(\beta), \ldots, x_\gamma(\beta), \ldots, x_N(\beta)\right]^T, \tag{2.29}$$

where $x_\gamma(\beta)$ with $\gamma = 1, 2, \ldots, N$ can be expressed as

$$x_\gamma(\beta) \in \begin{cases} \mathcal{X}_A, \gamma \notin I(\beta) \\ \mathcal{X}_B, \gamma \in I(\beta) \end{cases}. \tag{2.30}$$

Afterwards, all OFDM subblocks for U_A or U_B are concatenated together to generate an OFDM block, and then OFDM blocks of all users will be combined to form a single OFDM symbol $\mathbf{x}$. Owing to the following procedure similar to the OFDM-IM NOMA scheme, the detailed description is omitted here.

Moreover, the key advantages and disadvantages of OFDM-IM NOMA and DM-OFDM NOMA are summarized as Table 2.2. Similar to the previous description, the IUI and SE of OFDM-IM NOMA are dependent on the activated subcarriers during a transmission. Owing to these two frequency-domain NOMA techniques employed in the Section 2.3.2, there is no much concentration to compare their performance.

2.3.2 C-NOMA with Frequency-Domain IM

To resist the communication loss caused by long distance, the paradigm of cooperative NOMA (C-NOMA) was proposed in Ding et al. (2015). Motivated by the advantage of the IM-aided NOMA techniques, the IM concept has also been skillfully integrated with C-NOMA, formed as C-NOMA with generalized OFDM-IM (GCIM-NOMA) and C-NOMA with DM-OFDM-IM (CDM-NOMA) networks (Chen et al., 2020a,b).

As shown in Fig. 2.10, a downlink NOMA scenario based on the OFDM framework is considered, which consists of one BS and two users (cell-center user A and cell–edge user B). Meanwhile, given the similarity between GCIM-NOMA and CDM-NOMA, Fig. 2.10 involves system models of the two proposed schemes. For brevity, it is assumed that all nodes are configured with a single transmit/receive antenna and operate in a half-duplex mode. Moreover, the channel coefficients of links BS → A, BS → B, and A → B as $\mathbf{h}_A$, $\mathbf{h}_{B_1}$, and $\mathbf{h}_{B_2}$, where it is assumed that

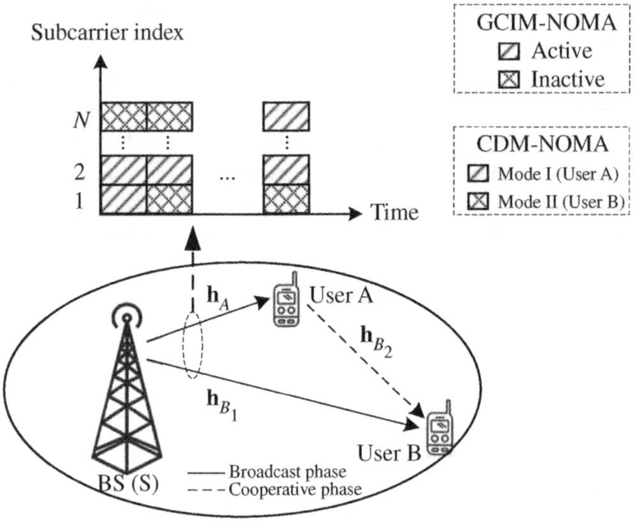

Figure 2.10 Schematic diagram of GCIM-NOMA and CDM-NOMA.

$||\mathbf{h}_A||^2 > ||\mathbf{h}_{B_1}||^2$. The complete communication protocol is divided into two phases (broadcast phase and cooperative phase) to illustrate.

2.3.2.1 Broadcast Phase

At the beginning of this phase, users information processed by the IM-based OFDM is broadcast to users A and B at the same frequency and time slot. The signal generation process are detailed in Sections 2.3.1.1 and 2.3.1.2. Here, it is assumed the transmitted vector is written as $\mathbf{x}$, created by (2.24) and (2.29), respectively, for GCIM-NOMA and CDM-NOMA.[2] When all processes, such as FFT and adding CP, are completed, the transmitted symbol at the BS can be successfully broadcast to users A and B.

At the receiving side, the FFT is applied after removing the CP. Therefore, the frequency-domain received signals at users A and B are given by

$$\mathbf{y}_k = \mathbf{X}\mathbf{h}_k + \mathbf{w}_k,\ k = A, B, \tag{2.31}$$

where $\mathbf{X} = \text{diag}(\mathbf{x})$ and $\mathbf{w}_k$ is the frequency-domain noise vector. User B retains the received signal $\mathbf{y}_B$ for demodulation in the follow-up phase, whereas user A recovers the transmitted signals via the ML detection described in (2.26) and then preserves transmitted symbol of user B for cooperative transmission in the next phase.

2 Notice that for the GCIM-NOMA, only **Case I**, as described in OFDM-IM NOMA, is considered in this subsection to construct $\mathbf{x}$ and operate the following detection process.

2.3.2.2 Cooperative Phase

In the cooperative phase, user A can utilize the decoded information belonging to U_B to regenerate a new OFDM-IM signal $\tilde{\mathbf{x}}_{B_2}$. Next, $\tilde{\mathbf{x}}_{B_2}$ is forwarded to user B with the full transmit power. The received signal at user B is similarly given by

$$\mathbf{y}_{B_2} = \widetilde{\mathbf{X}}_{B_2}\mathbf{h}_{B_2} + \mathbf{w}_{B_2}, \tag{2.32}$$

where $\widetilde{\mathbf{X}}_{B_2} = \text{diag}(\widetilde{\mathbf{x}}_{B_2})$, and $\mathbf{w}_{B_2}$ is the AWGN vector for the link A→B. Then, user B can make a decision based on the two observations, namely $\mathbf{y}_{B_1}$ and $\mathbf{y}_{B_2}$.

There are two detection methods to recover the information of user B, i.e., the ML detector and the low-complexity suboptimal ML detector. Due to the similar decoding procedure in the ML detector, we do not describe too much here. The low-complexity detector obtain the indices of the activated subcarriers $I_B(\beta)$ for the βth subblock at user B by greedy detector (GD) instead of ML, i.e., selecting L_B maximum combined energies,

$$\tilde{I}_B(\beta) = \underset{I_B(\beta)}{\arg\max}\left\{\mathcal{Y}_1 + \mathcal{Y}_2\right\}, \tag{2.33}$$

where we have

$$\begin{aligned}\mathcal{Y}_1 &= \left|y_{B_1,i_{B,1}(\beta)}(\beta)\right|^2 + \cdots \left|y_{B_1,i_{B,L_B}(\beta)}(\beta)\right|^2,\\ \mathcal{Y}_2 &= \left|y_{B_2,i_{B,1}(\beta)}(\beta)\right|^2 + \cdots \left|y_{B_2,i_{B,L_B}(\beta)}(\beta)\right|^2\end{aligned} \tag{2.34}$$

where $\mathbf{y}_{B_1} = \left[\mathbf{y}_{B_1}(1), \ldots, \mathbf{y}_{B_1}(\beta), \ldots, \mathbf{y}_{B_1}(g)\right]^T$ and $\mathbf{y}_{B_1}(\beta) = \left[y_{B_1,1}(\beta), \ldots, y_{B_1,N}(\beta)\right]$. Afterwards, the maximal-ratio combining (MRC) is employed to combine the constellation signals, followed by the ML detector to recover these information bits. Moreover, the computational complexity for the proposed schemes is illustrated in Table 2.3. It is clear that the complexity of the suboptimal detector is much lower than that of the ML detector, with a partial performance loss.

To comprehensively analyze the difference among the two proposed schemes and the conventional cooperative NOMA (C-NOMA) scheme, we further compare them according to the following simulation configurations, i.e.,

1. "GCIM-NOMA (4, 3, 8PSK, BPSK)" with the SE of users A and B equal to 2.25 and 1.25 bpcu, respectively, where $N = 4, L_B = 3, M_A = 8, M_B = 2$;

Table 2.3 The computational complexity of the ML and GD detector for the proposed schemes.

Schemes	ML detector	GD detector
GCIM-OFDM NOMA	$\mathcal{O}(2^{\lfloor \log_2(C(N,L_B)) \rfloor} {M_B}^{L_B})$	$\mathcal{O}(2^{\lfloor \log_2(C(N,L_B)) \rfloor} + {M_B}^{L_B})$
CDM-OFDM NOMA	$\mathcal{O}(2^{\lfloor \log_2(C(N,L_B)) \rfloor} {M_A}^{n-k} {M_B}^{L_B})$	$\mathcal{O}((2^{\lfloor \log_2(C(N,L_B)) \rfloor} + 1){M_B}^{L_B})$

2. "CDM-NOMA (4, 1, 8PSK)" with the SE of users A and B equal to 2.25 bpcu and 1.25 bpcu, respectively, $N = 4, L_B = 1, M_A = M_B = 8$.

where "C-NOMA (QPSK, BPSK)" is used as the benchmark with $M_A = 4$ and $M_B = 2$. Moreover, note that the optimal values of power coefficients are (0.29, 0.71) and (0.38, 0.52) for C-NOMA and GCIM-NOMA, respectively. For brevity, only the suboptimal ML detector is used in this simulation. As shown in Fig. 2.11, on one hand, we can find that CDM-NOMA exhibits the best bit error rate (BER) performance among all considered schemes. Specifically, compared with C-NOMA, SNR gains of about 3 and 2 dB are attained for users A and B, respectively, which is mainly due to the utilization of two independent information-bearing units and the elimination of error propagation of SIC. On the other hand, it is clear that there is no significant difference in the performance of GCIM-NOMA and C-NOMA at the high SNR region, which suggests that the limited transmission rate is the bottleneck of GCIM-NOMA.

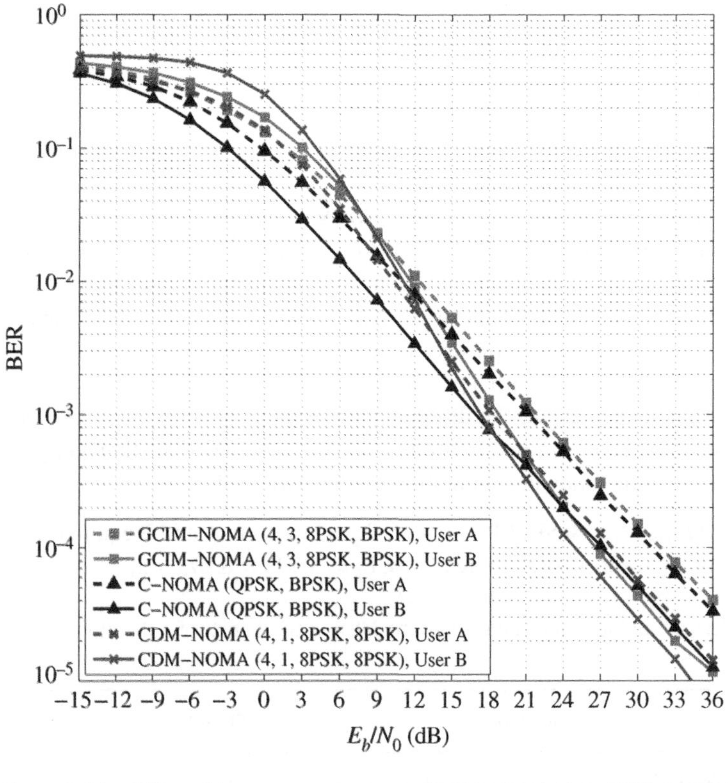

Figure 2.11 BER performance comparisons: "GCIM-NOMA (4, 3, 8PSK, BPSK)," "CDM-NOMA (4, 1, 8PSK)," and "C-NOMA (QPSK, BPSK)."

2.4 Code-Domain IM for NGMA

Code-domain IM (CIM) utilizes the indices of codes to convey information bits. The codes can be codewords in SCMA and spreading codes in MC-code division multiple access (MC-CDMA). In this part, we present two CIM-aied MA schemes, including CIM-SCMA and CIM-MC-CDMA.

2.4.1 CIM-SCMA

In contrast to conventional SCMA schemes that directly map the information bits of each user to a predefined codeword, CIM-SCMA conveys the data of each user not only by the codewords, but also by the codeword positions (Lai et al., 2020).

Figure 2.12 shows a system model of CIM-SCMA with K users. For each user, the information bits to be transmitted are divided into two parts. The first part, consisting of $\lfloor \log_2 (C(n,t)) \rfloor$ bits, is used to determine the active positions, namely

$$\mathcal{I} = \{i_1, \dots, i_t\}, \tag{2.35}$$

where $C(\cdot,\cdot)$ represents the binomial coefficient, $\lfloor \cdot \rfloor$ denotes the floor function, t is the number of active positions from n available positions, and $i_\beta \in \{1, \dots, n\}, \beta \in \{1, \dots, t\}$. The second part, consisting of $t \log_2(M)$bits, maps to t SCMA codewords, i.e., $\mathbf{c}_1, \dots, \mathbf{c}_t$, according to the corresponding N-dimension codebook, where M denotes the size of the codebook. The final transmitted nN-dimension codeword is constructed by putting $\mathbf{c}_\beta$ at the i_βth position and $\mathbf{0}$ at other inactive positions, where $\mathbf{0}$ is a N-dimension zero vector. Table 2.4 gives an example of mapping table for CIM-SCMA with $n = 4$ and $t = 2$.

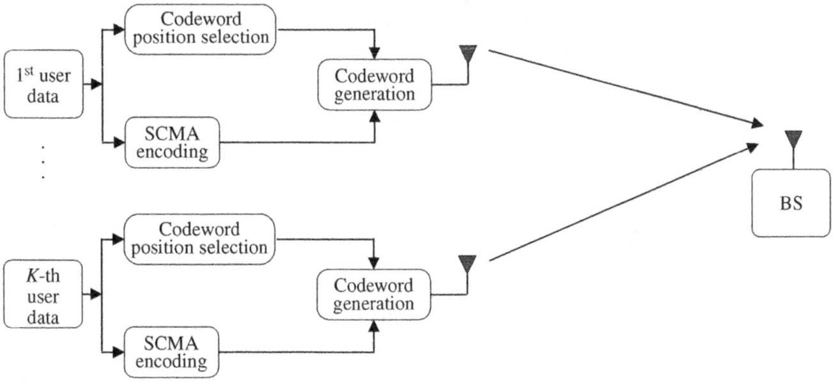

Figure 2.12 Block diagram of uplink CIM-SCMA.

Table 2.4 An example of mapping table for CIM-SCMA with $n = 4$ and $t = 2$.

Index bits	$\mathcal{I}$	Final codeword
[0 0]	$\{1,3\}$	$[\mathbf{c}_1^T, \mathbf{0}^T, \mathbf{c}_2^T, \mathbf{0}^T]^T$
[0 1]	$\{2,4\}$	$[\mathbf{0}^T, \mathbf{c}_1^T, \mathbf{0}^T, \mathbf{c}_2^T]^T$
[1 0]	$\{2,3\}$	$[\mathbf{0}^T, \mathbf{c}_1^T, \mathbf{c}_2^T, \mathbf{0}^T]^T$
[1 1]	$\{1,4\}$	$[\mathbf{c}_1^T, \mathbf{0}^T, \mathbf{0}^T, \mathbf{c}_2^T]^T$

The received signal at the BS can be expressed as

$$\mathbf{y} = \sum_{k=1}^{K} \text{diag}\left(\mathbf{h}_k\right) \mathbf{c}^k + \mathbf{w}, \tag{2.36}$$

where $\mathbf{y} = [y_1, \dots, y_{nN}]^T \in \mathbb{C}^{nN\times 1}$, $\mathbf{w} \in \mathbb{C}^{nN\times 1}$ is the AWGN vector, $\mathbf{h}_k = [h_{k,1}, \dots, h_{k,nN}]^T$ with $h_{k,l}, l = 1, \dots, nN$ being the channel coefficient from the kth user to the BS for the lth ORE, and $\mathbf{c}^k \in \mathbb{C}^{nN\times 1}$ denotes the transmitted codeword from the kth user. From (2.36), ML and MPA detectors can be used to recover the indices of the active positions and the corresponding codewords.

2.4.2 CIM-MC-CDMA

CIM-MC-CDMA combines the techniques of spread spectrum (SS) and IM under the framework of orthogonal frequency division multiplexing (OFDM) (Li et al., 2018). In this scheme, the information bits for each user are jointly conveyed by the indices of spreading codes and the conventional M-ary modulated symbols. Figure 2.13 shows a block diagram of CIM-MC-CDMA for K-user downlink transmission. As shown in Figure 2.13, the data for each user first enter an IM-OFDM-SS modulator, then the resulting symbols for all users are added together, and finally the sum is transmitted by a conventional OFDM transmitter. We next focus on the IM-OFDM-SS modulator, which is the same for all users.

IM-OFDM-SS is based on the framework of OFDM with N subcarriers. To perform IM efficiently, m information bits to be sent for each OFDM frame duration are equally separated into g blocks and IM is performed within $n = N/g$ subcarriers in a block-wise manner. Since the process in all blocks are the same and independent of each other, let us take the βth block as an example, where $\beta \in \{1, \dots, g\}$. In the βth block, the available $p = m/g$ bits are further divided into two parts. The first part consisting of $p_1 = \log_2(n/K)$ index bits is used to select the spreading code $\mathbf{c}_{i^{(\beta)}} \in \mathbb{C}^{n\times 1}$ from a predefined set $C = \{\mathbf{c}_1, \dots, \mathbf{c}_{n/K}\}$, where $i^{(\beta)} \in \{1, \dots, n/K\}$ is the index of the spreading code for the βth block. The second part with $p_2 = \log_2(M)$ symbol bits is mapped into an M-ary APM symbol $s^{(\beta)}$ with unit average power.

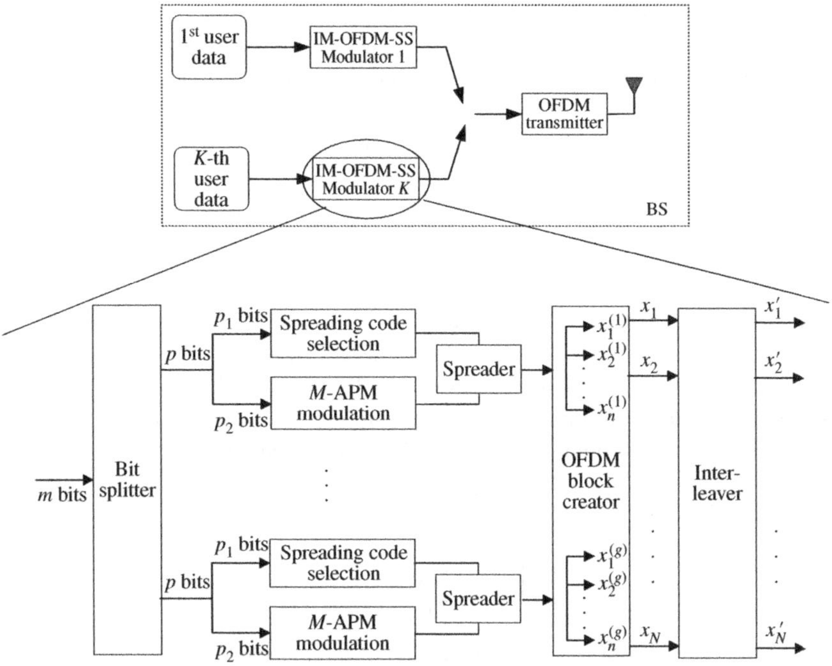

Figure 2.13 Block diagram of downlink CIM-MC-CDMA.

Then, the symbol $s^{(\beta)}$ is spread across n subcarriers by the selected spreading code $\mathbf{c}_{i^{(\beta)}}$, yielding

$$\begin{aligned}\mathbf{x}^{(\beta)} &= \left[x_1^{(\beta)}, \dots, x_n^{(\beta)}\right]^T = s^{(\beta)}\mathbf{c}_{i^{(\beta)}} \\ &= \left[s^{(\beta)}c_{i^{(\beta)},1}, \dots, s^{(\beta)}c_{i^{(\beta)},n}\right]^T,\end{aligned} \tag{2.37}$$

where $c_{i^{(\beta)},l}$, $l \in \{1, \dots, n\}$ is the lth element of $\mathbf{c}_{i^{(\beta)}}$. After obtaining $\mathbf{x}^{(\beta)}$ for all β, the OFDM block creator concatenates them, yielding the $N \times 1$ main OFDM block as follows:

$$\begin{aligned}\mathbf{x}_F &= \left[x_1, \dots, x_N\right]^T \\ &= \left[x_1^{(1)}, \dots, x_n^{(1)}, x_1^{(2)}, \dots, x_n^{(2)}, \dots, x_1^{(g)}, \dots, x_n^{(g)}\right]^T.\end{aligned} \tag{2.38}$$

To harvest the frequency diversity, an interleaver is adopted, which makes the symbols in the $\mathbf{x}^{(\beta)}$ spaced equally by g subcarriers, obtaining

$$\begin{aligned}\mathbf{x}'_F &= \left[x'_1, \dots, x'_N\right]^T \\ &= \left[x_1^{(1)}, \dots, x_1^{(g)}, x_2^{(1)}, \dots, x_2^{(g)}, \dots, x_n^{(1)}, \dots, x_n^{(g)}\right]^T.\end{aligned} \tag{2.39}$$

It should be noted that the set of spreading codes is different for different users, and there are no common spreading codes for any two sets. Such spreading code sets can be obtained by partitioning the columns of the $n \times n$ Hadamard matrix into K sets.

For the detection, without loss of generality, we focus on the first user, whose received signal on the lth subcarrier is

$$y_l = h_l \sum_{k=1}^{K} s^{(k)} c_{i^{(k)},l} + w_l, \quad l = 1, \dots, n, \tag{2.40}$$

where $s^{(k)}$ and $c_{i^{(k)},l}$ are the transmitted symbol and selected spreading code for the kth user on the lth subcarrier, respectively. h_l and w_l are the channel coefficient and noise on the lth subcarrier, respectively. From (2.40), both the optimal ML detection and single-user detection can be employed to detect the data.

The ML detector estimates the data of all K users jointly via

$$\left\{ \hat{s}^{(1)}, \hat{i}^{(1)}, \dots, \hat{s}^{(K)}, \hat{i}^{(K)} \right\} = \underset{\left\{ s^{(1)}, i^{(1)}, \dots, s^{(K)}, i^{(K)} \right\}}{\arg\min} \sum_{l=1}^{n} \left| y_l - h_l \sum_{k=1}^{K} s^{(k)} c_{i^{(k)},l} \right|^2 . \tag{2.41}$$

The computational complexity of (2.41) is of order $\sim \mathcal{O}((Mn/K)^K)$ per block in terms of complex multiplications, which grows exponentially with K.

To achieve low-complexity detection, we propose a single-user detector based on the MRC. For this detector, the τth equalizer of the first user for the lth subcarrier is given by

$$z_{\tau^{(1)},l} = \frac{h_l^* c_{\tau^{(1)},l}^*}{|h_l|^2 (K-1) + N_0}, \tag{2.42}$$

for $\tau = 1, 2, \dots, n/K$. Then, the estimate of the index of the spreading code for the first user can be obtained through

$$\hat{i}^{(1)} = \underset{\tau^{(1)}}{\arg\max} \left| \Delta_{\tau^{(1)}} \right|^2, \tag{2.43}$$

where $\Delta_{\tau^{(1)}} = \sum_{l=1}^{n} z_{\tau^{(1)},l} y_l$. Correspondingly, we have

$$\hat{s}^{(1)} = \underset{s}{\arg\min} \left| \sum_{l=1}^{n} z_{\hat{i}^{(1)},l} y_l - s \sum_{l=1}^{n} \frac{|h_l|^2}{|h_l|^2 (K-1) + N_0} \right|^2 . \tag{2.44}$$

As seen from (2.43) and (2.44), the computational complexity per user is of order $\mathcal{O}(M + n/K)$.

The key advantages and disadvantages of code-domain IM for NGMA are summarized in Table 2.5. Figure 2.14 shows the performance comparison between CIM-MC-CDMA and conventional MC-CDMA in the downlink at the overall SEs

Table 2.5 Advantages and disadvantages of code-domain IM for NGMA.

Schemes	Advantages	Disadvantages
CIM-SCMA	High SE	High detection complexity
CIM-MC-CDMA	Low detection complexity	Low SE

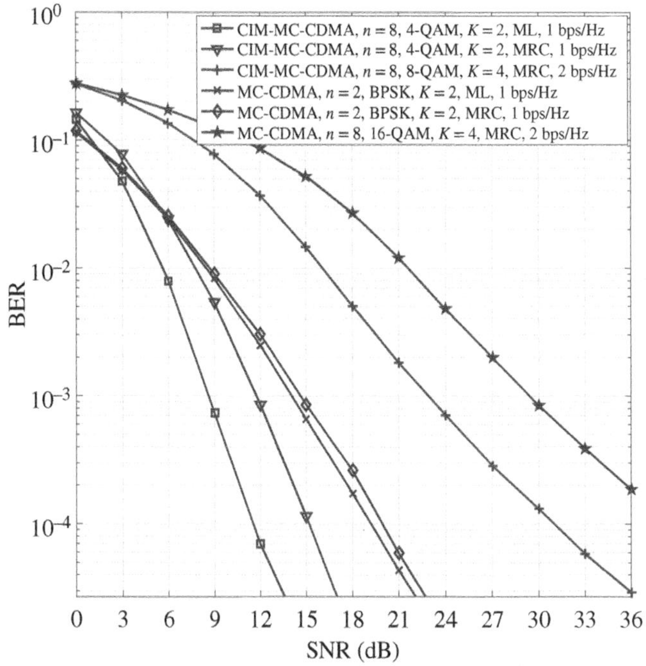

Figure 2.14 Performance comparison between CIM-MC-CDMA and conventional MC-CDMA in the downlink at the overall SEs of 1 and 2 bps/Hz.

of 1 and 2 bps/Hz. It can be seen from the figure that for both CIM-MC-CDMA and MC-CDMA, the MRC detectors suffer from performance loss compared with ML detectors due to the multiple access interference. On the other hand, at both SEs, CIM-MC-CDMA performs better than that of MC-CDMA in the SNR region of interest for ML and MRC detection.

2.5 Power-Domain IM for NGMA

Except for the modulation techniques in the space, frequency, and code domain for NGMA, a novel way of integrating NOMA with power-domain IM was proposed in Pei et al. (2022), where extra information bits are encoded into the power levels to improve the system SE. The resultant scheme is referred to as the NOMA scheme with information-guided power selection (PS-NOMA). This section presents the transmission model and decoding process of the PS-NOMA scheme, respectively.

2.5.1 Transmission Model

For ease of exposition, we first introduce the transmission model of the two-user PS-NOMA scheme, and then extend it to the multiuser case.

2.5.1.1 Two-User Case

A possible system structure of the two-user downlink PS-NOMA network is illustrated in Fig. 2.15, which consists of a BS, a near user (denoted by U_A), a far user (denoted by U_B). All the nodes are equipped with a single antenna. The channels of BS→ U_A and BS→ U_B are respectively denoted by h_A and h_B, which are assumed to follow $\mathcal{N}_c(0, \beta_A)$ and $\mathcal{N}_c(0, \beta_B)$. Notably, it is assumed that all the nodes are synchronous in both time and frequency, and perfect channel state information (CSI) is available to both users. Unlike conventional NOMA, an SE gain is obtained by selecting a power level (PL) from a set of N PLs at random at the transmitter where N is a power of two, so that the PL carries $\log_2(N)$ bits in addition to the amount

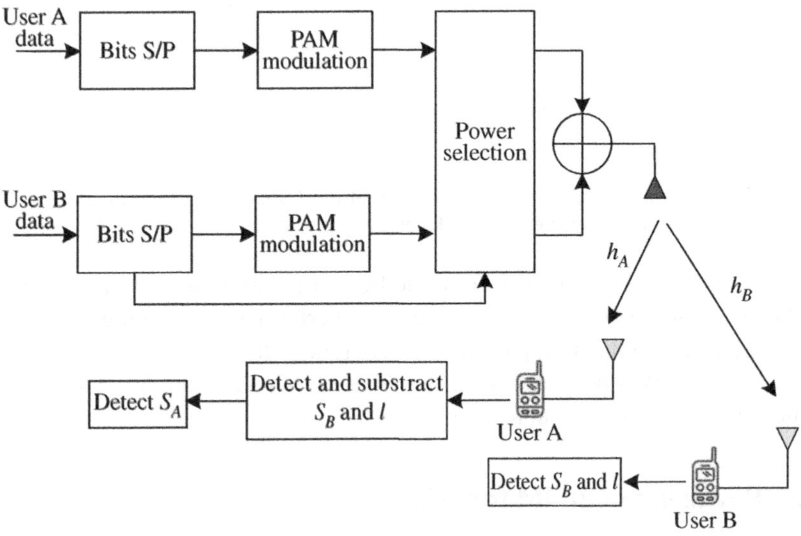

Figure 2.15 System model with two users, where U_A is the near user, and the other is the far user.

of information carried by the transmitted data symbols. Noticing that U_A is the near user and U_B is the far user, the power allocated to U_A should be smaller than that assigned to U_B considering users' fairness since $\beta_A \geq \beta_B$. In this manner, the transmit power matrix can be constructed as

$$\mathbf{P} = \begin{bmatrix} \sqrt{p_A(1)p_t}, & \sqrt{(1-p_A(1))p_t} \\ \sqrt{p_A(2)p_t}, & \sqrt{(1-p_A(2))p_t} \\ \vdots & \vdots \\ \sqrt{p_A(N)p_t}, & \sqrt{(1-p_A(N))p_t} \end{bmatrix}, \tag{2.45}$$

where $0 < p_A(l) < 0.5$ denotes the power allocation coefficient for U_A, $l \in \{1, \dots, N\}$, and p_t denotes the transmit power; the first and the second columns represent the power for U_A and U_B, respectively, and different rows represent different PLs. To ensure that the PLs are distinguishable, each PL is deliberately rotated with a certain angle, which is given by $\boldsymbol{\Theta} = \text{diag}(\exp(0), \exp(j\pi/\text{N}), \dots, \exp(j(\text{N}-1)\pi/\text{N}))$. Therefore, the rotated power matrix can be derived as

$$\mathbf{G} = \boldsymbol{\Theta}\mathbf{P} = \begin{bmatrix} \alpha_A(1), & \alpha_B(1) \\ \alpha_A(2), & \alpha_B(2) \\ \vdots & \vdots \\ \alpha_A(N), & \alpha_B(N) \end{bmatrix}, \tag{2.46}$$

where $\alpha_A(l) = \exp(j(l-1)\pi/N)\sqrt{p_A(l)p_t}$ and $\alpha_B(l) = \exp(j(l-1)\pi/N)\sqrt{(1-p_A(l))p_t}$.

Suppose that the BS chooses the PL $l \in \mathcal{N}$ ($\mathcal{N} = \{1, 2, \dots, N\}$). Subsequently, it conveys the superposition coded symbol

$$s_C = \alpha_A(l)s_A + \alpha_B(l)s_B \tag{2.47}$$

to U_A and U_B simultaneously, where $s_A \in \mathcal{S}_A$ and $s_B \in \mathcal{S}_B$ are the data symbols intended for U_A and U_B, respectively. It is assumed that s_A and s_B are M_A-ary pulse-amplitude modulation (PAM) and M_B-ary PAM symbols with $\mathbb{E}\{|s_A|^2\} = \mathbb{E}\{|s_B|^2\} = E_s = 1$;

$$\mathcal{S}_A = \{\pm d_A, \pm 3d_A, \dots, \pm(M_A - 1)d_A\} \tag{2.48}$$

and

$$\mathcal{S}_B = \{\pm d_B, \pm 3d_B, \dots, \pm(M_B - 1)d_B\}, \tag{2.49}$$

are the corresponding M_A and M_B points constellations, respectively, where $d_A = \sqrt{3/(M_A^2 - 1)}$ and $d_B = \sqrt{3/(M_B^2 - 1)}$ respectively represent half of the minimum distance between two adjacent points of the normalized M_A-PAM and M_B-PAM constellations. Denoting $\mathcal{M}_i = \{1, 2, \dots, M_i\}$ ($i \in \{A, B\}$), the signal constellation of s_C can be expressed as

$$\mathcal{S}_C = \{s_C | s_i = \mathcal{S}_i(k_i), \text{PL} = l, k_i \in \mathcal{M}_i, l \in \mathcal{N}, i \in \{A, B\}\}, \tag{2.50}$$

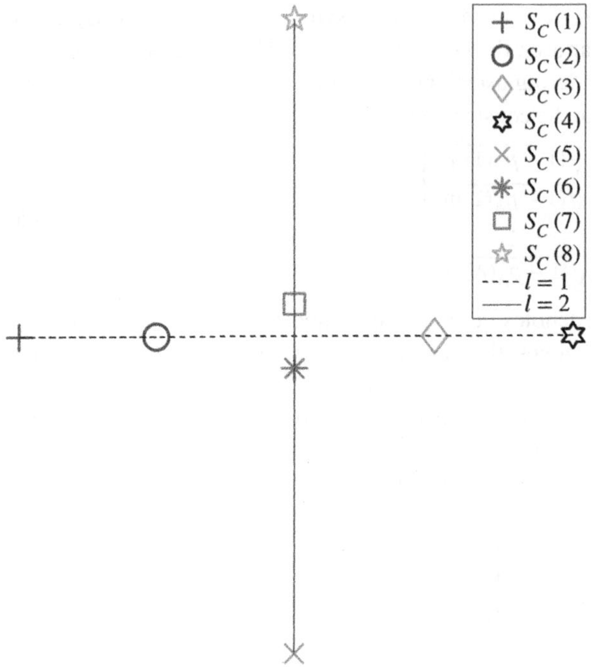

Figure 2.16 Constellation $S_C(i)$ ($i \in \{1, \dots, 8\}$) of superimposed signal transmitted with 2-PAM signals s_A as well as s_B and $N = 2$.

which is an irregular $M_A M_B N$-ary constellation. An example of $M_A = M_B = N = 2$ is shown in Fig. 2.16. The received signals at U_A and U_B can be given by

$$y_A = h_A s_C + n_A \tag{2.51}$$

and

$$y_B = h_B s_C + n_B, \tag{2.52}$$

respectively, where $n_A \sim \mathcal{N}_c(0, N_0)$ ($n_B \sim \mathcal{N}_c(0, N_0)$) indicates the AWGN at U_A (U_B) with noise variance N_0.

2.5.1.2 Multiuser Case

For the multi-user PS-NOMA, it is first assumed that there are $\mathcal{U}$ users (denoted by $U_1, \dots, U_{\mathcal{U}}$) to be served and N PLs to be selected. The channels BS→ U_i ($i \in \{1, \dots, \mathcal{U}\}$) is denoted by h_i, whose variance is β_i. Without loss of generality, the average powers of channels are sorted as $\beta_1 > \cdots > \beta_{\mathcal{U}}$. On the transmitter side, the rotation can be used to design the transmitted constellation. Specifically, the rotated power matrix can be written as

$$\mathbf{G} = \begin{bmatrix} \alpha_1(1), & \dots, & \alpha_{\mathcal{U}}(1) \\ \alpha_1(2), & \dots, & \alpha_{\mathcal{U}}(2) \\ \vdots & \vdots & \vdots \\ \alpha_1(N), & \dots, & \alpha_{\mathcal{U}}(N) \end{bmatrix}, \tag{2.53}$$

where $\alpha_i(l) = \exp(j(l-1)\pi/N)\sqrt{p_i(l)}$, $l \in \mathcal{N} = \{1, \dots, N\}$, $p_i(l)$ is the power for U_i in the lth PL, where $s_i \in \mathcal{S}_i$ with

$$\mathcal{S}_i = \{\pm d_i, \pm 3d_i, \dots, \pm(M_i - 1)d_i\}, \tag{2.54}$$

being the corresponding M_i-PAM constellation, in which $d_i = \sqrt{3/(M_i^2 - 1)}$ represents half of the minimum distance between two adjacent points of the normalized constellation. Consequently, the irregular $M_1 \times \cdots \times M_{\mathcal{U}} N$-ary signal constellation of s_C can be expressed as

$$\begin{aligned} \mathcal{S}_C = &\{s_C | s_i = \mathcal{S}_i(k_i), \mathrm{PL} = l, k_i \in \mathcal{M}_i, l \in \mathcal{N}, \\ &\mathcal{M}_i = \{1, 2, \dots, M_i\}, i \in \{1, \dots, \mathcal{U}\}\}. \end{aligned} \tag{2.55}$$

Finally, substituting (2.55) and the corresponding channel gain into (2.51) yields the received signal of U_i for $i \in \{1, \dots, \mathcal{U}\}$.

2.5.2 Signal Decoding

This subsection describes the decoding process of PS-NOMA. First, it is critical to clarify which user desires the information carried by the PL. For simplicity, it is assumed that the information of the PL is desired by U_B, such that the achievable rate of the far user can be further improved. In other scenarios of interest, it can also be used to improve the performance of U_A or both users. To enhance the performance of both users, it is reasonable to assume that both users share the information of the PL. For example, if there are two extra bits carried by four PLs in total, it is possible to allocate the first bit (corresponding to the first two PLs) to U_B, and the other bit to U_A (corresponding to the last two PLs).

Assume that both users can invoke ML detectors to decode their signals. In conventional NOMA scheme, whose PLs are fixed, the users already know the values of PLs when decoding. While in PS-NOMA, the PLs change over time and are unknown at the users. Hence, the conventional ML detector cannot be applied directly in PS-NOMA, and it is critical to decode PL at both users. In the design, U_A first decodes s_B and l; while for U_B, it detects s_B and l by treating s_A as noise. Herein, the joint constellation of s_B and PL are employed to help to decode. For example, in the case of $N = 2$ and 2-PAM s_B as shown in Fig. 2.17(b), a combined 4-ary constellation is formed, i.e.,

Algorithm 2.1 Detection Algorithm

1: **for** U_A **do**

2: Compare the received signal y_A with the joint constellation S_B^R:

$$S_B^R(\hat{i}) = \arg\min_i |y_A - h_A S_B^R(i)|^2, \tag{2.56}$$

where $i \in \{1, \dots, NM_B\}$, and $\hat{i}$ is the estimate of i at U_A.

3: Determine active PL $\hat{l}$ and the transmitted signal $\hat{s}_B$ by resolving $S_B^R(\hat{i})$, where $\hat{s}_B$ and $\hat{l}$ are the estimates of s_B and l at U_A, respectively.

4: Use SIC to decode s_A, whose ML detector can be expressed as

$$\hat{s}_A = \arg\min_{s_A} |y_A - h_A \alpha_B(\hat{l})\hat{s}_B - h_A \alpha_A(\hat{l}) s_A|^2, \tag{2.57}$$

where $\hat{s}_A$ is the estimate of s_A at U_A.

5: **end for**

6: **for** U_B **do**

7: Compare the received signal y_B with the joint constellation S_B^R:

$$S_B^R(\hat{i}) = \arg\min_i |y_B - h_B S_B^R(i)|^2, \tag{2.58}$$

where $i \in \{1, \dots, NM_B\}$, and $\hat{i}$ is the estimate of i at U_B.

8: Determine active PL $\hat{l}$ and the transmitted signal $\hat{s}_B$ by resolving $S_B^R(\hat{i})$, where $\hat{s}_B$ and $\hat{l}$ are the estimates of s_B and l at U_B, respectively.

9: **end for**

$$S_B^R(i) = \{s_B = \pm 1, l = 1, 2\}, i \in \{1, \dots, 4\}. \tag{2.59}$$

Explicitly, the information of s_B and PL is contained in each point of the constellation. If the received signal after equalization is close to $S_B^R(1)$, it is aware of the estimated $s_B = -1$ and $l = 1$. With these information, U_A then subtracts the estimated s_B and decodes its own signal, and this procedure can be seen as a common 2-PAM signal decoding as shown in Fig. 2.17(a). The detailed detection algorithm for both users is summarized in **Algorithm 2.1**.

For the multiuser case, U_i decodes the data symbol for U_j $(i < j < U)$ and the chosen PL sequentially through joint constellation like Fig. 2.17(b). For U_{U} (the farthest user), it just has to decode its own data symbol and PL. Specific decoding process is similar to **Algorithm 2.1**.

2.5.3 Performance Analysis

The advantages of PS-NOMA over the conventional NOMA can be described as the following two aspects: (i) better BER performance can be achieved without

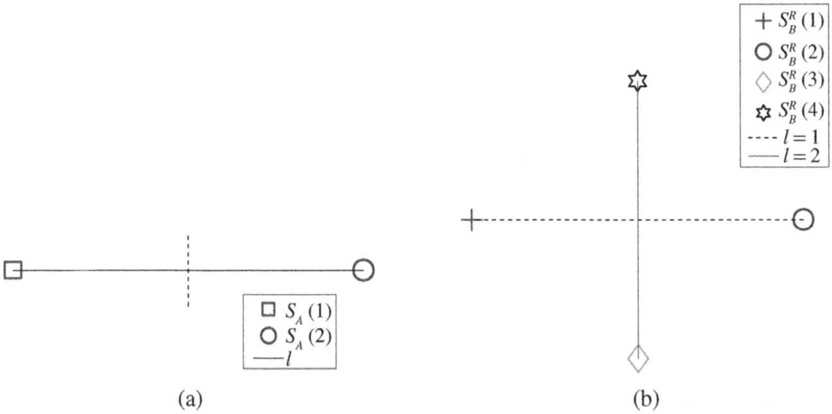

Figure 2.17 Constellations of the 2-PAM signals s_A and s_B: (a) constellations of s_A with $N = 2$; (b) constellations of s_B with $N = 2$, where $\mathcal{S}_B^R$ denotes the union constellation of s_B and PL.

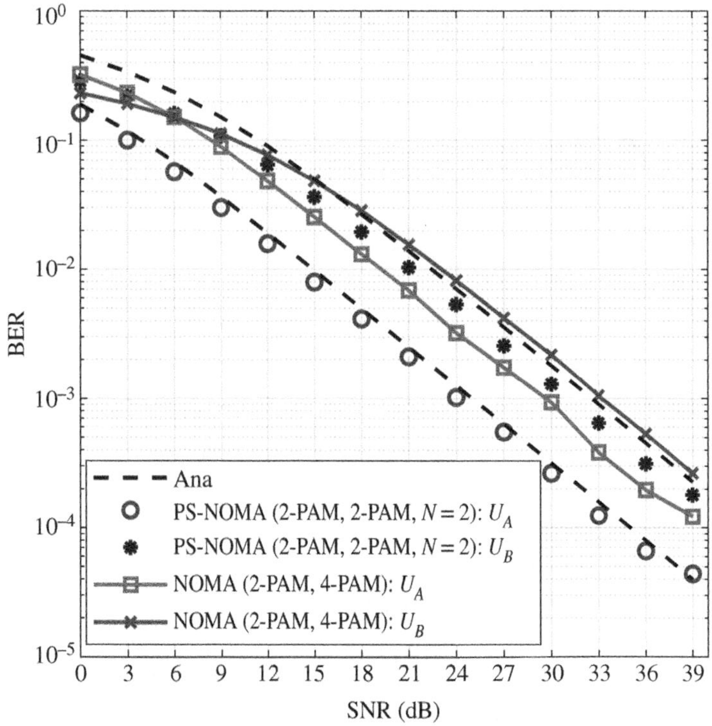

Figure 2.18 BER comparison between PS-NOMA and NOMA.

increasing constellation complexity; (ii) superior achievable rate performance and larger rate region can be attained. Correspondingly, the cost for these performance gains is the design difficulty for the power matrix and extra computational complexity. For clarity, in the sequel, computer simulations are provided to show the performance comparison between PS-NOMA and NOMA.

The BER performance of PS-NOMA and NOMA is compared, assuming that U_A and U_B in all considered schemes employ ML detection. Since the comparison of OMA and NOMA has been widely argued in previous works, the BER curves of OMA counterparts are not presented. In Fig. 2.18, we compare the BER performance of PS-NOMA (2-PAM, 2-PAM, $N = 2$) and NOMA (2-PAM, 4-PAM) with their theoretical BER results presented. Given fairness, it is assumed that the maximum spectral efficiency for both schemes are 3 bps/Hz, and both schemes have the same constellation order. It can be seen from Fig. 2.18 that the theoretical curves approximately match with their simulation counterparts. Besides, one can find that the BER performance of U_A is superior to that of U_B in both PS-NOMA and NOMA, which comes from the fact $\beta_A > \beta_B$. Although all the curves have the

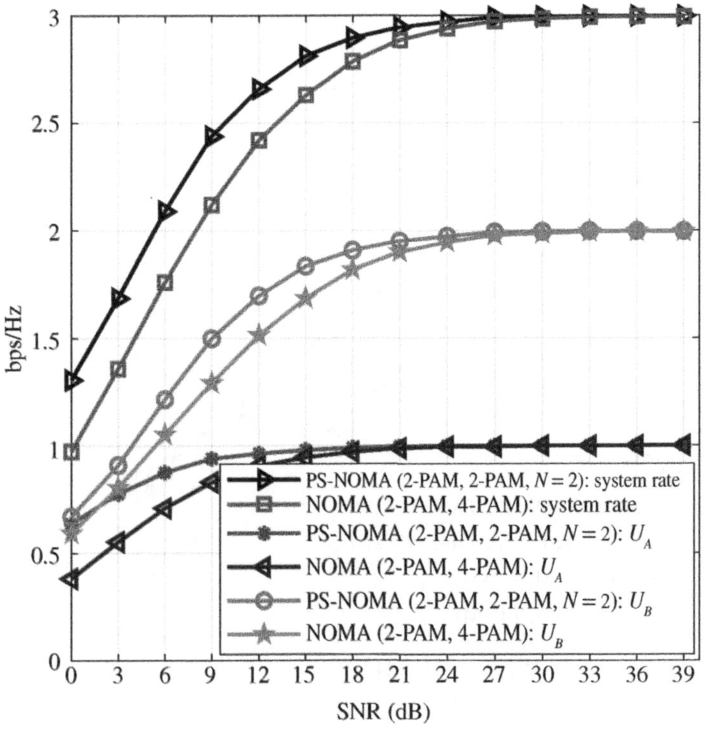

Figure 2.19 Achievable rate comparison between PS-NOMA and NOMA.

same constellation order as 8, the performances of PS-NOMA are always better than those of NOMA. This phenomenon tells us that applying PS can improve performance without increasing constellation complexity.

In Fig. 2.19, we present the curves of PS-NOMA versus NOMA. Compared to NOMA, PS-NOMA provides achievable rate gains for both U_A and U_B, and obtains the sum achievable rate enhancement in turn. This system performance improvement is more pronounced at low SNR, while the system gap narrows with increasing SNR. Since PS-NOMA and NOMA have the same constellation order, at high SNR, the achievable rate performances of them are equal. It is worth noting that the performance improvement of U_A is more obvious than that of U_B.

2.6 Summary

In this chapter, we explored the potential of space-domain, frequency-domain, code-domain, and power-domain IM techniques for NGMA primarily from transceiver design. Noticeably, the performance loss caused by the long transmission distance in NOMA or SCMA can be compensated by allocating the additional information bits created from the indices of transmit entities to the far user. Therefore, the IM-aided NGMA schemes not only can provide massive connectivity, but also can achieve highly spectrum-efficient and energy-efficient transmissions. We conclude that IM-aided NGMA schemes might be highly useful for next-generation wireless networks, thanks to their flexible structures and improved efficiency, and can be considered as strong candidates for the multiple-access techniques.

References

Mohammed Al-Ansi, Syed Alwee Aljunid, Essam Sourour, M S Anuar, and C B M Rashidi. Multi-RF and generalized single-RF combination models for spatial modulation and NOMA technologies. *IEEE Transactions on Vehicular Technology*, 71(7):7308–7324, 2022.

Abdullateef Almohamad, Mazen O Hasna, Saud Althunibat, and Khalid Qaraqe. A novel downlink IM-NOMA scheme. *IEEE Open Journal of the Communications Society*, 2:235–244, 2021.

Ibrahim Al-Nahhal, Octavia A Dobre, Ertugrul Basar, and Salama Ikki. Low-cost uplink sparse code multiple access for spatial modulation. *IEEE Transactions on Vehicular Technology*, 68(9):9313–9317, 2019.

Ibrahim Al-Nahhal, Octavia A Dobre, and Salama Ikki. On the complexity reduction of uplink sparse code multiple access for spatial modulation. *IEEE Transactions on Communications*, 68(11):6962–6974, 2020.

Emre Arslan, Ali Tugberk Dogukan, and Ertugrul Basar. Index modulation-based flexible non-orthogonal multiple access. *IEEE Wireless Communications Letters*, 9(11):1942–1946, Nov. 2020.

Ertugrul Basar, Umit Aygolu, Erdal Panayirci, and H Vincent Poor. Orthogonal frequency division multiplexing with index modulation. *IEEE Transactions on Signal Processing*, 61(22):5536–5549, Nov. 2013.

Ertugrul Basar, Miaowen Wen, Raed Mesleh, Marco Di Renzo, Yue Xiao, and Harald Haas. Index modulation techniques for next-generation wireless networks. *IEEE Access*, 5:16693–16746, 2017.

Yingyang Chen, Li Wang, Yutong Ai, Bingli Jiao, and Lajos Hanzo. Performance analysis of NOMA-SM in vehicle-to-vehicle massive MIMO channels. *IEEE Journal on Selected Areas in Communications*, 35(12):2653–2666, 2017.

Xuan Chen, Miaowen Wen, and Shuping Dang. On the performance of cooperative OFDM-NOMA system with index modulation. *IEEE Wireless Communications Letters*, 9(9):1346–1350, Sept. 2020a.

Xuan Chen, Miaowen Wen, Tianqi Mao, and Shuping Dang. Spectrum resource allocation based on cooperative NOMA with index modulation. *IEEE Transactions on Cognitive Communications and Networking*, 6(3):946–958, Sept. 2020b.

Xiang Cheng, Meng Zhang, Miaowen Wen, and Liuqing Yang. Index modulation for 5G: Striving to do more with less. *IEEE Wireless Communications*, 25(2):126–132, Apr. 2018.

Zhiguo Ding, Mugen Peng, and H Vincent Poor. Cooperative non-orthogonal multiple access in 5G systems. *IEEE Communications Letters*, 19(8):1462–1465, Aug. 2015.

Zijie Hong, Guoquan Li, Yongjun Xu, and Xiangyun Zhou. User grouping and power allocation for downlink NOMA-based quadrature spatial modulation. *IEEE Access*, 8:38136–38145, 2020.

M Hemanta Kumar, Sanjeev Sharma, M Thottappan, and Kuntal Deka. Precoded spatial modulation-aided cooperative NOMA. *IEEE Communications Letters*, 25(6):2053–2057, 2021.

Ke Lai, Jing Lei, Lei Wen, and Gaojie Chen. Codeword position index modulation design for sparse code multiple access system. *IEEE Transactions on Vehicular Technology*, 69(11):13273–13288, 2020.

Qiang Li, Miaowen Wen, H Vincent Poor, and Fangjiong Chen. Information guided precoding for OFDM. *IEEE Access*, 5:19644–19656, 2017.

Qiang Li, Miaowen Wen, Ertugrul Basar, and Fangjiong Chen. Index modulated OFDM spread spectrum. *IEEE Transactions on Wireless Communications*, 17(4):2360–2374, 2018.

Qiang Li, Miaowen Wen, Ertugrul Basar, H Vincent Poor, and Fangjiong Chen. Spatial modulation-aided cooperative NOMA: Performance analysis and comparative study. *IEEE Journal of Selected Topics in Signal Processing*, 13(3):715–728, 2019.

Jun Li, Shuping Dang, Yier Yan, Yuyang Peng, Saba Al-Rubaye, and Antonios Tsourdos. Generalized quadrature spatial modulation and its application to vehicular networks with NOMA. *IEEE Transactions on Intelligent Transportation Systems*, 22(7):4030–4039, 2020a.

Jun Li, Jia Hou, Lisheng Fan, Yier Yan, Xue-Qin Jiang, and Han Hai. NOMA-aided generalized pre-coded quadrature spatial modulation for downlink communication systems. *China Communications*, 17(11):120–130, 2020b.

Guoquan Li, Zijie Hong, Yu Pang, Yongjun Xu, and Zhengwen Huang. Resource allocation for sum-rate maximization in NOMA-based generalized spatial modulation. *Digital Communications and Networks*, 8(6):1077–1084, 2022.

Yusha Liu, Lie-Liang Yang, and Lajos Hanzo. Spatial modulation aided sparse code-division multiple access. *IEEE Transactions on Wireless Communications*, 17(3):1474–1487, 2017.

Yusha Liu, Lie-Liang Yang, Pei Xiao, Harald Haas, and Lajos Hanzo. Spatial modulated multicarrier sparse code-division multiple access. *IEEE Transactions on Wireless Communications*, 19(1):610–623, 2019.

Yusha Liu, Lie-Liang Yang, and Lajos Hanzo. Joint user-activity and data detection for grant-free spatial-modulated multi-carrier non-orthogonal multiple access. *IEEE Transactions on Vehicular Technology*, 69(10):11673–11684, 2020.

Yusha Liu, Luping Xiang, Lie-Liang Yang, and Lajos Hanzo. Space-time coded generalized spatial modulation for sparse code division multiple access. *IEEE Transactions on Wireless Communications*, 20(8):5359–5372, 2021.

Yuanwei Liu, Wenqiang Yi, Zhiguo Ding, Xiao Liu, Octavia A Dobre, and Naofal Al-Dhahir. Developing NOMA to next generation multiple access: Future vision and research opportunities. *IEEE Wireless Communications*, 29(6):120–127, Dec. 2022a.

Yuanwei Liu, Shuowen Zhang, Xidong Mu, Zhiguo Ding, Robert Schober, Naofal Al-Dhahir, Ekram Hossain, and Xuemin Shen. Evolution of NOMA toward next generation multiple access (NGMA) for 6G. *IEEE Journal on Selected Areas in Communications*, 40(4):1037–1071, Apr. 2022b.

Tianqi Mao, Zhaocheng Wang, Qi Wang, Sheng Chen, and Lajos Hanzo. Dual-mode index modulation aided OFDM. *IEEE Access*, 5:50–60, 2017.

Zhipeng Pan, Junshan Luo, Jing Lei, Lei Wen, and Chaojing Tang. Uplink spatial modulation SCMA system. *IEEE Communications Letters*, 23(1):184–187, 2018.

Xinyue Pei, Yingyang Chen, Miaowen Wen, Hua Yu, Erdal Panayirci, and H Vincent Poor. Next-generation multiple access based on NOMA with power level modulation. *IEEE Journal on Selected Areas in Communications*, 40(4):1072–1083, Apr. 2022.

Qintuya Si, Minglu Jin, Yunfei Chen, Nan Zhao, and Xianbin Wang. Performance analysis of spatial modulation aided NOMA with full-duplex relay. *IEEE Transactions on Vehicular Technology*, 69(5):5683–5687, 2020.

Armed Tusha, Seda Dogan, and Huseyin Arslan. A hybrid downlink NOMA with OFDM and OFDM-IM for beyond 5G wireless networks. *IEEE Signal Processing Letters*, 27:491–495, 2020.

Zhaocheng Wang and Jianfei Cao. NOMA-based spatial modulation. *IEEE Access*, 5:3790–3800, 2017.

Miaowen Wen, Ertugrul Basar, Qiang Li, Beixiong Zheng, and Meng Zhang. Multiple-mode orthogonal frequency division multiplexing with index modulation. *IEEE Transactions on Communications*, 65(9):3892–3906, Sept. 2017.

Miaowen Wen, Qiang Li, Ertugrul Basar, and Wensong Zhang. Generalized multiple-mode OFDM with index modulation. *IEEE Transactions on Wireless Communications*, 17(10):6531–6543, Oct. 2018.

Miaowen Wen, Beixiong Zheng, Kyeong Jin Kim, Marco Di Renzo, Theodoros A Tsiftsis, Kwang-Cheng Chen, and Naofal Al-Dhahir. A survey on spatial modulation in emerging wireless systems: Research progresses and applications. *IEEE Journal on Selected Areas in Communications*, 37(9):1949–1972, 2019.

Luping Xiang, Yusha Liu, Lie-Liang Yang, and Lajos Hanzo. Low complexity detection for spatial modulation aided sparse code division multiple access. *IEEE Transactions on Vehicular Technology*, 70(12):12858–12871, 2021.

Ping Yang, Yue Xiao, Ming Xiao, and Zheng Ma. NOMA-aided precoded spatial modulation for downlink MIMO transmissions. *IEEE Journal of Selected Topics in Signal Processing*, 13(3):729–738, 2019.

Xiangbin Yu, Jiali Cai, Mingfeng Xie, and Tao Teng. Ergodic rate analysis of massive spatial modulation MIMO system with NOMA in cooperative relay networks. *IEEE Systems Journal*, 16(3):4513–4524, 2021.

Caijun Zhong, Xiaoling Hu, Xiaoming Chen, Derrick Wing Kwan Ng, and Zhaoyang Zhang. Spatial modulation assisted multi-antenna non-orthogonal multiple access. *IEEE Wireless Communications*, 25(2):61–67, 2018.

3

NOMA Transmission Design with Practical Modulations

Tianying Zhong[1], *Yuan Wang*[1,2], *and Jiaheng Wang*[1,2]

[1] *National Mobile Communications Research Laboratory, Southeast University, Nanjing, China*
[2] *Pervasive Communication Research Center, Purple Mountain Laboratories, Nanjing, China*

3.1 Introduction

Non-orthogonal multiple access (NOMA) has been considered as a promising technology for the next-generation of wireless communication to support massive connectivity due to its non-orthogonal characteristics (Liu et al., 2017, Wong et al., 2017). Compared with the conventional orthogonal multiple access (OMA) scheme, NOMA can provide higher spectrum efficiency, better user fairness, and lower signaling cost (3GPP TR 38.812, 2018, Ding et al., 2017, Liu et al., 2022). In a power-domain multi-channel NOMA system,[1] the NOMA scheme is performed within each orthogonal channel[2] by employing superposition coding at the transmitter and successive interference cancellation (SIC) at the receiver, so that users' signals can be multiplexed in the power domain. Hence, the resource allocation scheme, including power allocation and channel assignment, is a critical issue to fulfill the benefits of NOMA.

Many different approaches have been proposed to optimize the power allocation and channel assignment of NOMA systems. In Wang et al. (2017), optimal power allocation is derived through exploring the convexity of the weighted sum rate maximization problem for NOMA. Zhu et al. (2017) develops power allocation schemes under various performance criteria and a channel assignment algorithm based on the deferred acceptance method. Branch and bound approach

1 According to the multiplexing method, NOMA can be divided into power-domain NOMA and code-domain NOMA. In this chapter, we focus on the former one. Hereafter, NOMA is used to refer to power-domain NOMA.
2 An orthogonal channel can be, e.g., a subcarrier in orthogonal frequency division multiple access (OFDMA) or a time slot in time division multiple access (TDMA).

Next Generation Multiple Access, First Edition.
Edited by Yuanwei Liu, Liang Liu, Zhiguo Ding, and Xuemin Shen.
© 2024 The Institute of Electrical and Electronics Engineers, Inc. Published 2024 by John Wiley & Sons, Inc.

is employed in Wei et al. (2017) to jointly allocate resources for minimizing the transmit power subject to the data rate and outage probability constraints. In Fang et al. (2016), energy-efficient resource allocation algorithms are developed by transforming and approximating the formulated problems into convex subproblems. A deep reinforcement learning framework is proposed in He et al. (2019) for channel assignment to maximize the sum rate or minimal rate. In Wang et al. (2019), the power allocation is predefined to follow the inverse proportional fairness criterion, while the channel assignment problem is formulated as a cooperative multi-agent game and is solved using deep deterministic policy gradient method. The power allocation of multiple-input multiple-output (MIMO)-NOMA is addressed through a communication deep neural network in Huang et al. (2020) to maximize the sum rate and energy efficiency. An energy and delay cost minimization problem of Internet-of-things systems is formulated in Wang et al. (2020), where the subcarrier allocation and task scheduling are solved based on matching theory and reinforcement learning. In Khairy et al. (2021), the authors optimize the altitudes of unmanned aerial vehicles (UAVs) and improve the channel access to maximize the sum rate via constrained deep reinforcement learning.

Most existing NOMA designs are based on the ideal information rate, which implies that the transmitted signals are Gaussian distributed and the SIC is perfect without any error propagation. In practice, however, decoding errors are inevitable and thus SIC is generally imperfect (Li et al., 2011, Narasimhan, 2005). More importantly, in practical systems, the transmitted signals are generally discrete symbols constrained to a signal constellation, e.g., pulse amplitude modulation (PAM) or quadrature amplitude modulation (QAM) symbols. The assumptions of Gaussian distributed signals and perfect SIC may lead to NOMA designs adverse to SIC and cause decoding failure.

This chapter focuses on the resource allocation of a multichannel NOMA system while taking into account practical modulations with imperfect SIC. We first briefly review the basic concepts of the downlink NOMA transmission and practical modulations in NOMA. Then, we introduce a performance metric, namely the *effective throughput*, to the NOMA system design, which takes into account both the data rate and the error performance. Both the exact expression and the lower bound of the effective throughput are derived. To maximize the effective throughput, we propose an analytical power allocation solution as well as an iterative joint resource allocation algorithm, which is shown to be efficient through simulations.

This chapter is organized as follows. Section 3.2 introduces the fundamentals of the multichannel NOMA system. In Section 3.3, the exact expression of effective throughput is provided. In Section 3.4, we formulate the effective throughput maximization problem, and then provide the power allocation strategy as well as the joint resource allocation scheme. Numerical results are presented in Section 3.5. Conclusions are drawn in Section 3.6.

3.2 Fundamentals

To start with, we review the basic concepts of the multichannel downlink NOMA transmission and the practical modulations for NOMA in this section.

3.2.1 Multichannel Downlink NOMA

We first introduce the system model of a typical multichannel downlink NOMA system. As depicted in Fig. 3.1, the single-antenna base station (BS) serves N single-antenna users[3] through K wireless channels. We denote the nth user by U_n $(n = 1, \dots, N)$ and the kth channel by C_k $(k = 1, \dots, K)$. The users are divided into K groups with each group assigned with a channel. Signals transmitted on different channels are assumed to have no interference with each other, while signals on the same channel are multiplexed in the power domain. We assume that the BS has full knowledge of the channel state information (CSI).

At the BS, after superposition coding, the transmit signal on channel C_k can be written as

$$x_k = \sum_{n=1}^{N} d_{k,n} \sqrt{p_{k,n}} s_n, \tag{3.1}$$

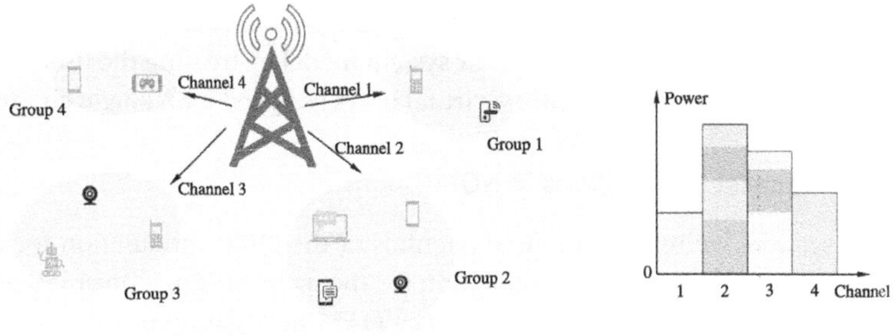

Figure 3.1 System model of a downlink multichannel NOMA system.

3 According to Ding et al. (2016), by carefully designing the precoding and detection matrices, the inter-cluster interference of MIMO–NOMA can be canceled completely. The MIMO–NOMA system can be reduced into several single-input single-output (SISO) NOMA systems. Thus, we focus on the single-antenna NOMA systems.

where $s_n \in \mathbb{C}$ is the intended signal of user U_n, $d_{k,n}$ is a binary variable and represents whether channel C_k is assigned to user U_n, i.e.,

$$d_{k,n} = \begin{cases} 0, & C_k \text{ is not assigned to } U_n, \\ 1, & C_k \text{ is assigned to } U_n. \end{cases} \tag{3.2}$$

Moreover, $p_{k,n} \geq 0$ is the power allocated to user U_n on channel C_k, subject to the total power constraint $\sum_{k=1}^{K} \sum_{n=1}^{N} d_{k,n} p_{k,n} \leq P$, where P is the total power budget. At the receiver, if assigned with channel C_k, user U_n receives the signal

$$y_{k,n} = g_{k,n} x_k + z_n, \tag{3.3}$$

where $g_{k,n} \in \mathbb{C}$ is the channel coefficient of channel C_k between the BS and user U_n, which includes both large-scale path loss and small-scale fading, and $z_n \in \mathbb{C}$ is the additive white Gaussian noise (AWGN) of user U_n, with zero mean and variance σ_n^2. To facilitate further analysis, we define the normalized channel gain as $h_{k,n} \triangleq |g_{k,n}|^2/\sigma_n^2 > 0$.

SIC is employed at the receiver to decode the intended signal s_n from $y_{k,n}$. For those users which are assigned with the same channel, each user decodes and removes the signals of the other users with higher power successively, and then decodes its own signal, while treating the signals of those users with lower power as interference. Considering the complexity of SIC, the decoding delay, and the co-channel interferences are proportional to the number of users on each channel (Zhu et al., 2017, Fang et al., 2016), we limit the number of users within each channel to be at most two. The total number of users N is less than or equal to $2K$.

We refer to the channels assigned to only one user as the single-user channels, the channels assigned to two users as the two-user channels, and the channels not being assigned to any users as the idle channels. Specifically, we assume that each user occupies one channel. The situation where each user can occupy more than one channel can be extended from our system model by treating the user assigned with multiple channels as multiple virtual users assigned with single channel.

3.2.2 Practical Modulations in NOMA

In this section, we review the fundamentals of the QAM modulation for downlink NOMA. Since the QAM modulation for the users on single-user channels is the same as that in OMA, we focus on the QAM modulation on two-user channels with the NOMA scheme. Specifically, we take the single channel case as an example where the BS serves two users U_1 and U_2 on the same channel. Note that the analysis of this case can be extended to all the two-user channels.

Suppose that U_n, $n \in \{1, 2\}$, employs M_n^2-QAM modulation, where M_1 and M_2 are both even integers and bits are Gray mapped to symbols with equal probabilities. The constellations of U_1 and U_2 are shown in Fig. 3.2(a) and (b), respectively.

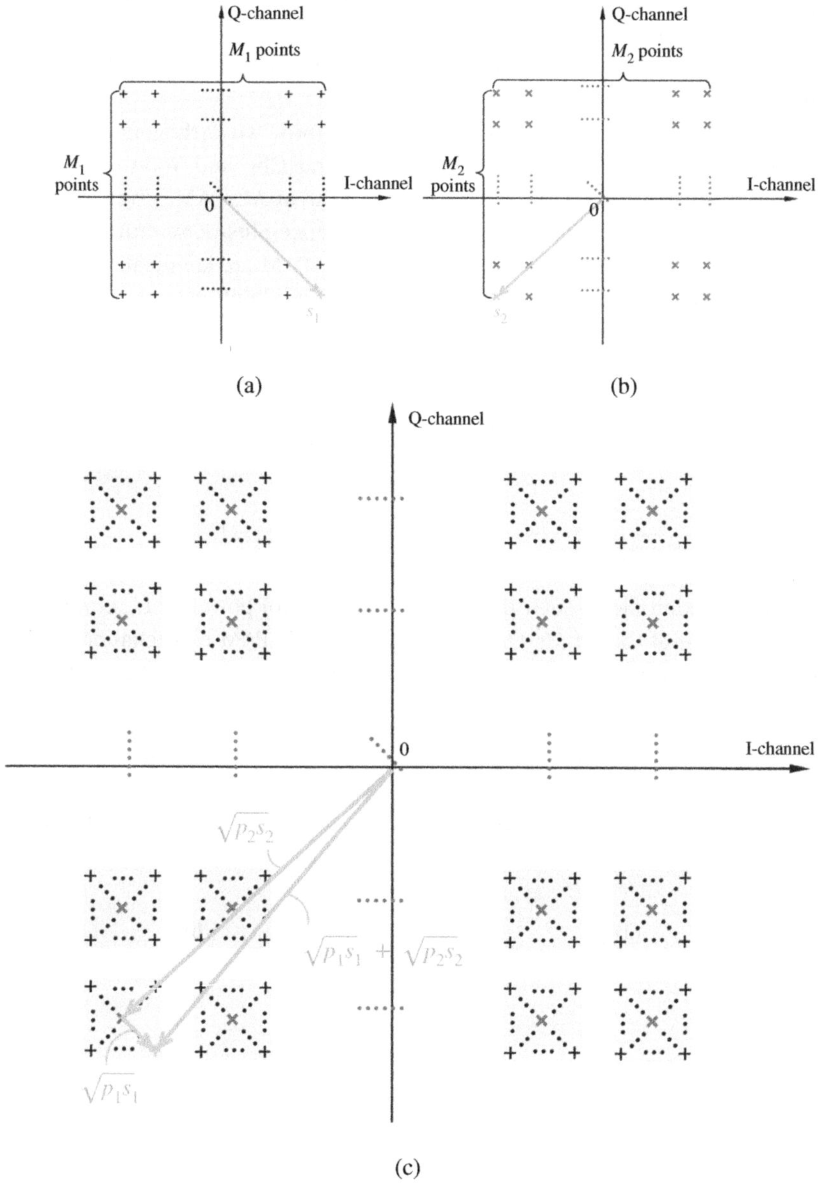

Figure 3.2 (a) The constellation of U_1 employing M_1^2-QAM; (b) The constellation of U_2 employing M_2^2-QAM; (c) The superposed constellation of M_1^2-QAM and M_2^2-QAM.

Suppose that U_1 transmits symbol s_1 with power p_1 and U_2 transmits symbol s_2 with power p_2. Then, the superposed signal $\sqrt{p_1}s_1 + \sqrt{p_2}s_2$ corresponds to a point in the superposed constellation shown in Fig. 3.2(c), which has $M_1^2M_2^2$ possible points.

An M_n^2-QAM constellation can be decoupled into two orthogonal M_n-PAM constellations. According to Goldsmith (2005) and Cho and Yoon (2002), the bit error rate (BER) of M_n^2-QAM is equal to that of M_n-PAM, given that the signal-to-noise ratios (SNRs) (or signal-to-interference-plus-noise ratios (SINRs)) and the bit mapping schemes of M_n^2-QAM and M_n-PAM are identical. Therefore, we next focus on investigating the BER of M_n-PAM modulation.

Specifically, assume that U_n employs M_n-PAM modulation and Gray mapping. The symbol of U_n is selected equiprobably from the set

$$\left\{(2m_n - M_n - 1)\sqrt{b_n/2},\ m_n = 1, 2, \ldots, M_n\right\}, \tag{3.4}$$

where

$$b_n = \frac{M_n}{\sum_{m_n=1}^{M_n}(2m_n - M_n - 1)^2} = \frac{3}{M_n^2 - 1}, \quad n \in \{1, 2\}, \tag{3.5}$$

is a normalization factor such that the average symbol power of U_n employing M_n-PAM is half of that of M_n^2-QAM, i.e., $1/2$. The M_n-PAM constellations of U_1 and U_2 are illustrated in Fig. 3.3(a) and (b), respectively. Suppose that U_1 transmits

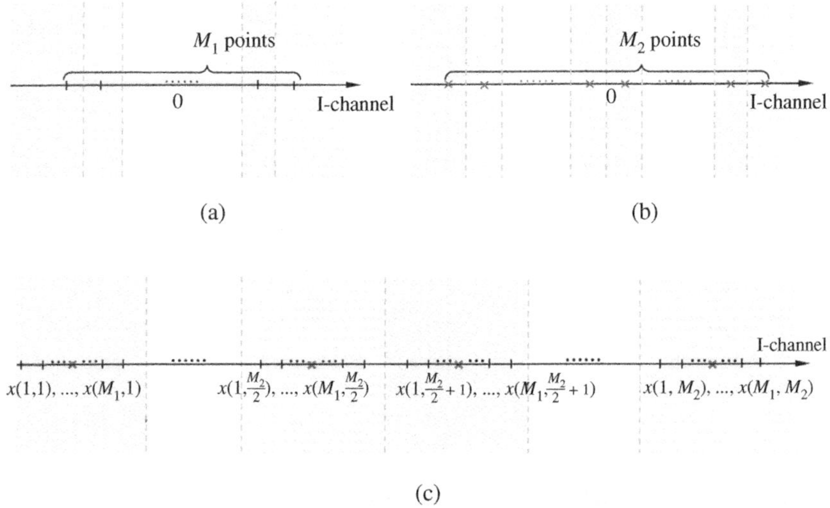

Figure 3.3 (a) The constellation of U_1 employing M_1-PAM; (b) The constellation of U_2 employing M_2-PAM; (c) The superposed constellation of M_1-PAM and M_2-PAM.

the m_1th symbol, $m_1 \in \{1, 2, \dots, M_1\}$, and U_2 transmits the m_2th symbol, $m_2 \in \{1, 2, \dots, M_2\}$. Then, the superposed symbol, denoted by $x(m_1, m_2)$, is given by

$$x(m_1, m_2) = \left(2m_1 - M_1 - 1\right)\sqrt{\frac{p_1 b_1}{2}} + \left(2m_2 - M_2 - 1\right)\sqrt{\frac{p_2 b_2}{2}}, \tag{3.6}$$

and there are $M_1 M_2$ such points in the superposed constellation as shown in Fig. 3.3(c). The symbol detection at the receivers is based on the minimum distance criterion. The corresponding decision regions are shown as shaded areas in Fig. 3.3.

3.3 Effective Throughput Analysis

In this section, we introduce the effective throughput as the performance metric to evaluate the impact of the imperfect SIC and practical modulations on the multichannel NOMA system. Then the exact expressions of the effective throughput are analyzed for the single-user channels and the two-user channels, respectively.

We first give the definition of the effective throughput. Suppose the bandwidth occupied by each channel is equal to B. Consider that user U_n assigned with channel C_k employs M_n^2-QAM. According to Goldsmith (2005), the symbol rate of user U_n is also equal to B. Thus, the effective symbol rate (which measures the correctly transmitted symbol rate (Sun et al., 2018, Polyanskiy et al., 2010)) of user U_n can be expressed as $B(1 - \epsilon_{k,n})$, where $\epsilon_{k,n}$ is the symbol error rate (SER) of user U_n on channel C_k. On this basis, we define the effective throughput of user U_n on channel C_k as

$$J_{k,n} \triangleq B(1 - \epsilon_{k,n})\log_2 M_n^2, \tag{3.7}$$

to measure the correctly transmitted bit rate.

In order to derive the exact expression of the effective throughput $J_{k,n}$, we need to determine $\epsilon_{k,n}$ of user U_n on channel C_k. According to Proakis and Salehi (2007), an M_n^2-QAM can be decoupled into two M_n-PAMs. When the SNRs and the bit mapping schemes of M_n^2-QAM and M_n-PAM are identical, the SER of the M_n^2-QAM can be expressed as

$$\epsilon_{k,n} = 1 - \left(1 - \phi_{k,n}\right)^2 = 2\phi_{k,n} - \phi_{k,n}^2, \tag{3.8}$$

where $\phi_{k,n}$ is the SER of M_n-PAM. Next, we determine the SER $\phi_{k,n}$ and the effective throughput $J_{k,n}$ of the single-user channels and the two-user channels, respectively.

3.3.1 Effective Throughput of the Single-User Channels

We consider a general NOMA case, where the number of users could be either even or odd, so there may exist single-user channels and two-user channels. Although the superposition coding is not employed in the single-user channels, the analysis of the single-user channel is necessary for NOMA design.

Suppose that channel C_k is a single-user channel. We denote the user assigned to channel C_k as $U_{k,0}$. According to Goldsmith (2005), the SER of user $U_{k,0}$, when adopting $M_{k,0}$-PAM, is

$$\phi_{k,0} = \frac{2(M_{k,0} - 1)}{M_{k,0}} Q\left(\sqrt{\frac{3h_{k,0}p_{k,0}}{M_{k,0}^2 - 1}}\right), \tag{3.9}$$

where $h_{k,0}$ and $p_{k,0}$ are the channel gain and the power of user $U_{k,0}$, respectively. Thus, the effective throughput $J_{k,0}$ is shown in the following result.

Proposition 3.1 The effective throughput of user $U_{k,0}$ on the single-user channel is

$$J_{k,0} = B\left(1 - \phi_{k,0}\right)^2 \log_2 M_{k,0}^2, \tag{3.10}$$

where $\phi_{k,0}$ is given in (3.9).

Proof: The expression in (3.10) can be obtained via (3.7) and (3.8). □

Therefore, the effective throughput for single-user channels can be given by $J_k = J_{k,0}$. Next, we provide the exact expression of the effective throughput for two-user channels.

3.3.2 Effective Throughput of the Two-User Channels

Suppose that channel C_k is a two-user channel, and then, the SER depends on the SIC decoding order. We denote the two users on channel C_k as users $U_{k,\text{I}}$ and $U_{k,\text{II}}$, where user $U_{k,\text{I}}$ has a better channel condition than user $U_{k,\text{II}}$, i.e., the channel gains satisfy $h_{k,\text{I}} \geq h_{k,\text{II}}$. Assume that $U_{k,i}$ ($i = \text{I}, \text{II}$) employs $M_{k,i}^2$-QAM and is allocated with the power $p_{k,i} \geq 0$. User $U_{k,\text{I}}$ decodes its signal directly, treating the signal of user $U_{k,\text{II}}$ as interference. User $U_{k,\text{II}}$ first decodes and cancels the signal of user $U_{k,\text{I}}$, and then decodes the signal of itself.

According to Wang et al. (2021), the SER of the strong channel user, i.e., $U_{k,\text{I}}$, when adopting $M_{k,\text{I}}$-PAM, is given by

$$\phi_{k,\text{I}} = \frac{1}{M_{k,\text{I}}M_{k,\text{II}}} \sum_{m=1}^{M_{k,\text{I}}} \sum_{j=1}^{M_{k,\text{II}}} \sum_{l=1}^{M_{k,\text{II}}} \left[f(m,1) Q\left(\sqrt{\frac{3h_{k,\text{I}}p_{k,\text{I}}}{M_{k,\text{I}}^2 - 1}} + 2(j-l)\sqrt{\frac{3h_{k,\text{I}}p_{k,\text{II}}}{M_{k,\text{II}}^2 - 1}}\right)\right.$$

$$+f(m, M_{k,\mathrm{I}})Q\left(\sqrt{\frac{3h_{k,\mathrm{I}}p_{k,\mathrm{I}}}{M_{k,\mathrm{I}}^2-1}}-2(j-l)\sqrt{\frac{3h_{k,\mathrm{I}}p_{k,\mathrm{II}}}{M_{k,\mathrm{II}}^2-1}}\right)\Bigg]$$
$$\times\Bigg[1-f(l,1)Q\left((2m-M_{k,\mathrm{I}}-1)\sqrt{\frac{3h_{k,\mathrm{I}}p_{k,\mathrm{I}}}{M_{k,\mathrm{I}}^2-1}}+(2j+1-2l)\sqrt{\frac{3h_{k,\mathrm{I}}p_{k,\mathrm{II}}}{M_{k,\mathrm{II}}^2-1}}\right)$$
$$-f(l,M_{k,\mathrm{II}})Q\left((M_{k,\mathrm{I}}+1-2m)\sqrt{\frac{3h_{k,\mathrm{I}}p_{k,\mathrm{I}}}{M_{k,\mathrm{I}}^2-1}}+(2l-2j+1)\sqrt{\frac{3h_{k,\mathrm{I}}p_{k,\mathrm{II}}}{M_{k,\mathrm{II}}^2-1}}\right)\Bigg], \tag{3.11}$$

where $Q(x) = \frac{1}{\sqrt{2\pi}}\int_x^{+\infty}\exp\left(-\frac{t^2}{2}\right)dt$ and $f(\cdot,\cdot)$ is defined as

$$f(x,y) \triangleq \begin{cases} 0, & x = y, \\ 1, & x \neq y. \end{cases} \tag{3.12}$$

The SER of the weak channel user, i.e., user $U_{k,\mathrm{II}}$, when adopting $M_{k,\mathrm{II}}$-PAM, is

$$\phi_{k,\mathrm{II}} = \frac{2M_{k,\mathrm{II}}-2}{M_{k,\mathrm{I}}M_{k,\mathrm{II}}} \times \sum_{m=1}^{M_{k,\mathrm{I}}} Q\left((M_{k,\mathrm{I}}+1-2m)\sqrt{\frac{3h_{k,\mathrm{II}}p_{k,\mathrm{I}}}{M_{k,\mathrm{I}}^2-1}}+\sqrt{\frac{3h_{k,\mathrm{II}}p_{k,\mathrm{II}}}{M_{k,\mathrm{II}}^2-1}}\right). \tag{3.13}$$

With the expressions of $\phi_{k,\mathrm{I}}$ and $\phi_{k,\mathrm{II}}$, we can obtain the effective throughput $J_{k,\mathrm{I}}$ and $J_{k,\mathrm{II}}$ in the following result.

Proposition 3.2 The effective throughput of user $U_{k,i}$ ($i = \mathrm{I}, \mathrm{II}$) on the two-user channel is

$$J_{k,i} = B\left(1-\phi_{k,i}\right)^2 \log_2 M_{k,i}^2, \tag{3.14}$$

where $\phi_{k,\mathrm{I}}$ and $\phi_{k,\mathrm{II}}$ are given in (3.11) and (3.13), respectively.

Proof: With (3.7) and (3.8), the expression in (3.14) can be derived. □

Consequently, the effective throughput for two-user channels can be given by $J_k = J_{k,\mathrm{I}} + J_{k,\mathrm{II}}$. According to Wang et al. (2021), the expressions of the error probabilities match the simulated ones perfectly. Thus, our derived effective throughput can accurately measure the error propagation caused by imperfect SIC in NOMA.

According to Propositions 3.1 and 3.2, the effective throughput is a function of the bandwidth, the channel quality, the modulation scheme, and the transmit power. Compared with the ideal information rate, the effective throughput considers practical modulations, and thus, is more practical. Furthermore, in (3.14) and

(3.10), the terms $(1-\phi_{k,i})^2$ and $(1-\phi_{k,0})^2$ reflect the error performance, whereas the terms $B\log_2 M_{k,i}^2$ and $B\log_2 M_{k,0}^2$ reflect the bit rate. Consequently, the effective throughput takes into account both the reliability and transmission rate of the NOMA system.

3.4 NOMA Transmission Design

In this section, we focus on the effective throughput maximization of the multichannel NOMA system by formulating a joint resource allocation problem. Then, we provide an analytical power allocation solution as well as an efficient joint resource allocation algorithm via iterative optimization for the effective throughput maximization NOMA design.

3.4.1 Problem Formulation

From (3.7), the effective throughput is a function of the bandwidth, the modulation scheme, and the SER, which further depends on the channel quality and transmit power. Thus, the effective throughput maximization problem can be formulated as a joint power allocation, channel assignment, and modulation selection problem:

$$\mathcal{P}_1 : \underset{\{p_{k,n},\, d_{k,n},\, M_n\}}{\text{maximize}} \quad J \triangleq \sum_{k=1}^{K}\sum_{n=1}^{N} d_{k,n} J_{k,n} \tag{3.15a}$$

$$\text{subject to} \quad d_{k,n} \in \{0,1\}, \quad k=1,\ldots,K, \quad n=1,\ldots,N \tag{3.15b}$$

$$\sum_{k=1}^{K} d_{k,n} = 1, \quad n=1,\ldots,N \tag{3.15c}$$

$$\sum_{n=1}^{N} d_{k,n} \le 2, \quad k=1,\ldots,K \tag{3.15d}$$

$$\sum_{k=1}^{K}\sum_{n=1}^{N} d_{k,n} p_{k,n} \le P \tag{3.15e}$$

$$p_{k,n} \ge 0, \quad k=1,\ldots,K, \quad n=1,\ldots,N \tag{3.15f}$$

$$M_n \in \mathcal{M}, \quad n=1,\ldots,N, \tag{3.15g}$$

where J is the effective system throughput. Constraints (3.15c) and (3.15d) indicate that each user occupies one channel and each channel can be assigned to at most two users. Constraint (3.15e) is the total power constraint. $\mathcal{M}$ represents the set of all possible modulation orders.

Problem $\mathcal{P}_1$ is an mixed integer program (MIP), which is NP-hard. To find the globally optimal solution, one has to employ the exhaustive search that has a high computational complexity. In practice, an efficient way to solve an MIP is to tackle the continuous variables and the discrete variables alternatively and iteratively. In our case, problem $\mathcal{P}_1$ can be decomposed into two subproblems: (i) optimizing power allocation for fixed channel assignment and modulation schemes; (ii) optimizing channel assignment and modulation schemes for fixed power allocation. Consequently, the joint solution can be obtained by iteratively solving them. Next, we address the power allocation subproblem and then provide the joint resource allocation scheme by solving problem $\mathcal{P}_1$.

3.4.2 Power Allocation

Suppose that the channel assignment and modulation schemes are fixed. For convenience, we denote the index sets of the single-user channels and the two-user channels by $\mathcal{K}^{\dagger}$ and $\mathcal{K}^{\ddagger}$, respectively. Then, the power allocation optimization subproblem can be formulated as

$$\mathcal{P}_2 : \underset{\{p_{k,0},p_{k,\mathrm{I}},p_{k,\mathrm{II}}\geq 0\}}{\text{maximize}} \quad J = \sum_{k\in\mathcal{K}^{\dagger}} J_{k,0} + \sum_{k\in\mathcal{K}^{\ddagger}} \left(J_{k,\mathrm{I}} + J_{k,\mathrm{II}}\right) \tag{3.16a}$$

$$\text{subject to} \quad \sum_{k\in\mathcal{K}^{\dagger}} p_{k,0} + \sum_{k\in\mathcal{K}^{\ddagger}} \left(p_{k,\mathrm{I}} + p_{k,\mathrm{II}}\right) \leq P. \tag{3.16b}$$

Note that problem $\mathcal{P}_2$ is a nonconvex optimization problem. Next, we address the nonconvexity of problem $\mathcal{P}_2$ through decomposition and approximation.

Specifically, we define the effective throughput of channel C_k as

$$J_k \triangleq \begin{cases} J_{k,0}, & k \in \mathcal{K}^{\dagger}, \\ J_{k,\mathrm{I}} + J_{k,\mathrm{II}}, & k \in \mathcal{K}^{\ddagger}, \end{cases} \tag{3.17}$$

and introduce the power budget of channel C_k as

$$q_k \triangleq \begin{cases} p_{k,0}, & k \in \mathcal{K}^{\dagger}, \\ p_{k,\mathrm{I}} + p_{k,\mathrm{II}}, & k \in \mathcal{K}^{\ddagger}, \end{cases} \tag{3.18}$$

which satisfies $q_k \geq 0$ and $\sum_{k=1}^{K} q_k \leq P$. Consequently, problem $\mathcal{P}_2$ can be equivalently decomposed into a power budget allocation subproblem among channels:

$$\mathcal{P}_3 : \underset{\{q_k\geq 0\}}{\text{maximize}} \quad J = \sum_{k=1}^{K} J_k \tag{3.19a}$$

$$\text{subject to} \quad \sum_{k=1}^{K} q_k \leq P, \tag{3.19b}$$

and a series of power allocation subproblems within the two-user channels:

$$\mathcal{P}_{4,k}\,(k \in \mathcal{K}^{\ddagger}) : \underset{\{p_{k,\mathrm{I}},\,p_{k,\mathrm{II}} \geq 0\}}{\text{maximize}} \quad J_k = J_{k,\mathrm{I}} + J_{k,\mathrm{II}} \tag{3.20a}$$

$$\text{subject to} \quad p_{k,\mathrm{I}} + p_{k,\mathrm{II}} = q_k. \tag{3.20b}$$

In the following section, we provide a closed-form solution to problem $\mathcal{P}_{4,k}$, and show that problem $\mathcal{P}_3$ can be transformed into a convex problem and be solved analytically.

3.4.2.1 Power Allocation within Channels

Substituting $p_{k,\mathrm{II}}$ by $q_k - p_{k,\mathrm{I}}$, problem $\mathcal{P}_{4,k}$ can be equivalently reformulated as

$$\mathcal{P}_{5,k}\,(k \in \mathcal{K}^{\ddagger}) : \underset{\{p_{k,\mathrm{I}} \in [0, q_k]\}}{\text{maximize}} \quad J_k = J_{k,\mathrm{I}} + J_{k,\mathrm{II}}. \tag{3.21}$$

From Proposition 3.2, one can find that the expressions of $J_{k,\mathrm{I}}$ and $J_{k,\mathrm{II}}$ are nonlinear. The exhaustive search is required to find the optimal solution to $\mathcal{P}_{5,k}$. Next, we derive the lower bound of J_k as the approximation and obtain the near-optimal solution via maximizing the lower bound of J_k.

Proposition 3.3 The effective throughput J_k is lower bounded by

$$\begin{aligned} J_k \geq J_k^L = {} & 2B\log_2 M_{k,\mathrm{I}} + 2B\log_2 M_{k,\mathrm{II}} \\ & - \frac{8B(M_{k,\mathrm{I}}-1)\log_2 M_{k,\mathrm{I}}}{M_{k,\mathrm{I}}} Q\left(\sqrt{\frac{3h_{k,\mathrm{I}}p_{k,\mathrm{I}}}{M_{k,\mathrm{I}}^2-1}}\right) \\ & - \left(\frac{8B(M_{k,\mathrm{II}}-1)\log_2 M_{k,\mathrm{II}}}{M_{k,\mathrm{II}}} + \frac{12B(M_{k,\mathrm{I}}-1)(M_{k,\mathrm{II}}-1)\log_2 M_{k,\mathrm{I}}}{M_{k,\mathrm{I}}}\right) \\ & \times Q\left((1-M_{k,\mathrm{I}})\sqrt{\frac{3h_{k,\mathrm{II}}p_{k,\mathrm{I}}}{M_{k,\mathrm{I}}^2-1}} + \sqrt{\frac{3h_{k,\mathrm{II}}(q_k-p_{k,\mathrm{I}})}{M_{k,\mathrm{II}}^2-1}}\right). \end{aligned} \tag{3.22}$$

Proof: With the upper bounds of $\phi_{k,\mathrm{I}}$ and $\phi_{k,\mathrm{II}}$ provided in Wang et al. (2021), the lower bound in (3.22) can be derived. □

In Proposition 3.3, the lower bound of J_k is given in a simpler expression compared with the exact expression of J_k. In the following result, we take the lower bound as an approximation of J_k and obtain a closed-form power allocation scheme via maximizing the lower bound.

Theorem 3.1 For a sufficiently large power budget q_k, the power allocation within the two-user channels that maximizes the lower bound of J_k is given by

$$p_{k,\mathrm{I}}^{\star} = \frac{(M_{k,\mathrm{I}}^2-1)q_k}{(M_{k,\mathrm{I}}-1+t_k)^2(M_{k,\mathrm{II}}^2-1)+M_{k,\mathrm{I}}^2-1}, \tag{3.23}$$

$$p_{k,\mathrm{II}}^{\star} = q_k - p_{k,\mathrm{I}}^{\star}, \tag{3.24}$$

where $t_k \triangleq \sqrt{h_{k,\mathrm{I}}/h_{k,\mathrm{II}}} \geq 1$ is defined as the channel gain ratio of channel C_k.

Proof: The derivative of J_k^L with respect to $p_{k,\mathrm{I}}$ exists when $q_k \neq 0$ and $0 < p_{k,\mathrm{I}} < q_k$. Setting $\partial J_k^L / \partial p_{k,\mathrm{I}}$ to be equal to zero, we obtain

$$\begin{aligned}
&\left((1 - M_{k,\mathrm{I}})\sqrt{\frac{3h_{k,\mathrm{II}}\theta_k}{M_{k,\mathrm{I}}^2 - 1}} + \sqrt{\frac{3h_{k,\mathrm{II}}(1-\theta_k)}{M_{k,\mathrm{II}}^2 - 1}}\right)^2 - \frac{3h_{k,\mathrm{I}}\theta_k}{M_{k,\mathrm{I}}^2 - 1} \\
&= \frac{2}{q_k} \ln\left[\left(\frac{3(M_{k,\mathrm{II}} - 1)}{2t_k} + \frac{M_{k,\mathrm{I}}(M_{k,\mathrm{II}} - 1)\log_2 M_{k,\mathrm{II}}}{t_k M_{k,\mathrm{II}}(M_{k,\mathrm{I}} - 1)\log_2 M_{k,\mathrm{I}}}\right)\right. \\
&\quad \left. \times \left(M_{k,\mathrm{I}} - 1 + \sqrt{\frac{M_{k,\mathrm{I}}^2 - 1}{M_{k,\mathrm{II}}^2 - 1}}\sqrt{\frac{\theta_k}{1-\theta_k}}\right)\right].
\end{aligned} \tag{3.25}$$

where $t_k = \sqrt{h_{k,\mathrm{I}}/h_{k,\mathrm{II}}}$ is the channel gain ratio, $\theta_k \triangleq p_{k,\mathrm{I}}/q_k \in (0, 1)$. When q_k is sufficiently large, the right-hand side of (3.25) tends to zero. Thus, (3.25) can be approximated by

$$\left((1 - M_{k,\mathrm{I}})\sqrt{\frac{3h_{k,\mathrm{II}}\theta_k}{M_{k,\mathrm{I}}^2 - 1}} + \sqrt{\frac{3h_{k,\mathrm{II}}(1-\theta_k)}{M_{k,\mathrm{II}}^2 - 1}}\right)^2 - \frac{3h_{k,\mathrm{I}}\theta_k}{M_{k,\mathrm{I}}^2 - 1} = 0. \tag{3.26}$$

To determine the (negative or positive) sign of $(1 - M_{k,\mathrm{I}})\sqrt{\frac{3h_{k,\mathrm{II}}\theta_k}{M_{k,\mathrm{I}}^2-1}} + \sqrt{\frac{3h_{k,\mathrm{II}}(1-\theta_k)}{M_{k,\mathrm{II}}^2-1}}$, we define $\tilde{p}_{k,\mathrm{I}} \triangleq \frac{(M_{k,\mathrm{I}}+1)q_k}{(M_{k,\mathrm{I}}-1)(M_{k,\mathrm{II}}^2-1)+M_{k,\mathrm{I}}+1} \in (0, q_k)$. When $p_{k,\mathrm{I}} \leq \tilde{p}_{k,\mathrm{I}}$, we have $(1 - M_{k,\mathrm{I}})\sqrt{\frac{3h_{k,\mathrm{II}}\theta_k}{M_{k,\mathrm{I}}^2-1}} + \sqrt{\frac{3h_{k,\mathrm{II}}(1-\theta_k)}{M_{k,\mathrm{II}}^2-1}} \geq 0$ and Eq. (3.26) has a root

$$p_{k,\mathrm{I}}^{(1)} = \frac{(M_{k,\mathrm{I}}^2 - 1)q_k}{(M_{k,\mathrm{I}} - 1 + t_k)^2(M_{k,\mathrm{II}}^2 - 1) + M_{k,\mathrm{I}}^2 - 1} \in (0, \tilde{p}_{k,\mathrm{I}}). \tag{3.27}$$

When $p_{k,\mathrm{I}} > \tilde{p}_{k,\mathrm{I}}$, we have $(1 - M_{k,\mathrm{I}})\sqrt{\frac{3h_{k,\mathrm{II}}\theta_k}{M_{k,\mathrm{I}}^2-1}} + \sqrt{\frac{3h_{k,\mathrm{II}}(1-\theta_k)}{M_{k,\mathrm{II}}^2-1}} < 0$ and the number of solutions to (3.26) depends on the relationship between $M_{k,\mathrm{I}} - 1$ and t_k. If $M_{k,\mathrm{I}} - 1 \leq t_k$, then (3.26) has no root; otherwise, (3.26) has a root

$$p_{k,\mathrm{I}}^{(2)} = \frac{(M_{k,\mathrm{I}}^2 - 1)q_k}{(M_{k,\mathrm{I}} - 1 - t_k)^2(M_{k,\mathrm{II}}^2 - 1) + M_{k,\mathrm{I}}^2 - 1} \in (\tilde{p}_{k,\mathrm{I}}, q_k). \tag{3.28}$$

Therefore, if $M_{k,\mathrm{I}} - 1 \leq t_k$, then J_k^L is monotonically increasing when $0 < p_{k,\mathrm{I}} \leq p_{k,\mathrm{I}}^{(1)}$ and is decreasing when $p_{k,\mathrm{I}}^{(1)} < p_{k,\mathrm{I}} < q_k$, with the peak point $p_{k,\mathrm{I}}^{(1)}$. If $M_{k,\mathrm{I}} - 1 > t_k$, then J_k^L is monotonically increasing with respect to $p_{k,\mathrm{I}}$ when $0 < p_{k,\mathrm{I}} \leq p_{k,\mathrm{I}}^{(1)}$, is decreasing when $p_{k,\mathrm{I}}^{(1)} < p_{k,\mathrm{I}} \leq p_{k,\mathrm{I}}^{(2)}$, and is increasing again when $p_{k,\mathrm{I}}^{(2)} < p_{k,\mathrm{I}} < q_k$.

To pinpoint the optimum point, we next compare $J_k^L|_{p_{k,\mathrm{I}}=q_k}$ with $J_k^L|_{p_{k,\mathrm{I}}=p_{k,\mathrm{I}}^{(1)}}$, which are shown in (3.29) and (3.30), respectively.

$$\begin{aligned} J_k^L|_{p_{k,\mathrm{I}}=q_k} &= 2B\log_2 M_{k,\mathrm{I}} + 2B\log_2 M_{k,\mathrm{II}} \\ &\quad - \frac{8B(M_{k,\mathrm{I}}-1)\log_2 M_{k,\mathrm{I}}}{M_{k,\mathrm{I}}} Q\left(\sqrt{\frac{3h_{k,\mathrm{I}}q_k}{M_{k,\mathrm{I}}^2-1}}\right) - Q\left((1-M_{k,\mathrm{I}})\sqrt{\frac{3h_{k,\mathrm{II}}q_k}{M_{k,\mathrm{I}}^2-1}}\right) \\ &\quad \times \left(\frac{8B(M_{k,\mathrm{II}}-1)\log_2 M_{k,\mathrm{II}}}{M_{k,\mathrm{II}}} + \frac{12B(M_{k,\mathrm{I}}-1)(M_{k,\mathrm{II}}-1)\log_2 M_{k,\mathrm{I}}}{M_{k,\mathrm{I}}}\right) \\ &< 2B\log_2 M_{k,\mathrm{I}} + 2B\log_2 M_{k,\mathrm{II}} - \frac{4B(M_{k,\mathrm{II}}-1)\log_2 M_{k,\mathrm{II}}}{M_{k,\mathrm{II}}} \\ &\quad + \frac{6B(M_{k,\mathrm{I}}-1)(M_{k,\mathrm{II}}-1)\log_2 M_{k,\mathrm{I}}}{M_{k,\mathrm{I}}}. \end{aligned} \tag{3.29}$$

$$\begin{aligned} J_k^L|_{p_{k,\mathrm{I}}=p_{k,\mathrm{I}}^{(1)}} &= 2B\log_2 M_{k,\mathrm{I}} + 2B\log_2 M_{k,\mathrm{II}} \\ &\quad - Q\left(\sqrt{\frac{3h_{k,\mathrm{I}}q_k}{(M_{k,\mathrm{I}}-1+t_k)^2(M_{k,\mathrm{II}}^2-1)+M_{k,\mathrm{I}}^2-1}}\right) \\ &\quad \times \left(\frac{4B(M_{k,\mathrm{I}}-1)(3M_{k,\mathrm{II}}-1)\log_2 M_{k,\mathrm{I}}}{M_{k,\mathrm{I}}} + \frac{8B(M_{k,\mathrm{II}}-1)\log_2 M_{k,\mathrm{II}}}{M_{k,\mathrm{II}}}\right). \end{aligned} \tag{3.30}$$

We can find that $J_k^L|_{p_{k,\mathrm{I}}=p_{k,\mathrm{I}}^{(1)}}$ approaches $2B\log_2 M_{k,\mathrm{I}} + 2B\log_2 M_{k,\mathrm{II}}$ as q_k increases. Hence, for a sufficiently large power budget, we have $J_k^L|_{p_{k,\mathrm{I}}=p_{k,\mathrm{I}}^{(1)}} > J_k^L|_{p_{k,\mathrm{I}}=q_k}$, and thus, $p_{k,\mathrm{I}}^{(1)}$ is the maximum point of the lower bound of J_k. Denoting $p_{k,\mathrm{I}}^{(1)}$ by $p_{k,\mathrm{I}}^{\star}$, Theorem 3.1 is proven. □

In Theorem 3.1, we provide a closed-form power allocation scheme within the two-user channels, which can be determined by the modulation orders and the channel gain ratio. One can observe that more power should be allocated to a user when its modulation order increases or when its channel gain decreases. Numerical results demonstrate that the proposed power allocation scheme can well approximate the optimal one obtained via an exhaustive search.

3.4.2.2 Power Budget Allocation Among Channels

With the closed-form power allocation scheme within channels, the power budget allocation problem among channels, i.e., problem $\mathcal{P}_3$, can be simplified into

$$\mathcal{P}_6 : \underset{\{q_k \geq 0\}}{\text{maximize}} \quad \tilde{J} \triangleq \sum_{k=1}^{K} \tilde{J}_k \tag{3.31a}$$

$$\text{subject to} \quad \sum_{k=1}^{K} q_k \leq P, \tag{3.31b}$$

where, for the two-user channels ($k \in \mathcal{K}^{\ddagger}$), $\tilde{J}_k \triangleq J_k^L|_{p_{k,\mathrm{I}}=p^{\star}_{k,\mathrm{I}}, p_{k,\mathrm{II}}=p^{\star}_{k,\mathrm{II}}}$ is the lower bound of the effective throughput J_k. According to (3.22) and (3.23), we have

$$\begin{aligned}\tilde{J}_k = {} & 2B\log_2 M_{k,\mathrm{I}} + 2B\log_2 M_{k,\mathrm{II}} \\ & - Q\left(\sqrt{\frac{3h_{k,\mathrm{I}} q_k}{(M_{k,\mathrm{I}} - 1 + t_k)^2 (M_{k,\mathrm{II}}^2 - 1) + M_{k,\mathrm{I}}^2 - 1}}\right) \\ & \times \left(\frac{4B(M_{k,\mathrm{I}} - 1)(3M_{k,\mathrm{II}} - 1)\log_2 M_{k,\mathrm{I}}}{M_{k,\mathrm{I}}}\right. \\ & \left. + \frac{8B(M_{k,\mathrm{II}} - 1)\log_2 M_{k,\mathrm{II}}}{M_{k,\mathrm{II}}}\right), \quad k \in \mathcal{K}^{\ddagger}.\end{aligned} \tag{3.32}$$

For the single-user channels ($k \in \mathcal{K}^{\dagger}$), the effective throughput $J_{k,0}$ can be obtained via (3.10), which, however, is still complicated. According to Goldsmith (2005), $J_{k,0}$ has the lower bound

$$J_{k,0} = B(1 - \phi_{k,0})^2 \log_2 M_{k,0}^2 \geq B\left(1 - 2\phi_{k,0}\right) \log_2 M_{k,0}^2, \tag{3.33}$$

which is a tight approximation for a sufficiently large power budget. Thus, we use the lower bound of $J_{k,0}$ to approximate $\tilde{J}_k$ as

$$\begin{aligned}\tilde{J}_k & \triangleq B\left(1 - 2\phi_{k,0}\right) \log_2 M_{k,0}^2 \\ & = 2B\log_2 M_{k,0} - \frac{8B(M_{k,0} - 1)\log_2 M_{k,0}}{M_{k,0}} \times Q\left(\sqrt{\frac{3h_{k,0} p_{k,0}}{M_{k,0}^2 - 1}}\right), \quad k \in \mathcal{K}^{\dagger}.\end{aligned} \tag{3.34}$$

Combining (3.32) and (3.34), $\tilde{J}_k$ can be given by

$$\tilde{J}_k = \begin{cases} 2B\log_2 M_{k,0} - \alpha_k Q(\sqrt{\beta_k q_k}), & k \in \mathcal{K}^{\dagger}, \\ 2B\log_2 M_{k,\mathrm{I}} + 2B\log_2 M_{k,\mathrm{II}} & \\ \quad -\alpha_k Q(\sqrt{\beta_k q_k}), & k \in \mathcal{K}^{\ddagger}, \end{cases} \tag{3.35}$$

where

$$\alpha_k \triangleq \begin{cases} \frac{8B(M_{k,0}-1)\log_2 M_{k,0}}{M_{k,0}}, & k \in \mathcal{K}^{\dagger}, \\ \frac{4B(M_{k,\mathrm{I}}-1)(3M_{k,\mathrm{II}}-1)\log_2 M_{k,\mathrm{I}}}{M_{k,\mathrm{I}}} & \\ \quad + \frac{8B(M_{k,\mathrm{II}}-1)\log_2 M_{k,\mathrm{II}}}{M_{k,\mathrm{II}}}, & k \in \mathcal{K}^{\ddagger}, \end{cases} \tag{3.36}$$

$$\beta_k \triangleq \begin{cases} \frac{3h_{k,0}}{M_{k,0}^2-1}, & k \in \mathcal{K}^{\dagger}, \\ \frac{3h_{k,\mathrm{I}}}{(M_{k,\mathrm{I}}-1+t_k)^2(M_{k,\mathrm{II}}^2-1)+M_{k,\mathrm{I}}^2-1}, & k \in \mathcal{K}^{\ddagger}. \end{cases} \tag{3.37}$$

To solve problem $\mathcal{P}_6$, we first provide the following useful results.

Lemma 3.1 $\tilde{J}_k$ is monotonically increasing with respect to q_k for $k = 1, \ldots, K$.

Lemma 3.2 $\tilde{J}_k$ is a concave function of q_k for $k = 1, \ldots, K$.

Proof: Lemmas 3.1 and 3.2 can be proven via deriving the first-order and second-order partial derivatives of $\tilde{J}_k$ with respect to q_k. □

Lemma 3.1 complies with the intuition that the larger the power budget is, the higher the effective throughput will be. Lemma 3.2 indicates that problem $\mathcal{P}_6$ is a convex optimization problem, whose solution can be found efficiently via standard convex optimization tools, e.g., CVX. Nevertheless, we are able to provide an analytical solution to problem $\mathcal{P}_6$.

Theorem 3.2 The optimal solution to problem $\mathcal{P}_6$ is given by

$$q_k^{\star} = \frac{W_0\left(\frac{\alpha_k^2\beta_k^2}{8\pi\lambda}\right)}{\beta_k}, \tag{3.38}$$

where λ is chosen such that $\sum_{k=1}^{K} q_k^{\star} = P$ and $W_0(\cdot)$ is the principal branch of Lambert W function.

Proof: The optimization problem $\mathcal{P}_6$ is equivalent to the following problem:

$$\mathcal{P}_9 : \underset{\{q_k \geq 0\}}{\text{minimize}} \quad \sum_{k=1}^{K} \alpha_k Q\left(\sqrt{\beta_k q_k}\right) \tag{3.39a}$$

$$\text{subject to} \quad \sum_{k=1}^{K} q_k \leq P. \tag{3.39b}$$

The Lagrangian is given by

$$L(q_1,\dots,q_K,\mu)=\sum_{k=1}^{K}\alpha_k Q\left(\sqrt{\beta_k q_k}\right)+\mu\left(\sum_{k=1}^{K}q_k-P\right),\tag{3.40}$$

where μ is the Lagrange multiplier. Since $\mathcal{P}_9$ is a convex optimization problem, its optimal solution is characterized by satisfying Karush–Kuhn–Tucker (KKT) conditions:

$$\frac{\partial L}{\partial q_k}=\mu-\frac{\alpha_k\sqrt{\beta_k}}{2\sqrt{2\pi q_k}}\exp\left(-\frac{\beta_k q_k}{2}\right)=0,\tag{3.41}$$

$$\mu\left(\sum_{k=1}^{K}q_k-P\right)=0,\tag{3.42}$$

$$\sum_{k=1}^{K}q_k-P\le 0,\tag{3.43}$$

$$\mu\ge 0.\tag{3.44}$$

From (3.41), it is obvious that $\mu\neq 0$. Hence, due to (3.42) and (3.44), $\mu>0$ and $\sum_{k=1}^{K}q_k-P=0$ hold. To obtain the root of (3.41), we have

$$\begin{aligned}
(3.41)\Longleftrightarrow&\ \frac{2\mu\sqrt{2\pi q_k}}{\alpha_k\sqrt{\beta_k}}\exp\left(\frac{\beta_k q_k}{2}\right)=1\\
\Longleftrightarrow&\ \frac{8\pi\mu^2 q_k}{\alpha_k^2\beta_k}\exp\left(\beta_k q_k\right)=1\Longleftrightarrow\frac{\alpha_k^2\beta_k^2}{8\pi\mu^2}=\beta_k q_k\exp\left(\beta_k q_k\right)\\
\Longleftrightarrow&\ W_0\left(\frac{\alpha_k^2\beta_k^2}{8\pi\mu^2}\right)=\beta_k q_k\Longleftrightarrow q_k^\star=\frac{W_0\left(\frac{\alpha_k^2\beta_k^2}{8\pi\mu^2}\right)}{\beta_k}>0,
\end{aligned}\tag{3.45}$$

where condition 1 $\Longleftrightarrow$ condition 2 means that conditions 1 and 2 are sufficient and necessary conditions for each other. For convenience, define $\lambda\triangleq\mu^2$, and thus, Theorem 3.2 is proven. λ is chosen such that $\sum_{k=1}^{K}q_k^\star=P$, i.e.,

$$\sum_{k=1}^{K}\frac{W_0\left(\frac{\alpha_k^2\beta_k^2}{8\pi\lambda}\right)}{\beta_k}=P.\tag{3.46}$$

The left-hand side is a monotonically decreasing function of λ and has the range $(0,+\infty)$. Thus, Eq. (3.46) has a unique solution. □

Theorem 3.2 provides an analytical power budget allocation scheme among channels in waterfilling form. Since $q_k^\star$ in (3.38) is monotonically decreasing with respect to λ. Thus λ can be efficiently found via a one-dimensional search.

In Theorems 3.1 and 3.2, we provide the analytical power allocation expressions within and among channels, respectively, which are both determined by the modulation orders and the channel gains. The former one is given in closed form and the latter one is given in a waterfilling fashion. By combining Theorems 3.1 and 3.2, an efficient power allocation scheme can be analytically obtained for downlink multichannel NOMA systems.

3.4.3 Joint Resource Allocation

As discussed in Section 3.4.1, the joint resource allocation problem can be alternatively solved by the problem for power allocation and the problem for channel assignment and modulation selection. In Section 3.4.2, we proposed the power allocation scheme for fixed channel assignment and modulation schemes. Now, we consider the channel assignment and modulation selection for effective throughput maximization with given power allocation. In this case, problem $\mathcal{P}_1$ reduces to

$$\mathcal{P}_7 : \underset{\{d_{k,n}, M_n\}}{\text{maximize}} \quad J = \sum_{k=1}^{K} \sum_{n=1}^{N} d_{k,n} J_{k,n} \tag{3.47a}$$

$$\text{subject to} \quad \text{constraints (3.15b)–(3.15d), (3.15g).} \tag{3.47b}$$

It is a combinatorial problem whose optimal solution has to be found via exhaustive search. Nevertheless, we are able to develop an efficient discrete optimization methods based on matching theory and machine learning to determine the channel assignment and modulation selection, respectively, in Wang et al. (2022).

Naturally, a joint resource allocation scheme, as the solution to the original problem $\mathcal{P}_1$, can be achieved by iteratively optimizing power allocation, channel

---

Algorithm 3.1 Joint Resource Allocation Algorithm

1. **Initialization**:
   (a) Initialize the power allocation;
   (b) Set precision δ and the parameter N_{converge}.
2. **Repeat**
3. Obtain the channel assignment and the modulation scheme according to Wang et al. [2022].
4. Update the power allocation according to (3.23), (3.24) and (3.38);
5. **Until** the change of the effective throughput is less thanδ for consecutive N_{converge} iterations.
6. **Return** the optimal power allocation, channel assignment,and modulation scheme, and the corresponding effective throughput.

---

assignment, and modulation selection. Specifically, we first initialize the power allocation, e.g., equally allocate power to each channel. Then, the channel assignment, modulation schemes, and power allocation are iteratively optimized until convergence. The details are described in Algorithm 3.1.

3.5 Numerical Results

In this section, numerical results are presented to evaluate the aforementioned algorithms. We employ the channel model with the fading coefficient being a zero-mean unit-variance complex Gaussian variable and the path loss being $-(128.1 + 37.6 \log_{10} \tau_n)$ dB (3GPP TR 36.814, 2010), where τ_n is the distance between the BS and user U_n in kilometers. The noise power spectral density is assumed to be -174 dBm/Hz for all users.

In Fig. 3.4, we compare the proposed power allocation scheme in Theorem 3.1 with the optimal one obtained via exhaustive search by illustrating $p^{\star}_{k,\mathrm{I}}/p^{\star}_{k,\mathrm{II}}$ versus the power budget q_k. The distances between the BS and the users are 0.2 km for $U_{k,\mathrm{I}}$ and 0.3 km for $U_{k,\mathrm{II}}$. The results show that the proposed scheme well approximates the optimal one even if the power budget is not large.

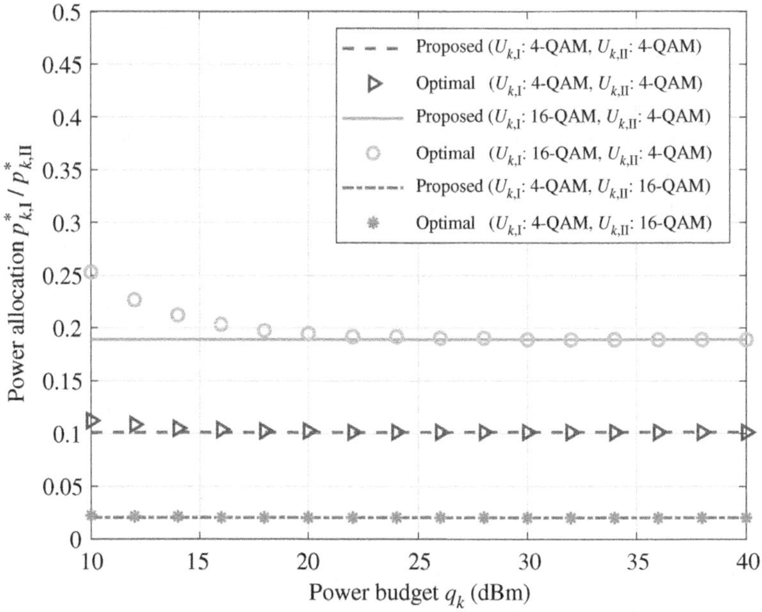

Figure 3.4 Comparison of the proposed and optimal power allocation schemes within channels.

Next, we evaluate the performance of the proposed joint resource allocation algorithm of NOMA. The total bandwidth is set as K (MHz). N users are located uniformly in a cell with the radius of 1 km and the BS is located in the center. Two users within a channel perform the NOMA scheme, while the users on different channels are assumed to be interference-free. The following effective throughput is obtained by averaging 100 randomly generated channel profiles.

For convenience, the proposed maximum effective throughput NOMA strategy is denoted by NOMA-MET. For comparison, six benchmarks are considered as follows:

(1) *NOMA-ES*: The exhaustive search (ES) scheme of NOMA.
(2) *NOMA-AM*: The modulation orders of all the users, including the users on the two-user channels, are determined separately via the adaptive modulation (AM) selection method in Wang et al. (2022).
(3) *NOMA-MSET*: The maximum short-packet effective throughput (MSET) NOMA scheme (Sun et al., 2018).
(4) *NOMA-UP*: The NOMA scheme based on the user pairing (UP) method where a near user and a far user are selected to share a channel (Qi et al., 2019).
(5) *NOMA-MSR*: The maximum sum rate (MSR) NOMA scheme where the residual interference caused by imperfect SIC is modeled as a continuously-valued Gaussian signal (Sun et al., 2017).
(6) *OMA*: The conventional OMA scheme where K (MHz) bandwidth is orthogonally allocated to N users without superposition (Sun et al., 2019).

In Fig. 3.5, we compare the effective throughput obtained via different algorithms versus the total power budget. One can find that the proposed NOMA-MET scheme outperforms the NOMA-AM, NOMA-MSET, NOMA-UP, NOMA-MSR, and OMA schemes, achieving almost the same performance as NOMA-ES. The NOMA-MSR scheme obtains a relatively lower effective throughput than other NOMA schemes when the total power budget is small.

Furthermore, Fig. 3.5 shows that as the total power budget increases, the effective throughput has an increasing trend due to lower error probabilities and higher modulation orders. Eventually, it reaches an upper bound. The upper bound of the effective throughput corresponds to the situation where all the users are error-free and the highest 256-QAM modulation is employed, which for NOMA is $\sum_{n=1}^{N} \log_2 256 = 8N = 40$ Mbps and for OMA is $\frac{K}{N} \sum_{n=1}^{N} \log_2 256 = 8K = 24$ Mbps, as shown in Fig. 3.5.

Figure 3.6 illustrates the effective throughput versus the number of users when the number of channels K is equal to 15 and the total power budget is 40 dBm (3GPP Report ITU-R M.2412-0, 2017). One can observe that the effective throughput of both NOMA and OMA schemes increases monotonically with respect to the

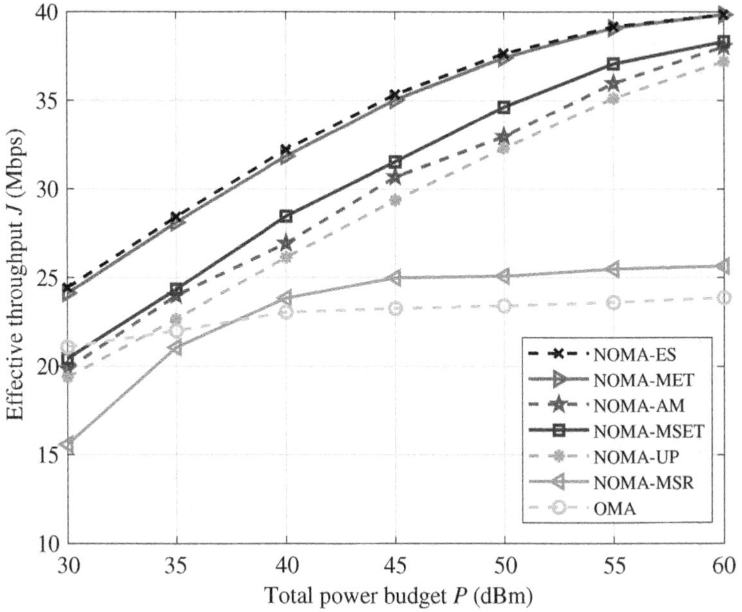

Figure 3.5 Comparison of the effective throughput versus the total power budget. We set $K = 3$ and $N = 5$.

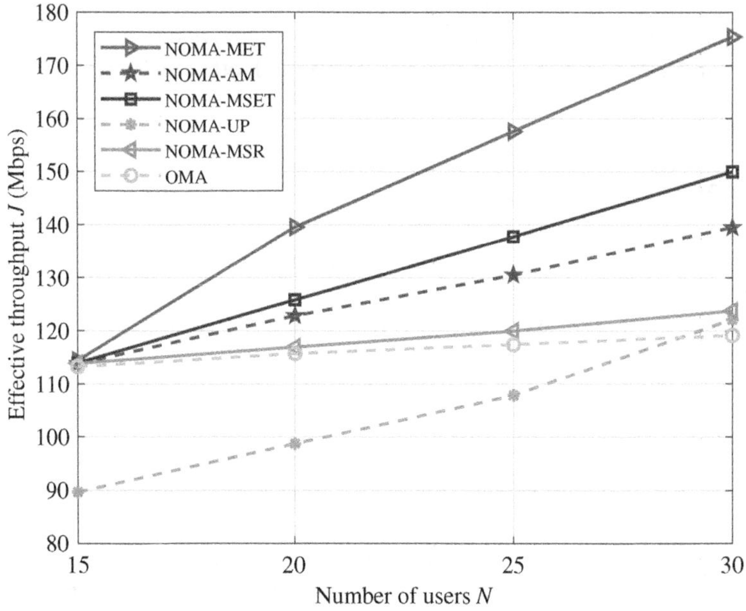

Figure 3.6 Comparison of the effective throughput obtained via different algorithms versus the number of users. We set $K = 15$ and $P = 40$ dBm.

number of users. When the number of users is equal to the number of channels, i.e., $N = K = 15$, the NOMA system degrades to an OMA system and NOMA-MET achieves similar performance to the NOMA-AM, NOMA-MSET, NOMA-MSR, and OMA schemes. As N increases, the superiority of the NOMA-MET scheme over OMA increases due to its overloading. The superiority of the NOMA-MET scheme over the NOMA-AM, NOMA-MSET, and NOMA-MSR schemes increases due to its near-optimal power allocation scheme and deep neural network (DNN)-based modulation selection scheme. The NOMA-UP scheme achieves low effective throughput due to its poor user pairing method.

3.6 Conclusion

In this chapter, we investigated the effective throughput maximization of downlink multichannel NOMA systems, considering practical QAM schemes with imperfect SIC. The exact and approximate expressions of the effective throughput were derived as functions of the transmit power, channel gain, and modulation order. We formulated a joint resource allocation problem to maximize the effective throughput. To address this problem, we developed an analytical power allocation scheme, including the closed-form power allocation within channels and the waterfilling-form power budget allocation among channels. Consequently, a joint resource allocation algorithm was proposed for the effective throughput maximization NOMA design. Numerical results were provided showing that the proposed NOMA scheme outperforms OMA and other existing NOMA schemes.

References

3GPP Report ITU-R M.2412-0. Guidelines for evaluation of radio interface technologies for IMT-2020. Oct. 2017.

3GPP TR 36.814. Further advancements for E-UTRA physical layer aspects (Release 9). Mar. 2010.

3GPP TR 38.812. Study on non-orthogonal multiple access (NOMA) for NR (Release 16). Dec. 2018.

Kyongkuk Cho and Dongweon Yoon. On the general BER expression of one- and two-dimensional amplitude modulations. *IEEE Transactions on Communications*, 50(7):1074–1080, Aug. 2002.

Zhiguo Ding, Fumiyuki Adachi, and H Vincent Poor. The application of MIMO to non-orthogonal multiple access. *IEEE Transactions on Wireless Communications*, 15(1):537–552, Jan. 2016.

Zhiguo Ding, Yuanwei Liu, Jinho Choi, Qi Sun, Maged Elkashlan, I Chih-Lin, and H Vincent Poor. Application of non-orthogonal multiple access in LTE and 5G networks. *IEEE Communications Magazine*, 55(2):185–191, Feb. 2017.

Fang Fang, Haijun Zhang, Julian Cheng, and Victor C M Leung. Energy-efficient resource allocation for downlink non-orthogonal multiple access network. *IEEE Transactions on Communications*, 64(9):3722–3732, Sept. 2016.

Andrea Goldsmith. *Wireless Communications*. Cambridge University Press, 2005.

Chaofan He, Yang Hu, Yan Chen, and Bing Zeng. Joint power allocation and channel assignment for NOMA with deep reinforcement learning. *IEEE Journal on Selected Areas in Communications*, 37(10):2200–2210, Oct. 2019.

Hongji Huang, Yuchun Yang, Zhiguo Ding, Hong Wang, Hikmet Sari, and Fumiyuki Adachi. Deep learning-based sum data rate and energy efficiency optimization for MIMO-NOMA systems. *IEEE Transactions on Wireless Communications*, 19(8):5373–5388, Aug. 2020.

Sami Khairy, Prasanna Balaprakashy, Lin X Cai, and Yu Cheng. Constrained deep reinforcement learning for energy sustainable multi-UAV based random access IoT networks with NOMA. *IEEE Journal on Selected Areas in Communications*, 39(4):1101–1115, Apr. 2021.

Peng Li, Rodrigo C de Lamare, and Rui Fa. Multiple feedback successive interference cancellation detection for multiuser MIMO systems. *IEEE Transactions on Wireless Communications*, 10(8):2434–2439, 2011. doi: 10.1109/TWC.2011.060811.101962.

Yuanwei Liu, Zhijin Qin, Maged Elkashlan, Zhiguo Ding, Arumugam Nallanathan, and Lajos Hanzo. Nonorthogonal multiple access for 5G and beyond. *Proceedings of the IEEE*, 105(12):2347–2381, Dec. 2017.

Yuanwei Liu, Shuowen Zhang, Xidong Mu, Zhiguo Ding, Robert Schober, Naofal Al-Dhahir, Ekram Hossain, and Xuemin Shen. Evolution of NOMA toward next generation multiple access (NGMA) for 6G. *IEEE Journal on Selected Areas in Communications*, 40(4):1037–1071, Apr. 2022.

R. Narasimhan. Error propagation analysis of V-BLAST with channel-estimation errors. *IEEE Transactions on Communications*, 53(1):27–31, 2005. doi: 10.1109/TCOMM.2004.840670.

Yury Polyanskiy, H Vincent Poor, and Sergio Verdu. Channel coding rate in the finite blocklength regime. *IEEE Transactions on Information Theory*, 56(5):2307–2359, May 2010.

J G Proakis and Masoud Salehi. *Digital Communications*. 5th ed. McGraw-Hill Education, 2007.

Lin Qi, Mugen Peng, Yaqiong Liu, and Shi Yan. Advanced user association in non-orthogonal multiple access based fog radio access networks. *IEEE Transactions on Communications*, 67(12):8408–8421, Dec. 2019.

Yan Sun, Derrick Wing Kwan Ng, Zhiguo Ding, and Robert Schober. Optimal joint power and subcarrier allocation for full-duplex multicarrier non-orthogonal multiple access systems. *IEEE Transactions on Communications*, 65(3):1077–1091, Mar. 2017.

Xiaofang Sun, Shihao Yan, Nan Yang, Zhiguo Ding, Chao Shen, and Zhangdui Zhong. Short-packet downlink transmission with non-orthogonal multiple access. *IEEE Transactions on Wireless Communications*, 17(7):4550–4564, Jul. 2018.

Chengjian Sun, Changyang She, Chenyang Yang, Tony Q S Quek, Yonghui Li, and Branka Vucetic. Optimizing resource allocation in the short blocklength regime for ultra-reliable and low-latency communications. *IEEE Transactions on Wireless Communications*, 18(1):402–415, Jan. 2019.

Jiaheng Wang, Qian Peng, Yongming Huang, Huiming Wang, and Xiaohu You. Convexity of weighted sum rate maximization in NOMA systems. *IEEE Signal Processing Letters*, 24(9):1323–1327, Sept. 2017.

Shaoyang Wang, Tiejun Lv, and Xuewei Zhang. Multi-agent reinforcement learning-based user pairing in multi-carrier NOMA systems. In *Proceedings of the IEEE ICC Workshops*, pages 1–6, Shanghai, China, May 2019.

Kunlun Wang, Yong Zhou, Zening Liu, Ziyu Shao, Xiliang Luo, and Yang Yang. Online task scheduling and resource allocation for intelligent NOMA-based industrial Internet of Things. *IEEE Journal on Selected Areas in Communications*, 38(5):803–815, May 2020.

Yuan Wang, Jiaheng Wang, Derrick Wing Kwan Ng, Robert Schober, and Xiqi Gao. A minimum error probability NOMA design. *IEEE Transactions on Wireless Communications*, 20(7):4221–4237, Jul. 2021.

Yuan Wang, Jiaheng Wang, Vincent W S Wong, and Xiaohu You. Effective throughput maximization of NOMA with practical modulations. *IEEE Journal on Selected Areas in Communications*, 40(4):1084–1100, 2022. doi: 10.1109/JSAC.2022.3143244.

Zhiqiang Wei, Derrick Wing Kwan Ng, Jinhong Yuan, and Hui Ming Wang. Optimal resource allocation for power-efficient MC-NOMA with imperfect channel state information. *IEEE Transactions on Communications*, 65(9):3944–3961, Sept. 2017.

Vincent W S Wong, Robert Schober, Derrick Wing Kwan Ng, and Li-Chun Wang. *Key Technologies for 5G Wireless Systems*. Cambridge University Press, 2017.

Jianyue Zhu, Jiaheng Wang, Yongming Huang, Shiwen He, Xiaohu You, and Luxi Yang. On optimal power allocation for downlink non-orthogonal multiple access systems. *IEEE Journal on Selected Areas in Communications*, 35(12):2744–2757, Dec. 2017.

4

Optimal Resource Allocation for NGMA

Sepehr Rezvani[1,2,3] *and Eduard Jorswieck*[1]

[1] *Institute for Communications Technology, Technical University of Braunschweig, Braunschweig, Germany*
[2] *Department of Telecommunication Systems, Technical University of Berlin, Berlin, Germany*
[3] *Department of Wireless Communications and Networks, Fraunhofer Institute for Telecommunications Heinrich-Hertz-Institute, Berlin, Germany*

4.1 Introduction

The previous generations of cellular networks have adopted the orthogonal multiple access (OMA)-based schemes to completely eliminate co-channel interference by having orthogonal signals for different users (Tse and Viswanath, 2005). In the OMA techniques, the information of multiple users can be retrieved via the low-complexity single-user detection algorithms. Nevertheless, the number of supported users is limited by the number of available orthogonal resource blocks, which is small in practice (Ding et al., 2017). Consequently, it is difficult for the OMA techniques to support massive connectivity[1] with heterogeneous data rate demands, which are major requirements of the next-generation wireless networks.

To tackle the above challenges, non-orthogonal multiple access (NOMA) has been proposed, which is based on the linear superposition coding (SC) of multiple user signals at the transmitter side combined with multiuser detection algorithms, such as successive interference cancellation (SIC) at the receivers side (Saito et al., 2013). The main purpose of NOMA is to reap the benefits promised by information theory for the downlink and uplink transmissions, modeled by the broadcast channel (BC) and multiple access channel (MAC), respectively. The capacity region of degraded BCs and MACs has been established several decades ago, and it is proved that SC-SIC is capacity-achieving, thus the optimal transmission strategy in both cases (Cover, 1972, Bergmans, 1973, Gallager, 1974).

1 The massive connectivity is important for ensuring that the next-generation multiple access (NGMA) techniques can effectively support internet of things (IoT) functionalities (Du et al., 2021).

Next Generation Multiple Access, First Edition.
Edited by Yuanwei Liu, Liang Liu, Zhiguo Ding, and Xuemin Shen.
© 2024 The Institute of Electrical and Electronics Engineers, Inc. Published 2024 by John Wiley & Sons, Inc.

The SIC has been widely studied and implemented in practical scenarios, e.g., code-division multiple access (CDMA) (Patel and Holtzman, 1994), vertical-bell laboratories layered space-time (V-BLAST) (Wolniansky et al., 1998), and multiuser multiple-input multiple-output (MIMO) networks. The concept of NOMA has been considered in the third-generation partnership project (3GPP) long-term evolution advanced (LTE-A) standard, where NOMA is referred to as multiuser superposition transmission (MUST) (3rd, 2015). In the research community, NOMA has two major types: code-domain NOMA and power-domain NOMA (Vaezi et al., 2019). In this chapter, we focus on the power-domain NOMA technique, where the SC is performed on the power domain.

The SIC complexity is shown to be cubic in the number of multiplexed (NOMA) users. Moreover, performing SIC among a large number of users increases error propagation (Dai et al., 2015). In this way, single-carrier NOMA (SC-NOMA) with SC-SIC among all the users is still impractical. To overcome the above issues, the multicarrier technology is introduced on NOMA, called multicarrier NOMA (MC-NOMA), where the users are grouped into multiple NOMA clusters, each operating on an isolated resource block. MC-NOMA with disjoint cluster sets is based on the combination of frequency-division multiple access (FDMA)/time-division multiple access (TDMA) and SC-SIC, namely FDMA/TDMA-NOMA. Orthogonal frequency-division multiple access (OFDMA) can also be introduced on NOMA, namely OFDMA–NOMA or Hybrid-NOMA, where each user can belong to multiple NOMA clusters, thus occupying multiple resource blocks. Similar to OFDMA, OFDMA–NOMA benefits from the multiplexing gain in the fading channels. Besides, downlink transmission in cellular systems, which can be modeled as interfering BCs, has been carried out through single-cell processing, where each user receives data from only one cell, and the signals from neighboring (interfering) cells are fully treated as noise. In such systems, users associated with the same cell can form a NOMA cluster or multiple NOMA clusters when the multicarrier/TDMA technology is adopted.

Generally speaking, resource allocation optimization is a fundamental task of multiuser wireless networks. The signal-to-interference-plus-noise ratio (SINR) of users with lower decoding order in a NOMA cluster is highly affected by its co-channel interference, which should be carefully managed via an efficient design of intra-cluster power allocation. For MC-NOMA, an additional inter-cluster power allocation is also required to efficiently determine the NOMA clusters' power budget. Subchannel allocation, i.e., determining the set of NOMA clusters (also called user grouping) as well as assigning subchannels to the clusters, is another fundamental task. There are a number of works addressing optimal power allocation for the downlink single-cell single-/multi-carrier single-input single-output (SISO) NOMA systems with various objectives, such as maximizing users' sum-rate (Ali et al., 2016, 2018, Zhu et al., 2017, Rezvani et al., 2022a), maximizing energy efficiency (EE) (Zhu et al., 2017, Fang et al., 2017, Rezvani et al., 2022a), and max–min rate fairness (Zhu et al., 2017). In multi-cell NOMA with single-cell processing, inter-cell interference (ICI) might affect the decoding order among NOMA users in each cell. The ICI of users in each cell is affected by the total power consumption of each neighboring (interfering) base station (BS). Therefore, in addition to the intra-cell/NOMA cluster(s) power allocation, a joint optimization of power control and SIC decoding order is required to effectively manage ICI and users' achievable rate addressed in Fu et al. (2017), Yang et al. (2018), Sun et al. (2017), Zhao et al. (2017), Wang et al. (2019), and Rezvani et al. (2022b).

In this chapter, we study power allocation for downlink NOMA systems. At first, we formulate the closed-form expression of optimal power allocation for the generic single-cell SC-NOMA systems with different performance metrics. The results are then extended to the single-cell MC-NOMA systems. We also investigate the successful SIC conditions, joint power allocation, and SIC decoding order for the multi-cell NOMA systems with single-cell processing.

This chapter is organized as follows: Section 4.2 introduces the fundamentals of the generic K-user downlink NOMA system with optimal power allocation algorithms. Section 4.3 describes the generic MC-NOMA system with optimal power allocation algorithms. The multi-cell NOMA systems with single-cell processing are studied in Section 4.4. The numerical results are presented in Section 4.5. The concluding remarks are provided in Section 4.6.

4.2 Single-Cell Single-Carrier NOMA

Consider a SISO Gaussian BC, where a BS serves K users with the set $\mathcal{K} = \{1, \dots, K\}$ in a unit time slot of a quasi-static flat fading channel. We assume that the perfect channel state information (CSI) is available at the BS as well as users. After SC, the transmitted signal by the BS is formulated by $x = \sum_{k\in\mathcal{K}} \sqrt{p_k} s_k$, where $s_k \sim \mathcal{CN}(0, 1)$ and $p_k \geq 0$ are the modulated symbol from Gaussian codebooks, and transmit power of user $k \in \mathcal{K}$, respectively. The received signal at user k (before SIC) is given by

$$y_k = \underbrace{\sqrt{p_k} g_k s_k}_{\text{intended signal}} + \underbrace{\sum_{i\in\mathcal{K}\setminus\{k\}} \sqrt{p_i} g_k s_i}_{\text{co-channel interference}} + z_k, \tag{4.1}$$

where g_k is the (generally complex) channel gain from the BS to user k, and $z_k \sim \mathcal{CN}(0, \sigma_k^2)$ is the additive white Gaussian noise (AWGN). At the receivers' side, SIC is applied with the optimal channel-to-noise ratio (CNR)-based decoding order. Let $h_k = |g_k|^2/\sigma_k^2$, $\forall k \in \mathcal{K}$, denote the CNR of user k. Without loss of generality, assume that $h_K > h_{K-1} > \cdots > h_1$. The optimal decoding order is $K \to K-1 \to \cdots \to 1$, where $i \to j$ represents that user i fully decodes and then subtracts the signal of user j before decoding its desired signal. During SIC, each user k fully decodes and then cancels the signal of weaker users $k-1, \dots, 1$. In this way, the strongest user K with the highest decoding order, namely NOMA cluster-head user, does not experience any co-channel interference. During SIC, the SINR of user $i \in \mathcal{K}$ for decoding the desired signal of user k is $\gamma_{k,i} = \frac{p_k h_i}{\sum_{j=k+1}^{K} p_j h_i + 1}$. For convenience, let $\gamma_{k,k} \equiv \gamma_k = \frac{p_k h_k}{\sum_{j=k+1}^{K} p_j h_k + 1}$ represent the SINR of user k for decoding its own signal s_k. The achievable rate (bps/Hz) of user k after successful SIC based on the Shannon's capacity formula is given by

$$R_k(\mathbf{p}) = \min_{i=k,\dots,K} \left\{ \log_2 \left(1 + \gamma_{k,i}(\mathbf{p}) \right) \right\}, \tag{4.2}$$

where $\mathbf{p} = [p_k]_{1\times K}$ is the set of allocated powers. For each user pair $i, k \in \mathcal{K}$ with $h_i > h_k$, the condition $\gamma_{k,i}(\mathbf{p}) \geq \gamma_k(\mathbf{p})$, known as SIC necessary condition, which is equivalent to $p_k h_i \geq p_k h_k$, holds independent of $\mathbf{p}$. Accordingly, for any given $\mathbf{p}$, the achievable rate of each user $k \in \mathcal{K}$ is equal to its channel capacity, as formulated by

$$R_k(\mathbf{p}) = \log_2 \left(1 + \gamma_k(\mathbf{p}) \right). \tag{4.3}$$

In the following, we obtain the optimal power allocation strategies to minimize the total power consumption of the BS, maximize users' sum-rate, and maximize system EE subject to the individual minimum rate demand of users.

4.2.1 Total Power Minimization Problem

The total power minimization problem is formulated by

$$\min_{\mathbf{p}\geq 0} \sum_{k=1}^{K} p_k \tag{4.4a}$$

$$\text{s.t.} \quad \sum_{k=1}^{K} p_k \leq P^{\max}, \tag{4.4b}$$

$$R_k(\mathbf{p}) \geq R_k^{\min}, \quad \forall k \in \mathcal{K}, \tag{4.4c}$$

where (4.4b) and (4.4c) are the BS's maximum power and users' minimum rate constraints, respectively. Denoted by $P^{\max}$ and $R_k^{\min}$, the power budget of the BS and individual minimum rate demand of the user k, respectively.

Problem (4.4) is a linear program with affine objective function and constraints. At the optimal point $\boldsymbol{p}^*$, each user gets power to only maintain its minimum rate demand, i.e., $R_k(\boldsymbol{p}^*) = R_k^{\min}$, $\forall k \in \mathcal{K}$ (Rezvani et al., 2022b).[2]

Theorem 4.1 The optimal powers for the total power minimization problem (4.4) can be obtained in closed form as follows:

$$p_i^* = \beta_i \left(\prod_{j=i+1}^{K} (1 + \beta_j) + \frac{1}{h_i} + \sum_{j=i+1}^{K} \frac{\beta_j \prod_{k=i+1}^{j-1} (1 + \beta_k)}{h_j} \right), \quad \forall i = 1, \dots, K, \tag{4.5}$$

where $\beta_i = 2^{R_i^{\min}} - 1$, $\forall i = 1, \dots, K$.

Proof: The proof is provided in Rezvani et al. (2021a, Appendix C). □

The optimal value of (4.4) can be obtained in closed form as

$$Q^{\min}(\boldsymbol{p}^*) = \sum_{i=1}^{K} \beta_i \left(\prod_{j=i+1}^{K} (1 + \beta_j) + \frac{1}{h_i} + \sum_{j=i+1}^{K} \frac{\beta_j \prod_{k=i+1}^{j-1} (1 + \beta_k)}{h_j} \right). \tag{4.6}$$

Remark 4.1 Problem (4.4) is feasible if and only if $Q^{\min} \le P^{\max}$.

4.2.2 Sum-Rate Maximization Problem

The sum-rate maximization problem is formulated by

$$\max_{\boldsymbol{p} \ge 0} \sum_{k=1}^{K} R_k(\boldsymbol{p}) \quad \text{s.t. (4.4b), (4.4c).} \tag{4.7}$$

Problem (4.7) is convex with strictly concave objective function, and affine constraints. At the optimal point, the maximum power constraint is active, i.e., $\sum_{k=1}^{K} p_k^* = P^{\max}$. Moreover, each user with lower decoding order (weaker user), i.e., users $1, \dots, K-1$, gets power to only maintain its individual minimum rate demand, and only the cluster-head user deserves additional power. In other words, at the optimal point, we guarantee that $R_k(\boldsymbol{p}^*) = R_k^{\min}$, $\forall k = 1, \dots, K-1$, and $R_K(\boldsymbol{p}^*) \ge R_K^{\min}$ (Rezvani et al. 2021a, Appendix B).

Theorem 4.2 The optimal powers for the sum-rate maximization problem (4.7) can be obtained in closed form as follows:

2 The proof is provided in Rezvani et al. (2021a, Appendix C), a supplementary material of Rezvani et al. (2022b).

$$p_i^* = \beta_i \left(\prod_{j=1}^{i-1} \left(1 - \beta_j\right) P^{\max} + \frac{1}{h_i} - \sum_{j=1}^{i-1} \frac{\prod_{k=j+1}^{i-1} \left(1 - \beta_k\right) \beta_j}{h_j} \right),$$
$$\forall i = 1, \dots, K-1, \tag{4.8}$$

and

$$p_K^* = P^{\max} - \sum_{i=1}^{K-1} \beta_i \left(\prod_{j=1}^{i-1} \left(1 - \beta_j\right) P^{\max} + \frac{1}{h_i} - \sum_{j=1}^{i-1} \frac{\prod_{k=j+1}^{i-1} \left(1 - \beta_k\right) \beta_j}{h_j} \right), \tag{4.9}$$

where $\beta_i = \frac{2^{R_i^{\min}} - 1}{2^{R_i^{\min}}}$, $\forall i = 1, \dots, K-1$.

Proof: The proof is provided in Rezvani et al. (2021a, Appendix B). □

Equations (4.8) and (4.9) can be reformulated, respectively, as

$$p_k^* = \left(\beta_k \prod_{j=1}^{k-1} \left(1 - \beta_j\right) \right) P^{\max} + c_k, \quad \forall k = 1, \dots, K-1, \tag{4.10}$$

and

$$p_K^* = \left(1 - \sum_{i=1}^{K-1} \beta_i \prod_{j=1}^{i-1} \left(1 - \beta_j\right) \right) P^{\max} - \sum_{i=1}^{K-1} c_i, \tag{4.11}$$

where $\beta_k = \frac{2^{R_k^{\min}} - 1}{2^{R_k^{\min}}}$, $\forall\, k \in \mathcal{K}$, and $c_k = \beta_k \left(\frac{1}{h_k} - \sum_{j=1}^{k-1} \frac{\prod_{i=j+1}^{k-1} (1-\beta_i)\beta_j}{h_j} \right)$, $\forall\, k \in \mathcal{K}$. Equation (4.11) can be reformulated as $p_K^* = \alpha P^{\max} - C$, where $\alpha = \left(1 - \sum_{i=1}^{K-1} \beta_i \prod_{j=1}^{i-1} \left(1 - \beta_j\right)\right)$ and $C = \sum_{i=1}^{K-1} c_i$ are non-negative constants (Rezvani et al., 2022a, Rezvani, 2023). The optimal value of (4.7) can be obtained in closed form as

$$R_{\text{sum}}\left(\boldsymbol{p}^*\right) = \sum_{k=1}^{K-1} R_k^{\min} + \log_2\left(1 + \left(\alpha P^{\max} - C\right) h_K\right). \tag{4.12}$$

4.2.3 Energy-Efficiency Maximization Problem

The EE maximization problem is formulated by

$$\max_{\boldsymbol{p} \geq 0} \frac{\sum_{k \in \mathcal{K}} R_k(\boldsymbol{p})}{\sum_{k \in \mathcal{K}} p_k + P_{\text{C}}} \quad \text{s.t. (4.4b), (4.4c),} \tag{4.13}$$

where constant P_{C} is the circuit power consumption (Zappone and Jorswieck, 2015). Problem (4.13) is a concave–convex fractional program with a pseudo-concave objective function and affine constraints. Hence, the optimal solution of (4.13) can be obtained by using the well-known Dinkelbach algorithm (Dinkelbach, 1967). In this algorithm, we iteratively solve the following problem

$$\max_{p \geq 0} \sum_{k \in \mathcal{K}} R_k - \lambda \left(\sum_{k \in \mathcal{K}} p_k + P_C \right), \quad \text{s.t. (4.4b), (4.4c),} \tag{4.14}$$

where $\lambda \geq 0$ is the fractional parameter, and the objective function is strictly concave. It is proved that the optimal powers in Theorem 4.2 are valid for problem (4.14) (Rezvani et al., 2022a). In this way, (4.14) can be equivalently transformed to

$$\max_{\tilde{q}} \log_2 (1 + \tilde{q}H) - \lambda \tilde{q}, \quad \text{s.t. } \tilde{q} \in \left[\tilde{Q}^{\min}, \tilde{P}^{\max} \right], \tag{4.15}$$

where $\tilde{q} = q - \frac{C}{\alpha}$, in which $q = \sum_{k \in \mathcal{K}} p_k$, is the total power consumption of the BS. Moreover, $H = \alpha h_K$, $\tilde{P}^{\max} = P^{\max} - \frac{C}{\alpha}$, and $\tilde{Q}^{\min} = Q^{\min} - \frac{C}{\alpha}$ (Rezvani et al., 2022a). The objective function of (4.15) is concave. By taking the first derivative, the optimal $\tilde{q}^*$ can be obtained in closed form as

$$\tilde{q}^* = \min \left\{ \max \left\{ \tilde{Q}^{\min}, \frac{1}{\ln(2)\lambda} - \frac{1}{H} \right\}, \tilde{P}^{\max} \right\}. \tag{4.16}$$

Accordingly, at the tth Dinkelbach iteration, for the given $\lambda = \frac{\sum_{k \in \mathcal{K}} R_k(\boldsymbol{p}^{(t-1)})}{\sum_{k \in \mathcal{K}} p_k^{(t-1)} + P_C}$, the optimal powers of problem (4.14) can be obtained by (4.8) and (4.9), in which $P^{\max}$ is substituted with $q = \min \left\{ \max \left\{ \tilde{Q}^{\min}, \frac{1}{\ln(2)\lambda} - \frac{1}{H} \right\}, \tilde{P}^{\max} \right\} + \frac{C}{\alpha}$. It can be seen that the power constraint (4.4b) might not be active at the optimal point of (4.13).

4.2.4 Key Features and Implementation Issues

Here, we present some important key features of NOMA, and their applications to the future wireless networks.

4.2.4.1 CSI Insensitivity

Generally speaking, the multiplexed users in a NOMA cluster can be divided into two categories, as follows:

1. *Cluster-head (strongest) user* K: This user is a NOMA-interference-free user, which cancels the signal of all the other multiplexed users. The cluster-head user acts as an OMA user, whose signal-to-noise ratio (SNR), formulated by $\gamma_K = p_K h_K$, is highly sensitive to its CNR h_K.
2. *Lower decoding order (weaker) users* $1, \ldots, K-1$: The SINR of each weaker user $k < K$, formulated by $\gamma_k(\boldsymbol{p}) = \frac{p_k h_k}{\sum_{j=k+1}^{K} p_j h_k + 1}$, is affected by its received NOMA interference power due to treating the stronger user(s) signal as noise. When

the received NOMA interference power dominates the noise power at the weaker user k, i.e., $\sum_{j=k+1}^{K} p_j h_k \gg 1$, the SINR of user k can be approximated to

$$\tilde{\gamma}_k(\mathbf{p}) \approx \frac{p_k}{\sum_{j=k+1}^{K} p_j}, \tag{4.17}$$

which is insensitive to its CNR h_k.

By substituting the approximated SINR $\tilde{\gamma}_k(\mathbf{p}), \forall k = 1, \ldots, K-1$, to problem (4.7), the optimal power coefficients can be obtained as

$$\tilde{\alpha}_k^* \approx \left(\beta_k \prod_{j=1}^{k-1} (1-\beta_j)\right), \quad \forall k = 1 \ldots, K-1, \quad \tilde{\alpha}_K^* \approx \alpha = 1 - \sum_{i=1}^{K-1} \beta_i \prod_{j=1}^{i-1} (1-\beta_j). \tag{4.18}$$

According to (4.10), (4.11), and (4.18), the condition $\sum_{j=k+1}^{K} p_j h_k \gg 1$, $\forall k = 1, \ldots, K-1$, corresponds to $c_k \approx 0, \forall k = 1 \ldots, K-1$, and subsequently $C = \sum_{k=1}^{K-1} c_k \approx 0$. The approximated SINR in (4.17) is comprehensively analyzed and evaluated in Rezvani (2023). It is shown that for a large CNR region of weaker users, the received NOMA interference power dominates the noise power, thus the SINR of these users remains constant. This is due to the fact that a part of the superimposed signal at the weaker users is treated as noise after SIC, whose received power is typically larger than that of AWGN power. In this way, enhancing or degrading the received superimposed signal power would enhance or degrade both the desired and interference signal powers with the same factor, such that the SINR of weaker users remains constant. It is worth noting that γ_k is an increasing function of h_k for all the users, since $\frac{\partial \gamma_k}{\partial h_k} > 0$, $\forall k \in \mathcal{K}$, however, $\frac{\partial \gamma_k}{\partial h_k} \approx 0$, when $\sum_{j=k+1}^{K} p_j h_k \gg 1, \forall k < K$.

The above results are useful for the topics depending on user's CSI insensitivity in the NOMA systems. For instance, we can conclude that NOMA works well in the imperfect CSI/fast fading scenarios. Another relevant topic is the channel enhancement of NOMA users by means of intelligent reflecting surfaces (IRS) (Mu et al., 2021). The CSI insensitivity of weaker users can highly impact on the design and placement of IRSs, which can be considered as a future work.

4.2.4.2 Rate Fairness

The water filling algorithm verifies that at the optimal point of sum-rate maximization of OMA systems, the achievable rate of each OMA user is usually more than its minimum rate demand. In other words, each OMA user deserves additional power. In contrast to OMA, the optimality conditions of sum-rate maximization of NOMA verify that only the cluster-head user deserves additional power, so achieves additional rate. Hence, rate fairness is not guaranteed at the optimal point of the sum-rate maximization of NOMA systems. The latter fairness issue will be more crucial if no minimum rate constraint is imposed for the weaker users. In this case, no power will be allocated to these users. Generally speaking, rate fairness is essential and needs to be taken into account during power allocation among NOMA users. When all the weaker users have the same minimum rate demand $R_{\min}$, the approximated optimal power coefficients in (4.18) can be reformulated as (Rezvani et al., 2022b)

$$\tilde{\alpha}_k^* \approx \frac{2^{R_{\min}} - 1}{\left(2^{R_{\min}}\right)^k}, \quad \forall k = 1 \ldots, K-1, \quad \tilde{\alpha}_K^* \approx \frac{1}{\left(2^{R_{\min}}\right)^{K-1}}. \tag{4.19}$$

Equation (4.19) shows that the optimal power coefficients are highly heterogeneous in many cases, and equal power allocation is usually infeasible. Also, usually more power will be allocated to the weaker users at the optimal point (Rezvani et al., 2022a, 2022b). As a result, imposing the constraint "allocating more power to the weaker users" does not guarantee any rate fairness since the optimal point lies in this region, however, only the cluster-head user gets additional rate.[3] Note that the latter constraint is not necessary to guarantee successful SIC (enable NOMA), since due to the degradation of SISO Gaussian BCs with perfect CSI, the stronger users can successfully decode and cancel the signal of weaker users independent of power allocation among NOMA users.

The rate fairness of NOMA can be efficiently managed via the well-known fairness schemes, such as proportional fairness. However, in contrast to the OMA systems, where the weighted sum-rate is concave, this function is nonconcave for the NOMA users, making the problem nonconvex (Rezvani et al., 2022a). The optimal solution of the latter problem is still open (Salaün et al., 2020). A more efficient way is to keep the sum-rate maximization, while adding the weights to the minimum rate demand of weaker users (Rezvani et al., 2022a). In contrast to OMA, the users minimum rate demand in the NOMA sum-rate maximization problem highly affect the optimal power allocation and achievable rates. In particular, since each weaker user $k < K$ achieves its minimum rate demand $R_k^{\min}$ at the optimal, by increasing $R_k^{\min}$, $\forall k < K$, we guarantee that each weaker user k gets additional power. Besides, less power remains for the cluster-head user, decreasing the rate of the cluster-head user. The weighted minimum rate fairness problem of NOMA users can thus be formulated by

$$\max_{\boldsymbol{p} \geq 0} \sum_{k=1}^{K} R_k(\boldsymbol{p}) \quad \text{s.t. (4.4b)}, \quad R_k(\boldsymbol{p}) \geq \omega_k R_k^{\min}, \quad \forall k \in \mathcal{K}, \tag{4.20}$$

3 In many cases, imposing the constraint "allocating more power to the weaker users" does not impact on the users' sum-rate at the optimal point.

where $\omega_k \geq 1$, $\forall k \in \mathcal{K}\setminus\{K\}$, and $\omega_K = 1$. The optimal solution of (4.20) can be obtained by (4.8) and (4.9), in which $R_k^{\min}$ is substituted with $\omega_k R_k^{\min}$. Subsequently, at the optimal point, we guarantee that $R_k(\boldsymbol{p}^*) = \omega_k R_k^{\min}$, $\forall k = 1, \dots, K-1$, while the minimum rate demand of the cluster-head user does not impact on the optimal solution, since the rest of the power will always be allocated to this user. The maximum minimal rate (max–min fairness) of NOMA users[4] can thus be attained by performing the bisection method over r on the top of problem (4.20), where $\omega_k R_k^{\min} = r$, $\forall k \in \mathcal{K}\setminus\{K\}$.

4.3 Single-Cell Multicarrier NOMA

Assume that the multicarrier technology is introduced on top of the single-cell NOMA system in Section 4.2. In MC-NOMA, the total bandwidth W (Hz) is divided into N isolated subchannels with the set $\mathcal{N} = \{1, \dots, N\}$, where the bandwidth of each subchannel is $W_s = W/N$. In MC-NOMA, SC-SIC is applied to each subchannel with maximum number of multiplexed users $U^{\max}$. Denoted by ρ_k^n, the binary channel allocation indicator, where if subchannel n is assigned to user k, we have $\rho_k^n = 1$, and otherwise, $\rho_k^n = 0$. The set of users occupying subchannel n is denoted by $\mathcal{K}_n = \{k \in \mathcal{K} | \rho_k^n = 1\}$. The set of subchannels assigned to user $k \in \mathcal{K}$ is indicated by $\mathcal{N}_k = \{n \in \mathcal{N} | \rho_k^n = 1\}$. In FDMA–NOMA, each user belongs to only one NOMA cluster, thus we have $\mathcal{K}_n \cap \mathcal{K}_m = \emptyset, \forall n, m \in \mathcal{N}$, $n \neq m$, or equivalently, $|\mathcal{N}_k| = 1$, $\forall k \in \mathcal{K}$. In the following, we consider the more general case OFDMA–NOMA with $|\mathcal{N}_k| \geq 1$, $\forall k \in \mathcal{K}$. In MC-NOMA, we have $|\mathcal{K}_n| \leq U^{\max}$, $\forall n \in \mathcal{N}$. Each subchannel $n \in \mathcal{N}$ can be viewed as a SISO Gaussian BC. In the following, we use similar notations as in Section 4.2 with the additional index n corresponding to subchannel n. The achievable rate of user $k \in \mathcal{K}_n$ on subchannel n is formulated by $R_k^n(\boldsymbol{p}^n) = W_s \log_2\left(1 + \gamma_k^n(\boldsymbol{p}^n)\right)$, where $\gamma_k^n(\boldsymbol{p}^n) = \frac{p_k^n h_k^n}{\sum_{j \in \mathcal{K}_n, h_j^n > h_k^n} p_j^n h_k^n + 1}$ is the SINR of user k for decoding its own signal, p_k^n and h_k^n are the transmit power and CNR of user k on subchannel n (Rezvani et al., 2022a). In each subchannel n, the index of the cluster-head user is denoted by $\Phi_n = \operatorname{argmax}_{k \in \mathcal{K}_n} h_k^n$. In the following, we obtain the optimal power allocation strategies to minimize the total power consumption of the BS, maximize users' sum-rate, and maximize system EE subject to the individual minimum rate demand of users on the assigned subchannels.

4 In the max–min rate problem, the users' individual minimum rate constraint will be removed.

4.3.1 Total Power Minimization Problem

The total power minimization problem is formulated by

$$\min_{\boldsymbol{p}\geq 0} \sum_{n\in\mathcal{N}}\sum_{k\in\mathcal{K}_n} p_k^n \tag{4.21a}$$

$$\text{s.t. } R_k^n(\boldsymbol{p}^n) \geq R_{k,n}^{\min}, \quad \forall k\in\mathcal{K}, \quad n\in\mathcal{N}_k, \tag{4.21b}$$

$$\sum_{n\in\mathcal{N}}\sum_{k\in\mathcal{K}_n} p_k^n \leq P^{\max}, \tag{4.21c}$$

$$\sum_{k\in\mathcal{K}_n} p_k^n \leq P_n^{\text{mask}}, \quad \forall n\in\mathcal{N}, \tag{4.21d}$$

where (4.21b) is the per-subchannel users' minimum rate constraint, in which $R_{k,n}^{\min}$ is the individual minimum rate demand of user k on subchannel n. (4.21c) is the maximum cellular power constraint. Equation (4.21d) is the per-subchannel maximum power constraint, where P_n^{mask} denotes the maximum allowable power on subchannel n. The main problem (4.21) is a linear program, so is convex. The optimal powers for each subchannel n can be obtained by solving the SC-NOMA problem (4.4) (Rezvani et al., 2022a). Let us define $q_n = \sum_{k\in\mathcal{K}_n} p_k^n$ representing the power consumption of cluster (subchannel) n.

Theorem 4.3 The feasible set of (4.21) can be characterized as the intersection of $q_n \in \left[Q_n^{\min}, P_n^{\text{mask}}\right]$, $\forall n\in\mathcal{N}$, and maximum cellular power constraint $\sum_{n\in\mathcal{N}} q_n \leq P^{\max}$, where

$$Q_n^{\min} = \sum_{k\in\mathcal{K}_n} \beta_k^n \left(\prod_{\substack{j\in\mathcal{K}_n\\ h_j^n > h_k^n}} \left(1+\beta_j^n\right) + \frac{1}{h_k^n} + \sum_{\substack{j\in\mathcal{K}_n\\ h_j^n>h_k^n}} \frac{\beta_j^n \prod_{\substack{l\in\mathcal{K}_n\\ h_k^n<h_l^n<h_j^n}} \left(1+\beta_l^n\right)}{h_j^n} \right), \tag{4.22}$$

in which $\beta_k^n = 2^{(R_{k,n}^{\min}/W_s)} - 1$, $\forall n\in\mathcal{N}$, $k\in\mathcal{K}_n$.

Proof: The proof is provided in Rezvani et al. (2022a, Appendix A). □

Remark 4.2 Problem (4.21) is feasible if and only if $Q_n^{\min} \leq P_n^{\text{mask}}$, $\forall n\in\mathcal{N}$, and $\sum_{n\in\mathcal{N}} Q_n^{\min} \leq P^{\max}$.

4.3.2 Sum-Rate Maximization Problem

The sum-rate maximization problem is formulated by

$$\max_{\boldsymbol{p}\geq 0} \sum_{n\in\mathcal{N}}\sum_{k\in\mathcal{K}_n} R_k^n(\boldsymbol{p}^n) \quad \text{s.t. (4.21b)–(4.21d).} \tag{4.23}$$

For any given feasible q_n, the optimal intra-cluster powers for users in $\mathcal{K}_n$ can be obtained by (Rezvani et al. 2022a)

$$p_k^{n^*} = \left(\beta_k^n \prod_{\substack{j \in \mathcal{K}_n \\ h_j^n < h_k^n}} \left(1 - \beta_j^n\right) \right) q_n + c_k^n, \quad \forall k \in \mathcal{K}_n \backslash \{\Phi_n\}, \tag{4.24}$$

and

$$p_{\Phi_n}^{n^*} = \left(1 - \sum_{\substack{i \in \mathcal{K}_n \\ h_i^n < h_{\Phi_n}^n}} \beta_i^n \prod_{\substack{j \in \mathcal{K}_n \\ h_j^n < h_i^n}} \left(1 - \beta_j^n\right) \right) q_n - \sum_{\substack{i \in \mathcal{K}_n \\ h_i^n < h_{\Phi_n}^n}} c_i^n, \tag{4.25}$$

where $\beta_k^n = \frac{2^{\left(R_{k,n}^{\min}/W_s\right)} - 1}{2^{\left(R_{k,n}^{\min}/W_s\right)}}$ and $c_k^n = \beta_k^n \left(\frac{1}{h_k^n} - \sum_{\substack{j \in \mathcal{K}_n \\ h_j^n < h_k^n}} \frac{\prod_{\substack{l \in \mathcal{K}_n \\ h_j^n < h_l^n < h_k^n}} (1-\beta_l^n)\beta_j^n}{h_j^n} \right)$, $\forall n \in \mathcal{N},\ k \in \mathcal{K}_n$. Problem (4.23) can thus be equivalently transformed to

$$\max_{\tilde{\boldsymbol{q}}} \sum_{n \in \mathcal{N}} W_s \log_2 \left(1 + \tilde{q}_n H_n\right) \tag{4.26a}$$

$$\text{s.t.} \sum_{n \in \mathcal{N}} \tilde{q}_n = \tilde{P}^{\max}, \quad \tilde{q}_n \in \left[\tilde{Q}_n^{\min}, \tilde{P}_n^{\text{mask}}\right], \quad \forall n \in \mathcal{N}, \tag{4.26b}$$

where $\alpha_n = \left(1 - \sum_{\substack{i \in \mathcal{K}_n \\ h_i^n < h_{\Phi_n}^n}} \beta_i^n \prod_{\substack{j \in \mathcal{K}_n \\ h_j^n < h_i^n}} \left(1 - \beta_j^n\right)\right)$, $c_n = \sum_{\substack{i \in \mathcal{K}_n \\ h_i^n < h_{\Phi_n}^n}} c_i^n$, $H_n = \alpha_n h_{\Phi_n}^n$, $\tilde{Q}_n^{\min} = Q_n^{\min} - \frac{c_n}{\alpha_n}$, $\tilde{P}_n^{\text{mask}} = P_n^{\text{mask}} - \frac{c_n}{\alpha_n}$, $\forall n \in \mathcal{N}$, and $\tilde{P}^{\max} = P^{\max} - \sum_{n \in \mathcal{N}} \frac{c_n}{\alpha_n}$ (Rezvani et al., 2022a). The inter-cluster power allocation problem (4.26) is equivalent to the sum-rate maximization problem of FDMA systems with N virtual OMA users.[5] For more details, please see (Rezvani et al., 2022a). In this way, (4.26) can be solved by using the water filling algorithm, where the optimal $\tilde{\boldsymbol{q}}^*$ can be obtained by

$$\tilde{q}_n^* = \begin{cases} \frac{W_s/(\ln 2)}{\nu^*} - \frac{1}{H_n}, & \left(\frac{W_s/(\ln 2)}{\nu^*} - \frac{1}{H_n}\right) \in \left[\tilde{Q}_n^{\min}, \tilde{P}_n^{\text{mask}}\right], \\ 0, & \text{otherwise.} \end{cases} \tag{4.27}$$

The optimal ν^*, which is the optimal Lagrange multiplier corresponding to the maximum power constraint (4.26b), can be obtained by using the bisection method (Rezvani et al., 2022a). After finding $\tilde{\boldsymbol{q}}^*$, we obtain $\boldsymbol{q}^*$ by using $q_n^* = (\tilde{q}_n^* + \frac{c_n}{\alpha_n})$, $\forall n \in \mathcal{N}$. Then, we find the optimal intra-cluster power allocation by (4.24) and (4.25).

5 It is shown that each NOMA cluster operating on an isolated subchannel can be modeled as a virtual OMA user Rezvani et al. (2022a).

4.3.3 Energy-Efficiency Maximization Problem

The EE maximization problem is formulated by

$$\max_{\boldsymbol{p}\geq 0} \frac{\sum_{n\in\mathcal{N}}\sum_{k\in\mathcal{K}_n} R_k^n(\boldsymbol{p}^n)}{\sum_{n\in\mathcal{N}}\sum_{k\in\mathcal{K}_n} p_k^n + P_{\mathrm{C}}} \quad \text{s.t. (4.21b)–(4.21d).} \tag{4.28}$$

To solve (4.28), we employ the Dinkelbach algorithm, where we iteratively solve the following problem

$$\max_{\boldsymbol{p}\geq 0} \left(\sum_{n\in\mathcal{N}}\sum_{k\in\mathcal{K}_n} R_k^n(\boldsymbol{p}^n)\right) - \lambda\left(\sum_{n\in\mathcal{N}}\sum_{k\in\mathcal{K}_n} p_k^n + P_{\mathrm{C}}\right) \quad \text{s.t. (4.21b)–(4.21d).} \tag{4.29}$$

Similar to (4.26), problem (4.29) can be rewritten as (Rezvani et al., 2022a)

$$\max_{\tilde{\boldsymbol{q}}} \sum_{n\in\mathcal{N}} W_s \log_2\left(1+\tilde{q}_n H_n\right) - \lambda\left(\sum_{n\in\mathcal{N}} \tilde{q}_n\right) \tag{4.30a}$$

$$\text{s.t.} \sum_{n\in\mathcal{N}} \tilde{q}_n \leq \tilde{P}^{\max}, \quad q_n \in \left[\tilde{Q}_n^{\min}, \tilde{P}_n^{\text{mask}}\right], \quad \forall n \in \mathcal{N}. \tag{4.30b}$$

The equivalent FDMA problem (4.30) is convex and can be solved by using the Lagrange dual with subgradient method or interior point methods (IPMs).

Corollary 4.1 In contrast to the sum-rate maximization problem, the maximum power constraint of the EE maximization problem might not be active. At the optimal point of both the sum-rate and EE maximization problems, all the users with lower decoding order get power to only maintain their minimum rate demands, and only the cluster-head users achieve additional rate.

4.3.4 Key Features and Implementation Issues

According to (4.27), the heterogeneity of inter-cluster power allocation for MC-NOMA is typically low, specifically for the high SNR regions of virtual OMA users, similar to the FDMA systems. For more details, please see Rezvani et al. (2022a). The equal inter-cluster power allocation has thus a near-optimal performance, in contrast to the equal intra-cluster power allocation, which is mostly infeasible.

The theoretical insights presented in Section 4.2.4 hold for each subchannel of the MC-NOMA systems, since each subchannel can be viewed as a SISO Gaussian BC. Another application of the CSI insensitivity of weaker users is for user grouping of MC-NOMA, or generally subchannel allocation. In the CNR region, where the received NOMA interference power dominates the noise power, the SINR of weaker users approximately remains constant over different subchannels. Hence, the most important factor for user grouping is to preferably increase the CNR of cluster-head users, while keeping the CNR of weaker users in the region, where the received NOMA interference power dominates the noise power. Further enhancing the CNR of weaker users does not significantly affect their SINR; thus, many user grouping strategies between the cluster-head and weaker users might have similar performance. It can be concluded that the CNR difference among the users might not be a good factor for user grouping. A more effective factor would be user grouping based on users' individual minimum rate demand, which significantly impacts the remaining power for the cluster-head users, and can be considered as a future work.

The rate fairness of MC-NOMA can be efficiently tuned by considering the weighted minimum rate fairness scheme among NOMA users within each subchannel, and the weighted sum-rate of different NOMA clusters, modeled as virtual OMA users, which is comprehensively discussed in Rezvani (2023) and Rezvani et al. (2022a).

4.4 Multi-cell NOMA with Single-Cell Processing

Consider the downlink transmission of a single-carrier multi-cell NOMA system with single-cell processing. In this network, the information of each user is available at only one cell; thus, users associated with the same cell form a NOMA cluster while the signals from other cells are fully treated as noise. The set of all BSs, and users served by BS b are indicated by $\mathcal{B}$ and $\mathcal{K}_b$, respectively. In the following, we use similar notations as in Section 4.2 with the additional index b corresponding to cell b.

In general, there are $|\mathcal{K}_b|!$ possible SIC decoding orders for users in $\mathcal{K}_b$. The set of all possible SIC decoding orders in cell b is denoted by $\boldsymbol{\pi}_b$, such that $|\boldsymbol{\pi}_b| = |\mathcal{K}_b|!$. The lth SIC decoding order in cell b is denoted by $\boldsymbol{\pi}_b(l)$. According to (4.2), the achievable rate of user $k \in \mathcal{K}_b$ after successful SIC is obtained by (Rezvani et al., 2022b, Rezvani, 2023)

$$\tilde{R}_{b,k}\left(\boldsymbol{p}, \boldsymbol{\pi}_b(l)\right) = \min_{i \in \{k\} \cup \Phi^{(l)}_{b,k}} \left\{ \log_2 \left(1 + \frac{p_{b,k} h_{b,i}}{\sum_{j \in \Phi^{(l)}_{b,k}} p_{b,j} h_{b,i} + \left(I_{b,i}\left(\boldsymbol{p}_{-b}\right) + \sigma^2_{b,i} \right)} \right) \right\}, \tag{4.31}$$

where $\Phi^{(l)}_{b,k}$ is the set of users in $\mathcal{K}_b$ with higher decoding order than user $k \in \mathcal{K}_b$ based on decoding order $\boldsymbol{\pi}_b(l)$, and $I_{b,k}(\boldsymbol{p}_{-b}) = \sum_{\substack{j \in \mathcal{B} \\ j \neq b}} \sum_{i \in \mathcal{K}_j} p_{j,i} h_{j,b,k}$ is the received ICI at user $k \in \mathcal{K}_b$, in which $h_{j,b,k}$ is the channel gain from BS j to user $k \in \mathcal{K}_b$. Moreover, $\boldsymbol{p} = [p_{b,k}], \forall b \in \mathcal{B}, k \in \mathcal{K}_b$, is the power allocation matrix of all the users, in which $\boldsymbol{p}_b$ is the bth row of $\boldsymbol{p}$ indicating the power allocation vector of users in cell b. Let us define the power consumption portion of BS b as $\alpha_b \in [0,1]$ such that $\sum\limits_{k\in\mathcal{K}_b} p_{b,k} = \alpha_b P_b^{\max}$. The received ICI power at user $k \in \mathcal{K}_b$ can be rewritten as $I_{b,k}(\boldsymbol{\alpha}_{-b}) = \sum_{\substack{j\in\mathcal{B}\\ j\neq b}} \alpha_j P_j^{\max} h_{j,b,k}$.

At each user $k \in \mathcal{K}_b$, the term $(I_{b,k}(\boldsymbol{\alpha}_{-b}) + \sigma_{b,k}^2)$ can be viewed as an equivalent noise power. Hence, in multi-cell NOMA with single-cell processing, the optimal decoding order for the user pair $k, i \in \mathcal{K}_b$ is $i \to k$ if and only if $\tilde{h}_{b,i}(\boldsymbol{\alpha}_{-b}) \geq \tilde{h}_{b,k}(\boldsymbol{\alpha}_{-b})$, where $\tilde{h}_{b,l}(\boldsymbol{\alpha}_{-b}) = \frac{h_{b,l}}{I_{b,l}(\boldsymbol{\alpha}_{-b})+\sigma_{b,l}^2}$, $l = k, i$. For more details, please see Rezvani et al. (2021a, Appendix A). In other words, the channel-to-interference-plus-noise ratio (CINR)-based decoding order is optimal in multi-cell NOMA.[6]

Remark 4.3 Assume that the CINR-based decoding order, denoted by $\boldsymbol{\pi}_b(l)$, is adopted in cell b for the given $\boldsymbol{\alpha}_{-b}$. For each user pair $k, i \in \mathcal{K}_b$ with $i \to k$, the inequality

$$\log_2\left(1 + \frac{p_{b,k}h_{b,k}}{\sum_{j\in\Phi_{b,k}^{(l)}} p_{b,j}h_{b,k} + (I_{b,k}(\boldsymbol{\alpha}_{-b}) + \sigma_{b,k}^2)}\right) \leq \log_2\left(1 + \frac{p_{b,k}h_{b,i}}{\sum_{j\in\Phi_{b,k}^{(l)}} p_{b,j}h_{b,i} + (I_{b,i}(\boldsymbol{\alpha}_{-b}) + \sigma_{b,i}^2)}\right), \tag{4.32}$$

which is equivalent to $\tilde{h}_{b,k}(\boldsymbol{\alpha}_{-b}) \leq \tilde{h}_{b,i}(\boldsymbol{\alpha}_{-b})$, holds independent of $\boldsymbol{p}_b$ (Rezvani et al., 2022b, Rezvani, 2023). According to (4.31), under the CINR-based decoding order $\boldsymbol{\pi}_b(l)$, each user $k \in \mathcal{K}_b$ achieves its channel capacity formulated by

$$R_{b,k}(\boldsymbol{p}, \boldsymbol{\pi}_b(l)) = \log_2\left(1 + \frac{p_{b,k}h_{b,k}}{\sum_{j\in\Phi_{b,k}^{(l)}} p_{b,j}h_{b,k} + (I_{b,k}(\boldsymbol{\alpha}_{-b}) + \sigma_{b,k}^2)}\right). \tag{4.33}$$

In multi-cell NOMA with single-cell processing, the decoding order in cell b needs to be jointly optimized with power control among the interfering cells, i.e., $\boldsymbol{\alpha}_{-b}$, while it is independent of intra-cell power allocation, i.e., $\boldsymbol{p}_b$. In the following, we consider two generic schemes for multi-cell NOMA systems as dynamic and static decoding orders. For each scheme, we propose optimal/suboptimal centralized power allocation strategy to minimize the total power consumption of the BS, maximize users' sum-rate, and maximize system EE subject to the individual minimum rate demand of users.

6 The term "interference" in "CINR" is referred to as ICI.

4.4.1 Dynamic Decoding Order

In this scheme, namely joint SIC ordering and power allocation (JSPA), we consider the optimal CINR-based decoding order for each cell requiring the joint optimization of SIC decoding order and power control among the cells.

4.4.1.1 Optimal JSPA for Total Power Minimization Problem

The JSPA total power minimization problem is formulated by

$$\min_{p\geq 0,\lambda} \sum_{b\in\mathcal{B}} \sum_{k\in\mathcal{K}_b} p_{b,k} \tag{4.34a}$$

$$\text{s.t.} \quad \sum_{k\in\mathcal{K}_b} p_{b,k} \leq P_b^{\max}, \tag{4.34b}$$

$$\sum_{l=1}^{|\mathcal{K}_b|!} \lambda_b^{(l)} R_{b,k}(\boldsymbol{p}, \boldsymbol{\pi}_b(l)) \geq R_{b,k}^{\min}, \quad \forall b\in\mathcal{B}, \quad k\in\mathcal{K}_b, \tag{4.34c}$$

$$\sum_{l=1}^{|\mathcal{K}_b|!} \lambda_b^{(l)} = 1, \quad \forall b\in\mathcal{B}, \quad \lambda_b^{(l)} \in \{0,1\}, \tag{4.34d}$$

where (4.34b) and (4.34c) are the per-BS maximum power and per-user minimum rate constraints, respectively. $P_b^{\max}$ denotes the maximum power of BS b, and $R_{b,k}^{\min}$ is the minimum rate demand of user $k\in\mathcal{K}_b$. The constraint (4.34d) represents that only one decoding order among all the possible decoding orders can be selected for each cell, where the binary decoding order indicator $\lambda_b^{(l)} = 1$ if the decoding order $\boldsymbol{\pi}_b(l)$ is chosen, and otherwise, $\lambda_b^{(l)} = 0$. Moreover, $\boldsymbol{\lambda} = [\lambda_b^{(l)}], \forall b \in \mathcal{B}, l\in\{1,\dots,|\mathcal{K}_b|!\}$, is the set of all decoding order indicators in the network.

It is verified that under the CINR-based decoding order, the ICI in the centralized total power minimization problem (4.34) verifies the basic properties of the standard interference function (Fu et al., 2017). Hence, the optimal JSPA can be obtained by using the well-known Yates power control framework (Yates, 1995). In this algorithm, we iteratively minimize the power consumption of each BS such that the power consumption of other cells is given. Hence, at each cell b, we first update the CINR-based decoding order for the given $\boldsymbol{\alpha}_{-b}$. Then, we solve the single-cell NOMA power minimization problem (4.4), whose optimal powers are formulated in (4.5), such that the channel gains are normalized by "ICI+AWGN," i.e., h_i is replaced with $\tilde{h}_{b,i}(\boldsymbol{\alpha}_{-b}) = \frac{h_{b,i}}{I_{b,i}(\boldsymbol{\alpha}_{-b}) + \sigma_{b,i}^2}$. For more details, please see Rezvani et al. (2021a, Subsection III-A1). For any finite initial point $\boldsymbol{p}^{(0)}$, this iterative distributed algorithm converges to a unique point, which is the globally optimal solution. Moreover, it is proved that the optimal $\boldsymbol{\alpha}^*$ in the total power minimization problem is a component-wise minimum (Fu et al., 2017). It means that for any feasible $\hat{\boldsymbol{\alpha}} = [\hat{\alpha}_b], \forall b\in\mathcal{B}$, in (4.34), it can be guaranteed that $\alpha_b^* \leq \hat{\alpha}_b$. Due to the existing ICI and/or BSs' power budget limitation to meet the users' minimum rate demand, problem (4.34) could be infeasible and described as follows:

1. *ICI*: Problem (4.34) is infeasible when the maximum power constraint (4.34b) is removed. This corresponds to the infeasibility of (4.34), which can be determined by the Perron–Frobenius eigenvalues of the matrices arising from the power control subproblems (Fu et al., 2017, Theorem 8).
2. *Limited power budget*: Problem (4.34) is infeasible while (4.34) without (4.34b) is feasible. As a result, (4.34) is infeasible only because of the lack of power resources to meet the minimum rate constraints in (4.34c).

4.4.1.2 Optimal JSPA for Sum-Rate Maximization Problem

The JSPA sum-rate maximization problem is formulated by

$$\max_{\boldsymbol{p}\geq 0,\boldsymbol{\lambda}} \sum_{b\in\mathcal{B}}\sum_{l=1}^{|\mathcal{K}_b|!}\lambda_b^{(l)}\sum_{k\in\mathcal{K}_b}R_{b,k}(\boldsymbol{p},\boldsymbol{\pi}_b(l)) \qquad \text{s.t. (4.34b)–(4.34d).} \tag{4.35}$$

The ICI in the centralized sum-rate maximization problem does not verify the basic properties of the standard interference function. As a result, Yates power control framework does not guarantee the global optimality for the sum-rate maximization problem. It is shown that the sum-rate function in multi-cell NOMA is nonconcave in powers, due to existing ICI, which makes the centralized sum-rate maximization problem nonconvex and strongly nondeterministic polynomial time (NP)-hard (Rezvani et al., 2022b). For any given feasible $\boldsymbol{\alpha}$, the centralized problem (4.35) can be decoupled into B single-cell NOMA sub-problems, whose globally optimal power allocation can be obtained by (4.8) and (4.9), where h_i is replaced with $\tilde{h}_{b,i}(\boldsymbol{\alpha}_{-b}) = \frac{h_{b,i}}{I_{b,i}(\boldsymbol{\alpha}_{-b})+\sigma_{b,i}^2}$ and $P^{\max}$ is replaced with $\alpha_b P_b^{\max}$. As a result, the globally optimal solution of problem (4.35) can be obtained by performing a greedy search on $\boldsymbol{\alpha}$. For more details, please see Rezvani et al. (2022b).

4.4.1.3 Optimal JSPA for EE Maximization Problem

The JSPA EE maximization problem is formulated by

$$\max_{\boldsymbol{p}\geq 0,\boldsymbol{\lambda}} \frac{\sum_{b\in\mathcal{B}}\sum_{l=1}^{|\mathcal{K}_b|!}\lambda_b^{(l)}\sum_{k\in\mathcal{K}_b}R_{b,k}(\boldsymbol{p},\boldsymbol{\pi}_b(l))}{\sum_{b\in\mathcal{B}}\sum_{k\in\mathcal{K}_b}p_{b,k}+P_{\text{C}}} \qquad \text{s.t. (4.34b)–(4.34d),} \tag{4.36}$$

where constant P_{C} is the total system's circuit power consumption. According to $\sum_{k\in\mathcal{K}_b}p_{b,k}=\alpha_b P_b^{\max}$, problem (4.36) can be equivalently transformed to

$$\max_{\boldsymbol{p}\geq 0,\boldsymbol{\lambda},\boldsymbol{\alpha}} \frac{\sum_{b\in\mathcal{B}}\sum_{l=1}^{|\mathcal{K}_b|!}\lambda_b^{(l)}\sum_{k\in\mathcal{K}_b}R_{b,k}(\boldsymbol{p},\boldsymbol{\pi}_b(l))}{\sum_{b\in\mathcal{B}}\alpha_b P_b^{\max}+P_{\text{C}}}$$

$$\text{s.t.} \quad \text{(4.34c), (4.34d),} \tag{4.37a}$$

$$\sum_{k\in\mathcal{K}_b}p_{b,k}=\alpha_b P_b^{\max}, \quad \forall b\in\mathcal{B}, \quad \alpha_b\in[0,1], \quad \forall b\in\mathcal{B}. \tag{4.37b}$$

For any given feasible $\boldsymbol{\alpha}$, the objective function (4.37a) can be equivalently transformed to maximizing users' sum-rate as $\max_{\boldsymbol{p}\geq 0,\lambda} \sum_{b\in\mathcal{B}} \sum_{l=1}^{|\mathcal{K}_b|!} \lambda_b^{(l)} \sum_{k\in\mathcal{K}_b} R_{b,k}(\boldsymbol{p}, \boldsymbol{\pi}_b(l))$. Subsequently, the centralized problem (4.37) can be decoupled into B single-cell NOMA sub-problems, whose globally optimal power allocation can be obtained by (4.8) and (4.9), where h_i is replaced with $\tilde{h}_{b,i}(\boldsymbol{\alpha}_{-b}) = \frac{h_{b,i}}{I_{b,i}(\boldsymbol{\alpha}_{-b})+\sigma_{b,i}^2}$ and $P^{\max}$ is replaced with $\alpha_b P_b^{\max}$. As a result, our proposed algorithm for solving (4.35) can be modified to find the optimal solution of problem (4.36).

The theoretical insights presented in Section 4.2.4 hold for each cell of multi-cell NOMA with single-cell processing under the optimal CINR-based decoding order. In this system, the ICI is treated as AWGN, such that the equivalent noise "ICI+AWGN" is typically larger than that of single-cell NOMA systems. As a result, for the region that the equivalent noise "ICI+AWGN" dominates the received NOMA interference power, corresponding to the strong ICI scenarios, the SINR approximation in (4.17) might not work well. However, via the existing ICI management techniques, such as power control and/or coordinated multipoint (Shin et al., 2017, Rezvani et al., 2021b), we expect that the NOMA interference dominates "ICI+AWGN" improving the performance of the SINR approximation in (4.17).

4.4.2 Static Decoding Order

In this scheme, we assume a fixed and potentially suboptimal decoding order λ, e.g., CNR-based decoding order, during power control among the cells. In this line, the inequality (4.32) will be violated if the fixed decoding order does not follow the CINR-based decoding order for some $\boldsymbol{\alpha}_{-b}$. Hence, for any fixed λ_b, it is required to guarantee successful SIC at users in cell b during the optimization of $\boldsymbol{\alpha}_{-b}$. The possible ways to guarantee successful SIC are as follows (Rezvani et al., 2022b):

1. *Fixed-rate-region power allocation (FRPA)*: In this scheme, $\boldsymbol{\alpha}_{-b}$ is limited by imposing (4.32) such that the CINR of users with higher decoding order remains larger than that of users with lower decoding order.
2. *Joint rate and power allocation (JRPA)*: In this scheme, the achievable rate of users with lower decoding order is limited by the rate of users with higher decoding order. In other words, we consider the complete achievable rate region of users formulated in (4.31).

4.4.2.1 Optimal FRPA for Total Power Minimization Problem

Assume that the potentially suboptimal decoding order $\boldsymbol{\pi}_b(l)$ is chosen. In the FRPA scheme, we impose the SIC constraint (4.32) on power control among the cells, which can be rewritten as

$$\frac{h_{b,k}}{I_{b,k}(\boldsymbol{\alpha}_{-b}) + \sigma_{b,k}^2} \leq \frac{h_{b,i}}{I_{b,i}(\boldsymbol{\alpha}_{-b}) + \sigma_{b,i}^2}, \quad \forall b \in \mathcal{B}, \quad k, i \in \mathcal{K}_b, \quad i \in \Phi_{b,k}^{(l)}. \tag{4.38}$$

The constraints in (4.38) for cell b limit the feasible region of power control $\boldsymbol{\alpha}_{-b}$ such that the predefined decoding order follows the CINR-based decoding order, so it remains optimal. Subsequently, each user achieves its channel capacity for decoding its own signal given by (4.33). The FRPA total power minimization problem is formulated by

$$\min_{\boldsymbol{p} \geq 0} \sum_{b \in \mathcal{B}} \sum_{k \in \mathcal{K}_b} p_{b,k} \quad \text{s.t. } (4.34\text{b}), (4.34\text{c}), (4.38). \tag{4.39}$$

Constraint (4.38) can be easily transformed to a linear form (Rezvani et al., 2022b). Therefore, problem (4.39) can be transformed to a linear program which can be solved by using the Dantzig's simplex method.

4.4.2.2 Optimal FRPA for Sum-Rate Maximization Problem

The FRPA sum-rate maximization problem is formulated as

$$\max_{\boldsymbol{p} \geq 0} \sum_{b \in \mathcal{B}} \sum_{k \in \mathcal{K}_b} R_{b,k}(\boldsymbol{p}) \quad \text{s.t. } (4.34\text{b}), (4.34\text{c}), (4.38). \tag{4.40}$$

For any fixed $\boldsymbol{\pi}_b(l)$, the FRPA problem (4.40) is identical to the JSPA problem (4.35) with additional SIC constraint (4.38) (Rezvani et al., 2021a, Remark 3). Hence, for any given $\boldsymbol{\alpha}$ in (4.40), the optimal powers of problem (4.40) can be obtained by using (4.8) and (4.9) (Rezvani et al., 2021a, Remark 4). Therefore, similar to the JSPA problem (4.35), the optimal solution of (4.40) can be obtained by performing a greedy search on $\boldsymbol{\alpha}$, however, under the SIC condition (4.38) (Rezvani et al., 2022b).

For some channel conditions, the CNR-based decoding order is optimal for a user pair independent of ICI, thus independent of BSs' power consumption portion in $\boldsymbol{\alpha}$.

Theorem 4.4 *SIC sufficient condition.* For each user pair $k, i \in \mathcal{K}_b$ with $\frac{h_{b,k}}{\sigma_{b,k}^2} \leq \frac{h_{b,i}}{\sigma_{b,i}^2}$, if

$$\frac{h_{b,i}}{\sigma_{b,i}^2} - \frac{h_{b,k}}{\sigma_{b,k}^2} \geq \sum_{j \in \mathcal{Q}_{b,k,i}} \frac{P_j^{\max}}{\sigma_{b,k}^2 \sigma_{b,i}^2} \left(h_{j,b,i} h_{b,k} - h_{j,b,k} h_{b,i}\right), \tag{4.41}$$

where $\mathcal{Q}_{b,k,i} = \left\{ j \in \mathcal{B} \backslash \{b\} | \frac{h_{b,i}}{h_{b,k}} < \frac{h_{j,b,i}}{h_{j,b,k}} \right\}$, the decoding order $i \to k$ is optimal.

Proof: The proof is provided in Rezvani et al. (2021a, Appendix E). □

If the SIC sufficient condition (4.41) is fulfilled for the user pair $k, i \in \mathcal{K}_b$, we have to adopt $i \to k$, otherwise, (4.38) makes the FRPA problem (4.40) infeasible for any $\boldsymbol{\alpha}$. The applications of Theorem 4.4 are discussed in Rezvani et al. (2022b).

4.4.2.3 Optimal FRPA for EE Maximization Problem

The FRPA EE maximization problem is formulated as

$$\max_{\boldsymbol{p}\geq 0} \frac{\sum_{b\in\mathcal{B}}\sum_{k\in\mathcal{K}_b} R_{b,k}(\boldsymbol{p}, \boldsymbol{\pi}_b(l))}{\sum_{b\in\mathcal{B}}\sum_{k\in\mathcal{K}_b} p_{b,k} + P_{\mathrm{C}}} \quad \text{s.t. (4.34b), (4.34c), (4.38).} \tag{4.42}$$

Similar to (4.37), we reformulate problem (4.42) as

$$\max_{\boldsymbol{p}\geq 0, \boldsymbol{\alpha}} \frac{\sum_{b\in\mathcal{B}}\sum_{k\in\mathcal{K}_b} R_{b,k}(\boldsymbol{p}, \boldsymbol{\pi}_b(l))}{\sum_{b\in\mathcal{B}} \alpha_b P_b^{\max} + P_{\mathrm{C}}} \quad \text{s.t. (4.34c), (4.37b), (4.38).} \tag{4.43}$$

For any given feasible $\boldsymbol{\alpha}$, the centralized problem (4.43) can be decoupled into B single-cell NOMA sum-rate maximization sub-problems. Hence, our proposed solution for (4.40) can be modified to solve (4.43).

4.4.2.4 Optimal JRPA for Total Power Minimization Problem

Assume that the potentially suboptimal decoding order $\boldsymbol{\pi}_b(l)$ is predefined. In the JRPA scheme, we allow users with lower decoding order to operate in lower rate than their channel capacity for decoding their own signal to guarantee successful SIC at users with higher decoding order. As a result, we consider the complete achievable rate region of users formulated in (4.31) instead of imposing the SIC constraint (4.32). The FRPA scheme is indeed a special case of JRPA; thus, the rate region of FRPA is a subset of the rate region of JRPA (Rezvani et al., 2022b). In other words, for any fixed and potentially suboptimal decoding order, JRPA always outperforms FRPA in terms of users' achievable rate.

The JRPA total power minimization problem is formulated by

$$\min_{\boldsymbol{p}\geq 0} \sum_{b\in\mathcal{B}} \sum_{k\in\mathcal{K}_b} p_{b,k} \quad \text{s.t. (4.34b),} \tag{4.44a}$$

$$\tilde{R}_{b,k}(\boldsymbol{p}, \boldsymbol{\pi}_b(l)) \geq R_{b,k}^{\min}, \quad \forall b \in \mathcal{B}, \quad k \in \mathcal{K}_b, \tag{4.44b}$$

where $\tilde{R}_{b,k}(\boldsymbol{p}, \boldsymbol{\pi}_b(l))$ in (4.44b) is based on (4.31). The minimum rate constraint (4.44b) can be equivalently transformed to an affine form (Rezvani et al., 2021a). Hence, (4.44) can be equivalently transformed to a linear program which can be solved by using the Dantzig's simplex method.

4.4.2.5 Suboptimal JRPA for Sum-Rate Maximization Problem

The JRPA sum-rate maximization problem is formulated as

$$\max_{\boldsymbol{p}\geq 0} \sum_{b\in\mathcal{B}} \sum_{k\in\mathcal{K}_b} \tilde{R}_{b,k}(\boldsymbol{p}, \boldsymbol{\pi}_b(l)) \quad \text{s.t. (4.34b), (4.44b).} \tag{4.45}$$

Although the feasible set of (4.45) is affine, so is convex, the objective function is non-differentiable and nonconcave, due to the minimum term and ICI, respectively. To make (4.45) more tractable, we define $r_{b,k}$ as the adopted rate of user $k \in \mathcal{K}_b$ such that $r_{b,k} \leq \log_2\left(1 + \frac{p_{b,k}h_{b,i}}{\sum_{j\in\Phi_{b,k}^{(l)}} p_{b,j}h_{b,i}+(I_{b,i}+\sigma_{b,i}^2)}\right)$, $\forall b \in \mathcal{B},\ k, i \in \mathcal{K}_b,\ i \in \{k\} \cup \Phi_{b,k}^{(l)}$. In this way, (4.45) can be rewritten as (Rezvani et al., 2022b)

$$\max_{\boldsymbol{p}\geq 0,\ \boldsymbol{r}\geq 0} \sum_{b\in\mathcal{B}} \sum_{k\in\mathcal{K}_b} r_{b,k} \quad \text{s.t. (4.34b)}, \tag{4.46a}$$

$$r_{b,k} \geq R_{b,k}^{\min}, \forall b \in \mathcal{B}, \quad k \in \mathcal{K}_b, \tag{4.46b}$$

$$r_{b,k} \leq \log_2\left(1 + \frac{p_{b,k}h_{b,i}}{\sum_{j\in\Phi_{b,k}^{(l)}} p_{b,j}h_{b,i} + (I_{b,i} + \sigma_{b,i}^2)}\right),$$
$$\forall b \in \mathcal{B}, \quad k, i \in \mathcal{K}_b, \quad i \in \{k\} \cup \Phi_{b,k}^{(l)}, \tag{4.46c}$$

where $\boldsymbol{r} = [r_{b,k}], \forall b \in \mathcal{B}, k \in \mathcal{K}_b$. The constraints in (4.46c) are still nonconvex. To this end, we can employ the sequential program to find a suboptimal solution for problem (4.46). For more details, please see Rezvani et al. (2022b, Appendix B).

4.4.2.6 Suboptimal JRPA for EE Maximization Problem

The JRPA EE maximization problem is formulated as

$$\max_{\boldsymbol{p}\geq 0} \frac{\sum_{b\in\mathcal{B}}\sum_{k\in\mathcal{K}_b} \tilde{R}_{b,k}(\boldsymbol{p}, \boldsymbol{\pi}_b(l))}{\sum_{b\in\mathcal{B}}\sum_{k\in\mathcal{K}_b} p_{b,k} + P_{\text{C}}} \quad \text{s.t. (4.34b), (4.44b).} \tag{4.47}$$

Similar to (4.46), problem (4.47) can be equivalently transformed to

$$\max_{\boldsymbol{p}\geq 0,\ \boldsymbol{r}\geq 0} \frac{\sum_{b\in\mathcal{B}}\sum_{k\in\mathcal{K}_b} r_{b,k}}{\sum_{b\in\mathcal{B}}\sum_{k\in\mathcal{K}_b} p_{b,k} + P_{\text{C}}} \quad \text{s.t. (4.34b), (4.46b), (4.46c).} \tag{4.48}$$

The objective function of (4.48) is a linear fractional function, which is nonconcave in general. Moreover, (4.46c) is nonconvex. Let us define $p_{b,k} = e^{\tilde{p}_{b,k}}$. Problem (4.48) can be rewritten as

$$\max_{\tilde{\boldsymbol{p}},\ \boldsymbol{r}\geq 0} \frac{\sum_{b\in\mathcal{B}}\sum_{k\in\mathcal{K}_b} r_{b,k}}{\sum_{b\in\mathcal{B}}\sum_{k\in\mathcal{K}_b} e^{\tilde{p}_{b,k}} + P_{\text{C}}} \quad \text{s.t.} \quad \text{(4.46b)}, \tag{4.49a}$$

$$\sum_{k\in\mathcal{K}_b} e^{\tilde{p}_{b,k}} \leq P_b^{\max}, \quad \forall b \in \mathcal{B}, \tag{4.49b}$$

$$\ln\left(2^{r_{b,k}} - 1\right) + \ln\left(\sum_{j\in\Phi_{b,k}^{(l)}} e^{\tilde{p}_{b,j}} h_{b,i} + \left(\sum_{\substack{j\in\mathcal{B}\\ j\neq b}}\left(\sum_{m\in\mathcal{K}_j} e^{\tilde{p}_{j,m}}\right) h_{j,b,i} + \sigma_{b,i}^2\right)\right)$$
$$- \left(\tilde{p}_{b,k} + \ln\left(h_{b,i}\right)\right) \leq 0, \quad \forall b \in \mathcal{B}, \quad k, i \in \mathcal{K}_b, \quad i \in \{k\} \cup \Phi_{b,k}^{(l)}. \tag{4.49c}$$

Constraint (4.46b) is affine, so is convex. Constraint (4.49b) is also convex. However, (4.49c) is nonconvex. On the left-hand side of (4.49c), the first term $\ln(2^{r_{b,k}}-1)$ is strictly concave on $r_{b,k}$, which makes (4.49) nonconvex. To this end, we apply the iterative sequential program. At each iteration t, we approximate the term $g_{b,k}\left(r_{b,k}^{(t)}\right)=\ln\left(2^{r_{b,k}^{(t)}}-1\right)$ to its first-order Taylor series expansion around $r_{b,k}^{(t-1)}$ obtained from the prior iteration $(t-1)$ as follows (Rezvani et al., 2022b, Appendix B):

$$\hat{g}_{b,k}\left(r_{b,k}^{(t)}\right)=g_{b,k}\left(r_{b,k}^{(t-1)}\right)+g'_{b,k}\left(r_{b,k}^{(t-1)}\right)\left(r_{b,k}^{(t)}-r_{b,k}^{(t-1)}\right), \tag{4.50}$$

where $g'_{b,k}\left(r_{b,k}^{(t-1)}\right)=\frac{2^{r_{b,k}^{(t-1)}}}{2^{r_{b,k}^{(t-1)}}-1}$. Hence, at each iteration t, we solve the following approximated problem

$$\max_{\tilde{\boldsymbol{p}},\,\boldsymbol{r}\geq 0}\ \frac{\sum_{b\in\mathcal{B}}\sum_{k\in\mathcal{K}_b} r_{b,k}}{\sum_{b\in\mathcal{B}}\sum_{k\in\mathcal{K}_b} e^{\tilde{p}_{b,k}}+P_{\mathrm{C}}}$$

$$\text{s.t.}\quad (4.46\text{b}),\ (4.49\text{b}), \tag{4.51a}$$

$$\hat{g}_{b,k}\left(r_{b,k}\right)+\ln\left(\sum_{j\in\Phi_{b,k}^{(l)}} e^{\tilde{p}_{b,j}}h_{b,i}+\left(\sum_{\substack{j\in\mathcal{B}\\ j\neq b}}\left(\sum_{m\in\mathcal{K}_j} e^{\tilde{p}_{j,m}}\right)h_{j,b,i}+\sigma_{b,i}^2\right)\right)$$
$$-\left(\tilde{p}_{b,k}+\ln\left(h_{b,i}\right)\right)\leq 0,\quad \forall b\in\mathcal{B},\quad k,i\in\mathcal{K}_b,\quad i\in\{k\}\cup\Phi_{b,k}^{(l)}. \tag{4.51b}$$

The approximated constraint (4.51b) is convex, since $\hat{g}_{b,k}\left(r_{b,k}\right)$ is affine, so is convex, and log-sum-exp is convex (Boyd and Vandenberghe, 2009). Accordingly, the feasible set of problem (4.51) is convex. The objective function (4.51a) is an affine-convex fractional function, so it can be defined as a concave–convex fractional function which is pseudoconcave. Therefore, the optimal solution of (4.51) can be obtained via the Dinkelbach algorithm, where we iteratively solve the following problem

$$\max_{\tilde{\boldsymbol{p}},\,\boldsymbol{r}\geq 0}\ \sum_{b\in\mathcal{B}}\sum_{k\in\mathcal{K}_b} r_{b,k}-\lambda\left(\sum_{b\in\mathcal{B}}\sum_{k\in\mathcal{K}_b} e^{\tilde{p}_{b,k}}+P_{\mathrm{C}}\right)$$

$$\text{s.t.}\quad (4.46\text{b}),\ (4.49\text{b}),\ (4.51\text{b}), \tag{4.52a}$$

where $\lambda\geq 0$ is the fractional parameter, and the objective function (4.52a) is concave. Problem (4.52) is convex, and can be solved by using the convex solvers, such as the Lagrange dual with subgradient method and IPMs.

4.5 Numerical Results

In this section, we evaluate the performance of approximated optimal powers in (4.18), SC-NOMA versus FDMA–NOMA versus FDMA, and different multi-cell NOMA schemes as FRPA, JRPA, and JSPA via the Monte Carlo simulations. In the following, the outage probability is obtained by dividing the number of infeasible case instances by the total number of channel realizations. Subsequently, the sum-rate and EE are set to zero when outage occurs.

4.5.1 Approximated Optimal Powers

The performance gap between the approximated and exact closed form of optimal powers formulated in (4.10), (4.11), and (4.18) is investigated in Fig. 4.1 for different number of NOMA users M and minimum rate demands $R_{\min}$.

The system settings for Fig. 4.1 are presented in Rezvani et al. (2021a, Subsection IV-F). As is shown, more minimum rate demands for the weaker users results in larger gap, since it would cause lower NOMA interference at weaker users. However, the average sum-rate gap is less than 1.7% and 0.7% for the AWGN powers less than −94 dBm in the macro-cell and femto-cell systems, respectively. The gap will be very low and negligible for lower number of users even at low CNR regimes, verifying the high accuracy of the approximated powers in (4.18), caused by the high insensitivity of users' SINR to their CNR.

4.5.2 SC-NOMA versus FDMA–NOMA versus FDMA

The performances of SC-NOMA (fully SC-SIC), FDMA–NOMA with different maximum number of multiplexed users on each subchannel, denoted by $U^{\max}$,

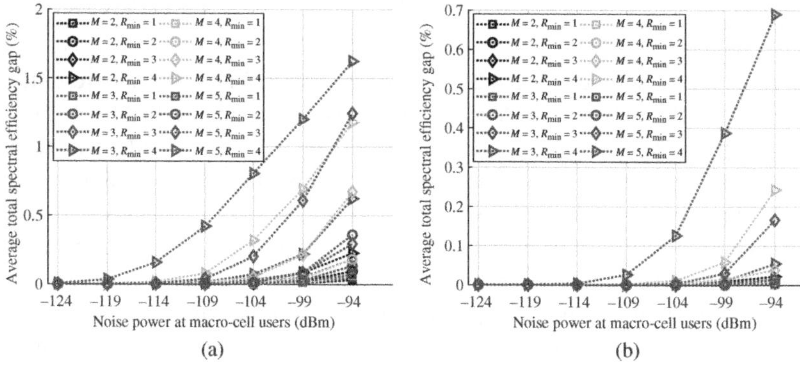

Figure 4.1 The performance gap between the approximated and exact closed form of optimal powers. Source: Rezvani et al., 2021a/IEEE. (a) Average sum-rate gap versus AWGN power at macro-cell users. (b) Average sum-rate gap versus AWGN power at femto-cell users.

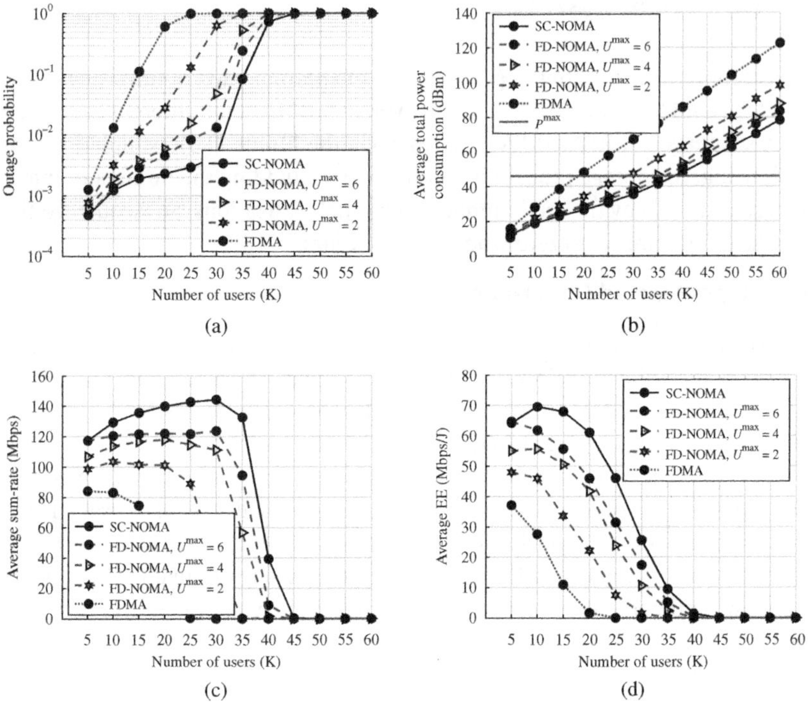

Figure 4.2 Performance comparison between FDMA, FDMA–NOMA with different U^{max}, and SC-NOMA. Source: Rezvani et al., 2022a/IEEE. (a) Outage probability versus number of users for $R_k^{min} = 3$ Mbps, $\forall k \in \mathcal{K}$. (b) Average total power consumption versus number of users for $R_k^{min} = 3$ Mbps, $\forall k \in \mathcal{K}$. (c) Average sum-rate versus number of users for $R_k^{min} = 3$ Mbps, $\forall k \in \mathcal{K}$. (d) Average EE versus number of users for $R_k^{min} = 3$ Mbps, $\forall k \in \mathcal{K}$.

and FDMA are illustrated in Fig. 4.2. The system settings for Fig. 4.2 are presented in Rezvani et al. (2022a, Table V). For simplicity, let us denote FDMA–NOMA with maximum X multiplexed users on each subchannel as X-NOMA. As is shown, there exist large performance gaps in terms of outage probability, BSs minimum power consumption, users' sum-rate, and system EE between FDMA and 2-NOMA. Significant gains can also be attained by multiplexing more than two users; however, the gaps between X-NOMA and $(X + 1)$-NOMA will be highly decreased after multiplexing more than $X = 4$ users on each subchannel. In other words, the numerical results show that multiplexing more than four users on each subchannel merely improves the system performance.

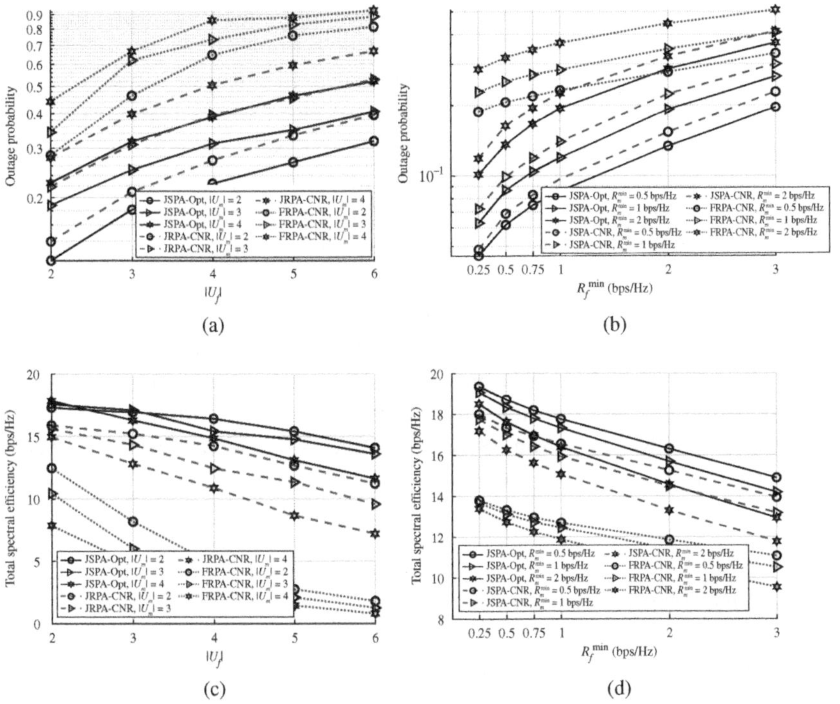

Figure 4.3 The outage probability and average sum-rate of the JSPA, JRPA, and FRPA schemes for different number of multiplexed users and minimum rate demands. Source: Rezvani et al., 2022b/IEEE. (a) Outage probability versus order of NOMA cluster for $R_m^{\min} = R_f^{\min} = 1$ bps/Hz. (b) Outage probability versus users minimum rate demand for second-order NOMA clusters. (c) Average sum-rate versus order of NOMA cluster for $R_m^{\min} = R_f^{\min} = 1$ bps/Hz. (d) Average sum-rate versus users minimum rate demand for second-order NOMA clusters.

4.5.3 Multi-cell NOMA: JSPA versus JRPA versus FRPA

Figure 4.3 evaluates the performance of the JSPA, JRPA, and FRPA algorithms in terms of outage probability and average sum-rate. For the FRPA and JRPA schemes, we adopt the CNR-based decoding order, which is potentially suboptimal.

The system settings for Fig. 4.3 are presented in Rezvani et al. (2022b, Table II). From the simulations, we observe that when at least one user pair within a cell does not satisfy the SIC sufficient condition in Theorem 4.4, so the CNR-based decoding order is not optimal, outage occurs with the FRPA scheme, due to imposing the SIC condition (4.38) on power control among the cells. In other words, in the latter case, the SIC condition (4.38) seriously limits the feasible set such that it usually makes the problem infeasible. In this way, FRPA is mostly infeasible for larger order of NOMA clusters (see Fig. 4.3(a)), where the channel gain difference among users will be lowered, and subsequently, the number of user pairs which do not satisfy the SIC sufficient condition in Theorem 4.4 increases. From Fig. 4.3(b), we observe that FRPA has also high outage probability when the order of NOMA clusters is low, even at lower minimum rate demands, verifying the inefficiency of imposing the SIC condition (4.38) on power control when the potentially suboptimal CNR-based decoding order is applied. The large performance gap between JRPA and FRPA verifies the importance of considering complete rate region, thus allowing users to operate in lower rates than their channel capacity, instead of imposing SIC condition (4.38) for the CNR-based decoding order. By comparing JSPA and JRPA, we observe that JRPA has a near-optimal performance in many cases.

4.6 Conclusions

In this chapter, we investigated resource allocation optimization for downlink NOMA systems. For SC-NOMA, we obtained the closed form of optimal power allocation for maximizing users' sum-rate and minimizing BS's power consumption under the users' individual minimum rate demand. Moreover, the EE maximization problem is solved by using the Dinkelbach algorithm, where in each iteration, the intra-cluster power allocation is obtained in closed form. The results are then extended to the MC-NOMA systems, where MC-NOMA is transformed into a virtual FDMA system. We showed that the SINR of users with lower decoding order is highly insensitive to the users' CNR. Also, we developed the weighted minimum rate fairness scheme to tackle the fairness issue of NOMA systems. For multi-cell NOMA with single-cell processing, we addressed the relation between ICI and SIC decoding order. In this line, we addressed power allocation with dynamic CINR-based decoding order, and joint power and rate allocation for any prefixed suboptimal decoding order. Numerical results show a significant performance gap between FDMA and FDMA–NOMA with the lowest (two) multiplexed users on each subchannel. For multi-cell NOMA, we observed that imposing the SIC condition usually makes the power allocation problem infeasible when the CNR-based decoding order is not optimal. Besides, we observed that allowing users to operate in lower rates than their channel capacity has a near-optimal performance for the suboptimal CNR-based decoding order.

Acknowledgments

The work of Sepehr Rezvani was supported in part by the German Research Foundation (DFG) under Grant JO 801/24-1, and the Federal Ministry of Education and Research (BMBF, Germany) in the program of "Souverän. Digital. Vernetzt." joint project 6G-RIC, project identification number: 16KISK020K and 16KISK030. The work of Eduard Jorswieck was supported in part by the Federal Ministry of Education and Research (BMBF, Germany) in the program of "Souverän. Digital. Vernetzt." joint project 6G-RIC, project identification number: 16KISK031.

References

3rd Generation Partnership Project (3GPP). *Study on Downlink Multiuser Superposition Transmission for LTE*. TSG RAN Meeting 67, RP150496, Mar. 2015.

M D Shipon Ali, Hina Tabassum, and Ekram Hossain. Dynamic user clustering and power allocation for uplink and downlink non-orthogonal multiple access (NOMA) systems. *IEEE Access*, 4:6325–6343, Aug. 2016. doi: 10.1109/ACCESS.2016.2604821.

Md Shipon Ali, Ekram Hossain, Arafat Al-Dweik, and Dong In Kim. Downlink power allocation for CoMP-NOMA in multi-cell networks. *IEEE Transactions on Communications*, 66(9):3982–3998, Sept. 2018. doi: 10.1109/TCOMM.2018.2831206.

P Bergmans. Random coding theorem for broadcast channels with degraded components. *IEEE Transactions on Information Theory*, 19(2):197–207, Mar. 1973. doi: 10.1109/TIT.1973.1054980.

Stephen Boyd and Lieven Vandenberghe. *Convex Optimization*. 7th ed. Cambridge University Press, 2009.

T M Cover. Broadcast channels. *IEEE Transactions on Information Theory*, 18(1):2–14, Jan. 1972. doi: 10.1109/TIT.1972.1054727.

Linglong Dai, Bichai Wang, Yifei Yuan, Shuangfeng Han, I Chih-lin, and Zhaocheng Wang. Non-orthogonal multiple access for 5G: Solutions, challenges, opportunities, and future research trends. *IEEE Communications Magazine*, 53(9):74–81, 2015. doi: 10.1109/MCOM.2015.7263349.

Zhiguo Ding, Xianfu Lei, George K Karagiannidis, Robert Schober, Jinhong Yuan, and Vijay K Bhargava. A survey on non-orthogonal multiple access for 5G networks: Research challenges and future trends. *IEEE Journal on Selected Areas in Communications*, 35(10):2181–2195, Oct. 2017. doi: 10.1109/JSAC.2017.2725519.

W Dinkelbach. On nonlinear fractional programming. *Management Science*, 13(7):492–498, Mar. 1967.

Jianbo Du, Wenhuan Liu, Guangyue Lu, Jing Jiang, Daosen Zhai, F Richard Yu, and Zhiguo Ding. When mobile-edge computing (MEC) meets nonorthogonal multiple access (NOMA) for the Internet of Things (IoT): System design and optimization. *IEEE Internet of Things Journal*, 8(10):7849–7862, May 2021. doi: 10.1109/JIOT.2020.3041598.

Fang Fang, Haijun Zhang, Julian Cheng, Sébastien Roy, and Victor C M Leung. Joint user scheduling and power allocation optimization for energy-efficient NOMA systems with imperfect CSI. *IEEE Journal on Selected Areas in Communications*, 35(12):2874–2885, Dec. 2017. doi: 10.1109/JSAC.2017.2777672.

Yaru Fu, Yi Chen, and Chi Wan Sung. Distributed power control for the downlink of multi-cell NOMA systems. *IEEE Transactions on Wireless Communications*, 16(9):6207–6220, Sept. 2017. doi: 10.1109/TWC.2017.2720743.

R G Gallager. Capacity and coding for degraded broadcast channels. *Problemy Peredachi Informatsii*, 10(3):185–193, 1974.

Xidong Mu, Yuanwei Liu, Li Guo, Jiaru Lin, and Naofal Al-Dhahir. Capacity and optimal resource allocation for IRS-assisted multi-user communication systems. *IEEE Transactions on Communications*, 69(6):3771–3786, 2021. doi: 10.1109/TCOMM.2021.3062651.

P Patel and J Holtzman. Analysis of a simple successive interference cancellation scheme in a DS/CDMA system. *IEEE Journal on Selected Areas in Communications*, 12(5):796–807, Jun. 1994. doi: 10.1109/49.298053.

Sepehr Rezvani. *Resource Allocation Optimization in Future Wireless Networks.* Mitteilungen aus dem Institut für Nachrichtentechnik der Technischen Universität Braunschweig, volume 76. Düren: Shaker Verlag, 2023. ISBN: 9783844091731. doi: 10.2370/9783844091731.

Sepehr Rezvani, Eduard A Jorswieck, Nader Mokari, and Mohammad R Javan. Optimal SIC ordering and power allocation in downlink multi-cell NOMA systems. *arXiv:2102.05015v3 [cs.IT]*, Sept. 2021a.

Sepehr Rezvani, Nader Mokari Yamchi, Mohammad Reza Javan, and Eduard Axel Jorswieck. Resource allocation in virtualized CoMP-NOMA HetNets: Multi-connectivity for joint transmission. *IEEE Transactions on Communications*, 69(6):4172–4185, 2021b. doi: 10.1109/TCOMM.2021.3067700.

Sepehr Rezvani, Eduard A Jorswieck, Roghayeh Joda, and Halim Yanikomeroglu. Optimal power allocation in downlink multicarrier NOMA systems: Theory and fast algorithms. *IEEE Journal on Selected Areas in Communications*, 40(4):1162–1189, Apr. 2022a. doi: 10.1109/JSAC.2022.3143237.

Sepehr Rezvani, Eduard Axel Jorswieck, Nader Mokari Yamchi, and Mohammad Reza Javan. Optimal SIC ordering and power allocation in downlink multi-cell NOMA systems. *IEEE Transactions on Wireless Communications*, 21(6):3553–3569, Jun. 2022b. doi: 10.1109/TWC.2021.3120325.

Yuya Saito, Yoshihisa Kishiyama, Anass Benjebbour, Takehiro Nakamura, Anxin Li, and Kenichi Higuchi. Non-orthogonal multiple access (NOMA) for cellular future radio access. In *Proceedings of the IEEE 77th Vehicular Technology Conference (VTC Spring)*, pages 1–5, Jun. 2013. doi: 10.1109/VTCSpring.2013.6692652.

Lou Salaün, Marceau Coupechoux, and Chung Shue Chen. Joint subcarrier and power allocation in NOMA: Optimal and approximate algorithms. *IEEE Transactions on Signal Processing*, 68:2215–2230, 2020. doi: 10.1109/TSP.2020.2982786.

Wonjae Shin, Mojtaba Vaezi, Byungju Lee, David J Love, Jungwoo Lee, and H Vincent Poor. Non-orthogonal multiple access in multi-cell networks: Theory, performance, and practical challenges. *IEEE Communications Magazine*, 55(10):176–183, Oct. 2017. doi: 10.1109/MCOM.2017.1601065.

Yan Sun, Derrick Wing Kwan Ng, Zhiguo Ding, and Robert Schober. Optimal joint power and subcarrier allocation for full-duplex multicarrier non-orthogonal multiple access systems. *IEEE Transactions on Communications*, 65(3):1077–1091, 2017. doi: 10.1109/TCOMM.2017.2650992.

David Tse and Pramod Viswanath. *Fundamentals of Wireless Communication.* Cambridge University Press, 2005. doi: 10.1017/CBO9780511807213.

Mojtaba Vaezi, Zhiguo Ding, and H Vincent Poor. *Multiple Access Techniques for 5G Wireless Networks and Beyond.* 1st ed. Switzerland: Springer Cham, 2019. ISBN 978-3-319-92090-0. doi: 10.1007/978-3-319-92090-0.

K Wang, Y Liu, Z Ding, A Nallanathan, and M Peng. User association and power allocation for multi-cell non-orthogonal multiple access networks. *IEEE Transactions on Wireless Communications*, 18(11):5284–5298, Nov. 2019. doi: 10.1109/TWC.2019.2935433.

P W Wolniansky, G J Foschini, G D Golden, and R A Valenzuela. V-BLAST: An architecture for realizing very high data rates over the rich-scattering wireless channel. In *Proceedings of the URSI International Symposium on Signals, Systems, and Electronics. Conference Proceedings (Cat. No.98EX167)*, pages 295–300, Oct. 1998. doi: 10.1109/ISSSE.1998.738086.

Zhaohui Yang, Cunhua Pan, Wei Xu, Yijin Pan, Ming Chen, and Maged Elkashlan. Power control for multi-cell networks with non-orthogonal multiple access. *IEEE Transactions on Wireless Communications*, 17(2):927–942, Feb. 2018. doi: 10.1109/TWC.2017.2772824.

R D Yates. A framework for uplink power control in cellular radio systems. *IEEE Journal on Selected Areas in Communications*, 13(7):1341–1347, Sept. 1995. doi: 10.1109/49.414651.

Alessio Zappone and Eduard Jorswieck. Energy efficiency in wireless networks via fractional programming theory. *Foundations and Trends in Communications and Information Theory*, 11(3–4):185–396, 2015. doi: 10.1561/0100000088.

Jingjing Zhao, Yuanwei Liu, Kok Keong Chai, Arumugam Nallanathan, Yue Chen, and Zhu Han. Spectrum allocation and power control for non-orthogonal multiple access in HetNets. *IEEE Transactions on Wireless Communications*, 16(9):5825–5837, Sept. 2017. doi: 10.1109/TWC.2017.2716921.

Jianyue Zhu, Jiaheng Wang, Yongming Huang, Shiwen He, Xiaohu You, and Luxi Yang. On optimal power allocation for downlink non-orthogonal multiple access systems. *IEEE Journal on Selected Areas in Communications*, 35(12):2744–2757, Dec. 2017. doi: 10.1109/JSAC.2017.2725618.

5

Cooperative NOMA

Yao Xu[1], Bo Li[2], Nan Zhao[3], Jie Tang[4], Dusit Niyato[5], and Kai-Kit Wong[6]

[1]*School of Electronic and Information Engineering, Nanjing University of Information Science and Technology, Nanjing, China*
[2]*Communication Research Center, Harbin Institute of Technology, Harbin, China*
[3]*School of Information and Communication Engineering, Dalian University of Technology, Dalian, China*
[4]*School of Electronic and Information Engineering, South China University of Technology, Guangzhou, China*
[5]*School of Computer Science and Engineering, Nanyang Technological University, Singapore, Singapore*
[6]*Department of Electronic and Electrical Engineering, University College London, London, UK*

5.1 Introduction

Cooperative non-orthogonal multiple access (CNOMA) can utilize relay forwarding and physical layer resource non-orthogonal multiplexing to achieve higher spectral efficiency and broader coverage simultaneously, thereby meeting the increasingly growing communication performance demands for future wireless communication networks. CNOMA was proposed by Ding et al. (2015) and the influence of relay selection on CNOMA was further investigated in Ding et al. (2016). Subsequently, CNOMA was extensively studied. For instance, CNOMA was integrated with simultaneous wireless information and power transfer (Liu et al., 2016), multiple-input single-output (Zhao et al., 2019), coordinated direct and relay transmission (CDRT) (Kim and Lee, 2015, Xu et al., 2021b), cognitive radio (Lv et al., 2017), and physical layer security (Cao et al., 2019).

Device-to-device (D2D) communication enables direct communication between devices without the need to go through a base station, thus D2D-assisted CNOMA technology can leverage the benefits of direct communication provided by D2D to further enhance spectral efficiency and device access density (Zhang et al., 2017, Kim et al., 2018, Xu et al., 2019, 2021c, Li et al., 2022, Kader et al., 2021, Zou et al., 2020). However, D2D-assisted CNOMA requires the cooperative procedure and the introduction of D2D may cause mutual interference between D2D users and cellular users, thus indirectly burying the high spectral efficiency potential

Next Generation Multiple Access, First Edition.
Edited by Yuanwei Liu, Liang Liu, Zhiguo Ding, and Xuemin Shen.
© 2024 The Institute of Electrical and Electronics Engineers, Inc. Published 2024 by John Wiley & Sons, Inc.

of non-orthogonal multiple access (NOMA). Meanwhile, in the D2D-assisted CNOMA system, the low adaptability caused by the fixed transmission strategy can decrease the reliability of the cell-edge user.

To address the aforementioned issue, device-to-multi-device (D2MD) can be applied to the CNOMA system, and the transmission scheme should be carefully designed. D2MD is a one-to-many D2D communication type, and one D2D transmitter can simultaneously broadcast data to several D2D receivers in a D2MD cluster, which is a critical 3GPP D2D communication scenario (Mach et al., 2015). In this chapter, we investigate a D2MD assisted CNOMA (D2MD-CNOMA) system, and present an adaptive aggregate transmission scheme using dynamic superposition coding, predesigning the decoding orders and prior information cancellation. The analysis of other performance metrics for the adaptive aggregate transmission scheme can be found in Xu et al. (2022).

Notations: In this chapter, the operator $\mathbb{E}(\cdot)$ is to get the expectation value; the operation $\Xi' \Rightarrow \Xi(z \to z')$ denotes that Ξ' can be obtained by replacing z with z'; "$\to$" means "approaching," and "$\sim$" means "be proportional to."

5.2 System Model for D2MD-CNOMA

5.2.1 System Configuration

Consider a downlink D2MD-CNOMA system shown in Fig. 5.1, where the base station can directly communicate with two cell-center users (i.e., U_1 and U_2) while

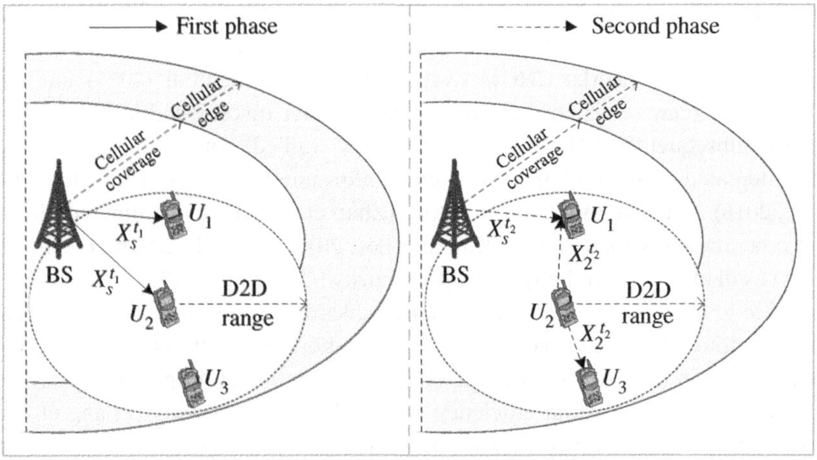

Figure 5.1 System model. An illustration of the proposed adaptive aggregate transmission scheme for the D2MD-CNOMA system with one base station, two cell-center users, and one cell-edge user.

serving a cell-edge user (i.e., U_3) via a decode-and-forward relaying user (i.e., U_2) due to the significant obstacle between the base station and U_3. Specifically, the system model consists of a D2MD cluster, in which U_2 is a potential D2D transmitter, and U_1 and U_3 are within the proximity detection region of U_2. All nodes have a single antenna and operate in half-duplex mode. The system model can be extended to the massive-user or multicell scenarios via the user pairing and hybrid multiple access in Xu et al. (2021a).

Hereafter, let subscripts s, 1, 2, and 3 denote the base station, U_1, U_2, and U_3, respectively. Let d_{xy} represent the distance between x and y, where $x, y \in \{s, 1, 2, 3\}$, $x \neq y$. In this chapter, the distance is set such that $d_{s1} < d_{s2}$ and $d_{23} < d_{21}$. Note that the system model can be used in practical applications if a feasibility-check procedure is performed to ensure the above distance conditions, and it can also be easily extended to other distance assumptions via downlink NOMA.

5.2.2 Channel Model

Assume that each channel link is subject to independent Nakagami-m block fading, and all the receiver nodes suffer from the additive white Gaussian noise with zero mean and variance N_0. The block fading means the channel changes over different blocks while remaining constant within one block. The channel coefficient, channel gain, and fading parameter between x and y are denoted by h_{xy}, λ_{xy} and m_{xy}, respectively. The expectation of λ_{xy} is $\Omega_{xy} = G_0 d_{xy}{}^{-\alpha_0}$, where α_0 and G_0 represent the path loss exponent and the channel gain for reference distance (i.e., 1 m), respectively.

Moreover, if the channel coefficient h_{xy} follows Nakagami-m distribution with an integer fading parameter m_{xy}, the cumulative distribution function and probability density function of the channel gain λ_{xy} exhibiting Gamma distribution can be expressed as

$$F_{\lambda_{xy}}(z) = 1 - \exp\left(-\frac{zm_{xy}}{\Omega_{xy}}\right) \sum_{i=0}^{m_{xy}-1} \frac{1}{i!} \left(\frac{zm_{xy}}{\Omega_{xy}}\right)^i, \tag{5.1}$$

$$f_{\lambda_{xy}}(z) = \frac{m_{xy}{}^{m_{xy}} z^{m_{xy}-1}}{\Omega_{xy}{}^{m_{xy}} \Gamma(m_{xy})} \exp\left(-\frac{m_{xy} z}{\Omega_{xy}}\right). \tag{5.2}$$

5.3 Adaptive Aggregate Transmission

To offload the data traffic and improve the spectral efficiency, we designed an adaptive transmission scheme for the D2MD-CNOMA system. One transmission block is divided into two consecutive and equal phases to complete downlink NOMA and D2D multicast transmissions, as detailed below.

5.3.1 First Phase

In the first phase (i.e., t_1), the base station broadcasts the signal $X_s^{t_1} = \sum_{i=1}^{3} \sqrt{a_i P_s} x_i$ to U_1 and U_2 using power-domain superposition coding, where P_s is the transmit power of the base station, the signal x_i contains the required information for U_i, $i \in \{1, 2, 3\}$, and a_i represents the corresponding power allocation coefficient. Since the condition $d_{s1} < d_{s2}$ holds and U_3 is a cell-edge user, the power allocation coefficient should satisfy $\sum_{i=1}^{3} a_i = 1$ and $a_3 > a_2 > a_1$. The received signals at U_j, $j \in \{1, 2\}$, can be written as

$$Y_j^{t_1} = h_{sj} X_s^{t_1} + n_j^{t_1}, \tag{5.3}$$

where $n_i^{t_k} \sim \mathcal{CN}(0, N_0)$, $k \in \{1, 2\}$, denotes the complex additive white Gaussian noise at the receiver U_i in t_k.

The cell-center user U_2 first decodes x_3 by treating x_1 and x_2 as noise, and then cancels x_3 and performs successive interference cancellation (SIC) to decode x_2 by treating x_1 as noise. Note that, based on downlink NOMA, U_2 does not need to decode x_1 in the first phase. Meanwhile, the cell-center user U_1 uses SIC to decode x_3, x_2, and x_1 sequentially. Therefore, the received signal-to-interference-and-noise ratios at U_j, $j \in \{1, 2\}$, for decoding x_3 and x_2 in t_1 can be respectively expressed as

$$\gamma_{j,x_3}^{t_1} = \frac{\lambda_{sj} a_3 \rho_s}{\lambda_{sj} \left(a_1 + a_2\right) \rho_s + 1}, \tag{5.4}$$

$$\gamma_{j,x_2}^{t_1} = \frac{\lambda_{sj} a_2 \rho_s}{\lambda_{sj} \left(a_1 + \kappa_{j,1}^{t_1} a_3\right) \rho_s + 1}, \tag{5.5}$$

where $\rho_s = \frac{P_s}{N_0}$ is the transmit signal-to-noise ratio, and the fractional error factor $\kappa_{j,1}^{t_1}$ is the residual interference levels for imperfect SIC. Specifically, a factor value of zero indicates that perfect SIC is performed, while a factor value of one represents that no SIC is performed. Moreover, the received signal-to-interference-and-noise ratio at U_1 for decoding x_1 in t_1 can be expressed as

$$\gamma_{1,x_1}^{t_1} = \frac{\lambda_{s1} a_1 \rho_s}{\lambda_{s1} \left(\kappa_{1,2}^{t_1} a_2 + \kappa_{1,3}^{t_1} a_3\right) \rho_s + 1}, \tag{5.6}$$

where $\kappa_{1,2}^{t_1}$ and $\kappa_{1,3}^{t_1}$ are the fractional error factors.

5.3.2 Second Phase

Since U_1 and U_2 cannot always decode x_3 successfully in the first phase, we propose an adaptive transmission scheme to improve the spectral efficiency. Moreover, the predetermined decoding order is designed to reduce the decoding complexity while guaranteeing the reliability of the cell-edge user.

Let the binary numbers A_1 and A_2 represent the decoding results for x_3 at U_1 and U_2 in t_1, respectively, where $A_j = 0$ and $A_j = 1, j \in \{1, 2\}$, denote failed decoding and successful decoding, respectively. A two-bit binary number $A = A_1A_2$ indicates the combination of two outage events. For instance, "$A = 3$ (decimal form)" denotes the event that both U_1 and U_2 can decode x_3 successfully in the first phase.

During the second phase (i.e., t_2), the base station transmits $X_s^{t_2} = \sqrt{P_s}x_1'$ to U_1, where x_1' is a new downlink signal required by U_1. Meanwhile, U_2 attempts to forward x_3 to U_3 as a decode-and-forward relay while acting as a D2D transmitter to communicate with U_1 and U_3 via NOMA-based D2D multicast communication. Specifically, the user U_2 broadcasts a superposed signal $X_2^{t_2}$ for case A, and $X_2^{t_2}$ can be expressed as

$$X_2^{t_2} = \sqrt{b_1^A P_u} x_{d_1} + \sqrt{b_2^A P_u} x_{d_3} + \sqrt{b_3^A P_u} x_3, \tag{5.7}$$

where x_{d_1} and x_{d_3} are the desired D2D signal for U_1 and U_3, respectively, P_u is the transmit power of U_2, and $b_i^A, i \in \{1, 2, 3\}$ is the power allocation coefficient with $\sum_{i=1}^{3} b_i^A = 1$. Due to $d_{23} < d_{21}$ and the cooperative transmission of x_3, the power allocation coefficients should satisfy $1 > b_1^A > b_3^A > b_2^A > 0$ via downlink NOMA when U_2 can decode x_3 successfully in the first phase (i.e., $A = 1, 3$). Conversely, if U_2 cannot decode x_3 in the first phase (i.e., $A = 0, 2$), the power allocation coefficient b_3^A in (5.7) should be set to $b_3^A = 0$. Meanwhile, based on downlink NOMA, the power allocation coefficients b_1^A and b_2^A should satisfy $1 > b_1^A > b_2^A > 0$. Therefore, the received signals at U_1 and U_3 in the second phase can be respectively expressed as

$$\begin{aligned} Y_{1,A}^{t_2} &= h_{21}\left(\sqrt{b_1^A P_u} x_{d_1} + \sqrt{b_2^A P_u} x_{d_3} + \sqrt{\varpi^A b_3^A P_u} x_3\right) \\ &\quad + h_{s1}\sqrt{P_s}x_1' + n_1^{t_2}, \end{aligned} \tag{5.8}$$

$$Y_{3,A}^{t_2} = h_{23}\left(\sqrt{b_1^A P_u} x_{d_1} + \sqrt{b_2^A P_u} x_{d_3} + \sqrt{b_3^A P_u} x_3\right), \tag{5.9}$$

where the flag value $\varpi^A \in \{0, 1\}$ indicates whether the interference term $\sqrt{b_3^A P_u} x_3$ in (5.8) exists.

When both U_1 and U_2 can decode x_3 successfully in the first phase (i.e., $A = 3$), U_1 can estimate $\sqrt{b_3^A P_u}$ in the second phase and use the decoded x_3 to cancel the interference $\sqrt{b_3^A P_u} x_3$ in (5.8). Therefore, the flag value ϖ^A should satisfy $\varpi^0 = \varpi^2 = \varpi^3 = 0$ and $\varpi^1 = 1$.

The received signals in (5.8) and (5.9) contain multiple signal components, which means that the diversity and uncertainty of decoding orders at U_1 and U_3 in the second phase may cause high complexity in practical implementation. To cope with this issue, we propose a predetermined decoding order strategy via power allocation coefficient design. Specifically, the D2D transmission is typically over the short distance, thus the condition $\Omega_{s1} < \Omega_{21}$ is more general than $\Omega_{s1} > \Omega_{21}$ in practice. In the proposed strategy, U_2 first attains statistical channel information Ω_{s1} and Ω_{s1} by using a two-step channel estimation (Lv et al., 2020). Then, the power allocation coefficient is designed as $b_1^A > \frac{\Omega_{s1}}{\Omega_{21}\alpha} > \max\{b_2^A, b_3^A\}$ at U_2, where $\alpha \triangleq \frac{P_u}{P_s}$ is a scale factor. Therefore, U_1 can use SIC to decode x_{d_1} and x_1' sequentially by treating the non-decoded signals as noise, while the decoding order at U_3 in the second phase is predetermined as x_{d_1}, x_3, and x_{d_3}. The received signal-to-interference-and-noise ratios for U_1 to decode x_{d_1} and x_1' in case A can be respectively expressed as

$$\gamma_{1,x_{d_1}}^{t_2,A} = \frac{\lambda_{21} b_1^A \alpha \rho_s}{\lambda_{s1}\rho_s + \lambda_{21}\left(b_2^A + \varpi^A b_3^A\right)\alpha\rho_s + 1}, \tag{5.10}$$

$$\gamma_{1,x_1'}^{t_2,A} = \frac{\lambda_{s1}\rho_s}{\lambda_{21}\left(\kappa_{1,1}^{t_2,A} b_1^A + b_2^A + \varpi^A b_3^A\right)\alpha\rho_s + 1}, \tag{5.11}$$

where $\kappa_{1,1}^{t_2,A}$ is a fractional error factor.

Moreover, the received signal-to-interference-and-noise ratios for U_3 to decode x_{d_1}, x_3, and x_{d_3} in case A can be expressed as

$$\gamma_{3,x_{d_1}}^{t_2,A} = \frac{\lambda_{23} b_1^A \alpha \rho_s}{\lambda_{23}\left(b_2^A + b_3^A\right)\alpha\rho_s + 1}, \tag{5.12}$$

$$\gamma_{3,x_3}^{t_2,A} = \frac{\lambda_{23} b_3^A \alpha \rho_s}{\lambda_{23}\left(\kappa_{3,1}^{t_2,A} b_1^A + b_2^A\right)\alpha\rho_s + 1}, \tag{5.13}$$

$$\gamma_{3,x_{d_3}}^{t_2,A} = \frac{\lambda_{23} b_2^A \alpha \rho_s}{\lambda_{23}\left(\kappa_{3,2}^{t_2,A} b_1^A + \kappa_{3,3}^{t_2,A} b_3^A\right)\alpha\rho_s + 1}, \tag{5.14}$$

where $\kappa_{3,1}^{t_2,A}$, $\kappa_{3,2}^{t_2,A}$, and $\kappa_{3,3}^{t_2,A}$ represent fractional error factors.

Based on (5.6), (5.11), and (5.14), the achievable rate for the data streams of x_1, x_1', and x_{d_3} can be respectively expressed as

$$C_{x_1} = \frac{1}{2}\log_2\left(1 + \gamma_{1,x_1}^{t_1}\right), \tag{5.15}$$

$$C_{x_1'}^A = \frac{1}{2}\log_2\left(1 + \gamma_{1,x_1'}^{t_2,A}\right), \tag{5.16}$$

$$C_{x_{d_3}}^A = \frac{1}{2}\log_2\left(1 + \gamma_{3,x_{d_3}}^{t_2,A}\right). \tag{5.17}$$

To perform SIC, U_1 and U_3 need to decode x_2 and x_{d_1} in t_1 and t_2, respectively. Therefore, based on (5.5), (5.10), and (5.12), the achievable rate for the data streams of x_2 and x_{d_1} can be respectively expressed as

$$C_{x_2} = \frac{1}{2}\log_2\left(1 + \min\left\{\gamma_{1,x_2}^{t_1}, \gamma_{2,x_2}^{t_1}\right\}\right), \tag{5.18}$$

$$C_{x_{d_1}}^{A} = \frac{1}{2}\log_2\left(1 + \min\left\{\gamma_{1,x_{d_1}}^{t_2,A}, \gamma_{3,x_{d_1}}^{t_2,A}\right\}\right). \tag{5.19}$$

The first-link and second-link achievable rates of x_3 at U_2 and U_3 are $\frac{1}{2}\log_2\left(1 + \gamma_{2,x_3}^{t_1}\right)$ and $\frac{1}{2}\log_2\left(1 + \gamma_{3,x_3}^{t_2,A}\right)$, respectively. Meanwhile, the achievable rate of x_3 at U_1 in the first phase is $\frac{1}{2}\log_2\left(1 + \gamma_{1,x_3}^{t_1}\right)$. The achievable rate of cooperative decode-and-forward relaying is dominated by the worst link, and U_1 must decode x_3 for SIC in the first phase. Therefore, using (5.4) and (5.13), we can express the achievable rate for x_3 as

$$C_{x_3}^{A} = \frac{1}{2}\log_2\left(1 + \min\left\{\gamma_{1,x_3}^{t_1}, \gamma_{2,x_3}^{t_1}, \gamma_{3,x_3}^{t_2,A}\right\}\right). \tag{5.20}$$

5.4 Performance Analysis

This section investigates the performance for the adaptive aggregate transmission scheme over Nakagami-m fading channels. Due to the adaptive characteristics, the corresponding performance analysis is quite difficult. In view of this, we use classification discussion and approximation method to obtain closed form of various metrics for proving the effectiveness of the proposed scheme and the scarcity of analysis. Specifically, the closed-form expressions for the outage probability of the proposed scheme are derived when the target data rates are fixed beforehand to satisfy the quality of service. Conversely, we use the ergodic sum capacity to evaluate the performance of the proposed scheme if the target rates change dynamically according to the channel qualities. The following derivations can provide guidance for system optimization.

5.4.1 Outage Probability

The event that the receiver node y can decode the signal $\bar{y}$ successfully in the first phase is denoted by $E_{y,\bar{y}}^{t_1} \triangleq \left\{\frac{1}{2}\log(1 + \gamma_{y,\bar{y}}^{t_1}) > R_{\bar{y}}\right\}$. Similarly, the inequality $E_{z,\bar{z}}^{t_2,A} \triangleq \left\{\frac{1}{2}\log(1 + \gamma_{z,\bar{z}}^{t_2,A}) > R_{\bar{z}}\right\}$ represents that the signal $\bar{z}$ can be decoded at the receiver node z in t_2 for case A, where $y \in \{1, 2\}$, $\bar{y} \in \{x_1, x_2, x_3\}$, $z \in \{2, 3\}$, $\bar{z} \in \left\{x_3, x_1', x_{d_1}, x_{d_3}\right\}$, and $R_{\bar{y}}$ and $R_{\bar{z}}$ are the target data rates of $\bar{y}$ and $\bar{z}$, respectively. The complement sets of $E_{y,\bar{y}}^{t_1}$ and $E_{z,\bar{z}}^{t_2,A}$ are denoted as $\bar{E}_{y,\bar{y}}^{t_1}$ and $\bar{E}_{z,\bar{z}}^{t_2,A}$, respectively. Moreover, let $\varphi_{\bar{y}} \triangleq 2^{R_{\bar{y}}} - 1$ and $\varphi_{\bar{z}} \triangleq 2^{R_{\bar{z}}} - 1$.

Before decoding x_1, U_1 must sequentially decode x_3 and x_2 first. Therefore, we can use (5.4)–(5.6) to calculate the outage probability for U_1 to decode x_1 in the first phase as

$$\begin{aligned} P_{1,\text{out}}^{t_1,x_1} &= 1 - \Pr\left(E_{1,x_3}^{t_1} \cap E_{1,x_2}^{t_1} \cap E_{1,x_1}^{t_1}\right) \\ &= 1 - \Pr\left(\lambda_{s1} > \max\{\phi_1, \phi_{2,1}, \phi_3\}\right) \\ &= F_{\lambda_{s1}}\left(\tilde{\phi}_1\right), \end{aligned} \tag{5.21}$$

where $\phi_1 = \frac{\varphi_{x_1}}{\rho_s\left(a_1 - \left(\kappa_{1,2}^{t_1}a_2 + \kappa_{1,3}^{t_1}a_3\right)\varphi_{x_1}\right)}$, $\phi_{2,j} = \frac{\varphi_{x_2}}{\rho_s\left(a_2 - \left(a_1 + \kappa_{j,1}^{t_1}a_3\right)\varphi_{x_2}\right)}$, $j \in \{1,2\}$, $\phi_3 = \frac{\varphi_{x_3}}{\rho_s\left(a_3 - (a_1 + a_2)\varphi_{x_3}\right)}$, and $\tilde{\phi}_1 = \max\left\{\phi_1, \phi_{2,1}, \phi_3\right\}$. Note that the conditions $\phi_1 > 0$, $\phi_{2,j} > 0$, and $\phi_3 > 0$ should be guaranteed via power allocation coefficient design to implement NOMA successfully in practice.

Since U_2 must adopt SIC to decode x_2 after decoding x_3, we can use (5.4) and (5.5) to calculate the outage probability for U_2 to decode x_2 during the first phase as

$$\begin{aligned} P_{2,\text{out}}^{t_1,x_2} &= 1 - \Pr\left(E_{2,x_3}^{t_1} \cap E_{2,x_2}^{t_1}\right) \\ &= 1 - \Pr\left(\lambda_{s2} > \max\left\{\phi_{2,2}, \phi_3\right\}\right) \\ &= F_{\lambda_{s2}}\left(\tilde{\phi}_2\right), \end{aligned} \tag{5.22}$$

where $\tilde{\phi}_2 = \max\{\phi_{2,2}, \phi_3\}$.

Based on the proposed scheme, U_3 may decode x_3 successfully in two cases (i.e., $A = 1, 3$). Therefore, the outage probability for U_3 to decode x_3 in the second phase can be expressed as

$$P_{3,\text{out}}^{t_2,x_3} = 1 - \sum_{\bar{k}=1,3} \mathbb{P}_{3,\bar{k}}^{t_2,x_3}, \tag{5.23}$$

where $\mathbb{P}_{3,\bar{k}}^{t_2,x_3} = \Pr(A = \bar{k}) \Pr(E_{3,x_{d_1}}^{t_2,\bar{k}} \cap E_{3,x_3}^{t_2,\bar{k}} | A = \bar{k})$. Using $\gamma_{1,x_3}^{t_1}, \gamma_{2,x_3}^{t_1}, \gamma_{3,x_{d_1}}^{t_2,1}$, and $\gamma_{3,x_3}^{t_2,1}$, we can calculate $\mathrm{P}_{3,1}^{t_2,x_3}$ as

$$\begin{aligned} \mathbb{P}_{3,1}^{t_2,x_3} &= \Pr\left(\lambda_{s1} < \phi_3, \lambda_{s2} > \phi_3, \lambda_{23} > \tilde{\phi}_3^1\right) \\ &= F_{\lambda_{s1}}(\phi_3)\left(1 - F_{\lambda_{s2}}(\phi_3)\right)\left(1 - F_{\lambda_{23}}\left(\tilde{\phi}_3^1\right)\right), \end{aligned} \tag{5.24}$$

where $\phi_4^A = \frac{\varphi_{x_{d_1}}}{\alpha\rho_s\left(b_1^A - (b_2^A + b_3^A)\varphi_{x_{d_1}}\right)}$, $\phi_6^A = \frac{\varphi_{x_3}}{\alpha\rho_s\left(b_3^A - \left(\kappa_{3,1}^{t_2,A}b_1^A + b_2^A\right)\varphi_{x_3}\right)}$, and $\tilde{\phi}_3^A = \max\left\{\phi_4^A, \phi_6^A\right\}$. The inequalities $\phi_4^A > 0$ and $\phi_6^A > 0$ should be satisfied to apply NOMA successfully. Otherwise, the outage probability for U_3 to decode x_3 is always one. Similarly, based on $\gamma_{1,x_3}^{t_1}, \gamma_{2,x_3}^{t_1}, \gamma_{3,x_{d_1}}^{t_2,3}$, and $\gamma_{3,x_3}^{t_2,3}$, the non-outage probability $\mathbb{P}_{3,3}^{t_2,x_3}$ can be calculated as

$$\mathbb{P}_{3,3}^{t_2,x_3} = \Pr\left(\lambda_{s1} > \phi_3, \lambda_{s2} > \phi_3, \lambda_{23} > \tilde{\phi}_3^3\right)$$
$$= \left(1 - F_{\lambda_{s1}}(\phi_3)\right)\left(1 - F_{\lambda_{s2}}(\phi_3)\right)\left(1 - F_{\lambda_{23}}\left(\tilde{\phi}_3^3\right)\right). \tag{5.25}$$

Substituting (5.24) and (5.25) into (5.23), we obtain $P_{3,\text{out}}^{t_2,x_3}$.

According to the proposed scheme, U_1 may decode x_{d_1} successfully in four cases (i.e., $A = 0, 1, 2, 3$). Therefore, the outage probability for U_1 to decode x_{d_1} during the second phase can be expressed as

$$P_{1,\text{out}}^{t_2,x_{d_1}} = 1 - \sum_{\bar{k}=0}^{3} \mathbb{P}_{1,\bar{k}}^{t_2,x_{d_1}}, \tag{5.26}$$

where $\mathbb{P}_{1,\bar{k}}^{t_2,x_{d_1}} = \Pr(A = \bar{k}) \Pr(E_{1,x_{d_1}}^{t_2,\bar{k}} | A = \bar{k})$. For simplicity, applying the equation (3.351.1) in Gradshteyn and Ryzhik (2007) to the following integral, we define

$$\mathcal{L}_1(\eta^*, \eta, \mu) \triangleq \int_0^{\eta^*} z^\eta \exp(-\mu z) dz$$
$$= \frac{\eta!}{\mu^{\eta+1}} - \exp(-\eta^* \mu) \sum_{\varepsilon_1=0}^{\eta} \frac{\eta! \eta^{*\varepsilon_1}}{\varepsilon_1! \mu^{\eta-\varepsilon_1+1}}. \tag{5.27}$$

Based on (5.1) and (5.27), we can use $\gamma_{1,x_3}^{t_1}, \gamma_{2,x_3}^{t_1}, \gamma_{1,x_{d_1}}^{t_2,0}$, and the binomial expansion of $(x + \rho_s^{-1})^i = \sum_{i_1=0}^{i} \binom{i}{i_1} \rho_s^{i_1 - i} x^{i_1}$ to calculate $\mathbb{P}_{1,0}^{t_2,x_{d_1}}$ as

$$\mathbb{P}_{1,0}^{t_2,x_{d_1}} = \Pr\left(\lambda_{s1} < \phi_3, \lambda_{s2} < \phi_3, \lambda_{21} > \lambda_{s1}\rho_s\phi_5^0 + \phi_5^0\right)$$
$$= \mathcal{L}_1\left(\phi_3, i_1 + m_{s1} - 1, \frac{m_{s1}}{\Omega_{s1}} + \frac{m_{21}\rho_s\phi_5^0}{\Omega_{21}}\right) F_{\lambda_{s2}}(\phi_3)\, \Delta_i^0, \tag{5.28}$$

where

$$\phi_5^A = \frac{\varphi_{x_{d_1}}}{\alpha\rho_s(b_1^A - (b_2^A + \varpi^A b_3^A)\varphi_{x_{d_1}})}, \tag{5.29a}$$

$$\Delta_i^A = \frac{m_{s1}{}^{m_{s1}}}{\Omega_{s1}{}^{m_{s1}}\Gamma(m_{s1})} \sum_{i=0}^{m_{21}-1} \sum_{i_1=0}^{i} \binom{i}{i_1} \frac{\rho_s{}^{i_1}}{i!}$$
$$\times \left(\frac{m_{21}\phi_5^A}{\Omega_{21}}\right)^i \exp\left(-\frac{m_{21}\phi_5^A}{\Omega_{21}}\right), \tag{5.29b}$$

where $\Gamma(\cdot)$ is the Gamma function. Note that the condition $\phi_5^A > 0$ should be met to implement NOMA successfully in practice. Following the same steps in (5.28), we can obtain $\mathbb{P}_{1,1}^{t_2,x_{d_1}}$ as

$$\begin{aligned}\mathbb{P}_{1,1}^{t_2,x_{d_1}} &= \Pr\left(\lambda_{s1} < \phi_3, \lambda_{s2} > \phi_3, \lambda_{21} > \lambda_{s1}\rho_s\phi_5^1 + \phi_5^1\right) \\ &= \mathcal{L}_1\left(\phi_3, i_1 + m_{s1} - 1, \frac{m_{s1}}{\Omega_{s1}} + \frac{m_{21}\rho_s\phi_5^1}{\Omega_{21}}\right)\left(1 - F_{\lambda_{s2}}\left(\phi_3\right)\right)\Delta_i^1. \end{aligned} \tag{5.30}$$

Moreover, using $\gamma_{1,x_3}^{t_1}, \gamma_{2,x_3}^{t_1}, \gamma_{1,x_{d_1}}^{t_2,2}$, and binomial expansion, we can calculate $\mathbb{P}_{1,2}^{t_2,x_{d_1}}$ as

$$\begin{aligned}\mathbb{P}_{1,2}^{t_2,x_{d_1}} &= \Pr\left(\lambda_{s1} > \phi_3, \lambda_{s2} < \phi_3, \lambda_{21} > \lambda_{s1}\rho_s\phi_5^2 + \phi_5^2\right) \\ &= \mathcal{L}_2\left(\phi_3, i_1 + m_{s1} - 1, \frac{m_{s1}}{\Omega_{s1}} + \frac{m_{21}\rho_s\phi_5^2}{\Omega_{21}}\right)F_{\lambda_{s2}}\left(\phi_3\right)\Delta_i^2, \end{aligned} \tag{5.31}$$

where the function $\mathcal{L}_2$ follows the equation (3.351.2) in Gradshteyn and Ryzhik (2007) with

$$\begin{aligned}\mathcal{L}_2\left(C, \eta, \mu\right) &\triangleq \int_C^{\infty} z^{\eta}\exp\left(-\mu z\right)dz \\ &= \exp\left(-C\mu\right)\sum_{\epsilon_1=0}^{\eta}\frac{\eta!C^{\epsilon_1}}{\epsilon_1!\mu^{\eta-\epsilon_1+1}}. \end{aligned} \tag{5.32}$$

Similar to (5.31), we can use $\gamma_{1,x_3}^{t_1}, \gamma_{2,x_3}^{t_1}, \gamma_{1,x_{d_1}}^{t_2,3}$, and binomial expansion to obtain $\mathbb{P}_{1,3}^{t_2,x_{d_1}}$ as

$$\begin{aligned}\mathbb{P}_{1,3}^{t_2,x_{d_1}} &= \Pr\left(\lambda_{s1} > \phi_3, \lambda_{s2} > \phi_3, \lambda_{21} > \lambda_{s1}\rho_s\phi_5^3 + \phi_5^3\right) \\ &= \mathcal{L}_2\left(\phi_3, i_1 + m_{s1} - 1, \frac{m_{s1}}{\Omega_{s1}} + \frac{m_{21}\rho_s\phi_5^3}{\Omega_{21}}\right)(1 - F_{\lambda_{s2}}(\phi_3))\Delta_i^3. \end{aligned} \tag{5.33}$$

Combining (5.26), (5.28), (5.30), (5.31), and (5.33), we obtain $P_{1,\text{out}}^{t_2,x_{d_1}}$.

Theorem 5.1 ***(Outage Probability for $P_{1,out}^{t_2,x_1'}$)*** The outage probability for U_1 to decode x_1' in the second phase can be written as

$$P_{1,\text{out}}^{t_2,x_1'} = 1 - \sum_{\bar{k}=0}^{3}\mathbb{P}_{1,\bar{k}}^{t_2,x_1'}, \tag{5.34}$$

where

$$\begin{aligned}\mathbb{P}_{1,0}^{t_2,x_1'} = \Bigg\{ & \Delta_{j_1}^0\left(\mathcal{L}_1\left(\phi_3, j_1 + m_{s1} - 1, \frac{m_{21}\phi_5^0\rho_s}{\Omega_{21}} + \frac{m_{s1}}{\Omega_{s1}}\right)\right. \\ & \left. - \mathcal{L}_1\left(\xi_1^0, j_1 + m_{s1} - 1, \frac{m_{21}\phi_5^0\rho_s}{\Omega_{21}} + \frac{m_{s1}}{\Omega_{s1}}\right)\right) \\ & - \Delta_{j_2}^0\left(\mathcal{L}_1\left(\phi_3, j_2 + m_{s1} - 1, \frac{m_{21}\rho_s\phi_7^0}{\Omega_{21}\varphi_{x_1'}} + \frac{m_{s1}}{\Omega_{s1}}\right)\right.\end{aligned}$$

$$-\mathcal{L}_1\left(\xi_1^0, j_2+m_{s1}-1, \frac{m_{21}\rho_s\phi_7^0}{\Omega_{21}\varphi_{x_1'}}+\frac{m_{s1}}{\Omega_{s1}}\right)\Bigg)\Bigg)\Bigg\}$$

$$\times F_{\lambda_{s2}}\left(\phi_3\right)\delta\left(\phi_7^0-\varphi_{x_1'}\phi_5^0\right), \tag{5.35a}$$

$$\mathbb{P}_{1,1}^{t_2,x_1'}=\mathbb{P}_{1,0}^{t_2,x_1'}\left(\xi_1^0\to\xi_1^1, \phi_7^0\to\phi_7^1, \phi_5^0\to\phi_5^1\right)\frac{\left(1-F_{\lambda_{s2}}\left(\phi_3\right)\right)}{F_{\lambda_{s2}}\left(\phi_3\right)}, \tag{5.35b}$$

$$\mathbb{P}_{1,2}^{t_2,x_1'}=F_{\lambda_{s2}}\left(\phi_3\right)\delta\left(\phi_7^2-\varphi_{x_1'}\phi_5^2\right)\left\{\Delta_{j_1}^2\mathcal{L}_2\left(\xi_2^2, m_{s1}+j_1-1, \frac{m_{21}\rho_s\phi_5^2}{\Omega_{21}}+\frac{m_{s1}}{\Omega_{s1}}\right)\right.$$

$$\left.-\Delta_{j_2}^2\mathcal{L}_2\left(\xi_2^2, m_{s1}+j_2-1, \frac{m_{21}\rho_s}{\Omega_{21}}\frac{\phi_7^2}{\varphi_{x_1'}}+\frac{m_{s1}}{\Omega_{s1}}\right)\right\}, \tag{5.35c}$$

$$\mathbb{P}_{1,3}^{t_2,x_1'}=\mathbb{P}_{1,2}^{t_2,x_1'}\left(\xi_2^2\to\xi_2^3, \phi_5^2\to\phi_5^3, \phi_7^2\to\phi_7^3\right)\frac{\left(1-F_{\lambda_{s2}}\left(\phi_3\right)\right)}{F_{\lambda_{s2}}\left(\phi_3\right)}, \tag{5.35d}$$

where

$$\phi_7^A=\frac{1}{(\kappa_{1,1}^{t_2,A}b_1^A+b_2^A+\varpi^A b_3^A)\alpha\rho_s}, \tag{5.36a}$$

$$\xi_1^A=\min\left\{\frac{\left(\phi_5^A+\phi_7^A\right)\varphi_{x_1'}}{(\phi_7^A-\varphi_{x_1'}\phi_5^A)\rho_s}, \phi_3\right\}, \tag{5.36b}$$

$$\xi_2^A=\max\left\{\frac{(\phi_5^A+\phi_7^A)\varphi_{x_1'}}{(\phi_7^A-\varphi_{x_1'}\phi_5^A)\rho_s}, \phi_3\right\}, \tag{5.36c}$$

$$\Delta_{j_1}^A=\frac{{m_{s1}}^{m_{s1}}}{{\Omega_{s1}}^{m_{s1}}\Gamma(m_{s1})}\exp\left(-\frac{m_{21}\phi_5^A}{\Omega_{21}}\right)$$

$$\times\sum_{j=0}^{m_{21}-1}\sum_{j_1=0}^{j}\binom{j}{j_1}\left(\frac{m_{21}\phi_5^A}{\Omega_{21}}\right)^j\frac{(\rho_s)^{j_1}}{j!}, \tag{5.36d}$$

$$\Delta_{j_2}^A=\frac{{m_{s1}}^{m_{s1}}}{{\Omega_{s1}}^{m_{s1}}\Gamma(m_{s1})}\exp\left(-\frac{m_{21}\phi_7^A}{\Omega_{21}}\right)$$

$$\times\sum_{j=0}^{m_{21}-1}\sum_{j_2=0}^{j}(-1)^{j-j_2}\binom{j}{j_2}\left(\frac{m_{21}\phi_7^A}{\Omega_{21}}\right)^j\left(\frac{\rho_s}{\varphi_{x_1'}}\right)^{j_2}\Big/ j!, \tag{5.36e}$$

and the step function $\delta(z)$ takes the value of zero for $z \leq 0$ and one for $z > 0$.

Proof: See Appendix 5.A. □

The outage probability for U_3 to decode x_{d_3} in the second phase should be discussed in four cases (i.e., $A = 0, 1, 2, 3$), and the corresponding expression can be written as

$$P_{3,\mathrm{out}}^{t_2,x_{d_3}} = 1 - \sum_{\bar{k}=0}^{3} \mathbb{P}_{3,\bar{k}}^{t_2,x_{d_3}}, \tag{5.37}$$

where $\mathbb{P}_{3,\bar{k}}^{t_2,x_{d_3}} = \Pr(A = \bar{k}) \Pr(E_{3,x_{d_1}}^{t_2,\bar{k}} \cap E_{3,x_{d_3}}^{t_2,\bar{k}} | A = \bar{k})$. When the events that both U_1 and U_2 cannot decode x_3 in the first phase occur (i.e., $A = 0$), we can use $\gamma_{1,x_3}^{t_1}, \gamma_{2,x_3}^{t_1}$, $\gamma_{3,x_{d_1}}^{t_2,0}$, and $\gamma_{3,x_{d_3}}^{t_2,0}$ to calculate $\mathbb{P}_{3,0}^{t_2,x_{d_3}}$ as

$$\begin{aligned} \mathbb{P}_{3,0}^{t_2,x_{d_3}} &= \Pr\left(\lambda_{s1} < \phi_3, \lambda_{s2} < \phi_3, \lambda_{23} > \tilde{\phi}_4^0\right) \\ &= F_{\lambda_{s1}}\left(\phi_3\right) F_{\lambda_{s2}}\left(\phi_3\right)\left(1 - F_{\lambda_{23}}\left(\tilde{\phi}_4^0\right)\right), \end{aligned} \tag{5.38}$$

where $\tilde{\phi}_4^A = \max\left\{\phi_4^A, \phi_8^A\right\}$ and $\phi_8^A = \frac{\varphi_{x_{d_3}}}{\alpha\rho_s\left(b_2^A - \left(\kappa_{3,2}^{t_2,A} b_1^A + \kappa_{3,3}^{t_2,A} b_3^A\right)\varphi_{x_{d_3}}\right)}$. Similarly, the condition $\phi_8^A > 0$ should be met, otherwise the outage probability for U_3 to decode x_{d_3} is always one.

If U_1 cannot decode x_3 in the first phase while U_2 can decode x_3 successfully (i.e., $A = 1$), the non-outage probability for U_3 to decode x_{d_3} during the second phase can be calculated as

$$\begin{aligned} \mathbb{P}_{3,1}^{t_2,x_{d_3}} &= \Pr\left(\lambda_{s1} < \phi_3, \lambda_{s2} > \phi_3, \lambda_{23} > \tilde{\phi}_5^1\right) \\ &= F_{\lambda_{s1}}\left(\phi_3\right)\left(1 - F_{\lambda_{s2}}\left(\phi_3\right)\right)\left(1 - F_{\lambda_{23}}\left(\tilde{\phi}_5^1\right)\right), \end{aligned} \tag{5.39}$$

where $\tilde{\phi}_5^A = \max\left\{\phi_4^A, \phi_6^A, \phi_8^A\right\}$.

Similarly, using $\gamma_{1,x_3}^{t_1}, \gamma_{2,x_3}^{t_1}, \gamma_{3,x_{d_1}}^{t_2,2}$, and $\gamma_{3,x_{d_3}}^{t_2,2}$, we can calculate $\mathbb{P}_{1,2}^{t_2,x_{d_3}}$ as

$$\begin{aligned} \mathbb{P}_{3,2}^{t_2,x_{d_3}} &= \Pr\left(\lambda_{s1} > \phi_3, \lambda_{s2} < \phi_3, \lambda_{23} > \tilde{\phi}_4^2\right) \\ &= \left(1 - F_{\lambda_{s1}}\left(\phi_3\right)\right) F_{\lambda_{s2}}\left(\phi_3\right)\left(1 - F_{\lambda_{23}}\left(\tilde{\phi}_4^2\right)\right). \end{aligned} \tag{5.40}$$

When the events that both U_1 and U_2 can decode x_3 successfully during the second phase occur (i.e., $A = 3$), the non-outage probability for U_3 to decode x_{d_3} can be calculated by using $\gamma_{1,x_3}^{t_1}, \gamma_{2,x_3}^{t_1}, \gamma_{3,x_{d_1}}^{t_2,3}, \gamma_{3,x_3}^{t_2,3}$, and $\gamma_{3,x_{d_3}}^{t_2,3}$ as

$$\begin{aligned} \mathbb{P}_{3,3}^{t_2,x_{d_3}} &= \Pr\left(\lambda_{s1} > \phi_3, \lambda_{s2} > \phi_3, \lambda_{23} > \tilde{\phi}_5^3\right) \\ &= \left(1 - F_{\lambda_{s1}}\left(\phi_3\right)\right)\left(1 - F_{\lambda_{s2}}\left(\phi_3\right)\right)\left(1 - F_{\lambda_{23}}\left(\tilde{\phi}_5^3\right)\right). \end{aligned} \tag{5.41}$$

Substituting (5.38)–(5.41) into (5.37), we obtain $P_{3,\mathrm{out}}^{t_2,x_{d_3}}$.

5.4.2 Ergodic Sum Capacity

When the target data rates vary dynamically according to the channel qualities, the ergodic sum capacity for the proposed scheme can be expressed as

$$\overline{C}_{\text{sum}} = \sum_{\forall \dot{x} \in \dot{X}} \overline{C}_{\dot{x}} + \sum_{\forall \ddot{x} \in \ddot{X}} \overline{C}^{3}_{\ddot{x}}, \tag{5.42}$$

where $\overline{C}_{\dot{x}} = \mathbb{E}[C_{\dot{x}}]$, $\overline{C}^{3}_{\ddot{x}} = \mathbb{E}[C^{3}_{\ddot{x}}]$, $\dot{X} \in \{x_1, x_2\}$, and $\ddot{X} \in \{x_3, x_{d_1}, x_{d_3}, x'_1\}$.

For simplicity, we first provide some defined functions and integral calculation equations before deriving the ergodic capacity. Applying the equation (3.353.5) in Gradshteyn and Ryzhik (2007) to the following integral, we have

$$\begin{aligned} \mathcal{L}_3(\eta, \mu) &\triangleq \int_0^\infty \frac{z^\eta}{1+z} \exp(-\mu z) dz \\ &= \sum_{\kappa=1}^{\eta} (-1)^{\eta-\kappa} (\kappa-1)! \mu^{-\kappa} + (-1)^{\eta-1} \exp(\mu) \text{Ei}(-\mu), \end{aligned} \tag{5.43}$$

where $\text{Ei}(\cdot)$ is the exponential integral function. Using the variable substitution, binomial expansion, the equations (3.351.2) and (3.351.4) in Gradshteyn and Ryzhik (2007), we can define the following integral as

$$\begin{aligned} \mathcal{L}_4(\eta, \overline{\eta}, \eta^*, \mu) &\triangleq \int_0^\infty \frac{z^\eta}{(z+\eta^*)^{\overline{\eta}}} \exp(-\mu z)\, dz \\ &= \sum_{\eta_1=0}^{\eta} \binom{\eta}{\eta_1} (-\eta^*)^{\eta-\eta_1} \exp(\mu \eta^*) \\ &\times \begin{cases} \exp(-\mu\eta^*) \sum\limits_{\eta_2=0}^{\eta_1-\overline{\eta}} \dfrac{(\eta_1-\overline{\eta})!(\eta^*)^{\eta_2}}{\eta_2! \mu^{\eta_1-\overline{\eta}-\eta_2+1}}, & \eta_1 - \overline{\eta} \geq 0, \\ \dfrac{-(-\mu)^{\overline{\eta}-\eta_1-1} \text{Ei}(-\mu\eta^*)}{(\overline{\eta}-\eta_1-1)!} + \dfrac{\exp(-\mu\eta^*)}{(\eta^*)^{\overline{\eta}-\eta_1-1}} & \\ \times \sum\limits_{\eta_3=0}^{\overline{\eta}-\eta_1-2} \dfrac{(-\mu\eta^*)^{\eta_3} (\overline{\eta}-\eta_1-\eta_3-2)!}{(\overline{\eta}-\eta_1-1)!}, & \eta_1 - \overline{\eta} < 0, \end{cases} \end{aligned} \tag{5.44}$$

where η^* denotes a constant. Moreover, we give an integral calculation equation

$$\int_0^\infty \ln(z+1) f_Z(z) dz = \int_0^\infty \frac{1-F_Z(z)}{1+z} dz. \tag{5.45}$$

Let $X_1 \triangleq \lambda_{s1}(a_1 + \kappa^{t_1}_{1,2} a_2 + \kappa^{t_1}_{1,3} a_3)\rho_s$ and $X_2 \triangleq \lambda_{s1}(\kappa^{t_1}_{1,2} a_2 + \kappa^{t_1}_{1,3} a_3)\rho_s$. Using (5.6), (5.15), and (5.45), we have

$$\overline{C}_{x_1} = \frac{1}{2\ln 2} \left\{ \int_0^\infty \frac{1-F_{X_1}(x)}{1+x} dx - \int_0^\infty \frac{1-F_{X_2}(x)}{1+x} dx \right\}$$

$$=\frac{1}{2\ln 2}\left\{\Delta_{d_1}\mathcal{L}_3\left(d_1,\frac{m_{s1}}{\Omega_{s1}\left(a_1+\kappa_{1,2}^{t_1}a_2+\kappa_{1,3}^{t_1}a_3\right)\rho_s}\right)\right.$$
$$\left.-\Delta_{d_2}\mathcal{L}_3\left(d_2,\frac{m_{s1}}{\Omega_{s1}\left(\kappa_{1,2}^{t_1}a_2+\kappa_{1,3}^{t_1}a_3\right)\rho_s}\right)\right\}, \tag{5.46}$$

where the equation $\mathcal{L}_3(\eta,\infty)=0$ holds, $F_{X_1}(x)=F_{\lambda_{s1}}\left(\frac{x}{\left(a_1+\kappa_{1,2}^{t_1}a_2+\kappa_{1,3}^{t_1}a_3\right)\rho_s}\right)$, $F_{X_2}(x)=F_{\lambda_{s1}}\left(\frac{x}{\left(\kappa_{1,2}^{t_1}a_2+\kappa_{1,3}^{t_1}a_3\right)\rho_s}\right)$, $\Delta_{d_1}=\sum_{d_1=0}^{m_{s1}-1}\left(\frac{m_{s1}}{\Omega_{s1}\left(a_1+\kappa_{1,2}^{t_1}a_2+\kappa_{1,3}^{t_1}a_3\right)\rho_s}\right)^{d_1}/d_1!$, and $\Delta_{d_2}=\sum_{d_2=0}^{m_{s1}-1}\left(\frac{m_{s1}}{\Omega_{s1}\left(\kappa_{1,2}^{t_1}a_2+\kappa_{1,3}^{t_1}a_3\right)\rho_s}\right)^{d_2}/d_2!$.

For mathematical tractability, we assume that U_1 and U_2 suffer the same residual interference as x_3 (i.e., $\kappa_{1,1}^{t_1}=\kappa_{2,1}^{t_1}$) when U_1 and U_2 perform imperfect SIC to decode x_3 in the first phase. Let $Y\triangleq\min(\lambda_{s1},\lambda_{s2})$, $Y_1\triangleq Y(a_1+a_2+\kappa_{1,1}^{t_1}a_3)\rho_s$, and $Y_2\triangleq Y(a_1+\kappa_{1,1}^{t_1}a_3)\rho_s$. Using the order statistic, we can obtain the cumulative distribution function of Y as

$$F_Y(y)=1-\left(1-F_{\lambda_{s1}}(y)\right)\left(1-F_{\lambda_{s2}}(y)\right)$$
$$=1-\Delta_e y^{e_1+e_2}\exp\left(-\left(\frac{m_{s1}}{\Omega_{s1}}+\frac{m_{s2}}{\Omega_{s2}}\right)y\right), \tag{5.47}$$

where $\Delta_e=\sum_{e_1=0}^{m_{s1}-1}\sum_{e_2=0}^{m_{s2}-1}\left(\frac{m_{s1}}{\Omega_{s1}}\right)^{e_1}\left(\frac{m_{s2}}{\Omega_{s2}}\right)^{e_2}/e_1!/e_2!$. Furthermore, the cumulative distribution functions of Y_1 and Y_2 can be written as $F_{Y_1}(y)=F_Y\left(\frac{y}{(a_1+a_2+\kappa_{1,1}^{t_1}a_3)\rho_s}\right)$ and $F_{Y_2}(y)=F_Y\left(\frac{y}{(a_1+\kappa_{1,1}^{t_1}a_3)\rho_s}\right)$, respectively. By using (5.5), (5.18), (5.45), and (5.47), the achievable data rate $\overline{C}_{x_2}$ can be calculated as

$$\overline{C}_{x_2}=\frac{1}{2\ln 2}\left\{\int_0^\infty\frac{1-F_{Y_1}(y)}{1+y}dy-\int_0^\infty\frac{1-F_{Y_2}(y)}{1+y}dy\right\}$$
$$=\frac{1}{2\ln 2}\left\{\Delta_{e_1}\mathcal{L}_3\left(e_1+e_2,\Theta_1\right)-\Delta_{e_2}\mathcal{L}_3\left(e_1+e_2,\Theta_2\right)\right\}, \tag{5.48}$$

where $\Delta_{e_1}=\Delta_e\left(\left(a_1+a_2+\kappa_{1,1}^{t_1}a_3\right)\rho_s\right)^{-e_1-e_2}$, $\Theta_1=\frac{m_{s1}/\Omega_{s1}+m_{s2}/\Omega_{s2}}{\left(a_1+a_2+\kappa_{1,1}^{t_1}a_3\right)\rho_s}$, $\Delta_{e_2}=\Delta_e\left(\left(a_1+\kappa_{1,1}^{t_1}a_3\right)\rho_s\right)^{-e_1-e_2}$, and $\Theta_2=\frac{m_{s1}/\Omega_{s1}+m_{s2}/\Omega_{s2}}{\left(a_1+\kappa_{1,1}^{t_1}a_3\right)\rho_s}$.

The ergodic capacity of x_3 is determined by $\bar{U} \triangleq \min\{\bar{U}_1, \bar{U}_2, \bar{U}_3\}$, where $\bar{U}_1 \triangleq \gamma_{1,x_3}^{t_1}$, $\bar{U}_2 \triangleq \gamma_{2,x_3}^{t_1}$, and $\bar{U}_3 \triangleq \gamma_{3,x_3}^{t_2,3}$. The cumulative distribution function of $\bar{U}_i$, $i \in \{1, 2, 3\}$ is given by

$$F_{\bar{U}_i}(u) = \begin{cases} \overline{F}_{\bar{U}_i}(u), & u < \theta_i, \\ 1, & u \geq \theta_i, \end{cases} \tag{5.49}$$

where $\overline{F}_{\bar{U}_1}(u) = F_{\lambda_{s1}}\left(\frac{u}{(a_3-(a_1+a_2)u)\rho_s}\right)$, $\overline{F}_{\bar{U}_2}(u) = F_{\lambda_{s2}}\left(\frac{u}{(a_3-(a_1+a_2)u)\rho_s}\right)$, $\overline{F}_{\bar{U}_3}(u) = F_{\lambda_{23}}\left(\frac{u}{(b_3^3-(\kappa_{3,1}^{t_2,3}b_1^3+b_2^3)u)\alpha\rho_s}\right)$, $\theta_1 = \theta_2 = \frac{a_3}{a_1+a_2}$, and $\theta_3 = \frac{b_3^3}{\kappa_{3,1}^{t_2,3}b_1^3+b_2}$. Using (5.49) and order statistic, we have

$$\begin{aligned} F_{\bar{U}}(u) &= 1 - \Pi_{i=1}^{3}(1 - F_{\bar{U}_i}(u)) \\ &= \begin{cases} 1 - \Pi_{i=1}^{3}(1 - \overline{F}_{\bar{U}_i}(u)), & u < \tilde{\theta}, \\ 1, & u \geq \tilde{\theta}, \end{cases} \end{aligned} \tag{5.50}$$

where $\tilde{\theta} = \min\{\theta_1, \theta_3\}$. Based on (5.20), (5.45), and (5.50), we can use Gaussian–Chebyshev quadrature to approximate $\overline{C}_{x_3}^3$ as

$$\begin{aligned} \overline{C}_{x_3}^3 &= \frac{1}{2\ln 2}\int_0^{\tilde{\theta}} \frac{\Pi_{i=1}^{3}\left(1 - \overline{F}_{\bar{U}_i}(u)\right)}{1+u} du \\ &\approx \frac{\tilde{\theta}\pi}{4\ln 2}\sum_{n_1=1}^{N_1} \frac{\sqrt{1-\psi_{n_1}^2}}{N_1} \frac{\Pi_{i=1}^{3}\left(1 - \overline{F}_{\bar{U}_i}(q_{n_1})\right)}{1+q_{n_1}}, \end{aligned} \tag{5.51}$$

where $q_{n_1} = \frac{(1+\psi_{n_1})\tilde{\theta}}{2}$, $\psi_{n_1} = \cos\left(\frac{2n_1-1}{2N_1}\pi\right)$, and N_1 is a complexity-accuracy trade-off parameter.

Let $V \triangleq \gamma_{1,x_1'}^{t_2,3}$. The cumulative distribution function of V is $F_V(v) = 1 - \Delta_f (v + F_1)^{-f_1-m_{21}} v^f \exp\left(-\frac{m_{s1}}{\Omega_{s1}\rho_s} v\right)$ by using (5.1), (5.11), and the binomial expansion, where $\Delta_f = \frac{m_{21}{}^{m_{21}}}{\Omega_{21}{}^{m_{21}}\Gamma(m_{21})} \sum_{f=0}^{m_{s1}-1} \sum_{f_1=0}^{f} \binom{f}{f_1} \frac{(f_1+m_{21}-1)!}{f!} \left(\frac{m_{s1}}{\Omega_{s1}}\right)^{f-f_1-m_{21}} \times \left(\kappa_{1,1}^{t_2,3} b_1^3 + b_2^3\right)^{-m_{21}} \alpha^{-m_{21}} \rho_s^{f_1-f}$ and $F_1 = \frac{m_{21}\Omega_{s1}}{m_{s1}\Omega_{21}\left(\kappa_{1,1}^{t_2,3}b_1^3+b_2^3\right)\alpha}$. Thus, using (5.16), (5.45)–(5.47), and the partial fraction decomposition, we can calculate the closed-form expression of $\overline{C}_{x_1'}^3$ as

$$\begin{aligned} \overline{C}_{x_1'}^3 &= \frac{\Delta_f}{2\ln 2}\int_0^{\infty} \frac{v^f}{(v+F_1)^{f_1+m_{21}}(1+v)} \exp\left(-\frac{m_{s1}}{\Omega_{s1}\rho_s} v\right) dv \\ &= \frac{\Delta_f}{2\ln 2}\left\{\sum_{f_2=1}^{f_1+m_{21}} -\frac{1}{(F_1-1)^{f_2}} \mathcal{L}_4\left(f, \bar{f}, F_1, \frac{m_{s1}}{\Omega_{s1}\rho_s}\right)\right. \\ &\quad \left. + \frac{1}{(F_1-1)^{f_1+m_{21}}} \mathcal{L}_3\left(f, \frac{m_{s1}}{\Omega_{s1}\rho_s}\right)\right\}, \end{aligned} \tag{5.52}$$

where $\bar{f} = f_1 + m_{21} - f_2 + 1$.

Let $W_1 \triangleq \gamma_{3,x_{d_1}}^{t_2,3}$, $W_2 \triangleq \gamma_{1,x_{d_1}}^{t_2,3}$, and $W \triangleq \min\{W_1, W_2\}$. Similar to (5.50), the cumulative distribution function of W can be expressed as

$$F_W(w) = 1 - \Pi_{i=1}^{2}(1 - F_{W_i}(w))$$
$$= \begin{cases} 1 - \Pi_{i=1}^{2}(1 - \overline{F}_{W_i}(w)), & w < \dfrac{b_1^3}{b_2^3 + b_3^3}, \\ 1, & w \geq \dfrac{b_1^3}{b_2^3 + b_3^3}, \end{cases} \tag{5.53}$$

where $\overline{F}_{W_1}(w) = F_{\lambda_{23}}\left(\frac{w}{\alpha\rho_s(b_1^3-(b_2^3+b_3^3)w)}\right)$, $\overline{F}_{W_2}(w) = 1 - \Delta_g(w)G(w)^{-g_1-m_{s1}}$, $G(w) = \frac{m_{21}\rho_s w}{(b_1^3-b_2^3w)\alpha\rho_s\Omega_{21}} + \frac{m_{s1}}{\Omega_{s1}}$, $\Delta_g(w) = \sum_{g=0}^{m_{21}-1}\sum_{g_1=0}^{g}\frac{m_{s1}{}^{m_{s1}}}{\Omega_{s1}{}^{m_{s1}}\Gamma(m_{s1})g!}(\frac{m_{21}}{\Omega_{21}})^g \times \binom{g}{g_1}\rho_s{}^{g_1}(g_1 + m_{s1} - 1)!\left(\frac{w}{(b_1^3-b_2^3w)\alpha\rho_s}\right)^g \exp\left(-\frac{m_{21}w}{(b_1^3-b_2^3w)\alpha\rho_s\Omega_{21}}\right)$. Therefore, we can use (5.19), (5.45), (5.53), and Gaussian–Chebyshev quadrature to obtain an approximation of $\overline{C}_{x_{d_1}}^3$ as

$$\overline{C}_{x_{d_1}}^3 \approx \frac{b_1^3\pi}{4(b_2^3 + b_3^3)\ln 2}\sum_{n_2=1}^{N_2}\frac{\sqrt{1-\psi_{n_2}^2}}{N_2}\frac{\Pi_{i=1}^{2}(1 - \overline{F}_{W_i}(q_{n_2}))}{1+q_{n_2}}, \tag{5.54}$$

where $q_{n_2} = \frac{(1+\psi_{n_2})b_1^3}{2(b_2^3+b_3^3)}$, $\psi_{n_2} = \cos\left(\frac{2n_2-1}{2N_2}\pi\right)$, and N_2 denotes a complexity-accuracy trade-off parameter.

Following the same steps in (5.46), we can use $\gamma_{3,x_{d_3}}^{t_2,3}$ to calculate $\overline{C}_{x_{d_3}}^3$ as

$$\overline{C}_{x_{d_3}}^3 = \frac{1}{2\ln 2}\left\{\Delta_{h_1}\mathcal{L}_3\left(h_1, \frac{m_{23}}{\Omega_{23}\left(b_2^3 + \kappa_{3,2}^{t_2,3}b_1^3 + \kappa_{3,3}^{t_2,3}b_3^3\right)\alpha\rho_s}\right)\right.$$
$$\left. -\Delta_{h_2}\mathcal{L}_3\left(h_2, \frac{m_{23}}{\Omega_{23}\left(\kappa_{3,2}^{t_2,3}b_1^3 + \kappa_{3,3}^{t_2,3}b_3^3\right)\alpha\rho_s}\right)\right\}, \tag{5.55}$$

where $\Delta_{h_1} = \sum_{h_1=0}^{m_{23}-1}\frac{1}{h_1!}\left(\frac{m_{23}}{\Omega_{23}(b_2^3+\kappa_{3,2}^{t_2,3}b_1^3+\kappa_{3,3}^{t_2,3}b_3^3)\alpha\rho_s}\right)^{h_1}$ and $\Delta_{h_2} = \sum_{h_2=0}^{m_{23}-1}\frac{1}{h_2!}\times\left(\frac{m_{23}}{\Omega_{23}(\kappa_{3,2}^{t_2,3}b_1^3+\kappa_{3,3}^{t_2,3}b_3^3)\alpha\rho_s}\right)^{h_2}$.

Combining (5.42), (5.46), (5.48), (5.51), (5.52), (5.54), and (5.55), we obtain $\overline{C}_{\text{sum}}$.

5.5 Numerical Results and Discussion

In this section, the performance of the adaptive aggregate transmission scheme (i.e., the Prop shown in the figures) is investigated through extensive Monte Carlo simulations. Specifically, we compare the outage probability and ergodic sum capacity of the adaptive aggregate transmission scheme with the D2D aided CDRT (D-CDRT) scheme in Zou et al. (2020) under the same simulation parameter settings. In the simulations, we consider two distance settings, i.e., Case I: $d_{s1} = 80\,\text{m}$, $d_{s2} = 90\,\text{m}$, $d_{21} = 15\,\text{m}$, and $d_{23} = 5\,\text{m}$; Case II: $d_{s1} = 50\,\text{m}$, $d_{s2} = 55\,\text{m}$, $d_{21} = 10\,\text{m}$, and $d_{23} = 5\,\text{m}$. The channel gain for reference distance, the path loss exponent, the noise power, and the scale factor are set to $G_0 = -40\,\text{dB}$, $\alpha_0 = 2.7$, $N_0 = -110\,\text{dBm}$, and $\alpha = 0.05$, respectively. Moreover, the target data rates are set to $R_{x_1} = R_{x_2} = R_{x_3} = R_{x_{d_3}} = 0.3\,\text{bit/s/Hz}$ and $R'_{x_1} = R_{x_{d_1}} = 0.1\,\text{bit/s/Hz}$. The power allocation coefficients are $a_1 = 0.01$, $a_2 = 0.09$, and $a_3 = 0.9$, and the power allocation coefficient combinations (A, b_1^A, b_2^A, b_3^A) are set to $(0, 0.99, 0.01, 0)$, $(1, 0.9, 0.01, 0.09)$, $(2, 0.99, 0.01, 0)$, and $(3, 0.9, 0.01, 0.09)$. For simplicity, we assumed that all the fractional error factors (e.g., $\kappa_{1,2}^{t_1}$ and $\kappa_{1,1}^{t_2,A}$) equal κ_0 for imperfect SIC.

To implement NOMA successfully, the D-CDRT scheme requires power control at the base station. Since $d_{s1} < d_{s2}$ and $d_{23} < d_{21}$, the power allocation coefficients of the D-CDRT are set as $\alpha_{e1} = a_2 + a_3$, $\alpha_c = a_1$, $\alpha_d = b_1^3 + b_3^3$, and $\alpha_{e2} = b_2^3$, while the corresponding power ratio is assumed to be $\beta = \frac{\alpha_d P_u}{2P_s}$ for fair comparisons.

5.5.1 Outage Probability

Figure 5.2 compares the outage probabilities for U_1 and U_3 to decode x_1 and x_3 in t_1 and t_2 (i.e., $P_{1,\text{out}}^{t_1,x_1}$ and $P_{3,\text{out}}^{t_2,x_3}$) between the adaptive aggregate transmission scheme and D-CDRT scheme. In this figure, the accurate theoretical results for the adaptive aggregate transmission scheme are in good agreement with the simulation ones. Figure 5.2 illustrates that the adaptive aggregate transmission scheme achieves the same outage probability for U_1 in t_1 as the D-CDRT scheme. This is because the decoding procedure and parameter settings (i.e., the target data rates, power allocation coefficients, and fractional error factors) of the adaptive aggregate transmission scheme are consistent with the D-CDRT. Although the decoding process of x_1 is affected by the residual interference caused by imperfect SIC, this process will not face the adjacent-channel interference. Therefore, the outage probability corresponding to x_1 has no error floors for large P_s. Unlike the stationary transmission scheme of the D-CDRT, the adaptive aggregate transmission scheme designs a predetermined decoding order strategy, thus U_3 can receive x_3 in t_2 as long as U_2 can decode x_3 successfully in t_1. Based on this, the adaptive aggregate transmission scheme can achieve a lower $P_{3,\text{out}}^{t_2,x_3}$ than that of the D-CDRT,

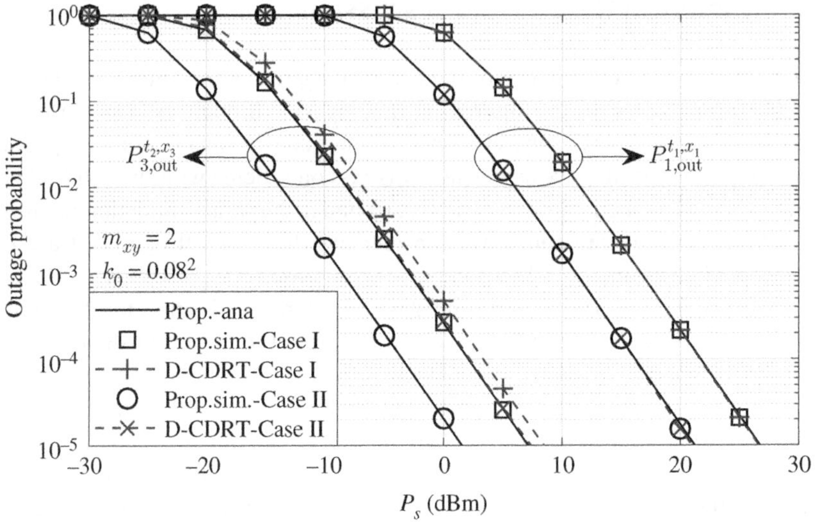

Figure 5.2 Outage probability for U_1 and U_3 to decode x_1 and x_3 in t_1 and t_2 (i.e., $P_{1,out}^{t_1,x_1}$ and $P_{3,out}^{t_2,x_3}$), respectively.

and the performance superiority becomes more evident in Case II. The adaptive aggregate transmission scheme has better robustness for facing imperfect SIC than that of the D-CDRT.

Figure 5.3 shows the outage probabilities for U_1 to decode x_1' and x_{d_1} in t_2, (i.e., $P_{1,out}^{t_2,x_1'}$ and $P_{1,out}^{t_2,x_{d_1}}$), respectively. Since U_1 suffers the inter-user interference from U_2 in t_2, both the adaptive aggregate transmission and D-CDRT schemes have error floors for the outage probability related to x_1' for large P_s. Specifically, when imperfect SIC is performed, the adaptive aggregate transmission scheme has a lower error floor for large P_s than the D-CDRT due to the aggregation transmission. It is worth noting that the decoding process of x_{d_1} for the adaptive aggregate transmission and D-CDRT schemes will not suffer the residual interference, thus imperfect SIC has no impact on $P_{1,out}^{t_2,x_{d_1}}$. Due to the interference from the base station, $P_{1,out}^{t_2,x_{d_1}}$ is subject to error floors for large P_s. Meanwhile, the adaptive aggregate transmission scheme achieves a higher error floor of $P_{1,out}^{t_2,x_{d_1}}$ for larger P_s because a portion of the transmit power at U_2 is allocated to x_{d_3}. Conversely, the adaptive aggregate transmission scheme can attain lower $P_{1,out}^{t_2,x_{d_1}}$ for small P_s, which benefits from the design of adaptive transmission.

Figure 5.4 depicts the outage probabilities for U_2 and U_3 to decode x_2 and x_{d_3} in t_1 and t_2 (i.e., $P_{2,out}^{t_1,x_2}$ and $P_{3,out}^{t_2,x_{d_3}}$), respectively. In this figure, the correctness of the theoretical analysis for the adaptive aggregate transmission scheme is verified

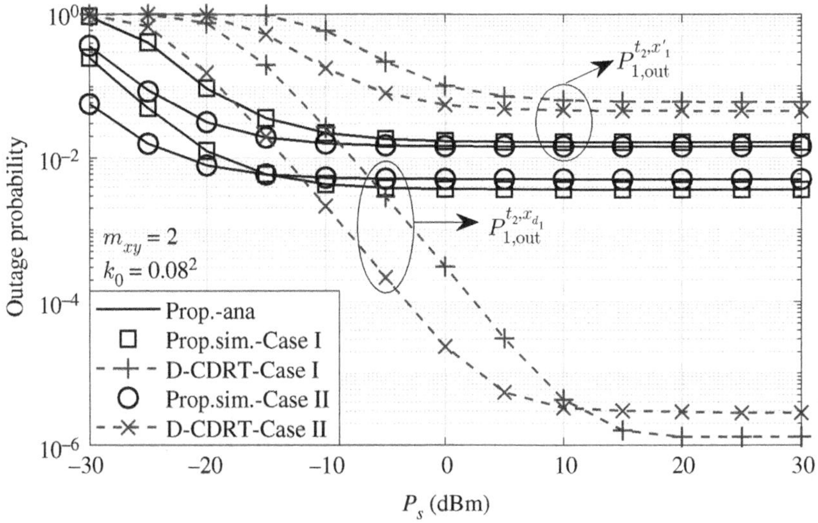

Figure 5.3 Outage probability for U_1 to decode x'_1 and x_{d_1} in t_2 (i.e., $P_{1,out}^{t_2,x'_1}$ and $P_{1,out}^{t_2,x_{d_1}}$), respectively.

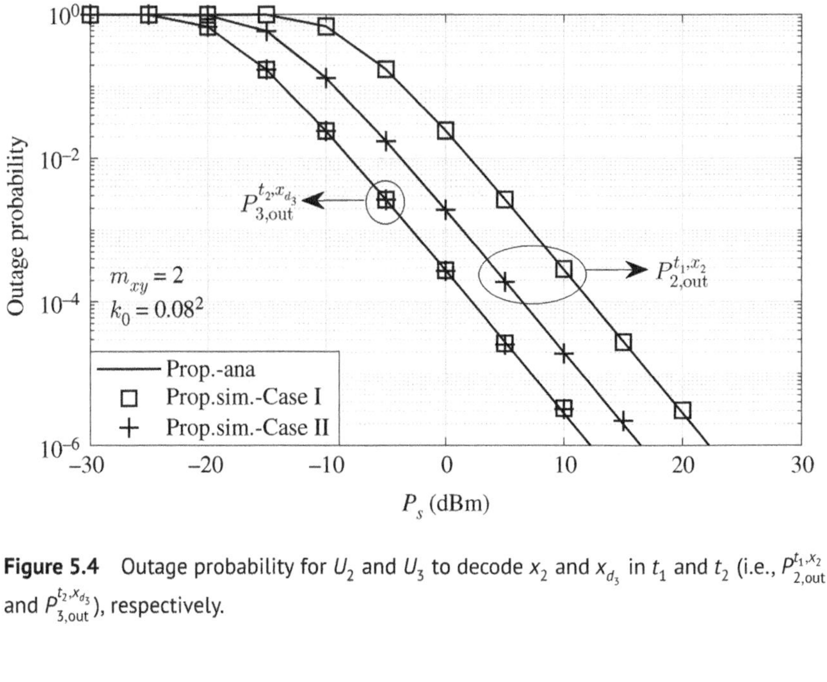

Figure 5.4 Outage probability for U_2 and U_3 to decode x_2 and x_{d_3} in t_1 and t_2 (i.e., $P_{2,out}^{t_1,x_2}$ and $P_{3,out}^{t_2,x_{d_3}}$), respectively.

through simulations. The proposed scheme can use the same amount of time resource to transmit two extra data streams (i.e., x_2 and x_{d_3}) compared with the D-CDRT scheme. From Fig. 5.4, we observe that $P_{2,\text{out}}^{t_1,x_2}$ in Case II is lower than that in Case I. This is because Case II has a larger λ_{s2} than that in Case I. The lines for $P_{3,\text{out}}^{t_2,x_{d_3}}$ in Case I almost coincide with that in Case II under imperfect SIC, because both Cases I and II have the same d_{23}. Moreover, both $P_{2,\text{out}}^{t_1,x_2}$ and $P_{3,\text{out}}^{t_2,x_{d_3}}$ have no error floors.

5.5.2 Ergodic Sum Capacity

Figure 5.5 shows the relationship between the ergodic sum capacity and the transmit power of the base station. To investigate the ergodic sum capacity difference between NOMA-based schemes (i.e., the adaptive aggregate transmission scheme, D-CDRT (Zou et al., 2020), and CDRT (Kim and Lee, 2015)) and orthogonal multiple access (OMA) schemes, we design two benchmarks using time division multiple access (TDMA) (i.e., OMA-I and OMA-II) for fair comparisons. Specifically, OMA-I divides the whole transmission period into seven phases and adopts TDMA to transmit six signals of the adaptive aggregate transmission scheme. Similarly, OMA-II completes the transmission of four signals in the D-CDRT scheme by using five phases. In Fig. 5.5, the theoretical analysis results of ergodic sum capacity for the adaptive aggregate transmission scheme are perfectly matched with the corresponding simulation ones. The

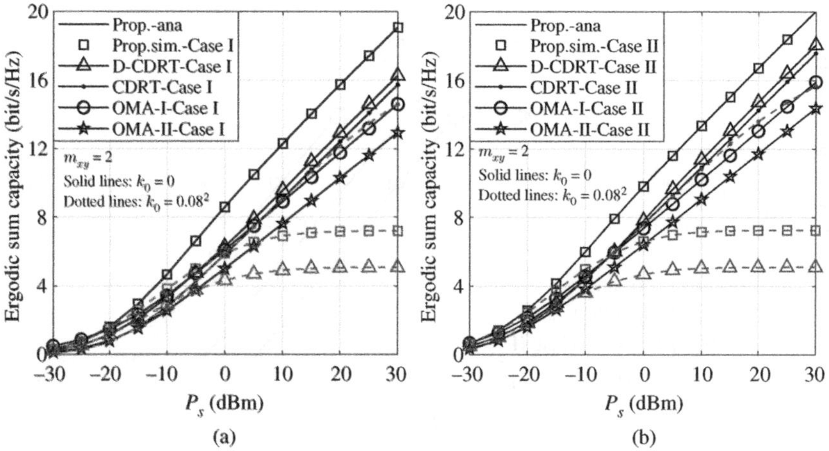

Figure 5.5 Comparison of Ergodic sum capacity among the adaptive aggregate transmission scheme, D-CDRT, CDRT, OMA-I, and OMA-II. (a) Case I and (b) Case II.

adaptive aggregate transmission scheme can realize the aggregation transmission of multiple data streams and outperforms the D-CDRT scheme in terms of ergodic sum capacity under both perfect and imperfect SIC. When perfect SIC is performed, the adaptive aggregate transmission scheme and the D-CDRT scheme can achieve better ergodic sum capacity compared with CDRT and OMA-based schemes. Instead, the opposite results can be observed in the high transmit power region for imperfect SIC. This is because imperfect SIC significantly affects the ergodic sum capacity of the adaptive aggregate transmission scheme and the D-CDRT schemes, resulting in the existence of capacity ceiling for large P_s. Moreover, since Case II has better channel qualities than Case I, the achievable ergodic sum capacity in Case II is superior to that in Case I.

Overall, the proposed adaptive aggregate transmission scheme can achieve better outage performance of the cell-edge user and spectral efficiency than the D-CDRT scheme. When the level of residual interference is low, the proposed scheme can also attain superior spectral efficiency than the OMA schemes.

5.A Appendix

5.A.1 Proof of Theorem 5.1

In the second phase, U_1 can adopt SIC to decode x_1' after decoding x_{d_1}. Based on the proposed scheme, the outage probability for U_1 to decode x_1' should be discussed in four cases (i.e., $A = 0, 1, 2, 3$) and given in (5.34).

If both U_1 and U_2 cannot decode x_3 successfully in the first phase (i.e., $A = 0$), U_2 only transmits x_{d_1} and x_{d_3} via superposition coding in the second phase. Based on the proposed predetermined decoding order strategy, U_2 can use SIC to decode x_1' after removing x_{d_1}. In this case, we can use $\gamma_{1,x_3}^{t_1}$, $\gamma_{2,x_3}^{t_1}$, $\gamma_{1,x_{d_1}}^{t_2,0}$ and $\gamma_{1,x_1'}^{t_2,0}$ to calculate the non-outage probability for U_1 to decode x_1' as

$$\begin{aligned}\mathbb{P}_{1,0}^{t_2,x_1'} &= \Pr\left(\gamma_{1,x_3}^{t_1} < \varphi_{x_3}, \gamma_{2,x_3}^{t_1} < \varphi_{x_3}, \gamma_{1,x_{d_1}}^{t_2,0} > \varphi_{x_{d_1}}, \gamma_{1,x_1'}^{t_2,0} > \varphi_{x_1'}\right) \\ &= \Pr\left(\xi_1^0 < \lambda_{s1} < \phi_3, \lambda_{s1}\frac{\rho_s\phi_7^0}{\varphi_{x_1'}} - \phi_7^0 > \lambda_{21} > \lambda_{s1}\rho_s\phi_5^0 + \phi_5^0\right) \\ &\quad \times F_{\lambda_{s2}}(\phi_3), \end{aligned} \tag{5.56}$$

where $\xi_1^A = \min\left\{\frac{(\phi_5^A+\phi_7^A)\varphi_{x_1'}}{(\phi_7^A-\varphi_{x_1'}\phi_5^A)\rho_s}, \phi_3\right\}$, $\phi_3 = \frac{\varphi_{x_3}}{\rho_s\left(a_3-(a_1+a_2)\varphi_{x_3}\right)}$, $\phi_5^A = \frac{\varphi_{x_{d_1}}}{\alpha\rho_s} \times \frac{1}{(b_1^A-(b_2^A+\varpi^A b_3^A)\varphi_{x_{d_1}})}$, and $\phi_7^A = \frac{1}{(\kappa_{1,1}^{t_2,A} b_1^A+b_2^A+\varpi^A b_3^A)\alpha\rho_s}$. The condition $\frac{\rho_s\phi_7^0}{\varphi_{x_1'}} - \phi_7^0 > \lambda_{s1}\rho_s\phi_5^0 + \phi_5^0$ should be satisfied, otherwise $\mathbb{P}_{1,0}^{t_2,x_1'}$ is always zero. Therefore,

$\mathbb{P}_{1,0}^{t_2,x_1'}$ in (5.56) can be rewritten as

$$\mathbb{P}_{1,0}^{t_2,x_1'} = \int_{\xi_1^0}^{\phi_3} \int_{\rho_s\phi_5^0 x+\phi_5^0}^{\rho_s\phi_7^0 x/\varphi_{x_1'} -\phi_7^0} f_{\lambda_{s1}}(x) f_{\lambda_{21}}(y)dxdy \times F_{\lambda_{s2}}(\phi_3)\delta\left(\phi_7^0 - \varphi_{x_1'}\phi_5^0\right), \tag{5.57}$$

The step function $\delta(z)$ is zero and one for $z \leq 0$ and $z > 0$, respectively. Using (5.27) and some integral calculations, we can calculate $\mathbb{P}_{1,0}^{t_2,x_1'}$ as

$$\begin{aligned}\mathbb{P}_{1,0}^{t_2,x_1'} = &\left\{ \Delta_{j_1}^0 \left(\mathcal{L}_1 \left(\phi_3, j_1 + m_{s1} - 1, \frac{m_{21}\phi_5^0\rho_s}{\Omega_{21}} + \frac{m_{s1}}{\Omega_{s1}} \right) \right.\right.\\ &\left. -\mathcal{L}_1 \left(\xi_1^0, j_1 + m_{s1} - 1, \frac{m_{21}\phi_5^0\rho_s}{\Omega_{21}} + \frac{m_{s1}}{\Omega_{s1}} \right)\right)\\ &- \Delta_{j_2}^0 \left(\mathcal{L}_1 \left(\phi_3, j_2 + m_{s1} - 1, \frac{m_{21}\rho_s\phi_7^0}{\Omega_{21}\varphi_{x_1'}} + \frac{m_{s1}}{\Omega_{s1}} \right)\right.\\ &\left.\left. -\mathcal{L}_1 \left(\xi_1^0, j_2 + m_{s1} - 1, \frac{m_{21}\rho_s\phi_7^0}{\Omega_{21}\varphi_{x_1'}} + \frac{m_{s1}}{\Omega_{s1}} \right)\right)\right\}\\ &\times F_{\lambda_{s2}}(\phi_3)\,\delta\left(\phi_7^0 - \varphi_{x_1'}\phi_5^0\right), \end{aligned} \tag{5.58}$$

where $\Delta_{j_1}^A = \frac{{m_{s1}}^{m_{s1}}}{{\Omega_{s1}}^{m_{s1}}\Gamma(m_{s1})} \exp\left(-\frac{m_{21}\phi_5^A}{\Omega_{21}}\right) \sum_{j=0}^{m_{21}-1} \sum_{j_1=0}^{j} \binom{j}{j_1} \left(\frac{m_{21}\phi_5^A}{\Omega_{21}}\right)^j \frac{(\rho_s)^{j_1}}{j!}$ and $\Delta_{j_2}^A = \frac{{m_{s1}}^{m_{s1}}}{{\Omega_{s1}}^{m_{s1}}\Gamma(m_{s1})} \exp\left(-\frac{m_{21}\phi_7^A}{\Omega_{21}}\right) \sum_{j=0}^{m_{21}-1} \sum_{j_2=0}^{j} \frac{(-1)^{j-j_2}}{j!} \binom{j}{j_2} \left(\frac{m_{21}\phi_7^A}{\Omega_{21}}\right)^j \left(\frac{\rho_s}{\varphi_{x_1'}}\right)^{j_2}$.

Similarly, if $A = 1$, U_2 will broadcast the combination of x_3, x_{d_1} and x_{d_3} in the second phase. Therefore, U_1 decodes x_{d_1} and x_1' sequentially by treating x_{d_3} and x_3 as noise. Based on this, using $\gamma_{1,x_3}^{t_1}$, $\gamma_{2,x_3}^{t_1}$, $\gamma_{1,x_{d_1}}^{t_2,1}$ and $\gamma_{1,x_1'}^{t_2,1}$, we can write $\mathbb{P}_{1,1}^{t_2,x_1'}$ for $A = 1$ as

$$\begin{aligned}\mathbb{P}_{1,1}^{t_2,x_1'} &= \Pr\left(\gamma_{1,x_3}^{t_1} < \varphi_{x_3}, \gamma_{2,x_3}^{t_1} > \varphi_{x_3}, \gamma_{1,x_{d_1}}^{t_2,1} > \varphi_{x_{d_1}}, \gamma_{1,x_1'}^{t_2,1} > \varphi_{x_1'}\right)\\ &= \Pr\left(\xi_1^1 < \lambda_{s1} < \phi_3, \lambda_{s1}\frac{\rho_s\phi_7^1}{\varphi_{x_1'}} - \phi_7^1 > \lambda_{21} > \lambda_{s1}\rho_s\phi_5^1 + \phi_5^1\right)\\ &\quad\times\left(1 - F_{\lambda_{s2}}(\phi_3)\right)\delta(\phi_7^1 - \varphi_{x_1'}\phi_5^1)\\ &= \mathbb{P}_{1,0}^{t_2,x_1'}\left(\xi_1^0 \to \xi_1^1, \phi_7^0 \to \phi_7^1, \phi_5^0 \to \phi_5^1\right)\frac{\left(1 - F_{\lambda_{s2}}(\phi_3)\right)}{F_{\lambda_{s2}}(\phi_3)}. \end{aligned} \tag{5.59}$$

When $A = 2$, U_1 can decode x_3 successfully while U_2 cannot decode x_3 in the first phase. Therefore, U_2 transmits the combination of x_{d_1} and x_{d_3} in the second phase. In this case, U_1 decodes x_{d_1} and x_1' sequentially by treating x_{d_3} as noise.

Based on this, we can use $\gamma_{1,x_3}^{t_1}$, $\gamma_{2,x_3}^{t_1}$, $\gamma_{1,x_{d_1}}^{t_2,2}$ and $\gamma_{1,x_1'}^{t_2,2}$ to calculate the non-outage probability for U_1 to decode x_1' as

$$\begin{aligned}
\mathbb{P}_{1,2}^{t_2,x_1'} &= \Pr\left(\gamma_{1,x_3}^{t_1} > \varphi_{x_3}, \gamma_{2,x_3}^{t_1} < \varphi_{x_3}, \gamma_{1,x_{d_1}}^{t_2,2} > \varphi_{x_{d_1}}, \gamma_{1,x_1'}^{t_2,2} > \varphi_{x_1'}\right) \\
&= \Pr\left(\lambda_{s1} > \phi_3, \lambda_{s1}\frac{\rho_s\phi_7^2}{\varphi_{x_1'}} - \phi_7^2 > \lambda_{21} > \lambda_{s1}\rho_s\phi_5^2 + \phi_5^2\right) \\
&\quad \times F_{\lambda_{s2}}(\phi_3)\delta\left(\phi_7^2 - \varphi_{x_1'}\phi_5^2\right),
\end{aligned} \tag{5.60}$$

Similar to (5.57), we can use (5.32) and some integral calculations to rewrite (5.60) as

$$\begin{aligned}
\mathbb{P}_{1,2}^{t_2,x_1'} &= \left\{\Delta_{j_1}^2\mathcal{L}_2\left(\xi_2^2, m_{s1} + j_1 - 1, \frac{m_{21}\rho_s\phi_5^2}{\Omega_{21}} + \frac{m_{s1}}{\Omega_{s1}}\right)\right. \\
&\quad \left. -\Delta_{j_2}^2\mathcal{L}_2\left(\xi_2^2, m_{s1} + j_2 - 1, \frac{m_{21}\rho_s}{\Omega_{21}}\frac{\phi_7^2}{\varphi_{x_1'}} + \frac{m_{s1}}{\Omega_{s1}}\right)\right\} \\
&\quad \times F_{\lambda_{s2}}\left(\phi_3\right)\delta\left(\phi_7^2 - \varphi_{x_1'}\phi_5^2\right),
\end{aligned} \tag{5.61}$$

where $\xi_2^A = \max\left\{\frac{(\phi_5^A+\phi_7^A)\varphi_{x_1'}}{(\phi_7^A-\varphi_{x_1'}\phi_5^A)\rho_s}, \phi_3\right\}$.

Similarly, U_1 can decode x_{d_1} and x_1' sequentially in the second phase by treating x_{d_3} as noise when both U_1 and U_2 can decode x_3 successfully (i.e., $A = 3$) in the first phase. This is because U_1 can use the known signal of x_3 to remove the corresponding interference caused by x_3. Following the same steps in (5.61), we can use $\gamma_{1,x_3}^{t_1}$, $\gamma_{2,x_3}^{t_1}$, $\gamma_{1,x_{d_1}}^{t_2,3}$ and $\gamma_{1,x_1'}^{t_2,3}$ to obtain $\mathbb{P}_{1,3}^{t_2,x_1'}$ as

$$\begin{aligned}
\mathbb{P}_{1,3}^{t_2,x_1'} &= \Pr\left(\gamma_{1,x_3}^{t_1} > \varphi_{x_3}, \gamma_{2,x_3}^{t_1} > \varphi_{x_3}, \gamma_{1,x_{d_1}}^{t_2,3} > \varphi_{x_{d_1}}, \gamma_{1,x_1'}^{t_2,3} > \varphi_{x_1'}\right) \\
&= \Pr\left(\xi_1^1 < \lambda_{s1} > \phi_3, \lambda_{s1}\frac{\rho_s\phi_7^3}{\varphi_{x_1'}} - \phi_7^3 > \lambda_{21} > \lambda_{s1}\rho_s\phi_5^3 + \phi_5^3\right) \\
&\quad \times \left(1 - F_{\lambda_{s2}}(\phi_3)\right)\delta(\phi_7^3 - \varphi_{x_1'}\phi_5^3) \\
&= \mathbb{P}_{1,2}^{t_2,x_1'}\left(\xi_2^2 \to \xi_2^3, \phi_5^2 \to \phi_5^3, \phi_7^2 \to \phi_7^3\right)\frac{\left(1 - F_{\lambda_{s2}}\left(\phi_3\right)\right)}{F_{\lambda_{s2}}\left(\phi_3\right)}.
\end{aligned} \tag{5.62}$$

Substituting (5.58), (5.59), (5.61) and (5.62) into (5.34), we obtain $P_{1,\text{out}}^{t_2,x_1'}$.

The analytical results in Theorem5.1 can be used to minimize the outage probability for U_1 to decode x_1' by formulating the optimization problem and designing the power allocation coefficient.

References

Yang Cao, Nan Zhao, Gaofeng Pan, Yunfei Chen, Lisheng Fan, Minglu Jin, and Mohamed-Slim Alouini. Secrecy analysis for cooperative NOMA networks with multi-antenna full-duplex relay. *IEEE Transactions on Communications*, 67(8):5574–5587, 2019. doi: 10.1109/TCOMM.2019.2914210.

Zhiguo Ding, Mugen Peng, and H Vincent Poor. Cooperative non-orthogonal multiple access in 5G systems. *IEEE Communications Letters*, 19(8):1462–1465, 2015. doi: 10.1109/LCOMM.2015.2441064.

Zhiguo Ding, Huaiyu Dai, and H Vincent Poor. Relay selection for cooperative NOMA. *IEEE Wireless Communications Letters*, 5(4):416–419, 2016. doi: 10.1109/LWC.2016.2574709.

I S Gradshteyn and I M Ryzhik. *Table of Integrals, Series, and Products*. Burlington, MA: Academic Press, 2007.

Md Fazlul Kader, S M Riazul Islam, and Octavia A Dobre. Simultaneous cellular and D2D communications exploiting cooperative uplink NOMA. *IEEE Communications Letters*, 25(6):1848–1852, 2021. doi: 10.1109/LCOMM.2021.3062111.

Jung-Bin Kim and In-Ho Lee. Non-orthogonal multiple access in coordinated direct and relay transmission. *IEEE Communications Letters*, 19(11):2037–2040, 2015. doi: 10.1109/LCOMM.2015.2474856.

Jung-Bin Kim, In-Ho Lee, and JunHwan Lee. Capacity scaling for D2D aided cooperative relaying systems using NOMA. *IEEE Wireless Communications Letters*, 7(1):42–45, 2018. doi: 10.1109/LWC.2017.2752162.

Runzhou Li, Peilin Hong, Kaiping Xue, and Te Yang. A cooperative D2D content sharing scheme using NOMA under social ties. *IEEE Communications Letters*, 26(6):1433–1437, 2022. doi: 10.1109/LCOMM.2021.3097762.

Yuanwei Liu, Zhiguo Ding, Maged Elkashlan, and H Vincent Poor. Cooperative non-orthogonal multiple access with simultaneous wireless information and power transfer. *IEEE Journal on Selected Areas in Communications*, 34(4):938–953, 2016. doi: 10.1109/JSAC.2016.2549378.

Lu Lv, Jian Chen, Qiang Ni, and Zhiguo Ding. Design of cooperative non-orthogonal multicast cognitive multiple access for 5G systems: User scheduling and performance analysis. *IEEE Transactions on Communications*, 65(6):2641–2656, 2017. doi: 10.1109/TCOMM.2017.2677942.

Lu Lv, Hai Jiang, Zhiguo Ding, Long Yang, and Jian Chen. Secrecy-enhancing design for cooperative downlink and uplink NOMA with an untrusted relay, 2020.

Pavel Mach, Zdenek Becvar, and Tomas Vanek. In-band device-to-device communication in OFDMA cellular networks: A survey and challenges. *IEEE Communications Surveys and Tutorials*, 17(4):1885–1922, 2015. doi: 10.1109/COMST.2015.2447036.

Yao Xu, Gang Wang, Bo Li, and Shaobo Jia. Performance of D2D aided uplink coordinated direct and relay transmission using NOMA. *IEEE Access*, 7:151090–151102, 2019. doi: 10.1109/ACCESS.2019.2946421.

Yao Xu, Julian Cheng, Gang Wang, and Victor C M Leung. Coordinated direct and relay transmission for multiuser networks: NOMA or hybrid multiple access? *IEEE Wireless Communications Letters*, 10(5):976–980, 2021a. doi: 10.1109/LWC.2021.3052894.

Yao Xu, Julian Cheng, Gang Wang, and Victor C M Leung. Adaptive coordinated direct and relay transmission for NOMA networks: A joint downlink-uplink scheme. *IEEE Transactions on Wireless Communications*, 20(7):4328–4346, 2021b. doi: 10.1109/TWC.2021.3058122.

Yao Xu, Bo Li, Nan Zhao, Yunfei Chen, Gang Wang, Zhiguo Ding, and Xianbin Wang. Coordinated direct and relay transmission with NOMA and network coding in Nakagami-*m* fading channels. *IEEE Transactions on Communications*, 69(1):207–222, 2021c. doi: 10.1109/TCOMM.2020.3025555.

Yao Xu, Jie Tang, Bo Li, Nan Zhao, Dusit Niyato, and Kai-Kit Wong. Adaptive aggregate transmission for device-to-multi-device aided cooperative NOMA networks. *IEEE Journal on Selected Areas in Communications*, 40(4):1355–1370, 2022. doi: 10.1109/JSAC.2022.3143267.

Zhengquan Zhang, Zheng Ma, Ming Xiao, Zhiguo Ding, and Pingzhi Fan. Full-duplex device-to-device-aided cooperative nonorthogonal multiple access. *IEEE Transactions on Vehicular Technology*, 66(5):4467–4471, 2017. doi: 10.1109/TVT.2016.2600102.

Nan Zhao, Wei Wang, Jingjing Wang, Yunfei Chen, Yun Lin, Zhiguo Ding, and Norman C Beaulieu. Joint beamforming and jamming optimization for secure transmission in MISO-NOMA networks. *IEEE Transactions on Communications*, 67(3):2294–2305, 2019. doi: 10.1109/TCOMM.2018.2883079.

Lisha Zou, Jian Chen, Lu Lv, and Bingtao He. Capacity enhancement of D2D aided coordinated direct and relay transmission using NOMA. *IEEE Communications Letters*, 24(10):2128–2132, 2020. doi: 10.1109/LCOMM.2020.3000996.

6

Multi-scale-NOMA: An Effective Support to Future Communication–Positioning Integration System

Lu Yin, Wenfang Guo, and Tianzhu Song

School of Electronic Engineering, Beijing University of Posts and Telecommunications, Beijing, China

6.1 Introduction

In recent years, the exponential growth of communication technology has spurred the emergence of novel intelligent industries, such as autonomous driving, Industrial Internet of Things (IIoT) and intelligent transportation systems, etc., which rely heavily on the availability of highly precise location information (Butt et al., 2021, Pan et al., 2022). Location-based services (LBS) provide users with an array of time and space-based services through the utilization of positioning and navigation technology. To efficiently support the growing demand for LBS, high-performance positioning becomes necessary, especially in positioning accuracy, reliability, and service range (Yin et al., 2018).

Global navigation satellite system (GNSS) is widely preferred by various industries due to its extensive coverage, accurate positioning performance, and cost-efficiency. GNSS has demonstrated excellent positioning accuracy in the open environment. However, in complex scenarios like urban canyons, indoor and underground environments, the GNSS's positioning performance is compromised by signal blockage, poor satellite geometry, and multipath effects (Pan et al., 2022, Saleh et al., 2022, Bai et al., 2022, Chen et al., 2022). Furthermore, GNSS signals are vulnerable to jamming and spoofing (Bai et al., 2022). To address these issues, it is essential to explore techniques for detecting and eliminating interference. Additionally, integrating GNSS with multi-sensor technologies such as inertial navigation systems (INS), light detection, and ranging (LiDAR) (Bai et al., 2022).

However, the integrated navigation system with satellite navigation as its fundamental component still faces challenges in providing the high-precision position services in scenarios where the GNSS signal remains unavailable for an extended

Next Generation Multiple Access, First Edition.
Edited by Yuanwei Liu, Liang Liu, Zhiguo Ding, and Xuemin Shen.
© 2024 The Institute of Electrical and Electronics Engineers, Inc. Published 2024 by John Wiley & Sons, Inc.

period. For example, when operating in standalone mode without recalibration from other sources (such as GNSS), the INS inherently accumulates errors, resulting in significant positioning errors. Therefore, INS can only serve as a short-term solution for bridging GNSS outages. In addition, the performance of perception systems, including but not limited to vision cameras, LiDAR, and radars, is highly dependent on environmental and weather conditions, which can decrease their stability and reliability (Saleh et al., 2022).

Wireless communication networks have the advantages of extensive coverage, high reliability, and cost-efficiency, making them a promising alternative to satellite positioning. Nowadays, communication–positioning integration technology, which integrates both communication and positioning functions in a single system, has become a research hotspot. In fact, cellular networks (1G–5G), satellite communication systems, Wi-Fi, bluetooth, ultra-wide band (UWB), and broadcasting systems can all provide positioning functions. Among them, wide-area communication systems, such as cellular networks(1G–3G), satellite communication systems, and broadcast systems, suffer from poor positioning performance due to a lack of specialized design for positioning signal. Local positioning technologies such as Wi-Fi, Bluetooth, and UWB provide high accuracy but are limited in range, effectiveness, and reliability (Chen et al., 2022). A detailed comparison of these technologies is provided in this article (Luo et al., 2017). 4G and 5G have taken positioning performance into account in signal design, and designed signals specifically for positioning, such as positioning reference signal (PRS), realizing the coexistence of communication signals and positioning signals, while the positioning accuracy is also improved.

Especially in the 5G era, benefiting from large bandwidth, high cell densities, and multiple-input–multiple-output (MIMO), it is possible to achieve submeter accuracy positioning, as demonstrated in various studies (Saleh et al., 2022, Bai et al., 2022, Li et al., 2023, Ko et al., 2022). However, due to the limitation of resources and hardware, 5G PRS has not been well used in fact. The subsequent Section 6.2 details the integration process between cellular networks and positioning function from loosely coupled systems to tightly coupled systems and introduces 5G PRS.

6.2 Positioning in Cellular Networks

In the 1970s, the concept of the cellular network was introduced, which opened up the possibility of using cellular networks for positioning. In 1996, the E911 Act promulgated by the US Federal Communications Commission further promoted the introduction of positioning function in the communication system. Subsequently, 2G began to develop positioning standards that were further refined in the 3G era. Despite these promising changes, the 1G–3G eras relied on pilot or control signals in the communication process to perform positioning, resulting in limited accuracy with the maximum positioning accuracy of 3G reaching only tens of meters.

In 2009, 3GPP RAN1#56b conference reached a consensus on broadcasting PRS for the first time in the 4G networks, which marked the formal integration of positioning systems in cellular networks at the signal level. The PRS is a group of Gold pseudo-random sequences modulated by quadrantal phase shift keying (QPSK). During resource mapping, PRS is arranged according to the comb structure in the frequency domain, which occupies only a certain number of OFDM symbols in the time domain, but it does not support continuous broadcasting. Due to resource limitations such as bandwidth and time, achieving sub-meter-level accuracy in 4G is challenging.

The PRS in the 5G era has been further improved in three main aspects (3GPP, 2022). Firstly, the DL PRS adopts Gold sequence with 4096 sequences to maintain interference randomization among multiple transmission and receiving points (TRPs) in the downlink and ensure good sequence cross-correlation characteristics. This enables user equipment (UE) to detect DL PRS of multiple TRPs under different network deployments, facilitating multipoint positioning. Secondly, the DL PRS supports sending and receiving beam scanning, allowing the merging processing of DL PRS resources on multiple beams, which results in merging gain and improves positioning performance. Finally, the DL PRS supports flexible time-frequency domain resource allocation to meet the positioning accuracy requirements in various application scenarios, while avoiding resource waste.

The integrated design of communication and positioning at the signal level can significantly improve the degree of integration and positioning performance, which is beyond many positioning systems. While the design of 5G PRS has achieved flexible resource allocation and transmission cycle, it consumes a substantial amount of communication resources to achieve high-precision positioning performance. As a result, the deployment of PRS remains a challenge for cellular providers due to additional bandwidth requirements (Shamaei and Kassas, 2021). In addition, 5G PRS positioning requires specialized hardware equipment that supports the reception of 5G PRS, and the contradiction between single base station for communication and multi-base station for positioning has not been properly handled.

As a result, ongoing efforts are focused on exploring new solutions to improve the performance of the integrated communication–positioning system. The design of the communication–positioning integration waveform is crucial and must take into account several key aspects, including:

1. *Positioning performance*: The positioning components of the integration waveform should be continuous to enable the receiver to track the signal in the carrier tracking loop and code tracking loop accurately.
2. *Resource consumption*: The positioning components should occupy the least amount of bandwidth and time resources possible while maintaining accuracy to improve the system's resource utilization efficiency.
3. *Interference interaction*: The introduction of the positioning components into the communication system can result in three types of interference. The first type of interference is between the positioning components and the communication components. It is critical to minimize the interference between the communication and positioning components by ensuring that the positioning waveform consumes enough communication resources to meet the positioning requirements while maintaining the quality of service (QoS) for communication services. The second type of interference is the "near and far effect," which is caused by the difference in distance between the positioning user and base stations. The autocorrelation value of the signal broadcast by the far base station may be overwhelmed by the cross-correlation value of the signal from the near base station (Dev et al., 2020, Schloemann et al., 2016). This effect can negatively impact the demodulation of signals from distant base stations. The third type of interference arises from the inherent contradiction between the single base station connection of communication and the multi-base station solution of positioning. Users communicating with different base stations do not interfere with each other, but may be interfered by positioning components from different base stations.

6.3 MS-NOMA Architecture

As shown in Fig. 6.1, we propose a novel multiple access technique for the future communication–positioning integrated systems, called multi-scale non-orthogonal multi-access (MS-NOMA). MS-NOMA performs non-orthogonal superposition of positioning and communication components in the frequency domain and power domain, providing different services to users while ensuring minimum interaction. In addition, the positioning component is designed to be continuous with the low power and large bandwidth to reduce interference and improve positioning accuracy. Furthermore, to avoid "near and far interference," the transmission power allocated to each positioning user depends on their own channel state information.

Specifically, Δf_c and Δf_p represent the subcarrier spacing of the communication and positioning waveform, and has $\Delta f_p = G\Delta f_c$, $G \in \mathbb{N}_+$. If the total bandwidth of the system is defined as B, then there are maximum $N = B/\Delta f_c - 1$

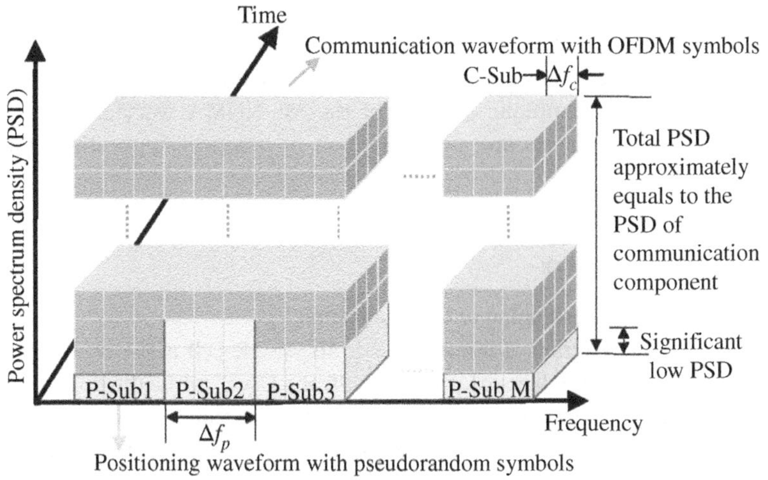

Figure 6.1 The MS-NOMA architecture.

and $M = B/\Delta f_p - 1$ sub-carriers for communication and positioning purposes, respectively. The power of the positioning users is much lower than that of the communication users, and the power sent to each positioning user is adjustable. For clear representation, define C-Sub/P-Sub and C-User/P-User as the abbreviations of communication/positioning sub-carrier and communication/positioning user, respectively. Subscript p and c represent the positioning and communication, respectively. $\|\cdot\|$ represents the Euclidean distance. The operator $\text{cov}(\cdot)$ represents the covariance. $\mathcal{M}$, $\mathcal{N}$, $\mathcal{K}$ and $\mathcal{K}^k$ represent the set $\{1, \dots, M\}$, $\{1, \dots, N\}$, $\{1, \dots, K\}$ and $\{1, \dots, k-1, k+1, \dots, K\}$, respectively.

6.4 Interference Analysis

As described in Section 6.3, the MS-NOMA waveform utilizes non-orthogonal frequency division multiplexing overlay transmission for both its communication and positioning components. While orthogonal frequency division multiplexing (OFDM) technology exhibits orthogonal characteristics, which prevent interference among users within the same system group, it does not prevent interference between users in different systems. To assess interference, we utilize the bit error rate (BER) to evaluate the interference of the positioning components on the communication components, and the code phase measurement error to evaluate the interference of the communication components on the positioning components.

We will analyse the interference of the MS-NOMA waveform in both single-cell network and multicell networks. In particular, The signal in the single-cell network will not be affected by signals from other base stations, thereby enabling a better reflection of the unique features of the MS-NOMA waveform itself. While the multicell networks are more common in practical applications for positioning.

6.4.1 Single-Cell Network

6.4.1.1 Interference of Positioning to Communication

Figure 6.2 shows the diagram of interference in the single-cell network. During demodulation, the C-user will be regarded as a "Strong User," while the P-User, as a "Weak User" due to its low power, so the P-User can be treated as noise during demodulation. Therefore, for C-User in the system, BER can be calculated by the following formula (Makki et al., 2019, Haci et al., 2017).

$$\mathrm{BER}^n = \alpha \mathrm{erfc}\left(\frac{\beta P_c T_c}{I^n + 2N_0}\right), \tag{6.1}$$

where n represents the index of C-User, satisfying $n \in \mathcal{N}$. α and β are determined by the modulation and coding schemes (Liu et al., 2016). P_c is the power of the communication component. T_c is the period of the communication symbol. N_0 is the environment noise's single-sided power spectral density (PSD). I^n represents

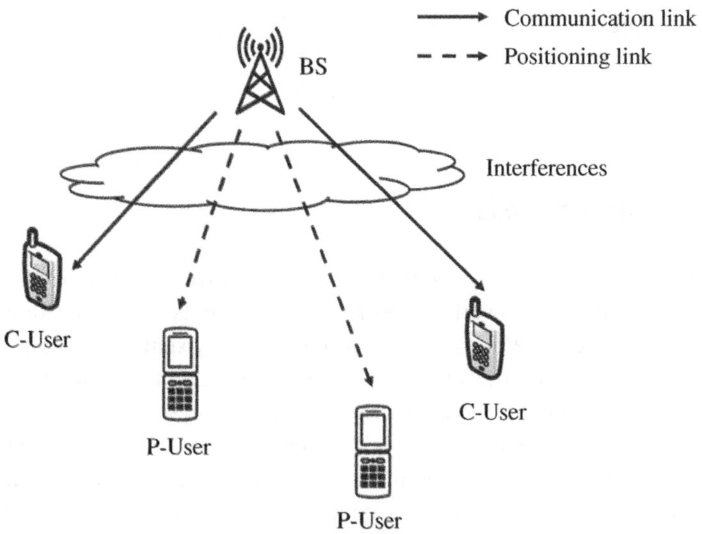

Figure 6.2 The diagram of interference in the single-cell network.

the interference of the positioning components to the nth C-User, which satisfies

$$I^n = \sum_{m \in \mathcal{M}} \bar{P}_p^m(n), \tag{6.2}$$

where m represents the index of P-User, satisfying $m \in \mathcal{M}$. $\bar{P}_p^m(n)$ is the power of the mth P-User over the nth C-User which satisfies

$$\bar{P}_p^m(n) = P_p^m G_p^m(n\Delta f_c) = P_p^m T_p \text{sinc}^2\left(m - \frac{n}{G}\right), \tag{6.3}$$

where P_p^m is the power of the positioning component. T_p is the period of the positioning symbol. $G_p^m(f) = T_p \text{sinc}^2\left[\left(f - m\Delta f_p\right) T_p\right]$ is the normalized PSD of the mth P-Sub.

So, we have

$$I^n = \sum_{m \in \mathcal{M}} P_p^m T_p \text{sinc}^2\left(m - \frac{n}{G}\right). \tag{6.4}$$

6.4.1.2 Interference of Communication to Positioning

The communication components carried by OFDM symbols can be regarded as powerful sources of interference for positioning components. Despite the relatively low power of positioning components, they can achieve high integration gain through time integration. If a positioning sequence uses a spreading code with length L, the presence of spreading gain can cause the positioning component to be stronger than the communication component, provided that the code length L is sufficiently long. Furthermore, positioning components are designed to be continuously broadcast, enabling the signal can be tracked through the code tracking and carrier tracking loops, which bring high-precision positioning results. In the signal bandwidth, the interference from communication components to the code tracking loop is treated as thermal noise, and the resulting code phase estimation error can be expressed as (Betz and Kolodziejski, 2009)

$$\left(\sigma_\rho^m\right)^2 = \frac{a\int_{B_0-B_{\text{fe}}/2}^{B_0+B_{\text{fe}}/2}\left[N_0 + G_s^m\left(f + m\Delta f_p\right)\right] G_p^m\left(f + m\Delta f_p\right)\sin^2\left(\pi f D T_p\right) df}{P_p^m\left[2\pi\int_{B_0-B_{\text{fe}}/2}^{B_0+B_{\text{fe}}/2} f G_p^m\left(f + m\Delta f_p\right)\ \sin\left(\pi f D T_p\right) df\right]^2}, \tag{6.5}$$

where a is determined by the loop parameters. B_0 is the central frequency of MS-NOMA waveform. B_{fe} is the double-sided front-end bandwidth. D is the early-late spacing of delay locked loop (DLL). $G_s^m(f)$ is the PSD of the communication components received by P-User m which satisfies

$$G_s^m(f) = \sum_{n \in \mathcal{N}} P_c G_c^n(f), \tag{6.6}$$

where $G_c^n(f) = T_c \text{sinc}^2\left[\left(f - n\Delta f_c\right) T_c\right]$ is the normalized PSD of the nth C-Sub.

In order to simplify (6.5), define

$$A_0^m = \int_{B_0-B_{\mathrm{fe}}/2}^{B_0+B_{\mathrm{fe}}/2} f G_p^m\left(f+m\Delta f_p\right)\ \sin\left(\pi f D T_p\right) df, \tag{6.7}$$

$$A_1^m = \int_{B_0-B_{\mathrm{fe}}/2}^{B_0+B_{\mathrm{fe}}/2} N_0 G_p^m\left(f+m\Delta f_p\right) \sin^2\left(\pi f D T_p\right) df, \tag{6.8}$$

$$A_2^m = \int_{B_0-B_{\mathrm{fe}}/2}^{B_0+B_{\mathrm{fe}}/2} G_s^m\left(f+m\Delta f_p\right) G_p^m\left(f+m\Delta f_p\right) \sin^2\left(\pi f D T_p\right) df. \tag{6.9}$$

Then, (6.5) can be written as

$$\left(\sigma_\rho^m\right)^2 = \frac{a\left(A_1^m + A_2^m\right)}{(2\pi)^2 P_p^m \left(A_0^m\right)^2}. \tag{6.10}$$

Notice that there are multiple P-Users, i.e., the bandwidth of the positioning waveform for one P-User is much smaller than the total bandwidth B. Moreover, the front-end bandwidth is larger than B as well. Insufficient front-end bandwidth can result in a flattened DLL correlation peak, degrading the performance of the phase discriminator. So we have $B_{\mathrm{fe}} \gg \frac{2}{T_p}$. When $D \to 0$, $\sin\left(\pi f D T_p\right)$ in (6.7)–(6.9) can be replaced by Taylor expansion around 0. Then, we have (6.11)–(6.13)

$$A_0^m = \pi D T_p^2 \int_{-B_{\mathrm{fe}}/2}^{B_{\mathrm{fe}}/2} f^2 \mathrm{sinc}^2\left(f T_p\right) df = \frac{1}{2\pi} D B_{\mathrm{fe}}, \tag{6.11}$$

$$A_1^m = \pi D T_p N_0 A_0^m, \tag{6.12}$$

$$\begin{aligned} A_2^m &= D^2 T_p \sum_{n\in\mathcal{N}} P_c T_c \int_{-B_{\mathrm{fe}}/2}^{B_{\mathrm{fe}}/2} \mathrm{sinc}^2\left[\left(f+m\Delta f_p - n\Delta f_c\right) T_c\right] \sin^2\left(f T_p\right) df \\ &\overset{G\gg 1}{\approx} D^2 T_p T_c P_c \sum_{n\in\mathcal{N}} \sin^2\left[\pi\left(m - \frac{n}{G}\right)\right] \\ &\quad \times \int_{(Gm-n-1)\Delta f_c}^{(Gm-n+1)\Delta f_c} \mathrm{sinc}^2\left[\left(f+m\Delta f_p - n\Delta f_c\right) T_c\right] df \\ &\approx D^2 T_p P_c \sum_{n\in\mathcal{N}} \sin^2\left(\frac{n}{G}\pi\right). \end{aligned} \tag{6.13}$$

Define

$$(C/N_0)^m = P_p^m/N_0, \tag{6.14}$$

$$(\mathrm{CPR})^m = 2GP_c/P_p^m. \tag{6.15}$$

Notice that (6.14) as the carrier-to-noise ratio of P-Sub m, (6.15) as the equivalent communication-to-positioning ratio of communication component transmitted by base station (BS) to P-Sub.

Then, we have (6.16)

$$\left(\sigma_\rho^m\right)^2 \approx \frac{aT_p^2}{2}\left[\frac{1}{B_{\mathrm{fe}}T_p\left(C/N_0\right)^m} + \frac{B(\mathrm{CPR})^m}{2B_{\mathrm{fe}}^2}\right]. \tag{6.16}$$

Notice that the first item in (6.16) is caused by the noise, the second one is caused by the communication components from BS.

6.4.2 Multicell Networks

We now shift our focus from the single-cell network to the multicell networks. To illustrate this scenario, we consider a typical positioning scenario in the spatial domain involving four base stations, as depicted in Fig. 6.3.

For convenience, the instantaneous channel gains of MS-NOMA waveform in typical positioning scenario are listed in Table 6.1.

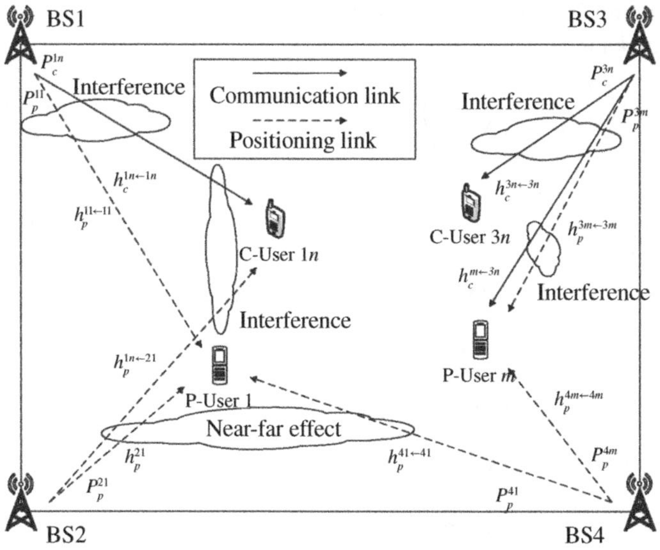

Figure 6.3 The diagram of interference in the multicell networks.

Table 6.1 Instantaneous channel gains of MS-NOMA waveform in 4-BSs scenario.

Instantaneous channel gain	Definition
h_c, h_p	Instantaneous channel gains of communication and positioning waveform, respectively
$h_c^{kn\leftarrow kn}$	h_c of sub-carrier kn received by the C-User kn
$h_p^{m\leftarrow km}$	h_p of sub-carrier km received by the P-User m
$h_c^{m\leftarrow kn}$	h_c of sub-carrier kn received by the P-User m
$h_p^{kn\leftarrow k'm}$	h_p of sub-carrier $k'm$ received by the C-User kn
$h_p^{m\leftarrow k'_k m}$	h_p of the strongest sub-carrier except the P-Sub km received by the P-User m
$h_c^{m\leftarrow k'}$	See (6.28)

6.4.2.1 Interference of Positioning to Communication

In contrast to the single-cell network, interference analysis in the multicell networks involves two differences. Firstly, we need to assume that the interference between communication components can be ideally eliminated (Maschietti et al., 2021). Secondly, the power allocation in each base station may differ, resulting in variations in the power of the positioning components across different base stations. Consequently, we must account for the interference from positioning components originating from multiple base stations.

The BER of C-Sub kn is

$$\text{BER}^{kn} = \alpha \text{erfc}\left(\frac{\beta |h_c^{kn\leftarrow kn}|^2 P_c T_c}{I^{kn} + 2N_0}\right), \tag{6.17}$$

where

$$I^{kn} = \sum_{k'\in\mathcal{K}} \sum_{m\in\mathcal{M}} \bar{P}_p^{kn\leftarrow k'm}, \tag{6.18}$$

where $\bar{P}_p^{kn\leftarrow k'm}$ is the power of the P-Sub $k'm$ received by C-User kn which satisfies

$$\begin{aligned}\bar{P}_p^{kn\leftarrow k'm} &= \left|h_p^{kn\leftarrow k'm}\right|^2 P_p^{k'm} G_p^m\left(n\Delta f_c\right)\\ &= \left|h_p^{kn\leftarrow k'm}\right|^2 P_p^{k'm} T_p \text{sinc}^2\left(m - \frac{n}{G}\right),\end{aligned} \tag{6.19}$$

where $P_p^{k'm}$ is the power of P-Sub $k'm$.

6.4.2.2 Interference of Communication to Positioning

DLL is used to track the signal, and the tracking/ranging accuracy is as (6.20). Compare (6.5) and (6.20), in the single-cell network, the impact of the cross-correlation will be vanished and the channel gains of the C-Subs and P-Subs will be equal.

$$\left(\sigma_\rho^{km}\right)^2 = \frac{a\int_{B_0-B_{\text{fe}}/2}^{B_0+B_{\text{fe}}/2}\left[N_0 + G_s^m\left(f+m\Delta f_p\right) + G_q^{km}\left(f+m\Delta f_p\right)\right]}{\left|h_p^{m\leftarrow km}\right|^2 P_p^{km}\left[2\pi\int_{B_0-B_{\text{fe}}/2}^{B_0+B_{\text{fe}}/2} fG_p^m\left(f+m\Delta f_p\right)\sin\left(\pi fDT_p\right)df\right]^2} \times G_p^m\left(f+m\Delta f_p\right)\sin^2\left(\pi fDT_p\right)df. \tag{6.20}$$

$G_s^m(f)$ is the PSD of the communication components received by P-User m which satisfies

$$G_s^m(f) = \sum_{k'\in\mathcal{K}}\sum_{n\in\mathcal{N}}\left|h_c^{m\leftarrow k'n}\right|^2 P_c G_c^n(f). \tag{6.21}$$

$G_q^{km}(f)$ is the PSD of the positioning components from other BSs, which satisfies

$$G_q^{km}(f) = \sum_{k'\in\mathcal{K}^k}\left|h_p^{m\leftarrow k'm}\right|^2 P_p^{k'm} G_p^m(f). \tag{6.22}$$

To simplify the formula (6.20), refer to formulas (6.7)–(6.9) defined in the analysis of the single-cell network and define

$$A_3^m = \int_{B_0-B_{\text{fe}}/2}^{B_0+B_{\text{fe}}/2} G_q^{km}\left(f+m\Delta f_p\right)G_p^m\left(f+m\Delta f_p\right)\sin^2\left(\pi fDT_p\right)df. \tag{6.23}$$

Then, (6.20) can be written as (6.24)

$$\left(\sigma_\rho^{km}\right)^2 = \frac{a\left(A_1^m + A_2^m + A_3^m\right)}{(2\pi)^2\left|h_p^{m\leftarrow km}\right|^2 P_p^{km}\left(A_0^m\right)^2}. \tag{6.24}$$

Then, by taking (6.21), (6.22) into (6.23) and rearranging items, we have (6.25)

$$A_3^m = \pi^2 D^2 T_p^4 \sum_{k'\in\mathcal{K}^k}\left|h_p^{m\leftarrow k'm}\right|^2 P_p^{k'm} \underbrace{\int_{-B_{\text{fe}}/2}^{B_{\text{fe}}/2} f^2\text{sinc}^4\left(fT_p\right)df}_{\bar{A}_3}, \tag{6.25}$$

where

$$\begin{aligned}\bar{A}_3 &\overset{B_{\text{fe}}\gg 2/T_p}{\approx} \int_{-\infty}^{\infty}\frac{\sin^4\left(\pi fT_p\right)}{\pi^4 f^2 T_p^4}df \\ &= \frac{1}{4\pi^4 T_p^4}\int_{-\infty}^{\infty}\left[\frac{4\sin^2\left(\pi fT_p\right)}{f^2} - \frac{\sin^2\left(2\pi fT_p\right)}{f^2}\right]df \\ &= \frac{1}{2\pi^2 T_p^3}.\end{aligned} \tag{6.26}$$

By taking (6.11)–(6.13) and (6.25)–(6.26) into (6.24), we have (6.27)

$$\begin{aligned}\left(\sigma_\rho^{km}\right)^2 &\approx \frac{aT_p}{2}\left[\frac{N_0}{B_{\rm fe}\left|h_p^{m\leftarrow km}\right|^2 P_p^{km}}+\frac{2P_c\sum_{k'\in\mathcal{K}}\sum_{n=1}^{N}\left|h_c^{m\leftarrow k'n}\right|^2\sin^2\left(\frac{n}{G}\pi\right)}{B_{\rm fe}^2\left|h_p^{m\leftarrow km}\right|^2 P_p^{km}}\right.\\&\left.+\frac{\sum_{k'\in\mathcal{K}^k}\left|h_p^{m\leftarrow k'm}\right|^2 P_p^{k'm}}{B_{\rm fe}^2\left|h_p^{m\leftarrow km}\right|^2 P_p^{km}}\right]\\&=\frac{aT_p^2}{2}\left[\frac{N_0}{B_{\rm fe}T_p\left|h_p^{m\leftarrow km}\right|^2 P_p^{km}}+\frac{BGP_c\sum_{k'\in\mathcal{K}}\left|h_c^{m\leftarrow k'}\right|^2}{B_{\rm fe}^2\left|h_p^{m\leftarrow km}\right|^2 P_p^{km}}\right.\\&\left.+\frac{\sum_{k'\in\mathcal{K}^k}\left|h_p^{m\leftarrow k'm}\right|^2 P_p^{k'm}}{B_{\rm fe}^2 T_p\left|h_p^{m\leftarrow km}\right|^2 P_p^{km}}\right],\end{aligned} \tag{6.27}$$

$$\left|h_c^{m\leftarrow k'}\right|^2=\frac{2}{N}\sum_{n\in\mathcal{N}}\left|h_c^{m\leftarrow k'n}\right|^2\sin^2\left(\frac{n}{G}\pi\right), \tag{6.28}$$

where (6.28) is defined as the normalized equivalent channel gain of the communication component transmitted by BS k' to the P-User m. And we can take (6.28) into (6.27) to simplify further.

In the meantime, define (6.29)–(6.31)

$$\left(C/N_0\right)^{km}=\left|h_p^{m\leftarrow km}\right|^2 P_p^{km}/N_0, \tag{6.29}$$

$$(\mathrm{CPR})^{km\leftarrow k'}=\frac{2G\left|h_c^{m\leftarrow k'}\right|^2 P_c}{\left|h_p^{m\leftarrow km}\right|^2 P_p^{km}}, \tag{6.30}$$

$$(\mathrm{PPR})^{km\leftarrow k'm}=\frac{\left|h_p^{m\leftarrow k'm}\right|^2 P_p^{k'm}}{\left|h_p^{m\leftarrow km}\right|^2 P_p^{km}}, \tag{6.31}$$

where (6.29) as the carrier-to-noise ratio of P-Sub km, (6.30) as the equivalent communication-to-positioning ratio of communication component transmitted by BS k' to P-Sub km, and (6.31) as the positioning-to-positioning ratio of P-Sub $k'm$ to P-Sub km. Then we have (6.32)

$$\left(\sigma_\rho^{km}\right)^2\approx\frac{aT_p^2}{2}\left[\frac{1}{B_{\rm fe}T_p\left(C/N_0\right)^{km}}+\frac{B\sum_{k'\in\mathcal{K}}(\mathrm{CPR})^{km\leftarrow k'}}{2B_{\rm fe}^2}+\frac{\sum_{k'\in\mathcal{K}^k}(\mathrm{PPR})^{km\leftarrow k'm}}{B_{\rm fe}^2 T_p}\right]. \tag{6.32}$$

Notice that the first item in (6.32) is caused by the noise, the second one is caused by the communication components from all BSs, and the third one is caused by the other BSs' positioning components. Because the positioning components are designed much weaker than communication ones in MS-NOMA, the interference from positioning components could be ignored. Then, (6.32) is further simplified as (6.33)

$$\left(\sigma_\rho^{km}\right)^2 \approx \frac{aT_p^2}{2}\left[\frac{1}{B_{\text{fe}}T_p\left(C/N_0\right)^{km}} + \frac{B\sum_{k'\in\mathcal{K}}(\text{CPR})^{km\leftarrow k'}}{2B_{\text{fe}}^2}\right]. \tag{6.33}$$

We define the ranging-factor $\left(\tilde{\sigma}_\rho^{km}\right)^2 = \left(\sigma_\rho^{km}\right)^2 P_p^{km}$ as (6.34) for later use. Notice that $\left(\tilde{\sigma}_\rho^{km}\right)^2$ has no relation to P_p^{km} which is a part of the denominator in (6.33).

$$\left(\tilde{\sigma}_\rho^{km}\right)^2 = \frac{aT_p^2}{2}\left[\frac{N_0}{B_{\text{fe}}T_p\left|h_p^{km}\right|^2} + \frac{BGP_c}{B_{\text{fe}}^2}\frac{\sum_{k'\in\mathcal{K}}\left|h_c^{m\leftarrow k'}\right|^2}{\left|h_p^{km}\right|^2} + \frac{\sum_{k'\in\mathcal{K}^k}\left|h_p^{k'm}\right|^2 P_p^{k'm}}{B_{\text{fe}}^2T_p\left|h_p^{km}\right|^2}\right]. \tag{6.34}$$

6.5 Resource Allocation

6.5.1 The Constraints

6.5.1.1 The BER Threshold Under QoS Constraint

To ensure the QoS of C-Users, the BERs of all C-Users should be limited under a certain threshold

$$\text{BER}^{kn} \le \Xi_{\text{th}}, \quad \forall k \in \mathcal{K}, \quad \forall n \in \mathcal{N}. \tag{6.35}$$

Then, by taking (6.17) to (6.35) and rearranging items, we have

$$\begin{aligned} I^{kn} &\le \frac{\beta\left|h_c^{kn\leftarrow kn}\right|^2 P_cT_c}{\text{erfc}^{-1}\left(\Xi_{\text{th}}/\alpha\right)} - 2N_0 \\ &\triangleq I_{\text{th}}^{kn}, \quad \forall k \in \mathcal{K}, \quad \forall n \in \mathcal{N}, \end{aligned} \tag{6.36}$$

where I_{th}^{kn} is defined as the interference threshold of C-User kn, which is determined by the QoS requirement Ξ_{th}.

6.5.1.2 The Total Power Limitation

The total transmit power is often limited. In MS-NOMA waveform, we have

$$\sum_{m\in\mathcal{M}} P_p^{km} + NP_c \le P_T^k, \quad \forall k \in \mathcal{K}, \tag{6.37}$$

where P_T^k is the total transmit power of BS k. Let's define the positioning power budget of BS k as $P_{\mathrm{th}}^k = P_T^k - NP_c$, then we have

$$\sum_{m \in \mathcal{M}} P_p^{km} \leq P_{\mathrm{th}}^k, \quad \forall k \in \mathcal{K}. \tag{6.38}$$

6.5.1.3 The Elimination of Near-Far Effect

To ensure that P-Users could receive as many positioning signals as possible, the power of the received positioning signals from different BSs must satisfy

$$\frac{|h_p^{m \leftarrow km}|^2 P_p^{km}}{|h_p^{m \leftarrow k'm}|^2 P_p^{k'm}} \geq \varrho\Omega, \quad \forall m \in \mathcal{M}, \quad \forall k \in \mathcal{K}, \quad \forall k' \in \mathcal{K}^k, \tag{6.39}$$

where Ω is the auto-correlation to cross-correlation ratio of positioning component, which is determined by the pseudorandom code and its length. ϱ is determined by the receiver's performance, which is usually larger than 1. For a particular P-Sub km, if the strongest cross-correlation satisfies (6.39), all k's in (6.39) will be satisfied. Therefore, (6.39) can be rewritten as

$$|h_p^{m \leftarrow km}|^2 P_p^{km} \geq \varrho\Omega |h_p^{m \leftarrow k_k' m}|^2 P_p^{k_k' m}, \quad \forall m \in \mathcal{M}, \quad \forall k \in \mathcal{K}, \tag{6.40}$$

where $k_k' m$ represents the index of the strongest sub-carrier except the P-Sub km received by the P-User m.

6.5.2 The Proposed Joint Power Allocation Model

Our objective is to achieve optimal positioning performance for all P-Users, with consideration for both accuracy and coverage, while meeting the QoS requirement and staying within the total transmit power budget. To simplify the calculation process, we represent the power allocation problem using the square of the horizontal positioning accuracy, denoted as $(\Psi^m)^2$, to model the power allocation problem. In Section 6.4.2.2, we have obtained the square of ranging error variance between BS k and P-User m, denoted as $\left(\sigma_\rho^{km}\right)^2$ (6.33), and the relationship between $\left(\sigma_\rho^{km}\right)^2$ and $(\Psi^m)^2$ can be determined as shown below.

Define $\varepsilon_\rho^m = \left[\varepsilon_\rho^{1m}, \varepsilon_\rho^{2m}, \ldots, \varepsilon_\rho^{km}\right]^T$ as the ranging errors of P-User m, where ε_ρ^{km} represents the ranging error between BS k and P-User m. Then, the positioning error of P-User m is (Lu et al., 2016)

$$\begin{aligned} \varepsilon_X^m &= \left[\left(G^m\right)^T G^m\right]^{-1} \left(G^m\right)^T \varepsilon_\rho^m \\ &= H^m \varepsilon_\rho^m, \end{aligned} \tag{6.41}$$

where

$$G^m = \begin{bmatrix} \iota_x^{1m} & \iota_y^{1m} & \iota_z^{1m} \\ \iota_x^{2m} & \iota_y^{2m} & \iota_z^{2m} \\ \cdots & \cdots & \cdots \\ \iota_x^{km} & \iota_y^{km} & \iota_z^{km} \end{bmatrix}, \tag{6.42}$$

$$\begin{cases} \iota_x^{km} = \dfrac{\left(x_p^m - x_b^k\right)}{\left\|X_b^k - X_p^m\right\|} \\ \iota_y^{km} = \dfrac{\left(y_p^m - y_b^k\right)}{\left\|X_b^k - X_p^m\right\|} \\ \iota_z^{km} = \dfrac{\left(z_p^m - z_b^k\right)}{\left\|X_b^k - X_p^m\right\|} \end{cases}, \tag{6.43}$$

where $X = [x, y, z]^T$ represents the coordinate, and X_p and X_b represents the position of P-User and BS, respectively. Because the ranging errors from the BSs are independent, their covariance matrix is diagonal under the assumption that the range measuring is unbiased

$$\begin{aligned} \left(\sigma_\rho^m\right)^2 &= \text{cov}\left(\varepsilon_\rho^m, \varepsilon_\rho^m\right) \\ &= \begin{bmatrix} \left(\sigma_\rho^{1m}\right)^2 & 0 & \cdots & 0 \\ 0 & \left(\sigma_\rho^{2m}\right)^2 & \cdots & 0 \\ \cdots & \cdots & \cdots & \cdots \\ 0 & 0 & \cdots & \left(\sigma_\rho^{km}\right)^2 \end{bmatrix}, \end{aligned} \tag{6.44}$$

where $\left(\sigma_\rho^{km}\right)^2 = \text{cov}\left(\varepsilon_\rho^{km}, \varepsilon_\rho^{km}\right)$ represents the ranging error variance of P-Sub km. Then, the covariance of the positioning error is

$$\begin{aligned} \left(\sigma_X^m\right)^2 &= \text{cov}\left(\varepsilon_X^m, \varepsilon_X^m\right) \\ &= H^m \left(\sigma_\rho^m\right)^2 \left(H^m\right)^T. \end{aligned} \tag{6.45}$$

The diagonal elements represent the positioning accuracy of each direction. Then, the horizontal positioning accuracy can be expressed as

$$\Psi^m = \sqrt{\sum_{k \in \mathcal{K}} \left\{ \left[\sum_{i=1}^{2} \left(\hbar_{ik}^m\right)^2 \right] \left(\sigma_\rho^{km}\right)^2 \right\}}, \tag{6.46}$$

where $\hbar_{ik}^m$s ($i \in \{1,2,3\}$) represent the elements of H^m. If the BSs are perfectly synchronized, the P-Users will use time-based algorithm to estimate their locations.

Then, the horizontal positioning accuracy of P-User m can be expressed as

$$\Psi^m = \sqrt{\sum_{k\in\mathcal{K}} \left(\lambda^{km}\sigma_\rho^{km}\right)^2}, \tag{6.47}$$

where $\lambda^{km} = \sqrt{\sum_{i=1}^{2} \left(\hbar_{ik}^{m}\right)^2}$ represents the geometric-dilution and the $\left(\sigma_\rho^{km}\right)^2$ in the multicell networks has been given in (6.33).

So, the average lower bound of horizontal positioning error for all P-Users in the network is minimized by finding the optimal power values $P_p^{km}, \forall m \in \mathcal{M}, \forall k \in \mathcal{K}$ under the given constraints. Considering the fact that the maximum negative value of a convex function is equivalent to its minimum. Then, the power allocation problem can be formulated as a convex optimization problem as follows

$$\begin{aligned}
\text{OP1 :}&\max_{P_p^{km}} -\frac{1}{M}\sum_{m\in\mathcal{M}} \left(\Psi^m\right)^2\\
&\text{s.t.}\, I^{kn} \le I_{\text{th}}^{kn}, \quad \forall n \in \mathcal{N}, \quad \forall k \in \mathcal{K}\\
&\sum_{m\in\mathcal{M}} P_p^{m\leftarrow km} \le P_{\text{th}}^{k}, \quad \forall k \in \mathcal{K}\\
&|h_p^{m\leftarrow km}|^2 P_p^{km} \ge \varrho\Omega |h_p^{m\leftarrow k_k' m}|^2 P_p^{k_k' m},\\
&\qquad \forall m \in \mathcal{M}, \quad \forall k \in \mathcal{K}.
\end{aligned} \tag{6.48}$$

6.5.3 The Positioning–Communication Joint Power Allocation Scheme

OP1 can be solved by the Lagrange duality method (Boyd and Vandenberghe, 2004). Then, the Lagrange dual function of OP1 is then given by

$$g(\mu, \nu, \beta) = \max_{P_p^{km}} \mathcal{L}\left(\left\{P_p^{km}\right\}, \mu, \nu, \beta\right), \tag{6.49}$$

where $\mathcal{L}\left(\left\{P_p^{km}\right\}, \mu, \nu, \beta\right)$ is the Lagrangian of OP1, $\mu = \left\{\mu^{kn}\right\} \succcurlyeq 0$, $\nu = \left\{\nu^{k}\right\} \succcurlyeq 0$ and $\beta = \left\{\beta^{km}\right\} \succcurlyeq 0$ are the matrices of dual variables associated with the corresponding constraints given in (6.36), (6.38), and (6.39). Then, the dual optimization problem can be formulated as

$$\begin{aligned}
&\min g(\mu, \nu, \beta)\\
&\text{s.t.}\ \mu \succcurlyeq 0,\ \nu \succcurlyeq 0,\ \beta \succcurlyeq 0.
\end{aligned} \tag{6.50}$$

It is easy to be proved that $\mathcal{L}\left(\left\{P_p^{km}\right\}, \mu, \nu, \beta\right)$ is linear in μ, ν, β for fixed P_p^{km}, and $g(\mu, \nu, \beta)$ is the maximum of linear function. Thus, the dual optimization problem is always convex. To solve this problem, the dual decomposition method introduced in Zhang (2008) is employed. For this purpose, we introduce a transformation $\sum_{n\in\mathcal{N}} = \sum_{m\in\mathcal{M}}\sum_{n\in\mathbb{N}_m}$ to decompose the Lagrange dual function to $K \times M$

independent sub-problems, where

$$\mathbb{N}_m = \{(2G-1)(m-1)+1, \dots, (2G-1)m\}. \tag{6.51}$$

Then, we have

$$\begin{aligned} g(\mu, \nu, \beta) &= \sum_{k\in\mathcal{K}} \left[g^k(\mu, \nu, \beta)\right] \\ &= \sum_{k\in\mathcal{K}} \left\{ \sum_{m\in\mathcal{M}} g^{km}(\mu, \nu, \beta) + \nu^k P_{\text{th}}^k \right\}, \end{aligned} \tag{6.52}$$

where

$$\begin{aligned} g^{km}(\mu, \nu, \beta) = \max_{P_p^{km}} &\left\{ -\frac{1}{M}\left(\lambda^{km}\sigma_\rho^{km}\right)^2 - \nu^k P_p^{km} + \sum_{n\in\mathbb{N}_m} \mu^{kn}\left(I_{\text{th}}^{kn} - I^{kn}\right) \right. \\ &\left. + \beta^{km}\left(|h_p^{m\leftarrow km}|^2 P_p^{km} - \varrho\Omega |h_p^{m\leftarrow k'_k m}|^2 P_p^{k'_k m}\right) \right\}. \end{aligned} \tag{6.53}$$

From (6.53), it is clear that we can decompose the Lagrange dual function $g^k(\mu, \nu, \beta)$ to M independent sub-problems by giving ν^k. Each of the sub-problems is given by

$$\text{OP2}: \max_{P_p^{km}} -\frac{1}{M}\left(\lambda^{km}\sigma_\rho^{km}\right)^2 - \nu^k P_p^{km} \tag{6.54}$$

$$\text{s.t.}\, I^{kn} \leq I_{\text{th}}^{kn}, \quad n \in \mathbb{N}_m, \tag{6.55}$$

$$|h_p^{m\leftarrow km}|^2 P_p^{km} \geq \varrho\Omega |h_p^{m\leftarrow k'_k m}|^2 P_p^{k'_k m}. \tag{6.56}$$

Similar to OP1, the dual problem of the sub-problems can be expressed as

$$\begin{aligned} &\min \tilde{g}^{km}\left(\tilde{\mu}^{kn}, \tilde{\beta}^{km}\right) \\ &\quad \text{s.t.}\, \tilde{\mu}^{kn} \geq 0, \quad \forall n \in \mathbb{N}_m, \\ &\qquad \tilde{\beta}^{km} \geq 0, \end{aligned} \tag{6.57}$$

where $\tilde{\mu}^{kn}$ and $\tilde{\beta}^{km}$ are the nonnegative dual variables for constraints (6.55) and (6.56), respectively.

The optimal power allocation solution $\tilde{P}_p^{km}$ of OP2 can be obtained by using the Karush–Kuhn–Tucker (KKT) conditions and the KKT conditions of OP2 can be written as

$$\sum_{n\in\mathbb{N}_m} \tilde{\mu}^{kn}\left(I_{\text{th}}^{kn} - I^{kn}\right) = 0 \tag{6.58}$$

$$\tilde{\beta}^{km}\left(|h_p^{m\leftarrow km}|^2 P_p^{km} - \varrho\Omega |h_p^{m\leftarrow k'_k m}|^2 P_p^{k'_k m}\right) = 0 \tag{6.59}$$

$$\frac{\partial \tilde{\mathcal{L}}\left(\left\{P_p^{km}\right\}, \tilde{\mu}^{kn}, \tilde{\beta}^{km}\right)}{\partial P_p^{km}} = 0. \tag{6.60}$$

It is obvious that the optimal solution $\tilde{P}_p^{km}$ satisfies (6.60). Thus, (6.60) can be simplified to (6.61).

$$\begin{aligned}\frac{\partial \tilde{\mathcal{L}}}{\partial P_p^{km}} &= \frac{-\frac{1}{M}\partial\left(\lambda^{km}\sigma_\rho^{km}\right)^2 - v^k P_p^{km}}{\partial P_p^{km}} + \frac{\partial \sum_{n\in\mathbb{N}_m} \tilde{\mu}^{kn}\left(I_{\text{th}}^{kn} - I^{kn}\right)}{\partial P_p^{km}} \\ &+ \frac{\partial\left\{\tilde{\beta}^{km}\left(|h_p^{m\leftarrow km}|^2 P_p^{km} - \varrho\Omega|h_p^{m\leftarrow k_k' m}|^2 P_p^{k_k' m}\right)\right\}}{\partial P_p^{km}} \\ &= -\frac{1}{M}\left(\frac{\lambda^{km}\tilde{\sigma}_\rho^{km}}{P_p^{km}}\right)^2 - v^k - \sum_{n\in\mathbb{N}_m}\tilde{\mu}^{kn}\underbrace{\frac{\partial I^{kn}}{\partial P_p^{km}}}_{J^{kn\leftarrow m}} + \tilde{\beta}^{km}|h_p^{m\leftarrow km}|^2. \end{aligned} \tag{6.61}$$

By taking (6.18) into (6.61), we have

$$J^{kn\leftarrow m} = \sum_{k'\in\mathcal{K}}\left|h_p^{kn\leftarrow k' m}\right|^2 T_p \text{sinc}^2\left(m - \frac{n}{G}\right). \tag{6.62}$$

Then, by setting (6.61) to 0, we can obtain the optimal power allocation solution as (6.63) shows

$$\begin{aligned}\tilde{P}_p^{km} &= \underbrace{\lambda^{km}}_{\text{geometric-dilution}} \times \underbrace{\tilde{\sigma}_\rho^{km}}_{\text{ranging-factor}} \\ &\times \underbrace{\left[M\left(\tilde{\beta}^{km}|h_p^{m\leftarrow km}|^2 - v^k - \sum_{n\in\mathbb{N}_m}\tilde{\mu}^{kn}J^{kn\leftarrow m}\right)\right]^{-1/2}}_{\text{constraint-scale}}. \end{aligned} \tag{6.63}$$

6.5.4 Remarks

From (6.63), it is observed that the optimal power allocation solution is determined by the geometric-dilution, ranging-factor and constraint-scale. It is necessary to have a clear understanding of these factors that affect the allocated power. The geometric-dilution λ^{km} associates with the relative positions between the P-User and all BSs. This means the power allocation procedure not only minimizes the ranging accuracy, but also considers the geometric distribution which affects the positioning accuracy as well.

The ranging-factor $\tilde{\sigma}_\rho^{km}$ reflects the ranging ability of P-User as (6.34) shows. If the loop parameters are fixed, $\tilde{\sigma}_\rho^{km}$ is determined by the channel gains of a certain P-User m, i.e., $\left|h_p^{km}\right|^2$, $\left|h_c^{m\leftarrow k'}\right|^2$ and $\left|h_p^{k'm}\right|^2$ which reflect the attenuation of the positioning component, the communication components and other P-Users' positioning components, respectively. If the positioning component's attenuation is large, i.e., $\left|h_p^{km}\right|^2$ is small, it will allocate stronger positioning power, and vice versa. Conversely, if the attenuation of communication components is large, i.e., $\left|h_c^{m\leftarrow k'}\right|^2$ is small, it will allocate weak positioning power because of the small interference from the communication components, and vice versa.

The constraint-scale reflects the impact of the constraints: $\tilde{\mu}^{kn}$ is the dual variable associated with the BER threshold of C-User kn. If C-User kn can accommodate a higher BER, $\tilde{\mu}^{kn}$ will be smaller, and thus result in a higher constraint-scale, and vice versa. In the extreme case that C-User kn cannot accommodate any additional interference, $\tilde{\mu}^{kn}$ will be infinite, and thus the constraint-scale will be zero, which indicates that the positioning component over C-User kn's band is not permitted. On the contrary, if C-User kn has no requirement on the BER, $\tilde{\mu}^{kn}$ will be zero, and thus the power-scale will be only determined by the other constraints.

$J^{kn\leftarrow m}$ is determined by the channel gains of positioning components $k'm$ ($k' \in \mathcal{K}$) at C-User kn as (6.62) shows. It is clear that a smaller $\sum_{k'\in\mathcal{K}}\left|h_p^{kn\leftarrow k'm}\right|^2$ will result in a higher constraint-scale. This is intuitively correct because the P-Users from all BSs will not cause too much interference when $\sum_{k'\in\mathcal{K}}\left|h_p^{kn\leftarrow k'm}\right|^2$ is small. In the real scenario, if $\sum_{k'\in\mathcal{K}}\left|h_p^{kn\leftarrow k'm}\right|^2 \to 0$, which means C-User kn is too far from all BSs to receive any communication/positioning component, the P-Users will not cause any interference to this C-User no matter how strong its transmit power is.

ν^k is the dual variable associated with the transmit power budget. A larger power budget results in a smaller ν^k, and thus results in a lower ranging error, and vice versa.

$\tilde{\beta}^{km}$ is a parameter related to the P-User's receiver performance. It reflects the influence of the cross-correlation (i.e., (6.39)) on the constraint-scale. There will be a larger $\tilde{\beta}^{km}$ with a smaller ϱ. Namely, if the receiver has a better anti-cross-correlation performance, there will be a higher ranging accuracy and better coverage, and vice versa.

6.6 Performance Evaluation

Firstly, we will assess the MS-NOMA waveform in the single-cell network, with emphasis on communication performance, ranging accuracy, and resource consumption. Subsequently, we will explore the positioning performance in a 4-BSs scenario. Simulation parameters are listed in Table 6.2.

Table 6.2 Simulation parameters.

Parameters	**Values**
Carrier frequency	3.5 GHz
Subcarrier frequency Δf_c	30 kHz
Number of positioning users M	20
Bandwidth B	20 MHz/50 MHz
Front-end bandwidth B_{fe}	2 B
Code loop noise bandwidth B_L	0.2 Hz
Predetection integration time T_{coh}	0.02 s
DLL' early-late spacing D	0.02 chip

6.6.1 Communication Performance

Referring to the preceding discussion in Section 6.4, BER is used to evaluate the interference of positioning components on communication components. Figures 6.4 and 6.5 illustrate the relationship between the average BER of communication users and $\frac{E_b}{N_0}$ and CPR from various perspectives. Where, $E_b = P_c T_c$

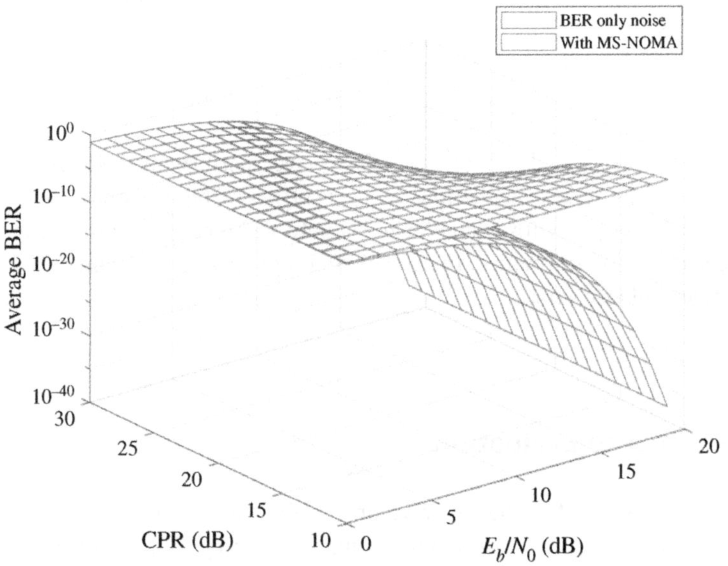

Figure 6.4 Average BER at (−37.5, 30) viewpoint.

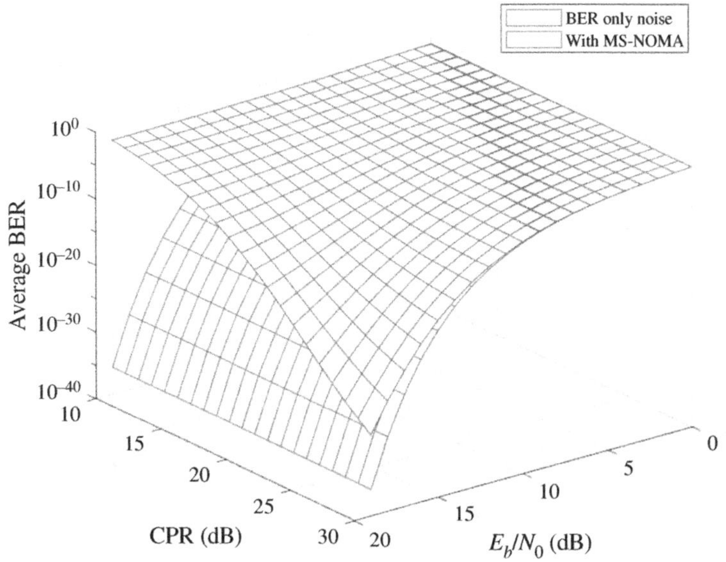

Figure 6.5 Average BER at (142.5, 30) viewpoint.

is the symbolic energy of communication component. Whether or not the MS-NOMA is used, the average BER of the communication component decreases with an increase in $\frac{E_b}{N_0}$ when the CPR is fixed and small. This reduction in BER can be attributed to the energy increment of communication component resulting from the increase in $\frac{E_b}{N_0}$. However, for MS-NOMA, when the $\frac{E_b}{N_0}$ is fixed and CPR increases, the average BER of the communication component decreases. This is because CPR represents the power ratio between the communication component and the positioning component, and as the $\frac{E_b}{N_0}$ is fixed and CPR increases, the power of the positioning component decreases, leading to a reduced impact on the BER.

An observation can be made that as the CPR becomes smaller, the BER tends to remain constant even when E_b/N_0 increases for MS-NOMA. This is because the interference caused by the positioning component dominates the BER performance rather than the environment noise (i.e., I^n is much larger than $2N_0$). As the power of positioning component becomes lower with larger CPR values, the BER decreases rapidly with increasing E_b/N_0 and becomes closer to that of the only existing noise scenario (CPR $= \infty$). This phenomenon can be observed more clearly in Fig. 6.5.

Figure 6.6 illustrates the relationship between the BER of a single communication user and the power of the positioning signals. Without power allocation, when the power of the positioning users are equal, the BER of each communication user

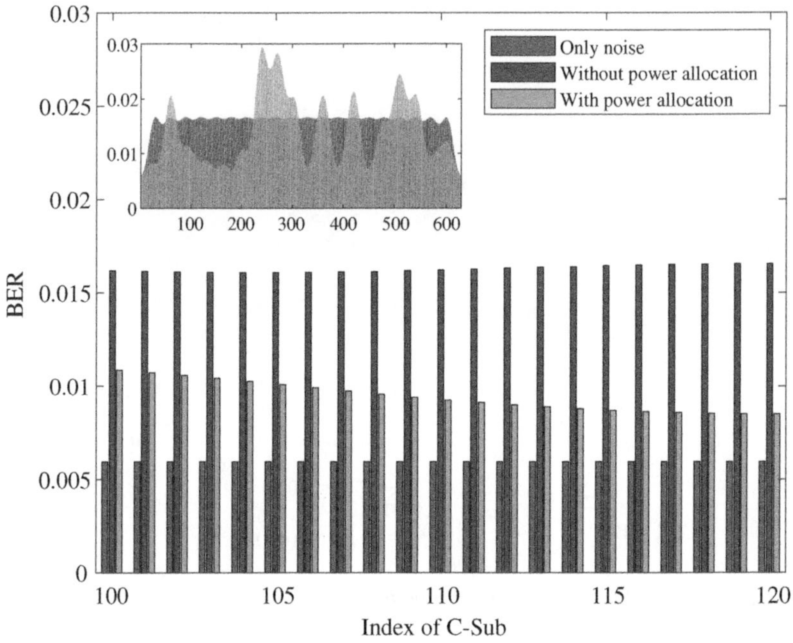

Figure 6.6 Example of BERs over C-Subs ($E_b/N_0 = 5$ dB, CPR = 15 dB).

is roughly the same. However, when power allocation is applied, the interference caused by the positioning component on the communication component varies. As expected, the BER of the communication user is lowest when there is only noise without any positioning component. Furthermore, it is evident that the BER of the communication user is closely associated with the power of the positioning component, so the power of the positioning component needs to be carefully set to ensure quality of service for each communication user.

6.6.2 Ranging Performance

We then examine the range measurement accuracy of the MS-NOMA waveform. The ranging accuracy of the MS-NOMA and PRS are compared. Where the lower bound of PRS is used as introduced in del Peral-Rosado et al. (2012).

Figure 6.7 shows the range measurement accuracy when $C/N_0 = 45\,\text{dB}\cdot\text{Hz}$, where superscript e and a represent the exact and approximate results, respectively. It is clear that the measurement errors of the MS-NOMA waveform are always smaller than the ones of PRS when CPR < 30 dB. And as CPR increased, the accuracy gap decreased. This is because, an increase in CPR means a decrease in the power of the positioning component for MS-NOMA, which will reduce

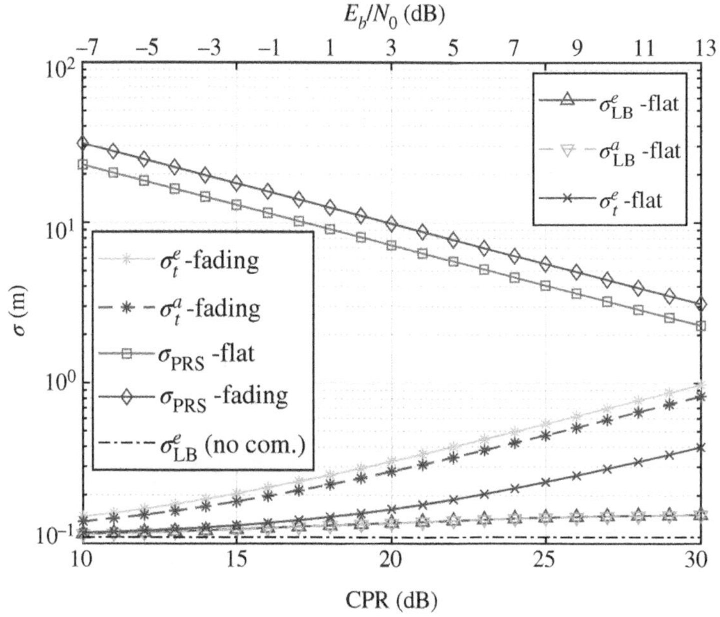

Figure 6.7 Range measurement accuracy ($C/N_0 = 45\,\text{dB} \cdot \text{Hz}$).

positioning accuracy. On the other hand, due to the influence of multipath, the accuracy of MS-NOMA in fading channels decreases with CPR increases. Figure 6.7 also confirms that the approximations of σ_ρ (see (6.16)) correspond to the exact one (see (6.5)) very well.

6.6.3 Resource Consumption of Positioning

Regardless of the type of communication–positioning integration signal used, the dedicated positioning waveform either consumes or interferes with the resources that are originally designated for communication. In this section, we compare the achievable positioning measurement frequency, resource element utilization, and energy consumption of MS-NOMA and PRS.

Some necessary parameters for this evaluation are defined as follows: Define E_c and E_p as the energy of communication and positioning components in a unit time, respectively. $E_{\text{total}} = E_c + E_p$ as the total energy.

6.6.3.1 Achievable Positioning Measurement Frequency

Define T_{meas} as the measuring period which means the receiver executes measuring every T_{meas} seconds. So, the positioning results update every T_{meas} seconds as well.

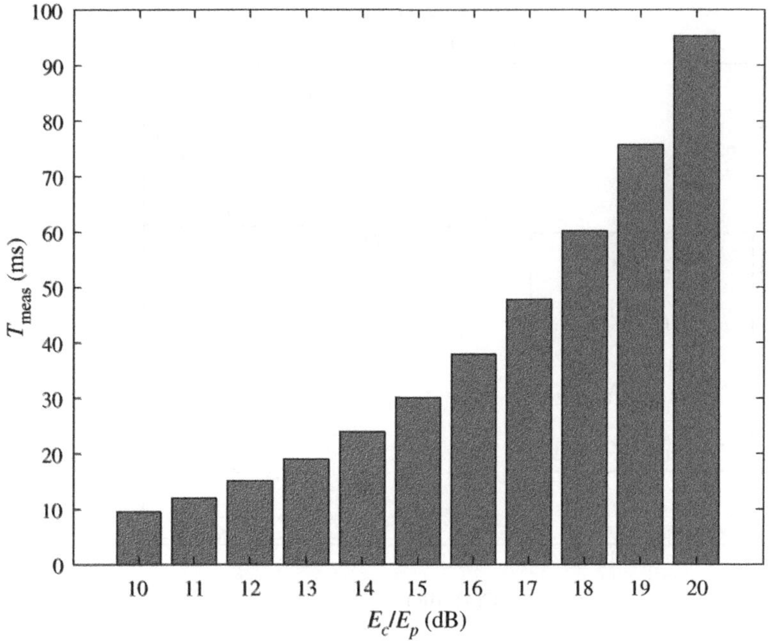

Figure 6.8 The relationship between T_{means} and E_c/E_p for PRS.

For MS-NOMA, its waveform is continuous and the receiver can perform measurements at any time within its calculation capacity. As a result, MS-NOMA has no relation to T_{meas}.

On the other hand, when using PRS for positioning, it is necessary to set the periodicity of the signal, denoted as T_{per}^{PRS}. Higher measurement frequencies, meaning lower T_{meas}, require more frequent broadcasting of PRS (i.e., lower T_{per}^{PRS} or setting T_{rep}^{PRS}). Therefore, PRS has a close relation to T_{meas} and the relationship between T_{meas} and E_c/E_p in PRS is shown in Fig. 6.8. It is clear that higher measuring frequency consumes more energy for positioning purpose (i.e., lower E_c/E_p) in PRS.

6.6.3.2 The Resource Element Consumption

For MS-NOMA, its unique waveform structure that allows positioning components to share band resources with communication components, without directly occupying resource elements (REs). However, the MS-NOMA can cause additional BER loss to the communication signals, which can be equivalent to occupying REs that should belong to communication. To evaluate its equivalent RE consumption, we introduce Eq. (6.64)

$$\Lambda_{\text{MS-NOMA}} = \frac{1}{N}\sum_{n} \text{BER}^{n} - \text{BER}_{0} \tag{6.64}$$

where, BER_0 is the BER without positioning component. Then, by replacing P_c and P_p with E_c and E_p in (6.64), we can obtain the $\Lambda_{\text{MS-NOMA}}$.

For PRS, it occupies the REs directly, so its RE consumption can be evaluated by

$$\Lambda_{\text{PRS}} = \frac{N_{\text{RE,PRS}}}{N_{\text{RE,total}}} \tag{6.65}$$

where, $N_{\text{RE,PRS}}$ is the number of REs allocated to PRS during each measuring period, $N_{\text{RE,total}}$ is the total number of REs allocated to both communication components and PRS, respectively. To ensure a fair comparison of the spreading gains between MS-NOMA and PRS, their code lengths should be the same during the integration time. So, we have

$$N_{\text{RE,PRS}} = \Delta f_p T_{\text{coh}}, \tag{6.66}$$

$$N_{\text{RE,total}} = N T_{\text{meas}} / T_c. \tag{6.67}$$

Then, (6.65) can be written as (6.68)

$$\Lambda_{\text{PRS}} = \left(\frac{N}{G} \frac{T_{\text{meas}}}{T_{\text{coh}}} \right)^{-1}. \tag{6.68}$$

Figure 6.9 shows the comparison of the RE consumption between MS-NOMA and PRS at different signal bandwidths. The results show that MS-NOMA consumes fewer REs than PRS in most cases. This indicates that, although the positioning component may interfere with communication components, its equivalent RE consumption is still smaller than PRS. As E_{total} increases, the value of $\Lambda_{\text{MS-NOMA}}$ slightly increases while Λ_{PRS} remains relatively constant. This is because the BER_0 decreases slightly in (6.64) when the power of communication component increases while Λ_{PRS} has no relation to E_{total} as (6.65) shows. On the other hand, Λ_{PRS} increases when T_{meas} decreases, whereas $\Lambda_{\text{MS-NOMA}}$ has little change. This finding is consistent with the previous discussion in Section 6.6.3.1. Therefore, PRS may not be suitable for frequent positioning scenarios such as automatic driving.

When the bandwidth B increases, it should be noted that MS-NOMA consumes fewer REs due to less positioning interference, as shown in (6.4) where T_p decreases. In contrast, Λ_{PRS} remains unaffected by the change in bandwidth, since B has no impact on the proportion of communication and positioning components in PRS.

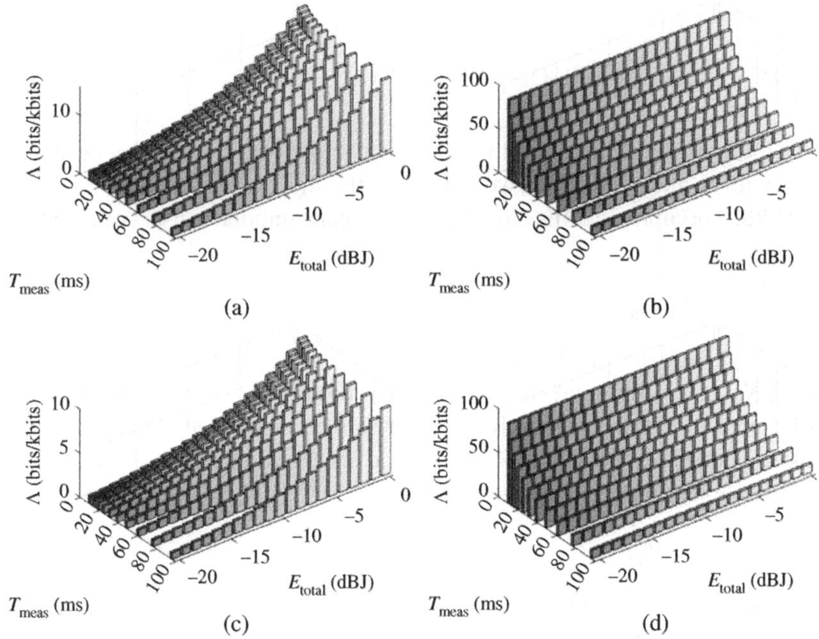

Figure 6.9 The resource element consumption. (a) MS-NOMA ($B = 20$ MHz). (b) PRS ($B = 20$ MHz). (c) MS-NOMA ($B = 50$ MHz). (d) PRS ($B = 50$ MHz).

6.6.3.3 The Power Consumption

For MS-NOMA, its unit energy is equivalent to its power due to its continuous nature, so we have

$$E_c = NP_c, \tag{6.69}$$

$$E_p = MP_p. \tag{6.70}$$

Then

$$\left(\frac{E_c}{E_p}\right)_{\text{MS-NOMA}} = \frac{N}{M}. \tag{6.71}$$

For PRS, P_p can be as strong as P_c. The energy of PRS is determined by the number of REs due to its potential discontinuity:

$$\left(\frac{E_c}{E_p}\right)_{\text{PRS}} = \frac{N_{\text{RE,com}}}{N_{\text{RE,PRS}}} \overset{E_c \gg E_p}{\approx} \frac{N_{\text{RE,total}}}{N_{\text{RE,PRS}}}. \tag{6.72}$$

Then

$$\left(\frac{E_c}{E_p}\right)_{\text{PRS}} = \frac{N}{G}\frac{T_{\text{meas}}}{T_{\text{coh}}}. \tag{6.73}$$

By giving certain E_{total} and $\frac{E_c}{E_p}$, P_c and P_p can be determined as analyzed above. Figure 6.10 shows the comparison of the ranging accuracy between MS-NOMA and PRS under the same E_{total} and $\frac{E_c}{E_p}$. It is evident that MS-NOMA consistently outperforms PRS in terms of ranging accuracy when consuming the same energy and bandwidth resources. When E_{total} increases, the ranging accuracy of both MS-NOMA and PRS improves. For MS-NOMA, as the $\frac{E_c}{E_p}$ is fixed, the bigger E_{total} means better the quality of positioning component and the higher the ranging accuracy. When $\frac{E_c}{E_p}$ increases and E_{total} is fixed, the ranging accuracy of MS-NOMA decreases gradually due to the proportion of positioning component decreases. On the other hand, since PRS and communication waveform are time division multiple access (TDMA) with equivalent power and different periods when E_c/E_p changes, the ranging accuracy remains constant, but E_c/E_p has an impact on the measuring frequency.

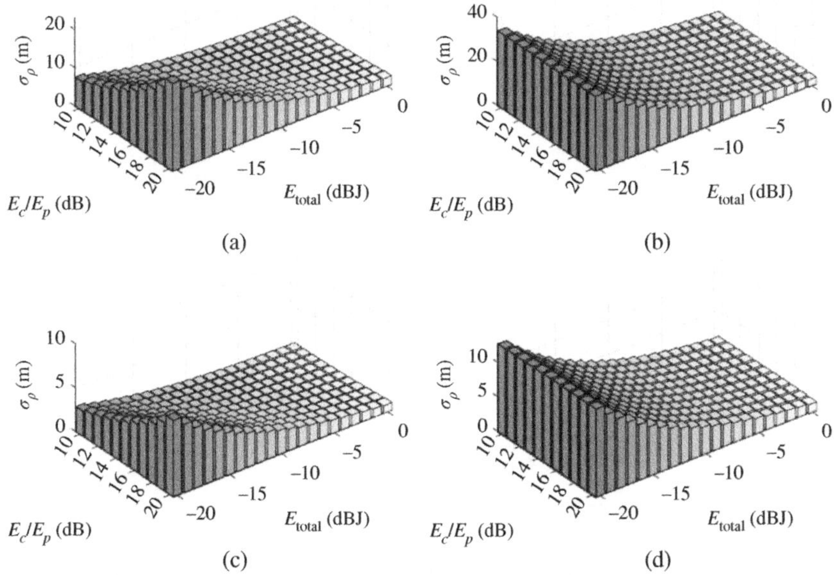

Figure 6.10 The energy consumption. (a) MS-NOMA ($B = 20$ MHz). (b) PRS ($B = 20$ MHz). (c) MS-NOMA ($B = 50$ MHz). (d) PRS ($B = 50$ MHz).

It is worth noting that even when the communication energy is 100 times higher than the positioning energy, MS-NOMA still achieves better ranging accuracy than PRS. Therefore, MS-NOMA consistently achieves the same ranging accuracy performance with lower energy consumption compared to PRS.

6.6.4 Positioning Performance

To assess the accuracy of the communication and positioning performances constrained positioning power allocation (CP4A) algorithm for the MS-NOMA waveform, we conducted simulations in a scenario with 4 fixed BSs positioned at (0,0), (0,200), (200,200), and (200,0), and 20 P-Users randomly placed within the coverage area. The simulations were conducted using the free space propagation model with 50 Monte Carlo runs.

6.6.4.1 Comparison by Using CP4A and the Traditional Method

Figures 6.11 and 6.12 demonstrate the positioning accuracy and coverage of the proposed CP4A algorithm and the traditional equal-power transmission strategy of MS-NOMA waveform. The positioning process requires a minimum of four

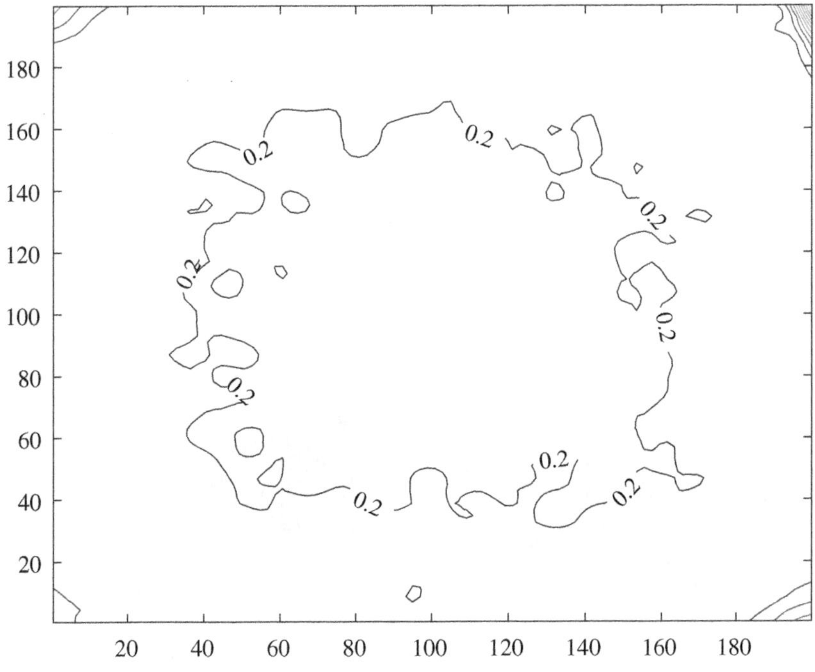

Figure 6.11 The proposed CP4A algorithm.

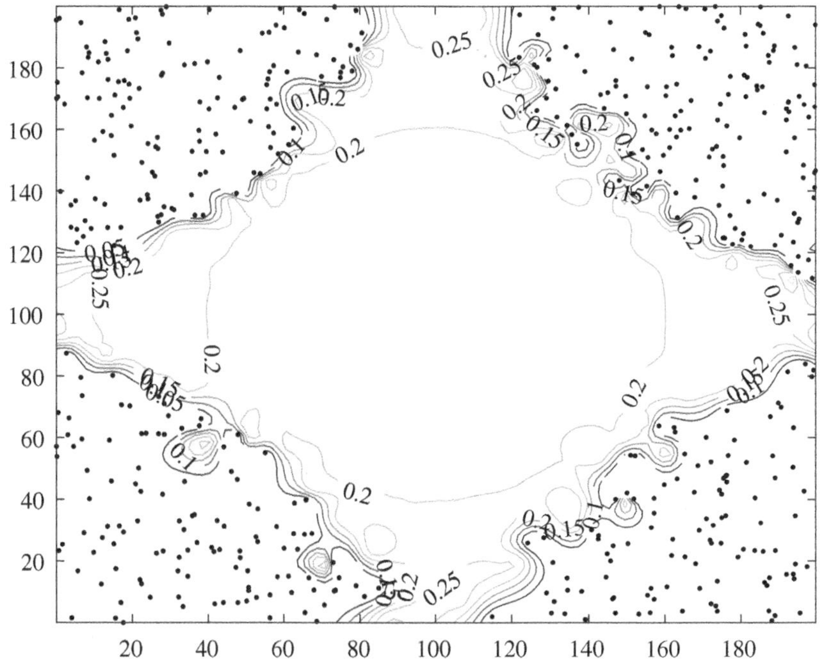

Figure 6.12 The traditional method without power allocation.

signals from the base stations. However, due to the "near and far interference" phenomenon, the signal from a distant base station may be overshadowed by a stronger signal from a nearby base station during demodulation, resulting in insufficient positioning components and no positioning result. In this experiment, if a P-User cannot obtain a positioning result, the positioning error is set to 0 and is noted with dark gray dots in the figure. As shown in Fig. 6.11, the proposed CP4A algorithm effectively reduces the "near and far effect," making all P-Users obtain a positioning result. In contrast, Fig. 6.12 indicates that a significant proportion of P-Users does not obtain a positioning result, resulting in reduced coverage of positioning.

Next, we will explore how the proposed CP4A algorithm allocates power to different positioning users to improve the positioning coverage through several specific parameters. As previously mentioned, the power allocated by each P-User for transmission depends on their individual channel state information. Figure 6.13 depicts the allocation of power, the channel gains $\left(\left|h_p^{m\leftarrow km}\right|^2\right)$, and the CPRs of the P-Users in a single simulation. To enhance clarity, the y-axis of the figure only represents the numerical values and does not include any units. In addition, the size of the channel gains is scaled down by a factor of 10 for display purposes. The

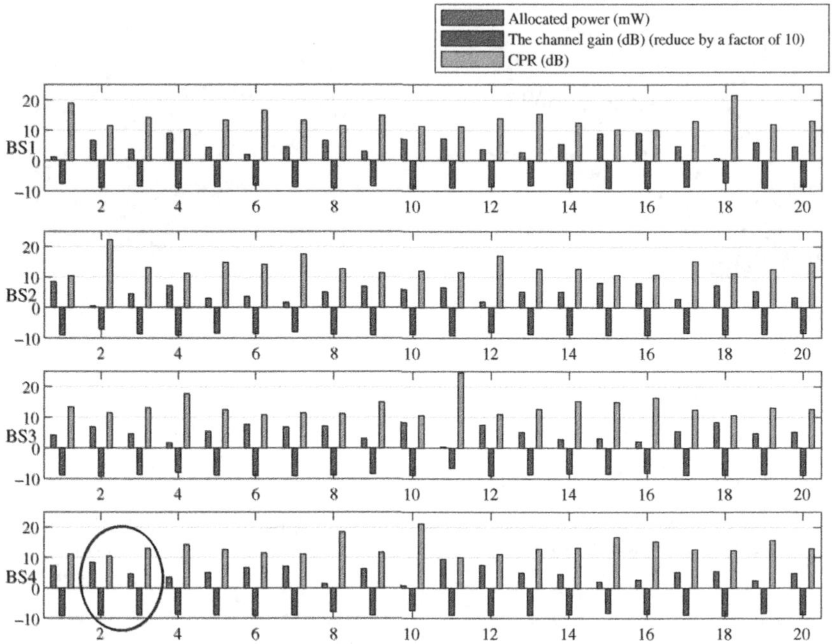

Figure 6.13 The allocated power and CPRs with different channel gains ($B = 50$ MHz).

figure highlights two significant features. Firstly, P-Users with poor channel states tend to be assigned higher transmission power for positioning. Secondly, the CPRs have a similar trend to the channel gains.

To reduce the impact of "near and far interference," it is preferable to assign higher power to P-Users with weaker channel states. However, the power allocation among P-Users is not solely determined by their channel states, but also by the geometric-dilution λ in (6.47), leading to different power for P-Users with similar channel gains. For instance, as highlighted by the dark gray circle in the Fig. 6.13, in BS4, the second P-User at coordinates (178,8) and the third P-User at coordinates (79,77) have similar channel gains, but the former will receive a higher power allocation due to its location at the edge of the coverage area.

Moreover, it is obvious that CPRs have similar tends with the channel gains. It does not seem correct as CPR is inversely proportional to $\left|h_p^{m\leftarrow km}\right|^2$ as defined in (6.30). Notice that worse channel states tend to allocate stronger positioning power. As a result, the denominator of CPR $\left(\text{i.e., } \left|h_p^{m\leftarrow km}\right|^2 P_p^{km}\right)$ will tend to be flat for all km. Then, CPR is approximately proportional to $\left|h_c^{m\leftarrow k'}\right|^2$. Because the channel gains of communication components are approximately equal to that of positioning components for P-User m $\left(\text{i.e., } \left|h_c^{m\leftarrow k'}\right|^2 \approx \left|h_p^{m\leftarrow km}\right|^2\right)$, it results in that

Table 6.3 The comparison between MS-NOMA and PRS.

Signal		MS-NOMA		PRS
Power allocation strategy		**CP4A**	**Equal power**	**–**
Positioning error	20 MHz	0.57 m	0.54 m	8.44 m
	50 MHz	0.21 m	0.20 m	3.17 m
Coverage	20 MHz	100%	50.8%	72.6%
	50 MHz	100%	50.5%	79.2%

CPRs have similar tends with $\left|h_p^{m\leftarrow km}\right|^2$. In addition, it is observed that the CPRs have weaker fluctuations than the channel gains, this reflects that the proposed CP4A algorithm weakens the effects of channel attenuation.

6.6.4.2 Comparision Between MS-NOMA and PRS

Table 6.3 and Fig. 6.14 present a comparison of the average positioning performance between MS-NOMA and PRS. The results demonstrate that MS-NOMA

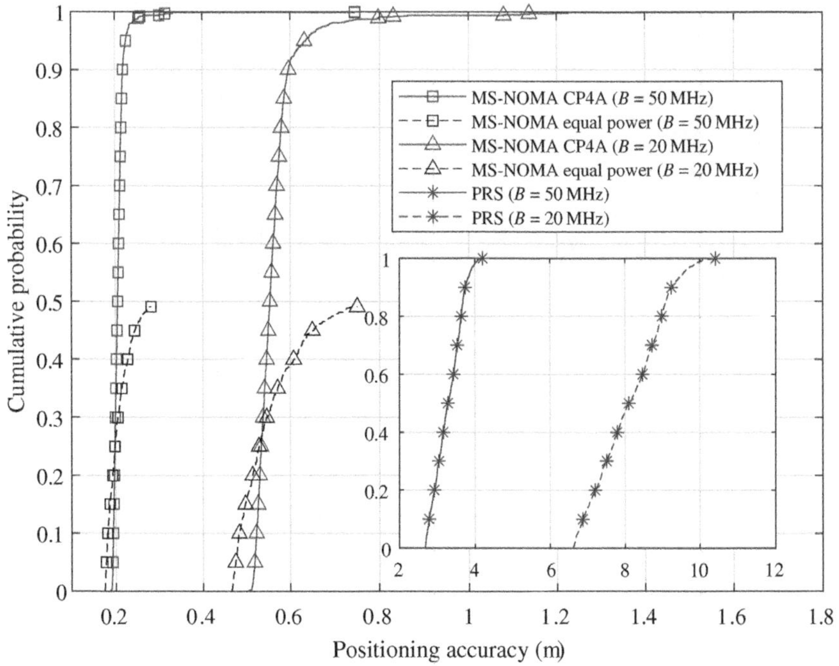

Figure 6.14 Detailed illustration of the positioning accuracy.

significantly outperforms PRS in terms of positioning accuracy, with an accuracy level below 1m. As can be seen from the Table 6.3, MS-NOMA waveform with CP4A algorithm can achieve 100% coverage of positioning, which is superior to using traditional power allocation algorithm and PRS for positioning.

As is shown in Fig. 6.14, the positioning error bounds of the proposed CP4A algorithm are higher than the one of the equal power allocation algorithm in some areas. This is mainly because of the "no result points" in Fig. 6.12. The proposed CP4A algorithm is able to achieve 100% positioning coverage, which considers the ranging errors and geometric-dilution jointly, and the geometric-dilution λ in peripheral area is usually larger than internal area which will further worsen the positioning accuracy. While the equal allocation algorithm has eliminated these edge points when calculating positioning accuracy because it cannot have positioning results.

Comparing the error bounds of MS-NOMA and PRS in Fig. 6.14, we can see that almost all of the error bounds of the 20 MHz MS-NOMA waveform are smaller than the minimum error bound of the 50 MHz PRS. The reasons for this are mainly twofold. Firstly, as can be seen from Fig. 6.15, the positioning error bounds of PRS are higher in the peripheral area than in the internal area due to the existence

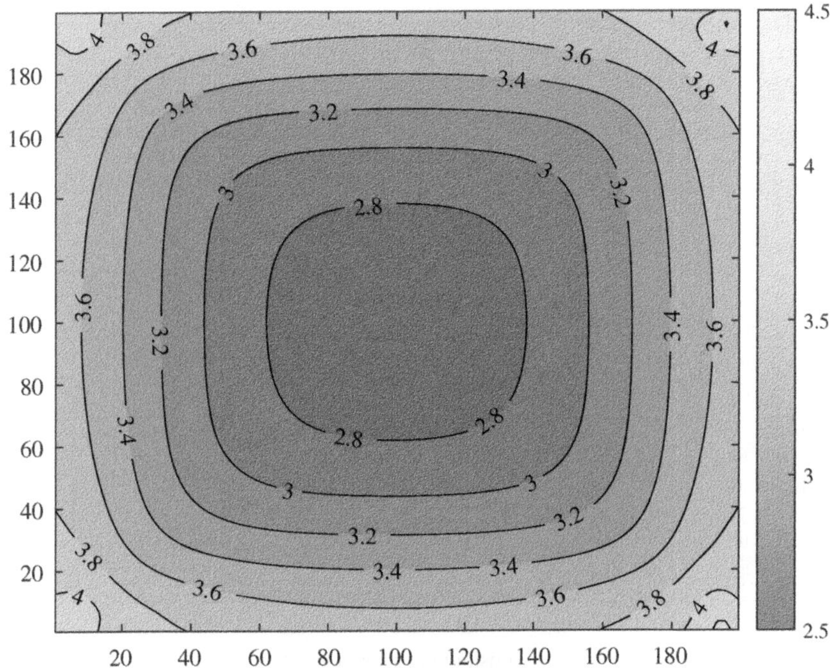

Figure 6.15 The positioning accuracy of PRS ($B = 50$ MHz).

of geometric dilution factor, and the poor positioning accuracy in the peripheral area will affect the overall positioning accuracy calculation. Secondly, as shown in Fig. 6.10, a considerable part of the ranging error variance of MS-NOMA at 20 MHz is smaller than that of 50 MHz PRS.

References

3GPP. TS 38.211 v17.1.0: NR physical channels and modulation. 2022.

L Bai, C Sun, A G Dempster, H Zhao, J W Cheong, and W Feng. GNSS-5G hybrid positioning based on multi-rate measurements fusion and proactive measurement uncertainty prediction. *IEEE Transactions on Instrumentation and Measurement*, 71:1–15, 2022. doi: 10.1109/TIM.2022.3154821.

J W Betz and K R Kolodziejski. Generalized theory of code tracking with an early-late discriminator Part I: Lower bound and coherent processing. *IEEE Transactions on Aerospace and Electronic Systems*, 45(4):1538–1556, 2009. doi: 10.1109/TAES.2009.5310316.

S P Boyd and L Vandenberghe. *Convex Optimization*. 2004.

M M Butt, A Pantelidou, and I Z Kovács. ML-assisted UE positioning: Performance analysis and 5G architecture enhancements. *IEEE Open Journal of Vehicular Technology*, 2:377–388, 2021.

L Chen, X Zhou, F Chen, L L Yang, and R Chen. Carrier phase ranging for indoor positioning with 5G NR signals. *IEEE Internet of Things Journal*, 9(13):10908–10919, 2022. doi: 10.1109/JIOT.2021.3125373.

C S G N Dev, L Pathak, G Ponnamareddy, and D Das. NRPos: A multi-RACH framework for 5G NR positioning. In *2020 IEEE 3rd 5G World Forum (5GWF)*, pages 25–30, 2020. doi: 10.1109/5GWF49715.2020.9221379.

H Haci, H Zhu, and J Wang. Performance of non-orthogonal multiple access with a novel asynchronous interference cancellation technique. *IEEE Transactions on Communications*, 65(3):1319–1335, 2017. doi: 10.1109/TCOMM.2016.2640307.

K Ko, W Ahn, and W Shin. High-speed train positioning using deep Kalman filter with 5G NR signals. *IEEE Transactions on Intelligent Transportation Systems*, 23(9):15993–16004, 2022. doi: 10.1109/TITS.2022.3146932.

Z Li, F Jiang, H Wymeersch, and F Wen. An iterative 5G positioning and synchronization algorithm in NLOS environments with multi-bounce paths. *IEEE Wireless Communications Letters*, page 1, 2023. doi: 10.1109/LWC.2023.3244575.

Y Liu, Z Ding, M Elkashlan, and H V Poor. Cooperative non-orthogonal multiple access with simultaneous wireless information and power transfer. *IEEE Journal on Selected Areas in Communications*, 34(4):938–953, 2016. doi: 10.1109/JSAC.2016.2549378.

Y Lu, D Zhongliang, Z Di, and H Enwen. Quality assessment method of GNSS signals base on multivariate dilution of precision. In *European Navigation Conference*, 2016.

J Luo, L Fan, and H Li. Indoor positioning systems based on visible light communication: State of the art. *IEEE Communications Surveys and Tutorials*, 19(4):2871–2893, 2017. doi: 10.1109/COMST.2017.2743228.

B Makki, T Svensson, M Coldrey, and M S Alouini. Finite block-length analysis of large-but-finite MIMO systems. *IEEE Wireless Communications Letters*, 8(1):113–116, 2019. doi: 10.1109/LWC.2018.2860017.

F Maschietti, G Fodor, D Gesbert, and P de Kerret. User coordination for fast beam training in FDD multi-user massive MIMO. *IEEE Transactions on Wireless Communications*, 20(5):2961–2976, 2021. doi: 10.1109/TWC.2020.3045922.

M Pan, P Liu, S Liu, W Qi, Y Huang, X You, X Jia, and X Li. Efficient joint DOA and TOA estimation for indoor positioning with 5G picocell base stations. *IEEE Transactions on Instrumentation and Measurement*, 71:1–19, 2022. doi: 10.1109/TIM.2022.3191705.

J A del Peral-Rosado, J A López-Salcedo, G Seco-Granados, F Zanier, and M Crisci. Joint channel and time delay estimation for LTE positioning reference signals. In *ESA Workshop on Satellite Navigation Technologies European Workshop on GNSS Signals and Signal Processing*, pages 1–8, Dec. 2012. doi: 10.1109/NAVITEC.2012.6423094.

S Saleh, A S El-Wakeel, and A Noureldin. 5G-enabled vehicle positioning using EKF with dynamic covariance matrix tuning. *IEEE Canadian Journal of Electrical and Computer Engineering*, 45(3):192–198, 2022. doi: 10.1109/ICJECE.2022.3187348.

J Schloemann, H S Dhillon, and R M Buehrer. A tractable analysis of the improvement in unique localizability through collaboration. *IEEE Transactions on Wireless Communications*, 15(6):3934–3948, 2016. doi: 10.1109/TWC.2016.2531646.

K Shamaei and Z M Kassas. Receiver design and time of arrival estimation for opportunistic localization with 5G signals. *IEEE Transactions on Wireless Communications*, 20(7):4716–4731, 2021. doi: 10.1109/TWC.2021.3061985.

L Yin, Q Ni, and Z Deng. A GNSS/5G integrated positioning methodology in D2D communication networks. *IEEE Journal on Selected Areas in Communications*, 36(2):351–362, 2018. doi: 10.1109/JSAC.2018.2804223.

R Zhang. Optimal power control over fading cognitive radio channel by exploiting primary user CSI. In *IEEE GLOBECOM 2008 -2008 IEEE Global Telecommunications Conference*, pages 1–5, 2008. doi: 10.1109/GLOCOM.2008.ECP.184.

7

NOMA-Aware Wireless Content Caching Networks

Yaru Fu[1], Zheng Shi[2], and Tony Q. S. Quek[3]

[1] *Department of Electronic Engineering and Computer Science, School of Science and Technology, Hong Kong Metropolitan University, Hong Kong, China*
[2] *School of Intelligent Systems Science and Engineering, Jinan University, Zhuhai, China*
[3] *Department of Information Systems Technology and Design, Information Systems Technology and Design, Singapore University of Technology and Design, Singapore, Singapore*

7.1 Introduction

In recent years, there has been a rapid growth in demand for data, largely due to the surge of data-hungry applications such as virtual reality (VR)/augmented reality (AR), ultra-high definition video, and mobile interactive games. These innovative applications are latency-sensitive and generate a significant amount of data, putting immense pressure on current networks due to scarce radio resources and limited backhaul capacity. The main issue is the heterogeneous demands of users; each individual has a unique content preference and will compete for communication resources if requesting content delivery simultaneously. According to Ericsson's mobility report (Ericsson, 2022), the mobile data traffic per month is forecast to increase to 160 Exabytes (EBs) in 2025 from 38 EBs each month in 2019, the majority of which will be video, pushing the endurance of our current cellular networks architecture to the limits. To address these significant challenges, wireless edge caching has been recognized as a promising solution. Specifically, by pre-fetching frequently accessed and reusable content and storing them at edge nodes like base stations (BSs), access points, and user devices, networks with low service latency and improved subscriber experiences can be achieved. This is because delivering contents from network edge nodes to proximal users experiences shorter transmission distance than that between the remote servers and the subscribers. Moreover, shorter transmission distance in general results in a higher bit rate, which in return reduces user's perceived latency (Liu et al., 2016, Chen et al., 2017, Zhang et al., 2019). In the meantime, non-orthogonal multiple access

Next Generation Multiple Access, First Edition.
Edited by Yuanwei Liu, Liang Liu, Zhiguo Ding, and Xuemin Shen.
© 2024 The Institute of Electrical and Electronics Engineers, Inc. Published 2024 by John Wiley & Sons, Inc.

(NOMA) has gathered significant research interest for its immense potential in enhancing cellular network's performance by facilitating the transmission of multiple users' signals on the same time or frequency resource block simultaneously. The superiority of NOMA compared to conventional orthogonal multiple access (OMA) systems in terms of spectrum efficiency (Zhao et al., 2020), power efficiency (Fu et al., 2017), and transmission efficiency (Shi et al., 2020) has been explicitly demonstrated. Hence, NOMA is widely perceived as the next-generation multiple access technology (Liu et al., 2017).

The integration of wireless edge caching and NOMA is a compelling research direction. For instance, the authors in Ding et al. (2018) proposed two caching schemes for a single-cell NOMA network and demonstrated that edge caching can improve NOMA system's coverage probability and spectrum efficiency. In addition, the outage performance of NOMA-assisted wireless content caching networks was analyzed in Zhao et al. (2018). It is shown that cache-enabled NOMA achieves superior outage performance than pure NOMA systems. Besides, the authors derived the optimal decoding order and studied the corresponding rate allocation for a cached-aided NOMA system with two users in Xiang et al. (2019), aiming to minimize the content transmission delay. Moreover, the authors in Doan et al. (2018) maximized the system's success probability via power allocation among clusters and users, wherein the success probability is defined as the probability that all mobile users can successfully decode their intended messages. Recently, the authors in Yang et al. (2020b, 2022) applied machine learning-enabled algorithms to solve the joint cache placement and transmission resource allocation problems for cache-aware NOMA systems.

However, due to the heterogeneity of users' preference distribution, the limited cache capacity goes far from satisfying most users' needs, which harms the benefits of wireless edge caching. To address this issue, it is of high necessity to fall back on some proactive mechanisms to further improve the caching performance. Thereof, recommendation has received a lot of attention due to its capacity of reshaping users' content request behaviors. The joint content caching and recommendation optimization for wireless edge caching systems was considered in Chatzieleftheriou et al. ((2019), Fu et al. 2021a, 2021b, 2021c), and Kastanakis et al. (2022). Specifically, in Chatzieleftheriou et al. (2019) and Fu et al. (2021a), the cache hit ratio maximization problem was studied by jointly optimizing the cache placement and recommendation decision. Thereof, alternating optimization algorithms were used to solve the mixed-integer optimization problems. Simulation results demonstrated that recommendation-aware caching achieves higher cache hit ratio than that of the caching systems without recommendation. In Fu et al. (2021b), authors investigated the joint coded-cache and recommendation optimization problem for wireless caching systems from a revenue maximization perspective. Here, content coding was used to correct the corrupted data, wherein the single-failure pattern was considered. Later on, the authors extended their idea to the multiple-failure case in Fu et al. (2021c). Recently, the authors in Kastanakis et al. (2022) performed real man assisted experiment to evaluate the performance of cache-enabled recommendation systems. Experimental results demonstrated that cache-aided recommendation scheme achieves an 8–10 times increase in terms of cache hit ratio compared to conventional caching systems.

Despite foregoing mentioned works validating the effectiveness of the interplay between content caching and recommendation, its successful implementation cannot be separated from content delivery, particularly when the time-varying wireless environment is taken into account. However, the joint study was overlooked by Chatzieleftheriou et al. (2019), Fu et al. (2021a, 2021b, 2021c), and Kastanakis et al. (2022). In this chapter, we aim to bridge this research gap by investigating the joint content caching, recommendation, and radio resource allocation problem for downlink NOMA networks with transmission mode selection at the BS. Our focus is on minimizing average latency, as future cellular systems are expected to support extensive intelligent applications with low latency requirements. For brevity, our main contributions are summarized as follows:

1. We present the average latency expression for cache-aware NOMA networks and demonstrate how it depends on factors such as cache placement, recommendation decisions, and resource management strategy for content delivery at the BS. Additionally, we discuss the importance of transmission mode selection at the BS for reducing latency. Specifically, we adopt multicasting for users requesting the same content, while applying NOMA for the remaining users.
2. To solve the nonconvex, multi-timescale, and mixed-integer average latency minimization problem, a *divide-and-rule* approach is developed. Firstly, we decompose the original problem into two subproblems: the short-term resource allocation problem at BS given caching and recommendation strategies, and the long-timescale joint caching and recommendation decision-making problem with a fixed resource management policy. Then, we solve each subproblem separately and optimize them alternately until convergence is reached.
3. Given recommendation and caching policies, the original optimization problem simplifies to a joint NOMA user pairing and inter-group power control problem. We transform the user pairing problem into a minimum weight perfect matching problem with respect to an undirected graph, which can be optimally solved by the enhanced Edmond's blossom algorithm with cubic time complexity. The intra-cluster power allocation is optimally determined as a byproduct. We then develop a time-efficient inter-group power control scheme. For the long-term optimization problem, we propose a joint dynamic programming and swap-then-compare enabled approach to determine the cache placement and personalized recommendation decision.

4. We mathematically examine the computational complexity and convergence performance for our designed joint optimization algorithm. Moreover, we perform extensive numerical simulations to evaluate the performance of our proposed solution against various benchmark schemes. Our performance metrics include the system's average latency and cache hit ratio, which are key indicators of the effectiveness of our algorithm. Simulation results demonstrate the superiority of our approach, providing compelling evidence for its potential as a practical solution for optimizing systems in real-world scenarios.

7.2 System Model

In this section, we first introduce the cache-enabled NOMA system model. Then, we elaborate on the users' content request model with recommendation and the channel model. Thereafter, we explicitly derive the system latency, showing how it can be jointly impacted by the cache decision, recommendation decision, and transmission-oriented NOMA user pairing and power control strategies.

7.2.1 System Description

A typical cache-aware NOMA system is considered, which contains one BS serving K users, whose index set is denoted by $\mathcal{K} = \{1, 2, \ldots, K\}$. The BS has the capability of proactive caching, which can store some popular content items in advance. Define $\overline{C}$ as the cache capacity of the BS. In total, there are I items in the system's content catalog. Define $\mathcal{I} = \{1, 2, \ldots, I\}$ as the set of content indices. Besides, for $i \in \mathcal{I}$, let B_i and $c_i \in \{0, 1\}$ be the data size (in bit) and the caching state of content i, respectively. In detail, $c_i = 1$ if and only if (iff) content i has been cached by the BS. As the cache capacity of the BS is limited by $\overline{C}$, we have

$$\sum_{i \in \mathcal{I}} c_i B_i \leq \overline{C}. \tag{7.1}$$

Considering the heterogeneity among users preference, we denote $a_{k,i}^{\text{pref}} \in [0, 1]$ as the inherent preference of user k to item i, which satisfies $\sum_{i \in \mathcal{I}} a_{k,i}^{\text{pref}} = 1$. For $k \in \mathcal{K}$, we define $\boldsymbol{a}_k = (a_{k,i}^{\text{pref}})_{i \in \mathcal{I}}$ as the preference information of user k. If the requested content items of the users are cached by the BS, the BS will deliver these contents to the corresponding users directly by adopting multicasting or NOMA transmission. Details will be introduced later. Given that some of the demanded items are not cached, the BS first fetches the un-cached items from the remote cloud or core network through backhaul link transmission and then transmits the requested contents to the corresponding users by different transmission modes.

7.2.2 Content Request Model

In conventional cache-enabled NOMA networks without recommendation, each user requests its demanded content items by $\boldsymbol{a}_k$. However, when incorporating recommendation algorithms, users' content request behavior changes. To capture this change, for $k \in \mathcal{K}$ and $i \in \mathcal{I}$, we define a binary variable $r_{k,i} \in \{0, 1\}$ as the recommendation decision of user k with respect to content i. More specifically, $r_{k,i} = 1$ iff content i has been recommended to user k. For brevity, let $\boldsymbol{r}_k = (r_{k,i})_{i \in \mathcal{I}}$ be the recommendation decision vector of user k, where $k \in \mathcal{K}$. In addition, we introduce the parameter $\overline{N}_k$ to represent the maximum number of recommended items that can be displayed on user k's device. On this ground, we have

$$\sum_{i \in \mathcal{I}} r_{k,i} = \overline{N}_k, \quad k \in \mathcal{K}.$$

To ensure alignment with best practices, we establish a minimum preference threshold, denoted as $\overline{\theta}_k$, which dictates that only content with a higher preference rating than the threshold can be recommended to user k. Moreover, in order to meet our recommendation quality standards, we impose the following constraint:

$$r_{k,i}(1 - a_{k,i}^{\text{pref}}) \leq (1 - \overline{\theta}_k), \quad k \in \mathcal{K}, \quad \text{and} \quad i \in \mathcal{I}. \tag{7.2}$$

With recommendations, the user's content request behavior will be jointly affected by its inherent preference $\boldsymbol{a}_k$ and the recommendation policy $\boldsymbol{r}_k$. Specifically, for $k \in \mathcal{K}$ and $i \in \mathcal{I}$, the request probability of user k toward content i is represented as $a_{k,i}^{\text{req}}$. Following the recommendation model described in Fu et al. (2021a), we have

$$a_{k,i}^{\text{req}} = b_k \frac{r_{k,i} a_{k,i}^{\text{pref}}}{\sum_{j \in \mathcal{I}} r_{k,j} a_{k,j}^{\text{pref}}} + (1 - b_k) \frac{(1 - r_{k,i}) a_{k,i}^{\text{pref}}}{\sum_{j \in \mathcal{I}} (1 - r_{k,j}) a_{k,j}^{\text{pref}}}, \tag{7.3}$$

wherein $b_k \in (0, 1)$ depicts the probability of user k accepting the recommendations.

7.2.3 Random System State

In this section, we provide a detailed explanation of the random system state, which comprises the random content request state and the random channel state. We define the content request state of user k as X_k, and based on our previous analysis, we know that $X_k \in \mathcal{I}$ and $p_{X_k}(x_k) = \Pr[X_k = x_k] = a_{k,x_k}^{\text{req}}$, where $x_k \in \mathcal{I}$ and a_{k,x_k}^{req} is defined in (7.3). Furthermore, we assume that the variables X_k for $k \in \mathcal{K}$ are mutually independent. Therefore, the random content requests of all users can be represented as $\mathbf{X} \triangleq (X_k)_{k \in \mathcal{K}} \in \mathcal{I}^K$. To describe the random channel state, we introduce $G_k \in \mathcal{G}$ as the random channel state of user k for $k \in \mathcal{K}$, where $\mathcal{G}$ represents the finite channel state set. The probability that the variable G_k takes the value of g_k is denoted as $p_{G_k}(g_k) = \Pr[G_k = g_k] \geq 0$, where $g_k \in \mathcal{G}$. It is worth noting that for $k \in \mathcal{K}$, the summation of probabilities $\sum_{g_k \in \mathcal{G}} p_{G_k}(g_k)$ equals to 1. Furthermore, we assume that G_k is mutually independent for $k \in \mathcal{K}$ and may have different probability mass functions. Based on these definitions, the random channel state can be expressed as $\mathbf{G} \triangleq (G_k)_{k \in \mathcal{K}} \in \mathcal{G}^K$. The vector $\mathbf{G}$ can be considered as the channel statistics in the model. Consequently, the proposed solutions in this chapter can be applied to practical systems as long as the channel statistics remain constant (Cui et al., 2017, Yang et al., 2021).

Thus far, we have introduced the random content request model and random channel state model. They jointly determine the random system state, which is defined as $(\mathbf{X}, \mathbf{G}) \in \mathcal{I}^K \times \mathcal{G}^K$. Likewise, we assume $\mathbf{X}$ is independent of $\mathbf{G}$, the probability of the random system state $(\boldsymbol{x}, \mathbf{g})$ is given below:

$$\Pr[(\mathbf{X}, \mathbf{G}) = (\boldsymbol{x}, \mathbf{g})] = \prod_{k \in \mathcal{K}} p_{X_k}(x_k) p_{G_k}(g_k) \triangleq p(\boldsymbol{x}, \mathbf{g}), \tag{7.4}$$

where $(\boldsymbol{x}, \mathbf{g}) \in \mathcal{I}^K \times \mathcal{G}^K$, $\boldsymbol{x} = (x_k)_{k \in \mathcal{K}} \in \mathcal{I}^K$, and $\mathbf{g} = (g_k)_{k \in \mathcal{K}} \in \mathcal{G}^K$.

7.2.4 System Latency Under Each Random State

In this section, we will examine the system latency for each random system state. To do so, we define $S_i(\mathbf{X})$ as the index set of users who require content i under the random content request state $\mathbf{X}$. The magnitude of $S_i(\mathbf{X})$ is denoted by $|S_i(\mathbf{X})|$. We then define the index sets of multicasting users and NOMA users as $\mathbb{S}'(\mathbf{X}) = (S_i(\mathbf{X}))_{i \in \mathcal{I}, |S_i(\mathbf{X})| \geq 2}$ and $\tilde{S}(\mathbf{X}) = \bigcup_{i \in \mathcal{I}, |S_i(\mathbf{X})| = 1} S_i(\mathbf{X})$, respectively. Note that if $|\tilde{S}(\mathbf{X})|$ is not even, an adjustment will be made. Specifically, we subtract from $\tilde{S}(\mathbf{X})$ the index of the user with the median value of channel gain, resulting in an updated set denoted as $\overline{S}(\mathbf{X})$.

$$\overline{S}(\mathbf{X}) = \tilde{S}(\mathbf{X}) \backslash \{k_m\}, \tag{7.5}$$

where k_m is the removed user whose channel gain has the intermediate value among that of all users in $\tilde{S}(\mathbf{X})$. On this basis, user k_m can be taken as a multicasting user. Thus, we update $\mathbb{S}'(\mathbf{X})$ as $\mathbb{S}(\mathbf{X})$, which is quoted below

$$\mathbb{S}(\mathbf{X}) = \mathbb{S}'(\mathbf{X}) \cup \{k_m\}. \tag{7.6}$$

In summary, based on whether $|\tilde{S}(\mathbf{X})|$ is an even number, we formulate the collection of index sets of multicasting users $\mathbb{S}(\mathbf{X})$ and the indices of NOMA users $\overline{S}(\mathbf{X})$ as

$$\mathbb{S}(\mathbf{X}) = \begin{cases} \mathbb{S}'(\mathbf{X}), & \text{if } |\tilde{S}(\mathbf{X})| \text{ is even,} \\ \mathbb{S}'(\mathbf{X}) \cup \{k_m\}, & \text{otherwise,} \end{cases} \tag{7.7}$$

and

$$\overline{\mathcal{S}}(\mathbf{X}) = \begin{cases} \tilde{\mathcal{S}}(\mathbf{X}), & \text{if } |\tilde{\mathcal{S}}(\mathbf{X})| \text{ is even}, \\ \tilde{\mathcal{S}}(\mathbf{X}) \backslash \{k_m\}, & \text{otherwise}, \end{cases} \tag{7.8}$$

respectively. we define $\mathcal{I}'(\mathbf{X})$ as the set of content indices demanded by users in $\mathbb{S}(\mathbf{X})$ and $\mathcal{I}''(\mathbf{X})$ as the index set of content requested by users in $\overline{\mathcal{S}}(\mathbf{X})$. The transmission strategy is as follows: for users in $\mathcal{S}_i(\mathbf{X})$, where $i \in \mathcal{I}'$, the BS transmits content i to them through multicasting. Meanwhile, for users in $\overline{\mathcal{S}}(\mathbf{X})$, downlink NOMA is applied. The performance comparison among the proposed hybrid scheme as well as the pure NOMA and pure multicasting transmission strategies will be presented in the section of numerical simulation.

Afterward, we will discuss the system latency. As stated in Section 7.2.1, for each content request, the BS will first retrieve the non-cached items from core networks through backhaul link transmission and then transmit the content items to the corresponding users. The associated backhaul latency, referred to as T_b, can be calculated as follows:

$$T_b = \frac{\sum_{i \in \mathcal{I}'(\mathbf{X})} (1 - c_i) B_i + \sum_{j \in \mathcal{I}''(\mathbf{X})} (1 - c_j) B_j}{R_b}, \tag{7.9}$$

in which R_b is the backhaul link rate.

Subsequently, we will analyze the latency in terms of transmission. Let $T_{t,M}$ denote the transmission latency for users who request the same items. It is important to note that the transmission latency for multicasting is determined by the worst-case latency among all users. However, since we primarily focus on optimizing the average delay, we adopt the summation of each user's latency. Based on this, we have:

$$T_{t,M} = \sum_{i \in \mathcal{I}'} \sum_{k \in \mathcal{S}_i(\mathbf{X})} \frac{B_i}{R_k}, \tag{7.10}$$

where R_k is the transmission data rate of user k, which is given below:

$$R_k = W \log_2 (1 + \frac{P'_i g_k}{N_k}),$$

in which W is the bandwidth. P'_i and N_k expresses the transmit power as regards multicasting content i and the power of the additive white Gaussian noise (AWGN) at user k, respectively.

Finally, we calculate the transmission latency for NOMA users, taking into account two crucial aspects for minimizing it, namely, user pairing and power allocation. For the sake of simplicity, we assume that two users are superimposed by NOMA, as that in Chen et al. (2016) and Sun et al. (2020). The set $\overline{\mathcal{S}}(\mathbf{X})$ consists of all NOMA users, and we construct L pairs denoted by $\mathcal{L} = \{1, 2, \dots, L\}$. It is easy to verify that $L = |\overline{\mathcal{S}}(\mathbf{X})|/2$. Furthermore, we define $y_{l,k} \in \{0, 1\}$ as the pairing indicator of user k in cluster l, i.e., $y_{l,k} = 1$ if user k is assigned to cluster l, where $k \in \overline{S}(\mathbf{X})$ and $l \in \mathcal{L}$. Thereby, the following constraint should be met,

$$\sum_{k \in \overline{S}} y_{l,k} = 1, \quad l \in \mathcal{L}, \quad k \in \overline{S}, \tag{7.11}$$

wherein $\mathbf{X}$ is ignored for brevity.

For ease of analysis, we define the strong user and the weak user in the lth pair as l_s and l_w, respectively. In addition, we denote the allocated transmit power for the two users in pair l as P_l''. Based on these definitions, we can state the following:

$$\begin{cases} R_{l_s} = W \log_2(1 + \frac{p_{l_s} g_{l_s}}{N_{l_s}}), \\ R_{l_w} = W \log_2 \left(1 + \frac{p_{l_w} g_{l_w}}{p_{l_s} g_{l_w} + N_{l_w}}\right). \end{cases} \tag{7.12}$$

where p_{l_s} and p_{l_w} is the transmit power assigned to users l_s and l_w, respectively, and satisfies the condition $P_l'' = p_{l_s} + p_{l_s}$. Meanwhile, N_{l_s} and N_{l_w} represent the powers of AWGN at these two users, respectively. With above mentioned definitions, we can express the transmission latency of users l_s and l_w as follows:

$$T_{l_s} = \frac{B_{l_s,i}}{R_{l_s}}, \quad \text{and} \quad T_{l_w} = \frac{B_{l_w,i}}{R_{l_w}}, \tag{7.13}$$

where $B_{l_s,i}$ and $B_{l_w,i}$ represent the data size of the content that is requested by users l_s and l_w, respectively. In addition, R_{l_s} and R_{l_w} are defined in (7.12). On these grounds, the total transmission latency for all NOMA users, defined as $T_{t,N}$, is quoted below:

$$T_{t,N} = \sum_{l \in \mathcal{L}} (T_{l_s} + T_{l_w}). \tag{7.14}$$

So far, we have examined both the backhaul latency and transmission latency (for both multicasting and NOMA users), which jointly affect the system's overall latency, denoted by T, which is expressed as follows:

$$T = T_b + T_{t,M} + T_{t,N}, \tag{7.15}$$

wherein T_b, $T_{t,M}$, and $T_{t,N}$ are defined in (7.9), (7.10), and (7.14), respectively. It is worth mentioning that T highly depends on the cache placement, recommendation decision, joint user pairing and power control scheme, as well as the system state, i.e., $(\mathbf{X}, \mathbf{G})$, although it is not explicitly shown in (7.15).

7.2.5 System's Average Latency

In Section 7.2.4, we discussed how to calculate the system latency for each random system state. In this section, we analyze how the average system latency is jointly determined by the cache placement, recommendation decision, and transmission-related resource management. For simplicity of notation, we define $\boldsymbol{c} = (c_i)_{i \in \mathcal{I}}$ and $\boldsymbol{r} = (\boldsymbol{r}_k)_{k \in \mathcal{K}}$ as the system's caching policy and recommendation strategy, respectively. In addition, let $\boldsymbol{y} = (y_{l,k})_{k \in \overline{\mathcal{S}}, l \in \mathcal{L}}$ be the NOMA user pairing scheme. Moreover, we denote the transmit power for the multicasting users and the NOMA users as $\boldsymbol{P}' = (P'_i)_{i \in \mathcal{I}'}$ and $\boldsymbol{P}'' = (p_{l_s}, p_{l_w})_{l \in \mathcal{L}}$, respectively. Furthermore, let $\boldsymbol{P} = (\boldsymbol{P}', \boldsymbol{P}'')$ be the system's power control strategy. In reality, the caching and recommendation policy do not change with the system state, i.e., $(\mathbf{X}, \mathbf{G})$. They will be updated once within a time period of several hours or a half day (Zhang et al., 2015). Thus, they are long-timescale variables. In contrast, the power control and user grouping schemes are crucially affected by the system state. On these grounds, we rewrite $\boldsymbol{P}$ and $\boldsymbol{y}$ as $\boldsymbol{P}(\mathbf{X}, \mathbf{G})$ and $\boldsymbol{y}(\mathbf{X}, \mathbf{G})$, respectively. With the aforementioned definitions, we can express the average system latency as follows:

$$\overline{T}(\mathbf{c}, \boldsymbol{r}, \boldsymbol{P}, \boldsymbol{y}) \triangleq \mathbb{E}[T(\mathbf{c}, \boldsymbol{r}, \boldsymbol{P}(\mathbf{X}, \mathbf{G}), \boldsymbol{y}(\mathbf{X}, \mathbf{G}))], \tag{7.16}$$

in which $\mathbb{E}$ depicts the expectation operation, which is taken over all the system states, i.e., $(\mathbf{X}, \mathbf{G}) \in \mathcal{I}^K \times \mathcal{G}^K$. In addition, $T(\boldsymbol{c}, \boldsymbol{r}, \boldsymbol{P}(\mathbf{X}, \mathbf{G}), \boldsymbol{y}(\mathbf{X}, \mathbf{G}))$ is defined in (7.15), from which we can observe that the recommendation decision, cache placement as well as transmission-oriented strategies (i.e., user pairing and power allocation) critically affect the system's average latency.

Our objective is to minimize the average latency of the cache-aware NOMA system, denoted as $\overline{T}(\boldsymbol{c}, \boldsymbol{r}, \boldsymbol{P}, \boldsymbol{y})$, while accounting for various practical constraints. These include the cache capacity budget at the BS, the total number of the recommendations per user, the recommendation quality requirement per user, the total transmit power constraint at the BS, as well as the user pairing-oriented constraint, i.e., each NOMA user can only be assigned in one cluster. It is important to note that the cache placement, recommendation decision, and user pairing variables are binary.

7.3 Algorithm Design

Upon analysis, it is apparent that the considered optimization problem is a nonlinear, mixed-integer programming problem with two timescales. Additionally, T and $\boldsymbol{a}_{k,i}^{\text{req}}$ are nonconvex functions with respect to $\boldsymbol{P}$ and $\boldsymbol{r}$, respectively, making the problem nonconvex and challenging to solve optimally. Given that the system's state significantly affects the strategies for user pairing and power allocation, it is crucial to perform these tasks for each $(\mathbf{X}, \mathbf{G})$. On the other hand, the cache placement and recommendation decision-making problem can be considered a long-timescale problem. To efficiently tackle the challenging optimization problem, we propose a divide-and-conquer approach. The details of this approach are expanded based on two aspects. First, we determine the optimal power allocation and user pairing scheme for each system state, given the content caching and user recommendation decisions (i.e., $\boldsymbol{c}$ and $\boldsymbol{r}$). This represents a short-term optimization problem. We then address the long-term optimization problem of cache placement and recommendation policy by utilizing the obtained user pairing and power allocation strategy for each system state. Afterwards, we propose an iterative optimization algorithm that alternately optimizes the two types of subproblems until convergence. By using this twofold analysis, we can effectively address the challenging optimization problem.

7.3.1 User Pairing and Power Control Optimization

Initially, we focus on optimizing user pairing for NOMA users. For a given power allocation of cluster l, denoted by P_l'', the optimal power allocation for the users in this cluster can be determined through a one-dimensional search. This ensures that the total latency of both users is minimized. The optimal transmit power of users l_s and l_w are denoted by $p_{l_s}^*$ and $p_{l_w}^*$, respectively. Besides, we define T_{l_s,l_w} as the corresponding latency of the two users in the lth cluster. T_{l_s,l_w} is calculated as $T_{l_s} + T_{l_w}$, where T_{l_s} and T_{l_w} are defined in (7.13). It is noteworthy that in this section, we assume that the recommendation and caching decisions, i.e., $(\boldsymbol{r}, \boldsymbol{c})$, are fixed. Hence, different resource management policies are required for varying $(\boldsymbol{r}, \boldsymbol{c})$.

Using the definitions mentioned earlier, we can determine the minimum transmit latency, denoted by $T_{k',k''}$, associated with any two users: a strong user denoted by k' and a weak user denoted by k''. To transform problem $\mathbb{P}'(1)$ into a minimum weight perfect matching problem regarding an undirected graph, we consider an undirected graph with $|\overline{S}|$ vertices, where each vertex corresponds to a different NOMA user. We calculate the minimum latency for every possible pair of users, given that they are paired in the same cluster. The latency for a pair of users can be considered as the weight of the edge that is determined by them. We construct an edge with non-negative weight of $T_{k',k''}$ between vertices k' and k'', for each pair of users. Since each cluster contains two users, there are $\binom{|\overline{S}|}{2}$ edges in total, and we denote the set of all edges as $\mathcal{U}$. A subset $U \subseteq \mathcal{U}$ is called a matching if each vertex $k \in \overline{S}$ has at most one incident edge in U. A matching U is called perfect if each node $k \in \overline{S}$ has exactly one incident edge in U. By the definition of perfect matching, we can transform the objective of latency minimization into finding a perfect matching whose sum of weights is the minimum.

Define U^* as the perfect matching with the minimum sum weight. For notation simplicity, we invoke the incident vector $h \in \{0,1\}^{\mathcal{U}}$ to represent matching $\mathcal{U}^\dagger \subseteq \mathcal{U}$. In addition, let $S^\dagger \subseteq \overline{S}$ be a subset of the vertices, and its boundary edges set is denoted by $f(S^\dagger) = \{(k', k'') \in \mathcal{U} | k' \in S^\dagger, k'' \in \mathcal{U} \backslash S^\dagger\}$. Moreover, it is stipulated that $f(k) = f(\{k\})$ for a single node, in which $k \in \overline{S}$. Furthermore, denote T_u as the total latency of edge u, where $u \in \mathcal{U}$. With above definitions and analysis, we mathematically formulate the minimum weight perfect matching problem as a 0–1 integer linear programming problem, referred to as $\mathbb{P}''(1)$, which is given below:

$$\underset{(h_u)_{u\in\mathcal{U}}}{\text{minimize}} \sum_{u\in\mathcal{U}} T_u h_u \qquad \mathbb{P}''(1)$$

$$\text{s.t. } C1^{\dagger} : \ h(f(k)) = 1, \quad \forall\, k \in \overline{\mathcal{S}},$$

$$C2^{\dagger} : \ h_u \in \{0,1\}, \quad \forall\, u \in \mathcal{U},$$

in which $C1^{\dagger}$ is used to restrict that each user is connected with one and only one incident edge. h_u in $C2^{\dagger}$ is the binary indicator with regard to edge u. More precisely, $h_u = 1$ iff edge u is accounted in the perfect matching. It is worth mentioning that the optimal solution of problem $\mathbb{P}''(1)$ can be obtained by extensive integer linear programming solvers at the cost of an exponentially increased time complexity (Berkelaar et al., 2011). Alternatively, the enhanced Edmond's blossom algorithm can also solve this problem, with a cubic computational complexity of $\mathcal{O}(|\overline{\mathcal{S}}|^3)$ (Thulasiraman and Swamy, 1991).

Up to this point, we have solved the problem of optimal user pairing and intra-cluster power control with fixed inter-group power allocation. Afterwards, we will address the issue of inter-group power control. To accomplish this, we employ a simple and intuitive method known as proportional power control (PPC), which allocates power to each multicasting and NOMA group based on the total number of users in the group. Namely,

$$P_i' = \frac{|\mathcal{S}_i|\overline{P}}{K} \quad \text{and} \quad P_l'' = \frac{2\overline{P}}{K}, \quad i \in \mathcal{I}', \quad l \in \mathcal{L}. \tag{7.17}$$

Before closing this section, we would like to propose a simple step to further improve the transmission latency of NOMA users. Note that with $\boldsymbol{P}$, the optimal pairing strategy for NOMA users is achieved. We use $\overline{g}_l$ as the average channel gain of the users in cluster l, i.e., $\overline{g}_l = \frac{g_{l_s}+g_{l_w}}{2}$. Then, we re-calculate the transmit power of cluster l as

$$\overline{P}_l'' = \frac{\overline{g}_l}{\sum_{l'\in\mathcal{L}}\overline{g}_{l'}} \frac{2L\overline{P}}{K}. \tag{7.18}$$

Using the adjusted value of $\overline{P}_l''$, we perform the optimal inter-cluster power control once again. It's worth noting that we'll only proceed with the aforementioned procedures if the latency performance can be further enhanced. If not, we'll maintain the original power control strategy, which is P_l''.

7.3.2 Cache Placement

This section focuses on optimizing cache placement with the information of personalized recommendation decision as well as the corresponding users pairing and power control policies. By leveraging this information, the cache decision-making problem at the BS can be simplified to minimizing $\overline{T}(\boldsymbol{c}, \boldsymbol{r}, \boldsymbol{P}, \boldsymbol{y})$, while taking into account the cache capacity budget requirement, i.e., $\sum_{i \in \mathcal{I}} c_i B_i \leq \overline{C}$. Thereof, only $\boldsymbol{c}$ is the binary variable to be determined. It is not difficult to check that the cache placement problem is a 0–1 Knapsack problem. Thereby, it can be optimally solved by the dynamic programming approach (DPA) in polynomial computational complexity (Willamson and Shmoys, 2011).

7.3.3 Recommendation Algorithm

We will now discuss the recommendation decision-making method. Given the caching policy as well as the user pairing and power allocation strategies, the objective function of the recommendation decision-making problem is to minimize $\overline{T}(\boldsymbol{c}, \boldsymbol{r}, \boldsymbol{P}, \boldsymbol{y})$ while satisfying constraints on the quantity and quality of recommendations, namely $\sum_{i \in \mathcal{I}} r_{k,i} = \overline{N}_k,\ k \in \mathcal{K}$ and $r_{k,i}(1 - a_{k,i}^{\text{pref}}) \leq (1 - \overline{\theta}_k),\ k \in \mathcal{K},\ i \in \mathcal{I}$, respectively. It is worth noting that $\boldsymbol{r}$ is the only variable in the recommendation decision-making optimization problem.

The recommendation problem remains a nonconvex integer programming problem because $a_{k,i}^{\text{req}}$ in $\overline{T}$ is nonconvex with respect to $\boldsymbol{r}$. To solve this problem efficiently, we have developed a swap-then-compare method (SCM). Specifically, given an initial recommendation state $\boldsymbol{r}$, the recommended items for user k are determined. For $k \in \mathcal{K}$, denote by $\mathcal{R}_k$ the index set of the recommended items to user k. Based on the recommendation quality threshold, the content that can be recommended to user k is obtained, whose index set is stipulated as $\mathcal{M}_k$. Moreover, we define $\overline{\mathcal{R}}_k = \mathcal{M}_k \backslash \mathcal{R}_k$ and let $\mathcal{Q}_k = \{(j_k, j_k') | j_k \in \mathcal{R}_k, j_k' \in \overline{\mathcal{R}}_k\}$, where $k \in \mathcal{K}$. Furthermore, denote by $\mathcal{Q} = \{\mathcal{Q}_k | k \in \mathcal{K}\}$. As the recommendation states of all users jointly determine the content request state of the system, the recommendation decisions of different users are no longer independent of each other, which is different from that in Fu et al. (2021a, 2021b, 2021c). The designed SCM works in an iterative manner. To elaborate, in each iteration, a recommended item j_i is randomly chosen from $\mathcal{R}_i$ and exchanged with another eligible content j_i' in $\overline{\mathcal{R}}_i$. If this results in a new recommendation policy that reduces the system latency, the exchange is approved. Otherwise, it is declined. These steps are repeated until further reductions in system latency are no longer possible.

7.3.4 Joint Optimization Algorithm and Property Analysis

Based on the analysis presented above, the solution to each subproblem is explicit. In this section, we introduce an iterative algorithm for jointly optimizing the different types of variables in the original minimization problem. We then analyze the convergence performance of the proposed methodology. To facilitate our analysis, we define some notations. Let $\boldsymbol{c}(t)$, $\boldsymbol{r}(t)$, $\boldsymbol{y}(t)$, and $\boldsymbol{P}(t)$ denote the cache placement decision, recommendation decision, NOMA user pairing strategy, and power control policy in the tth iteration, respectively. Our joint optimization scheme is derived from the alternating optimization approach, and proceeds as follows: given the long-term variables $\mathbf{c}(t)$ and $\boldsymbol{r}(t)$, we optimize $\boldsymbol{y}(t)$ and $\boldsymbol{P}(t)$ based on the algorithm designed in Section 7.3.1 under each system state. Upon obtaining the updated $\boldsymbol{y}(t)$ and $\boldsymbol{P}(t)$, we subsequently renew $\mathbf{c}(t)$ and $\boldsymbol{r}(t)$. These steps are repeated until the system performance cannot be further enhanced or the maximum iteration number N is reached.

Theorem 7.1 The convergence of the developed joint optimization method is ensured.

Proof: We can ensure convergence of our algorithm since each iteration results in a non-increased system latency and $\overline{T}$ is lower bounded by 0. □

The computational complexity as regards our developed approach is summarized in the following theorem.

Theorem 7.2 The worst-case computational complexity per iteration of the developed joint optimization method is $\max\{\mathcal{O}(KI^2), \mathcal{O}(K^3)\}$.

Proof: The time complexity per iteration of the joint optimization algorithm is dependent on the highest computational complexity of the three subproblems. The cache decision-making problem is solved by DPA with a time complexity of $\mathcal{O}(I^2)$, as discussed in Section 7.3.2. The computational complexity of the SCM-oriented recommendation optimization method is determined by the size of Q, with a highest complexity of $\mathcal{O}(KI^2)$. Additionally, as mentioned in Section 7.3.1, the transmission-related NOMA user pairing and power control scheme has a worst-case time complexity of K^3. These three aspects complete the proof. □

7.4 Numerical Simulation

In this section, we use extensive numerical results to validate the effectiveness of our developed joint content caching, recommendation, and transmission optimization algorithm for cache-enabled NOMA networks. The system parameters are set as follows: we consider a cache-enabled NOMA system with one BS serving four users. In total, there are 10 items in the content catalogue. The data size of each content follows a uniform distribution between 1 and 20. Each user's inherent preference distribution is modeled by a Zipf-like distribution with exponent

$\eta = 1$. For simplicity, we assume that each user is recommended by the same number of items. In addition, the recommendation acceptance ratio as well as the quality threshold of each user are identical. In other words, $\overline{N}_k = \overline{N}$, $b_k = b$, and $\overline{\theta}_k = \overline{\theta}$, respectively, where $k \in \mathcal{K}$. It is worth noting that the designed scheme is applicable to the scenarios where $\overline{N}_k$, b_k, and $\overline{\theta}_k$ are distinct among different users. Moreover, the system bandwidth W is set to be 1 MHz. The total transmit power of the BS is stipulated as $\overline{P} = 10\,\text{mW}$ and the noise power at each user is assumed to be $N_k = 1\,\text{mW}$ for $k \in \mathcal{K}$. Furthermore, the finite channel state for user k is assumed as follows: $\mathcal{G} = \{0.5, 1.5, 0.05, 0.15\}$, where $k \in \mathcal{K}$. Besides, for $k \in \{1,2\}, p_{G_k}(0.5) = 0.7015, p_{G_k}(1.5) = 0.2581, p_{G_k}(0.05) = 0, p_{G_k}(0.15) = 0$. While for $k \in \{3,4\}, p_{G_k}(0.5) = 0, p_{G_k}(1.5) = 0, p_{G_k}(0.05) = 0.7015, p_{G_k}(0.15) = 0.2581$. Here the setting of channel distribution follows the one in Cui et al. (2017). It should be noted that $\sum_{g_k \in \mathcal{G}} p_{G_k}(g_k) < 1$ in this setting as we only consider the dominated channel states for reducing computational complexity. For the cases with different system parameters, the developed joint optimization algorithm is still applicable with the information of channel statistics.

To provide a comprehensive performance comparison, the following benchmark schemes are taken into consideration:

- *TopHete*: In this scheme, the transmission-oriented resource allocation strategy is identical to the devised solution. However, top cache (Yang et al., 2020a) and heterogeneous top recommendation policies are leveraged. More specifically, the BS caches the most popular contents to fully use its cache capacity. In the meantime, each user is recommended by its top preferred content items.
- *TopHomo*: In this strategy, the cache decision and resource allocation are the same as that in TopHete. The difference lies in the recommendation mechanism. In contrast to the recommendation of the top-preferred content for each user, BS recommends the same top-ranked items to each user based on all users aggregated preference distribution (Fu et al., 2021a).
- *RandomP*: This scheme uses the same recommendation and caching strategies as that of the proposed algorithm. While the intrinsic user pairing is done in a random walk manner (Fu et al., 2019) and besides equal power control is performed for NOMA users.
- *PureNOMA*: This policy is identical to the proposed manner except that pure NOMA is used to do users' content delivery.
- *PureMC*: In this scheme, a pure multicasting transmission strategy is used. In addition, all the other aspects apply the same optimization solvers to our devised solutions.

Three different metrics, including the convergence performance, system's average latency and cache hit ratio, are used in this section.

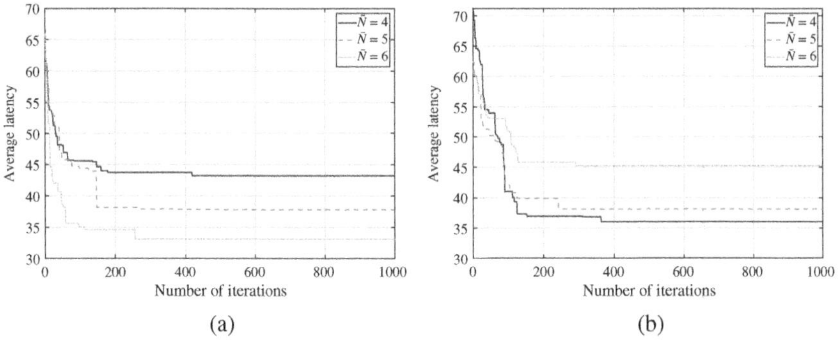

Figure 7.1 Convergence performance of the designed joint optimization algorithm. (a) $b = 0.1$ and (b) $b = 0.9$.

7.4.1 Convergence Performance

In this section, we depict the convergence performance of the developed joint optimization algorithm by considering different recommendation sizes, as shown in Fig. 7.1. For illustration, the backhaul link rate and the cache size of the BS are assigned with $R_b = 1$ bps/Hz and $\overline{C} = 50$, respectively. In addition, the recommendation acceptance ratios in Fig. 7.1(a) and (b) are set as $b = 0.1$ and $b = 0.9$, respectively. In the meantime, we use the system latency per iteration to indicate the speed of the convergence. In particular, it is seen from Fig. 7.1 that the average latency monotonically decreases with the number of iterations. The observations are consistent with the theoretical analysis in Section 7.3.4. Moreover, by comparing Fig. 7.1(a) and (b), it is found that higher recommendation acceptance ratio b would result in lower average latency under a small recommendation size. This justifies the efficiency of recommendation with regard to latency saving. Furthermore, for a large recommendation acceptance ratio b, the system latency is an increasing function of the recommendation size. This is due to the fact that plenty of recommendations will bring about the heterogeneity among users' behavior, such as, request probability. However, under low b, the system latency declines with the recommendation size. The reason is because lower recommendation acceptance ratio leads to heterogenous preference distribution. Therefore, the increase of the number of recommendations is beneficial to the improvement of edge caching in some sense.

7.4.2 System's Average Latency

In Fig. 7.2(a) and (b), we set the recommendation acceptance ratios as 0.1 and 0.9, respectively. From Fig. 7.2, it can be observed that the average latency of all the methods decreases with the Backhaul link rate R_b. In the meantime, the latency saving improvement becomes ignorable under high R_b. This is because the system

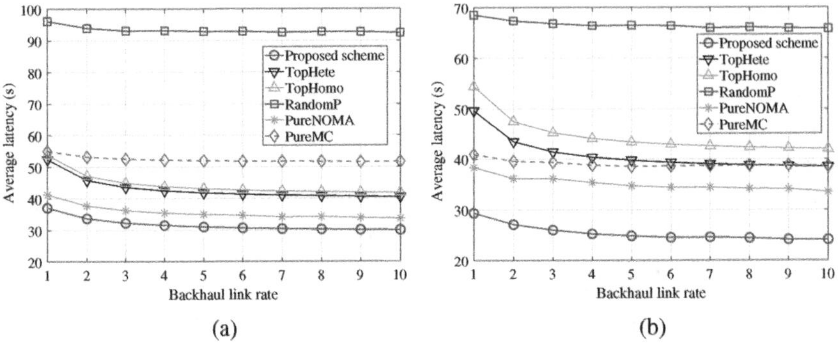

Figure 7.2 System's average latency versus Backhaul link rate, wherein $\overline{N} = 4$. (a) $b = 0.1$ and (b) $b = 0.9$.

average latency consists of the backhaul link delay and content transmission latency. Under a small value of R_b, the backhaul latency can be reduced by increasing R_b. This in turn significantly shortens the total latency. However, if R_b is high, constantly increasing the rate R_b cannot drastically reduces the latency introduced by backhaul transmission, which precludes further reduction of the total latency. It is not beyond our expectation that the proposed scheme exhibits the superior performance to the other baseline schemes. For example, if $R_b = 1$ and $b = 0.1$, the proposed scheme can conserve the average latency by 29.27%, 31.09%, 61.42%, 10.09%, and 32.46% by comparing to TopHete, TopHome, RandomP, PureNOMA, and PureMC, respectively. Besides, it is found that PureNOMA achieves the second-best performance and NOMA-oriented transmission is superior to OMA-aware systems.

In Fig. 7.3, we further examined the average latency of the six schemes versus the cache size, in which $b = 0.1$ and $b = 0.9$ are assumed in Fig. 7.3(a) and (b), respectively. It can be seen from Fig. 7.3 that the system latency of all the schemes is decreasing functions with respect to the cache size. Besides, our developed algorithm achieves the best performance, while the RandomP is unexceptionally the worst. It is worth noting that all the schemes except RandomP have identical performance. This reveals that the power control and user pairing are of pivotal importance for the reduction of the average system latency. Moreover, it is observed that the performance gap between TopHete and TopHome decreases with the cache size. This is due to the fact that the latency performance mainly depends on the resource management strategy in the case of sufficient caching resources. Furthermore, for a flat preference distribution, i.e., small b, the performance of TopHete has a comparable performance as TopHome. If b is high, the superiority of TopHete over TopHomo becomes remarkable. At last, it is also seen that our devised hybrid NOMA and multicasting transmission scheme performs better than the pure NOMA and pure multicasting strategies.

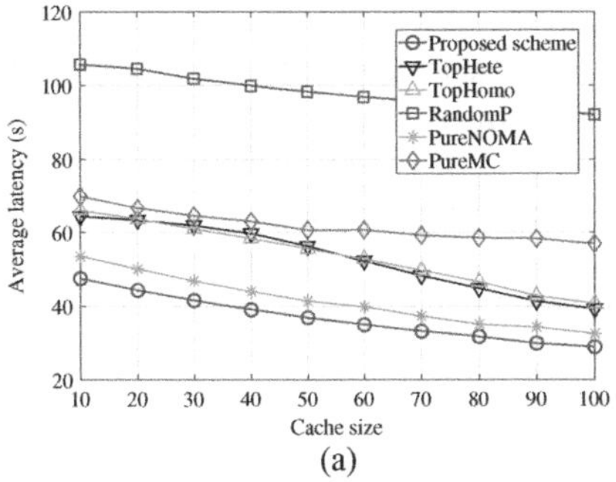

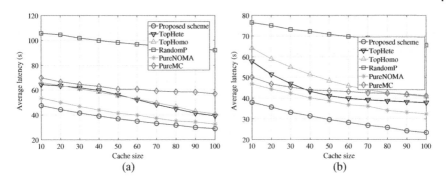

Figure 7.3 System's average latency versus Cache size, wherein $\overline{N} = 4$. (a) $b = 0.1$ and (b) $b = 0.9$.

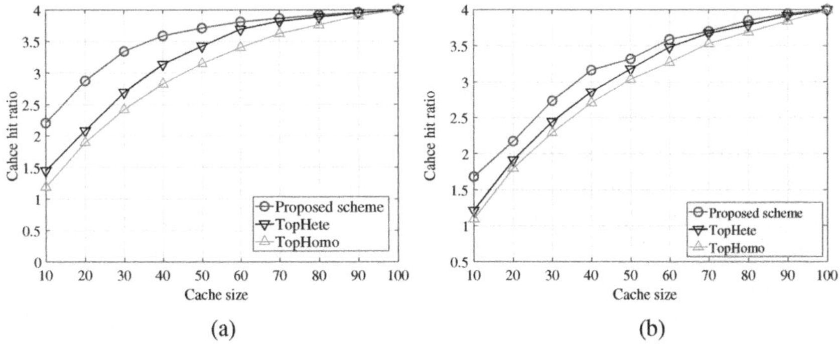

Figure 7.4 Cache hit ratio versus Cache size, wherein $b = 0.9$. (a) $\overline{N} = 4$ and (b) $\overline{N} = 6$.

7.4.3 Cache Hit Ratio

In the end, we investigate the cache hit ratio of our proposed joint optimization algorithm through comparison with some benchmarking schemes. A system cache hit ratio can be expressed as

$$\text{Cache hit ratio} = \sum_k \sum_i a_{k,i}^{\text{req}} c_i, \tag{7.19}$$

where the cache hit ratio of all users only relies on the recommendation decision and the cache placement strategy. Accordingly, we merely use two baseline schemes for comparisons, i.e., TopHete and TopHome, by considering different caching and recommendation policies.

Figure 7.4 compares the achievable cache hit ratio of the proposed scheme with the other two baseline schemes versus the cache size, where the recommendation acceptance ratio is assumed to be 0.9, the backhual transmission rate is set as $R_b = 1$, the recommendation sizes in Fig. 7.4(a) and (b) are set as $\overline{N} = 4$ and $\overline{N} = 6$, respectively. From the two figures, it can be seen that the cache hit ratio of all the schemes is an increasing function of the BS's cache capacity size, which justifies the significance of edge caching. Moreover, it is proved that our developed scheme attains the higher cache high ratio than the two baseline schemes, which reveals the effectiveness of our proposed scheme. It is also worth noticing that the proposed scheme achieves a more prominent improvement of the cache hit ratio for a low range cache size. This implies that the proposed scheme could be a promising solution to strike a balanced trade-off between the increased network data volume and the limited edge caching resources. Furthermore, by employing heterogeneous top recommendation, TopHete achieves superior performance than TopHomo. Lastly, by comparing Fig. 7.4(a) and (b), a large number of recommendations would impair the performance of edge caching, which has been explained in our foregoing analysis.

7.5 Conclusion

In this chapter, we studied the latency minimization problem for cache-aware NOMA systems with the consideration of recommendation. To achieve this goal, we first explicitly derived the expression of system's average latency with respect to content caching, recommendation, and transmission-oriented resource management strategies. Thereafter, we formulated the minimization problem mathematically, taking into account various practical constraints. To solve the formulated non-convex, mixed-integer, and multi-timescale optimization problem, a divided-and-rule manner was developed. Extensive simulation results validated that the designed joint optimization algorithm outperformed varied baseline schemes significantly in terms of system's latency and cache hit ratio. In future, we may consider the convergence among edge caching, computing, and advanced transmission technologies.

References

M Berkelaar, K Eikland, and P Notebaert. LP_solver-a mixed integer linear programming (MILP) solver. *version 5.5.2.0*, 2011.

L E Chatzieleftheriou, M Karaliopoulos, and I Koutsopoulos. Jointly optimizing content caching and recommendations in small cell networks. *IEEE Transactions on Mobile Computing*, 18(1):125–138, Jan. 2019.

Z Chen, Z Ding, P Xu, and X Dai. Optimal precoding for a QoS optimization problem in two-user MISO-NOMA downlink. *IEEE Communications Letters*, 20(6):1263–1266, Jun. 2016.

M Chen, M Mozaffari, W Saad, C Yin, M Debbah, and C S Hong. Caching in the sky: Proactive deployment of cache-enabled unmanned aerial vehicles for optimized quality-of-experience. *IEEE Journal on Selected Areas in Communications*, 35(5):1046–1061, May 2017.

Y Cui, W He, C Ni, C Guo, and Z Liu. Energy-efficient resource allocation for cache-assited mobile edge computing. In *IEEE Conference on Local Computer Networks (LCN)*, pages 1–9, Oct. 2017.

Z Ding, P Fan, George K Karagiannidis, R Schober, and H Vincent Poor. NOMA assisted wireless caching: Strategies and performance analysis. *IEEE Transactions on Communications*, 66(10):4854–4876, Oct. 2018.

K Doan, W Shin, M Vaezi, H Vincent Poor, and Tony Q S Quek. Optimal power allocation in cache-aided non-orthogonal multiple access systems. *IEEE International Conference on Communications Workshops (ICC Workshops)*, pages 1–6, May 2018.

Ericsson. The Ericsson mobility report. Jun. 2022. URL https://www.ericsson.com/49d3a0/assets/local/reports-papers/mobility-report/documents/2022/ericsson-mobility-report-june-2022.pdf.

Y Fu, Y Chen, and C Sung. Distributed power control for the downlink of multi-cell NOMA systems. *IEEE Transactions on Wireless Communications*, 16(9):6207–6220, Sept. 2017.

Y Fu, K Shum, C W Sung, and Y Liu. Optimal user pairing in cache-based NOMA systems with index coding. In *Proceedings of the IEEE International Conference on Communication (ICC)*, May 2019.

Y Fu, L Salaun, X Yang, W Wen, and Tony Q S Quek. Caching efficiency maximization for device-to-device communication networks: A recommend to cache approach. *IEEE Transactions on Wireless Communications*, 20(10):6580–6594, Oct. 2021a.

Y Fu, Q Yu, Tony Q S Quek, and W Wen. Revenue maximization for content-oriented wireless caching networks (CWCNs) with repair and recommendation considerations. *IEEE Transactions on Wireless Communications*, 20(1):284–298, Jan. 2021b.

Y Fu, Q Yu, Angus K Y Wong, Z Shi, H Wang, and Tony Q S Quek. Exploiting coding and recommendation to improve cache efficiency of reliability-aware wireless edge caching networks. *IEEE Transactions on Wireless Communications*, 20(11):7243–7256, Nov. 2021c.

S Kastanakis, P Sermpezis, V Kotronis, D Menasche, and T Spyropoulos. Network-aware recommendations in the wild: Methodology, realistic evaluations, experiments. *IEEE Transactions on Mobile Computing*, 21(7):2466–2479, Jul. 2022.

D Liu, B Chen, C Yang, and A F Molisch. Caching at the wireless edge: Design aspects, challenges, and future directions. *IEEE Wireless Communications*, 54(9):22–28, Sept. 2016.

Y Liu, Z Qin, M Elkashlan, Z Ding, A Nallanathan, and L Hanzo. Non-orthogonal multiple access for 5G and beyond. *Proceedings of the IEEE*, 105(12):2347–2381, Dec. 2017.

Z Shi, G Yang, Y Fu, H Wang, and S Ma. Achievable diversity order of downlink HARQ-aided NOMA systems. *IEEE Transactions on Vehicular Technology*, 69(1):471–487, Jan. 2020.

Z Sun, Y Jing, and X Yu. NOMA design with power-outage tradeoff for two-user systems. *IEEE Wireless Communications Letters*, 9(8):1278–1282, Aug. 2020.

K Thulasiraman and M N S Swamy. *Graphs: Theory and Algorithms*. A Wiley-Interscience Publication, 1991.

David P Willamson and David B Shmoys. *The Design of Approximation Algorithms*. Cambridge University Press, Electronic web edition, 2011.

L Xiang, Derrick Ng, X Ge, Z Ding, Vincent W S Wong, and R Schober. Cache-aided non-orthogonal multiple access: The two-user case. *IEEE Journal of Selected Topics in Signal Processing*, 13(3):436–451, Jun. 2019.

Z Yang, Y Liu, Y Chen, and L Jiao. Learning automata based Q-learning for content placement in cooperative caching. *IEEE Transactions on Communications*, 68(6):3667–3680, Jun. 2020a.

Z Yang, Y Liu, Y Chen, and Naofal A Dhahir. Cache-aided NOMA mobile edge computing: A reinforcement learning approach. *IEEE Transactions on Wireless Communications*, 19(10):6899–6915, Oct. 2020b.

X Yang, Y Fu, W Wen, Tony Q S Quek, and Z Fei. Mixed-timescale caching and beamforming in content recommendation aware Fog-RAN: A latency perspective. *IEEE Transactions on Wireless Communications*, 69(4):2427–2440, Apr. 2021.

Z Yang, Y Liu, Y Chen, and Joey T Zhou. Deep learning for latent events forecasting in content caching networks. *IEEE Transactions on Wireless Communications*, 21(1):413–428, Jan. 2022.

Y Zhang, M Zhang, Y Liu, T Chua, Y Zhang, and S Ma. Task-based recommendation on a web-scale. In *Proceedings of the IEEE International Conference on Big Data (Big Data)*, pages 827–836, Oct. 2015.

T Zhang, X Fang, Y Liu, G Y Li, and W Xu. D2D-enabled mobile user edge caching: A multi-winner auction approach. *IEEE Transactions on Vehicular Technology*, 68(12):12314–12328, Dec. 2019.

Z Zhao, M Xu, W Xie, Z Ding, and George K Karagiannidis. Coverage performance of NOMA in wireless caching networks. *IEEE Communications Letters*, 27(7):1458–1461, Jul. 2018.

N Zhao, Y Li, S Zhang, Y Chen, W Lu, J Wang, and X Wang. Security enhancement for NOMA-UAV networks. *IEEE Transactions on Vehicular Technology*, 69(4):3994–4005, Apr. 2020.

8

NOMA Empowered Multi-Access Edge Computing and Edge Intelligence

Yuan Wu[1,2], Yang Li[1], Liping Qian[3], and Xuemin Shen[4]

[1]*State Key Laboratory of Internet of Things for Smart City, University of Macau, Macao, China*
[2]*Zhuhai UM Science and Technology Research Institute, Zhuhai, China*
[3]*College of Information Engineering, Zhejiang University of Technology, Hangzhou, China*
[4]*Department of Electrical and Computer Engineering, University of Waterloo, Waterloo, Ontario, Canada*

8.1 Introduction

Mobile edge computing (MEC), a paradigm that deploys computation resources at the edge of radio access networks, has provided a promising approach for enabling various computation-intensive yet latency-sensitive applications in future wireless networks (Mao et al., 2017, Duan et al., 2023). Due to MEC, resource-constrained wireless devices can offload their tasks to nearby edge servers (ES) via short-distance radio transmissions, which thus can not only reduce the transmission latency but also save energy consumption compared to conventional cloud computing. The advantages of computation offloading in MEC have motivated lots of research efforts. For instance, in Chen et al. (2016) and Zheng et al. (2019), the authors studied the multi-user competitive computation offloading toward a single base station equipped with edge server. In Zhang et al. (2022a) and Liang et al. (2021), different multicell computation offloading schemes have been proposed, in which the edge-computing users can select one of the cells to offload their tasks. Moreover, the paradigm of cloud-edge collaborative computation offloading has raised much interest in recent years (Dai et al., 2018, Ren et al., 2019, Dou et al., 2023), in which edge server and cloud server can cooperatively provide the offloading services to the users. However, the explosive growth in wireless edge services and the growing number of mobile devices still lead to heavy pressure on providing timely and cost-efficient task offloading services.

Next Generation Multiple Access, First Edition.
Edited by Yuanwei Liu, Liang Liu, Zhiguo Ding, and Xuemin Shen.
© 2024 The Institute of Electrical and Electronics Engineers, Inc. Published 2024 by John Wiley & Sons, Inc.

The emerging multi-access MEC provides a promising solution for addressing the above issue (Taleb et al., 2017). Multi-access MEC enables the users to offload their tasks to different edge servers simultaneously by leveraging the feature of multi-homing in heterogeneous radio access networks. Many research efforts have been devoted to exploring the benefits of multi-access MEC and addressing the associated technical challenges, e.g., the energy consumption and the overall latency due to multi-access task offloading (Sheng et al., 2020, Wang et al., 2020a). Non-orthogonal multiple access (NOMA) has been envisioned as a spectrum-efficient multiple access scheme for enabling massive connectivity in future wireless networks (Liu et al., 2022, Wei et al., 2020, Liu et al., 2021, Wu et al., 2018b, Liu et al., 2019, Qian et al., 2017). In particular, the feature of multiple parallel transmissions with successive interference cancelation (SIC) for mitigating the co-channel interference fits the multi-access MEC very well. Exploiting NOMA, a group of users can simultaneously offload their workloads to an edge server over the same frequency channel, which thus improves the efficiency of task offloading transmission. There have been many recent studies focusing on the NOMA-aided multi-access MEC with its applications (Ding et al., 2019, Pham et al., 2020, Zhu et al., 2022a, Qian et al., 2021, Wu et al., 2018a).

In this chapter, we firstly review the recent advances in NOMA-empowered edge computing as well as edge intelligence. As a further step, we present a concrete design of latency-oriented NOMA-aided multi-access computation offloading. Specifically, we consider a hybrid NOMA and frequency division multiple access (FDMA) assisted dual computation offloading scheme with the objective of minimizing the overall latency for completing the users' tasks. Dual connectivity has been advocated as a promising approach for leveraging the multi-homing resources in heterogeneous small-cell networks (Rosa et al., 2016, Yilmaz et al., 2019). Due to dual connectivity, each user can execute a two-sided dual computation offloading, i.e., offload its workloads to an edge server connected to a small-cell base station (SBS) and the cloudlet server (CS) connected to the macro base station (MBS) simultaneously. To enable this dual computation offloading, we propose a hybrid NOMA-FDMA transmission scheme, in which the MBS utilizes FDMA to serve all users' offloading transmissions, and the SBS utilizes NOMA to serve a small number of users' offloading transmissions. Despite the flexible two-sided task offloading toward the edge server and cloudlet server, the dual computation offloading via hybrid NOMA–FDMA transmission is technically challenging. In particular, the users' co-channel interference in NOMA may degrade the performance of offloading transmission, which results in an important issue regarding how to properly group users into different NOMA groups for offloading their workloads. This issue becomes more complicated in dual computation offloading, since all users share the cloudlet server's computation resources, and thus, their dual offloading decisions are coupled. These challenges motivate our study in this chapter.

The remainder of this chapter is organized as follows. We review the related studies in Section 8.2. We present the system model and problem formulation in Section 8.3. We propose the algorithm for obtaining the optimal offloading solution in Section 8.4. We evaluate the algorithm performance in Section 8.5. We conclude this chapter and discuss the future directions in Section 8.6.

8.2 Literature Review

Achieving the benefits of MEC necessitates joint management of the task offloading scheduling and communication/computing resource allocation, which has attracted lots of research efforts (Chen et al., 2021). In Wang et al. (2020b), Wang et al. studied an intelligent dynamic computation offloading for smart Internet of Things systems. In Zhang et al. (2022b), Zhang et al. proposed a vehicular task offloading scheme that can distributively schedule the task offloading and allocate the edge resources. In Lim et al. (2021), Lim et al. investigated a joint resource allocation and incentive mechanism design for edge intelligence. Different performance matrices have been utilized to evaluate the joint computation offloading and resource allocation schemes. In Dai et al. (2018), Zhang et al. (2022a), and Huang et al. (2021), different approaches for energy minimization have been studied by jointly optimizing the multiuser task offloading and user association. In Chen et al. (2018) and Kuang et al. (2019), different approaches for optimizing the weighted sum of the energy consumption and task offloading latency have been proposed by jointly optimizing the offloading decision and the associated computation/communication resource allocation. Several works have considered the time-varying environments (e.g., the channel condition, the number of the arriving tasks, and the resource availability) and designed the corresponding dynamic edge resource allocation schemes (Qian et al., 2021, Wang et al., 2020b). To further improve the flexibility of computation offloading, there exist several studies investigating how to jointly exploit the resources at the edge servers and the cloudservers (Wang et al., 2019a). In particular, to facilitate a flexible usage of the computation resources at the edge servers and cloud (Zhang et al., 2020, El Haber et al., 2019), investigated a two-tier edge-cloud framework with the hierarchical offloading scheme, i.e., a mobile user can offload its computation workloads to an edge server that can further offload part of its received workloads to the cloud or other edge servers. Different from the hierarchical offloading scheme, the binary task offloading scheme has been investigated in Wang et al. (2019b), Du et al. (2017), and Zhao et al. (2019), where each task can be either processed locally or offloaded to the edge server.

Due to its advantage in enabling massive connectivity with ultra-high spectrum efficiency, NOMA has been regarded as one of the key technologies for the next-generation multiple access in wireless access networks (Liu et al., 2021, Wei et al., 2020, Liu et al., 2019, Xu et al., 2023, Wang et al., 2022). A comprehensive review of leveraging NOMA for future cellular networks has been presented in Liu et al. (2021). In Wei et al. (2020), Wei et al. investigated the ergodic sum-rate gain of NOMA over the uplink cellular networks and demonstrated the potential gain compared to orthogonal multiple access (OMA). In Liu et al. (2019), Liu et al. proposed a quality-of-service (QoS)-guaranteed resource allocation scheme for multi-beam satellite industrial Internet of Things by using NOMA. In Xu et al. (2023), Xu et al. proposed a downlink multi-antenna NOMA transmission framework with the concept of cluster-free SIC. In Wang et al. (2022), Wang et al. proposed a NOMA empowered integrated sensing and communication framework, where the superimposed NOMA communication signal is simultaneously exploited for target sensing. The benefit of NOMA has attracted many studies exploiting NOMA for task offloading (Yang et al., 2019, Ding et al., 2019, Wu et al., 2018a, 2019). In Yang et al. (2019), the authors exploited uplink NOMA to enable the multiuser' task offloading, with the objective of minimizing a total cost covering the completion time of the tasks and all users' total energy consumption. In Ding et al. (2019), a joint power and time allocation scheme has been proposed for users' computation offloading via NOMA. In Wu et al. (2018a), the authors studied a NOMA assisted multi-access computation offloading to minimize all users' task-completion latency. In Wu et al. (2019), the authors investigated a NOMA-assisted MEC for multiuser secure computation offloading in the presence of an eavesdropper. The small-cell dual connectivity provides a promising approach for realizing the multi-access edge computing. The dual connectivity allows each edge-computing user to be simultaneously associated with a MBS and a SBS. Thus, the user can flexibly offload its computational tasks toward the MBS and SBS (Centenaro et al., 2020). Such dual connectivity assisted MEC provides a two-side dual offloading to leverage the resources at the MBS and SBS and improves both the offloading efficiency as well as the resource utilization efficiency. In Cui and You (2021), the authors studied a dual connectivity assisted offloading scheme in which a wireless device connects with a SBS and a MBS simultaneously via OMA. Li et al. Li et al. (2021) proposed a dual connectivity assisted computation offloading via NOMA. In addition to solely rely on NOMA, there exist growing interest in exploring the hybrid NOMA and OMA for multiuser transmission. In Ding et al. (2018), a hybrid NOMA-OMA transmission scheme has been proposed for computation offloading. Due to the wide deployment and utilization of MEC, there has been a growing trend of providing machine learning and deep learning aided services in radio access networks, yielding the paradigm of so-called edge intelligence. In Dong et al.

(2020), a deep reinforcement learning approach based on self-imitative learning has been proposed to solve the NOMA-based energy-efficient task scheduling in vehicular edge computing networks. In Zhu et al. (2022b), a decentralized deep reinforcement learning framework has been proposed for multiple input multiple output NOMA vehicular edge computing networks. As an important paradigm of edge intelligence, federated learning (FL) provides a privacy-preserving scheme for realizing distributed machine learning and model training. The advantages of NOMA raise much attention in leveraging NOMA for edge intelligence and FL. In Wu et al. (2022), a NOMA assisted FL via wireless power transfer has been proposed. In Ni et al. (2022), a reconfigurable intelligent surface aided hybrid network that can integrate over-the-air FL and NOMA has been proposed. In Sun et al. (2020), NOMA has been exploited to facilitate efficient uplink FL updates. In Mo and Xu (2021), the authors proposed a NOMA-assisted federated edge learning system.

In this chapter, we demonstrate a design of dual connectivity enabled multiuser computation offloading (i.e., dual computation offloading), in which each edge-computing user is simultaneously associated with a SBS and the MBS, and thus can perform a two-sided task offloading to jointly utilize the computation resources at the SBS and those at the MBS. Based on this dual offloading model, we investigate the joint optimization of the edge-computing users' two-sided partial offloading, hybrid NOMA and FDMA transmission, and processing rate allocation to minimize the overall task-completion latency for all users. Part of the content of this chapter has been published in Li et al. (2022).

8.3 System Model and Formulation

8.3.1 Modeling of Two-Sided Dual Offloading

As shown in Fig. 8.1, we consider a two-tier radio access network with one MBS and a group of SBSs denoted by $\mathcal{K} = \{1, 2, \dots, K\}$. The MBS is equipped with a cloudlet server (CS) whose maximum processing rate is denoted by U_0 (in the following of this work, we use subscript "0" to denote the MBS). Each SBS k is equipped with an edge server (ES) whose maximum processing rate is denoted by U_k. Meanwhile, there exist a group of edge-computing users (EUs) denoted by $\mathcal{I} = \{1, 2, \dots, I\}$, with each EU i having a total computation workload S_i^{tot} to be completed. We use u_i^{loc} to denote EU i's local processing rate. With dual connectivity, each EU can execute the dual computation offloading to the MBS and one of the selected SBSs. Meanwhile, each SBS uses NOMA to accommodate the associated two EUs' task offloading.

We use $\Omega_k = \{i, j\}$ to denote the set of two selected EUs, which are associated with SBS k, namely, EU i and EU j form a NOMA group to offload their respective

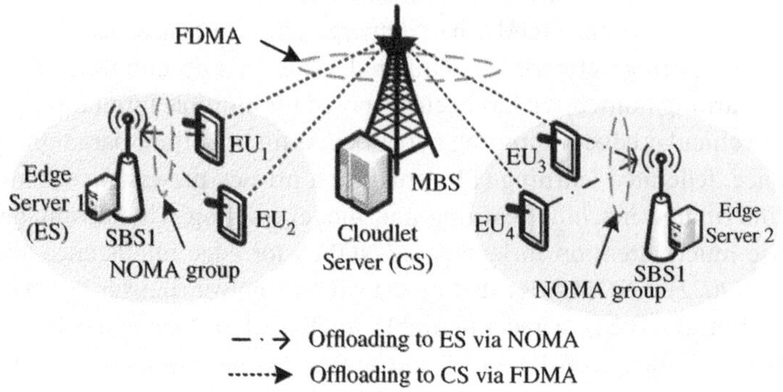

Figure 8.1 A two-tier radio access network with one MBS and two of SBSs. Each user offloads its computation workloads to the MBS and one of the SBSs.

workloads $s_{i,k} \in [0, S_i^{\text{tot}}]$ and $s_{j,k} \in [0, S_j^{\text{tot}}]$ to SBS k. Assuming that SBS k decodes EU j's offloaded data prior to EU i's offloaded data, we can express the offloading throughput from EU j to SBS k as

$$r_{j,k} = W_k \log_2 \left(1 + \frac{p_{j,k}G_{j,k}}{p_{i,k}G_{i,k} + N_k}\right), j \in \Omega_k, \tag{8.1}$$

and the offloading throughput from EU i to SBS k as

$$r_{i,k} = W_k \log_2 \left(1 + \frac{p_{i,k}G_{i,k}}{N_k}\right), i \in \Omega_k, \tag{8.2}$$

where parameter W_k denotes the channel bandwidth of SBS k. We use $p_{i,k}$ and $p_{j,k}$ to denote EU i's and EU j's transmission powers. Parameter $G_{j,k}$ denotes the channel power gain from EU j to SBS k, and $G_{i,k}$ denotes the channel power gain from EU i to SBS k. Parameter N_k denotes the power of the background noise. It is noticed that the uplink NOMA can provide an arbitrary decoding order. Thus, our modelings and proposed algorithms are also applicable to the other decoding order by changing the EUs' indices.

For each NOMA group Ω_k, we use t_k^{up} to denote the transmission duration, which thus yields the following constraints

$$r_{i,k}t_k^{\text{up}} \geq s_{i,k}, i \in \Omega_k, \tag{8.3}$$

$$r_{j,k}t_k^{\text{up}} \geq s_{j,k}, j \in \Omega_k. \tag{8.4}$$

With Eqs. (8.3) and (8.4), we can obtain the required minimum NOMA transmission powers of EU i and EU j as

$$p_{i,k}^{\text{req}}(s_{i,k}, t_k^{\text{up}}) = \frac{N_k}{G_{i,k}} \left(2^{\frac{s_{i,k}}{W_k t_k^{\text{up}}}} - 1 \right), i \in \Omega_k, \tag{8.5}$$

$$p_{j,k}^{\text{req}}(s_{i,k}, s_{j,k}, t_k^{\text{up}}) = \frac{N_k}{G_{j,k}} 2^{\frac{s_{i,k}}{W_k t_k^{\text{up}}}} \left(2^{\frac{s_{j,k}}{W_k t_k^{\text{up}}}} - 1 \right), j \in \Omega_k. \tag{8.6}$$

Each EU i can also offload its workloads denoted by $s_{i,0} \in [0, S_i^{\text{tot}}]$ to the MBS. The MBS uses FDMA to receive the EUs' offloaded data. Thus, we can express the offloading throughput from EU i to the MBS as

$$r_{i,0} = W_0 \log_2 \left(1 + \frac{P_{i,0} G_{i,0}}{N_0} \right), \forall i \in \mathcal{I}, \tag{8.7}$$

where parameter W_0 denotes the orthogonal channel bandwidth for each EU. Parameter $P_{i,0}$ denotes EU i's transmission power to the MBS, and $G_{i,0}$ denotes the channel gain from EU i to the MBS. We use t_0^{up} to denote the FDMA-transmission duration to the MBS. Thus, we have the following constraint

$$t_0^{\text{up}} r_{i,0} \geq s_{i,0}, \forall i \in \mathcal{I}. \tag{8.8}$$

The latency for completing EU i's workloads can be expressed as

$$t_i^{\text{ove}} = \max \left\{ \frac{S_i^{\text{tot}} - s_{i,k} - s_{i,0}}{u_i^{\text{loc}}}, t_k^{\text{up}} + \frac{s_{i,k}}{u_{i,k}}, t_0^{\text{up}} + \frac{s_{i,0}}{u_{i,0}} \right\}, \forall i \in \mathcal{I}, \tag{8.9}$$

where $u_{i,k}$ is the processing rate allocated by ES k for EU i, and $u_{i,0}$ is the processing rate allocated by the CS for EU i.

8.3.2 Overall Latency Minimization

We first consider that all EUs' NOMA groups $\Omega = \{\Omega_k\}_{k \in \mathcal{K}}$ are given and formulate an optimization problem to minimize all EUs' task-completion latency. To achieve this, we jointly optimize the offloading decisions $\mathbf{s} = \{\{s_{i,k}\}_{\forall i \in \mathcal{I}, k \in \mathcal{K}}, \{s_{i,0}\}_{\forall i \in \mathcal{I}}\}$, the transmission duration $\mathbf{t}^{\text{up}} = \{t_0^{\text{up}}, \{t_k^{\text{up}}\}_{\forall k \in \mathcal{K}}\}$, and the processing rate allocation $\mathbf{u} = \{\{u_{i,k}\}_{\forall i \in \mathcal{I}, k \in \mathcal{K}}, \{u_{i,0}\}_{\forall i \in \mathcal{I}}\}$. The details are as follows.

$$\text{(OLM):} \quad d_\Omega = \min_{\mathbf{s}, \mathbf{t}^{\text{up}}, \mathbf{u}} \max_{\forall i \in \mathcal{I}} \{t_i^{\text{ove}}\}$$

$$\text{Subject to:} \quad p_{i,k}^{\text{req}}(s_{i,k}, t_k^{\text{up}}) \leq P_i^{\max}, i \in \Omega_k, \tag{8.10}$$

$$p_{j,k}^{\text{req}}(s_{i,k}, s_{j,k}, t_k^{\text{up}}) \leq P_j^{\max}, j \in \Omega_k, \tag{8.11}$$

$$W_0 \log_2 \left(1 + \frac{P_{i,0} G_{i,0}}{N_0} \right) t_0^{\text{up}} \geq s_{i,0}, \forall i \in \mathcal{I}, \tag{8.12}$$

$$0 \leq s_{i,k} + s_{i,0} \leq S_i^{\text{tot}}, \forall i \in \Omega_k, \tag{8.13}$$

$$\sum_{i \in \Omega_k} u_{i,k} \leq U_k, \forall k \in \mathcal{K}, \tag{8.14}$$

$$\sum_{i \in \mathcal{I}} u_{i,0} \leq U_0, \tag{8.15}$$

Variables: $\mathbf{s}$, $\mathbf{t}^{\text{up}}$, and $\mathbf{u}$.

Constraints (8.10) and (8.11) ensure that in a NOMA group Ω_k (associated with SBS k), both EU i's and EU j's transmission powers cannot exceed their respective capacities denoted by $P_i^{\max}$ and $P_j^{\max}$. To solve Problem (OLM), which is a non-convex optimization problem, we adopt an approach of decomposition, i.e., separating Problem (OLM) into a top-problem for optimizing the overall latency, and the consequent subproblem for optimizing the EUs' two-sided offloading decisions, the computing resource and communication resource allocations.

Before decomposing Problem (OLM), we introduce a variable $d_\Omega^{\text{sub}} = \max_{\forall i \in \mathcal{I}}\{t_i^{\text{ove}}\}$ to denote the overall latency and rewrite the objective function of Problem (OLM) as minimizing d_Ω^{sub}, i.e., $d_\Omega = \min_{\mathbf{s},\mathbf{t}^{\text{up}},\mathbf{u}}\{d_\Omega^{\text{sub}}\}$. Correspondingly, each item in (8.9) should be no greater than d_Ω^{sub}, which leads to

$$S_i^{\text{tot}} - s_{i,k} - s_{i,0} \leq u_i^{\text{loc}} d_\Omega^{\text{sub}}, \forall i \in \Omega_k, k \in \mathcal{K}, \tag{8.16}$$

$$t_k^{\text{up}} u_{i,k} + s_{i,k} \leq u_{i,k} d_\Omega^{\text{sub}}, \forall i \in \Omega_k, k \in \mathcal{K}, \tag{8.17}$$

$$t_0^{\text{up}} u_{i,0} + s_{i,0} \leq u_{i,0} d_\Omega^{\text{sub}}, \forall i \in \mathcal{I}. \tag{8.18}$$

With (8.16)–(8.18), we can equivalently transform Problem (OLM) into

(OLM-E): $d_\Omega = \min_{\mathbf{s},\mathbf{t}^{\text{up}},\mathbf{u}} d_\Omega^{\text{sub}}$,

Subject to: constraints: (8.10)–(8.15), (8.16)–(8.18),

Variables: d_Ω^{sub}, $\mathbf{s}$, $\mathbf{t}^{\text{up}}$, and $\mathbf{u}$.

Problem (OLM) aims at finding the minimum feasible d_Ω^{sub}. Therefore, we decompose Problem (OLM-E) into two subproblems.

We firstly present the subproblem under a given d_Ω^{sub}. Assuming d_Ω^{sub} is given, we aim at finding the minimum total computation resources required by the CS, i.e.,

(OLM-E-Sub): $U_\Omega^{\text{req}} = \min_{\mathbf{s},\mathbf{t}_k^{\text{up}},\mathbf{u}} \sum_{i \in \mathcal{I}} u_{i,0}$

Subject to: constraints: (8.10)–(8.14), (8.16)–(8.18),

Variables: $\mathbf{s}$, $\mathbf{t}_k^{\text{up}}$, and $\mathbf{u}$,

where vector $\mathbf{t}_k^{\text{up}} = \{t_k^{\text{up}}\}_{\forall k \in \mathcal{K}}$.

We then present the top-problem that aims at finding the minimum d_Ω^{sub} and the corresponding t_0^{up} such that the U_Ω^{req} achieved from Problem (OLM-E-Sub) is no greater than U_0, i.e.,

$$\begin{aligned}
\text{(OLM-E-Top):} \quad & d_\Omega = \min\, d_\Omega^{\text{sub}} \\
\text{Subject to:} \quad & U_\Omega^{\text{req}} \leq U_0, \\
\text{Variables:} \quad & d_\Omega^{\text{sub}} \text{ and } t_0^{\text{up}}.
\end{aligned} \tag{8.19}$$

If U_Ω^{req} satisfies constraint (8.15), i.e., $U_\Omega^{\text{req}} \leq U_0$, the currently given d_Ω^{sub} is feasible for Problem (OLM-E). Otherwise, the currently given d_Ω^{sub} is infeasible.

The above decomposition approach can provide the following feature.

Proposition 8.1 The result of Problem (OLM-E-Sub) (i.e., U_Ω^{req}) is non-increasing with respect to d_Ω^{sub}.

Proof: Problem (OLM-E-Sub) aims at minimizing the required total computation resources under a given d_Ω^{sub}. Note that we define $d_\Omega^{\text{sub}} = \max_{i\in\mathcal{I}}\{t_i^{\text{ove}}\}$. When the currently given d_Ω^{sub} is less than the value of $\max_{i\in\mathcal{I}}\{t_i^{\text{ove}}\}$, there exist some EUs, which cannot complete their workloads within the given d_Ω^{sub}. In other words, if these EUs' workloads have to be completed within the given d_Ω^{sub}, more computation resources are required. Since the total allocated computation resources from the CS is unconstrained in Problem (OLM-E-Sub), U_Ω^{req} will increase. On the other hand, if d_Ω^{sub} is greater than $\max_{i\in\mathcal{I}}\{t_i^{\text{ove}}\}$, it means that the latency limit is still loose and the computation resources at the CS can be reduced. Since Problem (OLM-E-Sub) is to minimize the total computation resources required at the CS, U_Ω^{req} thus can be decreased. Therefore, U_Ω^{req} is non-increasing with respect to d_Ω^{sub}. This completes the proof. □

Proposition 8.1 indicates that U_Ω^{req} approaches the value of U_0 under the minimum d_Ω^{sub}. This feature enables us to design algorithms for solving Problem (OLM) in Section 8.4.

8.4 Algorithms for Optimal Offloading

We firstly propose a cell-based distributed decision (CDD) algorithm for Problem (OLM-E-Sub). With constraint (8.18), we can obtain the lower bound of $u_{i,0}$ as $u_{i,0} \geq \frac{s_{i,0}}{d_\Omega^{\text{sub}} - t_0^{\text{up}}}$. Since Problem (OLM-E-Sub) is to minimize the sum of $\{u_{i,0}\}_{\forall i\in\mathcal{I}}$, replacing $u_{i,0}$ with its lower bound will not affect the result. Thus, the equivalent form of the objective function of Problem (OLM-E-Sub) is $\sum_{i\in\mathcal{I}} \frac{s_{i,0}}{d_\Omega^{\text{sub}} - t_0^{\text{up}}}$. Similarly, with constraint (8.17), the lower bound of $u_{i,k}$ is $u_{i,k} \geq \frac{s_{i,k}}{d_\Omega^{\text{sub}} - t_k^{\text{up}}}$. Thus, constraint (8.14) can be transformed into

$$s_{i,k} + s_{j,k} \leq U_k(d_\Omega^{\text{sub}} - t_k^{\text{up}}),\ i,j \in \Omega_k. \tag{8.20}$$

Furthermore, to address constraint (8.11), we introduce a group of auxiliary variables $\{\beta_k\}_{k\in\mathcal{K}}$, where $\beta_k = \frac{P_j^{\max} G_{j,k}}{N_k} + 2^{\frac{s_{i,k}}{t_k^{\text{up}} W_k}}$. Then, Eq. (8.11) can be transformed into

$$2^{\frac{s_{i,k}+s_{j,k}}{t_k^{\text{up}} W_k}} \leq \beta_k, i, j \in \Omega_k, k \in \mathcal{K}. \tag{8.21}$$

Meanwhile, with Eq. (8.10), we can obtain

$$2^{\frac{s_{i,k}}{t_k^{\text{up}} W_k}} \leq 1 + \frac{P_i^{\max} G_{i,k}}{N_k}. \tag{8.22}$$

Moreover, we can obtain the viable interval of β_k as follows

$$\beta_k \in \left[1 + \frac{P_j^{\max} G_{j,k}}{N_k}, \frac{P_j^{\max} G_{j,k} + P_i^{\max} G_{i,k}}{N_k} + 1\right], \forall k \in \mathcal{K}. \tag{8.23}$$

After the above transformations, Problem (OLM-E-Sub) can be divided into K optimization problems, with each one for an individual cell to minimize the corresponding variable as

$$\begin{aligned}
\text{(Sub-}\beta\text{-cell):} \quad & U_{\Omega_k}^{\text{req}} = \min \frac{s_{i,0} + s_{j,0}}{d_{\Omega}^{\text{sub}} - t_0^{\text{up}}} \\
\text{Subject to:} \quad & \text{constraints: (8.12), (8.13), (8.16), (8.20), and (8.21)–(8.23),} \\
\text{Variables:} \quad & s_{i,0}, s_{i,k}, s_{j,0}, s_{j,k}, t_k^{\text{up}}, \text{and } \beta_k.
\end{aligned}$$

By solving Problem (Sub-β-cell) for each SBS, we can obtain the result of Problem (OLM-E-Sub). Since Problem (Sub-β-cell) is independent from each other, Problem (Sub-β-cell) for each SBS k can be solved in a parallel manner.

We can further provide the following feature to solve Problem (Sub-β-cell).

Proposition 8.2 Under the given $\{\beta_k, d_{\Omega}^{\text{sub}}, t_0^{\text{up}}\}$, Problem (Sub-$\beta$-cell) is a strictly convex problem with respect to variables $\{t_k^{\text{up}}, s_{i,k}, s_{i,0}, s_{j,k}, s_{j,0}\}$.

Proof: Except constraint (8.21), the objective function and the other constraints of Problem (Sub-β-cell) are convex when d_{Ω}^{sub} and t_0^{up} are given in advance. To address the coupling between β_k and t_k^{up}, we can enumerate β_k according to (8.23) and then constraint (8.21) becomes affine. Thus, Problem (Sub-β-cell) is a convex optimization problem under given β_k, d_{Ω}^{sub}, and t_0^{up}. This completes the proof. □

Exploiting Proposition 8.2, we propose the CDD algorithm to obtain the optimal solution of Problem (OLM-E-Sub). The pseudocode of our CDD algorithm is presented in Algorithm 8.1.

Algorithm 8.1 Cell-based Distributed Decision Algorithm (i.e., CDD algorithm) for Solving Problem (OLM-E-Sub)

Input: $d_{\Omega}^{\text{sub}}, t_0^{\text{up}}$.
Initialize: β_k's upper bound $\beta_k^{\text{ub}} = \frac{P_i^{\max} G_{i,k} + P_j^{\max} G_{j,k}}{N_0} + 1$, β_k's lower bound $\beta_k^{\text{lb}} = \frac{P_j^{\max} G_{j,k}}{N_0} + 1$.
Generate Problem (Sub-β-cell) for each cell and solve Problem (Sub-β-cell) by executing the following steps.
Update $\beta_k = \beta_k^{\text{ub}}$.
while $\beta_k \geq \beta_k^{\text{lb}}$ **do**
    Use the interior point method to solve Problem (Sub-β-cell) under β_k and obtain the optimal solutions $\{s_{i,k}^*, s_{j,k}^*, s_{i,0}^*, s_{j,0}^*, t_k^{\text{up}*}\} = \arg\min \frac{s_{i,0}+s_{j,0}}{d_{\Omega}^{\text{sub}} - t_0^{\text{up}}}$.
    if $N_k 2^{\frac{s_{i,k}^*}{W_k t_k^{\text{up}*}}} \left(2^{\frac{s_{j,k}^*}{W_k t_k^{\text{up}*}}} - 1 \right) \leq P_j^{\max} G_{j,k}$ **then**
        $\{s_{i,k}^*, s_{j,k}^*, s_{i,0}^*, s_{j,0}^*, t_k^{\text{up}*}\}$ is also the optimal solution for Problem (OLM-E-Sub), and compute $U_{\Omega_k}^{\text{req}} = \frac{s_{i,0}^* + s_{j,0}^*}{d_{\Omega}^{\text{sub}} - t_0^{\text{up}}}$.
        Break the WHILE-LOOP.
    else
        Update $\beta_k \leftarrow \beta_k - \Delta\beta_k$, where $\Delta\beta_k$ is the step size.
    end if
end while
Compute the result of Problem (OLM-E-Sub): $U_{\Omega}^{\text{req}} = \sum_{k \in \mathcal{K}} U_{\Omega_k}^{\text{req}}$.
Output: U_{Ω}^{req}.

With our CDD algorithm as a subroutine, we next propose a two-layer hybrid search (TLHS) algorithm for solving Problem (OLM-E-Top). Problem (OLM-E-Top) aims to find the minimum d_{Ω}^{sub} and the corresponding t_0^{up}. To this end, we propose a TLHS algorithm consisting of a bisection search on d_{Ω}^{sub} in the top layer and a linear search on t_0^{up} at the bottom layer. The pseudocode of our TLHS algorithm is shown in Algorithm 8.2.

Our CDD and TLHS algorithms solve Problem (OLM) under the given EUs' NOMA groups. By using CDD and TLHS algorithms as the subroutines, we can further optimize the EUs' NOMA groups to minimize overall latency. Specifically, we use a binary variable $a_{i,k}$ to denote the association between the EU i and SBS k, i.e., $a_{i,k} = 1$ meaning that EU i is associated with SBS k, and $a_{i,k} = 0$ meaning the opposite case. Since each EU is associated with one SBS, and two EUs form a

Algorithm 8.2 Two-layer Hybrid Search Algorithm (i.e., TLHS algorithm) for Solving Problem (OLM-E-Top)

Initialize: Initialize d_Ω^{sub}'s upper bound $d_\Omega^{\text{ub}} = \max\{S_i^{\text{tot}}/u_i^{\text{loc}}\}_{\forall i \in \mathcal{I}}$, d_Ω^{sub}'s lower bound $d_\Omega^{\text{lb}} = 0$, t_0^{up}'s lower bound $t_0^{\text{lb}} = 0$.
while $d_\Omega^{\text{lb}} < d_\Omega^{\text{ub}}$ **do**
  Update $d_\Omega^{\text{sub}} \leftarrow \frac{1}{2}(d_\Omega^{\text{lb}} + d_\Omega^{\text{ub}})$.
  Update t_0^{up}'s upper bound: $t_0^{\text{ub}} = d_\Omega^{\text{sub}}$.
  Update $t_0^{\text{up}} = t_0^{\text{lb}}$.
  while $t_0^{\text{up}} \leq t_0^{\text{ub}}$ **do**
    Solve Problem (OLM-E-Sub) by Algorithm 8.1 to obtain U_Ω^{req}.
    if $U_\Omega^{\text{req}} \leq U_0$ **then**
      Current d_Ω^{sub} and t_0^{up} are feasible for Problem (OLM).
      Update the lower bound of t_0^{up}, i.e., $t_0^{\text{lb}} = t_0^{\text{up}}$.
      Break Linear Search Loop.
    else
      Update $t_0^{\text{up}} \leftarrow t_0^{\text{up}} + \Delta t_0^{\text{up}}$, where Δt_0^{up} is the step size.
    end if
  end while
  if Finding the feasible t_0^{up} (i.e., $U_\Omega^{\text{req}} \leq U_0$) **then**
    Update d_Ω^{sub}'s upper bound, i.e., $d_\Omega^{\text{ub}} = d_\Omega^{\text{sub}}$.
  else
    Update d_Ω^{sub}'s lower bound, i.e., $d_\Omega^{\text{lb}} = d_\Omega^{\text{sub}}$.
  end if
end while
Find the minimum d_Ω^{sub}, then $d_\Omega = d_\Omega^{\text{sub}}$.
Output: d_Ω, t_0^{up}.

NOMA group, variable $a_{i,k}$ should satisfy

$$\sum_{k \in \mathcal{K}} a_{i,k} = 1, \forall i \in \mathcal{I}, \tag{8.24}$$

$$\sum_{i \in \mathcal{I}} a_{i,k} = 2, \forall k \in \mathcal{K}. \tag{8.25}$$

Thus, for SBS k, there exists $\Omega_k = \{i | a_{i,k} = 1, \forall i \in \mathcal{I}\}, \forall k \in \mathcal{K}$. We consider that the number of the EUs is twice than the number of the SBSs, i.e., $I = 2K$. Accordingly, the optimal grouping problem can be expressed as

$$\begin{aligned} \text{(Group):} \quad & \min_{\Omega} d_\Omega \\ \text{Subject to:} \quad & \text{constraints: (8.24), (8.25),} \\ \text{Variables:} \quad & \Omega = \{a_{i,k} \in \{0,1\}\}_{i \in \mathcal{I}, k \in \mathcal{K}}. \end{aligned}$$

Algorithm 8.3 CE-based Algorithm for Problem (Group)

1: **Initialize:** Initialize $\{\eta_{i,k}(0)\}_{\forall i \in \mathcal{I}, k \in \mathcal{K}}$, the learning rate $\alpha(0)$, and the stopping threshold ϵ_η.
2: Iteration $\tau = 1$.
3: **while** 1 **do**
4: According to the current $\{\eta_{i,k}(\tau)\}_{\forall i \in \mathcal{I}, k \in \mathcal{K}}$, randomly generate M feasible solutions $\{A_m(\tau)\}_{m=1,2,\ldots,M}$. Each $A_m(\tau)$ denotes a profile of $\{a^m_{i,k}(\tau)\}_{\forall i \in \mathcal{I}, k \in \mathcal{K}}$ which form a feasible solution for Problem (Group).
5: For each feasible solution $A_m(\tau)$, input the corresponding $\{a^m_{i,k}(\tau)\}_{\forall i \in \mathcal{I}, k \in \mathcal{K}}$ into Problem (OLM) and invoke TLHS algorithm to obtain the minimum overall latency $L(A_m(\tau)) = d_{A_m(\tau)}$.
6: Re-order $\{A_m(\tau)\}_{m=1,2,\ldots,M}$ according to the order from the smallest $L(A_m(\tau))$ to the biggest $L(A_m(\tau))$.
7: Let $\hat{L}(\tau) = L(A_{\lceil \rho M \rceil}(\tau))$ be the best ρ solutions' quantile, and calculate $\{C_{i,k}(\tau+1)\}_{\forall i \in \mathcal{I}, k \in \mathcal{K}}$ based on Eq. (8.26).
8: Update probability distributions $\eta_{i,k}(\tau+1) = (1-\alpha(\tau))\eta_{i,k}(\tau) + \alpha(\tau)C_{i,k}(\tau+1), \forall i \in \mathcal{I}, k \in \mathcal{K}$ for next round of iteration.
9: **if** $\sum_{i \in \mathcal{I}} \sum_{k \in \mathcal{K}} \eta_{i,k}(\tau+1) - \eta_{i,k}(\tau) \le \epsilon_\eta$ **then**
10: Determine the optimal Bernoulli-distribution of $a_{i,k}$, $\eta^*_{i,k} = \eta_{i,k}(\tau+1)$, $\forall i \in \mathcal{I}, k \in \mathcal{K}$.
11: Break the WHILE-LOOP.
12: **else**
13: Update $\tau = \tau + 1$.
14: **end if**
15: **end while**
16: Generate A^* according to the $\{\eta^*_{i,k}\}_{i \in \mathcal{I}, k \in \mathcal{K}}$.
17: **Output:** A^* is the optimal solution for Problem (Group).

We adopt a cross-entropy (CE)-based learning algorithm to solve the above binary programming problem (De Boer et al., 2005). In our CE-based learning algorithm, we treat the values of $\{a_{i,k}\}_{i \in \mathcal{I}, k \in \mathcal{K}}$ as random variables. Each $a_{i,k}$ follows a Bernoulli distribution of parameter $\eta_{i,k}$. To find the optimal solution of Problem (Group), we use the criterion of CE optimization to update $\{\eta_{i,k}\}_{i \in \mathcal{I}, k \in \mathcal{K}}$. The details are shown in Algorithm 8.3. In particular, in Step 7 of Algorithm 8.3, coefficient $C_{i,k}(\tau+1)$ is calculated based on the Kullback–Leibler cross-entropy approach (Abbas et al., 2017) as

$$C_{i,k}(\tau+1) = \frac{\sum_{m=1}^{M} \mathbb{I}(\{L(A_m(\tau)) \le \hat{L}(\tau)\})\mathbb{I}(\{a^m_{i,k}(\tau) = 1\})}{\sum_{m=1}^{M} \mathbb{I}(\{L(A_m(\tau)) \le \hat{L}(\tau)\})}, \tag{8.26}$$

where $\mathbb{I}(\cdot)$ is the indicator function. The iterative process in Algorithm 8.3 continues until the variation of each $\eta_{i,k}$ is sufficiently small. After its convergence, we can obtain the optimal solution A^* of Problem (Group) according to $\{\eta^*_{i,k}\}_{i\in\mathcal{I},k\in\mathcal{K}}$.

8.5 Numerical Results

This section presents the numerical results to validate our CDD algorithm and TLHS algorithm for solving Problem (OLM). Since our algorithms are applicable to arbitrary NOMA grouping, we use a fixed grouping scheme as $\Omega_k = \{2k-1, 2k\}, \forall k \in \mathcal{K}$ at first. We will further optimize the EUs' NOMA grouping later on. The MBS's channel bandwidth is set as $W_0 = 10$ Mbps, and each SBS's channel bandwidth is set as $W_k = 3$ Mbps, $\forall k \in \mathcal{K}$. The maximum transmission power of each EU is set as 0.5 W.

Figure 8.2 shows the results of our CDD algorithm for solving Problem (OLM-E-Sub) under different values of d_Ω^{sub}. In particular, we test our CDD algorithm under two scenarios, i.e., (i) $K = 3, I = 6$, and $U_0 = 10$ Gbps in Fig. 8.2(a), and (ii) $K = 4, I = 8$, and $U_0 = 40$ Gbps in Fig. 8.2(b). In each subplot, we vary d_Ω^{sub} and mark the value of U_0 (which is the horizontal line) and the corresponding intersection point where $U_\Omega^{\text{req}} = U_0$ holds. Both subplots in Fig. 8.2 show that U_Ω^{req} decreases as d_Ω^{sub} increases, which validates Proposition 8.1.

Figure 8.3 further illustrates the rationale behind our TLHS algorithm for solving Problem (OLM). In each subplot in Fig. 8.3, we demonstrate the variations in d_Ω^{sub} and U_Ω^{req} during the iterations of the bisection search in our TLHS algorithm. In particular, it can be observed that U_Ω^{req} gradually converges to U_0, at which d_Ω^{sub} converges to its correspondingly minimum value.

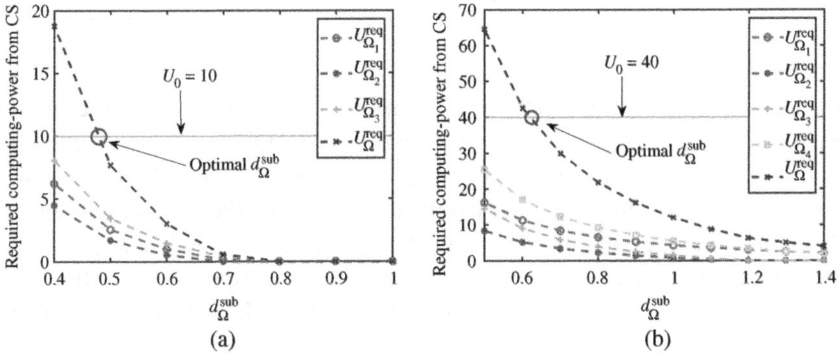

Figure 8.2 Illustration of our CDD algorithm: the optimal values of Problem (Sub-β-cell) and Problem (OLM-E-Sub) under different values of d_Ω^{sub}. (a) $K = 3, I = 6, U_0 = 10$ Gbps and (b) $K = 4, I = 8, U_0 = 40$ Gbps.

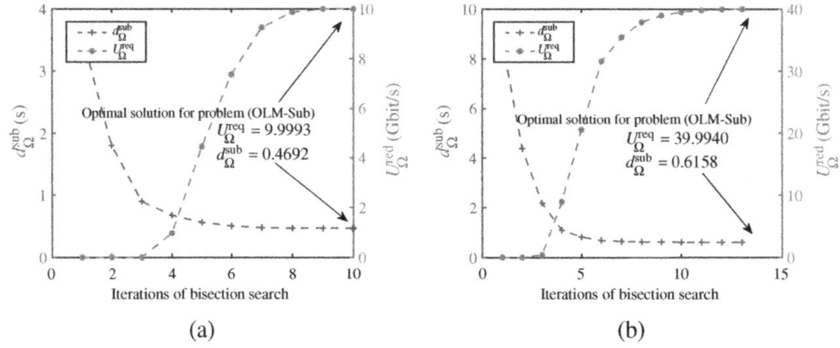

Figure 8.3 Illustration of our TLHS algorithm: variations of U_Ω^{req} and d_Ω^{sub} during the iterations of the bisection search. (a) $K = 3, I = 6, U_0 = 10$ Gbps and (b) $K = 4, I = 8, U_0 = 40$ Gbps.

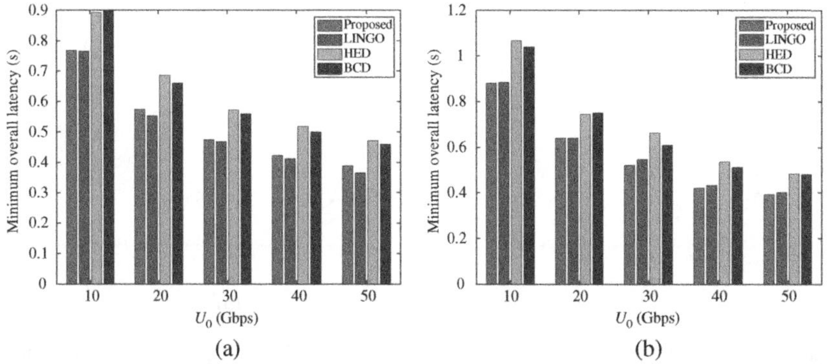

Figure 8.4 Performance comparisons among our proposed algorithm, LINGO, HED, and BCD. (a) $\{U_1, U_2, U_3\} = \{6, 4, 5\}$ Gbps and (b) $U_0 = 40$ Gbps.

For the purpose of comparison, we also use LINGO's global-solver (Schrage, 2006), block coordinate descent (BCD) (Yang et al., 2020), and a heuristic equal-division (HED) to solve our problem. Figure 8.4 shows the minimum overall latency achieved by these algorithms under different parameter-settings. The results in both subplots demonstrate that our proposed solution can achieve the almost same latency as the optimal solution provided by LINGO's global solver. Meanwhile, the latency achieved by our proposed solution is shorter than the other solutions provided by the BCD algorithm and HED algorithm. Figure 8.5 further illustrates the computation efficiency of TLHS algorithm (with CDD algorithm as its subroutine), in comparison with LINGO's global-solver, HED, and BCD schemes.

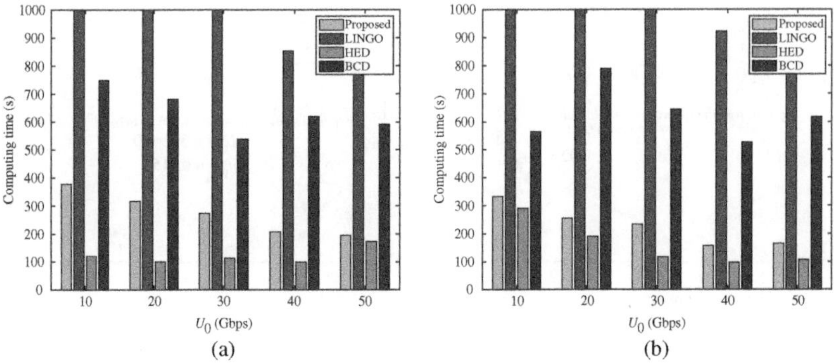

Figure 8.5 Computing time comparisons among our proposed algorithm, LINGO, HED, and BCD. (a) $\{U_1, U_2, U_3\} = \{6, 4, 5\}$ Gbps and (b) $U_0 = 40$ Gbps.

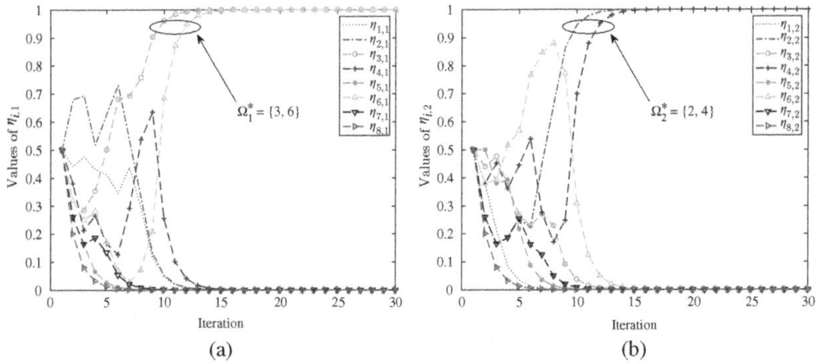

Figure 8.6 The convergence example of the EUs' probabilities (i.e., $\{\eta_{i,k}\}_{i\in\mathcal{I},k\in\mathcal{K}}$). (a) The values of $\{\eta_{i,1}\}_{i\in\mathcal{I}}$ and (b) the values of $\{\eta_{i,2}\}_{i\in\mathcal{I}}$.

To evaluate the performance of our proposed CE-based algorithm for optimizing the EUs' NOMA grouping, we set the parameters according to Hu et al. (2021), and test the CE-based learning algorithm on the scenario of $K = 4$ and $I = 8$. Figure 8.6 shows the convergence of all EUs' probabilities (i.e., $\{\eta_{i,k}\}_{i\in\mathcal{I},k\in\mathcal{K}}$) of choosing each individual SBS. We select the convergence of $\{\eta_{i,1}\}_{i\in\mathcal{I}}$ and $\{\eta_{i,2}\}_{i\in\mathcal{I}}$ for the sake of demonstration. Figure 8.6(a) shows the convergence of all EUs' Bernoulli-distribution parameters $\{\eta_{i,1}\}_{i\in\mathcal{I}}$ with respect to SBS 1. Specifically, the results show that both $\eta_{3,1}$ and $\eta_{6,1}$ converge to 1 while all the other $\{\eta_{i,1}\}_{i\neq 3,6}$ converge to zero. These results mean that EU 3 and EU 6 form a NOMA group to offload their tasks toward SBS 1.

Figure 8.7 finally shows the effectiveness of our CE-based algorithm in comparison with a randomized benchmark scheme. Figure 8.7(a) demonstrates the minimum latency under different values of U_0. Figure 8.7(b) further shows the

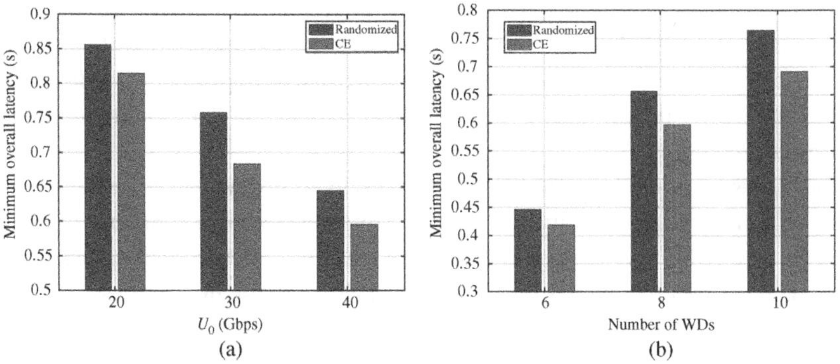

Figure 8.7 Illustration of the effectiveness of CE-based algorithm. (a) $K = 4, I = 8$ and (b) $U_0 = 40$ Gbps.

minimum latency under different numbers of the EUs. The results in both subplots in Figure 8.7 validate that our CE-based algorithm can effectively reduce the overall latency compared to the benchmark scheme.

8.6 Conclusion

In this chapter, we have studied the multiuser dual computation offloading via the hybrid NOMA–FDMA transmission, in which each EU can simultaneously offload its workloads to a cloudlet at the MBS and an ES at the SBS. We have formulated a joint optimization problem to minimize the overall latency for all EUs' to complete their tasks and proposed a layered and cell-based distributed algorithm for finding the optimal dual offloading solution. Based on this optimal dual offloading solution, we have further optimized the EUs' NOMA groups for their offloading to the SBSs by leveraging the CE aided learning algorithm. Numerical results have been provided to validate the effectiveness and efficiency of our proposed algorithms. Several future directions can be summarized as follows.

- With the growing interests in future hybrid terrestrial and non-terrestrial networks, e.g., integrated space-air-ground networks, we will further investigate the exploitation of NOMA empowered MEC for these advanced network architectures and study the corresponding joint resource allocation and task scheduling according to different service requirements, especially the significant latency due to the long-distance transmission as well as very limited energy capacity of the over-the-air devices.
- There are many emerging services such as digital twin (DT), Metaverse, and artificial intelligent generated content (AIGC) which are computation-intensive for

complicated model training and reasoning. Therefore, it is interesting to study how to leverage NOMA aided MEC for these services and address the associated technical challenges, e.g., design of cost-efficient approach for addressing the heavy computation-burdens due to deep learning model training and reasoning/prediction.

- The widespread deployment of edge servers in future networks raises growing concerns of data security and privacy. Motivated by these concerns, we will further study how to leverage the advanced security oriented techniques, e.g., Blockchain, for enhancing the security and privacy of the multi-access computation offloading services via distributed edge servers.

Acknowledgments

This work was supported in part by National Natural Science Foundation of China under Grants 62122069, 62072490, and 62071431, in part by FDCT-MOST Joint Project under Grant 0066/2019/AMJ, in part by Science and Technology Development Fund of Macau SAR under Grant 0162/2019/A3, in part by Research Grant of University of Macau under Grant MYRG2020-00107-IOTSC, in part by the Guangdong Basic and Applied Basic Research Foundation (2022A1515011287), in part by FDCT SKL-IOTSC(UM)-2021-2023, and in part by the Natural Sciences and Engineering Research Council of Canada.

References

Ali E Abbas, Andrea H Cadenbach, and Ehsan Salimi. A Kullback–Leibler view of maximum entropy and maximum log-probability methods. *Entropy*, 19(5):232, 2017.

Marco Centenaro, Daniela Laselva, Jens Steiner, Klaus Pedersen, and Preben Mogensen. Resource-efficient dual connectivity for ultra-reliable low-latency communication. In *2020 IEEE 91st Vehicular Technology Conference (VTC2020-Spring)*, pages 1–5, 2020.

Xu Chen, Lei Jiao, Wenzhong Li, and Xiaoming Fu. Efficient multi-user computation offloading for mobile-edge cloud computing. *IEEE/ACM Transactions on Networking*, 24(5):2795–2808, 2016.

Meng-Hsi Chen, Ben Liang, and Min Dong. Multi-user multi-task offloading and resource allocation in mobile cloud systems. *IEEE Transactions on Wireless Communications*, 17(10):6790–6805, 2018.

Xianfu Chen, Celimuge Wu, Zhi Liu, Ning Zhang, and Yusheng Ji. Computation offloading in beyond 5G networks: A distributed learning framework and applications. *IEEE Wireless Communications*, 28(2):56–62, 2021.

Haixia Cui and Fan You. User-centric resource scheduling for dual-connectivity communications. *IEEE Communications Letters*, 25(11):3659–3663, 2021.

Yueyue Dai, Du Xu, Sabita Maharjan, and Yan Zhang. Joint computation offloading and user association in multi-task mobile edge computing. *IEEE Transactions on Vehicular Technology*, 67(12):12313–12325, 2018.

Pieter-Tjerk De Boer, Dirk P Kroese, Shie Mannor, and Reuven Y Rubinstein. A tutorial on the cross-entropy method. *Annals of Operations Research*, 134(1): 19–67, 2005.

Zhiguo Ding, Derrick Wing Kwan Ng, Robert Schober, and H Vincent Poor. Delay minimization for NOMA-MEC offloading. *IEEE Signal Processing Letters*, 25(12): 1875–1879, 2018.

Zhiguo Ding, Jie Xu, Octavia A Dobre, and H Vincent Poor. Joint power and time allocation for NOMA–MEC offloading. *IEEE Transactions on Vehicular Technology*, 68(6):6207–6211, 2019.

Peiran Dong, Zhaolong Ning, Rong Ma, Xiaojie Wang, Xiping Hu, and Bin Hu. NOMA-based energy-efficient task scheduling in vehicular edge computing networks: A self-imitation learning-based approach. *China Communications*, 17(11):1–11, 2020.

Chenglong Dou, Ning Huang, Yuan Wu, and Tony Q S Quek. Energy-efficient hybrid NOMA-FDMA assisted distributed two-tier edge-cloudlet multi-access computation offloading. *IEEE Transactions on Green Communications and Networking*, 2023. doi: 10.1109/TGCN.2023.3248609.

Jianbo Du, Liqiang Zhao, Jie Feng, and Xiaoli Chu. Computation offloading and resource allocation in mixed fog/cloud computing systems with min-max fairness guarantee. *IEEE Transactions on Communications*, 66(4):1594–1608, 2017.

Sijing Duan, Dan Wang, Ju Ren, Feng Lyu, Ye Zhang, Huaqing Wu, and Xuemin Shen. Distributed artificial intelligence empowered by end-edge-cloud computing: A survey. *IEEE Communications Surveys & Tutorials*, 25(1):591–624, 2023.

Elie El Haber, Tri Minh Nguyen, and Chadi Assi. Joint optimization of computational cost and devices energy for task offloading in multi-tier edge-clouds. *IEEE Transactions on Communications*, 67(5):3407–3421, 2019.

Meixia Hu, Wei Wang, Wenchi Cheng, and Hailin Zhang. Initial probability adaptation enhanced cross-entropy-based tone injection scheme for PAPR reduction in OFDM systems. *IEEE Transactions on Vehicular Technology*, 70(7):6674–6683, 2021.

Xiaoyan Huang, Supeng Leng, Sabita Maharjan, and Yan Zhang. Multi-agent deep reinforcement learning for computation offloading and interference coordination in small cell networks. *IEEE Transactions on Vehicular Technology*, 70(9): 9282–9293, 2021.

Zhufang Kuang, Linfeng Li, Jie Gao, Lian Zhao, and Anfeng Liu. Partial offloading scheduling and power allocation for mobile edge computing systems. *IEEE Internet of Things Journal*, 6(4):6774–6785, 2019.

Changxiang Li, Hong Wang, and Rongfang Song. Intelligent offloading for NOMA-assisted MEC via dual connectivity. *IEEE Internet of Things Journal*, 8(4):2802–2813, 2021.

Yang Li, Yuan Wu, Minghui Dai, Bin Lin, Weijia Jia, and Xuemin Shen. Hybrid NOMA-FDMA assisted dual computation offloading: A latency minimization approach. *IEEE Transactions on Network Science and Engineering*, 9(5): 3345–3360, 2022.

Zezu Liang, Yuan Liu, Tat-Ming Lok, and Kaibin Huang. Multi-cell mobile edge computing: Joint service migration and resource allocation. *IEEE Transactions on Wireless Communications*, 20(9):5898–5912, 2021.

Wei Yang Bryan Lim, Jer Shyuan Ng, Zehui Xiong, Jiangming Jin, Yang Zhang, Dusit Niyato, Cyril Leung, and Chunyan Miao. Decentralized edge intelligence: A dynamic resource allocation framework for hierarchical federated learning. *IEEE Transactions on Parallel and Distributed Systems*, 33(3):536–550, 2021.

Xin Liu, Xiangping Bryce Zhai, Weidang Lu, and Celimuge Wu. QoS-guarantee resource allocation for multibeam satellite industrial Internet of Things with NOMA. *IEEE Transactions on Industrial Informatics*, 17(3):2052–2061, 2019.

Yuanwei Liu, Wenqiang Yi, Zhiguo Ding, Xiao Liu, Octavia Dobre, and Naofal Al-Dhahir. Application of NOMA in 6G networks: Future vision and research opportunities for next generation multiple access. *arXiv preprint arXiv:2103.02334*, 2021.

Yuanwei Liu, Shuowen Zhang, Xidong Mu, Zhiguo Ding, Robert Schober, Naofal Al-Dhahir, Ekram Hossain, and Xuemin Shen. Evolution of NOMA toward next generation multiple access (NGMA) for 6G. *IEEE Journal on Selected Areas in Communications*, 40(4):1037–1071, 2022.

Yuyi Mao, Changsheng You, Jun Zhang, Kaibin Huang, and Khaled B. Letaief. A survey on mobile edge computing: The communication perspective. *IEEE Communications Surveys & Tutorials*, 19(4):2322–2358, 2017.

Xiaopeng Mo and Jie Xu. Energy-efficient federated edge learning with joint communication and computation design. *Journal of Communications and Information Networks*, 6(2):110–124, 2021.

Wanli Ni, Yuanwei Liu, Zhaohui Yang, Hui Tian, and Xuemin Shen. Integrating over-the-air federated learning and non-orthogonal multiple access: What role can RIS play? *IEEE Transactions on Wireless Communications*, 21(12):10083–10099, 2022.

Quoc-Viet Pham, Hoang T Nguyen, Zhu Han, and Won-Joo Hwang. Coalitional games for computation offloading in NOMA-enabled multi-access edge computing. *IEEE Transactions on Vehicular Technology*, 69(2):1982–1993, 2020.

Liping Qian, Yuan Wu, Haibo Zhou, and Xuemin Shen. Joint uplink base station association and power control for small-cell networks with non-orthogonal multiple access. *IEEE Transactions on Wireless Communications*, 16(9):5567–5582, 2017.

Liping Qian, Yuan Wu, Fuli Jiang, Ningning Yu, Weidang Lu, and Bin Lin. NOMA assisted multi-task multi-access mobile edge computing via deep reinforcement learning for industrial Internet of Things. *IEEE Transactions on Industrial Informatics*, 17(8):5688–5698, 2021.

Jinke Ren, Guanding Yu, Yinghui He, and Geoffrey Ye Li. Collaborative cloud and edge computing for latency minimization. *IEEE Transactions on Vehicular Technology*, 68(5):5031–5044, 2019.

Claudio Rosa, Klaus Pedersen, Hua Wang, Per-Henrik Michaelsen, Simone Barbera, Esa Malkamäki, Tero Henttonen, and Benoist Sébire. Dual connectivity for LTE small cell evolution: Functionality and performance aspects. *IEEE Communications Magazine*, 54(6):137–143, 2016.

Linus E Schrage. *Optimization Modeling with LINGO*. Lindo System, 2006.

Min Sheng, Yanpeng Dai, Junyu Liu, Nan Cheng, Xuemin Shen, and Qinghai Yang. Delay-aware computation offloading in NOMA MEC under differentiated uploading delay. *IEEE Transactions on Wireless Communications*, 19(4):2813–2826, 2020.

Haijian Sun, Xiang Ma, and Rose Qingyang Hu. Adaptive federated learning with gradient compression in uplink NOMA. *IEEE Transactions on Vehicular Technology*, 69(12):16325–16329, 2020.

Tarik Taleb, Konstantinos Samdanis, Badr Mada, Hannu Flinck, Sunny Dutta, and Dario Sabella. On multi-access edge computing: A survey of the emerging 5G network edge cloud architecture and orchestration. *IEEE Communications Surveys & Tutorials*, 19(3):1657–1681, 2017.

Jianyu Wang, Jianli Pan, Flavio Esposito, Prasad Calyam, Zhicheng Yang, and Prasant Mohapatra. Edge cloud offloading algorithms: Issues, methods, and perspectives. *ACM Computing Surveys*, 52(1):1–23, Feb. 2019a.

Yue Wang, Xiaofeng Tao, Xuefei Zhang, Ping Zhang, and Y Thomas Hou. Cooperative task offloading in three-tier mobile computing networks: An ADMM framework. *IEEE Transactions on Vehicular Technology*, 68(3):2763–2776, 2019b.

Kunlun Wang, Yong Zhou, Zening Liu, Ziyu Shao, Xiliang Luo, and Yang Yang. Online task scheduling and resource allocation for intelligent NOMA-based industrial Internet of Things. *IEEE Journal on Selected Areas in Communications*, 38(5):803–815, 2020a.

Tian Wang, Yuzhu Liang, Yilin Zhang, Xi Zheng, Muhammad Arif, Jin Wang, and Qun Jin. An intelligent dynamic offloading from cloud to edge for smart IoT systems with big data. *IEEE Transactions on Network Science and Engineering*, 7(4):2598–2607, 2020b.

Zhaolin Wang, Yuanwei Liu, Xidong Mu, Zhiguo Ding, and Octavia A Dobre. NOMA empowered integrated sensing and communication. *IEEE Communications Letters*, 26(3):677–681, 2022.

Zhiqiang Wei, Lei Yang, Derrick Wing Kwan Ng, Jinhong Yuan, and Lajos Hanzo. On the performance gain of NOMA over OMA in uplink communication systems. *IEEE Transactions on Communications*, 68(1):536–568, 2020.

Yuan Wu, Kejie Ni, Cheng Zhang, Liping Qian, and Danny H K Tsang. NOMA-assisted multi-access mobile edge computing: A joint optimization of computation offloading and time allocation. *IEEE Transactions on Vehicular Technology*, 67(12):12244–12258, 2018a.

Yuan Wu, Liping Qian, Haowei Mao, Xiaowei Yang, Haibo Zhou, and Xuemin Shen. Optimal power allocation and scheduling for non-orthogonal multiple access relay-assisted networks. *IEEE Transactions on Mobile Computing*, 17(11):2591–2606, 2018b.

Wei Wu, Fuhui Zhou, Rose Qingyang Hu, and Baoyun Wang. Energy-efficient resource allocation for secure NOMA-enabled mobile edge computing networks. *IEEE Transactions on Communications*, 68(1):493–505, 2019.

Yuan Wu, Yuxiao Song, Tianshun Wang, Liping Qian, and Tony Q S Quek. Non-orthogonal multiple access assisted federated learning via wireless power transfer: A cost-efficient approach. *IEEE Transactions on Communications*, 70(4):2853–2869, 2022.

Xiaoxia Xu, Yuanwei Liu, Xidong Mu, Qimei Chen, and Zhiguo Ding. Cluster-free NOMA communications towards next generation multiple access. *IEEE Transactions on Communications*, 2023. doi: 10.1109/TCOMM.2023.3244926.

Zhaohui Yang, Cunhua Pan, Jiancao Hou, and Mohammad Shikh-Bahaei. Efficient resource allocation for mobile-edge computing networks with NOMA: Completion time and energy minimization. *IEEE Transactions on Communications*, 67(11):7771–7784, 2019.

Yang Yang, Marius Pesavento, Zhi-Quan Luo, and Björn Ottersten. Inexact block coordinate descent algorithms for nonsmooth nonconvex optimization. *IEEE Transactions on Signal Processing*, 68:947–961, 2020.

Osman N C Yilmaz, Oumer Teyeb, and Antonino Orsino. Overview of LTE-NR dual connectivity. *IEEE Communications Magazine*, 57(6):138–144, 2019.

Yongmin Zhang, Xiaolong Lan, Ju Ren, and Lin Cai. Efficient computing resource sharing for mobile edge-cloud computing networks. *IEEE/ACM Transactions on Networking*, 28(3):1227–1240, 2020.

Jian Zhang, Qimei Cui, Xuefei Zhang, Wei Ni, Xinchen Lyu, Miao Pan, and Xiaofeng Tao. Online optimization of energy-efficient user association and workload offloading for mobile edge computing. *IEEE Transactions on Vehicular Technology*, 71(2):1974–1988, 2022a.

Ke Zhang, Jiayu Cao, and Yan Zhang. Adaptive digital twin and multiagent deep reinforcement learning for vehicular edge computing and networks. *IEEE Transactions on Industrial Informatics*, 18(2):1405–1413, 2022b.

Junhui Zhao, Qiuping Li, Yi Gong, and Ke Zhang. Computation offloading and resource allocation for cloud assisted mobile edge computing in vehicular networks. *IEEE Transactions on Vehicular Technology*, 68(8):7944–7956, 2019.

Jianchao Zheng, Yueming Cai, Yuan Wu, and Xuemin Shen. Dynamic computation offloading for mobile cloud computing: A stochastic game-theoretic approach. *IEEE Transactions on Mobile Computing*, 18(4):771–786, 2019.

Bincheng Zhu, Kaikai Chi, Jiajia Liu, Keping Yu, and Shahid Mumtaz. Efficient offloading for minimizing task computation delay of NOMA-based multiaccess edge computing. *IEEE Transactions on Communications*, 70(5):3186–3203, 2022a.

Hongbiao Zhu, Qiong Wu, Xiao-Jun Wu, Qiang Fan, Pingyi Fan, and Jiangzhou Wang. Decentralized power allocation for MIMO-NOMA vehicular edge computing based on deep reinforcement learning. *IEEE Internet of Things Journal*, 9(14):12770–12782, 2022b.

9

Exploiting Non-orthogonal Multiple Access in Integrated Sensing and Communications

Xidong Mu, Zhaolin Wang, and Yuanwei Liu

School of Electronic Engineering and Computer Science, Queen Mary University of London, London, UK

9.1 Introduction

Recently, the concept of integrated sensing and communications (ISAC) has attracted significant attention from both industry and academia (Liu et al., 2022a, Zhang et al., 2022a, Tan et al., 2021). This is mainly driven by the unprecedented demands on sensing in next-generation wireless networks for supporting various "intelligence" enabled applications (Latva-aho and Leppanen, 2019). Among the family of sensing techniques, wireless radio sensing has been recognized as a promising sensing approach in 6G. For achieving high-quality and ubiquitous sensing, wireless radio sensing has evolved in some similar directions to wireless communications, including ultra-wideband, high radio frequencies, massive multiple-input multiple-output (MIMO), and ultra-dense networks (Liu et al., 2022a, Latva-aho and Leppanen, 2019). This thus opens up new exciting possibilities to integrate sensing and communication functionalities by sharing the same hardware platforms, radio resources, and signal processing pipelines, which motivates the research theme of ISAC. On the one hand, ISAC can enhance the resource efficiency of energy, spectrum, and hardware. For example, ISAC can equip the wireless communication networks with the sensing functionality without the need for significant changes to existing wireless infrastructures, i.e., leading to a relatively low cost. On the other hand, through deep integration and novel protocol designs, ISAC can further achieve mutualism benefits compared to the single-functional network (Liu et al., 2022a). Given the aforementioned potential advantages, ISAC can be extensively employed in future wireless networks, including but not limited to smart cities, industrial Internet-of-things (IoT), and smart homes.

Next Generation Multiple Access, First Edition.
Edited by Yuanwei Liu, Liang Liu, Zhiguo Ding, and Xuemin Shen.
© 2024 The Institute of Electrical and Electronics Engineers, Inc. Published 2024 by John Wiley & Sons, Inc.

Despite being promising for 6G, the strike of a good performance trade-off between the two functionalities is a challenging task when designing ISAC (Liu et al., 2022a). The intrinsic reason is that ISAC may suffer from severe inter-functionality interference due to the hardware platform and radio resource sharing. This calls for the development of efficient interference mitigation and resource management approaches. Reviewing the development road of wireless communications from 1G to 5G, multiple access (MA) techniques have long been one of the most fundamental enablers for efficiently accommodating multiple communication users via appropriate interference control and resource allocation (Liu et al., 2022b). Following the same path, this chapter aims to investigate the development of ISAC from the MA perspective with a particular focus on non-orthogonal multiple access (NOMA). By employing the superposition coding (SC) at the transmitter and the successive interference cancellation (SIC) at the receiver, NOMA allows multiple communication users to be well served over the same radio resources, thus significantly enhancing the connectivity and improving the resource efficiency (Liu et al., 2022b). The prominent features of NOMA inefficient interference management and flexible resource allocation match well with the requirements encountered in ISAC. Nevertheless, since NOMA is currently employed to serve users only for the communication functionality; utilizing NOMA in ISAC is not straightforward and requires re-designs. This motivates us to contribute to this chapter.

The organization of this chapter is as follows. Section 9.2 unveils the developing trend of ISAC with the idea from orthogonality to non-orthogonality and introduces the fundamental models of downlink and uplink ISAC as well as the corresponding design challenges due to the inter-functionality interference. Against these challenges, Section 9.3 proposes novel NOMA-empowered ISAC designs for both the downlink and uplink scenarios. To further elaborate it Section 9.4 introduces a case study of NOMA-empowered downlink ISAC and formulated the ISAC beamforming design problem. Section 9.5 developed an efficient double-layer penalty-based algorithm for addressing the challenging nonconvex beamforming design problem. Section 9.6 presents numerical results to demonstrate the effectiveness, and Section 9.7 concludes this chapter.

9.2 Developing Trends and Fundamental Models of ISAC

9.2.1 ISAC: From Orthogonality to Non-orthogonality

The main idea of ISAC is to facilitate both the sensing and communication functionalities in harmony via the same hardware platform. This is motivated by the fact that sensing and communication share some philosophy in common, i.e., relying on the exploitation of radio waves. Communication exploits the radio wave to convey information bits from the transmitter to the receiver, while sensing exploits the radio wave echo reflected by the target to analyze and obtain the desired sensing-related parameters (e.g., the location, speed, and shape of the target) (Zhang et al., 2021). The ultimate goal of ISAC is to go beyond the two separate functionalities and facilitate promising interplay between them for achieving mutualism benefits, namely communication-assisted sensing and sensing-assisted communication (Liu et al., 2022a). Nevertheless, on the road to ISAC, one of the most fundamental issues is how to effectively coordinate the mutual interference between sensing and communication. One straight solution is to accommodate the sensing and communication functionalities in the allotted *orthogonal* radio resources (e.g., in time-/frequency-/space-/code domains) (Zhang et al., 2022b, Shi et al., 2018, Chen et al., 2021), which can be generally termed as orthogonal-ISAC. Despite having a low implementation complexity and an interference-free sensing/communication process, the resultant spectrum- and energy efficiency of orthogonal-ISAC would be low since the radio resources have to be strictly orthogonally used.

To improve resource efficiency and promote the integration level, it is necessary to develop ISAC by sharing the same radio resources. In this case, since the sensing and communication functionalities have to be facilitated in a *non-orthogonal* manner with properly designed inter-functionality interference mitigation approaches, we term this as non-orthogonal-ISAC. Besides the improved resource efficiency, non-orthogonal-ISAC exhibits potential advantages, such as high compatibility and flexibility. For instance, non-orthogonal-ISAC enables communication (sensing) to be seamlessly integrated with sensing (communication) in the spectrum resource, which has been already occupied by sensing (communication) in history (Liu et al., 2022a). However, in order to fully reap the benefits provided by non-orthogonal-ISAC, efficient methods have to be developed for mitigating the inter-functionality interference. In the following, we introduce the fundamental models of downlink and uplink ISAC and identify the corresponding design challenge in each model if employing non-orthogonal-ISAC. Unless stated otherwise, we use "ISAC" to refer to "non-orthogonal-ISAC" in the remaining context.

9.2.2 Downlink ISAC

In order to better introduce the concept of ISAC, we consider a simple model, which includes one ISAC base station (BS), one sensing target, and one communication user. A typical downlink ISAC is illustrated in Fig. 9.1(a). On the one hand, the ISAC BS sends the sensing probing signal, which will be reflected by the target and returned back as the sensing echo. Thus, ISAC BS is able to estimate

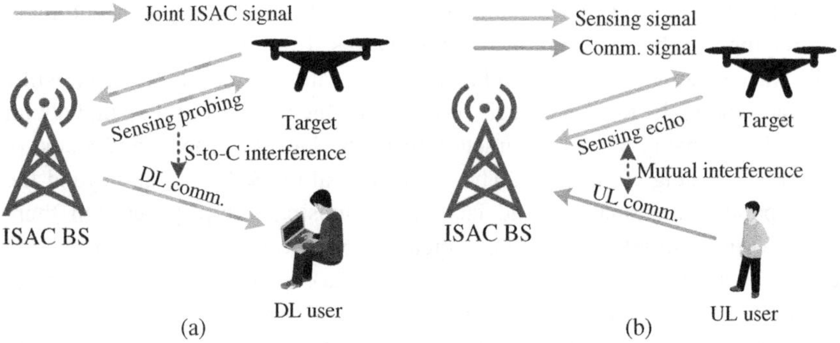

Figure 9.1 Illustration of (a) a downlink ISAC model and (b) an uplink ISAC model. In each model, there is one ISAC BS, one sensing target, and one communication user.

the relevant sensing parameters by analyzing the received sensing echo. On the other hand, the ISAC BS simultaneously sends the communication signal to the communication user. From the MA perspective, it can be observed that both the sensing and communication signals are transmitted by the ISAC BS, and the sensing parameter estimation and the information bit decoding are carried out at two different destinations (i.e., the ISAC BS for sensing and the downlink user for communication). Therefore, we classify this type of model as the downlink ISAC.

Recall the fact that sensing can be only carried out at the ISAC BS until receiving the echo of the transmit sensing waveform reflected by the target. In other words, there is no information contained in the transmit sensing waveform at first as that in the transmit communication waveform. Inspired by this observation, the key task in the downlink ISAC is the design of the joint S&C waveforms, which are capable of achieving both sensing and communication (Hua et al., 2021, Liu et al., 2018, 2020). Since the ISAC BS prior knows the communication information and the analysis of the sensing echo also does not focus on the information bits modulated on the joint S&C waveforms, there is in general no communication-to-sensing interference in the downlink ISAC. However, the reverse is not true. This is because, for well achieving sensing, the number of required joint S&C waveforms may be larger than the number of downlink communication streams (Liu et al., 2020, Hua et al., 2021). As a result, some of joint waveforms (termed as additional sensing waveforms) will lead to *sensing-to-communication interference* as shown in Fig. 9.1(a). Therefore, for downlink ISAC, on the one hand, how to efficiently exploit the limited DoFs of the joint S&C waveforms for facilitating the two functionality is one design challenge. On the other hand, how to mitigate the sensing-to-communication interference due to the employment of additional sensing waveforms is another major design challenge.

9.2.3 Uplink ISAC

We continue to introduce the fundamental model of the uplink ISAC, as shown in Fig. 9.1(b). In this case, the entire procedure of sensing remains unchanged, while the communication signal is uploaded by the uplink user to the ISAC BS. From the MA perspective, the sensing and communication signals are respectively transmitted by the ISAC BS and the uplink user, while the sensing parameter estimation and the information bit decoding are carried out at the same destination (ISAC BS). We classify this type of model as the uplink ISAC. It can be observed that since the sensing and communication signals are separately transmitted, there is no need to design the joint S&C waveforms in the uplink ISAC.

In the uplink ISAC, the ISAC BS has to analyze the sensing echo for obtaining the sensing results and decoding the communication signal for recovering the information message. Both of them are not prior known by the ISAC BS. When the ISAC BS processes the received mixed sensing and communication signals, the major design challenge in the uplink ISAC is how to mitigate the *mutual interference* between sensing and communication, as shown in Fig. 9.1(b).

9.3 Novel NOMA Designs in Downlink and Uplink ISAC

Against the above discussed challenges, in this section, we explore the possible applications of NOMA to enhance the performance of downlink and ISAC. On the one hand, for efficiently exploiting the limited DoFs in the downlink ISAC only employing joint S&C waveforms, we propose a novel NOMA-empowered downlink ISAC design to efficiently mitigate the conventional inter-user communication interference and thus provide more DoFs for striking a good sensing-versus-communication trade-off. On the other hand, to efficiently mitigate the mutual interference in the uplink ISAC, we propose a novel semi-NOMA-based uplink ISAC design.

9.3.1 NOMA-Empowered Downlink ISAC Design

In the NOMA-empowered downlink ISAC design, as shown in Fig. 9.2, the SC and SIC techniques are invoked for transmitting and detecting the communication signal for each user. Moreover, the superimposed communication signals are also exploited for target sensing. Therefore, the number of joint S&C waveforms is equal to the number of users, thus leading to no additional sensing-to-communication interference but providing limited DoFs for ISAC designs. The proposed NOMA-empowered downlink ISAC design can be regarded as a straightforward extension of employing NOMA from the conventional communication system to the ISAC system, i.e., the communication functionality

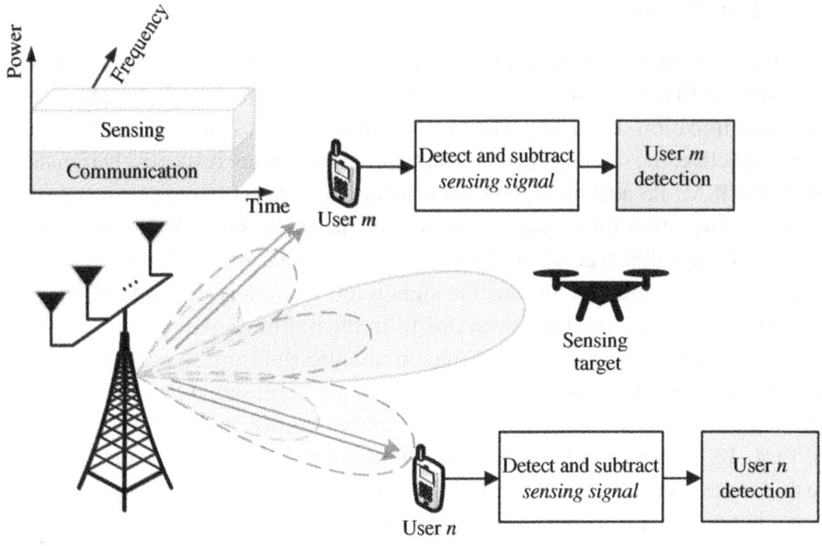

Figure 9.2 Illustration of the NOMA-empowered downlink ISAC design.

of ISAC is *empowered* by NOMA. Consequently, the NOMA-empowered downlink ISAC design inherits the key feature of conventional NOMA, namely the additional DoFs provided by SIC for inter-user interference cancellation. In conventional ISAC (Liu et al., 2018), the joint S&C waveform designs for mitigating the inter-user interference merely rely on the spatial DoFs. As a result, it may suffer from severe performance degradation when the spatial DoFs are deficient. Fortunately, this problem can be addressed by the proposed NOMA-empowered downlink ISAC with the aid of SIC. To further demonstrate the advantages of the proposed NOMA-empowered downlink ISAC design, we discuss it in both the overloaded and underloaded regimes.

In the overloaded regime, it is impossible to allocate at least one spatial DoF per communication user even without sensing. Therefore, the corresponding inter-user interference cannot be effectively suppressed in the spatial domain. In this case, the lack of efficient inter-user interference mitigation not only causes degraded communication performance but also further damages sensing performance given the limited DoFs in joint S&C waveform designs. As a remedy, NOMA-empowered downlink ISAC provides extra DoFs for inter-user interference mitigation via SIC and thus achieves enhanced sensing and communication performance than the conventional downlink ISAC designs.

In contrast to the overloaded regime, where the spatial DoFs are always insufficient, the deficiency of spatial DoFs in the underloaded regime may also occur in the following two cases. On the one hand, the correlated users' channels will directly cause the deficiency of spatial DoFs in the underloaded regime. On the other hand, in the sensing-prior design of ISAC, the joint S&C waveforms transmitted by the ISAC BS need to be constructed in some specific beam patterns for ensuring high-quality sensing, which might limit the spatial DoFs available for inter-user interference mitigation. In such cases, the proposed NOMA-empowered downlink ISAC design comes to the rescue. The details of the proposed NOMA-empowered downlink ISAC design will be further discussed in Sections 9.4–9.6.

9.3.2 Semi-NOMA-Based Uplink ISAC Design

In this section, we focus our attention on the uplink ISAC. As discussed in Section 9.2.3, the key problem in the uplink ISAC is how to efficiently mitigate the mutual interference between the sensing echo and the communication signal that collide at the ISAC BS. Recalling the fact that only the uplink communication signal conveys the information bits among the two types of signals. Based on this observation, we propose the semi-NOMA-based uplink ISAC design.

Before the proposed semi-NOMA-based uplink ISAC design, we first briefly review the employment of pure orthogonal multiple access (OMA) and NOMA schemes in the uplink ISAC. As shown in Fig. 9.3, let us take a basic uplink ISAC model as an example, where an ISAC BS aims to estimate the sensing-related parameters of one sensing target and recover the information messages uploaded by one communication user. On the one hand, in the conventional OMA-based uplink ISAC, the sensing echo and the communication signal can be allocated different orthogonal radio resources to facilitate each functionality in an interference-free manner, as illustrated in the top left of Fig. 9.3. As discussed in Section 9.2.1, the OMA-based uplink ISAC design may lead to low resource efficiency. On the other hand, in the conventional NOMA-based uplink ISAC, the sensing and communication functionalities to be realized via the fully shared radio resource, as illustrated in the top middle of Fig. 9.3. For mitigating the inter-functionality interference caused by resource sharing, similar to the uplink NOMA principle employed in the uplink communication system, the ISAC BS will successively process the two types of signals with the aid of SIC. In contrast to the conventional NOMA uplink communication, where the SIC decoding order can be flexibility designed among the mixed communication signals (Liu et al., 2022b), the conventional NOMA-based uplink ISAC design has to follow a fixed communication-to-sensing decoding order. This is because only the communication signal contains the information bits, which makes it possible to carry out SIC by first removing the communication signal from the received mixed sensing–communication signal. Then, the ISAC BS can analyze the sensing echo in an interference-free manner but have more available

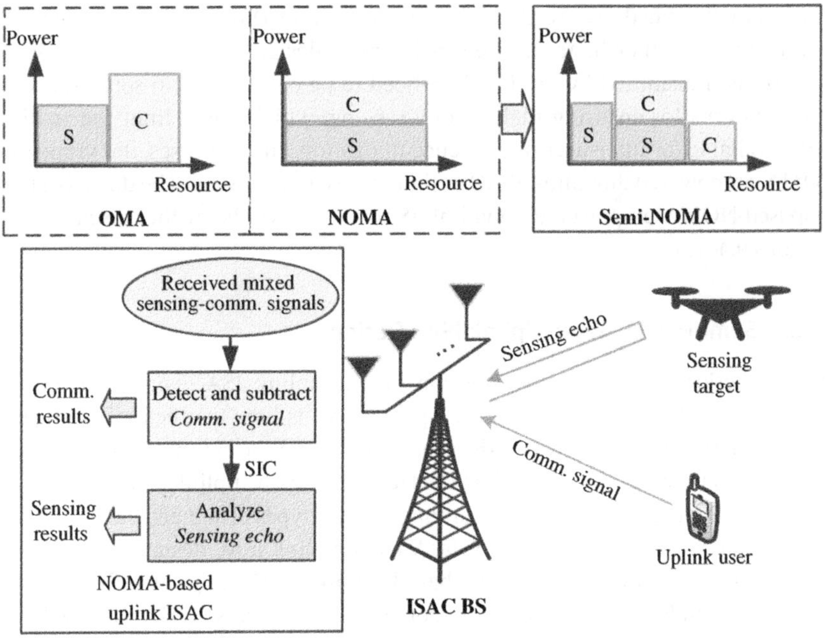

Figure 9.3 An uplink ISAC system with OMA, NOMA, and semi-NOMA schemes.

radio resources than the OMA-based uplink ISAC design. The procedure of mixed sensing–communication signal processing at the ISAC BS is illustrated at the bottom left of Fig. 9.3. However, one main drawback of the conventional NOMA-based uplink ISAC is that the fixed communication-to-sensing decoding order causes the uploaded information message to be always decoded against the sensing interference, thus resulting in a limited communication rate.

Against the pros and cons in the conventional OMA- and NOMA-based uplink ISAC, we propose a semi-NOMA-based uplink ISAC design, which not only unifies both the OMA-based and pure-NOMA-based uplink ISAC designs as special cases but also provides more flexible ISAC operations for well satisfying different objectives, e.g., sensing-prior design, communication-prior design, and sensing-versus-communication trade-off design. As shown at the top right of Fig. 9.3, the key idea behind the proposed semi-NOMA-based uplink ISAC design is to partition the total available radio resources into three orthogonal parts, namely the sensing-only resource block (S-RB), the communication-only RB (C-RB), and the mixed sensing–communication RB (S&C-RB). On the one hand, in the S&C-RB, the ISAC BS processes the received mixed sensing–communication signal following the principle of the pure-NOMA-based uplink ISAC design. On the other hand, the ISAC BS also processes the pure sensing echo/communication signal received from the S/C-RB in an interference-free manner as the OMA-based uplink ISAC design. Finally, the ISAC BS will combine the sensing/communication results obtained from the non-orthogonal S&C-RB and the orthogonal S/C-RB together. From the perspective of resource allocation, since only partial radio resources are shared among the sensing and communication functionalities with the employment of NOMA (except in the two extreme cases), we therefore term it the *semi-NOMA-based* uplink ISAC design.

The proposed semi-NOMA-based uplink ISAC design provides a general resource allocation framework for the uplink ISAC. By carefully optimizing the resource allocation among the three RBs, on the one hand, the semi-NOMA-based uplink ISAC design can be reduced to the OMA/pure-NOMA-based uplink ISAC design (e.g., make the S&C-RB null or make both the S- and C-RBs null). On the other hand, it can further facilitate diversified uplink ISAC operations according to changes in ISAC design objectives, which cannot be realized by either the OMA-based or the pure-NOMA-based uplink ISAC design. It is also worth noting that the allocation of the C-RB in the proposed semi-NOMA-based uplink ISAC design is of vital importance in improving the communication performance since the communication signal received by the C-RB is not sensing-interference-limited compared to that in the pure-NOMA-based uplink ISAC design. As a result, a better sensing-versus-communication trade-off can be achieved by the semi-NOMA-based uplink ISAC design. More details of the proposed semi-NOMA-based uplink ISAC design can be found in Zhang et al. (2023).

9.4 Case Study: System Model and Problem Formulation

9.4.1 System Model

In this section, we consider the NOMA-empowered downlink ISAC system as a case study, as illustrated in Fig. 9.2. In this system, there is a dual-functional BS equipped with an N-antennas uniform linear array (ULA), K single-antenna users, whose indices are collected in $\mathcal{K} = \{1, \dots, K\}$, and M sensing targets, whose indices are collected in $\mathcal{M} = \{1, \dots, M\}$.

9.4.1.1 Communication Model

As discussed in Section 9.3.1, in this system, NOMA is employed at the BS for serving multiple communication users. In particular, the BS transmits the superimposed signals of $\mathbf{w}_i s_i$ for $\forall i \in \mathcal{K}$ to all users (Liu et al., 2017), where $\mathbf{w}_i \in \mathbb{C}^{N\times 1}$

are beamformers for delivering the information symbol s_i to user i. Therefore, the received signal at user k is given by

$$y_k = \mathbf{h}_k^H \sum_{i\in\mathcal{K}} \mathbf{w}_i s_i + n_k = \sum_{i\in\mathcal{K}} \mathbf{h}_k^H \mathbf{w}_i s_i + n_k, \tag{9.1}$$

where $\mathbf{h}_k = \Lambda_k^{-1/2}\tilde{\mathbf{h}}_k, \forall k \in \mathcal{K}$ denotes the BS-user channel, $\Lambda_k^{-1/2}$ and $\tilde{\mathbf{h}}_k \in \mathbb{C}^{N\times 1}$ denote the large and small scale fading, respectively, and n_k denotes the circularly symmetric complex Gaussian noise with variance σ_n^2. We assume that the users' indexes are in increasing order with respect to their large-scale channel strength, i.e., $\Lambda_1^{-1} \leq \Lambda_2^{-1} \leq \cdots \leq \Lambda_K^{-1}$. Thus, user 1 is the weakest user while user K is the strongest user. In NOMA, user k first detects and removes the interference from all the weaker $j < k$ users by exploiting SIC, while treating the interference from all the stronger users $j > k$ as noise. Thus, the achievable rate of s_k after SIC at user k for $\forall k \in \mathcal{K}, k \neq K$ is

$$R_{k\to k} = \log_2\left(1 + \frac{|\mathbf{h}_k^H \mathbf{w}_k|^2}{\sum_{i\in\mathcal{K}, i>k} |\mathbf{h}_k^H \mathbf{w}_i|^2 + \sigma_n^2}\right). \tag{9.2}$$

However, the symbol s_k for user k also need to be decodable at user j, for $j > k$ and $\forall k \in \mathcal{K}, k \neq K$, to carry out SIC, yielding the following achievable rate

$$R_{k\to j} = \log_2\left(1 + \frac{|\mathbf{h}_j^H \mathbf{w}_k|^2}{\sum_{i\in\mathcal{K}, i>k} |\mathbf{h}_j^H \mathbf{w}_i|^2 + \sigma_n^2}\right). \tag{9.3}$$

Thus, the overall achievable rate of s_k for $\forall k \in \mathcal{K}, k \neq K$ is

$$R_k = \min\{R_{k\to k}, \ldots, R_{k\to K}\}. \tag{9.4}$$

At user K, the interference from all the other users is eliminated by SIC. Therefore, its achievable rate is given as

$$R_K = \log_2\left(1 + \frac{|\mathbf{h}_K^H \mathbf{w}_K|^2}{\sigma_n^2}\right). \tag{9.5}$$

The communication throughput is given by $R = \sum_{k\in\mathcal{K}} R_k$.

9.4.1.2 Sensing Model

In the ISAC system, the communication waveforms can be exploited to carry out target sensing, but need to satisfy the sensing requirements, which is equivalent to designing the covariance matrix of the transmitted signal (Stoica et al., 2007). The covariance matrix is given by

$$\mathbf{R}_{\mathbf{w}} = \sum_{i\in\mathcal{K}} \mathbf{w}_i \mathbf{w}_i^H. \tag{9.6}$$

With the prior information of targets, the objective of the sensing function is to maximize the effective sensing power, i.e., the power of probing signal in target directions (Stoica et al., 2007), which is given as

$$P(\theta_m) = \mathbf{a}^H(\theta_m)\mathbf{R}_{\mathbf{w}}\mathbf{a}(\theta_m), \tag{9.7}$$

where $\theta_m, \forall m \in \mathcal{M}$ are target directions and $\mathbf{a}(\theta_m)$ is the corresponding steering vector. In particular, $\mathbf{a}(\theta_m)$ is given by

$$\mathbf{a}(\theta_m) = [1, e^{j\frac{2\pi}{\lambda}d\sin(\theta_m)}, \dots, e^{j\frac{2\pi}{\lambda}d(N-1)\sin(\theta_m)}]^T, \tag{9.8}$$

where λ and d denote the carrier wavelength and antenna spacing, respectively. We assume that a similar level of sensing power is desired in the different target directions such that each target can be fairly tracked. Furthermore, the cross-correlation between transmitted signals at any two target directions θ_k and θ_p is expected to be low such that the sensing system can perform adaptive localization (Stoica et al. 2007). The cross-correlation is given by

$$C(\theta_k, \theta_p) = |\mathbf{a}^H(\theta_k)\mathbf{R}_{\mathbf{w}}\mathbf{a}(\theta_p)|, \quad \forall k \neq p \in \mathcal{M}. \tag{9.9}$$

The mean-squared cross-correlation of the $\frac{M^2-M}{2}$ pairs of sensing targets is given by

$$\overline{C} = \frac{2}{M^2 - M}\sum_{k=1}^{M-1}\sum_{p=k+1}^{M} C(\theta_k, \theta_p)^2. \tag{9.10}$$

Remark 9.1 *In contrast to the conventional ISAC system (Liu et al., 2018), where the inter-user interference is merely mitigated by exploiting spatial DoFs, the NOMA-empowered ISAC system further employs SIC for mitigating inter-user interference, see (9.2), (9.3), and (9.5). Therefore, when the available spatial DoFs are limited (e.g., the underloaded regime with highly correlated channels and overloaded regime), NOMA provides extra DoFs to guarantee the communication performance, which enables the feasibility of integrating the sensing function. This will be verified via the numerical results in Section 9.6.*

9.4.2 Problem Formulation

Given the NOMA-empowered downlink ISAC framework, we aim to maximize the weighted sum of communication throughput and effective sensing power, while satisfying the minimum communication rate of each user and sensing-specific requirements. The resultant optimization problem is formulated as

$$\max_{\mathbf{w}_k} \quad \rho_c \sum_{k\in\mathcal{K}} R_k + \rho_r \sum_{m\in\mathcal{M}} P(\theta_m) \tag{9.11a}$$

$$\text{s.t.}\quad R_k \geq R_{\min,k},\quad \forall k \in \mathcal{K}, \tag{9.11b}$$

$$|P(\theta_k) - P(\theta_p)| \leq P_{\text{diff}},\quad \forall k \neq p \in \mathcal{M}, \tag{9.11c}$$

$$\text{diag}\left(\sum_{i\in\mathcal{K}} \mathbf{w}_i\mathbf{w}_i^H\right) = \frac{P_t\mathbf{1}^{N\times 1}}{N}, \tag{9.11d}$$

$$\overline{C} \leq \xi, \tag{9.11e}$$

where $\rho_c \geq 0$ and $\rho_r \geq 0$ are the regularization parameters; by varying them we can obtain the performance trade-off between communication and sensing. Here, (9.11b) guarantees the minimum rate of each user and (9.11c) ensures the similar levels of sensing power in different target directions. The constraint (9.11d) is the constant per antenna constraint (Stoica et al., 2007), where P_t denotes the total transmit power. Finally, the constraint (9.11e) ensures a desired upper bound of the mean-squared cross-correlation. However, it is quite challenging to obtain the globally optimal solution for problem (9.11) due to the following reasons. On the one hand, the expression of achievable rate is neither convex nor concave, which makes the objective function non-concave and the constraint (9.11b) non-convex. On the other hand, the quadratic form of the covariance matrix makes the constraints (9.11c) and (9.11d) non-convex. In the following, we propose an efficient iterative algorithm to obtain a suboptimal solution by invoking successive convex approximation (Sun et al., 2017).

9.5 Case Study: Proposed Solutions

In this chapter, we develop an successive convex approximation (SCA)-based double-layer iterative algorithm to solve problem (9.11). Firstly, we define $\mathbf{W}_k \triangleq \mathbf{w}_k\mathbf{w}_k^H$, which satisfies $\mathbf{W}_k \succcurlyeq 0$, $\mathbf{W}_k = \mathbf{W}_k^H$, and $\text{rank}(\mathbf{W}_k) = 1$. Then, the problem (9.11) can be reformulated as

$$\max_{\gamma_k,\mathbf{W}_k}\quad f(\rho_c, \rho_r, \gamma_k, \mathbf{W}_k) = \rho_c \sum_{k\in\mathcal{K}} \gamma_k + \rho_r \sum_{m\in\mathcal{M}} P(\theta_m) \tag{9.12a}$$

$$\text{s.t.}\quad R_k \geq \gamma_k,\quad \forall k \in \mathcal{K}, \tag{9.12b}$$

$$\gamma_k \geq R_{\min,k},\quad \forall k \in \mathcal{K}, \tag{9.12c}$$

$$\mathbf{W}_k \succcurlyeq 0, \mathbf{W}_k = \mathbf{W}_k^H,\quad \forall k \in \mathcal{K}, \tag{9.12d}$$

$$\text{rank}(\mathbf{W}_k) = 1,\quad \forall k \in \mathcal{K}, \tag{9.12e}$$

$$\text{(9.11c)–(9.11e)}. \tag{9.12f}$$

Furthermore, we define $\mathbf{H}_k \triangleq \mathbf{h}_k\mathbf{h}_k^H$. Then, for $j \geq k$ and $k \in \mathcal{K}, k \neq K$, the constraint (9.12b) can be rewritten as

$$R_{k\to j} = \log_2\left(\sigma_n^2 + \sum_{i\in\mathcal{K},i\geq k} \mathrm{Tr}\left(\mathbf{H}_j\mathbf{W}_i\right)\right) \underbrace{-\log_2\left(\sigma_n^2 + \sum_{i\in\mathcal{K},i>k} \mathrm{Tr}\left(\mathbf{H}_j\mathbf{W}_i\right)\right)}_{F_{j,k}} \geq \gamma_k. \tag{9.13}$$

The non-convexity of this constraint lies in the second term $F_{j,k}$. To address this, we invoke the SCA. By using the first-order Taylor expansion at point $\left(\mathbf{W}_1^n, \ldots, \mathbf{W}_K^n\right)$, we have

$$\begin{aligned} F_{j,k} \geq \widehat{F}_{j,k} \triangleq &- \log_2\left(\sigma_n^2 + \sum_{i\in\mathcal{K},i>k} \mathrm{Tr}\left(\mathbf{H}_j\mathbf{W}_i^n\right)\right) \\ &- \frac{\sum_{i\in\mathcal{K},i>k} \mathrm{Tr}\left(\mathbf{H}_j\left(\mathbf{W}_i - \mathbf{W}_i^n\right)\right)}{\left(\sigma_n^2 + \sum_{i\in\mathcal{K},i>k} \mathrm{Tr}\left(\mathbf{H}_j\mathbf{W}_i^n\right)\right)\ln 2}. \end{aligned} \tag{9.14}$$

Then, we define

$$\widehat{R}_{k\to j} \triangleq \log_2\left(\sigma_n^2 + \sum_{i\in\mathcal{K},i\geq k} \mathrm{Tr}\left(\mathbf{H}_j\mathbf{W}_i\right)\right) + \widehat{F}_{j,k}, \tag{9.15}$$

which is a lower bound of $R_{k\to j}$. By exploiting it, the constraint (9.12b) can be approximated by $\widehat{R}_{k\to j} \geq \gamma_k$. Thus, problem (9.12) can be reformulated as

$$\max_{\gamma_k, \mathbf{W}_k} \quad f(\rho_c, \rho_r, \gamma_k, \mathbf{W}_k) \tag{9.16a}$$

$$\text{s.t.} \quad \widehat{R}_{k\to j} \geq \gamma_k, \, j \geq k, \quad \forall k \in \mathcal{K}, \quad k \neq K, \tag{9.16b}$$

$$R_K \geq \gamma_K, \tag{9.16c}$$

$$\text{(9.11c)–(9.11e), (9.12c)–(9.12e).} \tag{9.16d}$$

For this optimization problem, the non-convexity is only from the rank-one constraint (9.12e). Generally, the semidefinite relaxation (SDR) (Luo et al., 2010) is exploited to solve this problem by omitting the rank-one constraint. Then, the eigenvalues decomposition or Gaussian randomization is used to reconstruct the rank-one solution from the general-rank solution obtained by SDR, which may lead to a significant performance loss and not ensure the feasibility of the reconstructed matrix. To avoid these drawbacks, we attempt to transform the rank-one constraints to a penalty term in the objective function, which can also be solved by SCA. Toward this idea, we first introduce an equivalent equality constraint:

$$\|\mathbf{W}_k\|_* - \|\mathbf{W}_k\|_2 = 0, \quad k \in \mathcal{K}, \tag{9.17}$$

where $\|\cdot\|_*$ is the nuclear norm, which is the sum of singular values of the matrix, and $\|\cdot\|_2$ is the spectral norm, which is the largest singular values of the matrix. Thus, when the matrix $\mathbf{W}_k$ is a rank-one matrix, the equality (9.17) holds. Otherwise, as $\mathbf{W}_k$ is semidefinite, we must have that the sum of singular values is larger than the largest singular value, i.e., $\|\mathbf{W}_k\|_* - \|\mathbf{W}_k\|_2 > 0$. In order to obtain a rank-one matrix, we introduce a penalty term to the objective function based on (9.17), yielding

$$\max_{\gamma_k, \mathbf{W}_k} \quad f(\rho_c, \rho_r, \gamma_k, \mathbf{W}_k) - \frac{1}{\eta} \sum_{k \in \mathcal{K}} \left(\|\mathbf{W}_k\|_* - \|\mathbf{W}_k\|_2\right) \tag{9.18a}$$

$$\text{s.t.} \quad \text{(9.11c)–(9.11e), (9.12c), (9.12d), (9.12d), (9.16c).} \tag{9.18b}$$

However, the second term in the penalty term makes the objective not convex. By exploiting the first-order Taylor expansion at point $\mathbf{W}_k^n$, its upper bound of is given by

$$-\|\mathbf{W}_k\|_2 \leq \widehat{\mathbf{W}}_k^n \triangleq -\|\mathbf{W}_k^n\|_2 - \text{Tr}\left[\mathbf{v}_{\max,k}^n (\mathbf{v}_{\max,k}^n)^H \left(\mathbf{W}_k - \mathbf{W}_k^n\right)\right], \tag{9.19}$$

where $\mathbf{v}_{\max,k}^n$ is the eigenvector corresponding to the largest eigenvalue of $\mathbf{W}_k^n$. Thus, the problem (9.18) can be approximated by the following problem

$$\max_{\gamma_k, \mathbf{W}_k} \quad f(\rho_c, \rho_r, \gamma_k, \mathbf{W}_k) - \frac{1}{\eta} \sum_{k \in \mathcal{K}} \left(\|\mathbf{W}_k\|_* + \widehat{\mathbf{W}}_k^n\right) \tag{9.20a}$$

$$\text{s.t.} \quad \text{(9.11c)–(9.11e), (9.12c), (9.12b), (9.16b), (9.16c).} \tag{9.20b}$$

The problem (9.20) is a quadratic semidefinite program (QSDP), which can be efficiently solved by the standard interior point method,

It is worth noting that the choice of parameter η plays an important role in the objective function. If this parameter is chosen to be $\eta \to 0$ ($\frac{1}{\eta} \to \infty$), the rank of matrix $\mathbf{W}_k$ will be definitely one. Nevertheless, in this case, we cannot obtain a good solution regarding the maximization of throughput and effective sensing power, since the objective function is dominated by the penalty term. To tackle this, we can initialize a large η to obtain a good starting point for the throughput and the effective sensing power. Then, by gradually reducing η to a sufficiently small value via $\eta = \epsilon\eta, 0 < \epsilon < 1$, an overall suboptimal solution can be obtained. This procedure is terminated when the penalty term is sufficiently small, i.e., $\sum_{k \in \mathcal{K}} \left(\|\mathbf{W}_k\|_* - \|\mathbf{W}_k\|_2\right) \leq \varepsilon_2$. The overall algorithm to problem (9.11) is summarized in Algorithm 9.1. The complexity of this algorithm is $\mathcal{O}(I_o I_i (K^{6.5} N^{6.5} \log(1/e)))$, where I_o and I_i are the number of iterations of the outer and inner layers, e is the solution accuracy, and $\mathcal{O}(K^{6.5} N^{6.5} \log(1/e))$ is the complexity for solving the QSDP (9.20).

Algorithm 9.1 Proposed Double-Layer Penalty-Based Algorithm for Solving Problem (9.11)

Initialize the feasible $\mathbf{W}_k^0, \forall k \in \mathcal{K}$.
repeat
    $n \leftarrow 0$.
    repeat
        Update $\mathbf{W}_k^{n+1}$ by solving (9.20) with $\mathbf{W}_k^n, \forall k \in \mathcal{K}$.
        $n \leftarrow n + 1$.
    until the fractional reduction of the objective function value falls below a predefined threshold ε_1.
    $\mathbf{W}_k^0 \leftarrow \mathbf{W}_k^n, \forall k \in \mathcal{K}$.
    $\eta \leftarrow \epsilon\eta$.
until $\sum_{k\in\mathcal{K}} \left(\|\mathbf{W}_k^n\|_* - \|\mathbf{W}_k^n\|_2 \right) \leq \varepsilon_2$.

9.6 Case Study: Numerical Results

In this chapter, the numerical results are provided to demonstrate the characteristics of the NOMA-empowered downlink ISAC system. As shown in Fig. 9.4, we assume a BS equipped with a ULA with $N = 4$ antennas, serving $K = 2$ or 6 communication users and tracking $M = 2$ sensing targets in $\theta_1 = -40°$ and $\theta_2 = 40°$. The overall power budget is $P_t = 20$ dBm and the noise power at users is

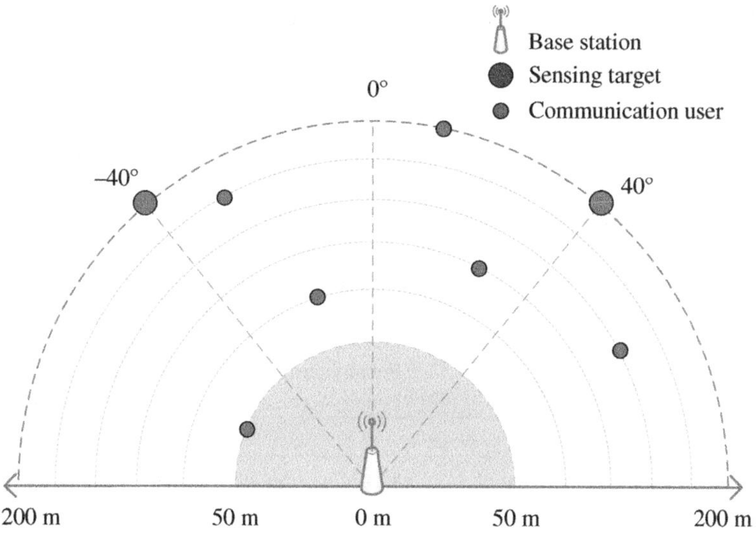

Figure 9.4 Simulation setup.

$\sigma_n^2 = -120$ dBm. The channels between BS and users are assumed to experience Rayleigh fading with the path loss of $\Lambda_k(\text{dB}) = 32.6 + 36.7\log_{10}(d_k)$. In particular, the path loss model is defined based on the 3GPP propagation environment [3GPP, 2010, Table B.1.2.1-1]. The fading model is given as follows (Kermoal et al., 2002):

$$\tilde{\mathbf{H}} = \mathbf{H}_w \mathbf{R}_{\tilde{\mathbf{H}}}^{1/2}, \tag{9.21}$$

where $\tilde{\mathbf{H}} = \left[\mathbf{h}_1/\|\mathbf{h}_1\|, \dots, \mathbf{h}_K/\|\mathbf{h}_K\|\right]$ and $\mathbf{H}_w$ is the normalized Rayleigh fading matrix satisfying $\mathbb{E}\left[\mathbf{H}_w^H \mathbf{H}_w\right] = \mathbf{I}$. The matrix $\mathbf{R}_{\tilde{\mathbf{H}}}$ is the spatial correlation matrix of $\tilde{\mathbf{H}}$. Its (i,j)th entry indicates the spatial correlation between the channels of user i and user j, the norm of which is $t^{|i-j|}$ for $t \in [0,1]$. The users are equally spaced between the distances of 50 and 200 m from BS. We set $R_{\min} = 1$ bit/s/Hz, $P_{\text{diff}} = 10$, and $\xi = 10$. The initial penalty factor of Algorithm 9.1 is set to $\eta = 10^5$. Finally, the convergence thresholds of inner and outer layers are set to $\epsilon_1 = 10^{-2}$ and $\epsilon_2 = 10^{-4}$.

9.6.1 Convergence of Algorithm 9.1

In Fig. 9.5, the convergence behavior of the proposed algorithm over one random channel realization is studied with $K = 6$, $t = 0$, $\rho_c = 10$, and $\rho_r = 1$. We can

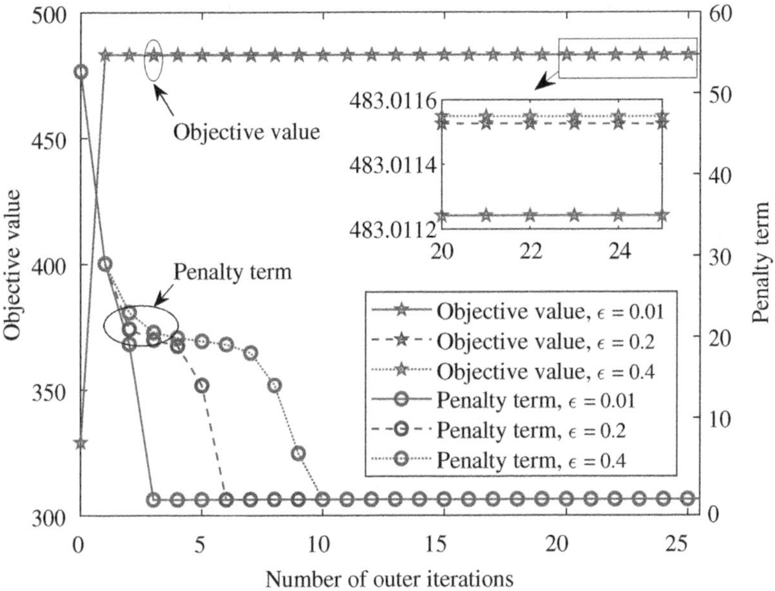

Figure 9.5 Convergence of Algorithm 9.1.

observe that for any values of the reduction factor ϵ, the objective value quickly converges to a stable value while the penalty term converges to almost zero after several outer iterations. It reveals that the proposed algorithm is capable of finding a feasible rank-one solution with high performance. Furthermore, it can be seen that as the value of ϵ becomes smaller, the proposed algorithm has a higher convergence speed while achieving a lower objective value, i.e., worse system performance, which is a trade-off. Thus, in the following simulation, we set $\epsilon = 0.2$, which achieves the suitable convergence speed and system performance simultaneously.

9.6.2 Baseline

For comparison, we consider the conventional ISAC system without the employment of NOMA (Liu et al., 2018), where the achievable rate at user k is given by

$$R_k^{\mathrm{b}} = \log_2\left(1 + \frac{|\mathbf{h}_k^H \mathbf{w}_k|^2}{\sum_{i\in\mathcal{K}, i\neq k}|\mathbf{h}_k^H \mathbf{w}_i|^2 + \sigma_n^2}\right). \tag{9.22}$$

The corresponding problem of maximizing the throughput $R^{\mathrm{b}} = \sum_{k\in\mathcal{K}} R_k^{\mathrm{b}}$ and effective sensing power at the sensing targets can be solved using Algorithm 9.1 with the interference term in (9.22).

In Fig. 9.6, we demonstrate the performance trade-off, i.e., the communication throughput versus the effective sensing power, of NOMA-empowered ISAC and the conventional ISAC. The results are obtained via Monte Carlo simulation over 400 random channel realizations. Two cases are considered, namely the underloaded regime ($K = 2$) and the overloaded regime ($K = 6$). As seen in Fig. 9.6, in both underloaded and overloaded regimes, the spatial factor has no effect

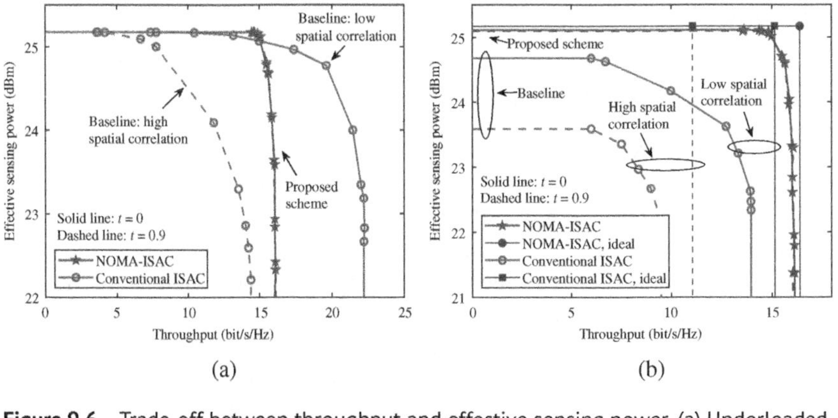

Figure 9.6 Trade-off between throughput and effective sensing power. (a) Underloaded $N = 4, K = 2$. (b) Overloaded $N = 4, K = 6$.

on the NOMA-empowered ISAC system. The reason is that the communication throughput is dominated by the strongest NOMA user. However, the performance of the conventional ISAC system is subject to the spatial factor, i.e., as the spatial correlation increases, the performance achievable area becomes smaller. Specifically, in the underloaded regime (Fig. 9.6(a)), the NOMA-empowered ISAC outperforms the conventional ISAC when the spatial correlation is high, but the result is the opposite when the spatial correlation is low. In the overloaded regime (Fig. 9.6(b)), compared to the conventional ISAC, the NOMA-empowered ISAC achieves considerable gain, which becomes even greater when the spatial correlation is high. This is because the inter-user interference cannot be well mitigated in the conventional ISAC system due to the limited spatial DoFs in the overloaded regime. Then, more resources are needed by the conventional ISAC to meet the communication requirements, thus leading to limited sensing performance. However, for the proposed NOMA-empowered ISAC, although the system is overloaded, the inter-user interference can still be mitigated by SIC, which provides more DoFs to be exploited for the target sensing. The above results underscore the importance of employing NOMA in the ISAC system when the communication system is overloaded or the channels are highly spatially correlated and verify Remark 9.1.

Furthermore, we also demonstrate the ideal ISAC system in Fig. 9.6(b), i.e., communication and sensing systems work independently, with no effect on each other. The communication throughput and effective sensing power in the ideal ISAC are obtained by removing the communication and sensing from problem (9.11), respectively. For the conventional ISAC system, it can be seen that there is a significant gap between the real case and the ideal case. However, for the NOMA-empowered ISAC system, there is only a slight performance gap and the performance upper bounds of communication and sensing can be nearly achieved simultaneously.

9.6.3 Transmit Beampattern

In Fig. 9.7, we present the obtained transmit beampattern by the proposed NOMA-empowered ISAC and the conventional ISAC over one random channel realization when the communication throughput is 13.5 bit/s/Hz. We set the spatial correlation factor as $t = 0$. Similarly, both underloaded ($K = 2$) and overloaded regimes ($K = 6$) are considered. It can be observed that in the underloaded regime, both NOMA-empowered ISAC and the conventional ISAC can achieve the dominant peak of the transmit beampattern in the directions of interest, i.e., $-40°$ and $40°$. In the overloaded regime, the dominant peaks can still be achieved by the proposed NOMA-empowered ISAC, while the conventional ISAC experiences severe power leakage in the undesired directions, leading to

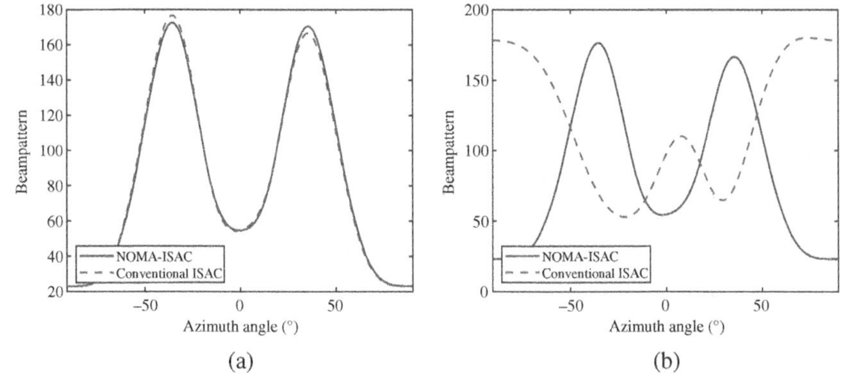

Figure 9.7 Obtained transmit beampattern by different schemes when the communication throughput is 13.5 bit/s/Hz. (a) Underloaded $N = 4, K = 2, t = 0$. (b) Overloaded $N = 4, K = 6, t = 0$.

significant sensing performance degradation. These results further emphasize the importance of NOMA in terms of guaranteeing the sensing performance when the ISAC system is overloaded and also verify Remark 9.1.

9.7 Conclusions

In this chapter, the development of ISAC has been investigated with NOMA from the MA perspective. Following the developing trend of non-orthogonal ISAC, the fundamental downlink and uplink ISAC models were introduced and the corresponding design challenges caused by the inter-functionality interference were also identified. To address these issues, novel designs, namely NOMA-empowered downlink ISAC and semi-NOMA-based uplink ISAC, were proposed. Moreover, the case study of NOMA-empowered downlink ISAC was provided, where a beamforming optimization problem was formulated to obtain the communication–sensing trade-off. This problem was solved by the proposed double-layer penalty-based algorithm. Numerical results indicated that the NOMA-empowered downlink ISAC framework outperforms the conventional one when the spatial DoFs are insufficient and can provide high quality communication and sensing functions simultaneously in the overloaded regime.

References

3GPP. *Further Advancements for E-UTRA Physical Layer Aspects (Release 9).* document 3GPP TS 36.814, Mar. 2010.

Xu Chen, Zhiyong Feng, Zhiqing Wei, Ping Zhang, and Xin Yuan. Code-division OFDM joint communication and sensing system for 6G machine-type communication. *IEEE Internet of Things Journal*, 8(15):12093–12105, 2021. doi: 10.1109/JIOT.2021.3060858.

Haocheng Hua, Jie Xu, and Tony Xiao Han. Optimal transmit beamforming for integrated sensing and communication. *arXiv preprint arXiv:2104.11871*, 2021.

J P Kermoal, L Schumacher, K I Pedersen, P E Mogensen, and F Frederiksen. A stochastic MIMO radio channel model with experimental validation. *IEEE Journal on Selected Areas in Communications*, 20(6):1211–1226, 2002. doi: 10.1109/JSAC.2002.801223.

Matti Latva-aho and Kari Leppanen. Key drivers and research challenges for 6G ubiquitous wireless intelligence. 2019.

Yuanwei Liu, Zhijin Qin, Maged Elkashlan, Zhiguo Ding, Arumugam Nallanathan, and Lajos Hanzo. Nonorthogonal multiple access for 5G and beyond. *Proceedings of the IEEE*, 105(12):2347–2381, 2017. doi: 10.1109/JPROC.2017.2768666.

Fan Liu, Longfei Zhou, Christos Masouros, Ang Li, Wu Luo, and Athina Petropulu. Toward dual-functional radar-communication systems: Optimal waveform design. *IEEE Transactions on Signal Processing*, 66(16):4264–4279, 2018. doi: 10.1109/TSP.2018.2847648.

Xiang Liu, Tianyao Huang, Nir Shlezinger, Yimin Liu, Jie Zhou, and Yonina C Eldar. Joint transmit beamforming for multiuser MIMO communications and MIMO radar. *IEEE Transactions on Signal Processing*, 68:3929–3944, 2020. doi: 10.1109/TSP.2020.3004739.

Fan Liu, Yuanhao Cui, Christos Masouros, Jie Xu, Tony Xiao Han, Yonina C Eldar, and Stefano Buzzi. Integrated sensing and communications: Toward dual-functional wireless networks for 6G and beyond. *IEEE Journal on Selected Areas in Communications*, 40(6):1728–1767, 2022a. doi: 10.1109/JSAC.2022.3156632.

Yuanwei Liu, Shuowen Zhang, Xidong Mu, Zhiguo Ding, Robert Schober, Naofal Al-Dhahir, Ekram Hossain, and Xuemin Shen. Evolution of NOMA toward next generation multiple access (NGMA) for 6G. *IEEE Journal on Selected Areas in Communications*, 40(4):1037–1071, 2022b. doi: 10.1109/JSAC.2022.3145234.

Zhi-Quan Luo, Wing-kin Ma, Anthony Man-cho So, Yinyu Ye, and Shuzhong Zhang. Semidefinite relaxation of quadratic optimization problems. *IEEE Signal Processing Magazine*, 27(3):20–34, 2010. doi: 10.1109/MSP.2010.936019.

Chenguang Shi, Fei Wang, Mathini Sellathurai, Jianjiang Zhou, and Sana Salous. Power minimization-based robust OFDM radar waveform design for radar and communication systems in coexistence. *IEEE Transactions on Signal Processing*, 66(5):1316–1330, 2018. doi: 10.1109/TSP.2017.2770086.

Petre Stoica, Jian Li, and Yao Xie. On probing signal design for MIMO radar. *IEEE Transactions on Signal Processing*, 55(8):4151–4161, 2007. doi: 10.1109/TSP.2007.894398.

Ying Sun, Prabhu Babu, and Daniel P Palomar. Majorization-minimization algorithms in signal processing, communications, and machine learning. *IEEE Transactions on Signal Processing*, 65(3):794–816, 2017. doi: 10.1109/TSP.2016.2601299.

Danny Kai Pin Tan, Jia He, Yanchun Li, Alireza Bayesteh, Yan Chen, Peiying Zhu, and Wen Tong. Integrated sensing and communication in 6G: Motivations, use cases, requirements, challenges and future directions. In *2021 1st IEEE International Online Symposium on Joint Communications & Sensing (JC&S)*, pages 1–6, 2021. doi: 10.1109/JCS52304.2021.9376324.

J Andrew Zhang, Fan Liu, Christos Masouros, Robert W. Heath, Zhiyong Feng, Le Zheng, and Athina Petropulu. An overview of signal processing techniques for joint communication and radar sensing. *IEEE Journal of Selected Topics in Signal Processing*, 15(6):1295–1315, 2021. doi: 10.1109/JSTSP.2021.3113120.

J Andrew Zhang, Md Lushanur Rahman, Kai Wu, Xiaojing Huang, Y Jay Guo, Shanzhi Chen, and Jinhong Yuan. Enabling joint communication and radar sensing in mobile networks—a survey. *IEEE Communications Surveys & Tutorials*, 24(1): 306–345, 2022a. doi: 10.1109/COMST.2021.3122519.

Qixun Zhang, Hongzhuo Sun, Xinye Gao, Xinna Wang, and Zhiyong Feng. Time-division ISAC enabled connected automated vehicles cooperation algorithm design and performance evaluation. *IEEE Journal on Selected Areas in Communications*, 40(7):2206–2218, 2022b. doi: 10.1109/JSAC.2022.3155506.

Chao Zhang, Wenqiang Yi, Yuanwei Liu, and Lajos Hanzo. Semi-integrated-sensing-and-communication (semi-ISaC): From OMA to NOMA. *IEEE Transactions on Communications*, 71(4):1878–1893, 2023. doi: 10.1109/TCOMM.2023.3241940.

Part II

Massive Access for NGMA

10

Capacity of Many-Access Channels

Lina Liu and Dongning Guo

Department of Electrical and Computer Engineering, Robert R. McCormick School of Engineering and Applied Science, Northwestern University, Evanston, IL, USA

10.1 Introduction

Massive machine-type communication (mMTC) is a type of communication service that offers scalable and efficient connectivity for a vast number of users. mMTC is particularly useful in the context of the Internet of Things (IoT) and the Internet of Everything (IoE). mMTC has some unique features, including the support of a large number of users per cell, small payload size per user, and sporadic data transmission. Additionally, mMTC users are often designed to have low power and low complexity, which makes them suitable for use cases with moderate data reliability and latency requirements.

This chapter focuses on uplink mMTC scenarios, where a large number of users in a cell may transmit data to a common receiver or base station. Traditional multiaccess schemes like frequency division multiple access (FDMA), time division multiple access (TDMA), code division multiple access (CDMA), and orthogonal frequency division multiple access (OFDMA) share the medium by partitioning in time, frequency or other signal dimensions, which helps to control interference through coordination. They were originally developed for voice and data communication, where transmissions are typically persistent over many slots, and a relatively small number of users share the radio resources in a cell at any point in time. These methods do not scale well for massive access. For instance, to negotiate a massive TDMA or FDMA schedule can result in a prohibitive latency overhead. A CDMA scheme cannot assign nearly orthogonal sequences to all users and would suffer from significant multiaccess interference when a large number of users are simultaneously active.

In contrast to the preceding coordinated approaches, the ALOHA protocol employs a contention-based method of sharing the medium. To recover from

Next Generation Multiple Access, First Edition.
Edited by Yuanwei Liu, Liang Liu, Zhiguo Ding, and Xuemin Shen.
© 2024 The Institute of Electrical and Electronics Engineers, Inc. Published 2024 by John Wiley & Sons, Inc.

collision, data is retransmitted after a random backoff. Only a small fraction of the time resources can be successfully used to communicate new data.

To design an efficient and scalable scheme for the uplink, we consider a baseband model of the system and draw tools from information theory to help understand the fundamental limits of the system, both in terms of the information capacity and the energy efficiency.

The multiaccess channel had its first coding theorem in Ahlswede (1971) and Liao (1972), and subsequently error exponents in Slepian and Wolf (1973) and Gallager (1985). These models typically assume a fixed (usually small) number of users and study fundamental limits as the coding block length approaches infinity. Alternative models with a very large number of users have been studied in, e.g., Gupta and Kumar (2000) and Guo and Verdú (2005). In such large-system analysis, the blocklength is still sent to infinity before the number of users is sent to infinity. These limits are not directly applicable to mMTC systems in which the number of users is comparable to or even exceed the blocklength.

Additionally, a typical mMTC device transmits small amounts of data intermittently. The unknown device activities are a major source of uncertainty for the receiver. Other practical issues that arise in mMTC applications, like finite payload size and energy concerns of users also pose questions beyond the scope of existing theoretical results.

In this chapter, we introduce the many-user information-theoretic regime, where the number of active users in a time slot and the coding blocklength simultaneously tend to infinity in some manner. A first such model was the many-access channel (MnAC) introduced in Chen and Guo (2013). This channel model, which is also referred to as the massive-access channel in literature, has since been studied by many authors (Chen et al., 2017, Polyanskiy, 2017, Wei et al., 2019, Robin and Erkip, 2021, Ravi and Koch, 2021, Gao et al., 2022). Until a penultimate result for arbitrary finite blocklength and number of users and given finite powers is obtained, taking various limits in the system parameters can yield analytical results that provide a good amount of practically relevant insights.

A related many-user setting, referred to as unsourced random access was studied in Polyanskiy (2017) and many follow-up works. In this model, all users employ a common codebook and the goal of the decoder is to put out all or most messages transmitted by the users irrespective of their sources.

The remainder of this chapter is organized as follows: In Section 10.2, we present a unified model for the MnAC framework. In Section 10.3, we investigate the capacity of MnAC with random user activities. In Section 10.4, we discuss the energy efficiency of MnAC by studying the minimum energy per bit and the capacity per unit-energy. In Section 10.5, we discuss the implications for the design and implementation of such systems, and propose open problems and future research directions.

10.2 The Many-Access Channel Model

Consider the MnAC framework with a single receiver and multiple transmitters (users). We assume a fully synchronous system and discuss asynchronous transmissions in Section 10.5. Let n denote the blocklength, i.e., the number of channel uses in a frame of transmission. The total number of users is denoted as ℓ_n, which is a function of n. In the baseband model, the received symbols in a block form a column vector of length n, which is represented by

$$\mathbf{Y} = \sum_{k=1}^{\ell_n} H_k \mathbf{S}_k(w_k) + \mathbf{Z}, \tag{10.1}$$

where w_k denotes the message of user k, $\mathbf{S}_k(w_k)$ denotes the corresponding n-symbol codeword of user k, H_k denotes the channel coefficient, and $\mathbf{Z}$ denotes additive white Gaussian noise (AWGN). Throughout the chapter, we use lower case letters (such as x) to denote a scalar, bold lower case letters ($\mathbf{x}$) to denote a column vector. The corresponding uppercase letters X and $\mathbf{X}$ denote the corresponding random scalar and random vector, respectively.

Since only the relative strengths of the signals and noise matter, we normalize all powers such that each element of the noise vector $\mathbf{Z}$ in (10.1) has unit variance. Specifically, if (10.1) represents a real-valued model, then the n entries of $\mathbf{Z}$ are independent and identically distributed (i.i.d.) standard normal random variables; if the model is complex-valued, then the n entries of $\mathbf{Z}$ are i.i.d. circularly symmetric complex Gaussian (CSCG) random variables with unit variance, whose distribution is denoted as $\mathcal{CN}(0, 1)$.

We assume each user has a message set of size M, so $w_k \in \{0, 1, \dots, M\}$. We use $w_k = 0$ to denote that user k is inactive and in this case, it transmits an all-zero codeword $\mathbf{s}_k(0) = \mathbf{0}$. We assume each user has one's own codebook, where each codeword of user k satisfies the power constraint

$$\frac{1}{n}\sum_{i=1}^{n} |s_{k,i}(w_k)|^2 \le P_k, \tag{10.2}$$

with P_k being user k's normalized power constraint.

An important measure of energy efficiency is the *energy per bit*. It often takes the normalized form of E_b/N_0, which is actually the (unitless) ratio of the energy per bit (E_b) and the noise power spectral density (N_0). It is important to relate E_b/N_0 to the aforementioned normalized power. In general, $\log_2 M$ bits are sent using n channel symbols. Without loss of generality, we assume the power constraint (10.2) is met with equality. With matched filtering and sampling at symbol rate, the signal energy per user symbol is $(\log_2 M)E_b/n$ where the variance of the additive noise is $N_0/2$ per dimension. In light of the normalized baseband model (10.1), we can express E_b/N_0 in terms of the power constraint (see also [Proakis et al., 1994, Chapter 7.5]):

$$\frac{E_b}{N_0} = \begin{cases} \frac{nP_k}{2\log_2 M}, & \text{if the model is real-valued,} \\ \frac{nP_k}{\log_2 M}, & \text{if the model is complex-valued.} \end{cases} \tag{10.3}$$

In subsequent sections of this chapter, we follow references such as (Polyanskiy, 2018) to refer to E_b/N_0 as the energy per bit with the understanding that it is normalized. The physical energy per bit in a given system is then easily obtained by multiplying the physical noise level N_0 in that system.

An equivalent matrix representation of the Gaussian MnAC (10.1) is often used in the literature. Let us assemble all codewords of user k, except for the first all-zero codeword, as a matrix $\underline{\mathbf{S}}_k = [\mathbf{S}_k(1), \dots, \mathbf{S}_k(M)]$ of dimension $n \times M$. Throughout the chapter, we use an underlined bold font to represent matrices. We use $\underline{\mathbf{S}} = [\underline{\mathbf{S}}_1, \dots, \underline{\mathbf{S}}_{\ell_n}]$ to denote the concatenation of the codebooks of all users, which is an $n \times (M\ell_n)$ matrix. The signal of user k can be represented as $\underline{\mathbf{S}}_k\mathbf{X}_k$, where the M-vector $\mathbf{X}_k = \mathbf{0}$ if user k is inactive, and $\mathbf{X}_k$ has a single nonzero entry at the w_kth element if user k's message is w_k. The channel model (10.1) can then be expressed as:

$$\mathbf{Y} = \underline{\mathbf{S}}\mathbf{U} + \mathbf{Z}, \tag{10.4}$$

where $\mathbf{U} = [\mathbf{U}_1^T, \mathbf{U}_2^T, \dots, \mathbf{U}_{\ell_n}^T]^T$, with $\mathbf{U}_k = H_k\mathbf{X}_k$, is an $M\ell_n$-dimensional vector representing the signals from all users.

The equivalent model (10.4) is closely related to compressed sensing problems (Donoho, 2006, Zhang et al., 2013, Aksoylar et al., 2016), albeit with a special sparsity pattern. This model also pushes the envelop of compressed sensing as the dimensions of $\underline{\mathbf{S}}$ and $\mathbf{U}$ can be extremely large, e.g., $M = 2^{100}$ if each user has 100 bits to transmit. Information-theoretic limits of exact support recovery was considered in Wainwright (2009), and stronger necessary and sufficient conditions have been derived subsequently (Fletcher et al., 2009, Wang et al., 2010, Rad, 2011).

10.3 Capacity of the MnAC

This section examines the MnAC model under the assumption that each active user is received at a fixed power as the blocklength tends to infinity, where different users may be received at different powers. For convenience, we focus on the real-valued channel model. We assume a block-fading model with known channel state information at the receiver (CSIR). Let n denote the blocklength. Let P_k denote user k's received signal-to-noise power ratio (SNR), which represents the combined effect of the transmit power, path loss, fading, and all other gains leading to the baseband model. Since only the product of all those coefficients matter, we can assume $H_k = 1, \forall k \in \{1, \dots, \ell_n\}$ without loss of generality. Under this assumption, $\mathbf{S}_k(w_k)$ denotes user k's codeword corresponding to the message $w_k \in \{0, 1, \dots, M\}$, which is subject to the power constraint (10.2).

We further assume that each user accesses the channel independently with an identical probability α_n during any given block. If user k is inactive, it is considered as transmitting an all-zero codeword with $w_k = 0$, and hence $\mathbf{s}_k(0) = \mathbf{0}$ regardless of the received SNR power ratio. On the other hand, if user k is active and transmits message $w_k > 0$, $\mathbf{S}_k(w_k)$ represents the corresponding codeword. The model (10.1) can be simplified to

$$\mathbf{Y} = \sum_{k=1}^{\ell_n} \mathbf{S}_k(w_k) + \mathbf{Z}. \tag{10.5}$$

In the equivalent model (10.4), the vector $\mathbf{U}$ reduces to $\mathbf{X}$ with

$$\mathbf{X}_k = \begin{cases} \mathbf{0}, & \text{with probability } 1 - \alpha_n, \\ \mathbf{e}_m, & \text{with probability } \frac{\alpha_n}{M}, m = 1, \dots, M, \end{cases} \tag{10.6}$$

where $\mathbf{e}_m$ is the binary column M-vector with a single 1 at the mth entry.

In the following, we present a sharp characterization of the capacity of Gaussian MnACs as well as the user identification cost. The capacity of the conventional multiaccess channel can be established using the fact that the user codewords and the received signal are jointly typical with high probability as the blocklength goes to infinity. This argument does not directly generalize to models where the number of users tends to infinity with the blocklength. While the empirical joint entropy of every fixed subset of user codewords and received signal converges to the corresponding joint entropy due to the law of large numbers, the asymptotic equipartition property is not guaranteed due to the exponential increase in the number of such subsets with the number of users (Chen et al., 2017).

10.3.1 The Equal-Power Case

In this section, we consider a base case where users have equal received power, i.e., $P_k = P$ for all $k \in 1, \dots, \ell_n$, and the same activity pattern α_n.

Definition 10.1 Let $\mathcal{S}_k$ and $\mathcal{Y}$ denote the input alphabet of user k and the output alphabet of the MnAC with random user activities, respectively. An (M, n) symmetric code with power constraint P for the MnAC channel $(\mathcal{S}_1 \times \mathcal{S}_2 \times \cdots \times \mathcal{S}_{\ell_n}, p_{Y|S_1,\dots,S_{\ell_n}}, \mathcal{Y})$ consists of the following mappings:

(1) The encoding functions $\mathcal{F}_k : \{0, 1, \dots, M\} \rightarrow \mathcal{S}_k^n$ for every user $k \in \{1, \dots, \ell_n\}$, which maps any message w to the codeword $\mathbf{s}_k(w) = [s_{k,1}(w), \dots, s_{k,n}(w)]^T$.

In particular, $\mathbf{s}_k(0) = \mathbf{0}$, for every k. Every codeword $\mathbf{s}_k(w)$ satisfies the power constraint:

$$\frac{1}{n}\sum_{i=1}^{n} s_{k,i}^2(w) \leq P. \tag{10.7}$$

(2) Decoding function $\mathcal{G} : \mathcal{Y}^n \rightarrow \{0, 1, \ldots, M\}^{\ell_n}$, which is a deterministic rule assigning a decision on the messages to each possible received vector.

The joint probability of error (JPE) of the (M, n) code is defined as:

$$\mathsf{P}_{e,J}^{(n)} = \mathsf{P}\{\mathcal{G}(\mathbf{Y}) \neq (W_1, \ldots, W_{\ell_n})\}, \tag{10.8}$$

where the messages $W_1, \ldots, W_{\ell_n}$ are independent, and for $k \in \{1, \ldots, \ell_n\}$, the message's distribution is

$$\mathsf{P}\{W_k = w\} = \begin{cases} 1 - \alpha_n, & w = 0, \\ \frac{\alpha_n}{M}, & w \in \{1, \ldots, M\}. \end{cases} \tag{10.9}$$

Definition 10.2 ***(Asymptotically Achievable Message Length)*** We say a positive nondecreasing sequence of message lengths $\{v(n)\}_{n=1}^{\infty}$ (in nats), or simply, $v(\cdot)$, is asymptotically achievable for the MnAC if there exists a sequence of $(\lceil \exp(v(n)) \rceil, n)$ codes according to Definition 10.1 such that the average error probability $\mathsf{P}_{e,J}^{(n)}$ given by (10.8) vanishes as $n \to \infty$.

Definition 10.3 ***(Symmetric Message-Length Capacity)*** For the MnAC channel described by (10.5), a positive nondecreasing function $B(n)$ of the blocklength n is said to be a symmetric message-length capacity of the MnAC channel if, for any $0 < \epsilon < 1$, $(1 - \epsilon)B(n)$ is an asymptotically achievable message length according to Definition 10.2, whereas $(1 + \epsilon)B(n)$ is not asymptotically achievable.

Here, a new notion of message length is introduced to characterize the channel capacity, which calculates the information (in nats) transmitted over the entire blocklength. For the special case of a (conventional) multiaccess channel, i.e., with fixed ℓ_n and $\alpha_n = 1$, the symmetric capacity $B(n)$ in Definition 10.3 is asymptotically linear in n, so that the asymptotic rate $\lim_{n\to\infty} B(n)/n$ is equal to the symmetric capacity of the multiaccess channel (in, e.g., bits per channel use). However, in the MnAC framework with random user activities, $B(n)/n$ is a function of n in general and may vanish in some cases. Therefore, the conventional notion of code rate is no longer an appropriate metric for evaluating the performance of such communication systems. By adopting the message length capacity notion, we can demonstrate that under certain conditions, we can still transmit an arbitrarily large amount of information over the entire blocklength. The ill-suited traditional notion of rate in the many-user regime is noted (for the Gaussian multiaccess channel) in Cover and Thomas [2006, pp. 546–547], "when the total number of senders is very large, so that there is a lot of interference, we can still send a total amount of information that is arbitrary large even though the rate per individual sender goes to 0." For the rest of this section, we use the term "capacity" specifically to refer to the message length capacity, as opposed to the conventional capacity.

Also note that the definition of symmetric capacity does not differentiate between all but the dominating term. As a result, the capacity expression is not unique. If $B(n)$ is a capacity, then any expression of the form $B(n) + o(B(n))$ is also a capacity.

Theorem 10.1 ***(Symmetric Capacity of the Gaussian MnAC)*** Suppose ℓ_n is nondecreasing with n and

$$\lim_{n\to\infty} \alpha_n = \alpha \in [0,1]. \tag{10.10}$$

Denote the average number of active users as

$$k_n = \alpha_n \ell_n. \tag{10.11}$$

Then, the symmetric message-length capacity $B(n)$ of the Gaussian MnAC (10.5), with every user's SNR constrained by P, is characterized as:

Case 1: ℓ_n and k_n are both unbounded, $k_n = O(n)$, and

$$\ell_n e^{-\delta k_n} \to 0 \tag{10.12}$$

for all $\delta > 0$: Let θ denote the limit of

$$\theta_n = \frac{2\ell_n h(\alpha_n)}{n \ln(1 + k_n P)}, \tag{10.13}$$

which may be ∞, where $h(p) = -p \ln p - (1-p)\ln(1-p)$ denotes the binary entropy function for $0 \le p \le 1$ with natural logarithm.

- If $\theta < 1$, then

$$B(n) = \frac{n}{2k_n} \ln(1 + k_n P) - \frac{h(\alpha_n)}{\alpha_n}. \tag{10.14}$$

- If $\theta > 1$, then a user cannot send even 1 bit reliably.
- If $\theta = 1$, then the message length $\frac{\epsilon n}{2k_n} \ln(1 + k_n P)$ is not achievable for any $\epsilon > 0$.

Case 2: ℓ_n is unbounded and k_n is bounded: $B(n)$ must be sublinear, i.e., the message length ϵn is not achievable for any $\epsilon > 0$.

Case 3: ℓ_n is bounded, i.e., $\ell_n = \ell < \infty$ for large enough n:

$$B(n) = \begin{cases} \frac{n}{2}\ln(1+P), & \text{if } \alpha = 0, \\ \frac{n}{2\ell}\ln(1+\ell P), & \text{if } \alpha > 0. \end{cases} \tag{10.15}$$

When ℓ_n is finite, the capacity result in Case 3 can be explained utilizing conventional multiaccess channel theory. The more significant finding is that Cases 1 and 2 can be achieved through a two-stage method involving user identification in the first stage and message decoding in the second stage.

Figure 10.1 illustrates the message length $B(n)$ (in bits) given by (10.14) with $P = 10$ (i.e., the SNR is 10 dB), $k_n = n/4$, and different scalings of the user number n. Evidently, $B(n)$ is sub-linear in n, and it depends on the scaling of k_n and n. Notably, when n grows too quickly (e.g., $\ell_n = n^3$), a user cannot transmit a single bit reliably.

As an important by-product in the proof of Theorem 10.1, the fundamental limits of user identification (without data transmission), where every user is active with certain probability, and the receiver aims to detect the set of active users, can be derived.

Let the total number of users be denoted as ℓ, and let other parameters depend on ℓ. To be specific, the probability of a user being active is denoted as α_ℓ, and the average number of active users is denoted as $k_\ell = \alpha_\ell \ell$. Suppose n_0 symbols

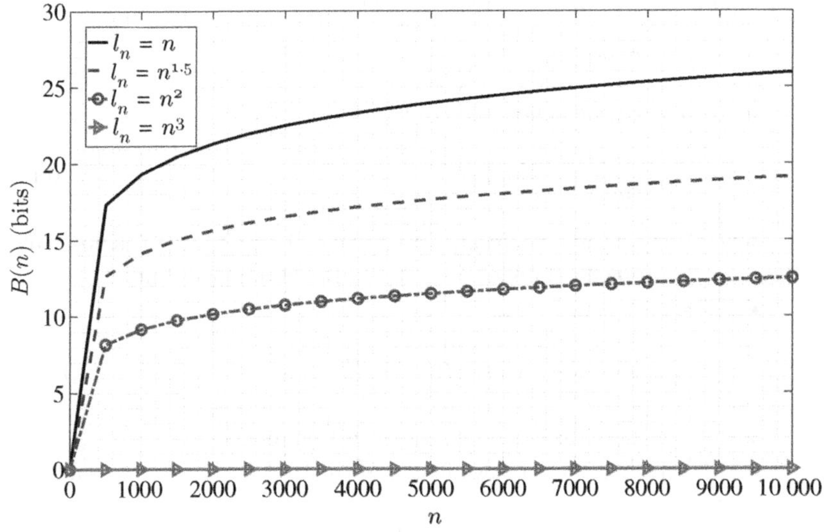

Figure 10.1 Plot of $B(n)$ given by (10.14), where $P = 10, k_n = n/4$.

are used for user identification purpose. We can modify the channel model in (10.4) to represent this problem. Since only signatures are transmitted, $M = 1$ in this context. Let $\underline{\mathbf{S}} = [\mathbf{S}_1, \ldots, \mathbf{S}_\ell]$, where $\mathbf{S}_j \in \mathbb{R}^{n_0}$ is the signature of user j. We also let $\mathbf{U} = \mathbf{X} \in \mathbb{R}^\ell$ be a random vector, which consists of i.i.d. Bernoulli entries with mean α_ℓ, and $\mathbf{Y}, \mathbf{Z} \in \mathbb{R}^\ell$.

The following theorem gives a sharp characterization of how many channel uses n_0 are needed for reliable identification.

Definition 10.4 ***(Minimum User Identification Cost)*** We say the identification is erroneous in case of any miss or false alarm. The minimum user identification cost is said to be $n(\ell)$ if $n(\ell) > 0$ and for every $0 < \epsilon < 1$, there exists a signature code of length $n_0 = (1+\epsilon)n(\ell)$ such that the probability of erroneous identification vanishes as $\ell \to \infty$, whereas the error probability is strictly bounded away from zero if $n_0 = (1-\epsilon)n(\ell)$.

Theorem 10.2 ***(Minimum Identification Cost Through the Gaussian MnAC)*** Let the total number of users be ℓ, where each user is active with the same probability. Suppose the average number of active users k_ℓ satisfies

$$\lim_{n\to\infty} \ell e^{-\delta k_\ell} = 0 \tag{10.16}$$

for all $\delta > 0$. Let

$$n(\ell) = \frac{\ell h(k_\ell/\ell)}{\frac{1}{2}\ln(1+k_\ell P)}. \tag{10.17}$$

The asymptotic identification cost is characterized as follows:

Case 1: As $k_\ell \to \infty$, $n(\ell)/k_\ell$ converges to a strictly positive number or goes to $+\infty$: The minimum user identification cost is $n(\ell)$.

Case 2: $\lim_{k_\ell\to\infty} n(\ell)/k_\ell = 0$: A signature of length $n_0 = \epsilon k_\ell$ yields vanishing error probability for any $\epsilon > 0$; on the other hand, if $n_0 < (1-\epsilon)n(\ell)$, then the identification error cannot vanish as $\ell \to \infty$.

The length of the signature matches the capacity penalty due to user activity uncertainty. Note that (10.16) implies $k_\ell \to \infty$ as $\ell \to \infty$. In the special case where $k_\ell = \lceil \ell^{1/d} \rceil$ for some $d > 1$, the minimum user identification cost is $n(\ell) = 2(d-1)k_\ell + o(k_\ell)$, which is linear in the number of active users. The minimum cost function $n(\ell)$ is illustrated in Fig. 10.2.

While a two-step approach consisting of active user detection followed by communication is asymptotically optimal when the payload size of each user grows to infinity logarithmically with the number of active users, using a partial resource to perform activity detection separately may be suboptimal when the payload of each user is finite (Wu et al., 2020).

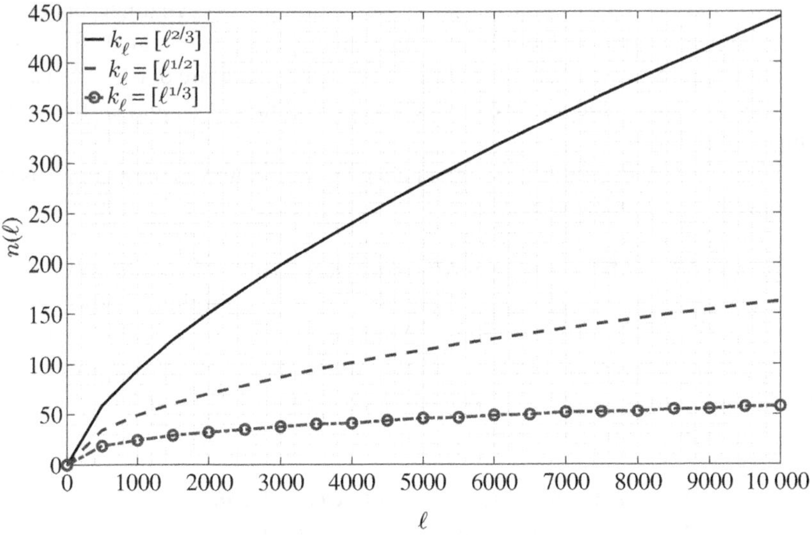

Figure 10.2 Plot of $n(\ell)$ specified in Theorem 10.2, where $P = 10$, i.e., SNR = 10 dB.

10.3.2 Heterogeneous Powers and Fading

In this section, we will generalize the characterization of the capacity region to include cases where users are received at different powers and/or have different activity patterns. Suppose ℓ_n users can be divided into a finite number of J groups, where group j consists of $\beta^{(j)}\ell_n$ users with $\sum_{i=1}^{J} \beta^{(j)} = 1$. We further assume every user in group j has the same received power constraint $P^{(j)}$ and transmits with probability $\alpha_n^{(j)}$. We refer to such MnAC with heterogeneous channel gains and activity patterns as the configuration $\left(\{\alpha^{(j)}\}, \{\beta^{(j)}\}, \{P^{(j)}\}, \ell_n\right)$. The error probability is defined as the probability that the receiver incorrectly detects the message of any user in the system. The problem is to determine the maximum achievable message length for users in each group such that the average error probability vanishes.

Definition 10.5 ***(Asymptotically Achievable Message Length Tuple)*** Consider a MnAC of configuration $\left(\{\alpha^{(j)}\}, \{\beta^{(j)}\}, \{P^{(j)}\}, \ell_n\right)$. A sequence of $\left(\lceil \exp\left(v^{(1)}(n)\right)\rceil, \ldots, \lceil \exp\left(v^{(J)}(n)\right)\rceil, n\right)$ code for this configuration consists of a $\left(\lceil \exp\left(v^{(j)}(n)\right)\rceil, n\right)$ symmetry code for every user in group j according to Definition 10.1, $j = 1, \ldots, J$. We say a message length tuple $\left(v^{(1)}(n), \ldots, v^{(J)}(n)\right)$ is asymptotically achievable if there exists a sequence of $\left(\lceil \exp\left(v^{(1)}(n)\right)\rceil, \ldots, \lceil \exp\left(v^{(J)}(n)\right)\rceil, n\right)$ codes such that the average error probability vanishes as $n \to \infty$.

Definition 10.6 ***(Capacity Region of the MnAC)*** Consider a MnAC of configuration $\left(\left\{\alpha^{(j)}\right\},\left\{\beta^{(j)}\right\},\left\{P^{(j)}\right\},\ell_n\right)$. The capacity region is the set of asymptotically achievable message length tuples. In particular, for every $\left(B^{(1)}(n),\ldots,B^{(J)}(n)\right)$ in the capacity region, if the users transmit with message length tuple $\left((1-\epsilon)B^{(1)}(n),\ldots,(1-\epsilon)B^{(J)}(n)\right)$, the average error probability vanishes as $n\to\infty$. If users transmit according to a message-length tuple outside the capacity region, then the communication cannot be reliable.

Theorem 10.3 Consider a MnAC of configuration $\left(\left\{\alpha^{(j)}\right\},\left\{\beta^{(j)}\right\},\left\{P^{(j)}\right\},\ell_n\right)$. Suppose $\ell_n\to\infty$ and for every $j\in\{1,\ldots,J\},\alpha_n^{(j)}\to\alpha^{(j)}\in[0,1]$. Let the average number of active users in group j be $k_n^{(j)}=\alpha_n^{(j)}\beta^{(j)}\ell_n=O(n)$, such that $\ell_n e^{-\delta k_n^{(j)}}\to 0$ for all $\delta>0$ and $j=1,\ldots,J$. Let $\theta_n^{(j)}$ be defined as

$$\theta_n^{(j)}=\frac{2\beta^{(j)}\ell_n h\left(\alpha_n^{(j)}\right)}{n\ln k_n^{(j)}},\tag{10.18}$$

and let $\theta^{(j)}$ denotes its limit. Suppose $\ln k_n^{(j_1)}/\ln k_n^{(j_2)}\to 1$ for any $j_1,j_2\in\{1,\ldots,J\}$. If $\sum_{j=1}^{J}\theta^{(j)}<1$, then the message length capacity region is characterized as

$$\sum_{j=1}^{J}k_n^{(j)}B^{(j)}(n)\le\frac{n}{2}\ln\left(\sum_{j=1}^{J}k_n^{(j)}\right)-\sum_{j=1}^{J}\beta^{(j)}\ell_n h\left(\alpha_n^{(j)}\right).\tag{10.19}$$

If $\sum_{j=1}^{J}\theta^{(j)}>1$, then some user cannot transmit a single bit reliably.

As far as the asymptotic message lengths are concerned, the impact of the received powers is inconsequential. This is in fact because $\ln\left(1+k_n^{(j)}P^{(j)}\right)=\ln\left(k_n^{(j)}\right)+o\left(k_n^{(j)}\right)$ as $k_n^{(j)}$ becomes large. Additionally, the only restriction on the messages is their weighted average, which differs from the classical multiaccess channel, where a separate upper bound typically applies to the sum rate of each subset of users.

As an achievable scheme, we can detect active users in each group and their transmitted messages in a time-division manner. Particularly, in the first stage, we let users in group 1 transmit the signatures before group 2, and so on. The signature length transmitted by users in group j is $n_0^{(j)}$. In the second stage, we let each group share the remaining time resource $n-\sum_{j=1}^{J}n_0^{(j)}$. users in group 1 transmit their message-bearing codewords before group 2, and so on. The time resource allocated to group j in the second stage is $\phi_j\left(n-\sum_{j=1}^{J}n_0^{(j)}\right)$, where $\phi_j\ge 0$ and $\sum_{j=1}^{J}\phi_j=1$. According to the group order, the receiver first identifies active users and then decodes the transmitted messages.

10.4 Energy Efficiency of the MnAC

Because many MTC devices are battery-driven, their energy efficiency is a crucial concern. In this section, we focus on two key energy-efficiency metrics: the minimum energy per bit and the capacity per unit-energy.

Under the assumption of fixed power regardless of blocklength, the MnAC capacity is given by Theorem 10.1 under the JPE criterion. Case 1 of the theorem is the most interesting, where the number of active users increases without bound. By (10.3), the energy per bit can be calculated as

$$\frac{E_b}{N_0} = \frac{nP}{2\log_2 eB(n)} \tag{10.20}$$

where the $\log_2 e$ factor converts nats to bits. If $\alpha_n = 1$, when all active users' identities are known, we have $\ell_n = k_n$, $h(\alpha_n) = 0$ and $\theta = 0$. According to (10.14), the minimum energy per bit is

$$\left(\frac{E_b}{N_0}\right)^* = \frac{k_n P}{\log_2(1 + k_n P)}. \tag{10.21}$$

When P is fixed as a constant and $k_n = O(n)$, $(E_b/N_0)^*$ goes to infinity as $n \to \infty$. This indicates that the system with fixed SNR is not energy efficient.

By allowing power to vary with blocklength and switching to per-user probability of error (PUPE) criterion, it is possible to achieve a finite energy per bit (Kowshik and Polyanskiy, 2021, Kowshik, 2022). We devote Section 10.4.1 to results in this energy-efficient setting.

The capacity per unit-energy, denoted as $\dot{C}$, represents the largest number of bits per unit-energy that can be transmitted reliably over a channel. It is inversely related to the energy per bit, as the rate per unit-energy is the reciprocal of the energy per bit. For a memoryless stationary channel, it has been shown in Verdú (1990) that the capacity per unit-energy is equal to

$$\dot{C} = \sup_{P>0} \frac{C(P)}{P}, \tag{10.22}$$

where $C(P)$ is the largest number of bits per channel use that can be transmitted reliably with average power per symbol not exceeding P. Note that in Verdú (1990) and Ravi and Koch (2021), the noise and power are unnormalized. In this chapter, we adopt the normalized noise and power to maintain consistency, as described in Section 10.2. The unit of $\dot{C}$ in this context is in bits.

In real-valued systems, the capacity per unit-energy for a single-user additive white Gaussian channel with unit variance is $\frac{\log_2 e}{2}$. In Verdú (1990), a two-user Gaussian multiple access channel was also studied, and it was demonstrated that both users can achieve the single-user capacity per unit-energy $\frac{\log_2 e}{2}$ by combining an orthogonal access scheme (through time-sharing) with orthogonal codebooks. This result can be directly generalized to any finite number of users. In

Section 10.4.2, we will demonstrate that it is also possible to achieve a single-user capacity per unit-energy in MnAC. However, the choice of error criteria can affect the results.

10.4.1 Minimum Energy per Bit for Given PUPE

For simplicity, let us assume that all users in an access point's service area are active. Let k_n denote the number of those users. (This is equivalent to letting $\alpha_n = 1$ in the model in Section 10.3, so that $k_n = \ell_n$.) To be consistent with results in the literature, we consider the complex-valued model with Rayleigh fading, so that (10.1) becomes

$$\mathbf{Y} = \sum_{k=1}^{k_n} H_k \mathbf{S}_k(w_k) + \mathbf{Z}, \tag{10.23}$$

where $H_1, H_2, \dots$ and all entries of $\mathbf{Z}$ are i.i.d. CSCG random variables of unit variance. We consider the linear scaling regime $k_n = \mu n$ as $n \to \infty$, where $\mu > 0$ is the user density.

Throughout this section, it is assumed that each user has only a finite number of bits to send. If one insists that almost all user messages be decoded correctly in the linear regime where $k_n = \mu n$, the energy required per bit also tends to infinity (Polyanskiy, 2018). Hence, we focus instead on the PUPE. By sacrificing a small fraction of users, the remaining users can communicate reliably using a small amount of energy per bit.

Definition 10.7 Let $\mathcal{S}_k$ and $\mathcal{Y}$ denote the input alphabet of user k and the output alphabet of the MnAC with fading, respectively. An (M, n, ϵ) code with power constraint P for the MnAC channel $(\mathcal{S}_1 \times \mathcal{S}_2 \times \cdots \times \mathcal{S}_{k_n}, P_{Y|S_1,\dots,S_{k_n}}, \mathcal{Y})$ consists of the following mappings:

(1) The encoding functions $\mathcal{F}_k : \{1, \dots, M\} \mapsto \mathcal{S}_k^n$ for every user $k \in \{1, \dots, k_n\}$, which maps any message w to the codeword $\mathbf{s}_k(w) = [s_{k1}(w), \dots, s_{kn}(w)]^T$. Every codeword $\mathbf{s}_k(w)$ satisfies the power constraint

$$\frac{1}{n}\sum_{i=1}^{n} s_{k,i}^2(w) \le P. \tag{10.24}$$

(2) Decoding function $\mathcal{G} : \mathcal{Y}^n \to \{1, \dots, M\}^{k_n}$, which is a deterministic rule assigning a decision on the messages to each possible received vector.

The average PUPE of the (M, n, ϵ) code satisfies

$$\mathsf{P}_{e,U}^{(n)} = \frac{1}{k_n}\sum_{k=1}^{k_n} \mathsf{P}\left\{W_k \neq (\mathcal{G}(\mathbf{Y}))_k\right\} \le \epsilon, \tag{10.25}$$

where the messages $W_1, \dots, W_{k_n}$ are chosen independently and uniformly, i.e., for $k \in \{1, \dots, k_n\}$, the message's distribution is

$$\mathsf{P}\{W_k = w\} = \frac{1}{M}, \quad w \in \{1, \dots, M\}. \tag{10.26}$$

Let the spectral efficiency be

$$S = \frac{k_n \log_2 M}{n} = \mu \log_2 M. \tag{10.27}$$

Hence, given S and μ, M is fixed. Let $P_{\text{tot}} = k_n P$ denote the total power, and $E = nP$ denote the total energy. The energy per bit can be represented as

$$\mathcal{E} = \frac{E_b}{N_0} = \frac{E}{\log_2 M} = \frac{P_{\text{tot}}}{S}. \tag{10.28}$$

To achieve a finite $\mathcal{E}$, it is necessary for the total power P_{tot} to also be finite, and hence the power P decays as $O(1/n)$.

Definition 10.8 Let $C_k = \{\mathbf{C}_k(1), \mathbf{C}_k(2), \dots, \mathbf{C}_k(M)\}$ be the codebook of user k, of size M. The power constraint is $\|\mathbf{C}_k(w_k)\|^2 \leq nP = \mathcal{E} \log_2 M, \forall k \in \{1, \dots, k_n\}, w_k \in \{1, \dots, M\}$. The collection of codebooks $\{C_k\}$ is called an $(n, M, \epsilon, \mathcal{E}, k_n)$ code if it satisfies the power constraint described before, and the PUPE is smaller than ϵ. Define the following fundamental limit:

$$\mathcal{E}^*(M, \mu, \epsilon) = \lim_{n \to \infty} \inf\{\mathcal{E} : \exists (n, M, \epsilon, \mathcal{E}, \mu n) - \text{code}\}. \tag{10.29}$$

The codebook C_k and the input alphabet $\mathcal{S}_k$ defers in the following way: the transmitted signal is $\mathbf{S}_k(w_k) = \mathbf{C}_k(w_k) 1\{\|\mathbf{C}_k(w_k)\|^2 \leq nP\}$, where $1\{\cdot\}$ is an indicator function.

The achievability bound on $\mathcal{E}$ without channel state information (CSI) can be derived by the asymptotic analysis of the scalar approximate message passing (AMP) algorithm (Bayati and Montanari, 2011) using a spacially coupled codebook design (Hsieh et al., 2021). Recall the compressed sensing model for the MnAC given by (10.4), where the codewords transmitted by the users are indicated by a block-sparse vector $\mathbf{U}$. Specifically, the support of $\mathbf{U}$ is denoted by $\mathbf{X} = [\mathbf{X}_1^T, \mathbf{X}_2^T, \dots, \mathbf{X}_{k_n}^T]^T \in \{0, 1\}^{Mk_n}$, where

$$\sum_{i=1}^{M} X_{k,i} = 1, \quad \text{almost surely.} \tag{10.30}$$

The goal is to obtain an estimate $\hat{\mathbf{X}}$ of $\mathbf{X}$ with a desired PUPE, which can be now represented as

$$\mathsf{P}_{e,U}^{(n)} = \frac{1}{k_n} \sum_{k=1}^{k_n} \mathsf{P}\left\{\hat{\mathbf{X}}_k \neq \mathbf{X}_k\right\}. \tag{10.31}$$

In spacially coupled coding schemes, the matrix of codebooks $\underline{\mathbf{S}} \in \mathbb{C}^{n \times Mk_n}$ is divided into A-by-D equally sized *blocks*, where each block matrix is of size $\frac{n}{A} \times \frac{Mk_n}{D}$. The entries within each block are i.i.d. Gaussian with zero mean and variance specified by the corresponding entry of a base matrix $\underline{\mathbf{Q}} \in \mathbb{C}^{A \times D}$ with nonnegative entries $Q_{a,d}$ such that $\sum_{a=1}^{A} Q_{a,d} = 1$ for all $d \in \{1, \ldots, D\}$. The design matrix $\underline{\mathbf{S}}$ is constructed by replacing each entry of the base matrix $Q_{a,d}$ with an $\frac{n}{A} \times \frac{Mk_n}{D}$ matrix whose entries are i.i.d. $\mathcal{CN}(0, EQ_{a,d}/(n/A))$. Thus, the design matrix $\underline{\mathbf{S}}$ has independent Gaussian entries given by

$$S_{i,j} \sim \mathcal{CN}\left(0, \frac{E}{n/A} Q_{\mathsf{a}(i),\mathsf{d}(j)}\right), \quad \text{for } i \in \{1, \ldots, n\}, \text{ and } j \in \{1, \ldots, Mk_n\}, \tag{10.32}$$

where the operators $\mathsf{a}(\cdot) : \{1, \ldots, n\} \rightarrow \{1, \ldots, A\}$ and $\mathsf{d}(\cdot) : \{1, \ldots, Mk_n\} \rightarrow \{1, \ldots, D\}$ map a particular row or column index to its corresponding *row block* or *column block* index. Consider a class of base matrices called (ϖ, Λ, ρ) base matrices.

Definition 10.9 A (ϖ, Λ, ρ) base matrix $\underline{\mathbf{Q}}$ is described by three parameters: coupling width $\varpi \geq 1$, coupling length $\Lambda \geq 2\varpi - 1$, and $\rho \in [0, 1)$, which specifies the fraction of "energy" allocated to the uncoupled entries in each column. The matrix has $A = \Lambda + \varpi - 1$ rows and $D = \Lambda$ columns, with each column containing ϖ identical non-zero entries in the band-diagonal. The (a, d)th entry of the base matrix, for $a \in \{1, \ldots, A\}, d \in \{1, \ldots, D\}$, is given by

$$Q_{a,d} = \begin{cases} \frac{1-\rho}{\varpi}, & \text{if } d \leq a \leq d + \varpi - 1, \\ \frac{\rho}{\Lambda - 1}, & \text{otherwise.} \end{cases} \tag{10.33}$$

Let $\tilde{\mu} = \frac{A}{D}\mu$ be the effective user density. Since usually $\varpi > 1$, it indicates that $\tilde{\mu} > \mu$.

To recover the support $\mathbf{X}$, we can use the block version of the AMP algorithm, whose analysis is well studied, e.g., Barbier and Krzakala (2017). However, evaluating the performance of block-AMP involves computing M-dimensional integrals, which become intractable for large M (e.g., $M = 2^{100}$). Alternatively, we can use the scalar AMP algorithm, whose asymptotic analysis only requires that the empirical distribution of entries of $\mathbf{U}$ be convergent (Bayati and Montanari, 2011), which is satisfied in the considered MnAC setting. Moreover, the empirical distribution of entries in $\mathbf{U}$ converges to the distribution of a random variable U_0, defined as

$$U_0 = \begin{cases} \mathcal{CN}(0, 1), & \text{with probability } \frac{1}{M}, \\ 0, & \text{with probability } 1 - \frac{1}{M}. \end{cases} \tag{10.34}$$

Thus, we can formulate the scalar problem as a scalar AWGN channel parameterized by the noise variance σ^2, given by

$$V_{\sigma^2} = U_0 + \sigma Z, \tag{10.35}$$

where $Z \sim \mathcal{CN}(0, 1)$ is independent of U_0. Let $X_0 = 1\{U_0 \neq 0\}$. The support estimation is based on the observation V_{σ^2}. In particular,

$$\hat{X}_0(\theta) = 1\{|V_{\sigma^2}|^2 > \theta\}, \tag{10.36}$$

where θ is a threshold parameter that can be optimized. In the AMP algorithm, we track the noise variance of the estimation in the scalar channel (10.35) through $\tau_d^{(t)}$, where the superscript (t) indicates the iteration index.

Define the replica potential as the following:

$$\mathcal{H}(\tau) = (\mu M) I(U_0; V_\tau) + \left(\ln \tau + \frac{1}{\tau E} - 1 \right), \tag{10.37}$$

where (U_0, V_τ) follows the joint distribution of U_0 and V_τ. We then define $\mathcal{M}$ as the maximum of the global minimizers of $\mathcal{H}$:

$$\mathcal{M}(\mu, E, M) = \max \left(\arg \min_{\tau > \frac{1}{E}} \mathcal{H}(\tau) \right). \tag{10.38}$$

The following Lemma holds.

Lemma 10.1 For any (ϖ, Λ, ρ) base matrix $\underline{\mathbf{Q}}$, and for each $d \in \{1, \dots, D\}$, $\tau_d^{(t)}$ is non-increasing in t and converges to a fixed point τ_d^{∞}. Furthermore, for any $\delta > 0$, there exists $\varpi_0 < \infty, \Lambda_0 < \infty$ and $\rho_0 > 0$ such that for all $\varpi > \varpi_0, \Lambda > \Lambda_0$, and $\rho < \rho_0$, the fixed points $\{\tau_d^{\infty} : d \in \{1, \dots, D\}\}$ satisfy

$$\tau_d^{\infty} \leq \tau^{(\infty)}(\tilde{\mu}) + \tilde{\mu} M \delta, \tag{10.39}$$

where $\tau^{(\infty)}(\tilde{\mu}) = \mathcal{M}(\tilde{\mu}, E, M)$.

It is worth noting that the fixed points of the spatially coupled system are at least as good as the uncoupled system (i.e., with $\underline{\mathbf{S}}$ having i.i.d entries) but with user density increased from μ to $\tilde{\mu}$. Taking limits as $\Lambda \to \infty$ and then $\varpi \to \infty$ yields $\tau^{(\infty)}(\tilde{\mu}) \to \tau^{(\infty)}(\mu)$, which is known as threshold saturation. We can utilize Lemma 10.1 to obtain the following achievability bound:

Theorem 10.4 ***(Scalar AMP Achievability, No CSI)*** Fix any $\mu > 0$, $E > 0$ and $M > 1$. For every $\mathcal{E} > \frac{E}{\log_2 M}$, there exist a sequence of $(n, M, \epsilon_n, \mathcal{E}, \mu n)$ (noCSI) codes such that an AMP decoder can achieve

$$\limsup_{n \to \infty} \epsilon_n \leq \pi^* \left(\tau^{(\infty)}(\mu), M \right), \tag{10.40}$$

where

$$\pi^*(\tau, M) = 1 - \frac{1}{1+\tau}\left((M-1)\left(\frac{1}{\tau}+1\right)\right)^{-\tau}, \tag{10.41}$$

$$\tau^{(\infty)}(\mu) \equiv \tau^{(\infty)}(\mu; E, M) = \mathcal{M}(\mu, E, M). \tag{10.42}$$

On the other hand, we can obtain the converse bounds on $\mathcal{E}$.

Theorem 10.5 ***[Converse]*** Let M be the codebook size. Given ϵ and μ, let $S = \mu \log_2 M$. Then assuming that the fading gain $|H|^2$ has a density with $\mathsf{E}[|H|^2] = 1$ and $\mathsf{E}[|H|^4] < \infty$, we have

$$\mathcal{E}^*(M, \mu, \epsilon) \geq \inf \frac{P_{\text{tot}}}{S}, \tag{10.43}$$

where infimum is taken over all $P_{\text{tot}} > 0$ that satisfies both

$$\theta S - \epsilon\mu \log_2(2^{S/\mu} - 1) - \mu h_2(\epsilon) \leq \log_2(1 + P_{\text{tot}}\alpha(1-\theta, 1)), \quad \forall \theta \in [0, 1], \tag{10.44}$$

$$\epsilon \geq 1 - \mathsf{E}\left[Q\left(Q^{-1}\left(\frac{1}{M}\right) - \sqrt{\frac{2P_{\text{tot}}}{\mu}|H|^2}\right)\right], \tag{10.45}$$

where $h_2(p) = -p\log_2 p - (1-p)\log_2(1-p)$ denotes the binary entropy function with a logarithm base of 2 for $0 \leq p \leq 1$, $\alpha(a, b)$ is given by

$$\alpha(a, b) = a\ln(a) - b\ln(b) + b - a, \tag{10.46}$$

obtained according to the distribution of H, and Q is the complementary cumulative distribution function of the standard normal distribution.

To be specific, bound (10.44) is obtained based on the Fano inequality, and bound (10.45) is the converse for a single user quasi-static fading channel.

In addition, bounds tighter than (10.44) can be derived if further assumptions are made on the codebook. For instance, assume that each codebook consists of i.i.d. entries of the form C/k_n where C is sampled from a distribution with zero mean and finite variance (P_{tot}), then the following converse bound can be obtained.

Theorem 10.6 Let M be the codebook size, and let μn users ($\mu < 1$) generate their codebooks independently with each code symbol i.i.d. of the form C/k_n where C is of zero mean and variance P_{tot}. Given ϵ and μ, let $S = \mu \log_2 M$. Then in order for the i.i.d. codebook to achieve PUPE ϵ with high probability, the energy per bit $\mathcal{E}$ should satisfy

$$\mathcal{E} \geq \inf \frac{P_{\text{tot}}}{S}, \tag{10.47}$$

where infimum is taken over all $P_{\text{tot}} > 0$ that satisfies

$$\ln M - \epsilon \ln(M-1) - h_2(\epsilon) \leq \left(M\mathcal{V}\left(\frac{1}{\mu M}, P_{\text{tot}}\right) - \mathcal{V}\left(\frac{1}{\mu}, P_{\text{tot}}\right)\right), \tag{10.48}$$

where $\mathcal{V}$ is given by

$$\mathcal{V}(r,\gamma) = r\ln(1+\gamma-\mathcal{F}(r,\gamma)) + \ln(1+r\gamma-\mathcal{F}(r,\gamma)) - \frac{\mathcal{F}(r,\gamma)}{\gamma}, \tag{10.49}$$

$$\mathcal{F}(r,\gamma) = \frac{1}{4}\left(\sqrt{\gamma(\sqrt{r}+1)^2+1} - \sqrt{\gamma(\sqrt{r}-1)^2+1}\right)^2. \tag{10.50}$$

We evaluate the bounds by fixing $M = 2^{100}$ and studying the trade-off between the minimum $\mathcal{E}^*$ and the user density μ under a target error probability $\mathsf{P}_{e,U}$. Figures 10.3 and 10.4 show the bounds for $\mathsf{P}_{e,U} = 0.1$ and $\mathsf{P}_{e,U} = 0.001$, respectively. For comparison, we also plot the bounds obtained through TDMA. Specifically, the frame of length n is divided equally among k_n users, so that each user transmits with a blocklength $1/\mu$. The smallest P_{tot} is computed to ensure the existence of a single user quasi-static AWGN code of rate S, blocklength $1/\mu$, and probability of error ϵ, using the bound from Yang et al. (2014).

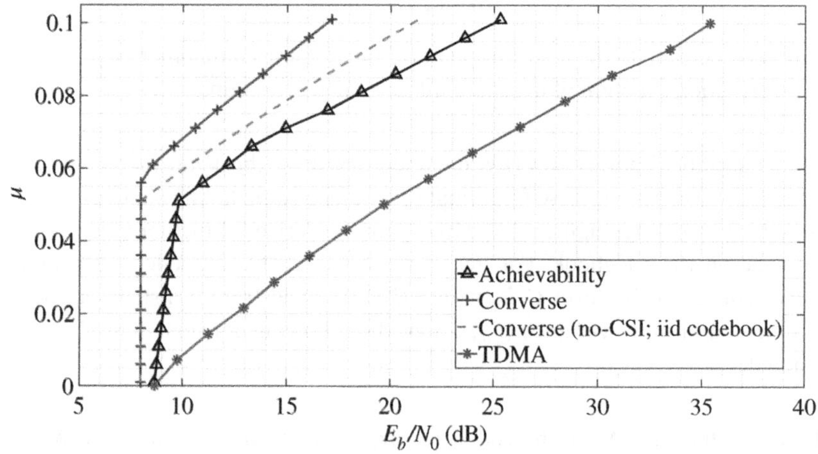

Figure 10.3 μ versus $\mathcal{E}$ for $\epsilon \leq 10^{-1}$, $M = 2^{100}$, i.e., each user transmits 100 bits.

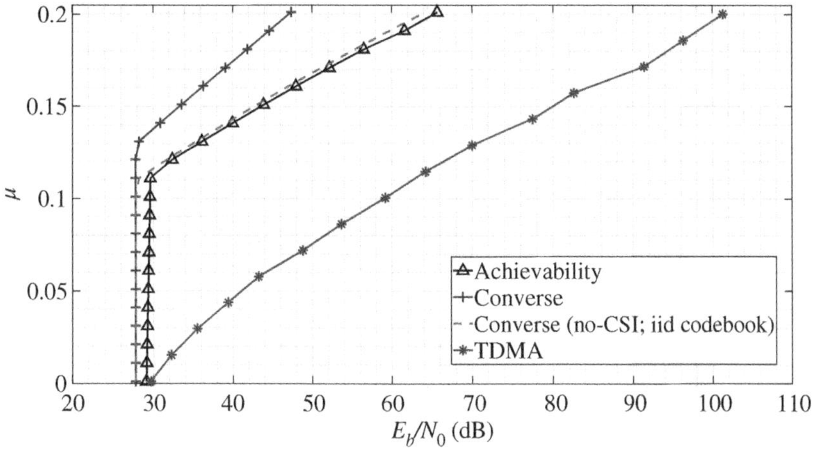

Figure 10.4 μ versus $\mathcal{E}$ for $\epsilon \le 10^{-3}$, $M = 2^{100}$, i.e., each user transmits 100 bits.

A significant observation is the possibility of almost perfect multiuser interference cancellation when user density is below a critical threshold. In particular, as μ increases from 0, the $\mathcal{E}^*$ is almost a constant, approximately equal to the single-user minimal energy per bit. In other words, $\mathcal{E}^*(\mu, \epsilon, k) \approx \mathcal{E}_{s.u.}(\epsilon, k)$ for $0 < \mu < \mu_{s.u.}$, where $\mathcal{E}_{s.u.}(\epsilon, k)$ is independent of μ and corresponds to the single-user minimal energy-per-bit for sending k bits with error ϵ, and $\mu_{s.u.}$ denotes the corresponding density threshold for achieving such $\mathcal{E}_{s.u.}$. As μ increases beyond $\mu_{s.u.}$, the tradeoff undergoes a "phase transition" and the energy per bit $\mathcal{E}^*$ exhibits a more familiar increase with μ. This behavior does not appear in standard schemes for multiple-access like TDMA, and is in fact related to the phase transitions in compressed sensing.

In the special case where each user has only one bit of information to send, the user may employ two carefully designed codewords that are unique to the user to represent 0 and 1, respectively. Alternatively, each user can use one's on/off activity to represent one bit of information, reducing the one-bit communication problem to a user identification problem. Evidently, the former scheme has a strictly larger design space and may achieve better error performance.

10.4.2 Capacity per Unit-Energy Under Different Error Criteria

In contrast to Sections 10.3 and 10.4.1, where either the transmit power or the message set size is fixed, we remove such constraints and allow both to vary with the blocklength in this section. We then investigate the capacity per unit-energy under different error criteria, i.e., JPE and PUPE.

Consider the model given in (10.5). We restrict ourselves to symmetric codes where users have the same message set size and energy constraint.

Definition 10.10 For $0 \leq \epsilon \leq 1$, an (n, M, E_n, ϵ) code for the Gaussian MnAC consists of:

(1) The encoding functions $\mathcal{F}_k : \{0, 1, \dots, M\} \to \mathcal{S}_k^n$ for every user $k \in \{1, \dots, \ell_n\}$, which maps any message w to the codeword $\mathbf{s}_k(w) = [s_{k1}(w), \dots, s_{kn}(w)]^T$. In particular, $\mathbf{s}_k(0) = \mathbf{0}$, for every k. Every codeword $\mathbf{s}_k(w)$ satisfies the energy constraint:

$$\sum_{i=1}^{n} s_{k,i}^2(w) \leq E_n. \tag{10.51}$$

(2) Decoding function $\mathcal{G} : \mathcal{Y}^n \to \{0, 1, \dots, M\}^{\ell_n}$, which is a deterministic rule assigning a decision on the messages to each possible received vector.

The messages $W_1, \dots, W_{\ell_n}$ are independent, and follow the distribution

$$\mathsf{P}\{W_k = w\} = \begin{cases} 1 - \alpha_n, & w = 0, \\ \frac{\alpha_n}{M}, & w \in \{1, \dots, M\}, \end{cases} \tag{10.52}$$

for $k \in \{1, \dots, \ell_n\}$. The probability of error satisfies

$$\mathsf{P}_{e,J}^{(n)} = \mathsf{P}\{\mathcal{G}(\mathbf{Y}) \neq (W_1, \dots, W_{\ell_n})\} \leq \epsilon, \tag{10.53}$$

if JPE criterion is employed, and

$$\mathsf{P}_{e,U}^{(n)} = \frac{1}{\ell_n} \sum_{k=1}^{\ell_n} \mathsf{P}\left\{W_k \neq (\mathcal{G}(\mathbf{Y}))_k\right\} \leq \epsilon, \tag{10.54}$$

if PUPE criterion is applied.

Definition 10.11 For a symmetric code, the rate per unit-energy $\dot{R}$ is said to be ϵ-achievable if for every $\delta > 0$ there exists an n' such that, if $n \geq n'$, then an (n, M, E_n, ϵ) code can be found whose rate per unit-energy satisfies $\frac{\log_2 M}{E_n} > \dot{R} - \delta$. Furthermore, $\dot{R}$ is said to be achievable if it is ϵ-achievable for all $0 < \epsilon < 1$. The capacity per unit-energy $\dot{C}$ is the supremum of all achievable rates per unit-energy. The ϵ-capacity per unit-energy $\dot{C}_\epsilon$ is the supremum of all ϵ-achievable rates per unit-energy.

In Verdú (1989), a rate per unit-energy $\dot{R}$ is said to be ϵ-achievable if, for every $\alpha > 0$, there exists an E' such that, if $E \geq E'$, then an (n, M, E, ϵ) code can be found whose rate per unit-energy satisfies $\frac{\log_2 M}{E_n} > \dot{R} - \delta$. This definition differs from Definition 10.11 in that the energy E is required to exceed a certain threshold rather than having a requirement on the blocklength n. However, it is worth noting that, for the capacity per unit-energy where a vanishing error probability is required, Definition 10.11 is equivalent to Verdú (1989)[Definition 2], since $\mathsf{P}_{e,J}^{(n)} \to 0$ and $\mathsf{P}_{e,U}^{(n)} \to 0$ only if $E_n \to \infty$ (see Lemmas 10.2 and 10.4).

In the following, we investigate the capacity per unit-energy for MnAC, examining both nonrandom and random access scenarios. Before presenting the results, we introduce the order notations that are applied throughout this section. Given two sequences of nonnegative real numbers, $\{a_n\}$ and $\{b_n\}$, we use the following notations: $a_n = O(b_n)$ if $\limsup_{n\to\infty} \frac{a_n}{b_n} < \infty$, $a_n = o(b_n)$ if $\lim_{n\to\infty} \frac{a_n}{b_n} = 0$, $a_n = \Omega(b_n)$ if $\liminf_{n\to\infty} \frac{a_n}{b_n} > 0$, and $a_n = \omega(b_n)$ if $\lim_{n\to\infty} \frac{a_n}{b_n} = \infty$. When $a_n = \Theta(b_n)$, it means that $a_n = O(b_n)$ and $a_n = \Omega(b_n)$.

We first assume that all users are active, i.e., $\alpha_n = 1$ and $\ell_n = k_n$.

Theorem 10.7 ***(No Random-Access Under JPE Criterion)*** Under the JPE criterion, the capacity per unit-energy of the Gaussian MnAC has the following behavior:

(1) If $k_n = o(n/\log_2 n)$, then $\dot{C}^J = \frac{\log_2 e}{2}$. Moreover, the capacity per unit-energy can be achieved by an orthogonal access scheme.
(2) If $k_n = \omega(n/\log_2 n)$, then $\dot{C}^J = 0$.
(3) If $k_n = \Theta(n/\log_2 n)$, then $0 < \dot{C}^J < \frac{\log_2 e}{2}$.

To achieve Part (1), we can employ an orthogonal access scheme, where the blocklength is divided equally among all users. In each time slot, only one user transmits while the others remain silent. The analysis is based on the point-to-point Gaussian channel with blocklength n/k_n. The power constraint then applies with $P = \frac{E_n}{n/k_n}$. It is shown that by choosing $E_n = c_n \ln n$ with $c_n = \ln\left(\frac{n}{k_n \ln n}\right)$, any rate per unit-energy $\dot{R} < \frac{\log_2 e}{2}$ is achievable, under the assumption that $k_n = o(n/\log_2 n)$. Note that the above choice implies that $E_n \to \infty$ while $P \to 0$ as n approaches infinity.

To prove Part (2), two lemmas that provide necessary conditions on the order of E_n are presented. Let the rate per unit-energy be $\dot{R} = \frac{\log_2 M}{E_n}$.

Lemma 10.2 If $M_n \geq 2$, then $\mathsf{P}_{e,J}^{(n)} \to 0$ only if $E_n \to \infty$.

According to this lemma, it can be shown that $\dot{R} > 0$ implies that

$$E_n = O(n/k_n). \tag{10.55}$$

Lemma 10.3 If $\dot{R} > 0$ and $k_n \geq 5$, then $\mathsf{P}_{e,J}^{(n)} \to 0$ only if $E_n = \Omega(\log_2 k_n)$.

Based on the two lemmas, no sequence $\{E_n\}$ can satisfy both conditions (10.55) and $E_n = \Omega(\log_2 k_n)$ if $k_n = \omega(n/\log_2 n)$.

Part (3) is analyzed through two sections $\dot{C} > 0$ and $\dot{C} < \frac{\log_2 e}{2}$. To demonstrate the former, we use the same orthogonal access scheme as in the proof of Part (1) and choose $E_n = c \log_2 n$ with a constant c. With the chosen E_n, $E_n \to \infty$ as $n \to \infty$ while $P = \Theta(1)$. On the other hand, we show the latter by demonstrating that $E_n = \Theta(\log_2 n)$.

Theorem 10.8 ***(No Random-Access Under PUPE Criterion)*** Under the PUPE criterion, the capacity per unit-energy of the Gaussian MnAC has the following behavior:

(1) If $k_n = o(n)$, then $\dot{C}^U = \frac{\log_2 e}{2}$. Furthermore, the capacity per unit-energy can be achieved by an orthogonal access scheme.
(2) If $k_n = \Omega(n)$, then $\dot{C}^U = 0$.

To prove Part (1), a similar argument as Lemma 10.2 is obtained:

Lemma 10.4 $\mathrm{P}_{e,U}^{(n)} \to 0$ only if $E_n \to \infty$.

In addition, any rate per unit-energy $\dot{R} < \frac{\log_2 e}{2}$ is achievable by an orthogonal access scheme where each user uses an orthogonal codebook of blocklength n/k_n. In proof of Part (2), it is further shown that to achieve a positive $\dot{R}$, it is required that $\frac{k_n E_n}{n}$, which is the total power of users, is bounded in n ($\frac{k_n E_n}{n} = \Theta(1)$).

Theorems 10.7 and 10.8 show that despite differences in the error criteria, the capacity per unit-energy exhibits similar behavior. In particular, there is a sharp transition between orders of growth of k_n where each user can achieve the single-user capacity per unit-energy $\frac{\log_2 e}{2}$, meaning that users can communicate as if they were free of interference, and orders of growth where no positive rate per unit-energy is feasible. Under the JPE criterion, the transition threshold separating these two regimes is at the order of growth $n/\log_2 n$. Under the PUPE criterion, the transition threshold shifts to the order of growth n. This shift occurs because necessary constraints for E_n, in order to achieve a vanishing error probability as $n \to \infty$, are different for the two error criteria, and the constraints required along with the JPE criterion are more stringent than those required along with the PUPE criterion. Additionally, while an orthogonal access scheme combined with orthogonal codebook is optimal for the PUPE criterion, it is suboptimal for the JPE criterion.

Let us now consider the case where a user's activation probability can be strictly smaller than 1. First, note that the case where k_n vanishes as $n \to \infty$ is not interesting. Indeed, this case only happens if $\alpha_n \to 0$. The probability that all users are inactive, given by $\left((1-\alpha_n)^{\frac{1}{\alpha_n}}\right)^{k_n}$, tends to one since $(1-\alpha_n)^{\frac{1}{\alpha_n}} \to 1/e$ and $k_n \to 0$. Therefore, if each user employs a code with $M = 2$ and $E_n = 0$ for all n, and the decoder always declares that all users are inactive, then the probability of error vanishes as $n \to \infty$, implying that $\dot{C} = \infty$. In the following, we avoid this trivial case and assume that ℓ_n and α_n are such that $k_n = \Omega(1)$, which implies that $\frac{1}{\alpha_n} = O(\ell_n)$.

Theorem 10.9 ***(Random MnAC Under JPE Criterion)*** Assume that $k_n = \Omega(1)$. Then the capacity per unit-energy of the Gaussian random MnAC has the following behavior:

(1) If $k_n \log_2 \ell_n = o(n)$, then $\dot{C}^J = \frac{\log_2 e}{2}$.
(2) If $k_n \log_2 \ell_n = \omega(n)$, then $\dot{C}^J = 0$.
(3) If $k_n \log_2 \ell_n = \Theta(n)$, then $0 < \dot{C}^J < \frac{\log_2 e}{2}$.

The achievability of Part (1) is shown using a non-orthogonal access scheme where all users employ different codebooks with codewords of length n. The first n_0 symbols of each codeword serve as the signature for user identification, and the remaining $n - n_0$ symbols are used for communication. The decoder performs a two-step decoding process, comprised of active user detection and message decoding. When $k_n \log_2 \ell_n = o(n)$, it can be shown that both the detection and decoding error probabilities tend to zero as $n \to \infty$.

The proof of Part (2) is similar to that of Part (2) of Theorem 10.7. First, it is shown that $\mathsf{P}_{e,J}^{(n)} \to 0$ only if $E_n = \Omega(\ell_n)$. Then, it is demonstrated that achieving $\dot{R} > 0$ requires that the total power of the active users, given by $k_n E_n/n$, is bounded as $n \to \infty$. Part (2) follows by noting that if $k_n \log_2 \ell_n = \omega(n)$, there is no sequence of E_n that can satisfy both conditions simultaneously.

When the PUPE criterion is employed, the analysis of capacity per unit-energy becomes trivial when $\alpha_n \to 0$ as $n \to \infty$. Specifically, if $\alpha_n \to 0$ as $n \to \infty$, then $\mathsf{P}\{W_k = 0\} \to 1$ for all $k = 1, \ldots, \ell_n$. As a result, if each user uses a code with $M = 2$ and $E_n = 0$ for all n, and the decoder always declares that all users are inactive, then the PUPE vanishes as $n \to \infty$. This implies that $\dot{C}^U = \infty$. To avoid this trivial case, we restrict our analysis to $\liminf_{n\to\infty} \alpha_n > 0$ in the following, which implies that k_n and ℓ_n are of the same order.

Theorem 10.10 ***(Random MnAC Under PUPE Criterion)*** If $\liminf_{n\to\infty} \alpha_n > 0$, then $\dot{C}^U$ has the following behavior:

(1) If $\ell_n = o(n)$, then $\dot{C}^U = \frac{\log_2 e}{2}$.
(2) If $\ell_n = \Omega(n)$, then $\dot{C}^U = 0$.

The achievability of Part (1) is proved using an orthogonal access scheme where each user uses an orthogonal codebook of blocklength n/ℓ_n. Out of these n/ℓ_n channel uses, the first one is used for sending a pilot signal to convey that the user is active, and the remaining channel uses are used to send the message.

The comparison of Theorem 10.10 with Theorem 10.8 reveals that the result remains unchanged regardless of the random access.

Theorem 10.9 demonstrates a sharp transition between orders of growth of k_n and ℓ_n, where interference-free communication is feasible and orders of growth where no positive rate per unit-energy is possible. This behavior is similar to that given in Theorem 10.7 when $\alpha_n = 1$. However, for a general α_n, the transition threshold depends on both ℓ_n and k_n. Nevertheless, if $\liminf_{n\to\infty} \alpha_n > 0$, then $k_n = \Theta(\ell_n)$, and the order of growth of $k_n \log_2 \ell_n$ coincides with that of both $k_n \log_2 k_n$ and $\ell_n \log_2 \ell_n$. As a result, the transition thresholds for both k_n and ℓ_n are also at $n/\log_2 n$, since $k_n \log_2 k_n = \Theta(n)$ is equivalent to $k_n = \Theta(n/\log_2 n)$.

As $\alpha_n \to 0$, the orders of growth of k_n and ℓ_n differ, and the transition threshold for ℓ_n is generally greater than $n/\log_2 n$. For example, when $\ell_n = n$ and $\alpha_n = \frac{1}{\sqrt{n}}$, we have $k_n \log_2 \ell_n = \sqrt{n} \log_2 n = o(n)$, and hence all active users can communicate without interference. This indicates that random user-activity enables interference-free communication at a growth rate beyond the $n/\log_2 n$ limit. Similarly, the transition threshold for k_n may be smaller than $n/\log_2 n$ as $\alpha_n \to 0$, although this only holds if ℓ_n grows superpolynomially with n. For instance, when $\ell_n = 2^n$ and $\alpha_n = \frac{\sqrt{n}}{2^n \log_2 n}$, we have $k_n = \frac{\sqrt{n}}{\log_2 n} = o(n/\log_2 n)$ and $k_n \log_2 \ell_n = \frac{n^{3/2}}{\log_2 n} = \omega(n)$, which implies that no positive rate per unit-energy is feasible. This further indicates that treating a random MnAC with an average number of k_n users as a nonrandom MnAC with k_n users may be overly optimistic.

In Section 10.4.1 and this section, we discuss two closely related metrics of energy efficiency for the MnAC. Here, we provide a brief discussion of these results. For ease of comparison, we assume $\alpha_n = 1$.

Lemmas 10.2 and 10.4 indicate that to achieve a vanishing probability of error, the total energy tends to infinity when $n \to \infty$. For a fixed and positive rate per unit-energy, this implies that the payload size of each user grows unbounded with the blocklength, i.e., $M \to \infty$ as $n \to \infty$. By not requiring the error probability to vanish as $n \to \infty$, finite payload is allowed in Section 10.4.1, which can be appealing from a practical perspective.

Section 10.4.1 established bounds on the minimum energy per bit, which suggest that interference-free communication is feasible when the user density μ is below a critical value. This observation is consistent with Theorems 10.7 and 10.8 (also Theorems 10.9 and 10.10). However, Theorems 10.7 and 10.8 imply that, irrespective of the value of μ and the error criteria considered, the capacity per unit-energy $\dot{C}$ is zero when the number of users grows linearly with n. This indicates that the minimum energy per bit is infinite under these conditions. In contrast, according to Theorem 10.4, the minimum energy per bit for a fixed error probability ϵ is finite, which means that the ϵ-capacity per unit-energy $\dot{C}_\epsilon$ is strictly positive. Thus, the capacity per unit-energy is strictly smaller than the ϵ-capacity per unit-energy. This suggests that, under the PUPE criterion, the strong converse does not hold. To have a positive ϵ-rate per unit-energy in this linear regime, however, it is necessary that the energy E_n and the payload $\log M$ are bounded in n (exactly as the setting in Section 10.4.1). Under the JPE criterion, on the other hand, the strong converse holds irrespective of the order of growth of k_n. To be specific, the following theorem holds:

Theorem 10.11 The ϵ-capacity per unit-energy $\dot{C}_\epsilon$ of the nonrandom MnAC under the JPE criterion has the following behavior:

(1) If $k_n = \omega(1)$ and $k_n = o(n/\log_2 n)$, then $\dot{C}_\epsilon^J = \frac{\log_2 e}{2}$ for every $0 < \epsilon < 1$.
(2) If $k_n = \omega(n/\log_2 n)$, then $\dot{C}_\epsilon^J = 0$ for every $0 < \epsilon < 1$.

10.5 Discussion and Open Problems

10.5.1 Scaling Regime

Sections 10.3 and 10.4.1 focus on the linear scaling regime of active users with respect to blocklength. Theorem 10.1 presents capacity results where the number of active users can scale as fast as linearly with the blocklength, i.e., $k_n = \mu n$ with $\mu > 0$. The same scaling regime is also considered in Theorems 10.4 and 10.5. This regime is particularly relevant in scenarios where the number of active users can exceed the blocklength, i.e., $\mu > 1$.

In contrast, in Section 10.4.2, it is demonstrated in Theorem 10.7 that, under the JPE criterion, a positive capacity per unit-energy can be achieved if $k_n = o(n/\log_2 n)$ and $k_n = \Theta(n/\log_2 n)$, and the single-user capacity per unit-energy can be achieved if $k_n = o(n/\log_2 n)$. This scaling regime is asymptotically below the linear regime as discussed in Section 10.3. It is then shown in Theorem 10.8 that, under the PUPE criterion, the single-user capacity per unit-energy can be achieved if $k_n = o(n)$ while no positive capacity per unit-energy is feasible if $k_n = \Omega(n)$.

All three settings investigate asymptotic results by assuming an infinite blocklength. The number of active users hence grow unbounded with the blocklength. However, the payload size of each user has different behaviors. In Section 10.4.1, the message set size M is a constant number, indicating a fixed and finite payload size. While in Section 10.3, the payload size of each user is given by the function $B(n)$ (nats), which depends on the blocklength n. As defined in Eq. (10.14), we have $B(n) = O(\ln n)$ since $k_n = O(n)$, which approaches infinity as $n \to \infty$. In Section 10.4.2, to achieve a vanishing error probability, the energy for transmission tends to infinity with blocklength n. When capacity per unit-energy is positive, this implies that the payload size grows unbounded with the blocklength.

In the context of IoT applications, short-packet transmission has become a critical feature in next-generation multiple access schemes. To enable efficient and reliable communication in such systems, it is important to focus more on research related to finite payload size. It is also essential to consider the finite blocklength regime, which takes into account not only latency constraints but also fading environments. Additionally, non-asymptotic regimes may provide more relevant and informative results in mMTC scenarios with sporadic transmissions and small payload sizes.

When the number of users is fixed, the achievability bound with finite blocklength for Gaussian MAC is derived in MolavianJazi and Laneman (2015), and improved in Yavas et al. (2021), where Gaussian random access channel is also investigated. When the total number of users can be comparable to the blocklength or even be unbounded, unsourced random access has been studied in the finite blocklength regime (Kowshik et al., 2020, Gao et al., 2022).

10.5.2 Some Practical Issues

Capacity results for the MnAC framework haven been discussed above, assuming fully synchronized receptions at the frame level. However, signals transmitted by users from different locations may reach the receiver with different delays, largely due to differing propagation delays. While proper synchronization can be implemented at the transmitters, such as using a common beacon to trigger transmissions, the relative delays cannot be entirely eliminated. Therefore, asynchronous reception at the frame level, or even at the symbol level, must be considered, which can present significant challenges for reliable and efficient multiaccess.

While frame-level asynchrony has little effect on the capacity region for conventional multiple access channels, it can significantly reduce the maximum achievable sum rate when the channel has memory (Cover et al., 1981, Hui and Humblet, 1985, Verdu, 1989). On the other hand, the impact of symbol-level asynchrony on the capacity region depends on the signature waveforms of transmitters (Verdú, 1989). When the number of users can grow with the blocklength, the capacity of the MnAC is investigate in (Shahi et al. 2022) with block asynchronism, where the start times of users are very large integer multiples of the blocklength, so that asymptotically user signals do not overlap. Despite these prior work, the capacity region of the asynchronous MnAC with frame-level or symbol-level asynchrony remains an open question.

Practical implementation of asynchronous multiaccess has also been studied in the literature. Asynchronous compressed sensing has been proposed, whose computational complexity is at least linear in the user population (Applebaum et al., 2012), and it is difficult to obtain a theoretical performance guarantee (Amalladinne et al., 2019). Asynchronous covariance-based methods have been designed for multiple-input multiple-output (MIMO) systems with few result on how the performance scale with the system size (Wang et al., 2022, Li et al., 2022). Asynchronous signaling scheme based on sparse OFDMA with sublinear codelength concerning the user population is proposed in Chen et al. (2022). The asynchronous massive access has also been studied in the context of unsourced random access under the PUPE metric, where it has been demonstrated that the performance loss in terms of energy per bit required for reliable communication due to asynchrony is small (Andreev et al., 2019).

Acknowledgments

This work was supported in part by the NSF under Grant No. 1910168 and SpectrumX, an NSF Spectrum Innovation Center (Grant No. 2132700).

References

Rudolf Ahlswede. Multi-way communication channels. In *Proceedings of the 2nd International Symposium Information Theory (Tsahkadsor, Armenian SSR), 1971*, pages 23–52. Publishing House of the Hungarian Academy of Sciences, 1971.

Cem Aksoylar, George K Atia, and Venkatesh Saligrama. Sparse signal processing with linear and nonlinear observations: A unified Shannon-theoretic approach. *IEEE Transactions on Information Theory*, 63(2):749–776, 2016.

Vamsi K Amalladinne, Krishna R Narayanan, Jean-Francois Chamberland, and Dongning Guo. Asynchronous neighbor discovery using coupled compressive sensing. In *ICASSP 2019-2019 IEEE International Conference on Acoustics, Speech and Signal Processing (ICASSP)*, pages 4569–4573. IEEE, 2019.

Kirill Andreev, Suhas S Kowshik, Alexey Frolov, and Yury Polyanskiy. Low complexity energy efficient random access scheme for the asynchronous fading MAC. In *2019 IEEE 90th Vehicular Technology Conference (VTC2019-Fall)*, pages 1–5. IEEE, 2019.

Lorne Applebaum, Waheed U Bajwa, Marco F Duarte, and Robert Calderbank. Asynchronous code-division random access using convex optimization. *Physical Communication*, 5(2):129–147, 2012.

Jean Barbier and Florent Krzakala. Approximate message-passing decoder and capacity achieving sparse superposition codes. *IEEE Transactions on Information Theory*, 63(8):4894–4927, 2017.

Mohsen Bayati and Andrea Montanari. The dynamics of message passing on dense graphs, with applications to compressed sensing. *IEEE Transactions on Information Theory*, 57(2):764–785, 2011.

Xu Chen and Dongning Guo. Gaussian many-access channels: Definition and symmetric capacity. In *2013 IEEE Information Theory Workshop (ITW)*, pages 1–5. IEEE, 2013.

Xu Chen, Tsung-Yi Chen, and Dongning Guo. Capacity of Gaussian many-access channels. *IEEE Transactions on Information Theory*, 63(6):3516–3539, 2017.

Xu Chen, Lina Liu, Dongning Guo, and Gregory W Wornell. Asynchronous massive access and neighbor discovery using OFDMA. *IEEE Transactions on Information Theory*, 69(4):2364–2384, 2022.

Thomas M Cover and Joy A Thomas. *Elements of Information Theory*. 2nd ed. Hoboken, NJ, USA: Wiley, 2006.

T Cover, R McEliece, and E Posner. Asynchronous multiple-access channel capacity. *IEEE Transactions on Information Theory*, 27(4):409–413, 1981.

David L Donoho. Compressed sensing. *IEEE Transactions on Information Theory*, 52(4):1289–1306, 2006.

Alyson K Fletcher, Sundeep Rangan, and Vivek K Goyal. Necessary and sufficient conditions for sparsity pattern recovery. *IEEE Transactions on Information Theory*, 55(12):5758–5772, 2009.

Robert Gallager. A perspective on multiaccess channels. *IEEE Transactions on Information Theory*, 31(2):124–142, 1985.

Junyuan Gao, Yongpeng Wu, Shuo Shao, Wei Yang, and H Vincent Poor. Energy efficiency of massive random access in MIMO quasi-static Rayleigh fading channels with finite blocklength. *IEEE Transactions on Information Theory*, 69(3):1618–1657, 2022.

Dongning Guo and Sergio Verdú. Randomly spread CDMA: Asymptotics via statistical physics. *IEEE Transactions on Information Theory*, 51(6):1983–2010, 2005.

Piyush Gupta and Panganmala R Kumar. The capacity of wireless networks. *IEEE Transactions on Information Theory*, 46(2):388–404, 2000.

Kuan Hsieh, Cynthia Rush, and Ramji Venkataramanan. Near-optimal coding for massive multiple access. In *2021 IEEE International Symposium on Information Theory (ISIT)*, pages 2471–2476. IEEE, 2021.

J Hui and P Humblet. The capacity region of the totally asynchronous multiple-access channel. *IEEE Transactions on Information Theory*, 31(2):207–216, 1985.

Suhas S Kowshik. Improved bounds for the many-user MAC. In *2022 IEEE International Symposium on Information Theory (ISIT)*, pages 2874–2879, 2022. doi: 10.1109/ISIT50566.2022.9834341.

Suhas S Kowshik and Yury Polyanskiy. Fundamental limits of many-user MAC with finite payloads and fading. *IEEE Transactions on Information Theory*, 67(9):5853–5884, 2021.

Suhas S Kowshik, Kirill Andreev, Alexey Frolov, and Yury Polyanskiy. Energy efficient coded random access for the wireless uplink. *IEEE Transactions on Communications*, 68(8):4694–4708, 2020.

Yang Li, Qingfeng Lin, Ya-Feng Liu, Bo Ai, and Yik-Chung Wu. Asynchronous activity detection for cell-free massive MIMO: From centralized to distributed algorithms. *IEEE Transactions on Wireless Communications*, 22(4):2477–2492, 2022.

H Liao. A coding theorem for multiple access communications. In *Proceedings of the International Symposium on Information Theory, Asilomar, CA, 1972*, 1972.

Ebrahim MolavianJazi and J Nicholas Laneman. A second-order achievable rate region for Gaussian multi-access channels via a central limit theorem for functions. *IEEE Transactions on Information Theory*, 61(12):6719–6733, 2015.

Yury Polyanskiy. A perspective on massive random-access. In *2017 IEEE International Symposium on Information Theory (ISIT)*, pages 2523–2527. IEEE, 2017.

Yury Polyanskiy. Information theoretic perspective on massive multiple-access. *Short Course (slides) Skoltech Inst. of Tech., Moscow, Russia*, 2018.

John G Proakis, Masoud Salehi, Ning Zhou, and Xiaofeng Li. *Communication Systems Engineering*, volume 2. New Jersey: Prentice Hall, 1994.

Kamiar Rahnama Rad. Nearly sharp sufficient conditions on exact sparsity pattern recovery. *IEEE Transactions on Information Theory*, 57(7):4672–4679, 2011.

Jithin Ravi and Tobias Koch. Scaling laws for Gaussian random many-access channels. *IEEE Transactions on Information Theory*, 68(4):2429–2459, 2021.

Jyotish Robin and Elza Erkip. Capacity bounds and user identification costs in Rayleigh-fading many-access channel. In *2021 IEEE International Symposium on Information Theory (ISIT)*, pages 2477–2482. IEEE, 2021.

Sara Shahi, Daniela Tuninetti, and Natasha Devroye. The strongly asynchronous massive access channel. *Entropy*, 25(1):65, 2022.

David Slepian and Jack Keil Wolf. A coding theorem for multiple access channels with correlated sources. *Bell System Technical Journal*, 52(7):1037–1076, 1973.

Sergio Verdú. The capacity region of the symbol-asynchronous Gaussian multiple-access channel. *IEEE Transactions on Information Theory*, 35(4):733–751, 1989.

Sergio Verdu. Multiple-access channels with memory with and without frame synchronism. *IEEE Transactions on Information Theory*, 35(3):605–619, 1989.

Sergio Verdú. On channel capacity per unit cost. *IEEE Transactions on Information Theory*, 36(5):1019–1030, 1990.

Martin J Wainwright. Information-theoretic limits on sparsity recovery in the high-dimensional and noisy setting. *IEEE Transactions on Information Theory*, 55(12):5728–5741, 2009.

Wei Wang, Martin J Wainwright, and Kannan Ramchandran. Information-theoretic limits on sparse signal recovery: Dense versus sparse measurement matrices. *IEEE Transactions on Information Theory*, 56(6):2967–2979, 2010.

Zhaorui Wang, Ya-Feng Liu, and Liang Liu. Covariance-based joint device activity and delay detection in asynchronous mMTC. *IEEE Signal Processing Letters*, 29:538–542, 2022.

Fan Wei, Yongpeng Wu, Wen Chen, Yanlin Geng, and Giuseppe Caire. On the fundamental limits of MIMO massive access communication. *arXiv preprint arXiv:1908.03298*, 2019.

Yongpeng Wu, Xiqi Gao, Shidong Zhou, Wei Yang, Yury Polyanskiy, and Giuseppe Caire. Massive access for future wireless communication systems. *IEEE Wireless Communications*, 27(4):148–156, 2020.

Wei Yang, Giuseppe Durisi, Tobias Koch, and Yury Polyanskiy. Quasi-static multiple-antenna fading channels at finite blocklength. *IEEE Transactions on Information Theory*, 60(7):4232–4265, 2014.

Recep Can Yavas, Victoria Kostina, and Michelle Effros. Gaussian multiple and random access channels: Finite-blocklength analysis. *IEEE Transactions on Information Theory*, 67(11):6983–7009, 2021.

Lei Zhang, Jun Luo, and Dongning Guo. Neighbor discovery for wireless networks via compressed sensing. *Performance Evaluation*, 70(7-8):457–471, 2013.

11

Random Access Techniques for Machine-Type Communication

Jinho Choi

School of Information Technology, Deakin University, Burwood, Victoria, Australia

11.1 Fundamentals of Random Access

11.1.1 Coordinated Versus Uncoordinated Transmissions

For multiuser communication systems, it is necessary to share a common radio resource. Suppose that multiple users communicate with a base station (BS). In a mobile cellular system, users transmit signals to the BS through uplink or multiple access channels, while the BS transmits signals to users through downlink or broadcast channels. A shared common radio resource can be divided into multiple channels to support multiple users. One approach is to divide the resource evenly among a predefined number of users, say M, which can be decided by the BS's capability in terms of the number of users' signals that it can handle simultaneously. Alternatively, the radio resource can be dynamically allocated according to the users' requirements to meet certain quality-of-service (QoS). Once the shared channel resource is divided, the BS can inform users of allocated channels so that users can transmit and receive signals through the allocated channels for uplink and downlink transmissions, respectively. The resulting approach is referred to as coordinated transmissions.

In coordinated transmissions, if users leave and join the system, the channel allocation has to be updated, which results in a system overhead. Thus, it is expected that users are connected to the system for a while with continuous transmissions. However, if users have a low-duty cycle and just a few packets to send when activated, such as a sensor, the overhead of coordinated transmissions would be overwhelming. In this case, it may be better to use uncoordinated transmissions, which do not require any channel allocation for users.

Next Generation Multiple Access, First Edition.
Edited by Yuanwei Liu, Liang Liu, Zhiguo Ding, and Xuemin Shen.
© 2024 The Institute of Electrical and Electronics Engineers, Inc. Published 2024 by John Wiley & Sons, Inc.

For uncoordinated transmissions, especially uplink transmissions, random access techniques can be used. In random access, a user can be activated when there are data packets to send to the BS. Since there is no allocated channel, the active user can transmit packets through a shared channel or one of shared channels. Thus, as long as no other users are transmitting packets, the user can successfully transmit packets through a shared channel without any interference. This means that a common channel can be shared by any number of users without channel allocation. This feature becomes useful when a large number of devices are to be connected, while transmitting data intermittently. As a result, random access is considered for machine-type communication (MTC), where it is expected to support a large number of devices with a limited spectrum.

11.1.2 Random Access Techniques

11.1.2.1 ALOHA Protocols

Suppose that there are multiple users and one receiver, which is a BS for uplink transmissions, in a random access system. ALOHA is a well-known random access scheme that allows multiple users to access a shared channel. The key operations of ALOHA are as follows.

1. Users send data packets when they have them.
2. The receiver sends feedback signals to users as follows: (i) a positive acknowledgment (ACK) is sent if the receiver succeeds to decode a data packet; or (ii) a negative acknowledgment (NACK) is sent if the receiver fails to decode a data packet (due to a collision when multiple users send packets at the same time).
3. The users receive NACK will try to send packets later.

There are two ALOHA protocols depending on whether users are time synchronized or not. In *pure ALOHA*, users are not synchronized. Thus, a collision happens even if only a small portion of a packet is overlapped with another packet. To avoid this problem, suppose that users are synchronized. In addition, thanks to synchronization, each user with data can start transmitting only at the beginning of a time slot, assuming that the packet size is equal to the length of the time slot. The resulting ALOHA protocol is referred to as *slotted-ALOHA* or S-ALOHA.

In order to find the throughput of pure ALOHA, we assume that (i) the length of packet, denoted by T, is the same for all users; and (ii) the number of active users to send packets follows a Poisson distribution. The average number of active users during T is denoted by G.

Suppose that a user transmits a packet at time t. This packet can be successfully transmitted if there are no other packets within the time window of $(t - T, t + T)$. Since the number of active users follows a Poisson distribution, the probability that there are no other active users is e^{-2G}, while the rate of transmission attempts by any active users within the unit time, T, is G. Thus, the throughput that is the average number of successful packet transmissions per unit time becomes

$$\eta_{\text{pure}}(G) = Ge^{-2G}. \tag{11.1}$$

To find the throughput of S-ALOHA, we assume that the length of slot is T and the length of packet is also T. Denote by K the number of active users within a slot, which follows a Poisson distribution. Let the average number of active users per unit time (i.e., T) be G. Then, the probability of successful packet transmissions per unit time becomes the probability that there is only one active user. Since the Poisson distribution with mean G is assumed, the probability of $K = 1$ becomes

$$\eta_S(G) = \Pr(K = 1) = Ge^{-G}, \tag{11.2}$$

which is the throughput of S-ALOHA. From (11.1) and (11.2), it is shown that the throughput of S-ALOHA is higher than that of pure-ALOHA. We can also compare them in terms of the peak throughput. The maxima of $\eta_{\text{pure}}(G)$ and $\eta_S(G)$ can be achieved when $G = \frac{1}{2}$ and 1, respectively. Thus, we have

$$\max_G \eta_{\text{pure}}(G) = \frac{e^{-1}}{2} < \max_G \eta_S(G) = e^{-1}, \tag{11.3}$$

which shows that the peak throughput of S-ALOHA is two-time higher than that of pure-ALOHA.

If a channel resource is divided into multiple blocks (in time or frequency), each resource block can be used for S-ALOHA and the overall system becomes *multichannel* S-ALOHA, where a user can choose one of multiple S-ALOHA systems.

11.1.2.2 CSMA

Like ALOHA, carrier-sense multiple access (CSMA) allows users to transmit their signals to a receiver without any coordination. When a user has a data packet to transmit, the user senses the channel prior to transmitting a data packet. There are two main protocols: CSMA with collision avoidance (CSMA/CA) and CSMA with collision detection (CSMA/CD). In CSMA/CD, a user should be able to stop transmitting signal as soon as a collision is detected. Thus, CSMA/CD is not used for wireless communication as a transmitter may not be able to detect any collision while transmitting signal. On the other hand, CSMA/CA is well-suited to wireless communication and has been adopted in WiFi.

While CSMA/CA outperforms S-ALOHA due to channel sensing capability, it may not be suitable for wireless systems with large propagation delays such as cellular systems. As a result, CSMA/CA is not extensively studied for MTC in cellular systems. The model of multichannel S-ALOHA becomes useful to analyze the performance of an MTC protocol as will be shown later.

11.1.3 Re-transmission Strategies

In random access, due to the nature of uncoordinated transmissions, packet collisions happen, and a user needs to retransmit collided packets according to a certain re-transmission strategy.

For S-ALOHA, a simple re-transmission strategy can be considered with a retransmission probability $f \in (0, 1)$. A user with a collided packet in slot t can retransmit it in slot $t + 1$ with probability f. In general, f is a control parameter (with a fixed f, it is known that all the transmitted packets will be collided once the backlog is large (Kelly and Yudovina, 2014)). Each user tries to estimate the size of backlog (i.e., the total number of packets awaiting re-transmissions) and decides f accordingly. For example, let $N_t = n$ be the size of backlog. Then, the average number of successfully transmitted packets without collision or throughput becomes the probability that there is exactly one transmission, which is given by $p_s(n) = nf(1 - f)^{n-1}$. Then, it can be shown that f maximizing $p_s(n)$ is $f^* = \frac{1}{n}$, which also leads to a throughput of $\left(1 - \frac{1}{n}\right)^{n-1} \to e^{-1}$ for a large n. In practice, each user needs to estimate N_t from the feedback information. Alternatively, the receiver (i.e., the BS) can estimate N_t and broadcast a control signal limiting user access according to N_t, which is typically used in MTC as an access-barring algorithm.

Another approach is based on the backoff interval. If a user has a collided packet, the user waits for a random period of time before re-transmission. For example, the re-transmission time is chosen uniformly at random within the backoff interval. Backoff algorithms usually adaptively change the backoff interval according to the history of feedback signals. A well-known example is an exponential backoff algorithm. In particular, in binary exponential backoff, each user doubles the backoff interval up to the maximum backoff interval after a collision occurs, and decreases the backoff interval to the minimum value after a successful transmission.

With more control and feedback signals, there can be better approaches to collision resolution in random access, e.g., tree algorithms. In the binary tree algorithm, the users involved in the collision in time slot t are divided into two groups. Each user can randomly choose one of the two groups (by flipping a coin). The users in the first group is to send their packets in time slot $t + 1$ and the users in the second group wait until all the users in the first group succeed to transmit their packets without collisions. Improved tree algorithms can also be found in Massey (1981) and, Yu and Giannakis (2007).

11.2 A Game Theoretic View

Random access can also be seen as a noncooperative game as multiple users compete for shared channel resources. In this section, we present a random access game.

11.2.1 A Model

Suppose that there are K users who are players in a random access game. The random access game can be seen a strategic form game with the following triplet:

1. $\mathcal{K} = \{1, \dots, K\}$: A set of users
2. $\mathcal{S}_k$: A set of strategies or actions of user k, $k \in \mathcal{K}$
3. $u_k : \mathcal{S} \to \mathbb{R}$: A payoff function of user $k \in \mathcal{K}$, where $\mathcal{S} = \prod_k \mathcal{S}_k$.

We assume that users/players always have data packets to send. Thus, $\mathcal{S}_k = \{0, 1\}$, where 0 and 1 represent the strategies of no transmission and transmission, respectively. For convenience, denote by $s_k \in \mathcal{S}_k$ the strategy taken by user k. The payoff function can be defined as

$$u_k(s_k, \mathbf{s}_{-k}) = \begin{cases} 1 - c, & \text{if } s_k = 1 \text{ and } s_{k'} = 0 \text{ for all } k' \in \mathcal{K}_{-k} \\ -c, & \text{if } s_k = 1 \text{ and } s_{k'} = 1 \text{ for any } k' \in \mathcal{K}_{-k} \\ 0, & \text{if } s_k = 0, \end{cases} \tag{11.4}$$

where $\mathbf{s}_{-k} = (s_1, \dots, s_{k-1}, s_{k+1}, \dots s_K)$, $\mathcal{K}_{-k} = \mathcal{K} \setminus \{k\}$, and $c > 0$ represents the cost of transmission. For example, let $K = 2$. Then, the payoff of user 1 becomes

$$u_1(0, 0) = 0,\ u_1(0, 1) = 0,\ u_1(1, 0) = 1 - c, \text{ and } u_1(1, 1) = -c.$$

The best response of player k for given $\mathbf{s}_{-k}$ is given by

$$\mathrm{BR}_k(\mathbf{s}_{-k}) = \operatorname*{argmax}_{s'_k \in \mathcal{S}_k} u_k(s'_k, \mathbf{s}_{-k}). \tag{11.5}$$

The Nash equilibrium (NE) is a solution concept (Fudenberg and Tirole, 1991), and a strategy profile $\mathbf{s} = (s_1, \dots, s_K)$ is an NE if it satisfies

$$u_k(s_k, \mathbf{s}_{-k}) \geq u_k(s'_k, \mathbf{s}_{-k}), \text{ for all } s'_k \in \mathcal{S}_k,\ k \in \mathcal{K}, \tag{11.6}$$

i.e., $\mathbf{s}$ is an NE if any player cannot have a higher payoff by unilaterally deviating from their strategies.

For the random access game with two players, $(s_1, s_2) = (0, 1)$ and $(1, 0)$ are NE for $0 < c < 1$. To see this, consider $(s_1, s_2) = (0, 1)$. It can be shown that

$$u_1(0, 1) = 0 \geq u_1(1, 1) = -c$$
$$u_2(1, 0) = 1 - c \geq u_2(0, 0) = 0,$$

which shows that $(s_1, s_2) = (0, 1)$ is an NE.

Each user can have a mixed strategy that is a distribution to randomize pure strategy, denoted by $\sigma_k = \{\sigma_k(s_k), s_k \in S_k\}$. Here, $\sigma_k(s_k)$ represents the probability that user k chooses s_k and $\sum_{s_k \in S_k} \sigma_k(s_k) = 1$, i.e., according to σ_k, user k can randomly choose one among available actions. Then, the payoff is given by

$$u_k(\sigma_1, \dots, \sigma_K) = \sum_{s_1 \in S_1} \cdots \sum_{s_K \in S_K} u_k(s_1, \dots, s_K)\sigma_1(s_1) \dots \sigma_K(s_K). \tag{11.7}$$

With mixed strategies, the best response is given by

$$\text{BR}_k(\sigma_{-k}) = \underset{\sigma_k \in \Delta_k}{\text{argmax}}\, u_k(\sigma_k, \sigma_{-k}), \tag{11.8}$$

where Δ_k represents the $(|S_k| - 1)$-dimensional simplex. Accordingly, $\sigma^* = (\sigma_1^*, \dots, \sigma_K^*)$ is called a mixed strategy NE if

$$u_k(\sigma_k^*, \sigma_{-k}^*) \geq u_k(\sigma_k, \sigma_{-k}^*), \text{ for all } \sigma_k \in \Delta_k,\ k \in \mathcal{K}. \tag{11.9}$$

Figure 11.1 shows the mixed strategy NE, $\sigma_k(1)$, as a function of c for different values of K, where $\sigma_k(1)$ is seen as the access or transmission probability. As the transmission cost, c, increases, $\sigma_k(1)$ decreases. It is also shown that $\sigma_k(1)$ decreases with K. That is, large numbers of users and/or high transmission cost do

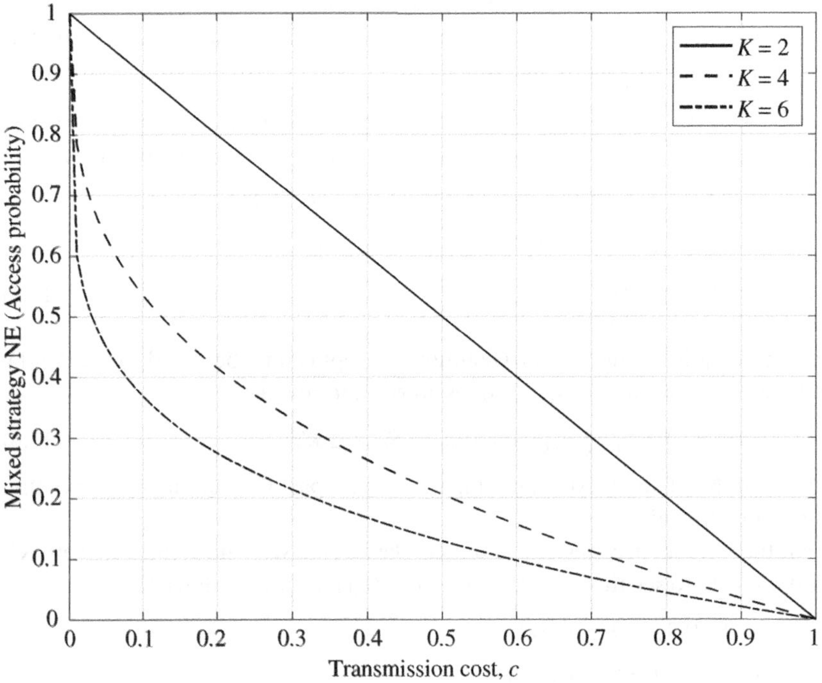

Figure 11.1 Mixed strategy NE as a function of c for different values of K.

not encourage users to transmit. This shows that appropriate access probabilities can be found through game theory.

11.2.2 Fictitious Play

In a noncooperative game including the random access game, users are expected to play according to one of equilibria. This means that each user may need to know the others' payoff functions and actions, while some information may not be available. For example, in the random access game, although we can assume that the same payoff function is used for all users, users may not know how many users are in the game, i.e., K (As shown in Fig. 11.1, since the mixed strategy NE depends on K, a user cannot find it without knowing K).

Fictitious play is an approach that allows each user to find a mixed strategy based on the best response in (11.8) under the assumption that the other users play according to fixed but unknown distributions. Suppose that user k observes the others' actions and can update a weight function $w_k^{(t)}(\mathbf{s}_{-k}) : \mathcal{S}_{-k} \to \mathbb{R}^+$, where $\mathcal{S}_{-k} = \prod_{l \neq k} \mathcal{S}_l$, as follows:

$$w_k^{(t)}(\mathbf{s}_{-k}) = w_k^{(t-1)}(\mathbf{s}_{-k}) + \begin{cases} 1, & \text{if } \mathbf{s}_{-k}^{(t-1)} = \mathbf{s}_{-k} \\ 0, & \text{otherwise,} \end{cases} \tag{11.10}$$

where $\mathbf{s}_{-k}^{(t-1)}$ represents the actions of the other users at time $t-1$. From the weight function, the empirical distribution can be obtained as follows:

$$p_k^{(t)}(\mathbf{s}_{-k}) = \frac{w_k^{(t)}(\mathbf{s}_{-k})}{\sum_{\mathbf{s}'_{-k} \in \mathcal{S}_{-k}} w_k^{(t)}(\mathbf{s}'_{-k})}, \ \mathbf{s}_{-k} \in \mathcal{S}_{-k}. \tag{11.11}$$

Using the empirical distribution of $\mathbf{s}_{-k}$, user k can find the following estimate of the payoff:

$$u_k(s_k, p_k^{(t)}) = \sum_{\mathbf{s}_{-k} \in \mathcal{S}_{-k}} u_k(s_k, \mathbf{s}_{-k}) p_k^{(t)}(\mathbf{s}_{-k}), \ s_k \in \mathcal{S}_k. \tag{11.12}$$

Finally, in fictitious play, based on the best response, the action can be taken as follows:

$$s_k^{(t)} = \underset{s_k \in \mathcal{S}_k}{\operatorname{argmax}}\, u_k(s_k, p_k^{(t)}). \tag{11.13}$$

Note that under fictitious play, $s_k^{(t)}$ does not necessarily converge in any sense. In particular, since users can update their weights as in (11.11), the assumption that the other users play according to fixed distributions is no longer valid. As a result, in general, it is difficult to guarantee convergence.

If a pure strategy NE exists, it is desirable that $s_k^{(t)}$ converges to s_k^*. However, some games may not have pure NEs, while all finite noncooperative games have mixed strategy NEs (Fudenberg and Tirole, 1991). Thus, it would be desirable that $s_k^{(t)}$ can converge to a distribution as follows:

$$\lim_{T\to\infty} \frac{\text{Number of times that } s_k^{(t)} = s_k,\ t = 1, \dots T}{T} = \sigma_k(s_k). \tag{11.14}$$

Under fictitious play, if $s_k^{(t)}$ can converge to a distribution, it is a mixed strategy NE (Fudenberg and Levine, 1998). For example, if the users have the identical payoff function, it is known that $s_k^{(t)}$ can converge to a mixed strategy NE. This implies that fictitious play can result in a mixed strategy NE when the payoff functions are the same for all the users in the random access game.

11.3 Random Access Protocols for MTC

In this section, we discuss two main random access protocols for MTC, namely 4-step and 2-step random access protocols, which are also called grant-based and grant-free random access protocols, respectively.

11.3.1 4-Step Random Access

The grant-based protocol for MTC in the 3rd generation partnership project (3GPP) standards is also called the random access channel (RACH) procedure. In the RACH procedure, uplink transmissions can be carried out based on four steps as illustrated in Fig. 11.2. The first step is random access to establish connection to the BS with a pool of preambles consisting of L preambles (or sequences), where an active device transmits a randomly selected a preamble on physical random access channel (PRACH). At the second step, the BS detects the preambles transmitted by active devices and sends responses. At the third step, the devices whose preambles are successfully detected transmit connection

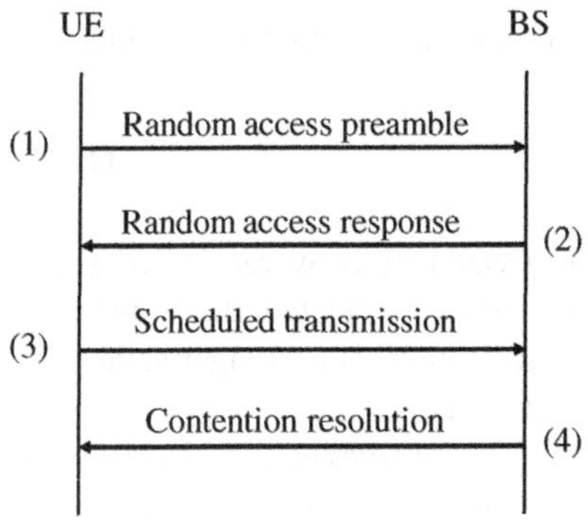

Figure 11.2 4-step random access protocol in MTC.

request messages (Msg3) including their device identities on granted physical uplink shared channel (PUSCH) resources, which is then followed by contention resolution at step 4.

There can be multiple devices that choose the same preamble and the BS can send the response to them so that all the devices transmit their packets through a granted PUSCH, which leads to packet collisions. Thus, the devices involved in a collision can only recognize the transmission failure due to the collision at step 4.

11.3.2 2-Step Random Access

In 5G, the four steps of the RACH procedure are reduced to two steps and the resulting protocol is called the 2-step random access protocol (3GPP, 2018) (Kim et al., 2021) or grant-free random access. In 2-step random access, unlike the RACH procedure, an active device does not wait for a response from the BS after transmitting a preamble, and immediately transmits a data packet. In other words, steps 1 and 3 of the RACH procedure are combined into step 1 of the 2-step random access protocol, while steps 2 and 4 become step 2. As a result, it is expected to shorten access delay, which is desirable for low latency applications, and improve the throughput so that more devices can be supported for massive MTC.

11.3.3 Analysis of 2-Step Random Access

To analyze 2-step random access, we can consider a slotted system, where a time slot is further divided into two sub-slots for preamble and data transmissions. We assume that there are L preambles. The second sub-slot for the data transmissions is further divided into L mini-slots. Each mini-slot is associated with one of L preambles. That is, if a user chooses preamble l, he/she transmits this preamble in the first sub-slot and transmits a data packet in the lth mini-slot of the second sub-slot. If the BS can detect the lth preamble from the signal received during the first sub-slot, it can expect to receive a data packet during the lth mini-slot of the second sub-slot from the user transmitting preamble l.

According to the model stated above, a user can succeed to transmit a data packet if the transmitted preamble is not collided, while users can choose any preambles uniformly at random. Thus, each preamble can be seen as a channel in multichannel S-ALOHA, and in order to find the throughput of 2-step random access, which is the average number of users transmitting data packets without collisions per slot, we can consider a model of multichannel S-ALOHA.

Let K_l denote the number of active users choosing channel (or preamble) l, while K denotes the total number of active users, i.e.,

$$K = K_1 + \cdots + K_L.$$

If K is assumed to be a Poisson random variable with mean G, it can be shown that each K_l is a Poisson random variable with mean $\frac{G}{L}$ (note that the sum of Poisson random variables is also a Poisson random variable). As a result, the throughput becomes

$$\begin{aligned}\eta_{2S}(G) &= \sum_{l=1}^{L} \Pr(K_l = 1) \\ &= \sum_{l=1}^{L} \frac{G}{L} e^{-\frac{G}{L}} \\ &= Ge^{-\frac{G}{L}} = L\left(\frac{G}{L}e^{-\frac{G}{L}}\right),\end{aligned} \tag{11.15}$$

since there should be only one active user choosing a given preamble to avoid preamble collision. The throughput is maximized when $G = L$ and the peak throughput is Le^{-1}. This shows that the peak throughput can increase linearly with the number of channels or preambles. However, the length of slot also increases linearly with L, meaning that the normalized throughput becomes independent of L.

11.3.4 Fast Retrial

As discussed earlier, in terms of the normalized throughput, multichannel S-ALOHA does not outperform single channel S-ALOHA. However, there are some aspects that multichannel ALOHA can be useful. One of them is fast retrial, which can shorten the access delay and can be applied to 2-step random access.

In (single channel) S-ALOHA, a user should not attempt re-transmission immediately upon receiving NACK feedback, because immediate re-transmissions by all the users involved in a collision will result in subsequent collisions. As a result, a user receiving NACK has to re-transmit after an arbitrary backoff time. However, if there are multiple channels, a user experiencing a collision may immediately attempt to send the collided packet, which may result in a short access delay.

In multichannel ALOHA, fast retrial (Choi et al., 2006) is a simple re-transmission strategy where a user with a collided packet can retransmit the collided packet immediately in the next time slot through a randomly selected channel. An example is shown in Fig. 11.3 with four channels (i.e., $L = 4$) and 3 users. In slot t, suppose that users 1 and 3 choose channel 1, which leads to collision. In the next time slot, user 1 chooses channel 2 and user 3 chooses channel 4, while a new user, i.e., user 3, chooses channel 1. Since all users choose different channels, they can succeed to transmit their packets without collision in time slot $t + 1$. Clearly, unlike single-channel S-ALOHA, immediate retransmissions in multichannel ALOHA do not necessarily result in subsequent packet collisions.

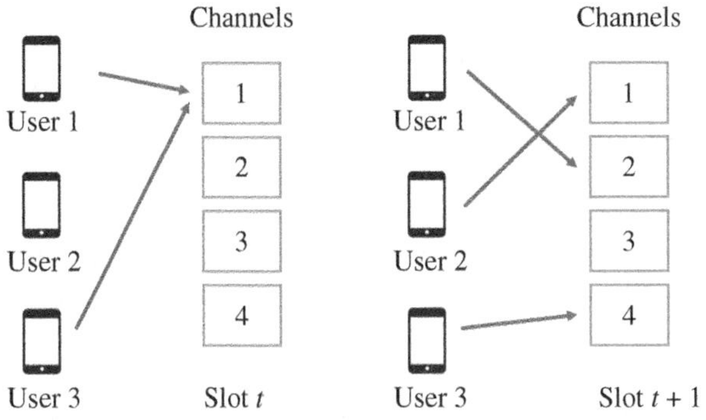

Figure 11.3 An illustration of fast retrial for multichannel ALOHA with 4 channels (i.e., $L = 4$) and 3 users.

As mentioned earlier, fast retrial explained above can also be applied to 2-step random access, where channels become preambles. In Choi (2021a), the performance of fast retrial for 2-step random access is analyzed in terms of stability as well as delay.

11.4 Variants of 2-Step Random Access

As discussed in Section 11.3.3, 2-step random access can be analyzed using the model of multichannel S-ALOHA. It was also shown that the throughput of 2-step random access is dependent on the number of preambles. Thus, in order to improve the throughput, it would be necessary to increase the number of preambles without any significant increase of radio resources. In this section, we present a few variants of 2-step random access that can improve the throughput by effectively increasing the number of preambles or exploiting the preamble transmissions when the BS is equipped with multiple antennas.

11.4.1 2-Step Random Access with MIMO

So far, a simple model is considered for the signal reception at a receiver, i.e., BS, in a time slotted system. That is, if there is only one packet transmitted within a given slot, it is assumed that the BS can successfully decode it. Otherwise, the BS fails to decode the signal.

We can consider a different model if the BS is equipped with multiple antennas. In particular, for a sufficiently large number of antennas, the channel hardening (i.e., the effect of small-scale fading is averaged out and users' channels behave deterministic like a wired channel) and favorable propagation (i.e., the propagation channels to different devices become orthogonal, which makes different users distinguishable in the space domain) can be exploited. The resulting system is referred to as massive MIMO. Thus, in massive MIMO, the BS can decode multiple users' packets transmitted through a shared resource block as long as their channels can be estimated. The structure of slot can also be modified. Recall that the slot is divided into two sub-slots. The second sub-slot is further divided into L mini-slots when massive MIMO is not considered. With massive MIMO, as mentioned earlier, multiple users' packets can be transmitted through a shared resource block, meaning that there is no need to divide the second sub-slot into L mini-slots.

For simplicity, we assume that the length of the first sub-slot for preamble transmissions is equal to the number of preambles, L, in unit time. The length of the second sub-slot for data transmission without massive MIMO, which is proportional to the number of preambles, is given by LQ, where Q is the length of data packet in unit time. Then, the length of slot without massive MIMO becomes

$$T_{\rm wo} = L(1+Q)\,(\text{in unit time}). \tag{11.16}$$

On the other hand, as mentioned earlier, with massive MIMO, all active users transmit their data packets through a shared resource block, which is the second sub-slot. Thus, the length of the second sub-slot becomes Q, and the length of slot becomes

$$T_w = L + Q\,(\text{in unit time}). \tag{11.17}$$

For convenience, denote by ρ the average number of active users per unit time. Then, without massive MIMO, the average number of active users per slot becomes $G_{\rm wo} = \rho T_{\rm wo} = \rho L(1+Q)$. With massive MIMO, we have $G_w = \rho T_w = \rho(L+Q)$. Then, from (11.15), the throughput with/without massive MIMO can be found as

$$\begin{aligned} \eta_{2S,\rm wo} &= G_{\rm wo} e^{-\frac{G_{\rm wo}}{L}} = \rho L(1+Q) e^{-\rho L(1+Q)} \\ \eta_{2S,w} &= G_w e^{-\frac{G_w}{L}} = \rho(L+Q) e^{-\rho\left(1+\frac{Q}{L}\right)}. \end{aligned} \tag{11.18}$$

Then, for a fair comparison of the performance of 2-step random access with/without massive MIMO, we consider the following normalized throughput with/without massive MIMO (in the number of bits per unit time):

$$\begin{aligned} \bar{\eta}_{2S,\rm wo} &= \frac{\eta_{2S,\rm wo} RQ}{T_{\rm wo}} = \rho RQ e^{-\rho(1+Q)} \\ \bar{\eta}_{2S,w} &= \frac{\eta_{2S,w} RQ}{T_w} = \rho RQ e^{-\rho\left(1+\frac{Q}{L}\right)}, \end{aligned} \tag{11.19}$$

where R represents the number of bits per unit time. Note that if the unit time can be the duration of data symbol in modulation, R becomes the number of bits per data symbol. From (11.19), the throughput ratio can be given by

$$\kappa = \frac{\bar{\eta}_{2S,w}}{\bar{\eta}_{2S,wo}} = e^{\rho Q\left(1-\frac{1}{L}\right)}, \tag{11.20}$$

which shows that the throughput gain due to massive MIMO increases with ρQ. In other words, the higher the traffic intensity (ρ) or the longer the data packets (Q), the higher the throughput gain, κ, by massive MIMO.

It is noteworthy that the normalized throughput with massive MIMO in (11.19) is bounded as follows:

$$\begin{aligned}\bar{\eta}_{2S,w} &= \frac{\rho QR\left(1+\frac{Q}{L}\right)}{1+\frac{Q}{L}} e^{-\rho\left(1+\frac{Q}{L}\right)} \\ &\leq \frac{LQR}{L+Q} e^{-1}.\end{aligned} \tag{11.21}$$

This shows that the normalized throughput can be maximized if $L = Q = \frac{T_w}{2}$, and this maximum linearly increases with T_w, i.e.,

$$\begin{aligned}\bar{\eta}_{2S,w} &\leq \frac{LQR}{L+Q} e^{-1} \\ &\leq \frac{Re^{-1}}{4} T_w.\end{aligned} \tag{11.22}$$

With massive MIMO, the performance is mainly decided by the number of preambles, and in order to achieve the maximum normalized throughput, a half of resource (i.e., T_w) has to be allocated to preamble transmission.

11.4.2 Sequential Transmission of Multiple Preambles

As discussed above, with massive MIMO, it is important to effectively increase the number of preambles. When orthogonal sequences are used for preambles, the length of preamble is proportional to the number of preambles. Thus, to increase the number of preambles without increasing the length, we can consider non-orthogonal sequences.

In Jiang et al. (2019), each user transmits multiple randomly selected (orthogonal) preambles, say N preambles. For convenience, the sequence of concatenated preambles is referred to as an extended preamble. Then, the length of the first sub-slot becomes NL, while the number of extended preambles becomes L^N. The extended preambles are not orthogonal. The highest cross-correlation between two different extended preambles is $\frac{N-1}{N}$. Despite the high cross-correlation of the extended preambles, if the BS is able to detect them thanks to the channel hardening and favorable propagation properties of massive MIMO (Jiang et al., 2019), the throughput can be significantly improved.

11.4.3 Simultaneous Transmission of Multiple Preambles

The length of extended preamble is N-time longer than that of preamble. As N increases, more resources are used. To avoid this, another approach can be considered. In Choi (2021b), super-positioned preambles are considered to effectively increase the number of preambles without increasing the length. Each user can choose N different preambles and transmit them simultaneously. The resulting preamble is a sum of N different preambles, which is called an S-preamble. Then, there are $\binom{L}{N}$ S-preambles when the size of the preamble pool is L. As N increases, there can be more S-preambles. However, as S-preambles have high cross-correlations, their detection becomes difficult although the BS is equipped with multiple antennas. In Choi (2021b), an approach that can effectively estimate the channel vectors when $N = 2$ was derived, but the approach for $N > 3$ is not yet known.

11.4.4 Preambles for Exploration

In 2-step random access, a preamble can be chosen and transmitted by multiple users. Denote by K_l the number of the users choosing and transmitting preamble l. In general, the preamble detection at the BS is to determine if $K_l = 0$ or $K_l > 0$ for all l. A bank of L correlators (each correlator is to measure the correlation with its associated preamble) is used for the preamble detection and the decision statistics is the energy of the correlator's output.

If the BS has multiple antennas, the BS becomes capable of not only detecting the presence of transmitted preambles, but also the numbers of users transmitting preambles, i.e., $K_l, l = 1, \dots, L$. Accordingly, the BS may send early feedback after the first sub-slot (i.e., after users' preamble transmissions). This early feedback, from user's perspective, can be viewed as a result of the exploration obtained by transmitting the preamble, and can help improve throughput.

To see how this early feedback can improve the throughput of 2-step random access, consider an example. As in Fig. 11.2, assume that $K = 3$ and $L = 4$. Active user 2 chooses channel 3 to transmit a preamble and two other active users (users 1 and 3) choose channel 4. From this, the feedback information from the BS to the active users is $\{K_1, K_2, K_3, K_4\} = \{0, 0, 1, 2\}$. The set of the users receiving $K_l = 1$ when they transmit preamble $l \in \{1, \dots, L\}$ is referred to as Group I, while the set of the users receiving $K_l > 1$ is referred to as Group II. In addition, the set of the preambles/uplink channels chosen by Group I is denoted by $\mathcal{S}$. Then, in this example, $\mathcal{S} = \{3\}$ and its complement becomes $\mathcal{S}^c = \{1, 2, 4\}$, while active user 2 belongs to Group I and active users 1 and 3 belong to Group II.

The users in Group I can proceed to transmit data packets through the channels corresponding to their transmitted preambles as there will not be other users

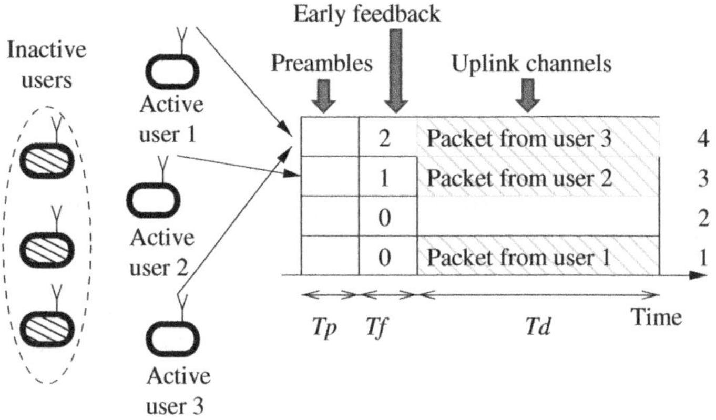

Figure 11.4 An example with $K = 3$ and $L = 4$, where active user 2 transmits preamble 3 and two other active users 1 and 3 transmit preamble 4.

transmitting. On the other hand, the users in Group II may not be able to transmit through the channels corresponding to their transmitted preambles due to possible collisions. Then, each user in Group II may choose a channel in $\mathcal{S}^c$. As shown in Fig. 11.4, users 1 and 3 in Group 2 reselect channels in $\mathcal{S}^c = \{1, 2, 4\}$, say channels 1 and 4, respectively, and they succeed to send their packets without collisions. This clearly shows that the early feedback (on preamble collisions) can help avoid data packet collisions and improve the throughput (Choi, 2020). In Burkov et al. (2022), a detailed analysis of this approach is presented.

11.5 Application of NOMA to Random Access

NOMA has been extensively studied as it can improve the spectral efficiency of cellular systems by exploiting the power domain (Ding et al., 2017). In MTC, in order to support more users/devices, it would also be natural to apply NOMA. In this section, we discuss the application of NOMA to random access and show how it can improve the throughput in MTC.

11.5.1 Power-Domain NOMA

In this subsection, we briefly explain the power-domain NOMA for uplink transmissions.

Suppose that there are K users and one receiver. Let P_k be the transmit power of user k and σ^2 the noise variance. For simplicity, we assume the additive white

Gaussian noise (AWGN) channel where the channel gains of users are normalized. Then, if user k is only a user transmitting in a slot, the signal-to-noise ratio (SNR) at the receiver is given by

$$\mathsf{SNR} = \frac{P_k}{\sigma^2}. \tag{11.23}$$

Let $\Gamma = \frac{P^*}{\sigma^2}$ be the SNR that allows the receiver to successfully decode the transmitted packet in a given slot, where P^* represents the minimum transmit power that meets the SNR threshold, Γ.

Suppose that two users, say users 1 and 2, transmit simultaneously with $P_k = P$, $k \in \{1, 2\}$, and the receiver is to decode the packet transmitted by user 1. Then, the resulting SNR becomes

$$\mathsf{SNR} = \frac{P_1}{P_2 + \sigma^2} = \frac{P}{P + \sigma^2} < \Gamma, \tag{11.24}$$

which means that the receiver is unable to decode the packet transmitted by user 1. Actually, the receiver fails to decode any packet due to the interfering signal from the other user.

In order to decode both the signals from users 1 and 2, different power levels can be used. Let $P_k \in \{P_H, P_L\}$, where $P_L < P_H$. Suppose that the receiver is to decode the signal associated with P_H, while the signal associated with P_L is regarded as the interfering signal. Let P_H and P_L be decided to satisfy the following inequalities:

$$\frac{P_H}{P_L + \sigma^2} \geq \Gamma$$
$$\frac{P_L}{\sigma^2} \geq \Gamma.$$

Then, the receiver is able to decode the signal associated with P_H and remove it from the received signal using successive interference cancelation (SIC). Then, the SNR becomes $\frac{P_L}{\sigma^2} \geq \Gamma$, which means that the receiver can also decode the signal associated with P_L. This shows that as long as two users choose different power levels, the receiver is able to decode all the signals successfully. With more power levels, the receiver is able to decode more signals that are transmitted simultaneously. In this case, each power level can be seen as a virtual channel for multiple access, which results in power-domain NOMA.

It is noteworthy that it is necessary to coordinate signal transmissions for power-domain NOMA by assigning different power levels to users. In Section 11.5.2, we show that power-domain NOMA can also be used for uncoordinated transmissions, i.e., random access.

11.5.2 S-ALOHA with NOMA

To apply the notion of power-domain NOMA to S-ALOHA, consider three different power levels as follows:

$$P_k \in \{0, P_L, P_H\}, \tag{11.25}$$

where P_H and P_L represent the high and low transmit power levels, respectively. Suppose that one of two non-zero power levels, i.e., P_H and P_L, can be equally chosen by an active user. Certainly, the receiver is able to decode a packet if there is one active user (as in S-ALOHA) regardless of the selection of transmit power level. Due to NOMA, the receiver can also decode up to two packets if two users are active simultaneously and each one choose a different power level.

In order to see the throughput improvement by NOMA, suppose that the number of active users follows a Poisson distribution with mean G. Then, the throughput becomes

$$\begin{aligned}\eta_{\text{noma}} &= \Pr(\text{one active user}) + \Pr(\text{two active users}) \underbrace{\frac{1}{2}}_{(a)} \underbrace{2}_{(b)} \\ &= Ge^{-G} + \frac{G^2}{2!}e^{-G},\end{aligned} \tag{11.26}$$

where *(a)* is the probability that one active user chooses P_H and the other active user chooses P_L and *(b)* is the number of successfully received packets, which is 2 as one transmits a packet with a transmit power of P_H and the other P_L. Recall that the throughput of S-ALOHA is $\eta_S(G) = Ge^{-G}$. Thus, by applying NOMA to S-ALOHA, additional throughput gain, which is $\frac{G^2}{2}e^{-G}$ according to (11.26), can be achieved. The resulting system is referred to as NOMA-ALOHA.

Figure 11.5 shows the throughput curves of pure ALOHA, S-ALOHA, and NOMA-ALOHA protocols as functions of G. NOMA-ALOHA outperforms the others at the expense of high transmit power. On the other hand, pure ALOHA has the worst performance but the least requirements (no synchronization required, no high transmit power required).

Ideally, a higher throughput can be achieved with more power levels as the probability of collisions decreases. However, transmit power increases exponentially as more power levels are required (Choi, 2017), meaning that the number of power levels has to be limited.

There is a similarity between NOMA and the capture effect in that it can increase throughput due to differences in received signal strength in the event of a collision. There can be one strong signal, while the other signals are relatively weak thanks to the near-far effect. In this case, the receiver can detect a strong signal even if

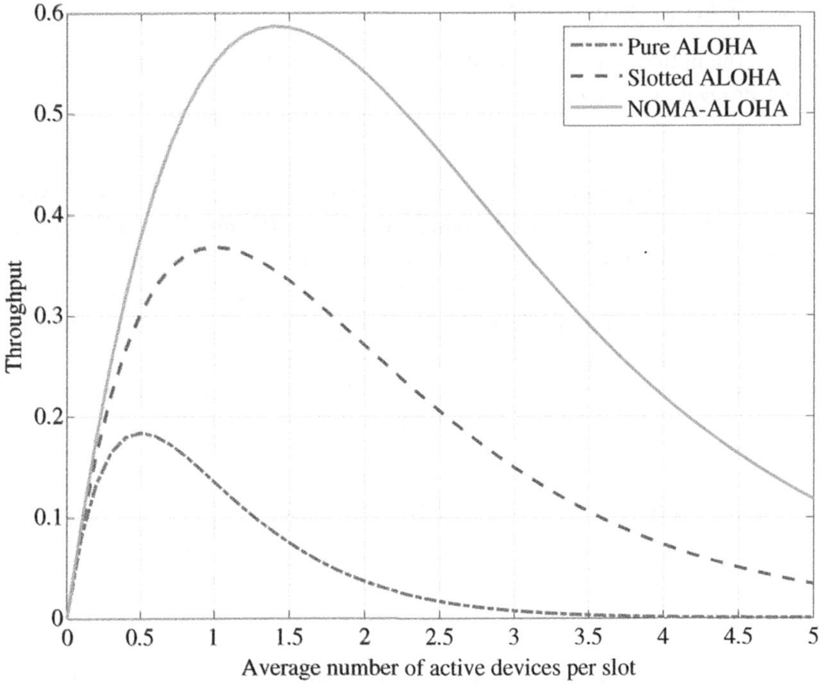

Figure 11.5 Throughput curves of pure ALOHA, S-ALOHA, and NOMA-ALOHA protocols as functions of G.

a collision occurs due to simultaneous transmission by multiple users, which is called the capture effect. However, if a tight power control policy is used in random access, the capture effect cannot be exploited. On the other hand, in power-domain NOMA, tight power control is required to perform SIC.

11.5.3 A Generalization with Multiple Channels

In wireless communication, the channel coefficient from a user to a receiver (i.e., a BS) is not fixed, but time-varying and can be modeled as a random variable. As a result, in order to apply power-domain NOMA, a tight power control policy is necessary. To see this, let h_k be the channel coefficient from user k to the receiver. If user k chooses power level P_H, then the transmit power, denoted by P_k, has to satisfy the following relation: $P_H = |h_k|^2 P_k$. In general, based on channel inversion power control, P_k is decided in NOMA-ALOHA as follows: $P_k = \frac{P}{|h_k|^2}$, $P \in \{0, P_L, P_H\}$.

In multichannel S-ALOHA, there are L channels and each channel may have a different coefficient, denoted by $h_{k,l}$. Then, when user k chooses channel l, P_k

becomes

$$P_k = \frac{P}{|h_{k,l}|^2}, \; P \in \{0, P_L, P_H\}. \tag{11.27}$$

Recall that each user chooses one of L channels uniformly at random in multichannel S-ALOHA. Now, in order to decrease the transmit power, a user can choose the channel associated with the largest channel coefficient (in magnitude). Thus, (11.27) can be replaced with

$$P_k = \min_l \frac{P}{|h_{k,l}|^2} = \frac{P}{\max_l |h_{k,l}|^2}, \; P \in \{0, P_L, P_H\}. \tag{11.28}$$

The resulting reduction in transmit power (i.e., $\frac{P}{\max_l |h_{k,l}|^2} \leq \frac{P}{|h_{k,l}|^2}$) can be viewed as the multichannel diversity gain in NOMA-ALOHA.

To find the throughput of multichannel NOMA-ALOHA, denote by K_H and K_L the numbers of the users transmitting with power levels P_H and P_L, respectively, and consider a user choosing channel l, where $l \in \{1, \dots, L\}$. When the user's transmit power is P_H, the packet can be successfully transmitted without collision if there is no other users with P_H and there is at most one user with P_L in channel l. Then, the corresponding probability is given by

$$P_{H,\text{nc}} = \left(1 - \frac{1}{L}\right)^{K_H - 1} \left(\binom{K_L}{0} \left(1 - \frac{1}{L}\right)^{K_L} + \binom{K_L}{1} \frac{1}{L} \left(1 - \frac{1}{L}\right)^{K_L - 1} \right).$$

When the transmit power is P_L, the probability of successful transmission becomes

$$P_{L,\text{nc}} = \left(1 - \frac{1}{L}\right)^{K_L - 1} \left(\binom{K_H}{0} \left(1 - \frac{1}{L}\right)^{K_H} + \binom{K_H}{1} \frac{1}{L} \left(1 - \frac{1}{L}\right)^{K_H - 1} \right).$$

As a result, the conditional throughput for given (K_H, K_L), becomes

$$\eta(K_H, K_L) = P_{H,\text{nc}} K_H + P_{L,\text{nc}} K_L. \tag{11.29}$$

If one of the two transmit powers, P_H and P_H, is equally chosen and the total number of active users follows a Poisson distribution with mean G, K_H and K_L are independent Poisson random variables with mean $\frac{G}{2}$. Then, after some manipulations, the average throughput is given by

$$\begin{aligned} \eta &= \mathbb{E}[\eta(K_H, K_L)] \\ &= G\left(1 + \frac{G}{2L}\right) e^{-\frac{G}{L}}. \end{aligned} \tag{11.30}$$

The throughput can be maximized when $G^* = \sqrt{2}L$, and the maximum throughput becomes

$$\eta^* = \max_G G\left(1 + \frac{G}{2L}\right) e^{-\frac{G}{L}} = Le^{-\sqrt{2}}(1 + \sqrt{2}). \tag{11.31}$$

We can see that the throughput of multichannel NOMA-ALOHA is higher than that of conventional multichannel ALOHA (without NOMA) in (11.15) by a factor of $1 + \frac{G}{2L}$. In terms of maximum throughput, multichannel NOMA-ALOHA offers $\frac{e^{-\sqrt{2}}(1+\sqrt{2})}{e^{-1}} \approx 1.6$ times higher performance than conventional multi-channel ALOHA.

11.5.4 NOMA-ALOHA Game

The NOMA-ALOHA game is a generalization of the random access game in Section 11.2. The set of actions of user k expands to $S_k = \{0, P_L, P_H\}$, and this is the set of power levels, from which a user choose one.

For multichannel S-ALOHA, assume that there are L orthogonal channels. In addition, for mixed strategies, denote by $a_{k,l}$ or $b_{k,l}$ the probability that user k transmits a packet with power level P_H or P_L through channel l, respectively. Since only one channel is chosen at a time, we have

$$a_k = \sum_{n=1}^{N} a_{k,l} \leq 1,\ b_k = \sum_{n=1}^{N} b_{k,l} \leq 1,$$
$$\text{and } a_k + b_k \leq 1,$$

while $1 - a_k + b_k$ is the probability that user k does not transmit. For convenience, let $\boldsymbol{\sigma}_{k,l} = (a_{k,l}, b_{k,l}, 1 - a_{k,l} - b_{k,l})$ and $\boldsymbol{\sigma}_k = [\boldsymbol{\sigma}_{k,1} \ \dots \ \boldsymbol{\sigma}_{k,L}]^{\mathrm{T}}$. Here, $\boldsymbol{\sigma}_k$ is the mixed strategy of user k. In addition, let Δ_k denote the set of all possible mixed strategy profiles for user k, $\boldsymbol{\sigma}_k$. Since all the users have the same set, we simply write $\Delta = \Delta_1 = \cdots \Delta_K$.

For given mixed strategies, $\{\boldsymbol{\sigma}_k\}$, the payoff of user k can be found as

$$\begin{aligned} u_k(\boldsymbol{\sigma}_k, \boldsymbol{\sigma}_{-k}) &= \sum_{n=1}^{N} (a_{k,l} q_{k,l}^H + b_{k,l} q_{k,l}^L) - c(1 - a_k - b_k) \\ &= \mathbf{q}_k^{\mathrm{T}} \mathbf{p}_k - c\mathbf{1}^{\mathrm{T}} \mathbf{p}_k, \end{aligned} \tag{11.32}$$

where $q_{k,l}^H = \prod_{i \neq k}(1 - a_{i,l})$, $q_{k,l}^L = \prod_{i \neq k}(1 - b_{i,l})$, $\mathbf{1} = [1 \ \dots \ 1]^{\mathrm{T}}$, and

$$\begin{aligned} \mathbf{p}_k &= [a_{k,1} \ \cdots \ a_{k,L} \ b_{k,1} \ \cdots \ b_{k,L}]^{\mathrm{T}} \\ \mathbf{q}_k &= [q_{k,1}^H \ \cdots \ q_{k,L}^H \ q_{k,1}^L \ \cdots \ q_{k,L}^L]^{\mathrm{T}}. \end{aligned} \tag{11.33}$$

Then, a mixed strategy NE, $\{\boldsymbol{\sigma}_k^*\}$ is to satisfy

$$u_k(\boldsymbol{\sigma}_k^*, \boldsymbol{\sigma}_{-k}^*) \geq u_k(\boldsymbol{\sigma}_k, \boldsymbol{\sigma}_{-k}^*) \text{ for all } \boldsymbol{\sigma}_k \in \Sigma. \tag{11.34}$$

Since the game with the payoff function in (11.32) is a symmetric game, fictitious play can be used to find a mixed strategy NE. However, as discussed in Section 11.5.3, the transmit power depends on the channel coefficients that might be taken into account the cost of transmission, c. In this case, c becomes a variable and different from a user to another, meaning that the resulting game is no longer a symmetric game. Thus, different approaches (e.g., multi-agent reinforcement learning algorithms) can be considered to find a solution that is not necessarily a mixed strategy NE.

11.6 Low-Latency Access for MTC

Massive MTC focuses on the connectivity that can support a large number of devices with limited spectrum, but generally does not guarantee low access latency. Thus, the approaches used for massive MTC may not be well-suited to mission-critical applications with specific low-latency requirements. In addition, random access requires a different design perspective when long propagation delay cannot be avoided in some cases (e.g., satellite communications). In this section, we discuss a few approaches that can be used when low-latency access is required and/or when propagation delay is long.

11.6.1 Long Propagation Delay

In general, the design of wireless communication systems is affected by signal propagation delay. For example, consider three different systems, namely wireless local area networks (WLAN), cellular systems, and low earth orbit (LEO) satellite systems, with nominal communication ranges, 10 m, 1 km, and 1000 km, respectively. Then, the propagation delays can be shown as follows:

$$\text{Propagation delay of WLAN} = 0.033\,\mu s$$
$$\text{Propagation delay of Cellular} = 3.333\,\mu s$$
$$\text{Propagation delay of LEO} = 3.333\,\text{ms}.$$

Without taking into account processing delay, the round trip time (RTT) can be estimated as twice the propagation delay. For low latency applications (that require a few msec latency), if the RTT is sufficiently short, retransmissions can be allowed, which might be the cases of WLAN and cellular systems. However, if the RTT is too long as in LEO satellite systems, it would be desirable to avoid retransmissions if possible. As a result, it might be desirable for MTC to design random access schemes to provide a reliable connectivity without retransmissions.

In satellite communications or any systems for low-latency applications with long propagation delays, the peak throughput of random access may not be a useful performance measure, because the system may experience frequent packet collisions when operating close to peak throughput. Instead, the probability of packet collisions can be a good measure and the system is expected to operate close to a certain low target packet collision probability (e.g., 10^{-3}).

Unlike the throughput, the packet collision probability is a performance measure from a user perspective. For multichannel S-ALOHA, the packet collision probability can be obtained as follows. Suppose that there are K users with packets to send. For a user choosing a specific channel among K users, the conditional packet collision probability is given by

$$\begin{aligned} P_{\text{col}}(K) &= \sum_{k=1}^{K-1} \left(\frac{1}{L}\right)^k \left(1 - \frac{1}{L}\right)^{K-1-k} \\ &= 1 - \left(1 - \frac{1}{L}\right)^{K-1}. \end{aligned} \tag{11.35}$$

If the number of active users, K, follows a Poisson distribution with mean G, the packet collision probability becomes

$$\begin{aligned} P_{\text{col}} &= \mathbb{E}[P_{\text{col}}(K) \,|\, K \geq 1] \\ &= 1 - \frac{1}{\Pr(K \geq 1)} \sum_{k=1}^{\infty} \left(1 - \frac{1}{L}\right)^{k-1} \frac{G^k e^{-G}}{k!} \\ &= 1 - \frac{e^{-G}}{\left(1 - \frac{1}{L}\right)(1 - e^{-G})} \left(e^{G\left(1-\frac{1}{L}\right)} - 1\right) \\ &= 1 - \frac{e^{-\frac{G}{L}} - e^{-G}}{\left(1 - \frac{1}{L}\right)(1 - e^{-G})}. \end{aligned} \tag{11.36}$$

In (11.36), the case that there is no user transmitting (i.e., $K = 0$) is excluded as there has to be at least one active user to find the packet collision probability as a user perspective performance measure. For a sufficiently small G (i.e., $G \ll 1$), it can be shown that

$$P_{\text{col}} \approx \frac{G}{2L}. \tag{11.37}$$

This clearly shows that more resources (i.e., channels) help reduce the packet collision probability. For example, in order to achieve a packet collision probability of 10^{-3} with one active device per unit time on average (i.e., $G = 1$), there should be $L = 500$ different channels if no retransmission strategies are used. Certainly, this is a case that requires more resources than one might think.

11.6.2 Repetition Diversity

In MTC with 2-step random access, for low-latency applications, it is expected to keep the packet collision probability low. In Choudhury and Rappaport (1983) and Östman et al. (2018), a simple diversity scheme for a low packet collision probability is considered. In this subsection, we slightly generalize this scheme.

Suppose that a user wishes to send M packets. As in Choudhury and Rappaport (1983) and Östman et al. (2018), each packet is sent multiple times, say D times, without any feedback. Hence, D is a design parameter. If one of D copies of a packet is received without collision, we assume that the BS can successfully decode the packet. The corresponding probability is given by

$$P_{\text{succ}}(1) = \sum_{d=1}^{D}(1 - P_{\text{col}})^d P_{\text{col}}^{D-d} = 1 - P_{\text{col}}^D. \tag{11.38}$$

Since there are M different packets to be sent by the user, the probability that all the packets are successfully received becomes $P_{\text{succ}}(M) = P_{\text{succ}}(1)^M = \left(1 - P_{\text{col}}^D\right)^M$, and the corresponding error or outage probability is given by

$$P_{\text{e}} = 1 - P_{\text{succ}}(M) = 1 - \left(1 - P_{\text{col}}^D\right)^M. \tag{11.39}$$

If P_{col} is sufficiently small, we have

$$P_{\text{e}} = 1 - (1 - MP_{\text{col}}^D + O(P_{\text{col}}^{2D})) \approx MP_{\text{col}}^D, \tag{11.40}$$

which indicates that as M increases, we need to increase D to keep P_{e} low.

It is noteworthy that since each user transmits a packet D times, the traffic intensity increases by a factor of D (i.e., G in (11.37) has to be replaced with GD). Thus, from (11.37), if $G \ll L$, we have

$$P_e \approx M\left(\frac{GD}{2L}\right)^D, \tag{11.41}$$

which shows that P_e can be minimized when $D^* = \frac{2L}{G}e^{-1} \approx 0.7358\frac{L}{G}$. In other words, D can be optimized for given traffic intensity to lower the probability of error. It can be further shown that

$$\min_D P_e \approx Mc^{\frac{L}{G}}, \tag{11.42}$$

where $c = e^{-2e^{-1}} \approx 0.4791$. Thus, with optimized D, when $G = 1$ and $M = 1$, $L = 10$ channels are enough to keep a probability of error below 10^{-3}. Note that without a diversity scheme, as discussed earlier, it was shown that $L = 500$ channels are required.

Figure 11.6 shows the probability of error as a function of G when $L = 4$ and $M = 4$. As expected, the probability of error is minimized when D is around $D^* \approx 29.44$ if $G = 0.1$.

In Casini et al. (2007), a random access frame that consists of multiple time slots, say N_{F} slots, is considered where each user sends multiple copies of a packet within a given frame. The resulting scheme is called contention resolution diversity S-ALOHA (CRDSA). In general, the number of copies that a user transmits has to be smaller than N_{F} to avoid collisions with packets sent by other users, but not too small to exploit the diversity gain. In addition, it is assumed that the receiver

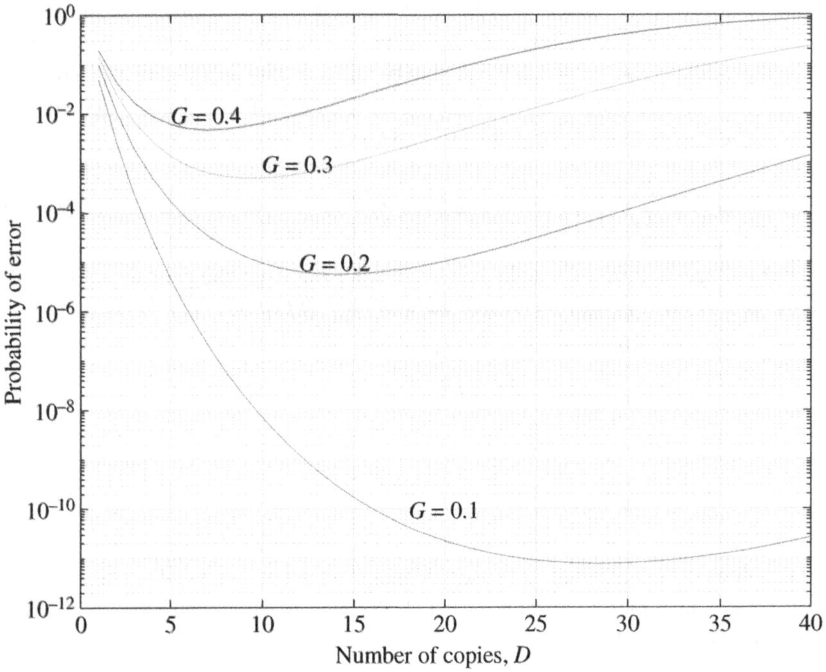

Figure 11.6 Probability of error as a function of G when $L = 4$ and $M = 4$.

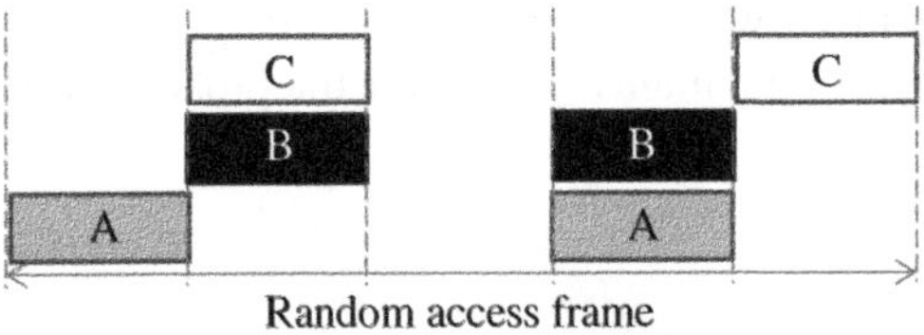

Figure 11.7 Contention resolution diversity S-ALOHA. Source: Adapted from Casini et al. (2007).

is able to perform SIC so that more packets can be decoded. For example, suppose that $N_F = 5$ and there are 3 users (users A, B, and C). Each user sends two copies of a packet as illustrated in Fig. 11.7. In the 2nd and 4th slots, two packets are collided. However, packets A and C can be decoded as they are not collided with others in the 1st and 5th slots, respectively. Once they are decoded, packet B can be decoded after removing packet C in the 2nd slot or packet A in the 4th slot using SIC, which can further decrease the probability of decoding error.

11.6.3 Channel Coding-Based Random Access

Since the repetition diversity scheme in Section 11.6.2 can be seen as a channel coding scheme at the packet level, other schemes based on channel coding techniques can also be applied to random access so that the probability of error decreases.

In Liva (2011), CRDSA is analyzed and optimized. In particular, the notion of graph-based coding Richardson and Urbanke, (2008) is applied to the design of CRDSA. In Choi and Ding (2021), network coding (Li et al., 2003) is applied to 2-step random access, and it is shown that a low error rate can be achieved without using SIC at the receiver. It is noteworthy that SIC needs additional requirements as a reliable channel estimation has to be carried out to reproduce the received signals for cancellation. Furthermore, in some cases, it may not be possible to use SIC. For example, in LEO satellite communications, the channel is highly time-varying. Thus, when user A transmits copies of a packet in the 1st and 4th slots as shown in Fig. 11.7, the channel coefficients can be different. This implies that even if the receiver, which is an LEO satellite, is able to decode packet A in the first slot, it may be unable to cancel the other copy of packet A in the 4th slot as its channel coefficient differs from that in the 1st slot and the received signal cannot be reproduced for SIC. Consequently, for highly time-varying channels, coded random access schemes that do not rely on SIC would be desirable.

In general, coded random access schemes can provide a low probability of error when the traffic intensity is low at the cost of encoding and decoding complexities at both transmitter (i.e., a user) and receiver.

References

3GPP. *Evolved Universal Terrestrial Radio Access (EUTRA) and Evolved Universal Terrestrial Radio Access Network (EUTRAN); Overall Description, TS 36.300 v.14.7.0.* 3rd Generation Partnership Project (3GPP), Jun. 2018.

Artem Burkov, Andrey Turlikov, and Roman Rachugin. Analyzing and stabilizing multichannel ALOHA with the use of the preamble-based exploration phase. *Information and Control Systems*, (5):49–59, Oct. 2022. doi: 10.31799/1684-8853-2022-5-49-59.

Enrico Casini, Riccardo De Gaudenzi, and Oscar Del Rio Herrero. Contention resolution diversity slotted ALOHA (CRDSA): An enhanced random access scheme for satellite access packet networks. *IEEE Transactions on Wireless Communications*, 6(4):1408–1419, 2007. doi: 10.1109/TWC.2007.348337.

J Choi. NOMA-based random access with multichannel ALOHA. *IEEE Journal on Selected Areas in Communications*, 35(12):2736–2743, Dec. 2017.

Jinho Choi. On improving throughput of multichannel ALOHA using preamble-based exploration. *Journal of Communications and Networks*, 22(5):380–389, 2020. doi: 10.1109/JCN.2020.000024.

Jinho Choi. On fast retrial for two-step random access in MTC. *IEEE Internet of Things Journal*, 8(3):1428–1436, 2021a. doi: 10.1109/JIOT.2020.3012449.

Jinho Choi. An approach to preamble collision reduction in grant-free random access with massive MIMO. *IEEE Transactions on Wireless Communications*, 20(3):1557–1566, 2021b. doi: 10.1109/TWC.2020.3034308.

Jinho Choi and Jie Ding. Network coding for K-repetition in grant-free random access. *IEEE Wireless Communications Letters*, 10(11):2557–2561, 2021. doi: 10.1109/LWC.2021.3107314.

Young-June Choi, Suho Park, and Saewoong Bahk. Multichannel random access in OFDMA wireless networks. *IEEE Journal on Selected Areas in Communications*, 24(3):603–613, Mar. 2006.

G Choudhury and S Rappaport. Diversity ALOHA - a random access scheme for satellite communications. *IEEE Transactions on Communications*, 31(3):450–457, 1983. doi: 10.1109/TCOM.1983.1095828.

Z Ding, Y Liu, J Choi, M Elkashlan, I Chih-Lin, and H V Poor. Application of non-orthogonal multiple access in LTE and 5G networks. *IEEE Communications Magazine*, 55(2):185–191, Feb. 2017.

Drew Fudenberg and David K Levine. *The Theory of Learning in Games*. Cambridge, MA: MIT Press, 1998.

Drew Fudenberg and Jean Tirole. *Game Theory*. Cambridge, MA: MIT Press, 1991.

Hao Jiang, Daiming Qu, Jie Ding, and Tao Jiang. Multiple preambles for high success rate of grant-free random access with massive MIMO. *IEEE Transactions on Wireless Communications*, 18(10):4779–4789, 2019. doi: 10.1109/TWC.2019.2929126.

F Kelly and E Yudovina. *Stochastic Networks*. Cambridge University Press, 2014.

J Kim, G Lee, S Kim, T Taleb, S Choi, and S Bahk. Two-step random access for 5G system: Latest trends and challenges. *IEEE Network*, 35(1):273–279, 2021. doi: 10.1109/MNET.011.2000317.

S-Y R Li, R W Yeung, and N Cai. Linear network coding. *IEEE Transactions on Information Theory*, 49(2):371–381, Feb. 2003.

Gianluigi Liva. Graph-based analysis and optimization of contention resolution diversity slotted ALOHA. *IEEE Transactions on Communications*, 59(2):477–487, 2011. doi: 10.1109/TCOMM.2010.120710.100054.

James L Massey. *Collision-Resolution Algorithms and Random-Access Communications*, pages 73–137. Vienna: Springer Vienna, 1981. ISBN 978-3-7091-2900-5. doi: 10.1007/978-3-7091-2900-54.

Johan Östman, Rahul Devassy, Guido Carlo Ferrante, and Giuseppe Durisi. Low-latency short-packet transmissions: Fixed length or HARQ? In *2018 IEEE Globecom Workshops (GC Wkshps)*, pages 1–6, 2018. doi: 10.1109/GLOCOMW.2018.8644397.

Tom Richardson and Rüdiger Urbanke. *Modern Coding Theory*. Cambridge University Press, 2008. doi: 10.1017/CBO9780511791338.

Y Yu and G B Giannakis. High-throughput random access using successive interference cancellation in a tree algorithm. *IEEE Transactions on Information Theory*, 53(12):2253–2256, Dec. 2007.

12

Grant-Free Random Access via Compressed Sensing: Algorithm and Performance

Yongpeng Wu, Xinyu Xie, Tianya Li, and Boxiao Shen

Department of Electronic Engineering,School of Electronic Information and Electrical Engineering, Shanghai Jiao Tong University, Shanghai, China

12.1 Introduction

The rapid evolution of the Internet-of-Things (IoT) has led to explosive growth in the number of devices. In the next 10 years, the number of IoT devices is predicted to reach hundreds of billions. For energy saving, these cheap devices stay silent most of the time and are randomly activated with low probability. Meanwhile, only small-size packets are delivered to the base station (BS).

Unfortunately, applying conventional access technologies tailored for human-type communications to this burgeoning scenario yields extremely low spectral efficiency since delicately arranging transmission resources to a massive number of users would result in prohibitive signaling overhead and long latency. To overcome these shortcomings, next generation multiple access schemes for massive machine-type communications (mMTC) prefer a protocol that allows the active users to access the network without any grant, hence the name *grant-free*. In the lack of pre-coordination, grant-free users usually send preambles to the BS to assist the upcoming data transmission. These pilot sequences are generally non-orthogonal to make full use of the limited time/frequency resources. Based on how they are allocated, we divide practical grant-free random access schemes into two categories. The first type, termed *sourced random access* (Liu and Yu, 2018, Ke et al., 2020), assigns pilots unique to the identity of the device, which serve as the training sequences for device activity detection (DAD) and channel estimation (CE). Another type, termed *unsourced random access* (URA) (Polyanskiy, 2017, Fengler et al., 2021), instead employs a common codebook for data encoding, i.e., the choices of codewords are irrelevant to the user identities. Accordingly, the receiver only acquires a list of transmitted messages with no obligation to link them to the transmitters.

Next Generation Multiple Access, First Edition.
Edited by Yuanwei Liu, Liang Liu, Zhiguo Ding, and Xuemin Shen.
© 2024 The Institute of Electrical and Electronics Engineers, Inc. Published 2024 by John Wiley & Sons, Inc.

In both scenarios, the sporadic traffic pattern of grant-free users attaches a sparse structure to the received signal. The recent development of *compressed sensing* (CS) suggests that a signal having a sparse representation can be reconstructed from a small set of measurements, which finds a position in the toolbox for solving the involving signal processing problems. In the remainder of this chapter, we will introduce a variety of CS algorithms for practical grant-free random access schemes. In Section 12.2, we propose a belief propagation (BP)-based algorithm for joint DAD, CE, and data decoding (DD) under a two-stage URA transmission framework. The multi-path channel sparsity is further exploited in Section 12.3, where a hybrid approximate message passing (AMP) algorithm is investigated for a channel-recovery-centric URA scheme. In Section 12.4, we focus on the emerging low earth orbit (LEO) satellite communications technology. A CS method combining AMP and sparse Bayesian learning (SBL) is developed for joint DAD and CE with the help of the orthogonal time-frequency space (OTFS) modulation to restrain the influence of high mobility.

12.2 Joint Device Detection, Channel Estimation, and Data Decoding with Collision Resolution for MIMO Massive Unsourced Random Access

In this section, we investigate a joint DAD, CE, and DD algorithm for multiple-input multiple-output (MIMO) massive URA. The data in the proposed framework is split into two parts. A portion of the data is coded by CS and the rest is low-density-parity-check (LDPC) coded, whereas the sparse interleave-division multiple access (IDMA) is exploited to reduce the multiuser interference. BP-based algorithms are built for the decoding of two transmission phases. Furthermore, we propose a collision resolution scheme referred to as *energy detection and sliding window protocol (ED-SWP)* to deal with the possible codeword collisions that are common in URA.

12.2.1 System Model and Encoding Scheme

In this section, we first introduce the system model of massive URA in MIMO systems. Then, the encoding framework is illustrated in detail, in which a portion of the data is coded by CS and the rest is LDPC coded.

12.2.1.1 System Model

Consider the uplink of a single-cell cellular network consisting of K_{tot} single-antenna devices, which are served by a BS equipped with M antennas.

The sporadic device activity is investigated for the massive access scenario, i.e., a small number, $K_a \ll K_{\text{tot}}$ of users are active within a coherence time. Each user has B bits of information to be coded and transmitted into a block-fading channel with L channel uses. Let $\mathbf{v}_k \in \{0,1\}^B$ denote device k's binary message and $f(\cdot) : \{0,1\}^B \rightarrow \mathbb{C}^L$ is some one-to-one encoding function. Typically, the corresponding received signal can be written as

$$\mathbf{Y} = \sum_{k \in \mathcal{K}_{\text{tot}}} \phi_k f(\mathbf{v}_k)\mathbf{h}_k^T + \mathbf{Z}, \tag{12.1}$$

where ϕ_k denotes the device activity indicator, which is one when user k is active or zero otherwise. $\mathbf{h}_k \in \mathbb{C}^{M\times 1}$ is the channel vector of device k, which is modeled as Rayleigh fading, i.e., $\mathbf{h}_k \triangleq \sqrt{\beta_k}\mathbf{g}_k$ and $\mathbf{g_k} \sim \mathcal{CN}(0, \mathbf{I}_M)$. β_k and $\mathbf{g}_k$ denote the path loss and small scaling fading, respectively. $\mathbf{Z} \in \mathbb{C}^{L\times M}$ is the additive white Gaussian noise (AWGN) matrix distributed as $\mathcal{CN}(0, \sigma^2\mathbf{I}_M)$.

The BS's task is to reproduce a list of transmitted messages $\mathcal{L}(\mathbf{Y})$ without identifying from whom they are sent, thereby leading to the so-called URA. The performance of a URA system is evaluated by the probability of missed detection and false alarm, denoted by p_{md} and p_{fa}, respectively, which are given by:

$$p_{\text{md}} = \frac{1}{K_a} \sum_{k \in \mathcal{K}_a} P\left(\mathbf{v}_k \notin \mathcal{L}(\mathbf{Y})\right), \tag{12.2}$$

$$p_{\text{fa}} = \frac{\left|\mathcal{L}(\mathbf{Y}) \setminus \left\{\mathbf{v}_k : k \in \mathcal{K}_a\right\}\right|}{|\mathcal{L}(\mathbf{Y})|}. \tag{12.3}$$

In this system, the code rate $R_c = B/L$ and the spectral efficiency $\mu = \frac{B\cdot K_a}{L\cdot M}$. Let P denote the power (per symbol) of each device, then the energy-per-bit E_b/N_0 is defined by $E_b/N_0 \triangleq \frac{LP}{2B}$.

12.2.1.2 Encoding Scheme

The entail data encoding process is in a hierarchical form. The B bits of information are first divided into two parts, namely, B_p and B_c bits with $B_p + B_c = B$. Typically, one would want $B_p \ll B_c$. The former B_p information bits are served as the preamble and coded by a CS-based encoder to pick codeword from the common codebook. In the receiver, the codewords are tasked to reconstruct part of the messages, the number of active devices, channel coefficients as well as interleaving patterns for the latter part. The remaining B_c bits are coded with LDPC codes. For clarity, we denote the former and latter encoding processes as CS and LDPC phases, respectively. Correspondingly, the total L channel uses are split into two segments of lengths L_p and L_c, respectively, with $L_p + L_c = L$. Since only a small fraction of data is CS coded and the rest is LDPC coded, the efficiency in our scheme is higher than those purely CS-coded schemes. We elaborate on these two encoding phases below.

CS Phase In the CS phase, each active user pick up a codeword from the share codebook $\mathbf{A} = [\mathbf{a}_1, \mathbf{a}_2, \dots, \mathbf{a}_{2^{B_p}}] \in \mathbb{C}^{L_p \times 2^{B_p}}$ according to the transmitted message, where the power of the codeword is fixed to $\|\mathbf{a}_i\|_2^2 = L_p$ by the power constraint. Specifically, let $\mathbf{v}_k^p$ denote the former B_p bits information and i_k is the corresponding decimal value (plus one), i.e., $i_k \in [1 : 2^{B_p}]$. Then, the coded sequence is obtained by taking the transpose of the i_kth column of $\mathbf{A}$, i.e., $\mathbf{a}_{i_k}^T$. Therefore, the corresponding received signal can be written as

$$\mathbf{Y} = \sum_{k \in \mathcal{K}_{\text{tot}}} \phi_k \mathbf{a}_{i_k} \mathbf{h}_k^T + \mathbf{Z} \triangleq \mathbf{A}\mathbf{\Phi}\tilde{\mathbf{H}} + \mathbf{Z}, \tag{12.4}$$

where $\mathbf{\Phi} = \text{diag}\{\phi_1, \dots, \phi_{2^{B_p}}\}$ denotes the diagonal binary selection matrix and $\phi_r = |\mathcal{K}_r|$ with $\mathcal{K}_r = \{k \in \mathcal{K}_a : i_k = r\}$ denoting the set of users selecting the rth codeword. Correspondingly, the effectively channel matrix $\tilde{\mathbf{H}} = [\tilde{\mathbf{h}}_1, \dots, \tilde{\mathbf{h}}_{2^{B_p}}]$ is given by $\tilde{\mathbf{h}}_r = \sum_{k \in \mathcal{K}_r} \mathbf{h}_k$. Intuitively, the non-zero elements of $\mathbf{\Phi}$ is no more than $K_a (K_a \ll B_p)$. As such, (12.4) can be formulated as the standard CS problem, where the matrix $\mathbf{\Phi}\tilde{\mathbf{H}}$ is row-sparse and can be recovered jointly.

LDPC Phase Likewise, let $\mathbf{v}_k^c$ denote the remaining B_c information bits of device k. The encoding process of $\mathbf{v}_k^c$ in the LDPC phase is as follows. Firstly, $\mathbf{v}_k^c$ is coded into $\mathbf{b}_k^c \in \{0, 1\}^{\tilde{L}_c}$ by the LDPC code with a rate $B_c/\tilde{L}_c$. Then, $\mathbf{v}_k^c$ is modulated to $\mathbf{s}_k^c$ and followed by zero padding, i.e., $\tilde{\mathbf{s}}_k^c = [\mathbf{s}_k^c, 0, \dots, 0] \in \mathbb{C}^{L_c \times 1}$. Consequently, $\tilde{\mathbf{s}}_k^c$ is permuted by the interleaver π_{i_k}, where i_k is the interleaving pattern obtained in the CS phase. Appending $\pi_{i_k}(\tilde{\mathbf{s}}_k^c)$ to the CS-coded message yields the final transmitted sequence $\mathbf{x}_k = \left[\mathbf{a}_{i_k}^T, \pi_{i_k}(\tilde{\mathbf{s}}_k^c)\right]^T$. And the received signal can be expressed as

$$\mathbf{Y} = \sum_{k \in \mathcal{K}_a} \mathbf{x}_k \mathbf{h}_{i_k}^T + \mathbf{Z}, \tag{12.5}$$

where $\mathbf{h}_{i_k}$ is assumed to follow independent quasi-static flat fading within the above two phases. At the receiver side, it is tasked to recover the message list $\mathcal{L}(\mathbf{Y})$ based on the received signal $\mathbf{Y}$, codebook $\mathbf{A}$, and the channel $\mathbf{h}_{i_k}$ estimated in the CS phase. We emphasize again that the BS only reproduces the transmitted messages without distinguishing the corresponding devices.

12.2.2 Collision Resolution Protocol

It is possible that two or more devices have the same preamble message, $\mathbf{v}^p$, which, although, may occur in a small probability. In this case, the collided devices will have the same interleaving pattern in the LDPC phase, which goes against the principles of the IDMA scheme. Moreover, they will choose the same codeword in the CS phase, leading to the received signal being

$$\mathbf{Y}_{\text{colli}} = \sum_{r \in \mathcal{R}} \mathbf{a}_r \sum_{j \in \mathcal{K}_r} \mathbf{h}_j^T + \mathbf{Z}, \tag{12.6}$$

where $\mathcal{R}$ and $\mathcal{K}_r$ denote the set of collided values and devices corresponding to the message index r, respectively. In this regard, the BS can only recover the superimposed channels of these devices, instead of their own, which will lead to the failure of the LDPC decoding.

In order to deal with the potential collisions, we develop a collision resolution protocol referred to as *ED-SWP*. The implementation of the protocol is twofold. On the one hand, the energy detection is utilized to figure out the collided user based on the energy of the estimated channels. On the other hand, the collided users leverage the sliding window protocol to retransmit new messages. As mentioned above, if a collision happens, the recovered channels of the collided devices will be a superposition of their own, that is

$$\hat{\mathbf{h}}_r = \sum_{k \in \mathcal{R}} \mathbf{h}_r^T + \mathbf{z}, \tag{12.7}$$

where $\mathbf{z} \sim \mathcal{CN}(0, \sigma_{\text{eff}}^2 \mathbf{I}_M)$ denotes the effective noise. And $\hat{\mathbf{h}}_r$ is distributed as $\hat{\mathbf{h}}_r \sim \mathcal{CN}(0, (|\mathcal{K}_r| + \sigma_{\text{eff}}^2)\mathbf{I}_M)$, which has a higher power than those without collision. Therefore, an effective way to detect collision is to perform energy detection on the estimated channel at the BS, i.e., $\epsilon_r = \mathbb{E}\left[\hat{\mathbf{h}}_r^H \hat{\mathbf{h}}_r\right]$. If ϵ_r is greater than a given threshold η, it is inferred that there are at least two devices that have the same preamble and thus the channels are superimposed. Since devices themselves have no information about whether they have been in a collision, the BS needs to feed this information back. Consequently, the BS will broadcast the collided index representations $\{r | \epsilon_r > \eta\}$ to all the devices for collision confirmation.

In order to get a new nonconflicting index representation, the collided devices will slide the window with length B_p bits forward within the total B bits to get new sequences, denoted by $\mathbf{v}_k^{p'}$. We denote B_0 as the sliding length which satisfies $0 < B_0 < B_p$. The reason behind $B_0 < B_p$ is that there should be a common part between the sequences before and after sliding the window, so as to splice back the sequences between different windows. After obtaining $\mathbf{v}_k^{p'}$, the CS-based encoder is again performed to selected a new codeword $\mathbf{a}_{r'}$. If the collision still exists, the window sliding will be executed again until the maximum number of retransmission is reached or no collision exists.

12.2.3 Decoding Scheme

The decoding process can be distilled into two key operations: the recovery of the preambles in the CS phase, and the LDPC decoding process combined with successive interference cancellation (SIC). Both are carried out with the low-complexity iterative message passing (MP) algorithm.

12.2.3.1 Joint DAD-CE Algorithm

According to (12.4), the recovery of the preambles and channels can be modeled as a standard CS problem, which can be resolved by the message passing (MP) algorithm, namely, the Joint-DAD-CE algorithm. Specifically, the activity ϕ_k and channel $\mathbf{h}_k$ are modeled as Bernoulli and Gaussian messages, respectively, which are referred to as variable nodes (VNs). While the received signal $\mathbf{Y}$ is modeled as the sum node (SN) and can be expressed as

$$y_{lm} = \sum_{i=1}^{K} A_{li}\phi_i h_{im} + n_{lm} = A_{lk}\phi_k h_{km} + \underbrace{\sum_{i\in\mathcal{K}\backslash k} A_{li}\phi_i h_{im} + n_{lm}}_{z_{lkm}}, \tag{12.8}$$

where the equivalent Gaussian noise $\mathbf{z}_{lk}$ is distributed to $\mathbf{z}_{lk} \sim \mathcal{CN}(\boldsymbol{\mu}_{z_{lk}}, \Sigma_{z_{lk}})$. The distribution of $\mathbf{z}_{ik}$ is determined by the messages of ϕ_k and $\mathbf{h}_k$ observed at VN. While the messages passed from SN to VN are given by

$$\mu^{\mathrm{SN}}_{lm\to km} = \mathbb{E}\left[h_{km}|y_{lm}, \mu_{z_{lkm}}, \Sigma_{z_{lk}}, \phi_k = 1\right] = (y_{lm} - \mu_{z_{lkm}})/A_{lk}, \tag{12.9}$$

$$\Sigma^{\mathrm{SN}}_{l\to k} = \mathrm{Var}\left[h_{km}|y_{lm}, \mu_{z_{lkm}}, \Sigma_{z_{lk}}, \phi_k = 1\right] = \Sigma_{z_{lk}}/|A_{lk}|^2, \tag{12.10}$$

$$p^{\mathrm{SN}}_{l\to k} = \frac{f(\mathbf{y}_l|\boldsymbol{\mu}'_{z_{lk}}, \Sigma'_{z_{lk}})}{f(\mathbf{y}_l|\boldsymbol{\mu}_{z_{lk}}, \Sigma_{z_{lk}}) + f(\mathbf{y}_l|\boldsymbol{\mu}'_{z_{lk}}, \Sigma'_{z_{lk}})}, \tag{12.11}$$

where $\boldsymbol{\mu}'_{z_{lk}} = A_{lk}\cdot\boldsymbol{\mu}^{\mathrm{VN}}_{k\to l} + \boldsymbol{\mu}_{z_{lk}}$, $\Sigma'_{z_{lk}} = |A_{lk}|^2\cdot\Sigma^{\mathrm{VN}}_{k\to l} + \Sigma_{z_{lk}}$. And $f(\mathbf{x}|\boldsymbol{\mu}, \Sigma)$ denotes the complex Gaussian distribution. Correspondingly, the messages at VN are updated by combining that from SN, which are expressed as

$$\boldsymbol{\mu}^{\mathrm{VN}}_{k\to l} = \Sigma^{\mathrm{VN}}_{k\to l}\cdot\left[(\Sigma^{\mathrm{pri}}_k)^{-1}\cdot\boldsymbol{\mu}^{pri}_k + \sum_{i\in\mathcal{L}\backslash l}(\Sigma^{\mathrm{SN}}_{i\to k})^{-1}\cdot\boldsymbol{\mu}^{\mathrm{SN}}_{i\to k}\right], \tag{12.12}$$

$$\Sigma^{\mathrm{VN}}_{k\to l} = \left[\sum_{i\in\mathcal{L}\backslash l}(\Sigma^{\mathrm{SN}}_{i\to k})^{-1} + (\Sigma^{\mathrm{pri}}_k)^{-1}\right]^{-1}, \tag{12.13}$$

$$p^{\mathrm{VN}}_{k\to l} = \frac{p_a\cdot\prod_{i\in\mathcal{L}\backslash l}p^{\mathrm{SN}}_{i\to k}}{p_a\cdot\prod_{i\in\mathcal{L}\backslash l}p^{\mathrm{SN}}_{i\to k} + (1-p_a)\cdot\prod_{i\in\mathcal{L}\backslash l}\left(1-p^{\mathrm{SN}}_{i\to k}\right)}, \tag{12.14}$$

where μ_k^{pri} and Σ_k^{pri} denote a prior mean and variance of $\mathbf{h}_k$, respectively. And p_a is the activated probability of users. Messages are updated iteratively between SN and VN until the stopping criterion is satisfied. The mean of $\mathbf{h}_k$ calculated at VN is the final CE output while the activity detection can be obtained by the hard decision of p_k^{VN}.

12.2.3.2 MIMO-LDPC-SIC Decoder

The estimated channel and activity obtained by the Joint DAD-CE algorithm can be applied in the LDPC phase directly, of which the received signal can be rewritten as follows.

$$\mathbf{Y}_c = \mathbf{Y}_{L_p+1:L,:} = \sum_{k \in \mathcal{K}_a} \pi_{i_k}\left(\tilde{\mathbf{s}}_k\right) \mathbf{h}_{i_k}^T + \mathbf{Z}_{L_p+1:L,:}, \tag{12.15}$$

where $\mathbf{Y}_c \in C^{L_c \times M}$ is the last L_c rows of $\mathbf{Y}$. After the recovery of interleaving patterns $\{i_k, k \in \hat{\mathcal{K}}_a\}$, the IDMA decoding process can be started. Firstly, an inverse operation called deinterleaving is performed on the received signal. Correspondingly, the iterative BP algorithm can be implemented for decoding the data embedded in the LDPC phase. Note that the SIC scheme is needed at the end of decoding, i.e., the signals of the users that are correctly decoded are canceled from $\mathbf{Y}_c$ in each iteration, which contributes to a decrement of the multiuser interference and the improved decoding performance.

Moreover, the overall performance can be further improved by jointly estimating the channel with the preambles as well as the correctly decoded LDPC codewords, which is called data-aided CE. Succinctly, the successfully decoded messages can be employed as soft pilots to conduct CE jointly with the real preambles, which brings to a substantial gains of CE. Accordingly, with the implementation of the SIC method, the residual signal of the incorrectly decoded messages can be obtained more accurately, contributing to the decoding performance enhancements.

12.2.4 Experimental Results

In this section, we assess the overall performance of the proposed framework with the metric defined in (12.2) and (12.3). The covariance-based maximum likelihood (CB-ML) proposed in Fengler et al. (2021) serves as the benchmark in this chapter. For the sake of fair comparison to the benchmarks, and isolating the fundamental aspects of the problem without additional model complication, we consider the flat path loss model in the simulation, i.e., the channel is i.i.d. Rayleigh fading model and the path loss is fixed to $\beta_k = 1$ in all schemes. We evaluate the performance of

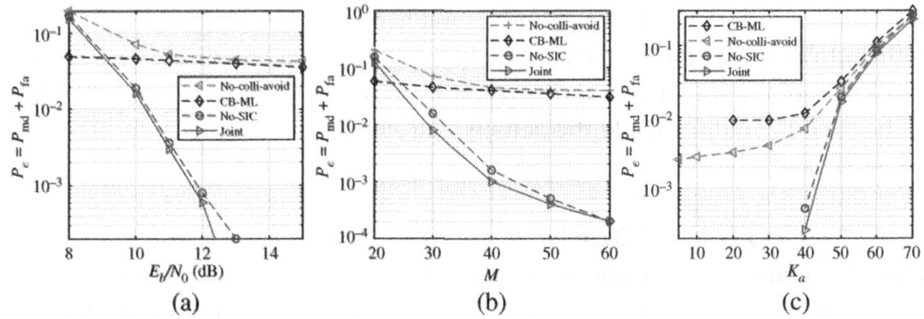

Figure 12.1 Performance of the proposed URA schemes versus different parameters. (a) As a function of E_b/N_0. (b) As a function of M. (c) As a function of K_a. Parameter settings: $E_b/N_0 = 10$ dB, $M = 30$, $K_a = 50$, $L = 1600$.

the proposed scheme compared with the state-of-the-art in terms of signal-to-noise ratio (SNR), antennas, and active users, as demonstrated in Fig. 12.1.

In Fig. 12.1, *No-colli-avoid*, *No-SIC*, and *Joint* denote the schemes without collision resolution, without SIC, and jointly estimating channels with preambles and data, respectively. Intuitively, the state-of-the-art method CB-ML suffers from high error floors, which stems from the poor parity check constraints. In contrast, with collision resolution, the proposed *Joint* and *No-SIC* schemes exhibit water-falling curves in terms of the error rate with respect to E_b/N_0 and K_a, while they gradually stabilize with respect to M. This is because the interference of devices cannot be reduced to infinitesimal by increasing M. As such, error still exits even for a large M. Moreover, since the CB-ML method employs the maximum likelihood estimation (MLE) as the inner decoding, it cannot exploit hardware parallelism implementation. On the contrary, thanking to the low-complexity linear BP structure, the proposed algorithm is capable for hardware parallelism with a lower computational burden. Altogether, the proposed algorithm exhibit remarkable performance enhancements with lower complexity.

12.3 Exploiting Angular Domain Sparsity for Grant-Free Random Access: A Hybrid AMP Approach

In this section, considering the structured channel sparsity, we exploit the possibility of reducing the required number of measurements of the CS recovery algorithm, which would save the time/frequency source in practical applications. Our approach to the activity detection and CE problem is through the generalized approximate message passing (GAMP) with a BP part for channel support estimation. Furthermore, we apply the proposed algorithm as the kernel of an uncoupled URA scheme, where the permutation ambiguity of the slot-wise codewords is resolved by channel clustering.

12.3.1 Sparse Modeling of Massive Access

Considering a single-cell network system consisting of massive single-antenna users and a BS equipped with a half-wave-spaced uniform planar array (UPA) with $M = M_v \times M_h$ elements, which arranges M_v and M_h antennas in the vertical and horizontal directions, respectively. Under the far-field assumption, the channel vector of the kth user is modeled as

$$\mathbf{h}_k = \sum_{p=1}^{P} g_{k,p} \mathbf{e}_h \left(\theta^h_{k,p}\right) \otimes \mathbf{e}_v \left(\theta^v_{k,p}\right) \in \mathbb{C}^M, \tag{12.16}$$

where $g_{k,p}$ is the gain of the pth physical path. Moreover, the vertical steering vector $\mathbf{e}_v$ and the horizontal steering vector $\mathbf{e}_h$ can be expressed as

$$\mathbf{e}_v \left(\theta^v_{k,p}\right) = \frac{1}{\sqrt{M_v}} \left[1, e^{-j\pi\theta^v_{k,p}}, \dots, e^{-j\pi(M_v-1)\theta^v_{k,p}}\right]^T \in \mathbb{C}^{M_v}, \tag{12.17}$$

$$\mathbf{e}_h \left(\theta^h_{k,p}\right) = \frac{1}{\sqrt{M_h}} \left[1, e^{-j\pi\theta^h_{k,p}}, \dots, e^{-j\pi(M_h-1)\theta^h_{k,p}}\right]^T \in \mathbb{C}^{M_h}, \tag{12.18}$$

where $\theta^v_{k,p} = \sin(\phi_{k,p})$ and $\theta^h_{k,p} = \cos(\phi_{k,p})\sin(\varphi_{k,p})$, with $\phi_{k,p}, \varphi_{k,p} \in [-\pi/2, \pi/2]$ the elevation and the horizontal angle of arrival (AoA), respectively. As the steering vectors are indeed discrete Fourier vectors, the angular domain representation $\mathbf{h}^a_k$ of $\mathbf{h}_k$ can be achieved by the standard spatial Fourier transformation, i.e.,

$$\mathbf{h}^a_k = \left(\mathbf{U}^H_h \otimes \mathbf{U}^H_v\right) \mathbf{h}_k = \mathbf{U}^H \mathbf{h}_k, \tag{12.19}$$

where $\mathbf{U}_h$ and $\mathbf{U}_v$ are discrete Fourier transform matrices and $\mathbf{U} \triangleq \mathbf{U}_h \otimes \mathbf{U}_v$ is a unitary matrix. The element of $\mathbf{h}^a_k$ captures the aggregated gain of the physical paths within the corresponding resolution bin. Due to insufficient scatterers in the realistic wireless communication environment, only a small number of bins have a high magnitude; they often appear in clusters due to the angular spread of scatterers.

Now, we can rewrite (12.4) to the equivalent received signal in the angular domain, expressed as

$$\widetilde{\mathbf{Y}} = \mathbf{Y}\mathbf{U}^* = \underbrace{\mathbf{A}\boldsymbol{\Phi}\widetilde{\mathbf{H}}\mathbf{U}^*}_{\triangleq \mathbf{X}} + \underbrace{\mathbf{Z}\mathbf{U}^*}_{\triangleq \widetilde{\mathbf{Z}}}. \tag{12.20}$$

As stated in Section 12.2.1.2, $\mathbf{X}$ is row sparse. Furthermore, profited from the sparse nature of the angular domain channel, the sparsity level is enhanced within each row.

12.3.2 Recovery Algorithm

To retrieve the sparse $\mathbf{X}$ from the received noisy superposition, we turn to the graphical models that desire a joint probability distribution $p(\mathbf{X}, \widetilde{\mathbf{Y}}) = p(\widetilde{\mathbf{Y}}|\mathbf{X})p(\mathbf{X})$. Focusing on encoding the sparsity properties in $p(\mathbf{X})$, we first assign a Bernoulli–Gaussian prior distribution to each element of $\mathbf{X}$, given by

$$p(x_{jm}|b_{jm}) = \mathcal{CN}(x_{jm}; 0, \sigma_x^2)\delta(b_{jm} - 1) + \delta(0)\delta(b_{jm} + 1), \tag{12.21}$$

where $\delta(\cdot)$ denotes the Dirac function and the binary state $b_{jm} = 1$ if x_{jm} takes nonzero value and $b_{jm} = -1$ otherwise. To further capture the underlying clustered support structure of the angular domain channel coefficients, we use a Markov random field (MRF) prior that models the distribution of the binary vector $\mathbf{b}_j = [b_{j1}, \dots, b_{jM}]$ as

$$\begin{aligned} p\left(\mathbf{b}_j\right) &\propto \exp\left(\sum_{m=1}^{M}\left(\frac{1}{2}\sum_{k\in\mathcal{R}_m}\beta_j b_{jk} - \alpha_j\right) b_{jm}\right) \\ &= \left(\prod_{m=1}^{M}\prod_{k\in\mathcal{R}_m}\exp\left(\beta_j b_{jm} b_{jk}\right)\right)^{\frac{1}{2}}\prod_{m=1}^{M}\exp\left(-\alpha_j b_{jm}\right), \end{aligned} \tag{12.22}$$

where $\mathcal{R}_m \subset \{1, \dots, M\}\backslash m$ is the set of related elements of index m. Moreover, α_j and β_j depict the correlation between entries of $\mathbf{b}_j$ and the sparsity of the signal, respectively. We use the posterior probability density for inference, which can be expressed as

$$\begin{aligned} p\left(\mathbf{B}, \mathbf{X}|\widetilde{\mathbf{Y}}\right) &\propto \underbrace{p\left(\widetilde{\mathbf{Y}}|\mathbf{B}, \mathbf{X}\right) p\left(\mathbf{X}|\mathbf{B}\right) p\left(\mathbf{B}\right)}_{\text{part (a)}} \\ &\propto \exp\left(-\frac{1}{\sigma^2}\|\widetilde{\mathbf{Y}} - \mathbf{A}\mathbf{X}\|_F^2\right)\prod_j \underbrace{p\left(\mathbf{x}_j|\mathbf{b}_j\right) p\left(\mathbf{b}_j\right)}_{\text{part (b)}}. \end{aligned} \tag{12.23}$$

Unfortunately, marginalizing a joint distribution with high dimensions for the maximum *a posteriori* estimation of (12.22) is computational prohibited.

To obtain a tractable proxy with low complexity, we resort to the tool of AMP. The message passing related to part (a) can be obtained straightly from GAMP, expressed as lines 4–11 of Algorithm 12.1. We refer interesting readers to for more details. We concentrate more on the message passing concerning the MRF support estimation module, whose factor graph is illustrated in Fig. 12.2, where $\gamma_{jm} = p(x_{jm}|b_{jm})$, $\eta_{jm}^{\alpha} = \exp(-\alpha_j b_{jm})$, $\eta_{jm,1}^{\beta} = \exp(\beta_j b_{jm} b_{jm_1})$. According to GAMP, the message from the VN x_{jm} to the factor node f_{jm} has a Gaussian approximation, i.e., $\nu_{x_{jm}\to g_{jm}} = \mathcal{CN}(x_{jm}; \hat{r}_{jm}, \mu_{jm}^{r})$. Then, the message from g_{jm} to b_{jm} can be calculated as

$$\nu_{g_{jm}\to b_{jm}} = \pi_{jm}\delta(b_{jm} - 1) + (1 - \pi_{jm})\delta(b_{jm} + 1), \tag{12.24}$$

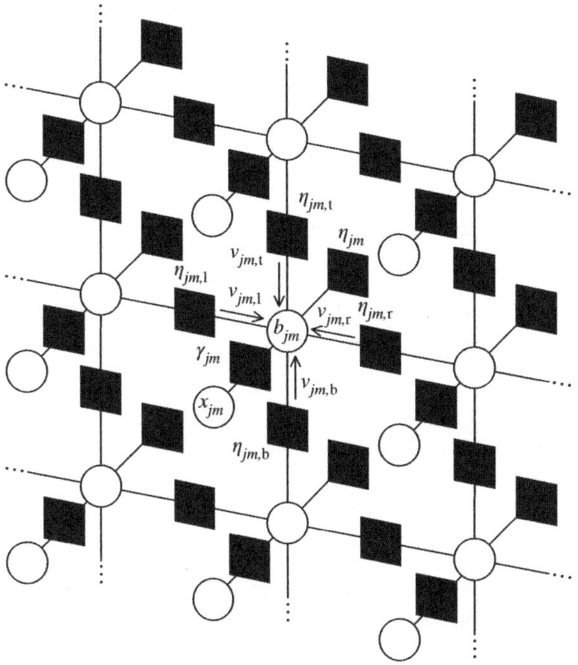

Figure 12.2 Factor graph for the MRF support structure.

with

$$\pi_{jm} = \left(1 + \frac{\mathcal{CN}\left(0; \hat{r}_{jm}, \mu_{jm}^{r}\right)}{\mathcal{CN}\left(0; \hat{r}_{jm}, \mu_{jm}^{r} + \sigma_x^2\right)} \right)^{-1}. \tag{12.25}$$

Each VN b_{jm} in MRF is linked to four nodes corresponding to the four adjutant antenna elements. We mark the messages form the left, right, top, and bottom by $\nu_{jm,\mathrm{l}}$, $\nu_{jm,\mathrm{r}}$, $\nu_{jm,\mathrm{t}}$, and $\nu_{jm,\mathrm{b}}$, respectively. We give the message update rule in example of $\nu_{jm,\mathrm{l}}$, which can be expressed as

$$\nu_{jm,\mathrm{l}} = \kappa_{jm,\mathrm{l}} \delta(b_{jm} - 1) + \left(1 - \kappa_{jm,\mathrm{l}}\right) \delta(b_{jm} + 1), \tag{12.26}$$

where

$$\kappa_{jm,\mathrm{l}} = \frac{\pi_{jm_q} \prod_{d \in \{\mathrm{l,t,b}\}} \kappa_{jm_\mathrm{l},d} e^{-\alpha+\beta} + \left(1 - \pi_{jm_q}\right) \prod_{d \in \{\mathrm{l,t,b}\}} \left(1 - \kappa_{jm_\mathrm{l},d}\right) e^{\alpha-\beta}}{\left(e^{\beta} + e^{-\beta}\right) \left(\pi_{jm_q} \prod_{d \in \{\mathrm{l,t,b}\}} \kappa_{jm_\mathrm{l},d} e^{-\alpha} + \left(1 - \pi_{jm_q}\right) \prod_{d \in \{\mathrm{l,t,b}\}} \left(1 - \kappa_{jm_\mathrm{l},d}\right) e^{\alpha}\right)}, \tag{12.27}$$

with $m_l = m - M_v$ the index of the left node. Hence, the message backward from b_{jm} to g_{jm} can be represented as

$$\nu_{b_{jm}\to g_{jm}} = \rho_{jm}\delta(b_{jm} - 1) + (1 - \rho_{jm})\delta(b_{jm} + 1), \tag{12.28}$$

with

$$\rho_{jm} = \frac{\prod_{d\in\{l,r,t,b\}} \kappa_{jm,d} e^{-\alpha}}{\prod_{d\in\{l,r,t,b\}} \kappa_{jm,d} e^{-\alpha} + \prod_{d\in\{l,r,t,b\}} (1 - \kappa_{jm,d}) e^{\alpha}}. \tag{12.29}$$

The parameter $\rho_{jm} \in (0, 1)$ as the output of the MRF module manifests as the soft support of x_{jm}. On such basis, the message from g_{jm} to x_{jm} is a Bernoulli–Gaussian distribution expressed as

$$\begin{aligned}\nu_{g_{jm}\to x_{jm}} &\propto \int_{b_{jm}} p(x_{jm}|b_{jm})\nu_{b_{jm}\to g_{jm}} \\ &= \rho_{jm}\mathcal{CN}(x_{jm}; 0, 1) + (1 - \rho_{jm})\delta(x_{jm}).\end{aligned} \tag{12.30}$$

Based on the graphic model, the approximate marginal posterior is the multiply of the forward and the backward messages, i.e.,

$$p(x_{jm}|\widetilde{\mathbf{Y}}) \propto \mathcal{CN}(x_{jm}; \hat{r}_{jm}, \mu^r_{jm}) \cdot \nu_{g_{jm}\to x_{jm}}. \tag{12.31}$$

Accordingly, the closed forms of the marginal posterior mean $\hat{x}_{jm}$ (which serves as the estimation of x_{jm}) and variance μ^x_{jm} of x_{jm} are given by

$$\hat{x}_{jm} = \mathbb{E}\left\{x_{jm}|\widetilde{\mathbf{Y}}; \hat{r}_{jm}, \mu^r_{jm}, \rho_{jm}, \sigma_x^2\right\} = \rho_{jm}\frac{\hat{r}_{jm}\sigma_x^2}{\mu^r_{jm} + \sigma_x^2}, \tag{12.32}$$

$$\begin{aligned}\mu^x_{jm} &= \mathrm{Var}\left\{x_{jm}|\widetilde{\mathbf{Y}}; \hat{r}_{jm}, \mu^r_{jm}, \rho_{jm}, \sigma_x^2\right\} \\ &= \rho_{jm}\frac{\mu^r_{jm}\sigma_x^2}{\mu^r_{jm} + \sigma_x^2} + \rho_{jm}(1 - \rho_{jm})|\hat{x}_{jm}|^2.\end{aligned} \tag{12.33}$$

The above message components are updated iteratively until the stopping criterion is satisfied. The proposed algorithm (summarized in Algorithm 12.1) can be seen as the hybrid of the GAMP for the linear reverse problem with the BP for the MRF-based support estimation, hence the name MRF-GAMP.

12.3.2.1 Application to Unsourced Random Access

To alleviate the computational burden, we take a divide-and-conquer strategy by uniformly dividing the long message into several fragments for slot-by-slot decoding and transmission. For simplicity here, we assume that no codeword is

---

Algorithm 12.1 MRF-GAMP

1: **Input:** $\widetilde{\mathbf{Y}}, \mathbf{A}, \tau, T_{\mathrm{mrf}}$

2: **Initialize:** $\forall l, m : \widehat{s}_{lm}(0) = 0$, $\forall j, m$: choose $\widehat{x}_{jm}(1)$, $\mu_{jm}^{x}(1)$, $\forall j, m, d : \kappa_{j'm_d} = 0.5$, $\forall j : \alpha_j = \beta_j = 0.4$

3: **for** $t = 1, 2, \ldots, T_{\max}$ **do**

4: $\quad \forall l, m : \mu_{lm}^{p}(t) = \sum_j |a_{lj}|^2 \mu_{jm}^{x}(t)$

5: $\quad \forall l, m : \widehat{p}_{lm}(t) = \sum_j a_{lj} \widehat{x}_{jm}(t) - \mu_{lm}^{p}(t) \widehat{s}_{lm}(t-1)$

6: $\quad \forall l, m : \mu_{lm}^{z}(t) = \frac{\mu_{lm}^{p}(t) \widetilde{y}_{lm} + \sigma^2 \widehat{p}_{lm}}{\mu_{lm}^{p} + \sigma^2}$

7: $\quad \forall l, m : \widehat{z}_{lm}(t) = \frac{\mu_{lm}^{p} \sigma^2}{\mu_{lm}^{p} + \sigma^2}$

8: $\quad \forall l, m : \mu_{lm}^{s}(t) = \frac{\mu_{lm}^{p}(t) - \mu_{lm}^{z}(t)}{\left[\mu_{lm}^{p}(t)\right]^2}$

9: $\quad \forall l, m : \widehat{s}_{lm}(t) = \frac{\widehat{z}_{lm}(t) - \widehat{p}_{lm}(t)}{\mu_{lm}^{p}(t)}$

10: $\quad \forall j, m : \mu_{jm}^{r}(t) = \left[\sum_l |a_{lj}|^2 \mu_{lm}^{s}(t)\right]^{-1}$

11: $\quad \forall j, m : \widehat{r}_{jm}(t) = \widehat{x}_{jm}(t) + \mu_{jm}^{r}(t) \sum_l a_{lj}^{*} \widehat{s}_{lm}(t)$

12: $\quad \forall j, m$: Compute input $\pi_{jm}(t)$ via (12.30)

13: $\quad$ **for** $t_{\mathrm{mrf}} = 1, 2, \ldots, T_{\mathrm{mrf}}$ **do**

14: $\quad\quad \forall j, m$: Update $\kappa_{jm}^{l}, \kappa_{jm}^{r}, \kappa_{jm}^{t}, \kappa_{jm}^{b}$ via (12.27)

15: $\quad$ **end for**

16: $\quad \forall j, m$: Compute output $\rho_{jm}(t)$ via (12.29)

17: $\quad \forall j, m : \widehat{x}_{jm}(t+1) = \mathbb{E}\left\{x_{jm} | \widetilde{\mathbf{Y}}\right\}$

18: $\quad \forall j, m : \mu_{jm}^{x}(t+1) = \mathrm{Var}\left\{x_{jm} | \widetilde{\mathbf{Y}}\right\}$

19: **end for**

20: **Output:** Estimated signal $\widehat{\mathbf{X}}$

---

reused by more than one user within the same slot, while a collision resolution mechanism can be found in detail in Xie et al. (2022). Accordingly, decoding takes two steps by first identifying the codewords, recovering the corresponding channels from every slot, and then combining codewords from each transmitter in sequence. The first CS decoding step employs the proposed MRF-GAMP as the decoder. A crucial observation to couple the slot-wise codewords is that the latent channel statistics of one transmitter (e.g., the AoA) remain almost unchanged for a relatively long period. Therefore, the message reconstruction procedure can be interpreted as a channel clustering process, and we slightly modify the K-means algorithm as the solution. Specifically, initialized by choosing K_a random cluster centers, the following two steps are repeatedly conducted until there are no further changes in the assignment: (i) *Assignment Step*: Assign K_a channels of one slot to K_a clusters simultaneously with the minimum aggregated channel-center distances. (ii) *Update Step*: Update each cluster center as the mean of its constituent data instances. Note that the assignment step is developed to meet the restriction that each cluster must consist of channels from different slots. The resultant assignment problem, equivalent to a minimum bipartite matching problem, can be solved by the Hungarian algorithm.

12.3.3 Experimental Results

Here, we conduct numerical experiments to support the results of the preceding parts. The UPA is of size 4×16, i.e., $M = 64$. We fix 50 active users whose channels have two physical paths with (ϕ, φ) uniformly chosen from $[-\pi/2, \pi/2] \times [-\pi/2, \pi/2]$ and $g \sim \mathcal{CN}(0, 1)$. For the URA transmission, the 96-bit messages are divided into 8 fragments each of length 12. Accordingly, the common codebook $\mathbf{A}$ is generated as an i.i.d. $L \times 2^{12}$ Gaussian matrix.

Figure 12.3 shows the dependency between the number of measurements L and the performance of activity detection and channel construction. The received SNR, defined as $\| \mathbf{AX} \|_F^2 \ / \ \| \widetilde{\mathbf{Z}} \|_F^2$, is set to be 0 dB. For comparison, we provide the multiple measurement vector (MMV) version of AMP. It can be observed that the proposed algorithm requires much fewer measurements than the MMV approach to achieve the targeted performance, which reveals the advantage of exploiting the enhanced angular domain sparsity.

Figure 12.4 demonstrates the required SNR of the designed sparsity-exploiting scheme to achieve $p_{\mathrm{md}} + p_{\mathrm{fa}} \leq 0.1$ with $L = 320$. Compared with the CB-ML scheme under the Rayleigh fading channel, the proposed method reveals higher energy efficiency and can accommodate more active devices.

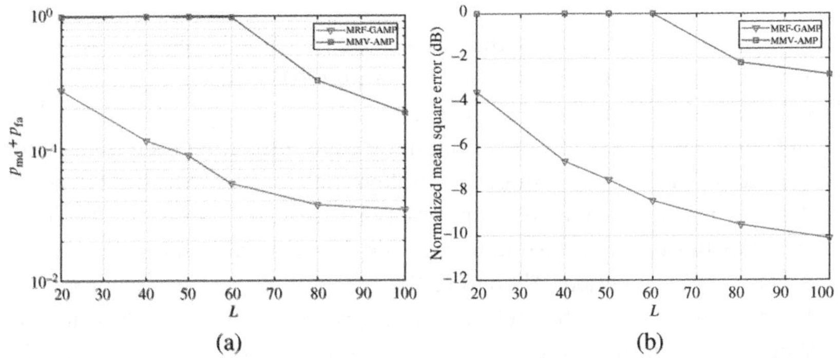

Figure 12.3 Performance of various algorithms with $K_a = 50$, $M = 64$, and SNR = 0 dB; (a) Detection error rate versus block-length; (b) Channel reconstruction error versus block-length.

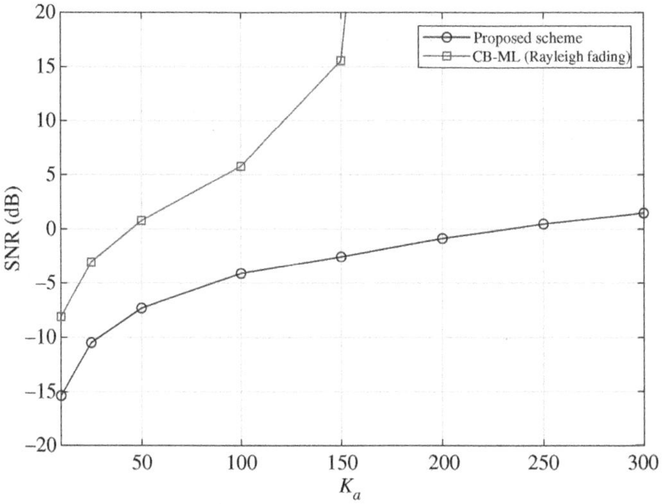

Figure 12.4 Minimum SNR required to achieve $p_{\text{md}} + p_{\text{fa}} \leq 0.1$ with different values of K_a. The total block-length is 3200 and $M = 64$.

12.4 LEO Satellite-Enabled Grant-Free Random Access

Satellite communication technology has been widely concerned since its inception, which will help to expand terrestrial communication networks due to its characteristics of large coverage and robustness for different geographical conditions. A lot of standardization works on satellite communications are currently underway, some of which hope to incorporate it into 5G and future networks. Among the application requirements of 5G, the mMTC is an important scenario and a basic part of IoT applications. A considerable part of the IoT devices is distributed in remote areas, such as deserts, oceans, and forests. However, the existing cellular communication networks are mainly deployed in places where populations are concentrated, and most of them are for communications between people. As a result, supporting remote mMTC is still a huge challenge for the existing communication facilities. Fortunately, satellite communications provide a very promising solution, especially those based on LEO satellites.

Over the past few years, bunches of methods have been proposed for CE and DAD in the grant-free random access systems, aiming at Rayleigh fading channel, Rician fading channel and Gaussian mixture model. However, the current works are based on the block fading channel which is constant during one transmission. The high mobility of LEO satellites inevitably leads to the rapid change of terrestrial-satellite link (TSL), and the large Doppler shift will also degrade the communication quality. Therefore, the current grant-free random access schemes can not be directly applied for the LEO satellite communications.

OTFS is a promising solution to tackle the time-variant channels (Surabhi et al., 2019), which converts the time-variant channels into the time-independent channels in the delay-Doppler domain. There has been a lot of literature focusing on the applications of OTFS. e.g., schemes for point-to-point communications (Hadani et al., 2017) and uplink transmissions (Ge et al., 2021). However, the grant-free random access scenario do not be considered sufficiently.

In this section, we investigate the joint CE and DAD in the LEO satellite communications with massive MIMO-OTFS, where the OTFS is adopted to suppress the doubly dispersive effect in the TSL link and the massive MIMO is integrated for improving the performance.

12.4.1 System Model

In this section, the basic system setup is first introduced. Then, we present the adopted channel model and input–output relationship of the MIMO-OTFS system when both the large differential delay and Doppler shift exist. Finally, we formulate the considered problem.

12.4.1.1 Channel Model

We consider the following channel response corresponding to the uth device in the delay-Doppler-space domain, i.e.,

$$\mathbf{h}_u(\tau, \nu) = \sum_{i=0}^{P} h_{u,i}\delta\left(\tau - \tau_{u,i}\right)\delta\left(\nu - \nu_{u,i}\right)\mathbf{v}(\psi_{u,i}, \varphi_{u,i}), \tag{12.34}$$

where $h_{u,i}$, $\nu_{u,i}$, $\tau_{u,i}$, $\psi_{u,i}$, and $\varphi_{u,i}$ denote the complex gain, differential Doppler shift, differential delay, azimuth angle, and elevation angle of the ith path, respectively, and $\mathbf{v}(\psi_{u,i}, \varphi_{u,i})$ is the steering vector of the antennas. The steering vector is the Kronecker product of the array response vectors $\mathbf{v}(\vartheta_{y_{u,i}})$ and $\mathbf{v}(\vartheta_{z_{u,i}})$ corresponding to the directions with respect to y- and z-axis, respectively, that is,

$$\mathbf{v}(\psi_{u,i}, \varphi_{u,i}) = \mathbf{v}(\vartheta_{y_{u,i}}) \otimes \mathbf{v}(\vartheta_{z_{u,i}}), \tag{12.35}$$

where $\vartheta_{y_{u,i}} = \sin\varphi_{u,i}\sin\psi_{u,i}$ and $\vartheta_{z_{u,i}} = \cos\varphi_{u,i}$ are the directional cosines along the y- and z-axis, respectively, and

$$\mathbf{v}(\vartheta) = \left[1 \exp\left\{-\bar{j}\pi\vartheta\right\} \dots \exp\left\{-\bar{j}\pi\left(N_y - 1\right)\vartheta\right\}\right]^T. \tag{12.36}$$

The Doppler and delay taps for the ith path of the uth device can be expressed as

$$\nu_{u,i} = \frac{k_{u,i} + \tilde{k}_{u,i} + b_{u,i}N}{NT_{\text{sym}}}, \quad \tau_{u,i} = \frac{l_{u,i} + c_{u,i}M}{M\Delta f}, \tag{12.37}$$

where N is the number of orthogonal frequency division multiplexing (OFDM) symbols, M is the number of subcarriers, and Δf is the subcarrier spacing; $l_{u,i} = 0, \dots, M-1$ and $k_{u,i} = \lceil -N/2 \rceil, \dots, \lceil N/2 \rceil - 1$ represent indexes of the delay tap and Doppler tap corresponding to the delay $\tau_{u,i}$ and Doppler $\nu_{u,i}$, respectively; $\tilde{k}_i \in (-\frac{1}{2}, \frac{1}{2}]$ corresponds to the fractional Doppler shift; T_{sym} is the time duration of a OFDM symbol with cyclic prefix (CP); $b_{u,i}$ and $c_{u,i}$ are integers, referred to as the outer Doppler tap and outer delay tap, respectively, which are non-zeros when differential Doppler $\nu_{u,i} > \Delta f$ and/or differential delay $\tau_{u,i} > T$.

12.4.1.2 Signal Modulation

The uth device employs the OFDM-based OTFS modulation to combat the doubly fading effect of the above channel. The detailed mathematical derivations of this modulation are omitted due to the limited spacing and interested readers can refer to Shen et al. (2022). We directly give the input–output relationship of the uth device in the delay-Doppler-angle domain at the $(a_z + a_y N_z)$th antenna as

$$
\begin{aligned}
Y^{\text{DDA}}[k, l, a_y, a_z] = & \sum_{u=0}^{U-1} \sum_{l'=0}^{M-1} \sum_{k'=\lceil -N/2 \rceil}^{\lceil N/2 \rceil - 1} \lambda_u \phi_u(l, l') H_u^{\text{DDA}}\left[k', l', a_y, a_z\right] \\
& \times X_u^{\text{DD}}\left[\left\langle k - k' \right\rangle_N, \left(l - l'\right)_M\right] + Z^{\text{DDA}}[k, l, a_y, a_z].
\end{aligned}
\tag{12.38}
$$

where $a_y = 0, \dots, N_y - 1$, $a_z = 0, \dots, N_z - 1$, $(\cdot)_M$ denotes mod M, $\langle x \rangle_N$ denotes $\left(x + \left\lfloor \frac{N}{2} \right\rfloor\right)_N - \left\lfloor \frac{N}{2} \right\rfloor$, Z_{n_y,n_z}^{DDS} is the noise, and the effective channel H_u^{DDS}. The effective channel can be expressed as

$$
\begin{aligned}
H_u^{\text{DDA}}[k, l, a_y, a_z] &= \frac{1}{\sqrt{N_y N_z}} \sum_{n_y=0}^{N_y-1} \sum_{n_z=0}^{N_z-1} H_{u,n_y,n_z}^{\text{DDS}}\left[k, l\right] e^{-\bar{j}2\pi\left(\frac{a_y n_y}{N_y} + \frac{a_z n_z}{N_z}\right)} \\
&= \sqrt{N_y N_z} \sum_{i=0}^{P} h_{u,i} e^{\bar{j}2\pi(M_{\text{cp}} - l)T_s \nu_{u,i}} \Pi_N(k - N T_{\text{sym}} \nu_{u,i}) \\
&\quad \times \Pi_{N_y}(a_y - N_y \Omega_{y_u}/2) \Pi_{N_z}(a_z - N_z \Omega_{z_u}/2) \delta\left(l T_s - \tau_{u,i}\right),
\end{aligned}
\tag{12.39}
$$

where $a_y = 0, \dots, N_y - 1$ and $a_z = 0, \dots, N_z - 1$ are indexes along the angular domain; $\Omega_{y_{u,i}} = \sin \varphi_u \sin \psi_u$ and $\Omega_{z_u} = \cos \varphi_u$ are the directional cosines along the y-axis and z-axis, respectively; $\Pi_N(x) \triangleq \frac{1}{N} \sum_{i=0}^{N-1} e^{-\bar{j}2\pi \frac{x}{N} i}$. Therefore, we can find in (12.39) that $H_u^{\text{DDA}}[k, l, a_y, a_z]$ has dominant elements only if $k \approx N T_{\text{sym}} \nu_{u,i}$, $l \approx \tau_{u,i} M \Delta f$, $a_y \approx N_y \Omega_{y_u}/2$, and $a_z \approx N_z \Omega_{z_u}/2$, which means that the channel in the delay-Doppler-angle domain has the 3D-structured sparsity (Shen et al., 2019).

12.4.1.3 Problem Formulation

To facilitate the following analysis, we rewrite (12.38) into the matrix form as

$$\mathbf{Y} = (\mathbf{\Phi} \odot \mathbf{X})\Lambda\tilde{\mathbf{H}} + \mathbf{Z}, \tag{12.40}$$

where the elements of $\mathbf{Z}$ are independent Gaussian noise, $\tilde{\mathbf{H}}$ is channel matrix, and the $(a_z + 1 + N_z a_y)$th column of $\mathbf{Y}$ is vec $\left(\mathbf{Y}_p^{\mathrm{DDA}}[:,:,a_y,a_z]\right)$. $\Phi = [\Phi_0, \dots, \Phi_{U-1}]$ contains the phase compensation matrices $\Phi_u \in \mathcal{C}^{M_\tau N \times UM_\tau N}$ of the uth device composed of $\phi_u(l,l)\mathbf{1}_N$ and $l = 0, \dots, M_\tau - 1$. $\mathbf{X}$ is the pilot matrix composed of the pilots of all the devices. Note that $\mathbf{X}_u$ is a doubly circulant matrix due to the 2D convolution in (12.38).

We denote the channel matrix $\Lambda\tilde{\mathbf{H}}$ as $\mathbf{H}$. Then, (12.40) can be written as

$$\mathbf{Y} = (\mathbf{\Phi} \odot \mathbf{X})\mathbf{H} + \mathbf{Z}. \tag{12.41}$$

Therefore, the joint DAD and CE in the massive MIMO-OTFS systems is a sparse signal reconstruction problem. However, since the phase compensation matrix $\mathbf{\Phi}$ is related to the channel, and $\mathbf{H}$ cannot be directly recovered with the unknown sensing matrix. Fortunately, when choosing large enough carrier spacing Δf and the number of OFDM symbols N in one OTFS frame, as shown in our simulations, the phase compensation matrix can be approximated by all-ones matrix $\mathbf{1}_{M_\tau N \times UM_\tau N}$ while with a little performance degradation. Therefore, (12.41) can be rewritten as

$$\mathbf{Y} \approx \mathbf{X}\mathbf{H} + \mathbf{Z}. \tag{12.42}$$

Note that the channel matrix $\mathbf{H}$ in the delay-Doppler-angle domain has different kinds of sparsities, compared with that in the previous literature (Hannak et al., 2015, Chen and Yu, 2017) which only focuses on the sparsity of the user or the sparsity in the angular domain. It is favorable for estimation algorithms capturing the 2D sparsity of $\mathbf{H}$.

12.4.2 Pattern Coupled SBL Framework

In this section, a 2D pattern coupled hierarchical prior is first proposed to capture the 2D burst block sparsity in the channel matrix. Then, the SBL framework is developed to estimate the channel matrix using the pilots and the received symbols. Finally, the covariance-free method is adopted to facilitate the computations of the SBL.

12.4.2.1 The Pattern-Coupled Hierarchical Prior

Considering that the sparsity patterns of nonzero coefficients are statistically dependent along the row and column. To explore this 2D sparsity, in the proposed prior, the precision of each coefficient involves the precisions of its neighborhoods. More precisely, the prior on the (i,j)th element of the channel matrix $\mathbf{H}$ is given by

$$p(h_{i,j}|\mathbf{A}, \mathbf{B}_{i,j}) = \mathcal{CN}\left(h_{i,j}|0, \left(\sum_{p=i-D}^{i+D}\sum_{q=j-D}^{j+D}\beta_{i,j,p,q}\alpha_{p,q}\right)^{-1}\right), \tag{12.43}$$

where $\mathbf{A}$ is the precision matrix whose (i,j)th element $\alpha_{i,j}$ is the local precision hyperparameter of $h_{i,j}$; $\mathbf{B}_{i,j} \in R^{UM_\tau N \times N_y N_z}$ is the pattern coupled hyperparameter matrix of $h_{i,j}$, and its (p,q)th element $\beta_{i,j,p,q}$ represents the relevance between $h_{i,j}$ and $h_{p,q}$, and $\beta_{i,j,p,q} \neq 0$ only if $p \in [i-D, i+D]$ and $q \in [j-D, j+D]$; D, called the range indicator, is a hyperparameter controlling the number of neighborhoods of $h_{i,j}$ in the consideration. Combined with (12.43), the prior distribution of $\mathbf{H}$ is given by

$$p(\mathbf{H}|\mathbf{A}, \mathbf{B}) = \prod_{i=0}^{UM_\tau N-1}\prod_{j=0}^{N_y N_z-1} p\left(h_{i,j}|\mathbf{A}, \mathbf{B}_{i,j}\right) = \prod_{j=0}^{N_y N_z-1}\mathcal{CN}(\mathbf{h}_j|0, \Upsilon_j), \tag{12.44}$$

where $\mathbf{B}$ contains all the pattern coupled hyperparameters, whose (i,j)th element is $\mathbf{B}_{i,j}$; $\mathbf{h}_j = [h_{0,j}, \dots, h_{UM_\tau N-1,j}]$ and represents the jth column of $\mathbf{H}$; $\Upsilon_j = \text{diag}(\gamma_{0,j}, \dots, \gamma_{UM_\tau N-1,j})$ is the covariance matrix of $\mathbf{h}_j$, and $\gamma_{i,j} = \left(\sum_{p=i-D}^{i+D}\sum_{q=j-D}^{j+D}\beta_{i,j,p,q}\alpha_{p,q}\right)^{-1}$. Then, following the conventional SBL, we adopt the Gamma distributions as the hierarchical prior for the local precision hyperparameter $\mathbf{A}$, i.e.,

$$p(\mathbf{A}|a, b) = \prod_{i=0}^{UM_\tau N-1}\prod_{j=0}^{N_y N_z-1}\Gamma(a)^{-1}b^a\alpha_{i,j}^a e^{-b\alpha_{i,j}}, \tag{12.45}$$

where $\Gamma(c) \triangleq \int_0^\infty t^{c-1}e^{-t}dt$. This hierarchical Bayesian modeling provides the analytically tractable solutions for the estimation of $\mathbf{H}$, and encourages the sparseness in the estimation since the overall prior of $\mathbf{H}$ (i.e., integrate out $\alpha_{i,j}$) is a Student-t distribution, which is sharply peaked at zero. In the conventional SBL, a and b are usually assigned by the small values (e.g., 10^{-4}), called the non-informative prior (Tipping, 2001).

12.4.2.2 SBL Framework

We now develop the SBL framework for the 2D burst block sparse signal recovery. Recalling that the noise obey a Gaussian distribution with the precision θ, and then the likelihood function is given by

$$p(\mathbf{Y}|\mathbf{H}, \theta) = \prod_{j=0}^{N_y N_z-1}\mathcal{CN}(\mathbf{y}_j|\mathbf{X}\mathbf{h}_j, \theta^{-1}\mathbf{I}_{UM_\tau N}), \tag{12.46}$$

where $\mathbf{y}_j = [y_{0,j}, \dots, y_{UM_\tau N-1}]$ is the jth column of $\mathbf{Y}$. According to the Bayes' rule, the posterior distribution of $\mathbf{H}$ is given as

$$p(\mathbf{H}|\mathbf{Y}, \mathbf{A}, \mathbf{B}, \theta) = \frac{p(\mathbf{Y}|\mathbf{H}, \theta)p(\mathbf{H}|\mathbf{A}, \mathbf{B})}{\int p(\mathbf{Y}|\mathbf{H}, \theta)p(\mathbf{H}|\mathbf{A}, \mathbf{B})d\mathbf{H}}. \tag{12.47}$$

Since in (12.47), the distributions appearing in the numerator are Gaussians, and the normalizing integral is a convolution of Gaussians. Thus, the posterior distribution of $\mathbf{H}$ is also Gaussian, and is given by

$$p(\mathbf{H}|\mathbf{Y}, \mathbf{A}, \mathbf{B}, \theta) = \prod_{j=0}^{N_y N_z - 1} \mathcal{CN}(\mathbf{h}_j|\mu_j, \Sigma_j), \tag{12.48}$$

where

$$\mu_j = \theta \Sigma_j \mathbf{X}^H \mathbf{y}_j, \Sigma_j = (\theta \mathbf{X}^H \mathbf{X} + \Upsilon_j^{-1})^{-1}. \tag{12.49}$$

Given a set of estimated hyperparameters $\{\mathbf{A}, \mathbf{B}, \theta\}$, the minimum mean square error (MMSE) estimation of $\hat{\mathbf{H}}$ is the mean of its posterior distribution, i.e.,

$$\hat{\mathbf{H}} = [\mu_0, \dots, \mu_{N_y N_z - 1}]. \tag{12.50}$$

Then, our problem reduces to estimating the set of hyperparameters. Following the conventional SBL, we place a Gamma hyperpriors over θ, i.e.,

$$p(\theta|r, s) = \Gamma(r)^{-1} s^r \theta^r e^{-s\theta}. \tag{12.51}$$

Then, the precision matrix $\mathbf{A}$ and the noise precision θ can be estimated by maximizing their log-posterior distribution with respect to $\mathbf{A}$ and θ, i.e.,

$$\hat{\mathbf{A}}, \hat{\theta} = \arg\max_{\mathbf{A}, \theta} \ \log p(\mathbf{A}, \theta|\mathbf{Y}, \mathbf{B}), \tag{12.52}$$

where the pattern coupled hyperparameter $\mathbf{B}$ is assumed to be known in this step.

Next, we adopt the expectation-maximization (EM) algorithm to approximate the optimal solutions in (12.52), where the channel matrix $\mathbf{H}$ is treated as the hidden variables. Then, the noise precision can be updated by

$$\theta^{t+1} = \hat{\theta} = \frac{M_\tau N N_y N_z + r}{\sum_{j=0}^{N_y N_z - 1} \mathrm{E}\left[\|\mathbf{y}_j - \mathbf{X}\mathbf{h}_j\|^2\right] + s}, \tag{12.53}$$

where

$$\mathrm{E}\left[\|\mathbf{y}_j - \mathbf{X}\mathbf{h}_j\|^2\right] = \|\mathbf{y}_j - \mathbf{X}\mu_j^t\|2 + (\theta^t)^{-1} \times \sum_{i=0}^{UM_\tau N-1} (1 - \Sigma_{i,j}^t (\gamma_{i,j}^t)^{-1}), \tag{12.54}$$

and $\gamma_{i,j}^t$ is the prior variance of $h_{i,j}$ in the tth iteration. In addition, the precision of the channel elements are updated by the following proposition (Shen et al., 2022).

Proposition 12.1 The range of the optimal (i,j)th element $\alpha^*_{i,j}$ in the $\mathbf{A}^{t+1}$ is given by

$$\left[\frac{a}{b+\omega^t_{i,j}}, \frac{a+(2D+1)^2}{b+\omega^t_{i,j}}\right], \tag{12.55}$$

where $\omega^t_{i,j} = \sum_{p=i-D}^{i+D} \sum_{q=j-D}^{j+D} \beta_{p,q,i,j} \mathrm{E}[|h_{pq}|^2]$. Therefore, a sub-optimal solution can be simply chosen as

$$\alpha^{t+1}_{i,j} = \hat{\alpha}_{i,j} = \frac{a}{b+\omega^t_{i,j}}. \tag{12.56}$$

Finally, the second moment $\mathrm{E}[|h_{pq}|^2] = |\mu^t_{p,q}|^2 + |\Sigma^t_{p,q}|^2$, where $\mu^t_{p,q}$ is the pth element of the posterior mean μ^t_q in the tth iteration, and $\Sigma^t_{p,q}$ is the (p,p)th entry of the covariance matrix Σ^t_q. Finally, we update the hyperparameters by the sparsity pattern of its neighborhoods, i.e.,

$$\beta_{i,j,p,q} = \begin{cases} S_{H_{p,q}}, & \text{if } S_{H_{i,j}=1}, \\ 1-S_{H_{p,q}}, & \text{otherwise.} \end{cases} \tag{12.57}$$

where

$$S_{H_{i,j}} = \mathbb{I}\left\{ \frac{\left|\mathbf{x}_i^{\mathrm{H}}\left(\theta^{-1}\mathbf{I}_{M_\tau N} + \mathbf{X}\Upsilon'_j\mathbf{X}^{\mathrm{H}}\right)^{-1}\mathbf{y}_j\right|^2}{\mathbf{x}_i^{\mathrm{H}}\left(\theta^{-1}\mathbf{I}_{M_\tau N} + \mathbf{X}\Upsilon'_j\mathbf{X}^{\mathrm{H}}\right)^{-1}\mathbf{x}_i} \geq \log\frac{1}{P_{\mathrm{fa}}} \right\}, \tag{12.58}$$

with $i = 0, \ldots, UM_\tau N - 1$, $j = 0, \ldots, N_y N_z - 1$, and $\mathbb{I}\{\cdot\}$ the indicator function which is 1 if the condition inside the braces is fulfilled and 0 otherwise. To facilitate the two-dimensional pattern coupled hierarchical prior with the sparse Bayesian learning and covariance-free method (TDSBL-CF) for solving the large-scale problems, the GAMP-based algorithm, called ConvSBL-GAMP, can be developed. The interested reader can refer to Shen et al. (2022).

12.4.3 Experimental Results

In this section, simulations are utilized to demonstrate the performance of the proposed algorithms. We consider the scenarios of the non-terrestrial networks recommended by the 3GPP (3rd Generation Partnership Project, 2019) and assume that the delay of each device is precompensated in the initial downlink synchronization so that the residual delay is in a small range. The detailed system parameters are summarized in Table 12.1. Here, we adopt the all-ones matrix to approximate the phase compensation matrix (see (12.41) and (12.42)) and set $D = s = a = 1, b = r = 10^{-4}$ and $P_{\mathrm{fa}} = 10^{-5}$ in our algorithms. Each user transmits

Table 12.1 Simulation Parameters.

Parameters	Values
Carrier frequency (GHz)	2
Subcarrier spacing (kHz)	330
Size of OTFS symbol (M, N)	(256, 15)
Length of CP	85
Number of LoS paths	1
Number of Non-LoS path	3
The residual delay (μs)	[0, 0.8]
Doppler shift (kHz)	[−41, 41]
Directional cosine along the y- and z-axis	[−1, 1]
Rician factor (dB)	5

Gaussian pilot matrix where entries in the pilot matrix are independent and obey $\mathcal{CN}(0, \frac{1}{M_\tau N})$. Moreover, the power of multi-path channel gain is normalized as 1, i.e., $\sum_{i=0}^{P} |h_{u,i}|^2 = 1$. Finally, we define the SNR as $\text{SNR} = 10\log_{10}\frac{\theta}{MN}$, where θ is the precision of the Gaussian noise.

To show the effectiveness of the proposed algorithms, we adopt generalized multiple measurement vector AMP (GMMV-AMP) (Ke et al., 2020) as benchmarks, which explores the sparsity in the angular domain. Besides, the normalized mean square error (NMSE) and the average activity error rate are adopted as the metrics for the CEs and DAD, respectively, given by

$$\text{NMSE} = \frac{\left\|\mathbf{H} - \hat{\mathbf{H}}\right\|_F^2}{\|\mathbf{H}\|_F^2}, \quad \text{AER} = \frac{1}{U}\sum_{u=0}^{U-1}\left|\lambda_u - \hat{\lambda}_u\right|. \tag{12.59}$$

Firstly, as shown in Fig. 12.5, we compare the performance for the CE between the proposed TDSBL-CF and ConvSBL-GAMP, where the ConvSBL-GAMP-1D (ConvSBL-GAMP-2D) represents that the ConvSBL-GAMP adopts the one (two) dimensional kernel $\mathcal{B}$. It is observed that ConvSBL-GAMP-2D outperforms the ConvSBL-GAMP-1D for different SNR values. For example, in order to achieve the same NMSE, the SNR required by ConvSBL-GAMP-2D is about 3 dB lower than that of the ConvSBL-GAMP-1D, since the 1D kernel only accounts for the sparsity in the angular domain while the 2D kernel can capture the 2D burst block sparsity. It is indicated the effectiveness of the proposed 2D kernel. Besides, the NMSE of the TDSBL-CF is lower than the ConvSBL-GAMP-2D when SNR is smaller than 10 dB, since the TDSBL-CF computes the posterior mean of the channel more precisely and its coupled hyperparameter can be adjusted automatically in each

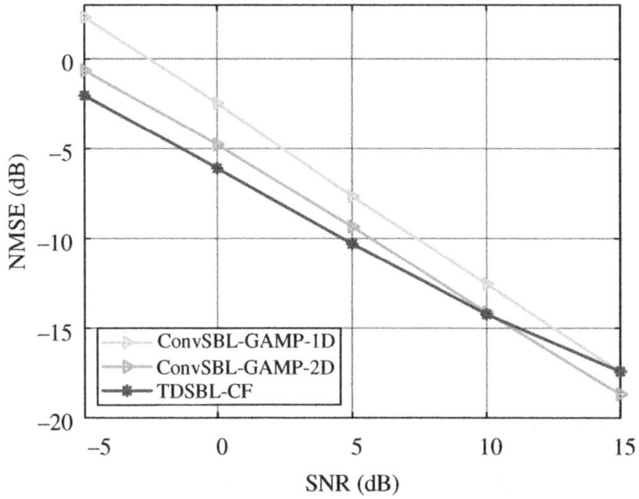

Figure 12.5 Performance comparison between the ConvSBL-GAMP and TDSBL-CF under different SNR values. ConvSBL-GAMP-1D (ConvSBL-GAMP-2D) represents the ConvSBL-GAMP with the one (two) dimensional kernel. $U = U_a = 2$, $N_y = N_z = 4$, and the pilot overhead is 0.3.

iteration while ConvSBL-GAMP only adopts the single fixed coupled parameter which is less flexible. In the following experiments, we show the performance comparison between the ConvSBL-GAMP and benchmarks for dealing with large dimensional problems, and omit the TDSBL-CF curve since its computation will be prohibitive in those scenarios.

Figure 12.6 shows the NMSE comparison between the ConvSBL-GAMP and GMMV-AMP. The number of the potential devices and active devices are 20 and 6, respectively, i.e., $U = 20$ and $U_a = 6$, and the numerical results are obtained by averaging over 1000 channel realizations. Here, we adopt the identity matrix to approximate the phase compensation matrix, and we also provide the performance comparison between the three algorithms when the phase compensation matrix is fully known (marked by "FULL"). It is observed that ConvSBL-GAMP outperforms the benchmark under different SNR values, even when that algorithm has the knowledge of the phase compensation matrix. For example, to achieve the NMSE of −15 dB, the SNR required by the ConvSBL-GAMP is about 5 dB lower than the GMMV-AMP, since the GMMV-AMP only captures the sparsity in the angular domain, while the ConvSBL-GAMP is able to deal with the sparsity in the angular domain and delay-Doppler domain. Therefore, utilizing the 2D burst block sparsity is favorable for the CE and our proposed algorithm can capture that feature better compared with the conventional

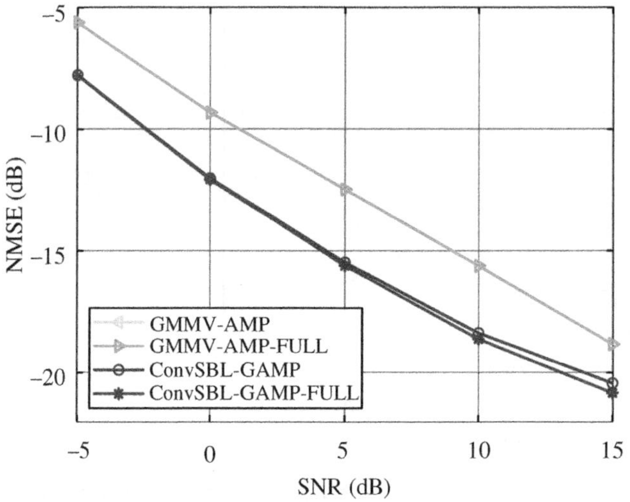

Figure 12.6 Performance comparison between the ConvSBL-GAMP and GMMV-AMP under different SNR values. $U = 20$, $U_a = 6$, $N_y = N_z = 4$, and the pilot overhead is 0.3.

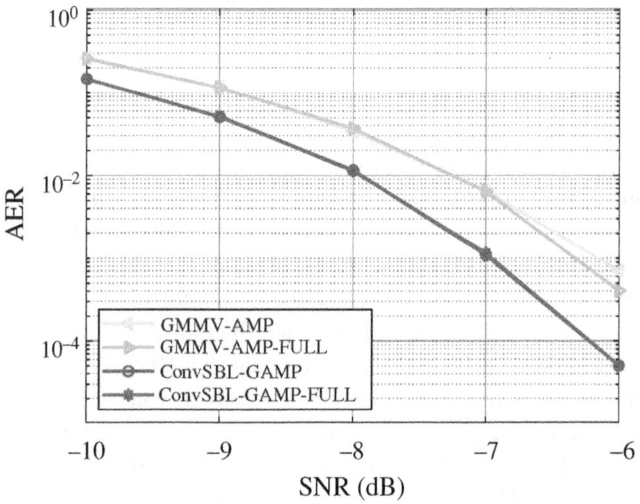

Figure 12.7 Performance comparison of DAD between the ConvSBL-GAMP and GMMV-AMP under different SNR values. $U = 20$, $U_a = 6$, the pilot overhead is 0.25, and $N_y = N_z = 4$.

algorithms. In addition, it is observed that the performance of the two algorithms only have marginal degradation when the phase compensation matrix is unknown.

Finally, we show in Fig. 12.7 the performance comparison for DAD between the GAMP algorithm combined with the sparse Bayesian learning and two-dimensional convolution (ConvSBL-GAMP) and GMMV-AMP under different SNR values. The detection threshold $\xi_{\text{th}} = 1$. There are 20 potential devices, six of which are active, i.e., $U = 20$ and $U_a = 6$. The number of antennas is set as $N_y = N_z = 4$, and the pilot overhead is 0.25. The numerical results here are obtained by averaging over 1000 channel realizations. It is observed that the error probability of the ConvSBL-GAMP is very close to zero when the SNR is larger than −6 dB and is lower than the other algorithm, which indicates that the ConvSBL-GAMP can successfully detect the active devices in the low SNR regime.

12.5 Concluding Remarks

This chapter investigated several CS approaches to enable grant-free random access. First, we developed a joint DAD, CE, and DD algorithm for MIMO massive URA. Based on the principle of the BP, the iterative MP algorithm was utilized to decode the twofold data, with substantial performance enhancements as well as complexity reduction. Then, we exploited the angular domain sparsity of the multipath channel and introduced the MRF model to capture the clustered channel sparsity in the framework of GAMP. The proposed algorithm further assisted an uncoupled URA scheme based on channel reconstruction and clustering, which has shown to be more energy efficient than state-of-the-art approaches. Finally, we studied the application of MIMO-OTFS for grant-free random access in LEO satellite communications. To exploit the 2D burst block sparsity in the delay-Doppler-angle domain, the GAMP algorithm combined with SBL and 2D convolution was proposed for joint DAD and CE. Simulation results demonstrated that the proposed algorithm could acquire accurate channel state information and device activity.

Acknowledgments

The work of Y. Wu is supported in part by the National Key R&D Program of under Grant China 2022YFB2902100, the Fundamental Research Funds for the Central Universities, National Science Foundation (NSFC) under Grant 62122052 and 62071289, 111 project BP0719010, and STCSM 22DZ2229005.

References

3rd Generation Partnership Project. 3rd generation partnership project; technical specification group radio access network; solutions for NR to support nonterrestrial networks (NTN) (Release 16). *3GPP TR 38.811V15.1.0*, 2019.

Z Chen and W Yu. Massive device activity detection by approximate message passing. In *2017 IEEE International Conference on Acoustics, Speech and Signal Processing (ICASSP)*, pages 3514–3518, New Orleans, LA, USA, 2017.

A Fengler, S Haghighatshoar, P Jung, and G Caire. Non-Bayesian activity detection, large-scale fading coefficient estimation, and unsourced random access with a massive MIMO receiver. *IEEE Transactions on Information Theory*, 67(5):2925–2951, 2021.

Y Ge, Q Deng, P C Ching, and Z Ding. OTFS signaling for uplink NOMA of heterogeneous mobility users. *IEEE Transactions on Communications*, 69(5):3147–3161, 2021.

R Hadani, S Rakib, M Tsatsanis, A Monk, A J Goldsmith, A F Molisch, and R Calderbank. Orthogonal time frequency space modulation. In *2017 IEEE Wireless Communications and Networking Conference (WCNC)*, pages 1–6, San Francisco, CA, USA, 2017.

G Hannak, M Mayer, A Jung, G Matz, and N Goertz. Joint channel estimation and activity detection for multiuser communication systems. In *2015 IEEE International Conference on Communication Workshop (ICCW)*, pages 2086–2091, London, UK, 2015.

M Ke, Z Gao, Y Wu, X Gao, and R Schober. Compressive sensing based adaptive active user detection and channel estimation: Massive access meets massive MIMO. *IEEE Transactions on Signal Processing*, 68:764–779, 2020.

L Liu and W Yu. Massive connectivity with massive MIMO – Part I: Device activity detection and channel estimation. *IEEE Transactions on Signal Processing*, 66(11):2933–2946, 2018.

Y Polyanskiy. A perspective on massive random-access. In *2017 IEEE International Symposium on Information Theory (ISIT)*, pages 2523–2527, 2017.

W Shen, L Dai, J An, P Fan, and R W Heath. Channel estimation for orthogonal time frequency space (OTFS) massive MIMO. *IEEE Transactions on Signal Processing*, 67(16):4204–4217, 2019.

B Shen, Y Wu, J An, C Xing, L Zhao, and W Zhang. Random access with massive MIMO-OTFS in LEO satellite communications. *IEEE Journal on Selected Areas in Communications*, 40(10):2865–2881, 2022.

G D Surabhi, R M Augustine, and A Chockalingam. Peak-to-average power ratio of OTFS modulation. *IEEE Communications Letters*, 23(6):999–1002, 2019.

M E Tipping. Sparse Bayesian learning and the relevance vector machine. *Journal of Machine Learning Research*, 1:211–244, 2001.

X Xie, Y Wu, J An, J Gao, W Zhang, C Xing, K-K Wong, and C Xiao. Massive unsourced random access: Exploiting angular domain sparsity. *IEEE Transactions on Communications*, 70(4):2480–2498, 2022.

13

Algorithm Unrolling for Massive Connectivity in IoT Networks

Yinan Zou, Yong Zhou, and Yuanming Shi

School of Information Science and Technology, ShanghaiTech University, Shanghai, China

13.1 Introduction

13.1.1 Massive Random Access

In order to support an enormous number of Internet of Things (IoT) devices to access wireless networks, massive random access has been identified as a typical use case of 5G (Sharma and Wang, 2019). Vast amount of data from IoT devices causes a heavy communication overhead in wireless network (Wu et al., 2020, Shi et al., 2023). In this scenario, it becomes imperative to devise a suitable access scheme that enables the multitude of IoT devices to connect to the wireless network and subsequently engage in intelligent applications (Letaief et al., 2021, Shi et al., 2020, Wang et al., 2021).

The conventional grant-based random access involves the following four steps (Hasan et al., 2013). First, each device transmits a pilot sequence to inform the base station (BS) that the device is active and has data to transmit. Second, the BS replies a response to the active device to enable its transmission. Third, the device dispatches a connection request to the BS, seeking resource allocation for its upcoming data transmission. Finally, if a pilot sequence is successfully detected, then the BS grants the connection request. However, if two or more active devices select an identical sequence, this results in collisions and consequently, network access failures. The impacted devices are required to initiate the procedure once again. A limitation of grant-based random access is the relatively high likelihood of sequence collision, mainly attributable to the substantial number of devices and the finite availability of orthogonal sequences, resulting in significant access latency. Furthermore, the four-step channel access leads to significant signaling overhead, thereby degrading the spectrum efficiency.

Next Generation Multiple Access, First Edition.
Edited by Yuanwei Liu, Liang Liu, Zhiguo Ding, and Xuemin Shen.
© 2024 The Institute of Electrical and Electronics Engineers, Inc. Published 2024 by John Wiley & Sons, Inc.

To enable low-overhead channel access, grant-free random access receives much attention, given its capability of enabling the direct transmission of pilot sequence and data from active devices to the BS without requiring explicit authorization (Liu et al., 2018). In particular, different from grant-based random access that adopts orthogonal pilot sequences, the pilot sequences utilized by grant-free random access are usually non-orthogonal and uniquely assigned to individual devices. If the device is active, it sends the assigned unique pilot sequence along with its data to the BS. After receiving the pilot sequences from the active devices, the BS conducts joint activity detection and channel estimation (JADCE).

13.1.2 Sparse Signal Processing

A significant characteristic of IoT data traffic is its sporadic nature, wherein only a small fraction of devices are active during any given time slot, while the majority of devices remain inactive to conserve energy. Due to this unique feature, JADCE problem can be formulated as multiple measurement vector (MMV) problems (Ziniel and Schniter, 2012). Numerous studies have been conducted with the objective of employing sparse signal processing techniques to effectively address MMV problems.

13.1.2.1 Bayesian Methods

As demonstrated in Chen et al. (2018b), approximate message passing (AMP) with vector denoiser and parallel AMP–MMV achieve almost the same performance, because both methods leverage channel statistical information for sparse signal recovery. With a large antenna array, vector AMP is capable of achieving perfect device detection, as illustrated in Liu and Yu (2018). In addition, generalized MMV-AMP is further developed in Ke et al. (2020) for effective JADCE, where spatial/angular domain channel models are adopted. Although AMP-based methods can efficiently tackle JADCE problems, they may fail to converge under non-Gaussian pilot sequence matrix.

13.1.2.2 Optimization-Based Methods

The mixed $\ell_{2,1}$-norm developed in He et al. (2018) can effectively exploit the intrinsic sparsity within the device state matrix which includes both device activity and channel conditions. The authors subsequently solve the formulated $\ell_{2,1}$-regularized problem by applying the alternating direction method of multipliers (ADMM) algorithm. A smoothed primal–dual algorithm proposed in Jiang et al. (2018) is capable of solving the high-dimensional JADCE problem. The authors also conduct theoretical analysis to characterize the trade-off between the estimation accuracy and the computational cost. For sparse and low-rank millimeter-wave (mmW)/Terahertz (Thz) channels, the authors in Shao et al.

(2021b) propose a massive access scheme based on multi-rank aware algorithms. By exploiting the low-rank property of received data matrix, the authors in Shao et al. (2019) proposed a dimension reduction scheme and further develop a Riemannian trust-region algorithm that is capable of efficiently tackling the JADCE problem. Nonetheless, the practical utility of optimization-based methods is hampered by their significant computational complexity.

13.1.2.3 Deep Learning-Based Methods

Recently, deep learning has emerged as a disruptive technology for the purpose of mitigating computational cost in wireless resource allocation (Zou et al., 2022, Wang et al., 2023). Besides, there are a lot of studies on utilizing deep learning technique for JADCE problems. Specifically, by exploiting the underlying Bernoulli–Gaussian mixture distribution, an adaptive-tuning learning algorithm is developed in Shao et al. (2021a) to tackle the massive device detection problem. In addition, auto-encoder-based methods developed in Cui et al. (2020) can also be utilized for the design of the pilot sequence matrix as well as the recovery of sparse channel matrix. The authors in Johnston and Wang (2022) develop two deep unfolding networks based on linearized ADMM and approximate AMP to tackle the JADCE problem for mmW channels. Moreover, the authors in Zhu et al. (2021) investigate asynchronous massive random access and design three learned AMP networks to balance the recoverability and computational complexity. However, these deep learning-based methods lack theoretical guarantees.

In this book chapter, we study grant-free massive access in an IoT network comprising one BS with multiple antennas and a massive number of IoT devices, where the IoT devices transmit non-orthogonal pilot sequences in a sporadic manner. To enhance the efficiency of channel access, we formulate the JADCE problem as a group–spare matrix recovery problem. By leveraging algorithm unrolling, we propose unrolled neural networks to tackle this problem. In addition, theoretical analysis is conducted to prove the coupling property of the weight matrix, which enables us to simply the neural network to reduce the number of the trainable parameters. Finally, we show that linear convergence can be achieved by the proposed unrolled neural networks.

13.2 System Model

Consider a single-cell IoT network, as shown in Fig. 13.1, where N single-antenna IoT devices are connected to a BS equipped with M antennas. Each device independently determines its active/inactive status in each time slot. We denote the channel response between device n and the BS as $\boldsymbol{h}_n \in \mathbb{C}^M$, which remains constant within a time slot but may vary across slots and devices in an independent

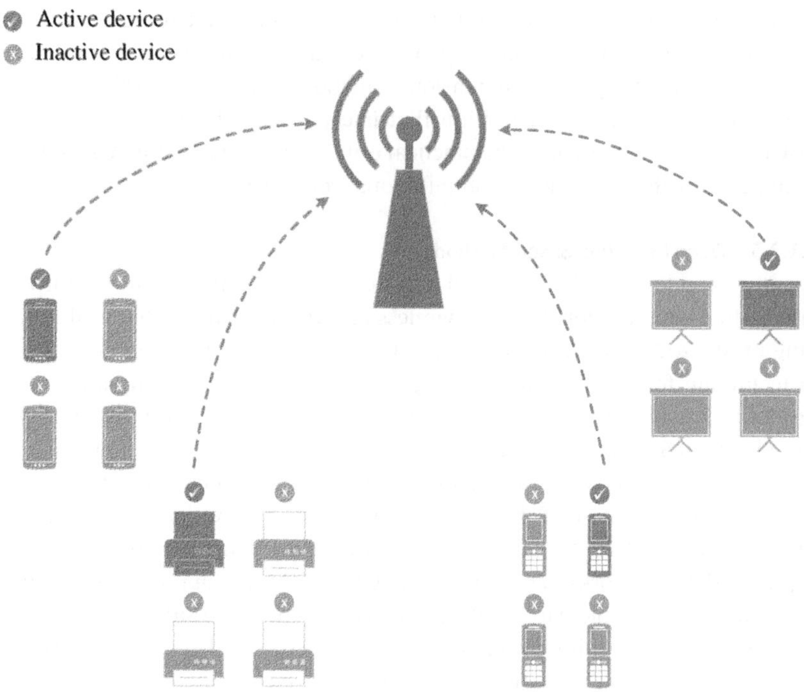

Figure 13.1 Illustration of an IoT network where only a small fraction of IoT devices is active.

manner. We denote a_n as the variable indicating the activity status of device n. Specifically, the binary variable a_n takes a value of 1 to indicate that the device n is active, while it takes a value of 0 if the device is inactive. By denoting $\boldsymbol{y}(\ell) \in \mathbb{C}^M$ as the signal received at the BS, we have

$$\boldsymbol{y}(\ell) = \sum_{n=1}^{N} \boldsymbol{h}_n a_n s_n(\ell) + \boldsymbol{z}(\ell), \quad \ell \in \mathcal{L}, \tag{13.1}$$

where $\mathcal{L} = \{1, \dots, L\}$, $s_n(\ell)$ denotes the ℓth symbol of device n's pilot sequence, L denotes the pilot length, and $\boldsymbol{z}(\ell)$ signifies the additive white Gaussian noise (AWGN). Note that the device number is generally much greater than the pilot length, i.e., $N \gg L$. Consequently, we allocate unique but non-orthogonal pilot sequence to different devices. By denoting $\boldsymbol{Y} = [\boldsymbol{y}(1), \dots, \boldsymbol{y}(L)]^{\mathrm{T}} \in \mathbb{C}^{L\times M}$, $\boldsymbol{A} = \mathrm{diag}(a_1, \dots, a_N) \in \mathbb{R}^{N\times N}$, $\boldsymbol{H} = [\boldsymbol{h}_1, \dots, \boldsymbol{h}_N]^{\mathrm{T}} \in \mathbb{C}^{N\times M}$, $\boldsymbol{Z} = [\boldsymbol{z}(1), \dots, \boldsymbol{z}(L)] \in \mathbb{C}^{L\times M}$, and $\boldsymbol{S} = [\boldsymbol{s}(1), \dots, \boldsymbol{s}(L)]^{\mathrm{T}} \in \mathbb{C}^{L\times N}$ with $\boldsymbol{s}(\ell) = [s_1(\ell), \dots, s_N(\ell)]^{\mathrm{T}} \in \mathbb{C}^N$, we have

$$\boldsymbol{Y} = \boldsymbol{SAH} + \boldsymbol{Z}, \tag{13.2}$$

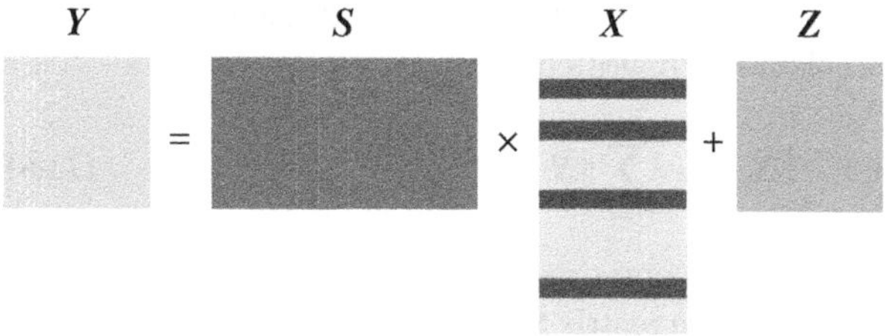

Figure 13.2 Illustration of the group-sparse matrix recovery problem.

Our objective is to detect activity matrix $\boldsymbol{A}$ and estimate channel matrix $\boldsymbol{H}$, referring to as JADCE (Liu and Yu, 2018).

We rewrite (13.2) as

$$\boldsymbol{Y} = \boldsymbol{S}\boldsymbol{X} + \boldsymbol{Z}, \tag{13.3}$$

where $\boldsymbol{X} = \boldsymbol{A}\boldsymbol{H} \in \mathbb{C}^{N\times M}$. Given that the device activity matrix $\boldsymbol{A}$ is a diagonal matrix with diagonal elements $\{a_1, \dots, a_N\}$, matrix $\boldsymbol{X}$ can be expressed as $\boldsymbol{X} = [a_1\boldsymbol{h}_1, \dots, a_N\boldsymbol{h}_N]^{\mathrm{T}}$. In the case where a particular device, i.e., device n, is not active, every element of the corresponding n-row of $\boldsymbol{X}$ is zero, indicating the group-row-sparse structure of matrix $\boldsymbol{X}$, as shown in Fig. 13.2.

The goal is to conduct JADCE, i.e., recover $\boldsymbol{X}$ from (13.3). In order to recover $\boldsymbol{X}$, we formulate the following matrix recovery problem

$$\mathcal{P} : \underset{\boldsymbol{X}\in\mathbb{C}^{N\times M}}{\text{minimize}} \ \frac{1}{2} \|\boldsymbol{Y} - \boldsymbol{S}\boldsymbol{X}\|_F^2 + \lambda F(\boldsymbol{X}), \tag{13.4}$$

where λ signifies the regularization parameter and $F(\boldsymbol{X})$ is the regularizer.

13.3 Learned Iterative Shrinkage Thresholding Algorithm for Massive Connectivity

13.3.1 Problem Formulation

To promote sparsity, we utilize the mixed ℓ_1/ℓ_2-norm, i.e., $\sum_i \|\boldsymbol{X}_{i,:}\|_2$ as the regularizer, and reformulate problem $\mathcal{P}$ as group least absolute shrinkage and selection operator (LASSO). We rewrite the complex matrix by splitting its real part and imaginary part and obtain the real-valued version of (13.4)

$$\tilde{\boldsymbol{Y}} = \tilde{\boldsymbol{S}}\tilde{\boldsymbol{X}} + \tilde{\boldsymbol{Z}} = \begin{bmatrix} \mathcal{R}\{\boldsymbol{S}\} & -\mathcal{I}\{\boldsymbol{S}\} \\ \mathcal{I}\{\boldsymbol{S}\} & \mathcal{R}\{\boldsymbol{S}\} \end{bmatrix} \begin{bmatrix} \mathcal{R}\{\boldsymbol{X}\} \\ \mathcal{I}\{\boldsymbol{X}\} \end{bmatrix} + \begin{bmatrix} \mathcal{R}\{\boldsymbol{Z}\} \\ \mathcal{I}\{\boldsymbol{Z}\} \end{bmatrix}, \tag{13.5}$$

where $\mathcal{R}(\cdot)$ and $\mathcal{I}(\cdot)$ represent the real part and imaginary part of a complex matrix, respectively. Therefore, problem $\mathcal{P}$ is rewritten as

$$\mathcal{P}_r : \underset{\tilde{X} \in \mathbb{R}^{2N \times M}}{\text{minimize}} \frac{1}{2} \| \tilde{Y} - \tilde{S}\tilde{X} \|_F^2 + \lambda \sum_{i=1}^{2N} \| \tilde{X}_{i,:} \|_2. \tag{13.6}$$

Consequently, the aim of JADCE becomes recovering the matrix $\tilde{X}$ according to the observed $\tilde{Y}$ and the pilot sequence matrix $\tilde{S}$.

Iterative shrinkage thresholding algorithm for group sparsity (ISTA-GS) can solve problem $\mathcal{P}_r$, whose iteration is

$$\tilde{X}^{k+1} = \eta_{\lambda/C}\left(\tilde{X}^k + \frac{1}{C}\tilde{S}^{\mathrm{T}}(\tilde{Y} - \tilde{S}\tilde{X}^k)\right), \tag{13.7}$$

where $\tilde{X}^k$ denotes the estimated matrix in the kth iteration, $\eta_{\lambda/C}(\cdot)$ denotes the multidimensional shrinkage thresholding operator and $1/C$ is the step size. The multidimensional shrinkage thresholding operator is applied to each row of the input

$$\eta_\theta(\tilde{X}[i,:]) = \max\{0, \|\tilde{X}[i,:]\|_2 - \theta\} \frac{\tilde{X}[i,:]}{\|\tilde{X}[i,:]\|_2}. \tag{13.8}$$

Although ISTA-GS can solve the problem, it suffers from an intrinsic trade-off between convergence rate and recovery performance due to the regularization parameter λ. Specifically, higher value of λ results in faster convergence rate, but at the cost of poorer recovery performance (Giryes et al., 2018). Furthermore, the selection of the step size $1/C$ is critical for determining the convergence rate. To address the convergence rate and recovery performance limitations, we shall propose an unrolled neural network framework based on ISTA-GS that learns the regularization parameter λ and step size $1/C$.

13.3.2 Unrolled Neural Networks

We present three unrolled neural networks based on ISTA-GS (Shi et al., 2022) to tackle problem $\mathcal{P}_r$ in this sub-section.

13.3.2.1 LISTA-GS

By following the idea of algorithm unrolling (Gregor and LeCun, 2010, Monga et al., 2021), we denote $W_1 = \frac{1}{C}\tilde{S}^T$, $\theta = \frac{\lambda}{C}$, and $W_2 = I - \frac{1}{C}\tilde{S}^T\tilde{S}$. Then, we rewrite (13.7) as

$$\tilde{X}^{k+1} = \eta_{\theta^k}(W_1\tilde{Y} + W_2\tilde{X}^k). \tag{13.9}$$

Scalar θ^k and matrices W_1 and W_2 are considered as trainable parameters. Consequently, (13.9) can be represented by a single-layer recurrent neural network

(RNN). By cascading K layer neural network, we can construct a K-layer RNN, which models the K iterations of ISTA-GS. Specifically, the K-layer RNN is

$$\tilde{\boldsymbol{X}}^{k+1} = \eta_{\theta^k}(\boldsymbol{W}_1\tilde{\boldsymbol{Y}} + \boldsymbol{W}_2\tilde{\boldsymbol{X}}^k),\ k \in \mathcal{K}, \tag{13.10}$$

where $\Theta = \{\boldsymbol{W}_1^k, \boldsymbol{W}_2^k, \theta^k\}_{k=0}^{K-1}$ are trainable parameters, and $\mathcal{K} = \{0,1,\dots,K-1\}$. The model given in (13.10) is referred to as learned ISTA-GS (LISTA-GS). Nevertheless, the number of trainable parameters of LISTA-GS is numerous, which brings much difficulty of training.

13.3.2.2 LISTA-GSCP

Because $\boldsymbol{W}_2^0 = (\boldsymbol{I} - \boldsymbol{W}_1^0\tilde{S})$ and coupling relationship in Chen et al. (2018a), we derive the following theorem. The necessary condition indicates properties of the trainable parameters if LISTA-GS is capable of recovering the group–sparse matrix.

Theorem 13.1 We assume that $\{\boldsymbol{W}_1^k, \boldsymbol{W}_2^k, \theta^k\}_{k=0}^{\infty}$ are given, where $\|\boldsymbol{W}_1^k\|_F \le C_{W_1}$ and $\|\boldsymbol{W}_2^k\|_F \le C_{W_2}$. The input to LISTA-GS is $\tilde{\boldsymbol{Y}}$ and the initial estimate $\tilde{\boldsymbol{X}}^0$ is $\boldsymbol{0}$. $\{\tilde{\boldsymbol{X}}^k\}_{k=1}^{\infty}$ is generated layer-by-layer by LISTA-GS. The following conditions should meet by $\{\boldsymbol{W}_1^k, \boldsymbol{W}_2^k, \theta^k\}_{k=0}^{\infty}$ if any group-row-sparse signal can be recovered without observation noise

$$\theta^k \to 0, \text{as}\, k \to \infty, \tag{13.11}$$

$$\boldsymbol{W}_2^k - (\boldsymbol{I} - \boldsymbol{W}_1^k\tilde{S}) \to \boldsymbol{0}, \text{as}\, k \to \infty. \tag{13.12}$$

Based on Theorem 13.1, we incorporate the coupling relationship between $\boldsymbol{W}_2^k$ and $\boldsymbol{W}_1^k$, i.e., $\boldsymbol{W}_2^k = (\boldsymbol{I} - \boldsymbol{W}_1^k\tilde{S})$, into LISTA-GS. By denoting $\boldsymbol{W}^k = (\boldsymbol{W}_1^k)^T$, we obtain LISTA-GS with coupling structure (LISTA-GSCP)

$$\tilde{\boldsymbol{X}}^{k+1} = \eta_{\theta^k}\left(\tilde{\boldsymbol{X}}^k + (\boldsymbol{W}^k)^T(\tilde{\boldsymbol{Y}} - \tilde{\boldsymbol{S}}\tilde{\boldsymbol{X}}^k)\right), k \in \mathcal{K}, \tag{13.13}$$

where $\Theta = \{\boldsymbol{W}^k, \theta^k\}_{k=0}^{K-1}$ are trainable parameters.

13.3.2.3 ALISTA-GS

Inspired by Liu and Chen (2019), the weight can be expressed as $\boldsymbol{W}^k = \gamma^k\boldsymbol{W}$ to further reduce the trainable parameter number. Before the training stage, the projected gradient descent (PGD) algorithm can be employed to address the following problem in order to obtain weight matrix $\boldsymbol{W}$

$$\underset{\boldsymbol{W}\in\mathbb{R}^{2L\times 2N}}{\text{minimize}} \quad \|\boldsymbol{W}^T\tilde{\boldsymbol{S}}\|_F^2 \tag{13.14}$$

$$\text{subject to} \quad \boldsymbol{W}[:,p]^T\tilde{\boldsymbol{S}}[:,p] = 1, \forall p \in [2N]. \tag{13.15}$$

Thus, we obtain the analytic LISTA-GS (ALISTA-GS)

$$\tilde{\boldsymbol{X}}^{k+1} = \eta_{\theta^k}\left(\tilde{\boldsymbol{X}}^k + \gamma^k \boldsymbol{W}^T(\tilde{\boldsymbol{Y}} - \tilde{\boldsymbol{S}}\tilde{\boldsymbol{X}}^k)\right), k \in \mathcal{K}, \tag{13.16}$$

where $\Theta = \{\theta^k, \gamma^k\}_{k=0}^{K-1}$ are trainable parameters. In addition, we can obtain matrix $\boldsymbol{W}$ by solving problem (13.14). In contrast to LISTA-GSCP, this particular network architecture places more focus on the selection of the learning step sizes.

13.3.3 Convergence Analysis

The convergence analytical results of ALISTA-GS and LISTA-GSCP are derived in this subsection. The structure of the unrolled neural networks adheres to that of ISTA-GS, enabling us to interpret the deep learning models within the context of optimization.

The convergence rates of ALISTA-GS and LISTA-GSCP are presented in the following theorems. Before establishing the convergence analysis, the ℓ_2-norm of each row of $\tilde{\boldsymbol{X}}^\natural$ and Frobenius norm of $\tilde{\boldsymbol{Z}}$ are assumed to be bounded by β and σ, respectively (Liu and Chen, 2019, Chen et al., 2018a). Moreover, the number of non-zero rows of $\tilde{\boldsymbol{X}}^\natural$ is bounded (Jiang et al., 2018), since each element of $\{a_1, \dots, a_n\}$ follows a Bernoulli distribution. For notational brevity, we replace $\tilde{\boldsymbol{X}}^k(\tilde{\boldsymbol{X}}^\natural, \tilde{\boldsymbol{Z}})$ with $\tilde{\boldsymbol{X}}^k$. We denote $\psi(\tilde{\boldsymbol{X}}) = [||\tilde{\boldsymbol{X}}[1,:]||_2, ||\tilde{\boldsymbol{X}}[2,:]||_2, \dots, ||\tilde{\boldsymbol{X}}[2N,:]||_2]^T$ and assume $\tilde{\boldsymbol{X}}^\natural$ and $\tilde{\boldsymbol{Z}}$ belonging to set $\mathcal{X}(\beta, s, \sigma) =:= \left\{(\tilde{\boldsymbol{X}}^\natural, \tilde{\boldsymbol{Z}}) |\ ||\tilde{\boldsymbol{X}}^\natural[p,:]||_2 \le \beta, \forall p, ||\tilde{\boldsymbol{X}}^\natural||_{2,0} \le s, ||\tilde{\boldsymbol{Z}}||_F \le \sigma\right\}$.

Theorem 13.2 We assume $\{\boldsymbol{W}^k, \theta^k\}_{k=0}^{\infty}$ are given. The input of LISTA-GSCP is $\tilde{\boldsymbol{Y}}$ and the initial point $\tilde{\boldsymbol{X}}^0$ is $\boldsymbol{0}$. We consider that $\{\tilde{\boldsymbol{X}}^k\}_{k=1}^{\infty}$ are generated by LISTA-GSCP. For all $(\tilde{\boldsymbol{X}}^\natural, \tilde{\boldsymbol{Z}}) \in \mathcal{X}(\beta, s, \sigma)$, if the sparsity level s is small enough, then the following error bound holds

$$||\tilde{\boldsymbol{X}}^k - \tilde{\boldsymbol{X}}^\natural||_F \le s\beta \exp(-ck) + C\sigma, \tag{13.17}$$

where constants $c > 0$ and $C > 0$ depend on $\tilde{\boldsymbol{S}}$ and s.

13.3.3.1 "Good" Parameters for Learning

We provide the definition of "good" parameters, which ensures its linear convergence rate.

Definition 13.1

(i) We normalize the columns of the mutual coherence of $\tilde{\boldsymbol{S}}$ as follows

$$\mu(\tilde{\boldsymbol{S}}) = \max_{\substack{p \neq q \\ 1 \le p,q \le 2N}} \left|(\tilde{\boldsymbol{S}}[:,p])^T \tilde{\boldsymbol{S}}[:,q]\right|. \tag{13.18}$$

(ii) We normalize the columns of the generalized mutual coherence of $\tilde{S} \in \mathbb{R}^{2L\times 2N}$ as follows

$$\tilde{\mu}(\tilde{S}) = \inf_{\substack{W\in\mathbb{R}^{2L\times 2N} \\ (W[:,p])^T \tilde{S}[:,p]=1,\forall p}} \left\{ \max_{\substack{p\neq q \\ 1\leq p,q\leq 2N}} \left|(W[:,p])^T\tilde{S}[:,q]\right| \right\}. \tag{13.19}$$

According to Lemma 1 in Chen et al. (2018a), matrix $W \in \mathbb{R}^{2L\times 2N}$ can be designed to reach the infimum in (13.19), i.e., $\mathcal{W}(\tilde{S}) \neq \emptyset$. As a result, a set of "good" weight matrices are defined as

$$\mathcal{W}(\tilde{S}) := \underset{W\in\mathbb{R}^{2L\times 2N}}{\operatorname{argmin}} \Big\{ \|W\|_{\max} | (W[:,p])^T \tilde{S}[:,p] = 1, \forall p,$$
$$\max_{\substack{p\neq q \\ 1\leq p,q\leq 2N}} \left|(W[:,p])^T\tilde{S}[:,q]\right| = \tilde{\mu}(\tilde{S}) \Big\}. \tag{13.20}$$

Then, we have the following definitions on "good" parameters.

Definition 13.2 $\Theta = \{W^k, \theta^k\}_{k=0}^{\infty}$ are "good" parameters if

$$W^k \in \mathcal{W}(\tilde{S}), \quad \theta^k = \tilde{\mu} \sup_{(\tilde{X}^\natural, \tilde{Z})} \|\tilde{X}^k - \tilde{X}^\natural\|_{2,1} + \sigma C_W, \forall k \in \mathbb{N}, \tag{13.21}$$

where $C_W = \max_{k\geq 0} \|W^k\|_{2,1}$ and $\tilde{\mu} = \tilde{\mu}(\tilde{S})$.

Subsequently, we prove the conclusion in Theorem 1.2 with the sequence of "good" parameters.

Similarly, we have the following theorem that presents the convergence of ALISTA-GS.

Theorem 13.3 With $\{\theta^k\}_{k=0}^{\infty}$ and $W \in \mathcal{W}(\tilde{S})$, if $s \leq (1 + 1/\tilde{\mu})/2$, then for all $(\tilde{X}^\natural, \tilde{Z}) \in \mathcal{X}(\beta, s, 0)$, and

$$\gamma^k \in (0, \frac{2}{1 + 2\tilde{\mu}s - \tilde{\mu}}), \tag{13.22}$$

$$\theta^k = \tilde{\mu}\gamma^k \sup_{\tilde{X}^\natural, \tilde{Z}} \|\tilde{X}^k - \tilde{X}^\natural\|_{2,1}, \tag{13.23}$$

then we obtain

$$\|\tilde{X}^k - \tilde{X}^\natural\|_F \leq s\beta \exp(-\sum_{\tau=0}^{k} c^\tau), \tag{13.24}$$

where $c^\tau = -\log(\gamma^\tau(2\tilde{\mu}s - \tilde{\mu}) + |1 - \gamma^\tau|)$.

13.4 Learned Proximal Operator Methods for Massive Connectivity

In this section, we further present multiple learned proximal operator methods (POMs) (Zou et al., 2021) that achieve better performance than different variants of LISTA methods presented in Section 13.3.

13.4.1 Problem Formulation

By considering the nonconvex regularizer such as minimax concave penalty (MCP) can better promote sparsity than ℓ_1-norm, we adopt MCP as the regularizer and formulate the following problem

$$\mathcal{P}_G : \underset{\tilde{X}\in\mathbb{R}^{2N\times M}}{\text{minimize}} \frac{1}{2} \| \tilde{Y} - \tilde{S}\tilde{X} \|_F^2 + \lambda G_\eta(\tilde{X}), \tag{13.25}$$

where $G_\eta(\tilde{X}) = \sum_{p,q} g_\eta(\tilde{X}[p,q])$ and η denotes the nonconvexity measure with

$$g_\eta(\tilde{X}[p,q]) = \begin{cases} |\tilde{X}[p,q]| - \eta \tilde{X}[p,q]^2, & \text{if } |\tilde{X}[p,q]| \le \dfrac{1}{2\eta}, \\ \dfrac{1}{4\eta_k}, & \text{if } |\tilde{X}[p,q]| > \dfrac{1}{2\eta}. \end{cases} \tag{13.26}$$

The POM can be used to tackle problem $\mathcal{P}_G$. The iteration update of POM is

$$\tilde{X}^{k+1} = P_{\lambda\gamma,G_\eta}\left(\tilde{X}^k + \gamma \tilde{S}^T(\tilde{Y} - \tilde{S}\tilde{X}^k)\right), \tag{13.27}$$

where $P_{\lambda\gamma,G_\eta}(\cdot)$ represents the proximal operator, γ denotes the step size, and $\tilde{X}^k$ denotes the estimate of matrix $\tilde{X}$ at iteration k. The proximal operator $P_{\lambda\gamma,G_\eta}(\cdot)$ is given by

$$P_{\theta,G_\eta}(U) = \arg\min_{X} \theta G_\eta(X) + \tfrac{1}{2} \|X - U\|_F^2 . \tag{13.28}$$

Note that (13.28) can be decomposed as follows

$$P_{\theta,G_\eta}(U)[p,g] = \arg\min_{X[p,q]} \theta g_\eta(X[p,q]) + \tfrac{1}{2}(X[p,q] - U[p,q])^2. \tag{13.29}$$

In the following, we derive the analytical solution to the univariate optimization problem.

Theorem 13.4 The analytical solution to $P_{\theta,G_\eta}(U)[p,q]$ is given by

$$P_{\theta,G_\eta}(U)[p,q] = \begin{cases} 0, & \text{if } |U[p,q]| \le \theta_k, \\ \dfrac{U[p,q] - \theta_k \operatorname{sign}(U[p,q])}{1 - 2\theta_k\eta_k}, & \text{if } \theta_k < |U[p,q]| \le \dfrac{1}{2\eta_k}, \\ U[p,q], & \text{if } |U[p,q]| > \dfrac{1}{2\eta_k}. \end{cases} \tag{13.30}$$

Proof: Note that $g_\eta(\cdot)$ is an even and differentiable function that satisfies the property $\lim_{z\to 0^+} \frac{g_\eta(z)}{z} = \alpha < +\infty$. Furthermore, both $g_\eta(\cdot)$ and $\frac{g_\eta(z)}{z}$ are monotonic non-decreasing and non-increasing functions over the intervals $[0, +\infty)$ and $(0, +\infty)$, respectively. By substituting (13.26) into (13.29), we obtain

$$P_{\theta,G_\eta}(\boldsymbol{U})[p,q] = \arg\min_{\boldsymbol{X}[p,q]} \theta_k \Bigg((|\boldsymbol{X}[p,q]| - \eta \boldsymbol{X}[p,q]^2)\mathbb{1}_{|\boldsymbol{X}[p,q]|\le\frac{1}{2\eta}}$$

$$(\boldsymbol{X}[p,q]) + \frac{1}{4\eta}\mathbb{1}_{|\boldsymbol{X}[p,q]|>\frac{1}{2\eta}}(\boldsymbol{X}[p,q]) \Bigg) + \frac{1}{2}(\boldsymbol{X}[p,q] - \boldsymbol{U}[p,q])^2. \tag{13.31}$$

In order to ensure that $P_{\theta,g_\eta}(\boldsymbol{U})[p,q]$ is continuously differentiable, it is necessary that η is less than $\frac{1}{2\theta}$. We denote $z(\boldsymbol{X}[p,q]) = \frac{(\boldsymbol{X}[p,q]-\boldsymbol{U}[p,q])^2}{2}$ and $(P_{\theta_k,g_{\eta_k}}(\boldsymbol{U}))[p,q] = \arg\min_{\boldsymbol{X}[p,q]} h(\boldsymbol{X}[p,q])$, where $h(\boldsymbol{X}[p,q])$ is

$$h(\boldsymbol{X}[p,q]) = \begin{cases} z(\boldsymbol{X}[p,q]) + \theta(\boldsymbol{X}[p,q] - \eta \boldsymbol{X}[p,q]^2), & \text{if} 0 \le \boldsymbol{X}[p,q] \le \dfrac{1}{2\eta}, \\ z(\boldsymbol{X}[p,q]) - \theta(\boldsymbol{X}[p,q] + \eta \boldsymbol{X}[p,q]^2), & \text{if} -\dfrac{1}{2\eta} \le \boldsymbol{X}[p,q] < 0, \\ z(\boldsymbol{X}[p,q]) + \dfrac{\theta}{4\eta}, & \text{otherwise.} \end{cases} \tag{13.32}$$

By setting $h'(\boldsymbol{X}[p,q]) = 0$, it can be deduced that $\boldsymbol{X}[p,q] = \frac{\boldsymbol{U}[p,q]-\theta}{1-2\theta\eta}$ if $\boldsymbol{X}[p,q] \in [0, \frac{1}{2\eta}]$, $\boldsymbol{X}_{p,q} = \frac{\boldsymbol{U}[p,q]+\theta}{1-2\theta\eta}$ if $\boldsymbol{X}[p,q] \in [-\frac{1}{2\eta}, 0)$, and $\boldsymbol{X}[p,q] = \boldsymbol{U}[p,q]$ otherwise. With aforementioned discussions, we have $\boldsymbol{X}[p,q] \in \{0, \frac{\boldsymbol{U}[p,q]-\theta}{1-2\theta\eta}, \frac{\boldsymbol{U}[p,q]+\theta}{1-2\theta\eta}, \boldsymbol{U}[p,q]\}$. Through simple mathematical manipulations, we have (13.30). □

While POM is able to address problem $\mathcal{P}_G$, it exhibits a sublinear convergence rate and requires a substantial number of iterations to converge. Therefore, we follow the idea of algorithm unrolling to unfold POM as a RNN.

13.4.2 Unrolled Neural Networks

13.4.2.1 LPOM-GS

By denoting $\boldsymbol{W}^k = \boldsymbol{I} - \gamma_k \tilde{\boldsymbol{S}}^T \tilde{\boldsymbol{S}}$ and $\boldsymbol{B}^k = \gamma_k \tilde{\boldsymbol{S}}^T$, we write (13.27) as

$$\tilde{\boldsymbol{X}}^{k+1} = P_{\lambda\gamma_k,G_\eta}\left(\boldsymbol{W}^k \tilde{\boldsymbol{X}}^k + \boldsymbol{B}^k \tilde{\boldsymbol{Y}}\right). \tag{13.33}$$

Through the technique of algorithm unrolling, we may view Eq. (13.33) as a single-layer neural network, where $\tilde{\boldsymbol{X}}^k$ and $\tilde{\boldsymbol{X}}^{k+1}$ are considered as the input and the output, $\boldsymbol{W}^k$ and $\boldsymbol{B}^k$ are interpreted as the weight parameters, and $P_{\lambda\gamma_k}$ is viewed as the activation function.

We formulate the K repetitions of Eq. (13.33) as a K-layer RNN by cascading multiple single-layer neural networks. Thus, we present the following unfolding neural network termed learned POM for group sparsity (LPOM-GS)

$$\tilde{\boldsymbol{X}}^{k+1} = P_{\theta_k, G_{\eta_k}}\left(\boldsymbol{W}^k\tilde{\boldsymbol{X}}^k + \boldsymbol{B}^k\tilde{\boldsymbol{Y}}\right), \quad k \in \mathcal{K}, \tag{13.34}$$

where $\theta_k = \lambda\gamma_k$. The trainable parameters of LPOM-GS are $\boldsymbol{\Theta} = \{\boldsymbol{W}^k, \boldsymbol{B}^k, \theta_k, \eta_k\}_{k=0}^{K-1}$.

13.4.2.2 LPOMCP-GS

The overfitting problem may occur due to the large number of trainable parameters in LPOM-GS. To lower the probability of overfitting, we leverage the interrelation between $\boldsymbol{W}^k$ and $\boldsymbol{B}^k$ to decrease the trainable parameter number. According to Chen et al. [2018a, Theorem 1], it has been established that $\boldsymbol{W}^k$ and $\boldsymbol{B}^k$ eventually exhibit the following coupling structure

$$\boldsymbol{W}^k = \boldsymbol{I} - \boldsymbol{B}^k\tilde{\boldsymbol{S}}. \tag{13.35}$$

With an objective of reducing the trainable parameter number in LPOM-GS, we incorporate the coupling structure described by Eq. (13.35). The resulting neural network structure, named LPOMCP-GS, is

$$\tilde{\boldsymbol{X}}^{k+1} = P_{\theta_k, G_{\eta_k}}\left(\tilde{\boldsymbol{X}}^k + \boldsymbol{B}^k(\tilde{\boldsymbol{Y}} - \tilde{\boldsymbol{S}}\tilde{\boldsymbol{X}}^k)\right), \quad k \in \mathcal{K}, \tag{13.36}$$

The trainable parameters of LPOMCP-GS are $\boldsymbol{\Theta} = \{\boldsymbol{B}^k, \theta_k, \eta_k\}_{k=0}^{K-1}$.

We demonstrate that $\tilde{\boldsymbol{X}}^k$ holds the no-false-positive property in the following theorem.

Theorem 13.5 The input of LPOMCP-GS is set to $\tilde{\boldsymbol{Y}} = \tilde{\boldsymbol{S}}\tilde{\boldsymbol{X}}^\natural + \tilde{\boldsymbol{Z}}$ and $\tilde{\boldsymbol{X}}^0 = \boldsymbol{0}$. The output of LPOMCP-GS is $\{\tilde{\boldsymbol{X}}^k\}_{k=1}^{\infty}$. We denote $\mathrm{supp}(\psi(\tilde{\boldsymbol{X}}^*))$ as S and $\|\boldsymbol{X}\|_{2,1} = \sum_n \|\boldsymbol{X}[n, :]\|_2$. If $\|\boldsymbol{B}^k\|_{2,1} \le \mu_B$, $\|\tilde{\boldsymbol{X}}^*[p, :]\|_2 \le \mu_x$, $\forall p$, $\|\tilde{\boldsymbol{Z}}\|_F \le \epsilon$, $|S| \le s$ and

$$\frac{1}{\alpha}\left(\phi \sup_{(\tilde{\boldsymbol{X}}^\natural, \tilde{\boldsymbol{Z}}) \in \mathcal{X}(\mu_x, s, \epsilon)} \|\tilde{\boldsymbol{X}}^k - \tilde{\boldsymbol{X}}^\natural\|_{2,1} + \mu_B\epsilon\right) = \theta_k \le \frac{1}{2\eta_k}, \tag{13.37}$$

$$\boldsymbol{B}^k \in \operatorname*{argmin}_{\boldsymbol{B}\in\mathbb{R}^{2N\times 2L}}\{\max \|\boldsymbol{B}\|_2 | \boldsymbol{B}[p, :]\tilde{\boldsymbol{S}}[:, p] = 1, \forall p, \max_{p\neq q} |\boldsymbol{B}[p, :]\tilde{\boldsymbol{S}}[:, q]| = \phi\}, \tag{13.38}$$

where $\phi = \inf_{\substack{\boldsymbol{B} \in \mathbb{R}^{2N\times 2L} \\ \boldsymbol{B}[p, :]\tilde{\boldsymbol{S}}[:, p] = 1, \forall p}} \max_{p\neq q} |\boldsymbol{B}[p, :]\tilde{\boldsymbol{S}}[:, q]|$, then

$$\mathrm{supp}(\psi(\tilde{\boldsymbol{X}}^k)) \subseteq \mathrm{supp}(\psi(\tilde{\boldsymbol{X}}^\natural)). \tag{13.39}$$

13.4.2.3 ALPOM-GS

As demonstrated by Liu and Chen (2019), the acquisition of weight matrix $\boldsymbol{B}^k = \gamma_k \boldsymbol{B}$ prior to the learning process leads to a reduction in trainable parameter number. Solving the following problem, the weight matrix $\boldsymbol{B}$ can be obtained

$$\underset{\boldsymbol{B}\in\mathbb{R}^{2N\times 2L}}{\text{minimize}} \quad \| \boldsymbol{B}\tilde{\boldsymbol{S}} \|_F^2 \tag{13.40}$$

$$\text{subject to} \quad \boldsymbol{B}[p,:]\tilde{\boldsymbol{S}}[:,p] = 1, \forall p \in [2N]. \tag{13.41}$$

The PGD method can be leveraged to solve problem (13.40) for obtaining matrix $\boldsymbol{B}$. Subsequently, we develop another unfolding structure, called ALPOM-GS, as follows

$$\tilde{\boldsymbol{X}}^{k+1} = P_{\theta_k, G_{\eta_k}}\left(\tilde{\boldsymbol{X}}^k + \gamma_k \boldsymbol{B}(\tilde{\boldsymbol{Y}} - \tilde{\boldsymbol{S}}\tilde{\boldsymbol{X}}^k)\right), \quad k \in \mathcal{K}, \tag{13.42}$$

where $\boldsymbol{\Theta} = \{\gamma_k, \theta_k, \eta_k\}_{k=0}^{K-1}$ are trainable parameters.

In the theorem below, we present the linear convergence of LPOMCP-GS. We define $\mathcal{X}(\mu_x, s, \epsilon) = \{(\tilde{\boldsymbol{X}}^\natural, \tilde{\boldsymbol{Z}}) | \ \|\tilde{\boldsymbol{X}}^\natural_{p,:}\|_2 \leq \mu_x, \forall p, \text{supp}(\psi(\tilde{\boldsymbol{X}}^\natural)) \leq s, \|\tilde{\boldsymbol{Z}}\|_F \leq \epsilon\}$.

Theorem 13.6 The input of LPOMCP-GS are $\tilde{\boldsymbol{Y}} = \tilde{\boldsymbol{S}}\tilde{\boldsymbol{X}}^\natural + \tilde{\boldsymbol{Z}}$ and $\tilde{\boldsymbol{X}}^0 = \boldsymbol{0}$. The output of LPOMCP-GS is $\{\tilde{\boldsymbol{X}}^k\}_{k=1}^\infty$. We denote $\text{supp}(\psi(\tilde{\boldsymbol{X}}^\natural))$ as S. For $\|\tilde{\boldsymbol{X}}^\natural[i,:]\|_2 \leq \mu_x, \forall i, \|\tilde{\boldsymbol{Z}}\|_F \leq \epsilon$, and $S \leq s$, we obtain

$$\|\tilde{\boldsymbol{X}}^k - \tilde{\boldsymbol{X}}^\natural\|_F \leq s\mu_x \ \exp(-c_1 k) + \epsilon c_2. \tag{13.43}$$

where $c_1 > 0$ and $c_2 > 0$.

13.5 Training and Testing Strategies

During the training phase, we adopt supervised learning to train the unrolled neural networks. By denoting $\tilde{\boldsymbol{Y}}_i$ and $\tilde{\boldsymbol{X}}_i^\natural$ as the ith data and the corresponding label, the training batch with P samples can be represented as $\{\tilde{\boldsymbol{X}}_i^\natural, \tilde{\boldsymbol{Y}}_i\}_{i=1}^P$. The following optimization problem is solved to train a k-layer network

$$\underset{\boldsymbol{\Theta}_{0:k-1}}{\text{minimize}} \sum_{i=1}^{P} \|\tilde{\boldsymbol{X}}^k(\Theta_{0:k-1}, \tilde{\mathbf{Y}}_{\mathbf{i}}, \tilde{\boldsymbol{X}}^0) - \tilde{\boldsymbol{X}}_i^\natural\|_F^2 . \tag{13.44}$$

Solving problem (13.44) often results in converging to a suboptimal local minimum (Borgerding et al., 2017). To this end, the layer-wise training approach partitions the training procedure (13.44) into two components, specifically (13.45) and (13.46). The objective of (13.45) is to find an appropriate initial solution for (13.46) in order to avoid entering the unfavorable local minima of (13.44).

Specifically, the trainable parameters in the first k layers (i.e., from layer 0 to layer $k-1$) are denoted as $\boldsymbol{\Theta}_{0:k-1}$. In addition, the trainable parameters in layer

$k-1$ are denoted by $\boldsymbol{\Theta}_{k-1}$. When training a k-layer neural network, we fix parameters $\boldsymbol{\Theta}_{0:k-2}$ that have already been trained. Subsequently, we utilize the Adam optimizer (Kabashima, 2003) with learning rate α_0 to learn parameters $\boldsymbol{\Theta}_{k-1}$ by solving problem (13.45)

$$\underset{\boldsymbol{\Theta}_{k-1}}{\text{minimize}} \sum_{i=1}^{P} \|\tilde{\boldsymbol{X}}^k\left(\Theta_{0:k-1}, \tilde{\boldsymbol{Y}}_i, \tilde{\boldsymbol{X}}^0\right) - \tilde{\boldsymbol{X}}^\natural\|_F^2, \tag{13.45}$$

where $\tilde{\boldsymbol{X}}^k(\Theta_{0:k-1}, \tilde{\boldsymbol{Y}}, \tilde{\boldsymbol{X}}^0)$ denotes the output of the k-layer network. Subsequently, leveraging the parameters $\boldsymbol{\Theta}_{k-1}$ obtained from (13.45) and utilizing the fixed parameters $\boldsymbol{\Theta}_{0:k-2}$ as the initial values, we adjust all parameters $\boldsymbol{\Theta}_{0:k-1}$ through the application of the Adam optimizer, employing a learning rate denoted as α_1. It is noteworthy that α_1 is set to be smaller than α_0. This optimization process is aimed at solving the following problem

$$\underset{\boldsymbol{\Theta}_{0:k-1}}{\text{minimize}} \sum_{i=1}^{P} \|\tilde{\boldsymbol{X}}^k\left(\Theta_{0:k-1}, \tilde{\boldsymbol{Y}}_i, \tilde{\boldsymbol{X}}^0\right) - \tilde{\boldsymbol{X}}^\natural\|_F^2 . \tag{13.46}$$

Upon successfully executing the procedure, the trainable parameters $\boldsymbol{\Theta}_{0:k-1}$ for the entire k-layer network are obtained, which will be utilized during the training of layer $(k+1)$.

After the training process for all K layers, all of the trainable parameters are obtained. These learned parameters can then be utilized to implement the proposed unrolled neural networks in the testing stage to conduct JADCE upon receipt of new signals.

13.6 Simulation Results

Under different parameter settings, we in this section assess the superiority of the proposed unrolled neural networks. In particular, the simulation settings and performance metrics are presented, followed by comparing the performance between the proposed algorithms and various baseline methods.

In simulations, the channel follows Rayleigh fading, and each entry of the noise matrix conforms to a complex Gaussian distribution with a mean of zero and a variance of σ^2. Moreover, the activity indicator is assumed to follow the Bernoulli distribution. Specifically, each device is active with probability 0.1 and non-active with probability 0.9. We define the signal-to-noise ratio (SNR) in the simulations as follows

$$\text{SNR} = \frac{\mathbb{E}[\|\boldsymbol{SX}\|_F^2]}{\mathbb{E}[\|\boldsymbol{Z}\|_F^2]}. \tag{13.47}$$

The training dataset $\{\tilde{\boldsymbol{X}}_i^{\natural}, \tilde{\boldsymbol{Y}}_i\}_{i=1}^{P}$ is obtained by converting matrices with complex values into matrices with real values using the transformation in Eq. (13.5). The training set comprises 51,200 data samples to train the model in the training stage, and the test set comprises 2048 data samples to test the trained model in the testing stage.

The proposed unrolled neural network is compared with the following baselines for group matrix recovery problem:

- *ISTA-GS* (Yuan and Lin, 2006): By replacing the soft-thresholding function in ISTA with multidimensional shrinkage thresholding function, ISTA-GS is the extension of ISTA to tackle MMV problems. The updated formula of ISTA-GS is given by (13.7).
- *FISTA-GS*: FISTA (Nesterov, 1983, Beck and Teboulle, 2009) are the accelerated version of ISTA. Correspondingly, FISTA-GS is the accelerated version of ISTA-GS.

To evaluate the matrix recovery performance, we utilize the normalized mean squared error (NMSE)

$$\mathrm{NMSE}(\tilde{\boldsymbol{X}}, \tilde{\boldsymbol{X}}^{\natural}) = 10\log_{10}\left(\frac{\mathbb{E}\,\|\tilde{\boldsymbol{X}} - \tilde{\boldsymbol{X}}^{\natural}\|_F^2}{\mathbb{E}\,\|\tilde{\boldsymbol{X}}^{\natural}\|_F^2}\right), \tag{13.48}$$

where $\tilde{\boldsymbol{X}}$ denotes the recovered matrix and $\tilde{\boldsymbol{X}}^{\natural}$ denotes the ground truth.

In the following, the pilot signature matrix $\boldsymbol{S}$ is Zadoff–Chu pilot sequence matrix (Chu, 1972) with each column of the pilot sequence matrix being normalized. Parameters L, N, and M are set to 125, 250, and 6, respectively. Besides, we set SNR to 40 dB. In Fig 13.3, ISTA-GS exhibits the trade-off between convergence rate and recovery performance. Specifically, a smaller λ produces a better performance but results in slower convergence rate, while a larger value of λ achieves faster convergence rate at the expense of recovery performance. Moreover, since LISTA-GS regards λ as the trainable parameter and adjusts it during training stage, LISTA-GS outperforms ISTA-GS with various values of λ.

In Fig. 13.4, LISTA-GS, LISTA-GSCP, and ALISTA achieve faster convergence rates and better performance than ISTA-GS and FISTA-GS. To achieve NMSE = −25 dB, the number of iterations required by FISTA-GS and LISTA-GS are 14 and 5, respectively. LISTA-GS also outperforms FISTA-GS in terms of NMSE performance. Moreover, we also observe that LISTA is capable of reaching convergence in few iterations.

The convergence performance of LPOM-GS and LISTA-GS are compared when M and SNR are set to 2 and dB, respectively. Results in Fig. 13.5 show that LPOM-GS, LPOMCP-GS, and ALPOM-GS achieve better performance than LISTA-GS. The proposed networks based on the POM improve the NMSE up to

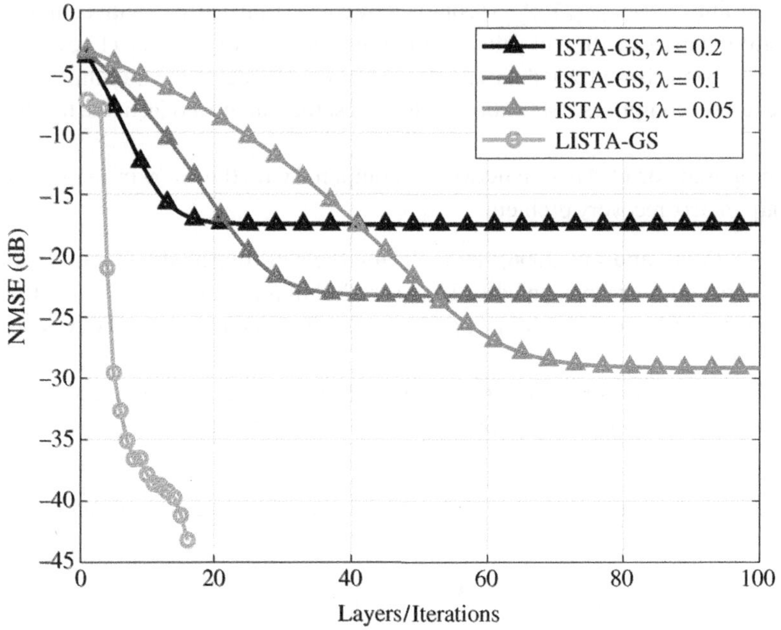

Figure 13.3 Performance comparison between LISTA-GS and ISTA-GS.

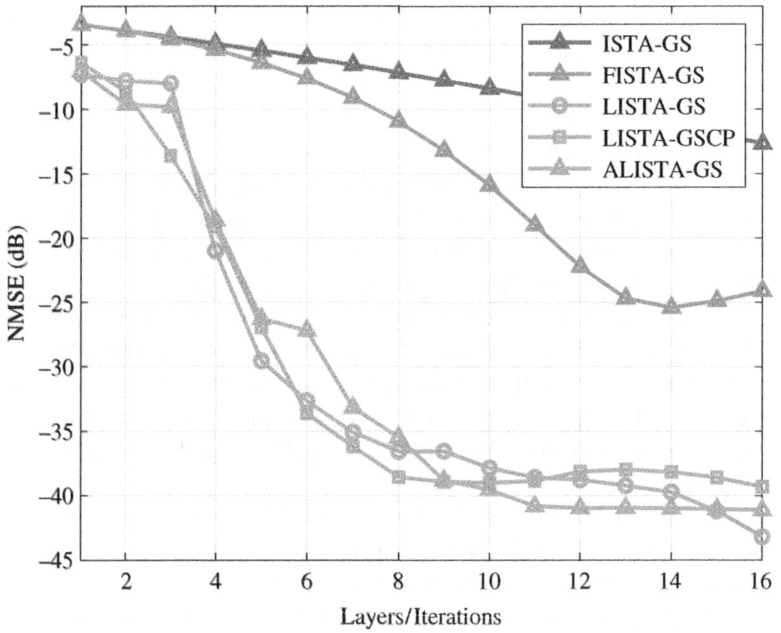

Figure 13.4 Convergence performance comparison among ISTA-GS, FISTA-GS, LISTA-GS, LISTA-GSCP, and ALISTA-GS.

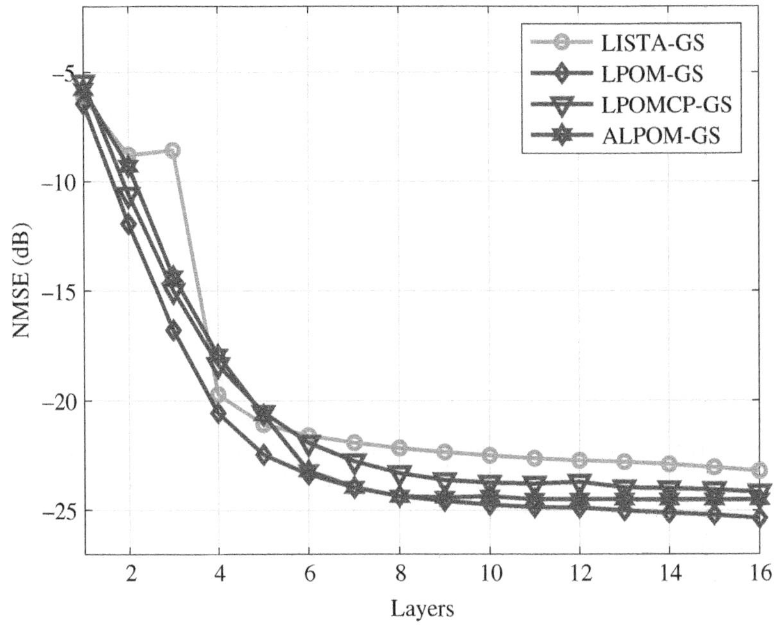

Figure 13.5 Convergence performance comparison among LISTA-GS, LPOM-GS, LPOMCP-GS, and ALPOM-GS.

8.7% over LISTA-GS by adopting MCP that has a stronger capability than the ℓ_1-norm to induce sparsity.

13.7 Conclusions

In this book chapter, we develop the unrolled neural networks that exhibit the linear convergence rate and achieve low computational complexity, thereby facilitating effective JADCE. In particular, we transform the iterative ISTA-GS algorithm into an unrolled RNN and fine-tune the parameters through end-to-end training in order to enhance the convergence rate. By considering the non-convex penalty has more capability than the ℓ_1-norm to induce sparsity, we then formulate a MCP-regularized problem. We also unroll the proximal operator method to an unrolled neural network to improve the recovery performance. Convergence analyses are given to show the linear convergence of the proposed unrolled neural networks. Simulation results demonstrate the efficiency of the proposed unrolled neural networks in terms of convergence rate and recovery performance for JADCE.

References

Amir Beck and Marc Teboulle. A fast iterative shrinkage-thresholding algorithm for linear inverse problems. *SIAM Journal on Imaging Sciences*, 2(1):183–202, 2009.

M Borgerding, P Schniter, and S Rangan. AMP-Inspired deep networks for sparse linear inverse problems. *IEEE Transactions on Signal Processing*, 65(16):4293–4308, Aug. 2017.

Xiaohan Chen, Jialin Liu, Zhangyang Wang, and Wotao Yin. Theoretical linear convergence of unfolded ISTA and its practical weights and thresholds. In *Proceedings Advances in Neural Information Processing Systems 31 (NeurIPS 2018)*, 2018a.

Zhilin Chen, Foad Sohrabi, and Wei Yu. Sparse activity detection for massive connectivity. *IEEE Transactions on Signal Processing*, 66(7):1890–1904, 2018b.

David Chu. Polyphase codes with good periodic correlation properties (Corresp.). *IEEE Transactions on Information Theory*, 18(4):531–532, 1972.

Ying Cui, Shuaichao Li, and Wanqing Zhang. Jointly sparse signal recovery and support recovery via deep learning with applications in MIMO-based grant-free random access. *IEEE Journal on Selected Areas in Communications*, 39(3):788–803, 2020.

Raja Giryes, Yonina C Eldar, Alex M Bronstein, and Guillermo Sapiro. Tradeoffs between convergence speed and reconstruction accuracy in inverse problems. *IEEE Transactions on Signal Processing*, 66(7):1676–1690, 2018.

Karol Gregor and Yann LeCun. Learning fast approximations of sparse coding. In *ICML'10: Proceedings of the 27th International Conference on International Conference on Machine Learning*, pages 399–406, 2010.

Monowar Hasan, Ekram Hossain, and Dusit Niyato. Random access for machine-to-machine communication in LTE-advanced networks: Issues and approaches. *IEEE Communications Magazine*, 51(6):86–93, 2013.

Qi He, Tony Q S Quek, Zhi Chen, Qi Zhang, and Shaoqian Li. Compressive channel estimation and multi-user detection in C-RAN with low-complexity methods. *IEEE Transactions on Wireless Communications*, 17(6):3931–3944, 2018.

Tao Jiang, Yuanming Shi, Jun Zhang, and Khaled B Letaief. Joint activity detection and channel estimation for IoT networks: Phase transition and computation-estimation tradeoff. *IEEE Internet of Things Journal*, 6(4):6212–6225, 2018.

Jeremy Johnston and Xiaodong Wang. Model-based deep learning for joint activity detection and channel estimation in massive and sporadic connectivity. *IEEE Transactions on Wireless Communications*, 21(11):9806–9817, 2022.

Yoshiyuki Kabashima. A CDMA multiuser detection algorithm on the basis of belief propagation. *Journal of Physics A: Mathematical and General*, 36(43):11111, Oct. 2003.

Malong Ke, Zhen Gao, Yongpeng Wu, Xiqi Gao, and Robert Schober. Compressive sensing-based adaptive active user detection and channel estimation: Massive access meets massive MIMO. *IEEE Transactions on Signal Processing*, 68:764–779, 2020.

Khaled B Letaief, Yuanming Shi, Jianmin Lu, and Jianhua Lu. Edge artificial intelligence for 6G: Vision, enabling technologies, and applications. *IEEE Journal on Selected Areas in Communications*, 40(1):5–36, 2021.

Jialin Liu and Xiaohan Chen. ALISTA: Analytic weights are as good as learned weights in Lista. In *Proceedings of the* International Conference on Learning Representations (ICLR), 2019.

Liang Liu and Wei Yu. Massive connectivity with massive MIMO—Part I: Device activity detection and channel estimation. *IEEE Transactions on Signal Processing*, 66(11):2933–2946, 2018.

Liang Liu, Erik G Larsson, Wei Yu, Petar Popovski, Cedomir Stefanovic, and Elisabeth De Carvalho. Sparse signal processing for grant-free massive connectivity: A future paradigm for random access protocols in the Internet of Things. *IEEE Signal Processing Magazine*, 35(5):88–99, 2018.

Vishal Monga, Yuelong Li, and Yonina C Eldar. Algorithm unrolling: Interpretable, efficient deep learning for signal and image processing. *IEEE Signal Processing Magazine*, 38(2):18–44, 2021.

Yurii Evgen'evich Nesterov. A method of solving a convex programming problem with convergence rate $O(1/k^2)$. In *Doklady Akademii Nauk*, volume 269, pages 543–547. Russian Academy of Sciences, 1983.

Xiaodan Shao, Xiaoming Chen, and Rundong Jia. A dimension reduction-based joint activity detection and channel estimation algorithm for massive access. *IEEE Transactions on Signal Processing*, 68:420–435, 2019.

Xiaodan Shao, Xiaoming Chen, Yiyang Qiang, Caijun Zhong, and Zhaoyang Zhang. Feature-aided adaptive-tuning deep learning for massive device detection. *IEEE Journal on Selected Areas in Communications*, 39(7):1899–1914, 2021a.

Xiaodan Shao, Xiaoming Chen, Caijun Zhong, and Zhaoyang Zhang. Exploiting simultaneous low-rank and sparsity in delay-angular domain for millimeter-wave/terahertz wideband massive access. *IEEE Transactions on Wireless Communications*, 21(4):2336–2351, 2021b.

Shree Krishna Sharma and Xianbin Wang. Toward massive machine type communications in ultra-dense cellular IoT networks: Current issues and machine learning-assisted solutions. *IEEE Communications Surveys & Tutorials*, 22(1):426–471, 2019.

Yuanming Shi, Kai Yang, Tao Jiang, Jun Zhang, and Khaled B Letaief. Communication-efficient edge AI: Algorithms and systems. *IEEE Communications Surveys & Tutorials*, 22(4):2167–2191, 2020.

Yandong Shi, Hayoung Choi, Yuanming Shi, and Yong Zhou. Algorithm unrolling for massive access via deep neural networks with theoretical guarantee. *IEEE Transactions on Wireless Communications*, 21(2):945–959, 2022. doi: 10.1109/TWC.2021.3100500.

Yuanming Shi, Yong Zhou, Dingzhu Wen, Youlong Wu, Chunxiao Jiang, and Khaled B Letaief. Task-oriented communications for 6G: Vision, principles, and technologies. *IEEE Wireless Communications*, 30(3):78–85, 2023. doi: 10.1109/MWC.002. 2200468.

Zhibin Wang, Jiahang Qiu, Yong Zhou, Yuanming Shi, Liqun Fu, Wei Chen, and Khaled B Letaief. Federated learning via intelligent reflecting surface. *IEEE Transactions on Wireless Communications*, 21(2):808–822, 2021.

Zixin Wang, Yong Zhou, Yinan Zou, Qiaochu An, Yuanming Shi, and Mehdi Bennis. A graph neural network learning approach to optimize RIS-assisted federated learning. *IEEE Transactions on Wireless Communications*, 22(9):6092–6106, 2023.

Yongpeng Wu, Xiqi Gao, Shidong Zhou, Wei Yang, Yury Polyanskiy, and Giuseppe Caire. Massive access for future wireless communication systems. *IEEE Wireless Communications*, 27(4):148–156, 2020.

Ming Yuan and Yi Lin. Model selection and estimation in regression with grouped variables. *Journal of the Royal Statistical Society Series B (Statistical Methodology)*, 68(1):49–67, 2006.

Weifeng Zhu, Meixia Tao, Xiaojun Yuan, and Yunfeng Guan. Deep-learned approximate message passing for asynchronous massive connectivity. *IEEE Transactions on Wireless Communications*, 20(8):5434–5448, 2021.

Justin Ziniel and Philip Schniter. Efficient high-dimensional inference in the multiple measurement vector problem. *IEEE Transactions on Signal Processing*, 61(2):340–354, 2012.

Yinan Zou, Yong Zhou, Yuanming Shi, and Xu Chen. Learning proximal operator methods for massive connectivity in IoT networks. In *2021 IEEE Global Communications Conference (GLOBECOM)*, pages 1–6, 2021. doi: 10.1109/GLOBECOM46510.2021.9685447.

Yinan Zou, Zixin Wang, Xu Chen, Haibo Zhou, and Yong Zhou. Knowledge-guided learning for transceiver design in over-the-air federated learning. *IEEE Transactions on Wireless Communications*, 22(1):270–285, 2022.

14

Grant-Free Massive Random Access: Joint Activity Detection, Channel Estimation, and Data Decoding

Xinyu Bian[1], Yuyi Mao[2], and Jun Zhang[1]

[1] *Department of Electronic and Computer Engineering, The Hong Kong University of Science and Technology, Hong Kong, China*
[2] *Department of Electronic and Information Engineering, The Hong Kong Polytechnic University, Hong Kong, China*

14.1 Introduction

The explosive growth of the Internet of Things (IoT) has given rise to an unprecedented demand for massive machine-type communications (mMTC). A distinctive feature of mMTC is that although a huge number of devices are connected to a base station (BS), only a small proportion of them are active for transmitting a short data packet at each time (Bockelmann et al., 2016), which entails innovative uplink access schemes for next-generation wireless networks (Liu et al., 2022). Specifically, the conventional uplink grant-based random access (RA) scheme shown in Fig. 14.1, where each user first initiates an RA procedure by transmitting a scheduling request to its serving BS and cannot start data transmission until the request is granted (Hasan et al., 2013), fails to provide scalable connectivity because of the long access latency and significant signaling overhead. On the contrary, grant-free RA, where users can directly transmit without approval from the BS (Chen et al., 2021), provides a promising alternative. Nevertheless, since the BS does not have knowledge on the set of active users in grant-free RA, user activity detection is essential in addition to channel estimation and data decoding. Besides, since users can only be assigned with non-orthogonal pilot sequences in grant-free massive RA, it is arduous to perform accurate user activity detection and channel estimation. Fortunately, thanks to the sparse traffic pattern of mMTC, the user activity detection and channel estimation problems for grant-free massive RA can be efficiently solved by many compressive sensing (CS) algorithms. Specifically, joint activity and data detection were performed for non-orthogonal multiple access

Next Generation Multiple Access, First Edition.
Edited by Yuanwei Liu, Liang Liu, Zhiguo Ding, and Xuemin Shen.
© 2024 The Institute of Electrical and Electronics Engineers, Inc. Published 2024 by John Wiley & Sons, Inc.

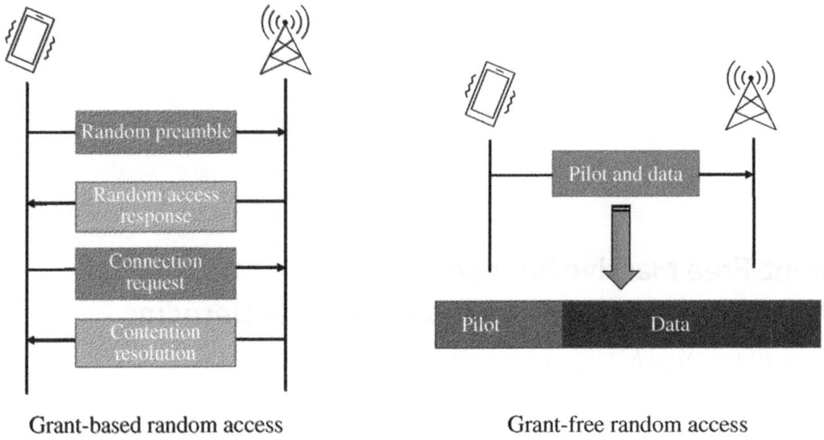

Figure 14.1 Grant-based and grant-free RA. With grant-based RA, each active user randomly picks a preamble from a predefined set of orthogonal sequences to make connections with the BS. Collisions may happen since multiple users may pick the same preamble. Once a collision is detected by the BS, the RA procedure restarts after timeout (Liu et al., 2018). In contrast, for grant-free RA, a dedicated non-orthogonal preamble is assigned to each device, and an active device directly transmits its pilot and data without needing approval from the BS.

systems in Wang et al. (2016) by applying the orthogonal matching pursuit (OMP) algorithm. A similar problem was tackled via approximate message passing (AMP) in Wei et al. (2017). These methods, however, assume perfect channel state information (CSI), which prompts the recent investigation on joint activity detection and channel estimation (JADCE). For example, an AMP algorithm with a minimum mean square error (MMSE) denoiser was proposed for JADCE in Chen et al. (2018). To improve the accuracy, channel sparsity from the spatial and angular domains was utilized in Ke et al. (2020).

The aforementioned JADCE algorithms mainly exploit the sparsity pattern in the received pilot signal, but the common sparsity embedded in both the received pilot and data signal can be further utilized. Du et al. first unveiled the benefits of such a common sparsity pattern in Du et al. (2018) for massive RA systems with a single-antenna BS under the framework of multiple measurement vector CS. An extension for multi-antenna BSs was conducted in Zou et al. (2020) via the bilinear generalized AMP (BiG-AMP) algorithm (Parker et al., 2014). However, these preliminary attempts were limited to uncoded transmissions and cannot take advantages of channel coding. In particular, the error detection mechanism of channel codes can help determine a subset of active users with high channel quality (Bian et al., 2021), and the soft decoding results that carry posterior information of the transmitted data symbols are valuable for improving the accuracy of activity detection, channel estimation, and multi-user detection. Nonetheless, the joint detection–decoding problem is computationally infeasible even for channel codes with a reasonable block length (Hochwald and Ten Brink, 2003). Hence, it is in urgent need to integrate a channel decoder with other key components in a massive RA receiver.

In this chapter, we first propose a turbo receiver for joint activity detection, channel estimation, and data decoding in grant-free massive RA, which iterates between a joint estimator and a channel decoder. To exploit the common sparsity pattern, the joint estimator for activity detection, channel estimation, and data symbol detection is developed by solving a bilinear inference problem. We derive the extrinsic information using outputs of the channel decoder, which serves as prior information for the next turbo iteration. To facilitate fast execution while retaining the performance gain of the iterative receiver, we further develop a side information (SI)-aided receiver that executes a sequential estimator and a channel decoder alternatively. Specifically, the sequential estimator for JADCE is developed based on the AMP algorithm, which processes only the received pilot signal and leaves data symbol detection to an MMSE equalizer. To leverage the common sparsity and channel decoding results, the estimates on whether a user is active or not are used as SI for the sequential estimator. Simulation results show that the turbo receiver significantly reduces the activity detection, channel estimation, and data decoding errors compared with the baseline schemes, while the SI-aided receiver maintains a noticeable performance improvement compared against a data-assisted design that only leverages the common sparsity, despite saving substantial execution time.

Notations: We use lower-case letters, bold-face lower-case letters, bold-face upper-case letters, and math calligraphy letters to denote scalars, vectors, matrices, and sets, respectively. The entry in the ith row and jth column of matrix $\mathbf{M}$ is denoted as m_{ij}, and the matrix transpose, complex conjugate, and conjugate transpose operators are denoted as $(\cdot)^T$, $(\cdot)^*$, and $(\cdot)^H$, respectively. Besides, $\mathbf{M}_{\setminus i,j}$ represents all the elements in matrix $\mathbf{M} \triangleq [m_{ij}]$ except m_{ij}. In addition, $\exp(\cdot)$ denotes the exponential function, $\delta(\cdot)$ denotes the Dirac delta function, $\lfloor \cdot \rfloor$ denotes the floor function, and $\mathcal{CN}(x; \mu, v)$ denotes the probability density function (PDF) of a complex Gaussian random variable x with mean μ and variance v. Wherever appropriate, $\mathcal{CN}(\boldsymbol{\mu}, \boldsymbol{\Sigma})$ also denotes the complex Gaussian distribution with mean $\boldsymbol{\mu}$ and covariance matrix $\boldsymbol{\Sigma}$.

14.2 System Model

We consider an uplink cellular system, where N single-antenna users are served by an M-antenna BS. At each time instant, K $(K \leq N)$ among the N users become active for short-packet transmission. Denote $u_n \in \{0,1\}$ as the user activity indicator, where $u_n = 1$ means user n is active and vice versa. The sets of system users and active users are represented by $\mathcal{N} \triangleq \{1,\dots,N\}$ and $\Xi \triangleq \{n \in \mathcal{N} | u_n = 1\}$, respectively, and the set of BS antennas is denoted as $\mathcal{M} \triangleq \{1,\dots,M\}$. To avoid the system from being overloaded, the number of BS antennas is assumed to be no less than the number of active users, i.e., $M \geq K$ (Wong et al., 2007).

We assume quasi-static block fading channels and each transmission block contains T symbol intervals. The uplink channel vector from user n to the BS is modeled as $\mathbf{f}_n = \sqrt{\beta_n}\boldsymbol{\alpha}_n$, where $\boldsymbol{\alpha}_n$ and β_n denote the small-scale and large-scale fading coefficients, respectively. We focus on Rayleigh fading channels, i.e., $\boldsymbol{\alpha}_n \sim \mathcal{CN}(\mathbf{0}, \mathbf{I}_M)$, and assume the users are static with $\{\beta_n\}$'s known by the BS (Chen et al., 2018). A grant-free RA scheme with two phases in each transmission block is adopted for uplink transmission. In particular, L symbols, denoted as $\mathbf{T}_p$, are reserved for pilot transmission in the first phase; Whereas $L_d \triangleq T - L$ symbols, denoted as $\mathbf{T}_d$, are used for payload delivery in the second phase. Since it is prohibitive to use orthogonal pilot sequences in mMTC, we assign the users with a set of non-orthogonal and unique pilot sequences $\{\mathbf{x}_{pn}\}$'s by sampling a complex Gaussian distribution, i.e., $\mathbf{x}_{pn} \triangleq [x_{n1},\dots,x_{nL}]$ with $x_{nl} \sim \mathcal{CN}(0,1)$, which achieves asymptotic orthogonality when L is sufficiently large. Define $\mathbf{X}_p \triangleq [\mathbf{x}_{p1},\dots,\mathbf{x}_{pN}]^T$ as the collection of pilot sequences.

In each transmission block, each active user has N_b payload bits, denoted as $\boldsymbol{b}_n \triangleq \left[b_{n1},\dots,b_{nN_b}\right]$, $n \in \Xi$, to transmit, which are encoded for error detection and correction. Following the long-term evolution (LTE) and 5G new radio (NR) standards, cyclic redundancy check (CRC) bits are generated and attached to the payload bits to form a code block. We represent the CRC generation and attachment procedures by function $\Upsilon: \{0,1\}^{N_b} \to \{0,1\}^{N_d}$, where N_d denotes the size of a code block. Thus, the code blocks of the active users can be expressed as $\boldsymbol{d}_n \triangleq \left[d_{n1},\dots,d_{nN_d}\right] = \Upsilon(\boldsymbol{b}_n)$, $n \in \Xi$. Each code block is encoded by a channel encoder $\Phi: \{0,1\}^{N_d} \to \{0,1\}^{N_c}$, and the coded bits can be represented as $\boldsymbol{c}_n \triangleq \left[c_{n1},\dots,c_{nN_c}\right] = \Phi(\boldsymbol{d}_n)$, $n \in \Xi$, where N_c is the number of coded bits and the code rate ϕ is defined as $\phi \triangleq \frac{N_d}{N_c}$.

The coded bits are then modulated to a set of constellation points $\mathcal{X}$ with normalized average power via an invertible mapping $\mu: \{0,1\}^{\log_2|\mathcal{X}|} \to \mathcal{X}$, i.e., for an arbitrary bit sequence with length $\log_2|\mathcal{X}|$, $\mu([c_1,\dots,c_{\log_2|\mathcal{X}|}]) = s$ if and only if $\mu^{-1}(s) = [c_1,\dots,c_{\log_2|\mathcal{X}|}]$, where $s \in \mathcal{X}$ is a constellation point. The modulated symbols for the active users are denoted as $\mathbf{x}_{dn} \triangleq \left[x_{n(L+1)},\dots,x_{nT}\right]$, $n \in \Xi$. We assume $N_c = L_d \log_2|\mathcal{X}|$ for simplicity and assign zero vectors to $\mathbf{x}_{dn}$ for the inactive users. Let $\mathbf{X}_d \triangleq [\mathbf{x}_{d1},\dots,\mathbf{x}_{dN}]^T$ denote the transmitted data symbols from all the users. The received signal of the transmission block $\tilde{\mathbf{Y}} \in \mathbb{C}^{M \times T}$ at the BS can be expressed as follows:

$$\tilde{\mathbf{Y}} \triangleq [\tilde{\mathbf{Y}}_p, \tilde{\mathbf{Y}}_d] = \sqrt{\gamma}\mathbf{H}\underbrace{[\mathbf{X}_p, \mathbf{X}_d]}_{\triangleq \mathbf{X}} + \underbrace{[\tilde{\mathbf{N}}_p, \tilde{\mathbf{N}}_d]}_{\triangleq \tilde{\mathbf{N}}}, \tag{14.1}$$

where γ is the uplink transmit power, $\tilde{\mathbf{Y}}_p \in \mathbb{C}^{M\times L}$ and $\tilde{\mathbf{Y}}_d \in \mathbb{C}^{M\times L_d}$ are the received pilot and data signal, respectively, and $\mathbf{H} \triangleq [\mathbf{h}_1, \dots, \mathbf{h}_N] \in \mathbb{C}^{M\times N}$ with $\mathbf{h}_n \triangleq u_n\mathbf{f}_n$ denotes the effective channel coefficient matrix. Besides, $\tilde{\mathbf{N}} = [\tilde{\mathbf{n}}_1, \dots, \tilde{\mathbf{n}}_T] \in \mathbb{C}^{M\times T}$ is the additive white Gaussian noise (AWGN) with zero mean and variance σ^2 for each element, where σ^2 is assumed known, and $\tilde{\mathbf{N}}_p \in \mathbb{C}^{M\times L}$ and $\tilde{\mathbf{N}}_d \in \mathbb{C}^{M\times L_d}$ are the noise of the received pilot and data signal, respectively. Define $\mathbf{Y} \triangleq \tilde{\mathbf{Y}}/\sqrt{\gamma}$, $\mathbf{Y}_p \triangleq \tilde{\mathbf{Y}}_p/\sqrt{\gamma}$, $\mathbf{Y}_d \triangleq \tilde{\mathbf{Y}}_d/\sqrt{\gamma}$, $\mathbf{N} \triangleq \tilde{\mathbf{N}}/\sqrt{\gamma}$, $\mathbf{N}_p \triangleq \tilde{\mathbf{N}}_p/\sqrt{\gamma}$, and $\mathbf{N}_d \triangleq \tilde{\mathbf{N}}_d/\sqrt{\gamma}$.

In the following, we will develop efficient algorithms to detect the active users, estimate their channel coefficients and the transmitted payload bits.

14.3 Joint Estimation via a Turbo Receiver

In this section, a turbo receiver is developed to jointly estimate the user activity, channel coefficients, and payload data of the active users. Although the turbo principle has achieved great success in conventional multiuser multiple-input and multiple-output (MIMO) systems (Haykin et al., 2004, Wautelet et al., 2004), its applications in grant-free massive RA are still unchartered due to the new component of user activity detection. Our design exploits the common sparsity pattern in both the received pilot and data signal, and takes advantages of the soft decoding information in order to optimize the activity detection and data reception performance.

14.3.1 Overview of the Turbo Receiver

Inspired by the *turbo decoding principle* (Berrou and Glavieux, 1996), the proposed turbo receiver iterates between a joint estimator and a channel decoder as shown in Fig. 14.2. Specifically, the joint estimator is designed based on the BiG-AMP algorithm (Parker et al., 2014) for user activity detection, channel estimation, and soft data symbol detection, where the posterior probabilities of the transmitted data symbols are estimated and converted as extrinsic information of the coded bits in each turbo iteration. On the other hand, the channel decoder is developed based on the belief propagation (BP) algorithm (Kschischang et al., 2001), which accepts the extrinsic information of the coded bits as input to generate their posteriors. The extrinsic LLRs of the coded bits, i.e., the logarithm of ratio between

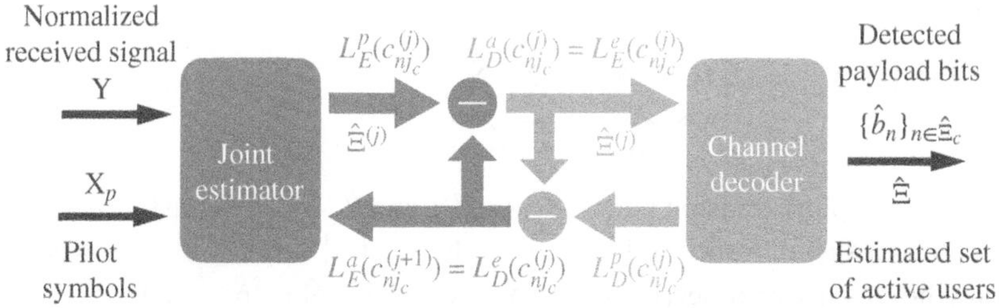

Figure 14.2 The proposed turbo receiver for massive RA. Source: Bian et al. (2023). Reproduced with permission of IEEE.

the probabilities that a coded bit is "0" or "1," are obtained accordingly and translated to priors of the transmitted data symbols for the use of the joint estimator in the next turbo iteration. After Q_1 rounds or when an exit condition is achieved, the turbo iteration terminates, and hard decision is performed to obtain the code block, followed by a CRC. Note that the turbo receiver is initiated by the AMP algorithm in Ke et al. (2020),[1] and its workflow is summarized in Algorithm 14.1 with details of the joint estimator and channel decoder are presented in Section 14.3.2 and 14.3.3.

14.3.2 The Joint Estimator

The joint estimator detects the set of active users, estimates their channel coefficients and the transmitted data symbols. Since the user activity pattern is encapsulated in $\mathbf{H}$, it remains to estimate the effective channel coefficients and soft data symbols. We resort to the MMSE estimators, which can be expressed for the effective channel coefficients and soft data symbols, respectively, as follows (Kay, 1993):

$$\hat{h}_{mn} \triangleq \mathbb{E}\left[h_{mn}|\mathbf{Y}\right] = \int h_{mn}p(h_{mn}|\mathbf{Y})dh_{mn}, \quad \forall m \in \mathcal{M}, \quad n \in \mathcal{N}, \tag{14.2}$$

$$\hat{x}_{nt} \triangleq \mathbb{E}\left[x_{nt}|\mathbf{Y}\right] = \sum x_{nt}p(x_{nt}|\mathbf{Y}), \quad \forall n \in \mathcal{N}, \quad t \in \mathbf{T}_d, \tag{14.3}$$

where $\hat{h}_{mn}$ ($\hat{x}_{nt}$) is the estimate of h_{mn} (x_{nt}), and $p(h_{mn}|\mathbf{Y})$ ($p(x_{nt}|\mathbf{Y})$) denotes the marginal posterior distribution of h_{mn} (x_{nt}) given the normalized received

1 With prior knowledge of the user active probability, the AMP algorithm (Ke et al., 2020) estimates the effective channel coefficients and their variances based on the received pilot signal, and a set of belief indicators $\{\tilde{\rho}_{mn}\}$'s are derived as the posterior probabilities of the effective channel coefficients to be nonzero. We term $\tilde{\rho}_{mn}$ as the *posterior sparsity level* of user n at the mth BS antenna, and define $\overline{\rho}_n \triangleq \frac{1}{M}\sum_{m\in\mathcal{M}}\tilde{\rho}_{mn}$ as the *average sparsity level* of user n, which is a reliable statistic of the activity status and updated iteratively by the joint estimator of the proposed turbo receiver.

Algorithm 14.1 The Proposed Turbo Receiver for Massive RA

Input: The normalized received signal $\mathbf{Y}$, pilot symbols $\mathbf{X}_p$, maximum number of iterations Q_1, and accuracy tolerance ϵ_1.

Output: The estimated set of active users $\hat{\Xi}$, the set of users that pass CRC $\hat{\Xi}_c$ and their detected payload bits $\{\hat{\boldsymbol{b}}_n\}$'s.

Initialize: $j \leftarrow 0$, $n \in \mathcal{N}$, $\hat{x}_{nt}^{(0)} \leftarrow 0$, $t \in \mathbf{T}_d$, $\lambda_n^{(0)} = \frac{K}{N}$, $n \in \mathcal{N}$, $\eta_{nt,s}^{(1)} \leftarrow \frac{1}{|\mathcal{X}|}$, $t \in \mathbf{T}_d$, $L_E^a\left(c_{nj_c}^{(1)}\right) \leftarrow 0, j_c = 1, \dots, N_c$.

1: Execute the AMP algorithm in Ke et al. (2020) to obtain the initial estimates of the effective channel coefficients $\{\hat{h}_{mn}^{(0)}\}$'s and their variances $\{V_{mn}^{h(0)}\}$'s, and the average sparsity levels $\{\bar{\rho}_n^{(0)}\}$'s.

2: **while** $j < Q_1$ and $\frac{\sum_{n,t}|\hat{x}_{nt}^{(j)} - \hat{x}_{nt}^{(j-1)}|^2}{\sum_{n,t}|\hat{x}_{nt}^{(j-1)}|^2} > \epsilon_1$ **do**

3: $j \leftarrow j + 1$

//The Joint Estimator//

4: Based on $\{\hat{h}_{mn}^{(j-1)}\}$'s, $\{V_{mn}^{h(j-1)}\}$'s, and $\{\bar{\rho}_n^{(j-1)}\}$'s, the joint estimator executes Algorithm 14.2 to estimate the set of active users $\hat{\Xi}^{(j)}$, the posterior probabilities that x_{nt} equals s, i.e., $\tilde{\eta}_{nt,s}^{(j)}$, $n \in \hat{\Xi}^{(j)}$, $t \in \mathbf{T}_d$, and the soft data symbols $\hat{x}_{nt}^{(j)}$, $t \in \mathbf{T}_d$.

5: Convert $\tilde{\eta}_{nt,s}^{(j)}$ to the posterior LLRs of coded bits $L_E^p\left(c_{nj_c}^{(j)}\right)$, $n \in \hat{\Xi}^{(j)}$ according to (14.14).

6: Calculate the extrinsic information $L_E^e\left(c_{nj_c}^{(j)}\right)$, $n \in \hat{\Xi}^{(j)}$ as input of the channel decoder $L_D^a\left(c_{nj_c}^{(j)}\right)$, $n \in \hat{\Xi}^{(j)}$ according to (14.15).

//The Channel Decoder//

7: Perform soft data decoding via a BP-based channel decoder and obtain the posterior LLRs of the coded bits $L_D^p\left(c_{nj_c}^{(j)}\right)$, $n \in \hat{\Xi}^{(j)}$.

8: Calculate the extrinsic information $L_D^e\left(c_{nj_c}^{(j)}\right)$, $n \in \hat{\Xi}^{(j)}$ via (14.17) as input of joint estimator $L_E^a\left(c_{nj_c}^{(j+1)}\right)$ for the next turbo iteration, and obtain the prior probabilities that x_{nt} equals s, i.e., $\eta_{nt,s}^{(j+1)}$, $t \in \mathbf{T}_d$ according to (14.19).

9: **end while**

10: Determine the set of active users $\hat{\Xi}$ as $\hat{\Xi}^{(j)}$.

11: Perform hard decision based on $L_D^p\left(c_{nj_c}^{(j)}\right)$ via (14.20) to obtain the code blocks $\hat{\boldsymbol{d}}_n$, $n \in \hat{\Xi}$.

12: Perform CRC to determine $\hat{\Xi}_c$ and detach the CRC bits from $\hat{\boldsymbol{d}}_n$ to obtain $\hat{\boldsymbol{b}}_n$, $n \in \hat{\Xi}_c$.

---

signal $\mathbf{Y}$. The marginal posterior distributions can be rewritten in terms of the joint posterior distribution $p(\mathbf{H},\mathbf{X}|\mathbf{Y})$ as follows:

$$p\left(h_{mn}|\mathbf{Y}\right) = \int_{\mathbf{H}_{\backslash m,n}} \sum_{\mathbf{X}} p(\mathbf{H},\mathbf{X}|\mathbf{Y})d\mathbf{H}, \tag{14.4}$$

$$p\left(x_{nt}|\mathbf{Y}\right) = \sum_{\mathbf{X}_{\backslash n,t}} \int_{\mathbf{H}} p(\mathbf{H},\mathbf{X}|\mathbf{Y})d\mathbf{H}, \tag{14.5}$$

where $p\left(\mathbf{H},\mathbf{X}|\mathbf{Y}\right)$ can be factorized via the Bayes' rule:

$$\begin{aligned} p(\mathbf{H},\mathbf{X}|\mathbf{Y}) &= \frac{p(\mathbf{Y}|\mathbf{H},\mathbf{X})p(\mathbf{H})p(\mathbf{X})}{p(\mathbf{Y})} \overset{(a)}{=} \frac{p(\mathbf{Y}|\mathbf{H},\mathbf{X})p(\mathbf{H}|\mathbf{U})p(\mathbf{U})p(\mathbf{X})}{p(\mathbf{Y})} \\ &\overset{(b)}{=} \frac{1}{p(\mathbf{Y})}\prod_{m=1}^{M}\prod_{t=1}^{T} p\left(y_{mt}|\sum_{n=1}^{N} h_{mn}x_{nt}\right) \\ &\times \prod_{n=1}^{N}\left[p\left(u_n\right)\prod_{m=1}^{M} p\left(h_{mn}|u_n\right)\prod_{t=1}^{T} p\left(x_{nt}\right)\right]. \end{aligned} \tag{14.6}$$

In (14.6), (a) holds since $p(\mathbf{H}) = p(\mathbf{H},\mathbf{U}) = p(\mathbf{H}|\mathbf{U})p(\mathbf{U})$, as the user activity pattern is deterministic given $\mathbf{H}$, and (b) is attributed to the conditional independence of random variables.

However, because of the high-dimensional integrals and summations, the marginal distributions in (14.4) and (14.5) are intractable. Fortunately, the factorization in (14.6) implies efficient approximations via the BP algorithm operating on factor graphs (Kschischang et al., 2001). As shown in Fig. 14.3, a factor graph consists of variable nodes (as indicated by circles), factor nodes (correspond to PDFs as indicated by squares), and edges connecting variable nodes and factor nodes. Messages are propagated in the factor graph and updated iteratively in

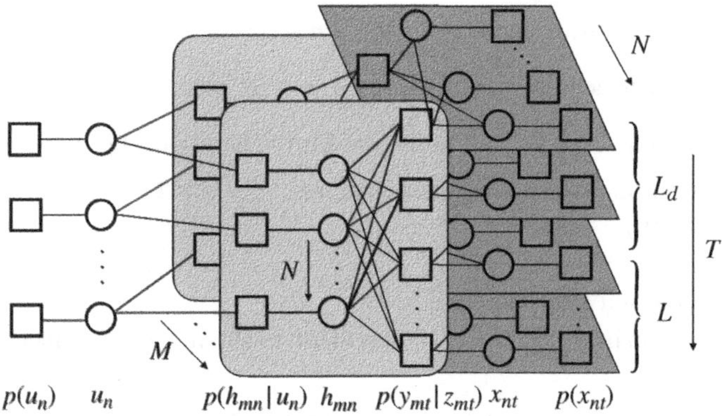

Figure 14.3 The factor graph of the joint posterior distribution $p(\mathbf{H},\mathbf{X}|\mathbf{Y})$, where $z_{mt} \triangleq \sum_{n\in\mathcal{N}} h_{mn}x_{nt}$. Source: Bian et al. (2023). Reproduced with permission of IEEE.

the formats of PDFs. In particular, the message from a factor node to a variable node is the integral of the product of that factor and messages from other adjacent variable nodes of that factor node, while the message from a variable node to a factor node is the product of messages from other adjacent factor nodes of that variable node. Besides, the posterior PDF of a variable is the product of messages from all its adjacent factor nodes.

Although the BP algorithm is efficient in calculating marginal distributions, computations of the high-dimensional integrals in BP still exhibit excessive complexity since N is very large in massive RA systems. Thus, we resort to the framework of AMP, a variant of BP that provides more tractable approximations for marginal distributions (Donoho et al., 2010), to develop the joint estimator with affordable complexity. Since the joint estimation of effective channel coefficients and soft data symbols belongs to a bilinear inference problem, the BiG-AMP algorithm (Parker et al., 2014) offers a viable solution by estimating $\mathbf{H}$ and $\mathbf{X}$ alternatively. Key steps of the BiG-AMP-based joint estimator are summarized in Algorithm 14.2, which is an iterative algorithm that estimates three sets of variables, including: (1) the linear mixing variables $\{z_{mt}\}$'s ($z_{mt} \triangleq \sum_{n\in\mathcal{N}} h_{mn}x_{nt}$); (2) the effective channel coefficients $\{h_{mn}\}$'s; and (3) the soft data symbols $\{x_{nt}\}$'s, as elaborated in the following.

(1) *Estimate the linear mixing variables*: In each iteration, the linear mixing variable z_{mt} is first estimated from y_{mt}. The basic principle is that the posterior probability can be derived for MMSE estimation by using the Bayes' rule if the prior distribution of a variable and its likelihood function are available. Since $y_{mt} = z_{mt} + n_{mt}$, the likelihood function is given as follows:

$$p\left(y_{mt}\Big|\sum_{n\in\mathcal{N}} h_{mn}x_{nt}\right) = \frac{\gamma}{\pi\sigma^2}\exp\left(-\frac{\gamma}{\sigma^2}\left|y_{mt} - \sum_{n\in\mathcal{N}} h_{mn}x_{nt}\right|^2\right). \tag{14.7}$$

The prior distribution of z_{mt} is approximated as a complex Gaussian distribution with mean $M^{p(j)}_{mt}(i)$ and variance $V^{p(j)}_{mt}(i)$ in the ith iteration of the joint estimator, as shown in Lines 3 and 4 in Algorithm 14.2, respectively, where the superscript "$^{(j)}$" denotes the turbo iteration index, $\hat{h}^{(j)}_{mn}(i-1)$ and $V^{h(j)}_{mn}(i-1)$ are the most updated estimate of the effective channel coefficient and its variance, $\hat{x}^{(j)}_{nt}(i-1)$ and $V^{x(j)}_{nt}(i-1)$ are the latest estimate of the soft data symbol and its variance, and $\hat{s}^{(j)}_{mt}(i-1)$ denotes the *scaled residual* of z_{mt}. Since the pilot symbols are known, for $t \in \mathbf{T}_p$, we have $\hat{x}^{(j)}_{nt}(i-1) = x_{nt}$ and $V^{x(j)}_{nt}(i-1) = 0$.

The posterior distribution $p(z_{mt}|y_{mt})$ can also be approximated by a complex Gaussian distribution with mean $\hat{z}^{(j)}_{mt}(i)$ and variance $V^{z(j)}_{mt}(i)$ given in Lines 5 and 6 of Algorithm 14.2, respectively, due to the fact that both the approximated prior distribution of z_{mt} and the likelihood function of z_{mt} are complex Gaussian. Note that the posterior mean of z_{mt} also gives the MMSE

Algorithm 14.2 The Joint Estimator Based on BiG-AMP

Input: The normalized received signal $\mathbf{Y}$, pilot symbols $\mathbf{X}_p$, the estimates of the likelihood that each user is active $\{\lambda_n^{(j)}\}$'s, the estimates of the effective channel coefficients $\{\hat{h}_{mn}^{(j-1)}\}$'s and their variances $\{V_{mn}^{h(j-1)}\}$'s, the prior probabilities that x_{nt} equals s, i.e., $\{\eta_{nt,s}^{(j)}\}$'s, the threshold of determining the active user θ, maximum number of iterations Q_2, and accuracy tolerance ϵ_2.

Output: The estimated set of active users $\hat{\Xi}^{(j)}$, and the posterior probabilities that x_{nt} equals s, i.e., $\tilde{\eta}_{nt,s}^{(j)}$, $n \in \hat{\Xi}^{(j)}$, $t \in \mathbf{T}_d$.

Initialize: $i \leftarrow 0, \hat{h}_{mn}^{(j)}(0) \leftarrow \hat{h}_{mn}^{(j-1)}, V_{mn}^{h(j)}(0) \leftarrow V_{mn}^{h(j-1)}, \hat{s}_{mt}^{(j)}(0) \leftarrow 0, \hat{x}_{nt}^{(j)}(0) \leftarrow 0, t \in \mathbf{T}_d$, $V_{nt}^{x(j)}(0) \leftarrow 1, t \in \mathbf{T}_d$.

while $i < Q_2$ and $\frac{\sum_{m,t}|\hat{z}_{mt}^{(j)}(i)-\hat{z}_{mt}^{(j)}(i-1)|^2}{\sum_{m,t}|\hat{z}_{mt}^{(j)}(i-1)|^2} > \epsilon_2$ **do**

$i \leftarrow i+1$

//Estimate the Linear Mixing Variable z_{mt}//

$\forall m,t: M_{mt}^{p(j)}(i) = \sum_n \hat{h}_{mn}^{(j)}(i-1)\hat{x}_{nt}^{(j)}(i-1) - \hat{s}_{mt}^{(j)}(i-1)$
$\sum_n \left(|\hat{x}_{nt}^{(j)}(i-1)|^2 V_{mn}^{h(j)}(i-1) + |\hat{h}_{mn}^{(j)}(i-1)|^2 V_{nt}^{x(j)}(i-1)\right)$

$\forall m,t: V_{mt}^{p(j)}(i) = \sum_n \left(|\hat{x}_{nt}^{(j)}(i-1)|^2 V_{mn}^{h(j)}(i-1) + |\hat{h}_{mn}^{(j)}(i-1)|^2 V_{nt}^{x(j)}(i-1)\right)$
$+\sum_n V_{mn}^{h(j)}(i-1)V_{nt}^{x(j)}(i-1)$

$\forall m,t: \hat{z}_{mt}^{(j)}(i) = \mathbb{E}\left[z_{mt}|M_{mt}^{p(j)}(i), V_{mt}^{p(j)}(i)\right] = \frac{y_{mt}V_{mt}^{p(j)}(i)+(\sigma^2/\gamma)M_{mt}^{p(j)}(i)}{(\sigma^2/\gamma)+V_{mt}^{p(j)}(i)}$

$\forall m,t: V_{mt}^{z(j)}(i) = \mathrm{Var}\left[z_{mt}|M_{mt}^{p(j)}(i), V_{mt}^{p(j)}(i)\right] = \frac{(\sigma^2/\gamma)V_{mt}^{p(j)}(i)}{(\sigma^2/\gamma)+V_{mt}^{p(j)}(i)}$

$\forall m,t: \hat{s}_{mt}^{(j)}(i) = \left(\hat{z}_{mt}^{(j)}(i) - M_{mt}^{p(j)}(i)\right)/V_{mt}^{p(j)}(i)$

$\forall m,t: V_{mt}^{s(j)}(i) = \left(1 - V_{mt}^{z(j)}(i)/V_{mt}^{p(j)}(i)\right)/V_{mt}^{p(j)}(i)$

//Estimate the Effective Channel Coefficients //

$\forall m,n: Q_{p,mn}^{h(j)}(i) = \left(\sum_{t\in\mathbf{T}_p}|x_{nt}|^2 V_{mt}^{s(j)}(i)\right)^{-1}$

$\forall m,n: P_{p,mn}^{h(j)}(i) = \hat{h}_{mn}^{(j)}(i-1) + Q_{p,mn}^{h(j)}(i)\sum_{t\in\mathbf{T}_p} x_{nt}^*\hat{s}_{mt}^{(j)}(i)$.

$\forall m,n: Q_{d,mn}^{h(j)}(i) = \left(\sum_{t\in\mathbf{T}_d}\left|\hat{x}_{nt}^{(j)}(i-1)\right|^2 V_{mt}^{s(j)}(i)\right)^{-1}$

$\forall m,n: P_{d,mn}^{h(j)}(i) = \hat{h}_{mn}^{(j)}(i-1)\left(1 - Q_{d,mn}^{h(j)}(i)\sum_{t\in\mathbf{T}_d} V_{nt}^{x(j)}(i-1)V_{mt}^{s(j)}(i)\right)$
$+Q_{d,mn}^{h(j)}(i)\sum_{t\in\mathbf{T}_d}\hat{x}_{nt}^{(j)*}(i-1)\hat{s}_{mt}^{(j)}(i)$

$\forall m,n: P_{mn}^{h(j)}(i) = \frac{P_{p,mn}^{h(j)}(i)Q_{d,mn}^{h(j)}(i)+P_{d,mn}^{h(j)}(i)Q_{p,mn}^{h(j)}(i)}{Q_{p,mn}^{h(j)}(i)+Q_{d,mn}^{h(j)}(i)}$

$\forall m,n: Q_{mn}^{h(j)}(i) = \frac{Q_{p,mn}^{h(j)}(i)Q_{d,mn}^{h(j)}(i)}{Q_{p,mn}^{h(j)}(i)+Q_{d,mn}^{h(j)}(i)}$

$$\forall m,n : K_{mn}^{(j)}(i) = \ln\left(\frac{\mathcal{CN}\left(0;P_{mn}^{h(j)}(i),Q_{mn}^{h(j)}(i)+\beta_n\right)}{\mathcal{CN}\left(0;P_{mn}^{h(j)}(i),Q_{mn}^{h(j)}(i)\right)}\right)$$

$$= \ln\left(\frac{Q_{mn}^{h(j)}(i)}{Q_{mn}^{h(j)}(i)+\beta_n}\right) + \frac{\left|P_{mn}^{h(j)}(i)\right|^2 \beta_n}{\left(Q_{mn}^{h(j)}(i)+\beta_n\right)Q_{mn}^{h(j)}(i)}$$

$$\forall m,n : L_{mn}^{(j)}(i) = \ln\left(\frac{\lambda_n^{(j)}}{1-\lambda_n^{(j)}}\right) + \sum_{k\in\mathcal{M}\setminus\{m\}}\left(K_{kn}^{(j)}(i)\right)$$

$$\forall m,n : \rho_{mn}^{(j)}(i) = \exp\left(L_{mn}^{(j)}(i)\right) \Big/ \left(1+\exp\left(L_{mn}^{(j)}(i)\right)\right)$$

$$\forall m,n : \mu_{mn}^{(j)}(i) = \beta_n P_{mn}^{h(j)}(i) \Big/ \left(\beta_n + Q_{mn}^{h(j)}(i)\right),$$

$$\tau_{mn}^{(j)}(i) = \beta_n Q_{mn}^{h(j)}(i) \Big/ \left(\beta_n + Q_{mn}^{h(j)}(i)\right)$$

$$\forall m,n : \tilde{\rho}_{mn}^{(j)}(i) = \rho_{mn}^{(j)}(i) \Big/ \left(\rho_{mn}^{(j)}(i) + \left(1-\rho_{mn}^{(j)}(i)\right)\cdot\right.$$

$$\left.\exp\left(-\ln\left(\frac{Q_{mn}^{h(j)}(i)}{Q_{mn}^{h(j)}(i)+\beta_n}\right) - \frac{|P_{mn}^{h(j)}(i)|^2\beta_n}{\left(Q_{mn}^{h(j)}(i)+\beta_n\right)Q_{mn}^{h(j)}(i)}\right)\right)$$

$$\forall m,n : \hat{h}_{mn}^{(j)}(i) = \mathbb{E}\left[h_{mn}|P_{mn}^{h(j)}(i), Q_{mn}^{h(j)}(i)\right] = \tilde{\rho}_{mn}^{(j)}(i)\mu_{mn}^{(j)}(i)$$

$$\forall m,n : V_{mn}^{h(j)}(i) = \mathrm{Var}\left[h_{mn}|P_{mn}^{h(j)}(i), Q_{mn}^{h(j)}(i)\right]$$

$$= \tilde{\rho}_{mn}^{(j)}(i)\left(\left|\mu_{mn}^{(j)}(i)\right|^2 + \tau_{mn}^{(j)}(i)\right) - \left|\hat{h}_{mn}^{(j)}(i)\right|^2$$

//Estimate the Soft Data Symbols//

$$\forall n,t\in\mathbf{T}_d : Q_{nt}^{x(j)}(i) = \left(\sum_m \left|\hat{h}_{mn}^{(j)}(i-1)\right|^2 V_{mt}^{s(j)}(i)\right)^{-1}$$

$$\forall n,t\in\mathbf{T}_d : P_{nt}^{x(j)}(i) = \hat{x}_{nt}^{(j)}(i-1)\left(1 - Q_{nt}^{x(j)}(i)\sum_m V_{mn}^{h(j)}(i-1)V_{mt}^{s(j)}(i)\right)$$

$$+Q_{nt}^{x(j)}(i)\sum_m \hat{h}_{mn}^{(j)*}(i-1)\hat{s}_{mt}^{(j)}(i)$$

$$\forall n,t\in\mathbf{T}_d : \tilde{\eta}_{nt,s}^{(j)}(i) = \frac{\eta_{nt,s}^{(j)}\mathcal{CN}\left(s;P_{nt}^{x(j)}(i),Q_{nt}^{x(j)}(i)\right)}{\sum_{s'\in\mathcal{X}}\eta_{nt,s'}^{(j)}\mathcal{CN}\left(s';P_{nt}^{x(j)}(i),Q_{nt}^{x(j)}(i)\right)}$$

$$\forall n,t\in\mathbf{T}_d : \hat{x}_{nt}^{(j)}(i) = \mathbb{E}\left[x_{nt}|P_{nt}^{x(j)}(i), V_{nt}^{x(j)}(i)\right] = \bar{\rho}_n^{(j-1)}\sum_{s\in\mathcal{X}}\tilde{\eta}_{nt,s}^{(j)}(i)s$$

$$\forall n,t\in\mathbf{T}_d : V_{nt}^{x(j)}(i) = \mathrm{Var}\left[x_{nt}|P_{nt}^{x(j)}(i), V_{nt}^{x(j)}(i)\right]$$

$$= \sum_{s\in\mathcal{X}}\tilde{\eta}_{nt,s}^{(j)}(i)\left|\bar{\rho}_n^{(j-1)}s - \hat{x}_{nt}^{(j)}(i)\right|^2$$

end while

Update $\bar{\rho}_n^{(j)} = \frac{1}{M}\sum_{m\in\mathcal{M}}\tilde{\rho}_{mn}^{(j)}$, $\hat{h}_{mn}^{(j)}$, and $V_{mn}^{h(j)}$ for the next turbo iteration.

Update $\lambda_n^{(j+1)} = \kappa\bar{\rho}_n^{(j)} + (1-\kappa)\lambda_n^{(j)}, \forall n\in\mathcal{N}$.

Determine the estimated set of active users as $\hat{\Xi}^{(j)} \triangleq \{n\in\mathcal{N} \mid \bar{\rho}_n^{(j)} \geq \theta\}$.

---

estimate $\hat{z}_{mt}^{(j)}(i)$. Besides, the *scaled residual* $\hat{s}_{mt}^{(j)}(i)$ of z_{mt} and the corresponding *inverse-residual-variance* $V_{mt}^{s(j)}(i)$ are updated in Lines 7 and 8, respectively, which are useful for approximating the likelihood functions of the effective channel coefficients and soft data symbols.

(2) *Estimate the effective channel coefficients*: The effective channel coefficients and their variances are estimated by incorporating both the received pilot and data signals. To approximate the posterior distribution of h_{mn} in the ith iteration of the joint estimator, the belief of variable node h_{mn} is first derived based on the BP algorithm as follows:

$$B_{h_{mn}}^{(j)}(i) = I_{f_{h_{mn}} \to h_{mn}}^{(j)}(i) \prod_{t \in \mathbf{T}_p} I_{f_{y_{mt}} \to h_{mn}}^{(j)}(i) \prod_{t \in \mathbf{T}_d} I_{f_{y_{mt}} \to h_{mn}}^{(j)}(i), \tag{14.8}$$

where $I_{f_{h_{mn}} \to h_{mn}}^{(j)}(i)$ denotes the message from factor node $p(h_{mn}|u_n)$ to variable node h_{mn} that serves as the prior distribution of h_{mn}, and $I_{f_{y_{mt}} \to h_{mn}}^{(j)}(i)$ represents the message from factor node $p(y_{mt}|z_{mt})$ to variable node h_{mn}. Thus, the term $\prod_{t \in \mathbf{T}_p} I_{f_{y_{mt}} \to h_{mn}}^{(j)}(i) \prod_{t \in \mathbf{T}_d} I_{f_{y_{mt}} \to h_{mn}}^{(j)}(i)$ can be interpreted as the likelihood function of h_{mn} in the ith iteration. Specifically, the term $\prod_{t \in \mathbf{T}_p} I_{f_{y_{mt}} \to h_{mn}}^{(j)}(i)$, which corresponds to the received pilot symbols, is approximated as a complex Gaussian PDF with mean $P_{p,mn}^{h(j)}(i)$ and variance $Q_{p,mn}^{h(j)}(i)$, and the term $\prod_{t \in \mathbf{T}_d} I_{f_{y_{mt}} \to h_{mn}}^{(j)}(i)$ that relates to the received data symbols, is approximated as another complex Gaussian PDF with mean $P_{d,mn}^{h(j)}(i)$ and variance $Q_{d,mn}^{h(j)}(i)$. Consequently, the term $\prod_{t \in \mathbf{T}_p} I_{f_{y_{mt}} \to h_{mn}}^{(j)}(i) \prod_{t \in \mathbf{T}_d} I_{f_{y_{mt}} \to h_{mn}}^{(j)}(i)$ is also approximated as a complex Gaussian PDF with mean $P_{mn}^{h(j)}(i)$ and variance $Q_{mn}^{h(j)}(i)$ given in Lines 13 and 14 of Algorithm 14.2, respectively. To further derive $B_{h_{mn}}^{(j)}(i)$, we obtain $I_{f_{h_{mn}} \to h_{mn}}^{(j)}(i)$ as follows:

$$I_{f_{h_{mn}} \to h_{mn}}^{(j)}(i) = \left(1 - \rho_{mn}^{(j)}(i)\right) \delta(h_{mn}) + \rho_{mn}^{(j)}(i) \mathcal{CN}(h_{mn}; 0, \beta_n), \tag{14.9}$$

where $\rho_{mn}^{(j)}(i)$ approximates the probability that h_{mn} is nonzero, and it is defined as the *sparsity level* of user n at the mth BS antenna. Note that estimates of the likelihood that each user is active or not in the considered transmission block, i.e., $\{\lambda_n^{(j)}\}$'s, which are updated in each turbo iteration for more accurate estimation of the BiG-AMP algorithm, are required for the calculations of $\{\rho_{mn}^{(j)}(i)\}$'s in Lines 15–17 of Algorithm 14.2. Therefore, the posterior distribution of h_{mn} is approximated in the ith iteration as follows:

$$\begin{aligned} r_{h_{mn}}^{(j)}(i) &= \frac{B_{h_{mn}}^{(j)}(i)}{\int B_{h_{mn}}^{(j)}(i) dh_{mn}} = \left(1 - \tilde{\rho}_{mn}^{(j)}(i)\right) \delta(h_{mn}) \\ &+ \tilde{\rho}_{mn}^{(j)}(i) \mathcal{CN}\left(h_{mn}; \mu_{mn}^{(j)}(i), \tau_{mn}^{(j)}(i)\right), \end{aligned} \tag{14.10}$$

where $\mu_{mn}^{(j)}(i)$ and $\tau_{mn}^{(j)}(i)$ are given in Line 18 of Algorithm 14.2, and $\tilde{\rho}_{mn}^{(j)}(i)$ presented in Line 19 is defined as the *posterior sparsity level* of user n at the mth BS antenna. Based on the posterior distribution, we obtain the MMSE estimate of the effective channel coefficient and its variance in Lines 20 and 21 of Algorithm 14.2, respectively.

(3) *Estimate the soft data symbols*: Since x_{nt} and h_{mn} are symmetry in the bilinear inference problem of the joint estimator, the conditional mean $P_{nt}^{x(j)}(i)$ and variance $Q_{nt}^{x(j)}(i)$ given x_{nt} are obtained in Lines 23 and 22 of Algorithm 14.2, respectively. Then, prior distributions of the transmitted data symbols can be estimated as follows:

$$p(x_{nt}) = I_{f_{x_{nt}} \to x_{nt}}^{(j)}(i) \approx \overline{\rho}_n^{(j-1)} \sum_{s \in \mathcal{X}} \eta_{nt,s}^{(j)} \delta(x_{nt} - s), t \in \mathbf{T}_d, \tag{14.11}$$

where $\overline{\rho}_n^{(j-1)} \triangleq \frac{1}{M} \sum_{m \in \mathcal{M}} \tilde{\rho}_{mn}^{(j-1)}$, and $\eta_{nt,s}^{(j)}$ denotes the probability that x_{nt} belongs to constellation point s. As will be introduced in Section 14.3.3, $\{\eta_{nt,s}^{(j)}\}$'s are obtained from the channel decoder in the last turbo iteration. Thus, the approximated posterior distributions of the transmitted data symbols in the ith iteration of the joint estimator can be expressed as follows:

$$r_{x_{nt}}^{(j)}(i) = \overline{\rho}_n^{(j-1)} \sum_{s \in \mathcal{X}} \tilde{\eta}_{nt,s}^{(j)}(i) \delta(x_{nt} - s), \quad t \in \mathbf{T}_d, \tag{14.12}$$

where $\tilde{\eta}_{nt,s}^{(j)}(i)$ denotes the posterior probability that x_{nt} belongs to constellation point s as derived using the Bayes' rule in Line 24 of Algorithm 14.2. The soft data symbols and the corresponding posterior variances are estimated via Lines 25 and 26, respectively.

Once the while loop of Algorithm 14.2 is terminated, $\overline{\rho}_n^{(j)}$, $\hat{h}_{mn}^{(j)}$ and $V_{mn}^{h(j)}$ are updated for the next turbo iteration. Accordingly, we update $\{\lambda_n^{(j)}\}$'s in Line 28 using the average sparsity level $\overline{\rho}_n^{(j)}$, i.e., $p(u_n = 1) \triangleq \lambda_n^{(j+1)} = \kappa \overline{\rho}_n^{(j)} + (1 - \kappa)\lambda_n^{(j)}$, $n \in \mathcal{N}$, where $\kappa \in [0, 1]$ is the *learning rate*. This operation avoids using the average sparsity levels exclusively to eliminate the potential estimation errors caused by inaccurate prior information of the BiG-AMP algorithm. The set of active users is determined as $\hat{\Xi}^{(j)} \triangleq \{n \in \mathcal{N} | \overline{\rho}_n^{(j)} \geq \theta\}$, where θ is an empirical threshold (Ke et al., 2020).

In order to minimize the data decoding error by eliminating some redundancy from the prior information of the coded bits, extrinsic information of the joint estimator is derived as input of the channel decoder. In particular, the posterior probability of a transmitted data symbols is translated to the posterior probabilities of the corresponding coded bits as follows:

$$p\left(c_{nj_c}^{(j)} = b | \mathbf{Y}\right) = \sum_{s \in \mathcal{X}_{j_c}^b} \tilde{\eta}_{nt,s}^{(j)}, \quad n \in \hat{\Xi}^{(j)}, \tag{14.13}$$

where $\hat{j}_c \triangleq \text{mod}\,(j_c, \log_2|\mathcal{X}|)$, $t = L + 1 + \left\lfloor \frac{j_c}{\log_2|\mathcal{X}|} \right\rfloor$, and $\mathcal{X}_l^b$ represents the set of constellation points with the lth position ($l = 0, \ldots, \log_2|\mathcal{X}| - 1$) of the corresponding bit sequence as b. For example, suppose the bit sequences "00," "01," "10," and "11" are modulated to constellation points s_0, s_1, s_2, and s_3, respectively, in quadrature phase shift keying (QPSK), we have $\mathcal{X}_0^0 = \{s_0, s_1\}$, $\mathcal{X}_0^1 = \{s_2, s_3\}$, $\mathcal{X}_1^0 = \{s_0, s_2\}$, and $\mathcal{X}_1^1 = \{s_1, s_3\}$. Then, the posterior probabilities of the coded bits are used to derive the posterior LLRs as follows:

$$L_E^p\left(c_{nj_c}^{(j)}\right) \triangleq \ln\left(\frac{p(c_{nj_c}^{(j)} = 0|\mathbf{Y})}{p(c_{nj_c}^{(j)} = 1|\mathbf{Y})}\right), \quad n \in \hat{\Xi}^{(j)}, \tag{14.14}$$

which are converted to the extrinsic information as defined below (Vucetic and Yuan, 2001):

$$L_E^e\left(c_{nj_c}^{(j)}\right) \triangleq L_E^p\left(c_{nj_c}^{(j)}\right) - L_E^a\left(c_{nj_c}^{(j)}\right), \quad n \in \hat{\Xi}^{(j)}. \tag{14.15}$$

The prior information $L_E^a\left(c_{nj_c}^{(j)}\right) \triangleq \ln\left(\frac{p(c_{nj_c}^{(j)}=0)}{p(c_{nj_c}^{(j)}=1)}\right)$ is obtained from the channel decoder in the last turbo iteration.

14.3.3 The Channel Decoder

The channel decoder determines the most probable code block for each estimated active user, which can be formulated as the following maximum *a posteriori* probability (MAP) estimation problem:

$$\hat{\mathbf{c}}_n = \arg \max_{\mathbf{c}_n \in \{0,1\}^{N_c}} p\left(\mathbf{c}_n \mid \{L_E^e(c_{nj_c}^{(j)})\}\right), \quad n \in \hat{\Xi}^{(j)}. \tag{14.16}$$

We adopt a BP-based channel decoder to solve (14.16), which is able to accept the extrinsic information of the coded bits derived from the joint estimator as input, and calculate the posterior LLRs of the coded bits $L_D^p(c_{nj_c}^{(j)})$, $n \in \hat{\Xi}^{(j)}$ as the soft decoding results. This can be achieved by a variety of off-the-shelf BP-based channel decoders, e.g., the decoders developed in Gallager (1962), Hocevar (2004) and Berrou and Glavieux (1996), Hagenauer et al. (1996) for low-density parity-check (LDPC) code and turbo code, respectively.

Similar to the joint estimator, extrinsic information of the channel decoder is derived as follows:

$$L_D^e\left(c_{nj_c}^{(j)}\right) \triangleq L_D^p\left(c_{nj_c}^{(j)}\right) - L_D^a\left(c_{nj_c}^{(j)}\right), \quad n \in \hat{\Xi}^{(j)}, \tag{14.17}$$

which is adopted as prior information $L_E^a(c_{nj_c}^{(j+1)})$, $n \in \hat{\Xi}^{(j)}$ for the use of the joint estimator in the next-turbo iteration. Therefore, the prior distribution of a coded bit is given as

$$p\left(c_{nj_c}^{(j)}\right) = \begin{cases} \dfrac{1}{1+\exp\left(L_D^e\left(c_{nj_c}^{(j)}\right)\right)}, & c_{nj_c}^{(j)} = 1, \\ \dfrac{\exp\left(L_D^e\left(c_{nj_c}^{(j)}\right)\right)}{1+\exp\left(L_D^e\left(c_{nj_c}^{(j)}\right)\right)}, & c_{nj_c}^{(j)} = 0, \end{cases} \quad n \in \hat{\Xi}^{(j)}, \tag{14.18}$$

and prior distributions of the transmitted data symbols can be estimated according to the following expression:

$$\eta_{nt,s}^{(j+1)} = \prod_{j_c=v_1}^{v_2} p\left(c_{nj_c}^{(j)}\right), \quad t \in \mathbf{T}_d, \quad n \in \hat{\Xi}^{(j)}. \tag{14.19}$$

where $v_1 \triangleq (t-L-1)\log_2|\mathcal{X}|$, $v_2 \triangleq (t-L)\log_2|\mathcal{X}|-1$, and $\mu([c_{nv_1},\dots,c_{nv_2}]) = s$. Note that for the users that are determined as inactive, we reuse the prior information of the transmitted data symbols from the last turbo iteration by setting $L_E^a\left(c_{nj_c}^{(j+1)}\right) = L_E^a\left(c_{nj_c}^{(j)}\right)$ and $\eta_{nt,s}^{(j+1)} = \eta_{nt,s}^{(j)}$, $n \in \mathcal{N}\backslash\hat{\Xi}^{(j)}$.

The values of $\{L_D^p(c_{nj_c})\}$'s are also utilized to obtain the code block $\hat{\boldsymbol{d}}_n$, $n \in \hat{\Xi}$ after the last turbo iteration by performing hard decision as follows:

$$\hat{d}_{nj_c} = \begin{cases} 0, & L_D^p(c_{nj_c}) \geq 0, \\ 1, & L_D^p(c_{nj_c}) < 0. \end{cases} \tag{14.20}$$

CRC is then performed for $\hat{\boldsymbol{d}}_n$, $n \in \hat{\Xi}$ and the CRC bits are detached to obtain the payload bits $\hat{\boldsymbol{b}}_n$ for the users that pass parity check, which is denoted as $\hat{\Xi}_c$ in Algorithm 14.1.

14.4 A Low-Complexity Side Information-Aided Receiver

Although the proposed turbo receiver effectively exploits the common sparsity pattern in the received pilot and data signal via BiG-AMP, the joint estimation of effective channel coefficients and soft data symbols is performed iteratively by incorporating all the received symbols in each turbo iteration, which incurs significant computation overhead even with a reasonable size of the payload data. In addition, in order to estimate the prior information, the channel decoder needs to be executed in each turbo iteration for all users in $\hat{\Xi}^{(j)}$ (See Line 7 of Algorithm 14.1), which brings additional computation overhead. Therefore, low-complexity receivers for massive RA are needed to leverage both the common sparsity pattern and channel decoding results more efficiently. In this section, we develop a low-complexity SI-aided receiver.

14.4.1 Overview of the SI-Aided Receiver

As shown in Fig. 14.4, the SI-aided receiver iterates between a sequential estimator and a channel decoder. The sequential estimator cascades the AMP algorithm (Ke et al., 2020) for JADCE, and an MMSE-based soft demodulator to compute the prior LLRs of the coded bits. A BP-based channel decoder is adopted to obtain the posterior LLRs of the coded bits similar as the turbo receiver, while hard decision is performed for parity check in each iteration. The SI, i.e., the estimates on whether a user is active, is updated jointly based on the average sparsity levels, the posterior LLRs of the coded bits, and the parity check results, which provides more precise prior knowledge for the sequential estimator to achieve more accurate JADCE (Ma et al., 2019). The workflow of the SI-aided receiver is summarized in Algorithm 14.3, whereas details of the sequential estimator and the channel decoder are introduced in Section 14.4.2, and design of the SI is elaborated in Section 14.4.3.

14.4.2 The Sequential Estimator and the Channel Decoder

The sequential estimator adopts the AMP algorithm (Ke et al., 2020) with the normalized received pilot signal $\mathbf{Y}_p$ to estimate the effective channel coefficients $\{\hat{h}_{mn}^{(j)}\}$'s and the set of active users $\hat{\Xi}^{(j)}$. We also derive the sparsity levels $\{\rho_{mn}^{(j)}\}$'s from the AMP algorithm following similar steps in Lines 15–17 of Algorithm 14.2. Soft data symbol detection is performed using an MMSE-based soft demodulator based on the results of JADCE. Specifically, the estimated data symbols $\hat{x}_{nt}$, $n \in \hat{\Xi}^{(j)}$, $t \in \mathbf{T}_d$ are obtained with the normalized received data signal $\mathbf{Y}_d$ via an

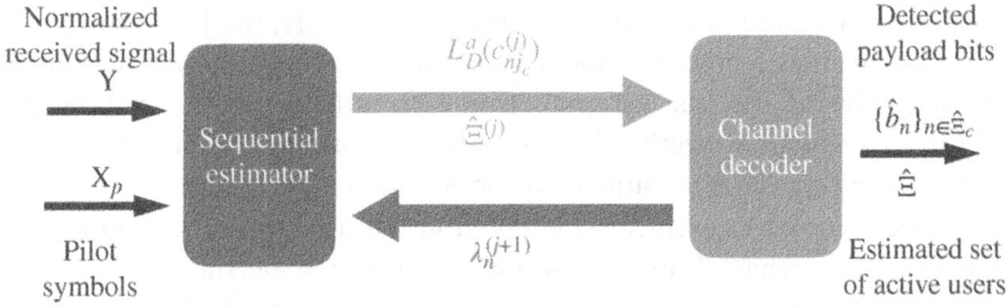

Figure 14.4 The proposed SI-aided receiver for massive RA. Source: Bian et al. (2023). Reproduced with permission of IEEE.

Algorithm 14.3 The Proposed SI-Aided Receiver for Massive RA

Input: The normalized received signal $\mathbf{Y}$, pilot symbols $\mathbf{X}_p$, maximum number of iterations Q_3, and accuracy tolerance ϵ_3.

Output: The estimated set of active users $\hat{\Xi}$, the set of users that pass CRC $\hat{\Xi}_c$ and their detected payload bits $\hat{\boldsymbol{b}}_n$.

Initialize: $j \leftarrow 0$, $\lambda_n^{(1)} \leftarrow \frac{K}{N}$, $n \in \mathcal{N}$, $\hat{x}_{nt}^{(0)} \leftarrow 0$, $t \in \mathbf{T}_d$, $\hat{\Xi}_c \leftarrow \emptyset$.

while $j < Q_3$ and $\frac{\sum_{n,t} |\hat{x}_{nt}^{(j)} - \hat{x}_{nt}^{(j-1)}|^2}{\sum_{n,t} |\hat{x}_{nt}^{(j-1)}|^2} > \epsilon_3$ **do**

  $j \leftarrow j + 1$

  //The Sequential Estimator//

  Execute the AMP algorithm (Ke et al. 2020) with the SI $\{\lambda_n^{(j)}\}$'s as the prior knowledge of the user activity to estimate the effective channel coefficients $\{\hat{h}_{mn}^{(j)}\}$'s and set of active users $\hat{\Xi}^{(j)}$.

  Estimate the transmitted data symbols $\hat{x}_{nt}^{(j)}$, $t \in \mathbf{T}_d$ via an MMSE equalizer, i.e., $\hat{\mathbf{X}}_{d,a}^{(j)} = ((\hat{\mathbf{H}}_a^{(j)})^H \hat{\mathbf{H}}_a^{(j)} + (\sigma^2/\gamma)\mathbf{I})^{-1} (\hat{\mathbf{H}}_a^{(j)})^H \mathbf{Y}_d$, where $\mathbf{H}_a^{(j)} \triangleq \left[\left\{\hat{\mathbf{h}}_k^{(j)}\right\}_{k \in \hat{\Xi}^{(j)}}\right]$ stacks the effective channel coefficients of all the estimated active users, and $\hat{x}_{nt}^{(j)}$ is the entry of $\hat{\mathbf{X}}_{d,a}^{(j)}$.

  Compute the prior LLRs of the coded bits as $L_D^a\left(c_{nj_c}^{(j)}\right)$, $n \in \hat{\Xi}^{(j)}$ via soft demodulation according to (14.21).

//The Channel Decoder//

  Perform soft data decoding via a BP-based channel decoder to obtain the posterior LLRs of the coded bits $L_D^p\left(c_{nj_c}^{(j)}\right)$, $n \in \hat{\Xi}^{(j)} \setminus \hat{\Xi}_c$.

  Perform hard decision, determine the set of users in $\hat{\Xi}^{(j)}$ that pass $\hat{\Xi}_c^{(j)}$ and obtain their payload bits $\{\hat{\boldsymbol{b}}_n\}$'s.

  $\hat{\Xi}_c \leftarrow \hat{\Xi}_c \bigcup \hat{\Xi}_c^{(j)}$

  Update the SI $\{\lambda_n^{(j+1)}\}$'s according to (14.22).

end while

MMSE equalizer. Then, the prior LLRs of the coded bits are derived via soft demodulation as follows:

$$L_D^a\left(c_{nj_c}^{(j)}\right) \triangleq \ln\left(\frac{p\left(c_{nj_c}^{(j)} = 0 | \hat{x}_{nt}^{(j)}\right)}{p\left(c_{nj_c}^{(j)} = 1 | \hat{x}_{nt}^{(j)}\right)}\right)$$

$$= \ln\left(\frac{\sum_{s \in \mathcal{X}_{\hat{j}_c}^0} \exp\left(-\gamma \|\hat{x}_{nt}^{(j)} - s\|_2^2/\sigma^2\right)}{\sum_{s \in \mathcal{X}_{\hat{j}_c}^1} \exp\left(-\gamma \|\hat{x}_{nt}^{(j)} - s\|_2^2/\sigma^2\right)}\right), \quad n \in \hat{\Xi}^{(j)}, \tag{14.21}$$

where $\hat{j}_c \triangleq \mathrm{mod}\,(j_c, \log_2|\mathcal{X}|)$ and $t = L + 1 + \left\lfloor \frac{j_c}{\log_2|\mathcal{X}|} \right\rfloor$.

With the knowledge of $\{L_D^a(c_{nj_c}^{(j)})\}$'s, the BP-based channel decoder calculates the posterior LLRs of the coded bits $L_D^P(c_{nj_c}^{(j)})$, $n \in \hat{\Xi}^{(j)} \backslash \hat{\Xi}_c$ and decides the code blocks according to (14.20). CRC is performed for all users in $\hat{\Xi}^{(j)} \backslash \hat{\Xi}_c$ to obtain their payload bits $\{\hat{\boldsymbol{b}}_n\}$'s. Note that channel decoding is not performed for users that have already passed the parity check, which differs the turbo receiver and helps to save the computations.

14.4.3 The Side Information

Since the AMP algorithm as implemented in the sequential estimator is based on the framework of Bayesian estimation (Kay, 1993), prior knowledge of the user activity, i.e., the SI for the sequential estimator $\{\lambda_n\}$'s, has significant impacts on the estimation accuracy. Therefore, in order to obtain more precise estimates through multiple iterations of Algorithm 14.3, we propose to update the SI by jointly utilizing the results of the sequential estimator and channel decoder, according to three different cases depending on the estimated set of active users and their parity check results via the following update rule:

$$\lambda_n^{(j+1)} = \begin{cases} 1, & n \in \hat{\Xi}_c, \\ \kappa_1 \overline{\rho}_n^{(j)} + \frac{1-\kappa_1}{N_c} \sum\limits_{j_c} \frac{\left|L_D^P(c_{nj_c}^{(j)})\right|}{1+\left|L_D^P(c_{nj_c}^{(j)})\right|}, & n \in \hat{\Xi}^{(j)} \backslash \hat{\Xi}_c, \\ \kappa_2 \overline{\rho}_n^{(j)} + (1-\kappa_2)\lambda_n^{(j)}, & n \in \mathcal{N} \backslash (\hat{\Xi}^{(j)} \cup \hat{\Xi}_c). \end{cases} \tag{14.22}$$

In particular, in the first case of (14.22), we set $\lambda_n^{(j+1)} = 1, n \in \hat{\Xi}_c$ since the users that have passed the parity check in the current or previous iterations can be safely determined as active. In the second case, the users that are estimated as active but fail to pass the parity check in the current iteration, i.e., $n \in \hat{\Xi}^{(j)} \backslash \hat{\Xi}_c$, are handled, where the average sparsity levels and the posterior LLRs of the coded bits are jointly utilized to update the SI since both of them are informative on users' activity. The term $\frac{1}{N_c} \sum_{j_c} \frac{\left|L_D^P(c_{nj_c}^{(j)})\right|}{1+\left|L_D^P(c_{nj_c}^{(j)})\right|} \in [0, 1)$ indicates the decoding reliability of user n as its complement, i.e., $\frac{1}{N_c} \sum_{j_c} \frac{1}{1+\left|L_D^P(c_{nj_c}^{(j)})\right|}$, provides an accurate estimate of the bit error rate (Goektepe et al., 2018). Parameter $\kappa_1 \in [0, 1]$ is an empirical weighting factor balancing the contributions of the channel estimation and data decoding results. The third case uses a similar methodology as that of the turbo receiver, where the SI is updated with the learning rate $\kappa_2 \in [0, 1]$ for users that neither pass the CRC in any iteration nor being determined as inactive in the current iteration.

To demonstrate the rationality of the SI update rule in (14.22), we provide numerical examples on the evolutions of $\{\lambda_n^{(j)}\}$'s, considering two scenarios with $K = 40$ and 80 in Fig. 14.5(a) and (b), respectively. From these figures, we see that

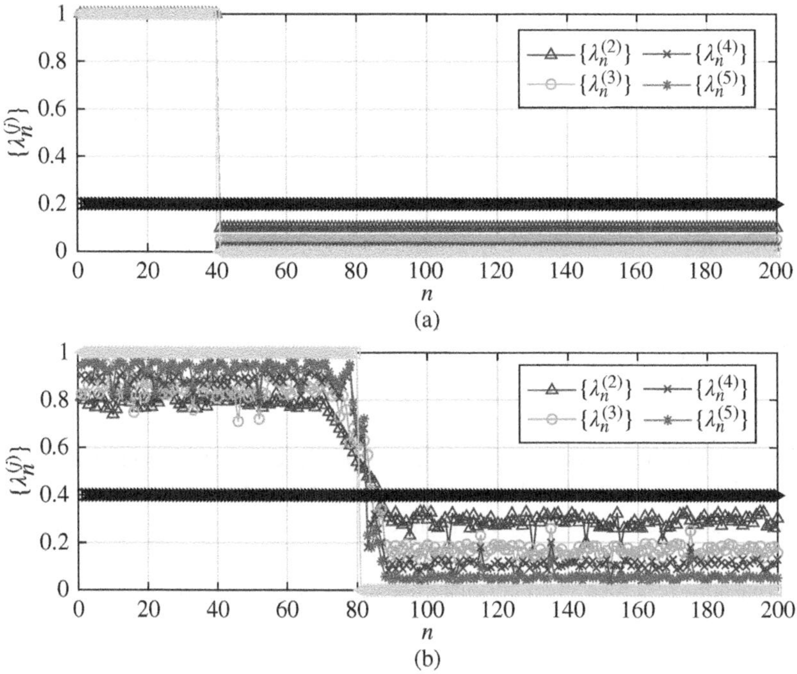

Figure 14.5 Illustrations on the SI update rule in (14.22). We set $N = 200$ and $\kappa_1 = \kappa_2 = 0.5$. The first K users are active in the transmission block. (a) $K = 40$ and (b) $K = 80$. Source: Bian et al. (2023). Reproduced with permission of IEEE.

Algorithm 14.3 iterates, the SI evolves from the initial values $\lambda_n^{(0)} = \frac{K}{N}$, $n \in \mathcal{N}$, to the perfect estimates, i.e., $\lambda_n^{(0)} = 1$, $n = 1, \ldots, K$ and $\lambda_n^{(0)} = 0$, $n = K + 1, \ldots, N$.

14.5 Simulation Results

14.5.1 Simulation Setting and Baseline Schemes

We simulate a single-cell cellular network, where 200 users are randomly distributed in a circle with a radius of 500 m centered at the 64-antenna BS. The path loss of user n is calculated as $\beta_n = -128.1 - 36.7\log_{10}(r_n)$ (dB), where r_n (km) is the distance to the BS. To achieve more stable convergence behavior, a damping factor $\omega \in (0, 1]$ (Parker et al., 2014) is applied to moderate the updates of M_{mt}^p, V_{mt}^p, $\hat{h}_{mn}$, and $\hat{x}_{nt}$. For instance, the damped version of the estimated soft data symbol can be expressed as $\overline{x}_{nt}(i) = \omega\hat{x}_{nt}(i) + (1 - \omega)\overline{x}_{nt}(i - 1)$, $t \in \mathbf{T}_d$. In particular, the mean and variance of z_{mt} in Lines 3 and 4 of Algorithm 14.2 are replaced with the

Table 14.1 Simulation parameters.

Parameters	Values	Parameters	Values	Parameters	Values
Code rate	1/2	L	50	L_d	150
Modulation	QPSK	N_d	150	N_c	300
CRC type	CRC-8	ω	0.6	θ	0.4
Channel coding	LDPC	Q_2	100	ϵ_1, ϵ_2	10^{-5}
Noise power density	−169 dBm/Hz	κ, κ_2	0.5	κ_1	0.5
Bandwidth	1 MHz	Q_1, Q_3	6	T	200
Transmit power	23 dBm	N_b	142	ϵ_3	10^{-5}

damped versions $\overline{M}^{p}_{mt}$ and $\overline{V}^{p}_{mt}$, respectively, while the damped versions of $\hat{h}_{mn}$ and $\hat{x}_{nt}$ are used in Lines 10–12 and Lines 22 and 23 of Algorithm 14.2. The simulation results are averaged over 10^5 independent channel realizations, and other critical simulation parameters are summarized in Table 14.1.

Two baseline schemes and a performance upper bound are adopted for comparisons:

- *Separate design (Ke et al., 2020)*: This can be viewed as an instance of the SI-aided receiver by setting $Q_3 = 1$, where the AMP algorithm is first performed for JADCE, and data symbols are then detected using an MMSE equalizer. The detected soft data symbols are converted to prior LLRs of the coded bits for data decoding via soft demodulation using (14.21).
- *Data-assisted design with BiG-AMP (Zou et al., 2020)*: This is a special case of the turbo receiver when $Q_1 = 1$, where the BiG-AMP algorithm is used for joint activity detection, channel estimation, and soft data symbol detection. The detected soft data symbols are converted to prior LLRs of the coded bits for data decoding using (14.21).
- *Turbo receiver with known user activity*: This scheme serves as a performance upper bound, where perfect knowledge of the user activity is assumed and consequently, channel estimation and data decoding are performed via the proposed turbo receiver by setting $\lambda_n^{(j)} = 1, n \in \Xi$, and $\lambda_n^{(j)} = 0, n \in \mathcal{N}\backslash\Xi$.

Note that all the simulated schemes adopt the same BP-based LDPC decoder (Gallager, 1962) for fair comparisons.

14.5.2 Results

The activity detection error probability (including the missed detection and false alarm probability) and the normalized mean square error (NMSE) of channel

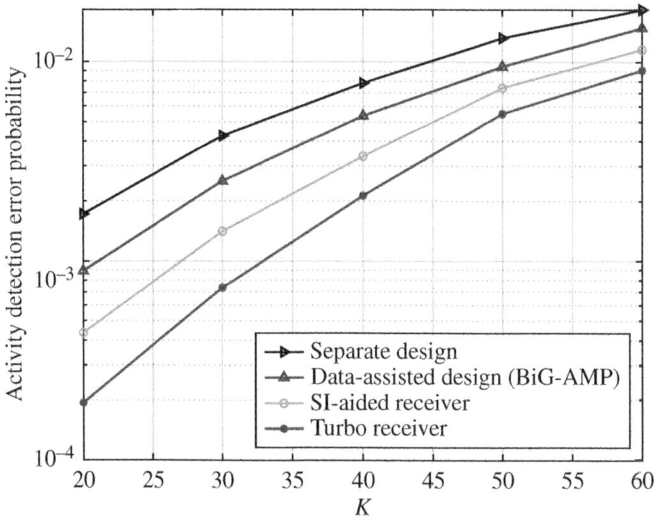

Figure 14.6 Activity detection error probability versus the number of active users. Source: Bian et al. (2023). Reproduced with permission of IEEE.

estimation are first evaluated in Figs. 14.6 and 14.7, respectively. Compared with the separate design, the data-assisted design achieves much lower activity detection and channel estimation errors, validating the benefits of incorporating the received data symbols. Besides, with the soft channel decoding results to refine the prior distributions of the transmitted data symbols through multiple turbo iterations, it is seen that the proposed turbo receiver significantly outperforms the data-assisted design. Moreover, despite with some performance loss compared with the turbo receiver, the low-cost SI-aided receiver secures noticeable performance improvement compared with the data-assisted design, which can be credited to the use of the customized SI as prior knowledge of the user activity.

Figure 14.8 shows the block error rate (BLER) of all the simulated schemes versus the number of active users. Similar to activity detection and channel estimation, the turbo receiver achieves the best BLER performance. Assuming the BLER requirement is 10^{-3}, the turbo receiver is able to support 40 active users while the separate design can only support 20, which is a remarkable 100% increase. Compared with the baseline schemes, it also greatly narrows the performance gap to the performance upper bound with perfect knowledge of the user activity, owing to the more accurate activity detection and channel estimation. Because of the same reason, the SI-aided receiver brings notable BLER reduction compared with the data-assisted design.

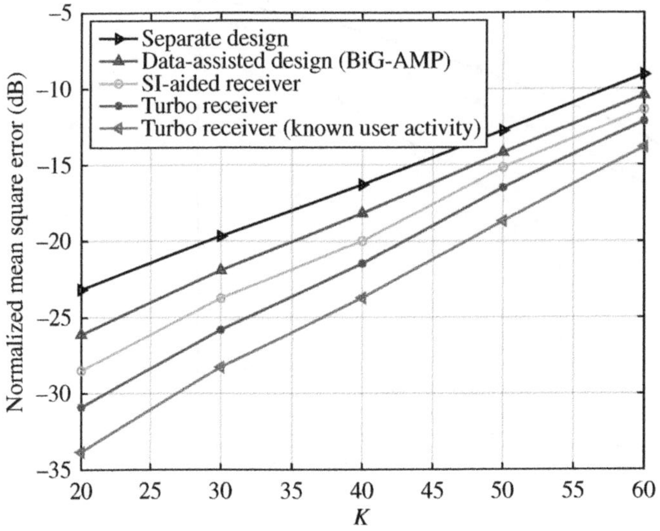

Figure 14.7 NMSE of channel estimation versus the number of active users. Source: Bian et al. (2023). Reproduced with permission of IEEE.

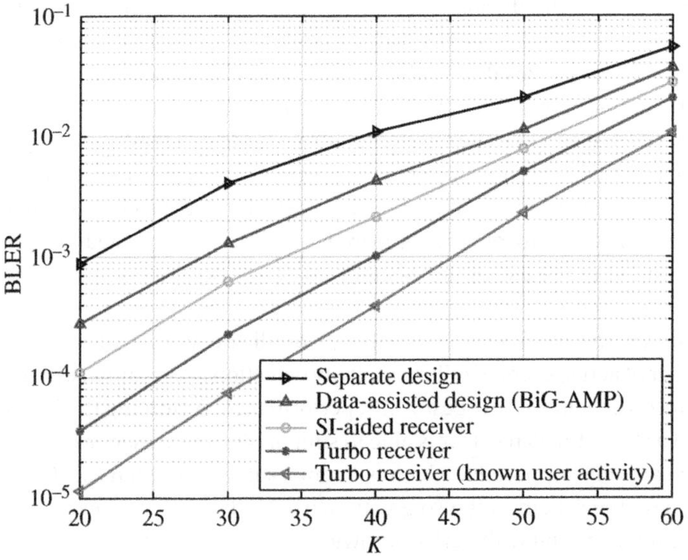

Figure 14.8 BLER versus the number of active users. Source: Bian et al. (2023). Reproduced with permission of IEEE.

Next, we investigate the impacts of the number of iterations, i.e., Q_1 for the turbo receiver and Q_3 for the SI-aided receiver, since both of them iterate between an estimator and a channel decoder in Fig. 14.9. By measuring their average execution time on the same computing server, the computation complexity of different schemes are compared. Considering that the average execution time is platform-specific, it is normalized with respect to that of the separate design. It is observed that the separate design has the lowest complexity but the highest BLER since it ignores both the common sparsity pattern and the information offered by the channel decoder. Besides, the performance achieved by both of the proposed receivers improves with the number of iterations, which again corroborates the effectiveness of the iterative estimation on the prior information of the user activity and the transmitted data symbols. However, such performance improvement is accompanied with increased computation complexity. Compared with the turbo receiver, the SI-aided receiver enjoys 66–74% average execution time reduction since the sequential estimator only processes the pilot signal for JADCE, and channel decoding is performed just for the users that have not passed the parity check. In addition, we notice that the major performance gains of the proposed receivers come from the first few iterations, e.g., seven in the considered scenario, and the subsequent iterations only contribute to marginal further improvement. Therefore, there is no need to execute a large number of

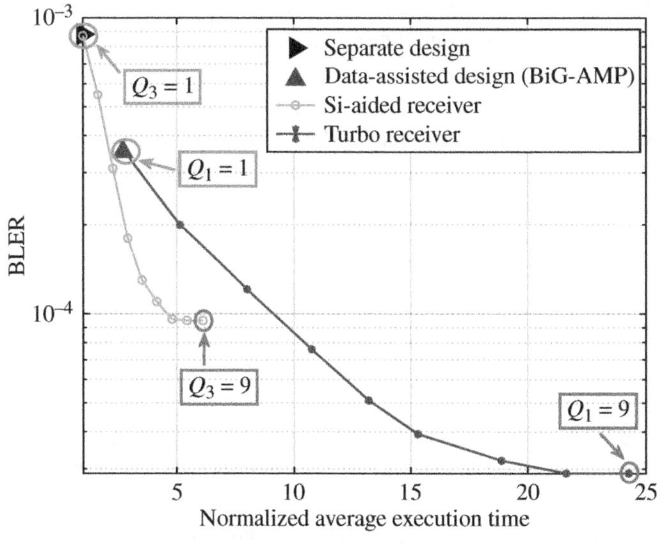

Figure 14.9 BLER versus the normalized average execution time ($K = 20$). Source: Bian et al. (2023) with IEEE permission.

iterations, and wise choices of hyper-parameters for the proposed receivers are critical to balance the performance gain and the computation cost.

14.6 Summary

This chapter investigated the joint design of user activity detection, channel estimation, and data decoding for grant-free massive RA. We leveraged the common sparsity pattern in the received pilot and data signal, together with the data decoding results for performance enhancement. A turbo receiver was first proposed under the BiG-AMP framework, which is further enhanced by the extrinsic information from the channel decoder. To reduce the complexity, we also developed a low-cost SI-aided receiver, where the SI is updated iteratively to take advantages of the common sparsity pattern and the channel decoding results. Simulation results demonstrated that substantial performance gains can be obtained with advanced receivers for a given uplink massive RA protocol. We refer the interested readers to Bian et al. (2023) for detailed derivations and proofs, complexity analysis, as well as more numerical evaluations of the proposed receivers.

References

C Berrou and A Glavieux. Near optimum error correcting coding and decoding: Turbo-codes. *IEEE Transactions on Communications*, 44:1261–1271, 1996. doi: 10.1109/26.539767.

X Bian, Y Mao, and J Zhang. Supporting more active users for massive access via data-assisted activity detection. In *Proceedings of IEEE International Conference on Communications (ICC)*, 2021. doi: 10.1109/ICC42927.2021.9500797.

X Bian, Y Mao, and J Zhang. Joint activity detection, channel estimation, and data decoding for grant-free massive random access. *IEEE Internet of Things Journal*, 2023. doi: 10.1109/JIOT.2023.3243947.

C Bockelmann, N Pratas, H Nikopour, K Au, T Svensson, C Stefanovic, P Popovski, and A Dekorsy. Massive machine-type communications in 5G: Physical and mac-layer solutions. *IEEE Communications Magazine*, 54:59–65, 2016. doi: 10.1109/MCOM.2016.7565189.

Z Chen, F Sohrabi, and W Yu. Sparse activity detection for massive connectivity. *IEEE Transactions on Signal Processing*, 66:1890–1904, 2018. doi: 10.1109/TSP.2018.2795540.

X Chen, D Ng, W Yu, E G Larsson, N Al-Dhahir, and R Schober. Massive access for 5G and beyond. *IEEE Journal on Selected Areas in Communications*, 39:615–637, 2021. doi: 10.1109/JSAC.2020.3019724.

D L Donoho, A Maleki, and A Montanari. Message passing algorithms for compressed sensing: I. Motivation and construction. In *Proceedings of IEEE Information Theory Workshop (ITW)*, 2010. doi: 10.1109/ITWKSPS.2010.5503193.

Y Du, B Dong, W Zhu, P Gao, Z Chen, X Wang, and J Fang. Joint channel estimation and multiuser detection for uplink grant-free NOMA. *IEEE Wireless Communications Letters*, 7:682–685, 2018. doi: 10.1109/LWC.2018.2810278.

R G Gallager. Low density parity check codes. *IRE Transactions on Information Theory*, 8:21–28, 1962. doi: 10.1109/TIT.1962.1057683.

B Goektepe, S Faehse, L Thiele, T Schierl, and C Hellge. Subcode-based early HARQ for 5G. In *Proceedings of IEEE International Conference on Communications (ICC)*, 2018. doi: 10.1109/ICCW.2018.8403491.

J Hagenauer, E Offer, and L Papke. Iterative decoding of binary block and convolutional codes. *IEEE Transactions on Information Theory*, 42:429–445, 1996. doi: 10.1109/18.485714.

M Hasan, E Hossain, and D Niyato. Random access for machine-to-machine communication in LTE-advanced networks: Issues and approaches. *IEEE Communications Magazine*, 51:86–93, 2013. doi: 10.1109/MCOM.2013.6525600.

S Haykin, M Sellathurai, Y de Jong, and T Willink. Turbo-MIMO for wireless communications. *IEEE Communications Magazine*, 42:48–53, 2004. doi: 10.1109/MCOM.2004.1341260.

D E Hocevar. A reduced complexity decoder architecture via layered decoding of LDPC codes. In *Proceedings of IEEE Workshop on Signal Processing Systems (SIPS)*, 2004. doi: 10.1109/SIPS.2004.1363033.

B M Hochwald and S Ten Brink. Achieving near-capacity on a multiple-antenna channel. *IEEE Transactions on Communications*, 51:389–399, 2003. doi: 10.1109/TCOMM.2003.809789.

Steven Kay. *Fundamentals of Statistical Signal Processing: Estimation Theory*. Englewood Cliffs, NJ: Prentice Hall, 1993.

M Ke, Z Gao, Y Wu, X Gao, and R Schober. Compressive sensing based adaptive active user detection and channel estimation: Massive access meets massive MIMO. *IEEE Transactions on Signal Processing*, 68:764–779, 2020. doi: 10.1109/TSP.2020.2967175.

F R Kschischang, B J Frey, and H A Loeliger. Factor graphs and the sum-product algorithm. *IEEE Transactions on Information Theory*, 47:498–519, 2001. doi: 10.1109/18.910572.

L Liu, E G Larsson, W Yu, P Popovski, C Stefanović, and E Carvalh. Sparse signal processing for grant-free massive connectivity: A future paradigm for random access protocols in the Internet of Things. *IEEE Signal Processing Magazine*, 35:88–99, 2018. doi: 10.1109/MSP.2018.2844952.

Y Liu, S Zhang, X Mu, Z Ding, R Schober, N Al-Dhahir, E Hossain, and X Shen. Evolution of NOMA toward next generation multiple access (NGMA). *IEEE*

Journal on Selected Areas in Communications, 40:1037–1071, 2022. doi: 10.1109/JSAC.2022.3145234.

A Ma, Y Zhou, C Rush, D Baron, and D Needell. An approximate message passing framework for side information. *IEEE Transactions on Signal Processing*, 67:1875–1888, 2019. doi: 10.1109/TSP.2019.2899286.

J T Parker, P Schniter, and V Cevher. Bilinear generalized approximate message passing –Part I: Derivation. *IEEE Transactions on Signal Processing*, 62:5839–5853, 2014. doi: 10.1109/TSP.2014.2357776.

B Vucetic and J Yuan. *Turbo Codes: Principles and Applications*. New York: Springer, 2001.

B Wang, L Dai, Y Zhang, T Mir, and J Li. Dynamic compressive sensing-based multi-user detection for uplink grant-free NOMA. *IEEE Communications Letters*, 20:2320–2323, 2016. doi: 10.1109/LCOMM.2016.2602264.

X Wautelet, A Dejonghe, and L Vandendorpe. MMSE-based fractional turbo receiver for space-time BICM over frequency-selective MIMO fading channels. *IEEE Transactions on Signal Processing*, 52:1804–1809, 2004. doi: 10.1109/TSP.2004.827198.

C Wei, H Liu, Z Zhang, J Dang, and L Wu. Approximate message passing-based joint user activity and data detection for NOMA. *IEEE Communications Letters*, 21:640–643, 2017. doi: 10.1109/LCOMM.2016.2624297.

K Wong, A Paulraj, and R Murch. Efficient high-performance decoding for overloaded MIMO antenna systems. *IEEE Transactions on Wireless Communications*, 6:1833–1843, 2007. doi: 10.1109/TWC.2007.360385.

Q Zou, H Zhang, D Cai, and H Yang. A low-complexity joint user activity, channel and data estimation for grant-free massive MIMO systems. *IEEE Signal Processing Letters*, 27:1290–1294, 2020. doi: 10.1109/LSP.2020.3008550.

15

Joint User Activity Detection, Channel Estimation, and Signal Detection for Grant-Free Massive Connectivity

Zhichao Shao[1], Shuchao Jiang[2], Chongbin Xu[2], Xiaojun Yuan[1], and Xin Wang[2]

[1]*National Key Laboratory of Wireless Communications, University of Electronic Science and Technology of China, Chengdu, China*
[2]*Key Laboratory for Information Science of Electromagnetic Waves (MoE), Department of Communication Science and Engineering, Fudan University, Shanghai, China*

15.1 Introduction

The explosive development of Internet of Things brings a serious challenge to the design of new random access mechanisms to accommodate massive device connections with an affordable system overhead (Hasan et al., 2013, Ghavimi and Chen, 2015, Chen et al., 2021). Extensive research effort has recently been devoted to addressing this massive connectivity problem (Ding et al., 2017, Liu et al., 2017, 2018, Dai et al., 2015, Wang et al., 2016, Wei et al., 2017, Liu and Yu, 2018, Du et al., 2018). Particularly, grant-free non-orthogonal multiple access (GF-NOMA) was proposed (Dai et al., 2015, Liu et al., 2018) to reduce the signaling overhead and enhance the access capability of massive connectivity, where a huge number of users with sporadic activity patterns are allowed to communicate with an access point (AP) non-exclusively without requiring a scheduling grant. Under this grant-free and non-orthogonal random access mechanism, the receiver at the AP needs to identify active users, estimate their channels, and detect their data as well.

Several works on GF-NOMA assumed perfect symbol-level and frame-level synchrony in sporadic transmission. Specifically, under the assumption of perfect channel state information (CSI) at the receiver (CSIR), joint active user identification and signal detection algorithms were developed Wang et al. (2016) and Wei et al. (2017). For systems without CSIR, (Liu and Yu, 2018, Hannak et al., 2015) established joint channel estimation and active user identification algorithms, followed by separated signal detection. In addition, joint channel and data estimation algorithms were developed for massive multiple-input

Next Generation Multiple Access, First Edition.
Edited by Yuanwei Liu, Liang Liu, Zhiguo Ding, and Xuemin Shen.
© 2024 The Institute of Electrical and Electronics Engineers, Inc. Published 2024 by John Wiley & Sons, Inc.

multiple-output (MIMO) systems (Zhang et al., 2018, Ding et al., 2019) and for single carrier systems (Sun et al., 2018). Recently, Du et al. (2018) proposed a joint channel estimation and multiuser detection algorithm, named block sparsity adaptive subspace pursuit (BSASP). This algorithm transfers the single measurement-vector compressive sensing (SMV-CS) problem to multiple measurement-vector compressive sensing (MMV-CS), and reconstructs the sparse signal by exploiting the inherent block sparsity of the channel. BSASP generally suffers from the non-orthogonality of the spreading matrix. This issue becomes more serious in massive connectivity case, since the non-orthogonal spreading sequence is usually inevitable in order to accommodate as many potential users as possible. In Wei et al. (2019), a message-passing-based joint channel estimation and data decoding algorithm was proposed for grant-free sparse code multiple access systems, where the messages were approximated by Gaussian distributions with minimized Kullback–Leibler divergence.

However, achieving and maintaining synchronization requires frequent coordination between the AP and all the potential users, which poses a heavy burden especially for low-cost devices. Allowing asynchronous transmissions at the user ends can relieve this burden, but the problem is how to handle this asynchrony at the receiver. A more attractive solution to the above problem is to design advanced signal processing techniques to handle the effect of asynchronous reception at the receiver. For MIMO systems, pioneering works in this direction can be found Ding et al. (2019), Liu and Liu (2021), Zhu et al. (2021), and Di Renna and de Lamare (2021). In particular, the authors in Ding et al. (2019) proposed an asynchronous massive connectivity scheme, in which users send short packets to a multi-antenna AP in a sporadic manner without frame/packet synchronization. To detect the packets from the received signals, a sliding window method was introduced and a turbo bilinear generalized approximate message passing (BiGAMP) algorithm was developed by exploiting the signal sparsity inherent in the sporadic arrivals of random short packets. Under the same frame-asynchrony assumption, the authors in Liu and Liu (2021) considered to detect the active users and estimate their delays and channels based on the preambles transmitted by active users. They cast such a task into a group least absolute shrinkage and selection operator (LASSO) problem and utilized a block coordinate descent algorithm to solve the problem. Additionally, the authors in Zhu et al. (2021) formulated the receiver design in the pilot-phase transmission as a sparse signal recovery problem with hierarchical sparsity including the user-level sparsity and the delay-level sparsity inherent in asynchronous massive connectivity. A learned approximate message passing (LAMP) scheme that combines deep learning and the approximate message passing (AMP) algorithm was then developed for joint user activity identification, delay detection, and channel estimation. Recently, the authors in Di Renna and de Lamare (2021) developed a message-scheduling generalized approximate message passing (GAMP) algorithm to achieve joint active device detection and channel estimation for asynchronous massive connectivity based on the pilot transmission. We emphasize that all these schemes (Ding et al., 2019, Liu and Liu, 2021, Zhu et al., 2021, Di Renna and de Lamare, 2021) still assumed perfect symbol-level synchrony in transmissions, which is difficult to achieve in an asynchronous system.

In this chapter, we develop receiver designs for both synchronous and asynchronous massive connectivity. In synchronous system, a joint user identification, channel estimation, and signal detection (JUICESD) algorithm based on message passing principles is proposed, where the joint detection problem can be represented as a complicated trilinear recovery problem. In asynchronous system, a turbo approximate message passing (TAMP) algorithm is proposed, where it resorts to sophisticated applications of message passing principles to iteratively identify active users, determine the time delays, estimate the channels, and detect the data symbols.

15.2 Receiver Design for Synchronous Massive Connectivity

In this section, all users are synchronized in frames, which can be achieved by the AP sending a beacon signal to initialize uplink transmissions.

15.2.1 System Model

Consider a typical massive connectivity scenario, in which a large number of single-antenna users with sporadic traffic communicate with a single-antenna AP. Based on the received signals, the AP is responsible for judging which users are active, estimating the channels of the active users, and recovering the signals transmitted by the active users.

15.2.1.1 Synchronous Uplink Transmission

We follow the spreading based NOMA schemes in Du et al. (2018) for grant-free transmissions. Specifically, each user k is assigned with a unique spreading sequence $\boldsymbol{a}_k = [a_{1,k}, \dots, a_{L,k}]^T$ as its signature, where L is the spreading length. For massive connectivity case, L can be much less than the total number of users K, and therefore orthogonal spreading sequence design is generally impossible. Here, we assume that the elements of each $\boldsymbol{a}_k$ are randomly and independently drawn from the Gaussian distribution $\mathcal{N}(0, 1/L)$.

Consider the transmission in a frame of T slots, where each slot consists of L transmission symbols corresponding to the length of the spreading sequence. The received signal can be modeled as (Du et al., 2018)

$$\boldsymbol{r}_t = \sum_{k=1}^{K} \boldsymbol{a}_k h_k u_k x_{k,t} + \boldsymbol{w}_t, \quad t = 1, \dots, T, \tag{15.1}$$

where K is the total number of users, h_k is the channel coefficient from user k to the AP, u_k is an indicator to represent the activity state of user k (with $u_k = 1$ meaning that user k is active and $u_k = 0$ otherwise), $x_{k,t}$ is the transmit signal of user k at time slot t, and each entry of $\boldsymbol{w}_t$ is the complex additive white Gaussian noise (AWGN) with mean zero and variance N_0. The user symbols $\{x_{k,t}\}$ are modulated by using a common signal constellation $\mathcal{S}$ with cardinality $|\mathcal{S}|$, i.e., each $x_{k,t}$ is independently and uniformly distributed over the signal constellation $\mathcal{S}$, where $p_{\mathcal{S}}(s) = \frac{1}{|\mathcal{S}|}\sum_{j=1}^{|\mathcal{S}|} \delta(s - s_j)$ with $\sum_{j=1}^{|\mathcal{S}|} \| s_j \|^2 / |\mathcal{S}| = 1$.

Block fading is assumed, i.e., h_k and u_k remain unchanged within each transmission frame. The channel coefficient h_k is modeled as $h_k = \sqrt{\beta_k}\alpha_k$, $\forall k$, where $\alpha_k \sim \mathcal{CN}(0, 1)$ denotes the Rayleigh fading component, and β_k denotes large scale fading component including path-loss and shadowing. Then, we have $h_k \sim p_{H_k}(h_k) = \mathcal{CN}(0, \beta_k)$, implying that the channels of different users are not necessarily identically distributed. Moreover, the activity indicator u_k is drawn from the Bernoulli distribution $p_u(u_k) = (1 - \lambda)\delta(u_k) + \lambda\delta(u_k - 1)$.

15.2.1.2 Problem Formulation

The received signal (15.1) is rewritten as

$$\boldsymbol{R} = [\boldsymbol{r}_1, \dots, \boldsymbol{r}_T] = \boldsymbol{AHUX} + \boldsymbol{W}, \tag{15.2}$$

where $\boldsymbol{A} = [\boldsymbol{a}_1, \dots, \boldsymbol{a}_K] \in \mathbb{R}^{L\times K}$, $\boldsymbol{H} = \text{diag}([h_1, \dots, h_K]^T) \in \mathbb{C}^{K\times K}$, $\boldsymbol{U} = \text{diag}([u_1, \dots, u_K]^T) \in \{0, 1\}^{K\times K}$, $\boldsymbol{X} = [x_{k,t}] \in \mathcal{S}^{K\times T}$, and $\boldsymbol{W} = [\boldsymbol{w}_1, \dots, \boldsymbol{w}_T] \in \mathbb{C}^{L\times T}$. For matrices and vectors, $(\cdot)^T$ denotes transpose, diag($\boldsymbol{b}$) is the diagonal matrix with the diagonal elements specified by $\boldsymbol{b}$.

Suppose that the AP jointly estimates ($\boldsymbol{H}$, $\boldsymbol{U}$, $\boldsymbol{X}$) by following the maximum *a posteriori* probability (MAP) principle. Conditioned on $\boldsymbol{R}$, the MAP estimate of ($\boldsymbol{H}$, $\boldsymbol{U}$, $\boldsymbol{X}$) is given by

$$\begin{aligned}
\left(\hat{\boldsymbol{H}}, \hat{\boldsymbol{U}}, \hat{\boldsymbol{X}}\right) &= \arg \max_{(\boldsymbol{H},\boldsymbol{U},\boldsymbol{X})} p_{H,U,X|R}(\boldsymbol{H}, \boldsymbol{U}, \boldsymbol{X}|\boldsymbol{R}) \\
&= \arg \max_{(\boldsymbol{H},\boldsymbol{U},\boldsymbol{X})} \exp\left(-\frac{\| \boldsymbol{R} - \boldsymbol{AHUX} \|_F^2}{N_0}\right) p_H(\boldsymbol{H}) p_U(\boldsymbol{U}) p_X(\boldsymbol{X}),
\end{aligned} \tag{15.3}$$

Problem (15.3) is nonconvex and generally difficult to solve. Message passing algorithms could provide possible solutions. However, the variables to be estimated in (15.3), i.e., $\boldsymbol{H}$, $\boldsymbol{U}$, and $\boldsymbol{X}$, are all coupled to form a trilinear function. Exact message passing based on the sum-product rule is too complicated to implement, while the existing low complexity AMP-type algorithms (Rangan, 2012, Rangan et al., 2016, Parker et al., 2014a,2014b) cannot be applied to the trilinear model in (15.2) directly. To bypass the dilemma, we next develop a low-complexity yet efficient iterative algorithm to solve the problem by a judicious design of the receiver structure and the message updates.

15.2.2 Proposed JUICESD Algorithm

To facilitate a low-complexity yet efficient solution to the joint detection problem in (15.3), we divide the whole detection scheme into two modules by introducing appropriate auxiliary variables, based on which we develop the proposed JUICESD algorithm.

15.2.2.1 JUICESD Algorithm Structure

Introduce the following auxiliary variables:

$$g_k \doteq h_k u_k, \quad \forall k, \tag{15.4}$$

$$y_{k,t} \doteq g_k x_{k,t}, \quad \forall k, t. \tag{15.5}$$

We refer to g_k and $y_{k,t}$ as the effective channel of user k and the effective signal of user k at time slot t, respectively. Based on the *a priori* distributions of h_k, u_k, and $x_{k,t}$, we obtain the *a priori* distributions of g_k and $y_{k,t}$ from (15.4) and (15.5) respectively as

$$p_{G_k}(g_k) = (1-\lambda)\delta(g_k) + \lambda p_{H_k}(g_k), \tag{15.6}$$

$$p_{y_{k,t}}(y_{k,t}) = (1-\lambda)\delta(y_{k,t}) + \frac{\lambda}{|S|}\sum_{j=1}^{|S|} p_{H_k}(y_{k,t}/s_j). \tag{15.7}$$

Rewrite the system model in (15.1) as

$$\boldsymbol{r}_t = \sum_{k=1}^{K} \boldsymbol{a}_k y_{k,t} + \boldsymbol{w}_t, \quad t = 1, \dots, T. \tag{15.8}$$

Clearly, the signal model (15.8) is linear in the auxiliary variables $\{y_{k,t}\}$ since $\{\boldsymbol{a}_k\}$ are known to the receiver. In addition, given $\{y_{k,t}\}$, the estimations of $\{g_k\}$ and $\{x_{k,t}\}$ are nonlinear yet decoupled for different k. These two properties are of importance to our algorithm design.

With (15.4)–(15.8), we can represent the system model by the factor graph in Fig. 15.1. The factor graph consists of two types of nodes:

- Variable nodes $\{g_k\}$,$\{x_{k,t}\}$, $\{y_{k,t}\}$, and $\{r_{l,t}\}$, depicted as white circles in Fig. 15.1, corresponding to the variables in (15.5) and (15.8), where $r_{l,t}$ is the lth entry of $\boldsymbol{r}_t$;

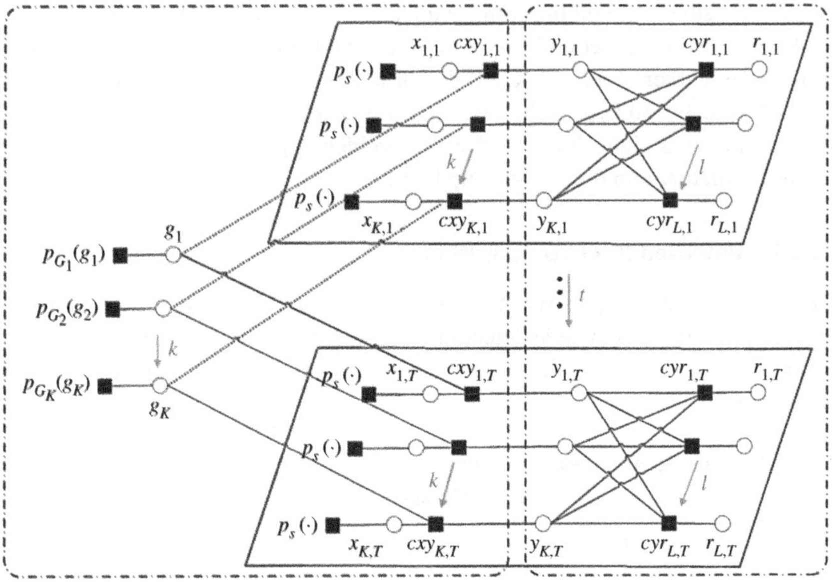

Figure 15.1 Factor graph representation of the considered synchronous system.

- Check nodes $\{p_{G_k}(g_k)\}$, $p_S(\cdot)$, $\{cxy_{k,t}\}$, and $\{cyr_{l,t}\}$, depicted as black boxes in Fig. 15.1, corresponding to the marginal *a priori* distributions of $\{g_k\}$, the marginal *a priori* distributions of $\{x_{k,t}\}$, the equality constraints in (15.5), and the equality constraints in (15.8), respectively.

A variable node is connected to a check node when the variable is involved in the check constraint.

As shown in Fig. 15.1, we divide the whole receiver structure into two modules, one for the multiuser signal model in (15.8) and the other for the nonlinear equality constraint in (15.5). We next outline their main functionalities.

The module on the right hand side of Fig. 15.1 focuses on the multiuser signal model in (15.8). With the messages of $\{y_{k,t}\}$ fed back from the other module as the *a priori*, the estimation of $\{y_{k,t}\}$ given $\boldsymbol{r}_t$ in (15.8) can be performed slot-by-slot. Hence, we refer to this module as slot-wise multiuser detection (SMD). The refined estimates of $\{y_{k,t}\}$ are forwarded to the other module. The detailed operations will be described later in Section 15.2.2.2.

The module on the left hand side of Fig. 15.1 is for the nonlinear constraints in (15.5). With the estimates of $\{y_{k,t}\}$ as input, the effective channels $\{g_k\}$ and the effective signals $\{y_{k,t}\}$ can be refined based on (15.5) by noticing that the effective channel of the same user remains unchanged in one frame. Thus, we refer to this module as combined signal and channel estimation (CSCE). The output of the CSCE is forwarded to the SMD module for further processing. The detailed operations of the CSCE will be specified in Section 15.2.2.3.

The above two modules are executed iteratively until convergence.

15.2.2.2 SMD Operation

The SMD module is to estimate $\{y_{k,t}\}$ based on the received signal model (15.8) and the messages of $\{y_{k,t}\}$ from the CSCE module. We next outline the SMD operation in each time slot t by following the GAMP algorithms (Rangan, 2012, Rangan et al., 2016). The detailed derivations are omitted for brevity.

(1) *Initialization*: Denote by $\{m_{cxy_{k,t}\to y_{k,t}}(y_{k,t})\}$ the messages of $\{y_{k,t}\}$ fed back from CSCE. (Specifically, they are passed from the check nodes $\{cxy_{k,t}\}$ as detailed in Section 15.2.2.3.) With no feedback from CSCE at the beginning, each $m_{cxy_{k,t}\to y_{k,t}}(y_{k,t})$ is initialized to $p_{y_{k,t}}(y_{k,t})$ in (15.7). The means $\{\hat{y}_{k,t}\}$ and variances $\{v_{y_{k,t}}\}$ of $\{y_{k,t}\}$ are calculated at variable nodes $\{y_{k,t}\}$.

(2) *Message update at check nodes* $\{cyr_{l,t}\}$: Based on the linear model $z_{l,t} = \sum_{k=1}^{K} a_{l,k}y_{k,t}$, the messages of $\{y_{k,t}\}$ are cumulated to obtain an estimate of $\{z_{l,t}\}$. With the "Onsager" correction applied, the messages of $\{z_{l,t}\}$ in the form of means $\{\hat{p}_{l,t}\}$ and variances $\{v_{p_{l,t}}\}$ are calculated as (Rangan, 2012)

$$v_{p_{l,t}} = \sum_{k=1}^{K} |a_{l,k}|^2 v_{y_{k,t}}, \quad \forall l, \tag{15.9}$$

$$\hat{p}_{l,t} = \sum_{k=1}^{K} a_{l,k}\hat{y}_{k,t} - v_{p_{l,t}}\hat{s}_{l,t}, \quad \forall l, \tag{15.10}$$

where initially we set $\hat{s}_{l,t} = 0$ for $\forall l$. Then, the means $\{\hat{z}_{l,t}\}$ and the variances $\{v_{z_{l,t}}\}$ are computed by using the observations $\{r_{l,t}\}$ as

$$v_{z_{l,t}} = \mathrm{Var}\{z_{l,t}|\hat{p}_{l,t}, v_{p_{l,t}}, r_{l,t}\}, \quad \forall l, \tag{15.11}$$

$$\hat{z}_{l,t} = \mathrm{E}\{z_{l,t}|\hat{p}_{l,t}, v_{p_{l,t}}, r_{l,t}\}, \quad \forall l, \tag{15.12}$$

where the mean $\mathrm{E}\{\cdot\}$ and variance $\mathrm{Var}\{\cdot\}$ operations are taken with respect to the *a posteriori* distribution of $z_{l,t}$ given the *a priori* distribution $z_{l,t} \sim \mathcal{CN}(\hat{p}_{l,t}, v_{p_{l,t}})$ and the observation $r_{l,t} = z_{l,t} + w_{l,t}$ with $w_{l,t}$ being the lth entry of $\boldsymbol{w}_t$. Lastly, the residual $\{\hat{s}_{l,t}\}$ and the inverse-residual-variances $\{v_{s_{l,t}}\}$ are computed by

$$v_{s_{l,t}} = \left(1 - v_{z_{l,t}}/v_{p_{l,t}}\right)/v_{p_{l,t}} \forall l, \tag{15.13}$$

$$\hat{s}_{l,t} = \left(\hat{z}_{l,t} - \hat{p}_{l,t}\right) / v_{p_{l,t}} \, \forall l. \tag{15.14}$$

(3) *Message update at variable nodes* $\{y_{k,t}\}$: With the residual $\{\hat{s}_{l,t}\}$ and inverse-residual-variances $\{v_{s_{l,t}}\}$, the messages of $\{y_{k,t}\}$ are computed in the form of means $\{\hat{r}_{k,t}\}$ and variances $\{v_{r_{k,t}}\}$ as

$$v_{r_{k,t}} = \left(\sum_{l=1}^{L} |a_{l,k}|^2 v_{s_{l,t}}\right)^{-1}, \quad \forall k, \tag{15.15}$$

$$\hat{r}_{k,t} = \hat{y}_{k,t} + v_{r_{k,t}} \sum_{l=1}^{L} a_{l,k} \hat{s}_{l,t}, \quad \forall k. \tag{15.16}$$

Then, the means and variances of $\{y_{k,t}\}$ are updated by

$$v_{y_{k,t}} = \text{Var}\{y_{k,t} | m_{cxy_{k,t} \to y_{k,t}}(y_{k,t}), \hat{r}_{k,t}, v_{r_{k,t}}\}, \quad \forall k, \tag{15.17}$$

$$\hat{y}_{k,t} = \text{E}\{y_{k,t} | m_{cxy_{k,t} \to y_{k,t}}(y_{k,t}), \hat{r}_{k,t}, v_{r_{k,t}}\}, \quad \forall k, \tag{15.18}$$

where the mean $\text{E}\{\cdot\}$ and variance $\text{Var}\{\cdot\}$ operations are taken with respect to the *a posteriori* distribution of $y_{k,t}$ given its *a priori* distribution $m_{cxy_{k,t} \to y_{k,t}}(y_{k,t})$ and the feedback message $y_{k,t} \sim \mathcal{CN}(\hat{r}_{k,t}, v_{r_{k,t}})$. The operations in (15.17) and (15.18) essentially give a nonlinear denoiser since they take the structure information of $y_{k,t}$ such as sparsity into account by regarding $m_{cxy_{k,t} \to y_{k,t}}(y_{k,t})$ as the *a priori* distribution. The above refined messages are used to update the messages of $\{z_{l,t}\}$ in the next iteration. The iteration continues until convergence or the maximum iteration number Q is reached. Finally, the messages of $\{y_{k,t}\}$ with mean $\{\hat{r}_{k,t}\}$ and variance $\{v_{r_{k,t}}\}$ are passed to the CSCE module.

15.2.2.3 CSCE Operation

With the messages $\{\hat{r}_{k,t}\}$ and $\{v_{r_{k,t}}\}$ from the SMD as input, the CSCE module in Fig. 15.1 deals with the nonlinear constraints in (15.5) to yield more accurate estimates of $\{y_{k,t}\}$. Since the constraints in (15.5) are decoupled for different users, we next present the message passing operations for each individual user as follows.

(1) *Messages passed from* $\{cxy_{k,t}\}$ *to* $\{g_k\}$: With the output of SMD, the message from $y_{k,t}$ to $cxy_{k,t}$ is given by a complex Gaussian distribution with mean $\hat{r}_{k,t}$ and variance $v_{r_{k,t}}$, i.e., $m_{y_{k,t} \to cxy_{k,t}}(y_{k,t}) = \mathcal{CN}(\hat{r}_{k,t}, v_{r_{k,t}})$. Then, from the sum-product rule, we obtain

$$\begin{aligned} m_{cxy_{k,t} \to g_k}(g_k) &= \int_{y_{k,t}, x_{k,t}} m_{y_{k,t} \to cxy_{k,t}}(y_{k,t}) p_S(x_{k,t}) \delta(y_{k,t} - g_k x_{k,t}) \\ &= \sum_{j=1}^{|S|} \frac{p_S(s_j)}{\pi v_{r_{k,t}} / \| s_j \|^2} \exp\left(-\frac{\| g_k - \hat{r}_{k,t}/s_j \|^2}{v_{r_{k,t}} / \| s_j \|^2}\right), \quad \forall t, \end{aligned} \tag{15.19}$$

where $p_S(x_{k,t}) = \frac{1}{|S|}\sum_{j=1}^{|S|}\delta(x_{k,t} - s_j)$ is the *a priori* distribution of the transmit signal $x_{k,t}$.

(2) *Messages passed from* $\{g_k\}$ *to* $\{cxy_{k,t}\}$: For the same channel coefficient g_k, we have T messages $m_{cxy_{k,t}\to g_k}(g_k)$, $t = 1, \ldots, T$. From the sum-product rule, we obtain

$$m_{g_k,t}(g_k) = \prod_{t'=1,t'\neq t}^{T} m_{cxy_{k,t'}\to g_k}(g_k) \tag{15.20}$$

$$= \prod_{t'=1,t'\neq t}^{T}\sum_{j=1}^{|S|}\frac{p_S(s_j)}{\pi v_{r_{k,t'}}/\parallel s_j\parallel^2}\exp\left(-\frac{\parallel g_k - \hat{r}_{k,t'}/s_j\parallel^2}{v_{r_{k,t'}}/\parallel s_j\parallel^2}\right), \quad \forall t,$$

$$m_{g_k\to cxy_{k,t}}(g_k) = m_{g_k,t}(g_k)p_{G_k}(g_k), \quad \forall t, \tag{15.21}$$

where $p_{G_k}(g_k)$ is the *a priori* distribution of the effective channel g_k given in (15.6).

(3) *Messages passed from* $\{cxy_{k,t}\}$ *to* $\{y_{k,t}\}$: With the message $m_{g_k\to cxy_{k,t}}(g_k)$ from node g_k, we obtain

$$m_{cxy_{k,t}\to y_{k,t}}(y_{k,t}) = \int_{g_k,x_{k,t}} m_{g_k\to cxy_{k,t}}(g_k)p_S(x_{k,t})\delta(y_{k,t} - g_k x_{k,t})$$

$$= \sum_{j=1}^{|S|} p_S(s_j)m_{g_k\to cxy_{k,t}}(y_{k,t}/s_j), \quad \forall t. \tag{15.22}$$

These messages are passed to the variable nodes $\{y_{k,t}\}$ to activate the next SMD operation. The SMD and CSCE operations iterate until convergence.

15.2.2.4 Overall Algorithm and Complexity Analysis

Based on the above described procedures, we outline the main steps of the proposed JUICESD algorithm in Algorithm 15.1, where Q is the maximum number of iterations allowed in the SMD module, and Q' is the maximum number of iterations between the SMD and CSCE modules.

The complexity of the JUICESD algorithm is described as follows. The algorithm consists of two modules: SMD and CSCE. The SMD operation is based on slot-wise GAMP; hence, its complexity is $\mathcal{O}(KLT)$. The complexity of the CSCE operation is dominated by (15.20). Recall that (15.20) is the product of $(T-1)$ $|S|$-component Gaussian mixtures. A direct evaluation of (15.20) results in a complexity of $\mathcal{O}(KT|S|^{T-1})$ for the CSCE. Consequently, the total complexity of JUICESD is $\mathcal{O}(KLT + KT|S|^{T-1})$.

15.2.3 Numerical Results

Numerical results are provided to verify the effectiveness of the proposed algorithms. The signal-to-noise ratio (SNR) is defined by $\text{SNR} = \frac{1}{N_0}$. The data

Algorithm 15.1 JUICESD Algorithm

Input: Received signal R, signature matrix A, signal distribution $p_S(s)$, channel distributions $\{p_{H_k}(\cdot)\}$, user activity probability λ.

for $q' = 1, 2, \dots, Q'$

  // SMD module

  Initialization: $\hat{y}_{k,t} = \mathrm{E}\{y_{k,t}\}, v_{y_{k,t}} = \mathrm{Var}\{y_{k,t}\}, \hat{s}_{l,t}^{(0)} = 0;$

  for $q = 1, 2, \dots, Q$

    Calculate $v_{z_{l,t}}$ and $\hat{z}_{l,t}$, $\forall l, t$, based on (15.9)–(15.12);

    Calculate $v_{y_{k,t}}$ and $\hat{y}_{k,t}$, $\forall k, t$, based on (15.13)–(15.18);

  end

  // CSCE module

  Calculate the message of g_k in each time slot using (15.19);

  Calculate the combined message of g_k using (15.20) and (15.21);

  Calculate the refined messages of $\{y_{k,t}\}$ using (15.22);

end

Output: Obtain the final estimates $\{\hat{h}_k = \hat{y}_{k,1}/s_p\}$, $\{\hat{x}_{k,t} = \hat{y}_{k,t}/\hat{h}_k\}$ and $\{\hat{u}_k = 1\}$ when the corresponding magnitude of $\{\hat{g}_k\}$ is larger than a predetermined threshold according to (15.4) and (15.5).

signals of all active users are quadrature phase shift keying (QPSK) modulated. In simulations, we mainly focus on the case that perfect power control is adopted to compensate for large scale fading such that $\beta_1 = \beta_2 = \cdots = \beta_K \equiv 1$.

We first define the activity error rate (AER) and the symbol error rate (SER) in the considered systems. For each frame, if the activities of all users are correct, the activity error is 0; otherwise, the activity error is 1. The AER is then obtained as the average of the activity errors over all simulated frames. The SER is calculated as follows. For inactive users, if its active state is judged correctly, all symbols are regarded as detected correctly; otherwise all incorrectly. For active users, we estimate their transmitted QPSK signals. A symbol is detected correctly only when both the user activity and the user symbol are judged correctly.

In Fig. 15.2, we compare the AER and SER performance of our proposed algorithms with the state-of-the-art BSASP algorithm in Du et al. (2018). For comparison, we also include the performance of the two-phase detection scheme (Hannak et al., 2015) in which the receiver first estimates the user activities and channel coefficients jointly, and then recovers the signals sent by the active users, both through AMP algorithms. From Fig. 15.2, we see that the two-phase detection scheme has a relatively high error floor while the BSASP algorithm works well. This demonstrates the benefit of JUICESD design. Furthermore, the proposed JUICESD algorithm outperforms BSASP by about 3–5 dB in terms of AER and SER. This is because BSASP suffers from non-orthogonal spreading

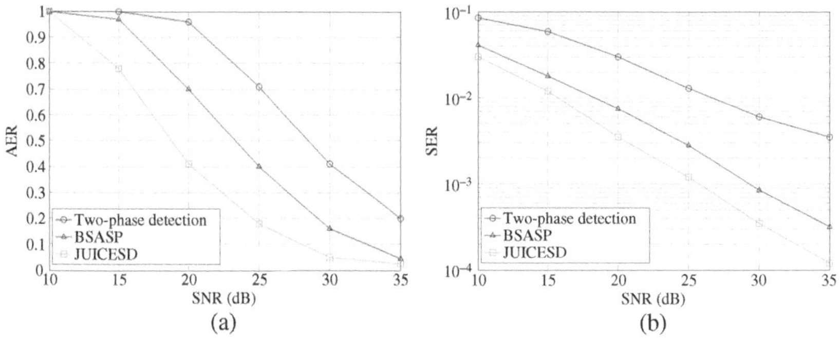

Figure 15.2 Performance versus SNR: $K = 200$, $L = 50$, $\lambda = 0.1$, $T = 7$. (a) AER versus SNR. (b) SER versus SNR.

matrix, while the proposed JUICESD algorithm relies on message passing principles to iteratively cancel/suppress the interference among users.

We next apply the two-phase detection scheme, and JUICESD to a much larger system with a relative higher computational cost. We set the number of potential users $K = 2000$, the length of spreading sequence $L = 500$. We use the OracleActivity-linear minimum mean squared error (LMMSE) (in which user activity information is assumed to be perfectly known while both channel and data are estimated based on LMMSE principles) and the OracleCSIR-AMP in Wei et al. (2017) (in which perfect CSIR is assumed and the user activity and data are jointly detected by using AMP) as baselines. Figure 15.3 shows the SER performance of the considered algorithms, where the proposed JUICESD algorithm outperforms the two-phase detection scheme. In addition, JUICESD almost approaches to OracleCSIR-AMP especially when the user activity probability λ is small. This implies that JUICESD can estimate the channel accurately through joint detection.

15.2.4 Summary

We proposed a novel JUICESD framework for the scenario of synchronous massive connectivity. Numerical results demonstrate that JUICESD achieves a significant performance gain over the state-of-the-art algorithms. This section is focused only on the single-antenna configuration at the receiver. Multi-antenna configuration will be an interesting extension to support more users and/or higher transmission rates. To exploit the potential structural information of the multi-antenna channels, such as sparsity in the angular domain, more advanced designs can be involved.

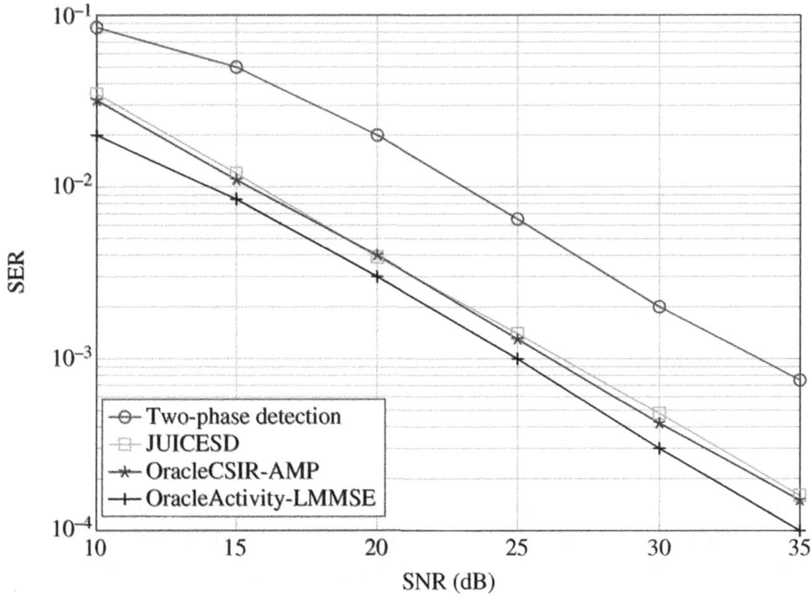

Figure 15.3 SER versus SNR: $K = 2000$, $L = 500$, $\lambda = 0.1$, $T = 7$.

15.3 Receiver Design for Asynchronous Massive Connectivity

In this section, all users transmit data to the AP at any time without the procedure of symbol-level or frame-level synchronization.

15.3.1 System Model

Consider a typical massive connectivity scenario, in which K single-antenna users with sporadic traffic distributed in a large area communicate with an M-antenna AP. Based on the received signals, the AP is responsible for judging which users are active, estimating the delays and channels of the active users, and recovering the signals transmitted by the active users.

15.3.1.1 Asynchronous Uplink Transmission

Let N denote the packet length (in symbols) of each active user, and $s_{k,n}$ the nth symbol transmitted by user k. Then the baseband transmit signal of user k can be expressed as

$$x_k(t) = \sum_{n=1}^{N} s_{k,n} g_T \left(t - nT_s\right), \tag{15.23}$$

where $s_{k,n}$ is chosen from a common signal constellation $\mathcal{S}$, $g_T(\cdot)$ is the normalized transmit shaping filter, and T_s is the symbol interval. We assume the data packets of each user are generated according to a Poisson point process with rate λ.

Suppose that the signal is transmitted over a flat fading channel. Let the path delay between the mth antenna on AP and user k be $\tau_{m,k}$, and the related amplitude gain as $a_{m,k}$. Then the channel impulse response between the mth antenna and user k $h_{m,k}(t)$ is $a_{m,k}\delta\left(t-\tau_{m,k}\right)$, where $a_{m,k} \sim \mathcal{CN}(0,\tilde{\beta}_k)$, and $\tilde{\beta}_k$ is the power gain of large-scale fading for user k. We assume that $\tau_{1,k}=\tau_{2,k}=\cdots=\tau_{M,k}\doteq\overline{\tau}_k$ since the size of antenna array is usually small.

With asynchronous transmission, the relative delay between the transmission time of user k and the reference clock at the AP is $\tilde{\tau}_k$. The received baseband signal at the mth antenna of AP can be expressed as

$$y_m(t)=\sum_{k=1}^{K} e^{-j2\pi f_c\tau_k} h_{m,k}(t) * x_k(t-\tilde{\tau}_k)+w_m(t), \tag{15.24}$$

where f_c is the carrier frequency, $\tau_k=\overline{\tau}_k+\tilde{\tau}_k$ is the effective delay between user k and the receiver, and $w_m(t)$ is the AWGN at the mth antenna with zero mean and two-sided power spectral density $N_0/2$.

Substituting (15.23) into (15.24), we obtain

$$y_m(t)=\sum_{k=1}^{K} h_{m,k}\sum_{n=1}^{N} s_{k,n}g_T\left(t-\tau_k-nT_s\right)+w_m(t), \tag{15.25}$$

where $h_{m,k}=a_{m,k}e^{-j2\pi f_c\tau_k}$ is the baseband equivalent channel coefficient. Under a block fading assumption, $h_{m,k}$ remains unchanged within each transmission packet.

Let $g_R(t)=g_T^*(t_0-t)$ be the impulse response of the matched filter at the receiver. The signal after matched filtering is

$$y'_m(t)=\sum_{k=1}^{K} h_{m,k}\sum_{n=1}^{N} s_{k,n}g\left(t-\tau_k-nT_s\right)+w'_m(t), \tag{15.26}$$

where $y'_m(t)=y_m(t)*g_R(t)$, $w'_m(t)=w_m(t)*g_R(t)$, and $g(t)=g_T(t)*g_R(t)$ is the impulse response of the effective shaping filter.

Without loss of generality, we assume $t_0=0$ and sample $y'_m(t)$ at lT_s with l being integer. Then the discrete received signal $y_{m,l}\triangleq y'_m(lT_s)$ can be modeled as

$$y_{m,l}=\sum_{k=1}^{K} h_{m,k}\sum_{n=1}^{N} s_{k,n}g\left((l-n)T_s-\tau_k\right)+w_{m,l}, \tag{15.27}$$

where $w_{m,l}=w'_m(lT_s)$ is the equivalent noise.

Notice that τ_k in (15.27) may not be an integer multiples of T_s. Thus, for a given n (i.e., the index of transmit symbol), $g\left((l-n)T_s-\tau_k\right)$ can be non-zero for some

l (i.e., the index of receive symbol) even if the effective shaping filter $g(t)$ meets the Nyquist criterion. This leads to inter-symbol interference (ISI) in the case of asynchronous transmission. Further notice that such ISI tap coefficients are determined by the corresponding user delays (through the effective shaping filer $g(t)$), which are continuously valued and unknown to the receiver. This ISI effect makes the receiver design of massive connectivity very challenging, as elaborated later in Section 15.3.3.

15.3.1.2 Problem Formulation

Since the received signal is frameless, we focus on a specific observation window $(0, LT_s]$. By selecting the length of the window and sliding the window properly, all transmit packets can be potentially recovered. Denote by K_a the number of active users in the window. Then (15.27) can be rewritten as

$$y_{m,l} = \sum_{k=1}^{K_a} h_{m,k} \sum_{n=1}^{N} s_{k,n} g\left((l-n)T_s - \tau_k\right) + w_{m,l}, \quad 1 \leq l \leq L. \tag{15.28}$$

Notice that in (15.28) we allow $\tau_k < 0$ to represent packets that are transmitted before the observation window. Define $g_{k,l,n}(\tau_k) \triangleq g\left((l-n)T_s - \tau_k\right)$, $\boldsymbol{g}_{k,l}(\tau_k) \triangleq [g_{k,l,1}(\tau_k), \ldots, g_{k,l,N}(\tau_k)]^T$, and

$$\boldsymbol{G}(\boldsymbol{\tau}) \triangleq \begin{pmatrix} \boldsymbol{g}_{1,1}(\tau_1) & \cdots & \boldsymbol{g}_{1,L}(\tau_1) \\ \boldsymbol{g}_{2,1}(\tau_2) & \cdots & \boldsymbol{g}_{2,L}(\tau_2) \\ \vdots & \ddots & \vdots \\ \boldsymbol{g}_{K_a,1}(\tau_{K_a}) & \cdots & \boldsymbol{g}_{K_a,L}(\tau_{K_a}) \end{pmatrix}. \tag{15.29}$$

Let $\boldsymbol{s}_k = [s_{k,1}, \ldots, s_{k,N}]^T$, $\boldsymbol{S}(k,:) = \left[\boldsymbol{0}_{1\times N(k-1)}\, \boldsymbol{s}_k^T\, \boldsymbol{0}_{1\times N(K_a-k)}\right]$, $k = 1, \ldots, K_a$, $\boldsymbol{Y} = [y_{m,l}]$, $\boldsymbol{H} = [h_{m,k}]$. Then (15.28) can be rewritten in a matrix form as

$$\boldsymbol{Y} = \boldsymbol{H}\boldsymbol{S}\boldsymbol{G}(\boldsymbol{\tau}) + \boldsymbol{W}. \tag{15.30}$$

The focus of this section is to design an efficient receiver to recover the transmit packets in the observation window. With the received signal model (15.30), the MAP probability estimate of $(\boldsymbol{H}, \boldsymbol{S}, \boldsymbol{\tau})$ can be obtained by

$$\begin{aligned}\left(\hat{\boldsymbol{H}}, \hat{\boldsymbol{S}}, \hat{\boldsymbol{\tau}}\right) &= \arg\max_{(\boldsymbol{H},\boldsymbol{S},\boldsymbol{\tau})} p_{\boldsymbol{H},\boldsymbol{S},\boldsymbol{\tau}|\boldsymbol{Y}}(\boldsymbol{H}, \boldsymbol{S}, \boldsymbol{\tau}|\boldsymbol{Y}) \\ &= \arg\max_{(\boldsymbol{H},\boldsymbol{S},\boldsymbol{\tau})} \exp\left(-\frac{\| \boldsymbol{Y} - \boldsymbol{H}\boldsymbol{S}\boldsymbol{G}(\boldsymbol{\tau}) \|_F^2}{N_0}\right) p_{\boldsymbol{H}}(\boldsymbol{H}) p_{\boldsymbol{S}}(\boldsymbol{S}) p_{\boldsymbol{\tau}}(\boldsymbol{\tau}). \end{aligned} \tag{15.31}$$

In (15.31), $\boldsymbol{Y}$ is tri-linear with respect to $\boldsymbol{H}$, $\boldsymbol{S}$, and $\boldsymbol{G}$. Furthermore, $\boldsymbol{G}$ is non-linear with $\boldsymbol{\tau}$. In general, (15.31) is difficult to solve. We next develop an efficient Bayesian receiver design to approach a viable solution for (15.31).

15.3.2 Extended Probability Model and Factor Graph Construction

To facilitate the Bayesian iterative receiver design, we first construct an extended probability model of the considered problem by introducing appropriate auxiliary variables. We then construct the factor graph accordingly.

15.3.2.1 Auxiliary Variables

The Effective Signal We combine $\boldsymbol{S}$ and $\boldsymbol{G}(\boldsymbol{\tau})$ to form an effective signal $\boldsymbol{X} = \mathbf{S}G(\boldsymbol{\tau})$, where the (k, l)th element of $\boldsymbol{X}$ is given by

$$x_{k,l} \triangleq \sum_{n=1}^{N} s_{k,n} g\left((l-n)T_s - \tau_k\right), \quad l = 1, \dots, L. \tag{15.32}$$

We refer to $x_{k,l}$ as the effective signal of user k at slot l. Then, the received signal can be rewritten in a bilinear form as

$$y_{m,l} = \sum_{k=1}^{K_a} h_{m,k} x_{k,l} + w_{m,l}, \quad l = 1, \dots, L. \tag{15.33}$$

Decomposition of Delay Due to the nonlinear constraint (15.32), it is difficult to estimate τ_k directly. We decompose τ_k into the symbol-level delay $\lfloor \tau_k \rfloor$ and the sub-symbol-level delay $\dot{\tau}_k$ according to the symbol interval T_s, i.e.,

$$\tau_k = \lfloor \tau_k \rfloor T_s + \dot{\tau}_k, \text{ with } 0 \le \dot{\tau}_k < T_s. \tag{15.34}$$

With this decomposition, the symbol-level delay $\lfloor \tau_k \rfloor$ determines the positions of the non-zero elements in $\boldsymbol{x}_k \triangleq [x_{k,1}, \dots, x_{k,L}]$ while the sub-symbol-level delay $\dot{\tau}_k$ determines the constellation information of the non-zero elements in $\boldsymbol{x}_k$.

Decomposition of Effective Signal Corresponding to the decomposition of delay τ_k, we also decompose the effective signal $x_{k,l}$ as

$$x_{k,l} \triangleq d_{k,l} \tilde{x}_{k,l}, \tag{15.35}$$

where $d_{k,l} \in \{0, 1\}$ with $d_{k,l} = 1$ if $x_{k,l} \neq 0$, and $d_{k,l} = 0$ otherwise. Correspondingly, $\tilde{x}_{k,l} = x_{k,l}$ for $d_{k,l} = 1$, and $\tilde{x}_{k,l}$ can take any value for $d_{k,l} = 0$. With this decomposition, the sparsity information and constellation information of $x_{k,l}$ can be processed through $d_{k,l}$ and $\tilde{x}_{k,l}$, respectively, which facilitates the design of Bayesian receiver detailed later.

Extended Transmit Constellation Recall that the observation window is longer than the transmit sequence (i.e., $L > N$). To facilitate the subsequent processing, we extend $\boldsymbol{s}_k$ to a length-L sequence $\tilde{\boldsymbol{s}}_k$ by adding appropriate zero elements. Focus on the active user k in the specific window $(0, LT_s]$, $\tilde{\boldsymbol{s}}_k$ is defined as[1]

$$\tilde{s}_{k,l} = \begin{cases} s_{k,l-\lfloor\tau_k\rfloor}, & \max\{1, 1+\lfloor\tau_k\rfloor\} \le l \le \min\{L, N+\lfloor\tau_k\rfloor\}, \\ 0, & \text{otherwise.} \end{cases} \tag{15.36}$$

Then (15.32) can be rewritten as

$$\begin{aligned} x_{k,l} &\triangleq \sum_{l'=1}^{L} \tilde{s}_{k,l'} g\left(\left(l - (l' - \lfloor\tau_k\rfloor)\right) T_s - \tau_k\right), \\ &= \sum_{l'=1}^{L} \tilde{s}_{k,l'} g\left((l-l')T_s - \dot{\tau}_k\right), \quad l = 1, \ldots, L, \end{aligned} \tag{15.37}$$

where $\tilde{s}_{k,l'} \in \mathcal{S}' = \{0, \mathcal{S}\}$.

15.3.2.2 Extended Probability Model Construction

Define $\tilde{\boldsymbol{X}} = [\tilde{x}_{k,l}]$, $\boldsymbol{D} = [d_{k,l}]$, $\tilde{\boldsymbol{S}} = [\tilde{s}_{k,l}]$ $\boldsymbol{S} = [s_{k,n}]$, $\lfloor\boldsymbol{\tau}\rfloor = \left[\lfloor\tau_1\rfloor, \ldots, \lfloor\tau_{K_a}\rfloor\right]^T$ and $\dot{\boldsymbol{\tau}} = \left[\dot{\tau}_1, \ldots, \dot{\tau}_{K_a}\right]^T$. Then the extended probability model of the considered receiver design problem can be written as

$$\begin{aligned} & p\left(\boldsymbol{H}, \boldsymbol{X}, \tilde{\boldsymbol{X}}, \boldsymbol{D}, \tilde{\boldsymbol{S}}, \boldsymbol{S}, \lfloor\boldsymbol{\tau}\rfloor, \dot{\boldsymbol{\tau}} | \boldsymbol{Y}\right) \overset{(a)}{\propto} p\left(\boldsymbol{Y}|\boldsymbol{H}, \boldsymbol{X}, \tilde{\boldsymbol{X}}, \boldsymbol{D}, \tilde{\boldsymbol{S}}, \boldsymbol{S}, \lfloor\boldsymbol{\tau}\rfloor, \dot{\boldsymbol{\tau}}\right) \\ &\quad \cdot p\left(\boldsymbol{H}|\boldsymbol{X}, \tilde{\boldsymbol{X}}, \boldsymbol{D}, \tilde{\boldsymbol{S}}, \boldsymbol{S}, \lfloor\boldsymbol{\tau}\rfloor, \dot{\boldsymbol{\tau}}\right) \quad p\left(\boldsymbol{X}|\tilde{\boldsymbol{X}}, \boldsymbol{D}, \tilde{\boldsymbol{S}}, \boldsymbol{S}, \lfloor\boldsymbol{\tau}\rfloor, \dot{\boldsymbol{\tau}}\right) p\left(\tilde{\boldsymbol{X}}|\boldsymbol{D}, \tilde{\boldsymbol{S}}, \boldsymbol{S}, \lfloor\boldsymbol{\tau}\rfloor, \dot{\boldsymbol{\tau}}\right) \\ &\quad \cdot p\left(\boldsymbol{D}|\tilde{\boldsymbol{S}}, \boldsymbol{S}, \lfloor\boldsymbol{\tau}\rfloor, \dot{\boldsymbol{\tau}}\right) p\left(\tilde{\boldsymbol{S}}|\boldsymbol{S}, \lfloor\boldsymbol{\tau}\rfloor, \dot{\boldsymbol{\tau}}\right) p\left(\boldsymbol{S}, \lfloor\boldsymbol{\tau}\rfloor, \dot{\boldsymbol{\tau}}\right) \\ &\overset{(b)}{=} p(\boldsymbol{Y}|\boldsymbol{H}, \boldsymbol{X}) p(\boldsymbol{H}) p(\boldsymbol{X}| \sim \mathbf{X}, \boldsymbol{D}) p\left(\tilde{\boldsymbol{X}}|\tilde{\boldsymbol{S}}, \dot{\boldsymbol{\tau}}\right) p(\boldsymbol{D}|\lfloor\boldsymbol{\tau}\rfloor) \cdot p\left(\tilde{\boldsymbol{S}}|\boldsymbol{S}, \lfloor\boldsymbol{\tau}\rfloor\right) p(\lfloor\boldsymbol{\tau}\rfloor) \\ &\quad \cdot p(\dot{\boldsymbol{\tau}}) p(\boldsymbol{S}) \\ &\overset{(c)}{=} \prod_{m,l} p\left(y_{m,l} \Big| \sum_{k=1}^{K_a} h_{m,k} x_{k,l}\right) \cdot \prod_{m,k} p\left(h_{m,k}\right) \cdot \prod_{k,l} p\left(x_{k,l}|d_{k,l}, \tilde{x}_{k,l}\right) p\left(\tilde{x}_{k,l}|\tilde{\boldsymbol{s}}_k, \dot{\tau}_k\right) p\left(d_{k,l}|\lfloor\tau_k\rfloor\right) \\ &\quad \cdot \prod_{k} p\left(\tilde{\boldsymbol{s}}_k|\boldsymbol{s}_k, \lfloor\tau_k\rfloor\right) p\left(\lfloor\tau_k\rfloor\right) p\left(\dot{\tau}_k\right) \cdot \prod_{k,n} p\left(s_{k,n}\right). \end{aligned} \tag{15.38}$$

where (a) is due to the Bayesian formula; (b) is due to the Markov property of involved variables; (c) is due to the (conditional) independence of elements in the matrices $\boldsymbol{Y}$ and $\boldsymbol{H}$, the conditional independence of elements in $\boldsymbol{x}_k$, $\tilde{\boldsymbol{x}}_k$ and $\boldsymbol{d}_k$, and the independence of different users.

In (15.38), $\{\tilde{x}_{k,l}\}$ depend on $\dot{\tau}_k$ and $\tilde{\boldsymbol{s}}_k$ where $\tilde{\boldsymbol{s}}_k$ is the extended transmit sequence. From (15.35), $\{\tilde{x}_{k,l}\}$ carry the non-zero elements in $\{x_{k,l}\}$ while $\{d_{k,l}\}$ carry the zero/non-zero positions of $\{x_{k,l}\}$ with $d_{k,l} = 1$ for non-zero position and $d_{k,l} = 0$

---

1 Due to the ISI effect incurred by asynchronous transmission, the transmit signals outside the observation window may also contribute to the received signals in the window. We will discuss the treatment of such boundary effect later in Section 15.3.3.

for zero position. Notice that $\tilde{x}_{k,l}$ corresponding to a zero element of $\{x_{k,l}\}$ can take any value since $d_{k,l} = 0$ in this case. To simplify the receiver design, we assume that $\{\tilde{x}_{k,l}\}$ are generated by an $\tilde{\mathbf{s}}_k$ sequence with independent and identically distributed elements drawn from S'.[2] Then the term $\prod_k p\left(\tilde{\mathbf{s}}_k|\mathbf{s}_k, \lfloor\tau_k\rfloor\right) \cdot \prod_{k,n} p\left(s_{k,n}\right)$ in (15.38) reduces to $\prod_{k,l} p\left(\tilde{s}_{k,l}\right)$, and (15.38) becomes

$$p\left(\boldsymbol{H}, \boldsymbol{X}, \boldsymbol{D}, \tilde{\boldsymbol{X}}, \tilde{\boldsymbol{S}}, \lfloor\boldsymbol{\tau}\rfloor, \dot{\boldsymbol{\tau}}|\boldsymbol{Y}\right) \propto \prod_{m,l} p\left(y_{m,l}|\sum_{k=1}^{K_a} h_{m,k}x_{k,l}\right) \cdot \prod_{m,k} p\left(h_{m,k}\right)$$
$$\cdot \prod_{k,l} p\left(x_{k,l}|d_{k,l}, \tilde{x}_{k,l}\right) p\left(\tilde{x}_{k,l}|\tilde{\mathbf{s}}_k, \dot{\tau}_k\right) p\left(d_{k,l}|\lfloor\tau_k\rfloor\right) p\left(\tilde{s}_{k,l}\right) \cdot \prod_k p\left(\lfloor\tau_k\rfloor\right) p\left(\dot{\tau}_k\right). \tag{15.39}$$

15.3.2.3 Factor Graph Construction

Based on (15.39), the factor graph of the proposed Bayesian receiver is shown in Fig. 15.4. The factor graph consists of two types of nodes:

- Variable nodes $\{h_{m,k}\}$, $\{x_{k,l}\}$, $\{d_{k,l}\}$, $\{\tilde{x}_{k,l}\}$, $\{\tilde{s}_{k,l}\}$, $\{\dot{\tau}_k\}$, and $\{\lfloor\tau_k\rfloor\}$ depicted as white circles in Fig. 15.4, corresponding to the variables in (15.39);

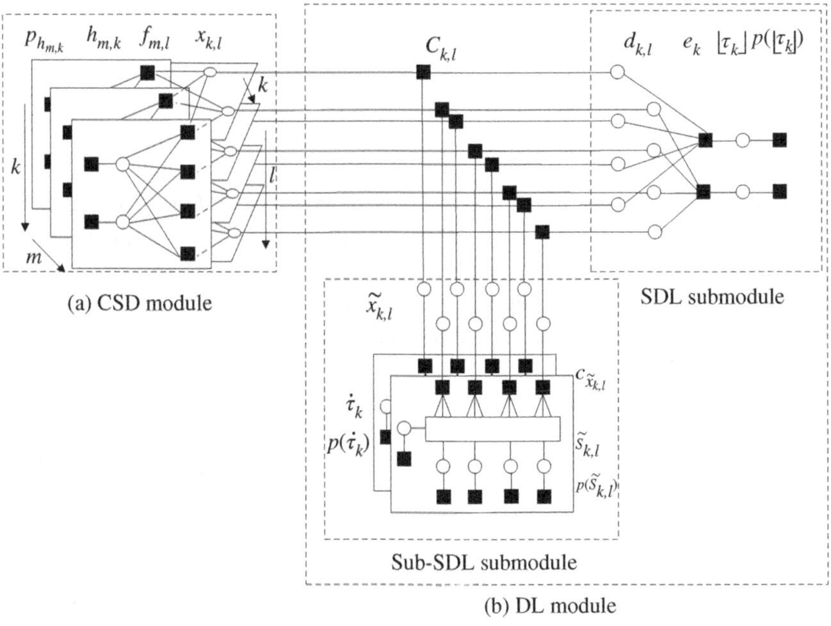

Figure 15.4 The factor graph for the Bayesian receiver with toy-problem parameters $M = 3$, $K_a = 2$, $L = 4$, $N = 2$. Subgraph (a) models the bilinear constraint in (15.33), subgraph (b) models the non-linear constraints in (15.34), (15.35), (15.36), and (15.37).

2 With some abuse of notation, we still use $\tilde{\mathbf{s}}_k$ for the decoupled sequence.

- Check nodes $\{p_{h_{m,k}}\}$, $\{p(\tilde{s}_{k,l})\}$, $\{p(\lfloor\tau_k\rfloor)\}$, $\{p(\dot{\tau}_k)\}$, $\{f_{m,l}\}$, $\{c_{k,l}\}$, $\{c_{\tilde{x}_{k,l}}\}$, and $\{e_k\}$ depicted as black boxes in Fig. 15.4, corresponding to the *a priori* distributions of $\{h_{m,k}\}$, the *a priori* distributions of $\{\tilde{s}_{k,l}\}$, the *a priori* distributions of $\{\lfloor\tau_k\rfloor\}$, the *a priori* distributions of $\{\dot{\tau}_k\}$, the conditional distributions $p(y_{m,l}|\sum_{k=1}^{K_a} h_{m,k}x_{k,l})$, $p(x_{k,l}|d_{k,l},\tilde{x}_{k,l})$, $p(\tilde{x}_{k,l}|\tilde{s}_k,\dot{\tau}_k)$, and $p\left(d_{k,l}|\lfloor\tau_k\rfloor\right)$, respectively.

With the factor graph in Fig. 15.4, a message passing-based receiver can be readily derived based on the sum-product rule. However, due to the fully asynchronous transmission, the distribution of effective signal $x_{k,l}$, which is the convolution of transmit constellation symbols and ISI tap coefficients determined by the unknown continuously-valued delay, is difficult to characterize. This imposes challenges on message calculations related to $x_{k,l}$, as well as the deconvolution of transmit symbols and user delay from $x_{k,l}$. To solve the problem, we next develop the proposed TAMP algorithm.

15.3.3 Proposed TAMP Algorithm

Based on the probability model and the factor graph in Section 15.3.2, we derive the TAMP algorithm as follows.

15.3.3.1 Structure of Bayesian Receiver

We divide the whole receiver into two modules, i.e., the channel-signal decomposition (CSD) module and the delay learning (DL) module, as outlined below.

CSD Module The CSD module represented by subgraph (a) in Fig. 15.4 focuses on the bilinear signal model in (15.33). With the *a priori* of $\{h_{m,k}\}$ and the messages of $\{x_{k,l}\}$ fed back from the DL module, the estimation of $\{h_{m,k}\}$ and $\{x_{k,l}\}$ can be refined based on $\{y_{m,l}\}$. The refined estimates of $\{x_{k,l}\}$ are forwarded to the DL module. More details are described in Section 15.3.3.2.

DL Module The DL module represented by subgraph (b) in Fig. 15.4 focuses on the non-linear models (15.34)–(15.37). With the messages of $\{x_{k,l}\}$ fed back from the CSD module and passed through the check nodes $\{c_{k,l}\}$ according to (15.35), the estimates of $\{d_{k,l}\}$ and $\{\tilde{x}_{k,l}\}$ are then refined respectively, based on which the messages of $\{x_{k,l}\}$ are updated and forwarded to the CSD module. More details are specified in Section 15.3.3.3.

The operations of the CSD module and DL module iterate until convergence and the transmit packets are then recovered.

15.3.3.2 Channel and Signal Decomposition Module

We propose to solve the problem by leveraging the near-optimal BiGAMP algorithm (Parker et al., 2014a,2014b). A brief explanation of the adopted BiGAMP algorithm is provided in the following. Denote by $\{m_{c_{k,l} \to x_{k,l}}(x_{k,l})\}$ the messages of $\{x_{k,l}\}$ fed back from the DL module. With the *a priori* distributions $\{p_{h_{m,k}}\}$ and the messages $\{m_{c_{k,l} \to x_{k,l}}(x_{k,l})\}$ available, BiGAMP operates as follows. First, the means and variances of $\{h_{m,k}\}$ and $\{x_{k,l}\}$ are computed for initialization. Then the messages from the variable nodes $\{h_{m,k}, x_{k,l}\}$ to the factor nodes $\{f_{m,l}\}$ are accumulated to obtain an estimate of $\boldsymbol{HX}$ with means $\{\overline{p}_{m,l}(i)\}$ and variance $\{\overline{v}^p_{m,l}(i)\}$. The adjusted means $\{\hat{p}_{m,l}(i)\}$ and variances $\{v^p_{m,l}(i)\}$ are obtained by applying the Onsager correction, and then are used to calculate the marginal means $\{\hat{q}_{m,k}(i)\}$ and variances $\{v^q_{m,k}(i)\}$ of $\{h_{m,k}\}$ as well as the marginal means $\{\hat{r}_{k,l}(i)\}$ and variances $\{v^r_{k,l}(i)\}$ of $\{x_{k,l}\}$. Finally, each pair of $\hat{q}_{m,k}(i)$ and $v^q_{m,k}(i)$ is merged with the *a priori* distribution $p_{h_{m,k}}$ to produce the *a posteriori* mean $\hat{h}_{m,k}(i+1)$ and variance $v^h_{m,k}(i+1)$ for $\{h_{m,k}\}$. A similar process is performed on each $x_{k,l}$. The above operations are iterated until convergence or the maximum number of iterations *ItNum* is achieved. Then the extrinsic messages of $\{x_{k,l}\}$, i.e., means $\{\hat{r}_{k,l}(ItNum)\}$ and variances $\{v^r_{k,l}(ItNum)\}$, are passed to the DL module.

15.3.3.3 Delay Learning Module

With the messages $\{\hat{r}_{k,l}\}$ and $\{v^r_{k,l}\}$ from the CSD module as input, the DL module deals with (15.34)–(15.37) to estimate the user delays and resolve the ISI caused by asynchrony.

Decomposition of DL Module Based on Fig. 15.4, we divide the DL module into three parts, i.e., the check nodes $\{c_{k,l}\}$, symbol-level delay learning (SDL), and sub-symbol-level delay learning (sub-SDL), respectively, for the decomposition/combination of $\{x_{k,l}\}$ into/from $\{d_{k,l}\}$ and $\{\tilde{x}_{k,l}\}$, the refined estimation of $\{d_{k,l}\}$ (and so the symbol-level delays $\{\lfloor \tau_k \rfloor\}$), and the refined estimation of $\{\tilde{x}_{k,l}\}$ (and so the sub-symbol-level delays $\{\dot{\tau}_k\}$ as well as the transmit symbols $\{\tilde{s}_{k,l}, s_{k,n}\}$). The detailed operations are derived as follows.

Check Nodes $\{c_{k,l}\}$ As illustrated by Fig. 15.5(a), there are three types of messages passed from $\{c_{k,l}\}$.

Messages passed from $\{c_{k,l}\}$ to $\{d_{k,l}\}$: With the message $m_{x_{k,l} \to c_{k,l}}(x_{k,l}) = \mathcal{CN}(\hat{r}_{k,l}, v^r_{k,l})$ passed from $x_{k,l}$ and the message $m_{\tilde{x}_{k,l} \to c_{k,l}}(\tilde{x}_{k,l})$ passed from $\tilde{x}_{k,l}$, the message passed from $\{c_{k,l}\}$ to $\{d_{k,l}\}$ is given by

$$m_{c_{k,l} \to d_{k,l}}(d_{k,l}) = m_{c_{k,l} \to d_{k,l}}(1)\delta(d_{k,l} - 1) + m_{c_{k,l} \to d_{k,l}}(0)\delta(d_{k,l}), \tag{15.40}$$

where

$$\begin{aligned} m_{c_{k,l} \to d_{k,l}}(1) &= \int_{\tilde{x}_{k,l}, x_{k,l} \neq 0} \left(m_{x_{k,l} \to c_{k,l}}(x_{k,l}) m_{\tilde{x}_{k,l} \to c_{k,l}}(\tilde{x}_{k,l}) \cdot \delta\left(x_{k,l} - d_{k,l}\tilde{x}_{k,l}\right) \right), \\ m_{c_{k,l} \to d_{k,l}}(0) &= 1 - m_{c_{k,l} \to d_{k,l}}(1). \end{aligned} \tag{15.41}$$

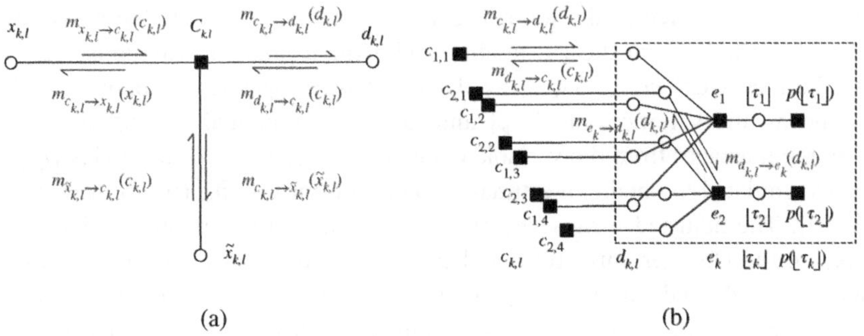

Figure 15.5 Illustration of message flow of check node $c_{k,l}$ and SDL module with toy-problem parameters $K_a = 2$, $L = 4$ and $N = 2$. (a) Message flow of check node $c_{k,l}$. (b) Message flow in SDL module.

Messages passed from $\{c_{k,l}\}$ *to* $\{\tilde{x}_{k,l}\}$: Given the messages $m_{x_{k,l}\to c_{k,l}}(x_{k,l})$ and $m_{d_{k,l}\to c_{k,l}}(d_{k,l})$, the messages $m_{c_{k,l}\to \tilde{x}_{k,l}}(\tilde{x}_{k,l})$ can be calculated accordingly. However, with $x_{k,l} = \tilde{x}_{k,l} d_{k,l}$ in (15.35), the inference of $\tilde{x}_{k,l}$ from $x_{k,l}$ is infeasible when $d_{k,l} = 0$. In order to bypass this difficulty, we follow the treatment in Ziniel and Schniter (2013) and Jiang et al. (2022) by regarding $d_{k,l} = 0$ as the limit of $d_{k,l} = \varepsilon$ with ε approaching 0, then the messages passed from $\{c_{k,l}\}$ to $\{\tilde{x}_{k,l}\}$ can be approximately obtained as

$$m_{c_{k,l}\to \tilde{x}_{k,l}}(\tilde{x}_{k,l}) \approx \mathcal{CN}\left(\hat{r}_{k,l}, v^r_{k,l}\right). \tag{15.42}$$

The specific calculations can be found in Ziniel and Schniter [2013].

Messages passed from $\{c_{k,l}\}$ *to* $\{x_{k,l}\}$: With the messages $m_{\tilde{x}_{k,l}\to c_{k,l}}(\tilde{x}_{k,l})$ and $m_{d_{k,l}\to c_{k,l}}(d_{k,l})$, the message passed from $\{c_{k,l}\}$ to $\{x_{k,l}\}$ is given by

$$m_{c_{k,l}\to x_{k,l}}(x_{k,l}) = \int_{d_{k,l},\tilde{x}_{k,l}} \left(m_{d_{k,l}\to c_{k,l}}(d_{k,l}) \cdot m_{\tilde{x}_{k,l}\to c_{k,l}}(\tilde{x}_{k,l}) \delta\left(x_{k,l} - d_{k,l}\tilde{x}_{k,l}\right)\right). \tag{15.43}$$

Recall that $d_{k,l} \in \{0, 1\}$ indicates whether $x_{k,l}$ is zero and $\tilde{x}_{k,l}$ is the non-zero value of $x_{k,l}$ (when $d_{k,l} = 1$). Hence, we further rewrite (15.43) as

$$m_{c_{k,l}\to x_{k,l}}(x_{k,l}) = m_{d_{k,l}\to c_{k,l}}(0)\delta(x_{k,l}) + m_{d_{k,l}\to c_{k,l}}(1) m_{\tilde{x}_{k,l}\to c_{k,l}}(x_{k,l}). \tag{15.44}$$

This activates the operation of CSD in the next iteration.

SDL SDL is used to update the active/inactive probability of the elements within $\boldsymbol{x}_k$ based on the block sparsity. Due to the ISI effect in the asynchronous case, the number of actual non-zero entries in $\boldsymbol{x}_k$ is larger than N, which we denote by $N + \triangle N$. For the typical raised cosine filter $g(t)$, $\triangle N$ can be infinite. Nevertheless, most of the energy of $g(t)$ is concentrated in a finite range; hence, we truncate and approximate $g(t)$ with a finite-length impulse $\tilde{g}(t)$ centered at t_0 with width $2N_gT_s$ as

$$\tilde{g}(t) = \begin{cases} g(t), & t_0 - N_gT_s < t \leq t_0 + N_gT_s, \\ 0, & \text{otherwise.} \end{cases} \tag{15.45}$$

Furthermore, we define $g'_{k,n} \triangleq \tilde{g}(nT_s - \dot{\tau}_k)$. For illustration, we give an example of $\tilde{g}(t)$ and $g'_{k,n}$ in Fig. 15.6 when $N_g = 2$.

It can be verified that $\triangle N = 2N_g - 1$ in the truncated case. Notice that the distributions of these $N + \triangle N$ non-zero elements are generally different due to the boundary effect. We simply ignore such difference in iteration and address them in the packet recovery stage in Section 15.3.3.4.

Also notice that the non-zero positions of $x_{k,l}$ are all adjacent and its number is at most $N + \triangle N$ (some non-zero positions may be outside the observation window). The message passing operations in SDL can be derived based on the sum-product algorithm, as illustrated in Fig. 15.5(b). The detailed derivations are as follows.

Messages passed from $\{d_{k,l}\}$ *to* $\{e_k\}$: With the messages $\{m_{c_{k,l}\to d_{k,l}}(d_{k,l})\}$ passed form $\{c_{k,l}\}$, the messages $m_{d_{k,l}\to e_k}(d_{k,l})$ can be obtained directly as

$$m_{d_{k,l}\to e_k}(d_{k,l}) = m_{c_{k,l}\to d_{k,l}}(d_{k,l}). \tag{15.46}$$

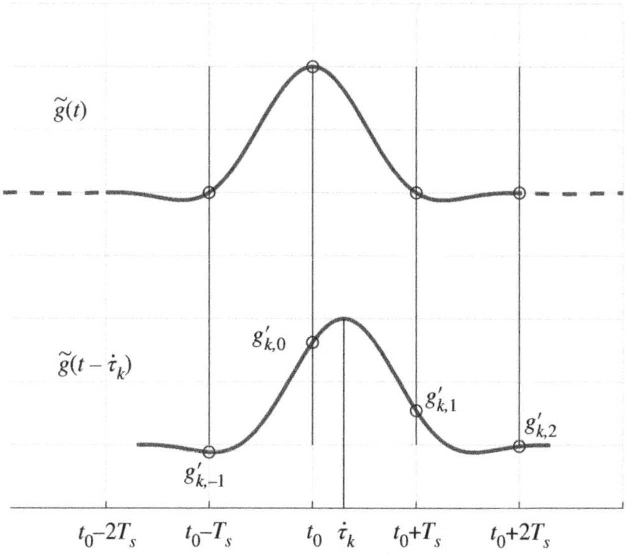

Figure 15.6 Example of truncated filter $\tilde{g}(t)$ and its sampling points at the receiver with asynchrony.

Messages passed from $\{e_k\}$ *to* $\{d_{k,l}\}$: For simplicity, we introduce $\triangle t_k$ as the index difference between the first non-zero entry of $\boldsymbol{x}_k$ and the beginning of the observation window $(0, LT_s]$, then $\lfloor \tau_k \rfloor = \triangle t_k + N_g - 1$. For active packets, we have $-(N + \triangle N - 1) \leq \triangle t_k \leq L - 1$ and $-N - N_g + 1 \leq \lfloor \tau_k \rfloor \leq L + N_g - 2$.

For any $k \in \{1, \ldots, K_a\}$, $\{d_{k,l}\}$ are constrained by

$$d_{k,l} = \begin{cases} 1, & \max\{\triangle t_k + 1, 1\} \leq l \leq \min\{\triangle t_k + N + \triangle N, L\}, \\ 0, & \text{otherwise}. \end{cases} \tag{15.47}$$

The constraint (15.47) is represented by the factor node e_k in the subgraph (b) in Fig. 15.4. Using the messages of $\{d_{k,l'}, l' = 1, \ldots, L, l' \neq l\}$, the distribution of $\triangle t_k$ is computed as

$$p^{(l)}_{\triangle t_k}(\triangle t_k) \propto \prod_{l'=1, l'\neq l}^{\max\{\triangle t_k+1,1\}-1} m_{c_{k,l'} \to d_{k,l'}}(0) \prod_{l'=\max\{\triangle t_k+1,1\}, l'\neq l}^{\min\{\triangle t_k+N+\triangle N, L\}} m_{c_{k,l'} \to d_{k,l'}}(1) \cdot \prod_{l'=\min\{\triangle t_k+N+\triangle N,L\}+1, l'\neq l}^{L} m_{c_{k,l'} \to d_{k,l'}}(0). \tag{15.48}$$

Then

$$p^{(l)}_{\lfloor \tau_k \rfloor}(\lfloor \tau_k \rfloor) \propto p^{(l)}_{\triangle t_k}(\lfloor \tau_k \rfloor - N_g + 1) p(\lfloor \tau_k \rfloor), \tag{15.49}$$

where $p(\lfloor \tau_k \rfloor)$ is the *a priori* distribution of $\lfloor \tau_k \rfloor$. With uniformly and randomly asynchronous transmissions, $p(\lfloor \tau_k \rfloor)$ is a uniform distribution function.

The message passed from e_k to each $d_{k,l}$ is given by

$$m_{e_k \to d_{k,l}}(d_{k,l}) = m_{e_k \to d_{k,l}}(1)\delta(d_{k,l} - 1) + m_{e_k \to d_{k,l}}(0)\delta(d_{k,l}), \tag{15.50}$$

where

$$\begin{aligned} m_{e_k \to d_{k,l}}(1) &= \sum_{\lfloor \tau_k \rfloor = l-N-N_g}^{l+N_g-2} p^{(l)}_{\lfloor \tau_k \rfloor}(\lfloor \tau_k \rfloor), \\ m_{e_k \to d_{k,l}}(0) &= 1 - m_{e_k \to d_{k,l}}(1). \end{aligned} \tag{15.51}$$

Messages passed from $\{d_{k,l}\}$ *to* $\{c_{k,l}\}$: The message in (15.50) is passed to $\{c_{k,l}\}$ directly, i.e.,

$$m_{d_{k,l} \to c_{k,l}}(d_{k,l}) = m_{e_k \to d_{k,l}}(d_{k,l}). \tag{15.52}$$

At the beginning, $m_{d_{k,l} \to c_{k,l}}(d_{k,l})$ can be initialized as (Ding et al., 2019)

$$\begin{aligned} m_{d_{k,l} \to c_{k,l}}(1) &= \frac{(2L + N + 2N_g - 2)(N + 2N_g - 1)}{2(N + 2N_g - 2 + L)L}, \\ m_{d_{k,l} \to c_{k,l}}(0) &= 1 - m_{d_{k,l} \to c_{k,l}}(1). \end{aligned} \tag{15.53}$$

Sub-SDL With the messages $\{m_{c_{k,l}\to\tilde{x}_{k,l}}(\tilde{x}_{k,l})\}$ as input, sub-SDL is used to refine the estimates of $\{\tilde{x}_{k,l}\}$ by estimating the sub-symbol level delay as well as the transmit symbols based on (15.37). From (15.37), this is essentially a blind deconvolution problem. To solve the problem, we derive a Bayesian recursive method to estimate the sub-symbol level delay and utilize the BCJR algorithm (Bahl et al., 1974) to resolve the ISI, as detailed below.

Messages calculation in sub-SDL: Using the messages of $\{m_{c_{k,l}\to\tilde{x}_{k,l}}(\tilde{x}_{k,l})\}$, we first calculate the *a posteriori* distribution of $\{\tilde{x}_{k,l}\}$ as

$$\begin{aligned} p'_{\tilde{x}_{k,l}}(\tilde{x}_{k,l}) &= \Pr\{\tilde{x}_{k,l}|\{m_{c_{k,l}\to\tilde{x}_{k,l}}(\tilde{x}_{k,l})\}\} \\ &= \int_{\dot{\tau}_k}\sum_{\{\tilde{s}_{k,l}\}} \Pr\{\tilde{x}_{k,l},\tilde{s}_{k,l},\dot{\tau}_k|\hat{\boldsymbol{r}}_k,\boldsymbol{v}_k^r\} \\ &\propto \int_{\dot{\tau}_k}\sum_{\{\tilde{s}_{k,l}\}} \Pr\{\tilde{x}_{k,l},\tilde{s}_{k,l}|\dot{\tau}_k,\hat{\boldsymbol{r}}_k,\boldsymbol{v}_k^r\}\Pr\{\hat{\boldsymbol{r}}_k,\boldsymbol{v}_k^r|\dot{\tau}_k\}\Pr\{\dot{\tau}_k\}. \end{aligned} \tag{15.54}$$

Next we derive $\Pr\{\tilde{x}_{k,l},\tilde{s}_{k,l}|\dot{\tau}_k,\hat{\boldsymbol{r}}_k,\boldsymbol{v}_k^r\}$ and $\Pr\{\hat{\boldsymbol{r}}_k,\boldsymbol{v}_k^r|\dot{\tau}_k\}$, respectively. We resort to Bahl-Cocke-Jelinek-Raviv (BCJR) algorithm (Bahl et al., 1974) to compute $\Pr\{\tilde{x}_{k,l},\tilde{s}_{k,l}|\dot{\tau}_k,\hat{\boldsymbol{r}}_k,\boldsymbol{v}_k^r\}$. To this end, $S_{k,l}=[\tilde{s}_{k,l-(2N_g-2)},\dots,\tilde{s}_{k,l}]\in\mathcal{S}'^{(2N_g-1)}$ is denoted as the state of user k at time l.

Denote by $S_{k,l-1}$ and $S_{k,l}$ the states of user k at time $l-1$ and l respectively, the convolution output of user k at time l for specific $\dot{\tau}_k$ can be modeled as

$$\tilde{x}_{k,l} = \sum_{n=1}^{2N_g-1}\tilde{s}_{k,l-n}g'_{k,f_{\mathrm{ISI}}(n)} + \tilde{s}_{k,l}g'_{k,N_g}, \tag{15.55}$$

where $f_{\mathrm{ISI}}(n)$ is the nth element in the vector $[N_g-1,N_g-2,\dots,-N_g+1]$, i.e., $\tilde{x}_{k,l}$ is the convolution of the ISI coefficients and the symbols consisting of the previous state $S_{k,l-1}$ and the newly arrived symbol $\tilde{s}_{k,l}$ in the current state $S_{k,l}$. Then, we obtain the conditional probability $\Pr\{S_{k,l-1},S_{k,l}|\dot{\tau}_k,\hat{\boldsymbol{r}}_k,\boldsymbol{v}_k^r\}$ according to the BCJR algorithm. Consequently, we obtain $\Pr\{\tilde{x}_{k,l},\tilde{s}_{k,l}|\dot{\tau}_k,\hat{\boldsymbol{r}}_k,\boldsymbol{v}_k^r\}=\Pr\{S_{k,l-1},S_{k,l}|\dot{\tau}_k,\hat{\boldsymbol{r}}_k,\boldsymbol{v}_k^r\}$.

The rest is to derive

$$\Pr\{\hat{\boldsymbol{r}}_k,\boldsymbol{v}_k^r|\dot{\tau}_k\} = \sum_{\tilde{\mathbf{s}}_k}\Pr(\hat{\boldsymbol{r}}_k,\boldsymbol{v}_k^r,\tilde{\mathbf{s}}_{\mathbf{k}}|\dot{\tau}_k) = \sum_{\tilde{\mathbf{s}}_k}\Pr(\hat{\boldsymbol{r}}_k,\boldsymbol{v}_k^r|\tilde{\mathbf{s}}_k,\dot{\tau}_k)\Pr(\tilde{\mathbf{s}}_k|\dot{\tau}_k). \tag{15.56}$$

Finally, $p'_{\tilde{x}_{k,l}}(\tilde{x}_{k,l})$ can be obtained according to (15.54)–(15.56).

Messages passed from $\{\tilde{x}_{k,l}\}$ *to* $\{c_{k,l}\}$: From the sum-product rule, the messages $m_{\tilde{x}_{k,l}\to c_{k,l}}(\tilde{x}_{k,l})$ can be obtained as

$$m_{\tilde{x}_{k,l}\to c_{k,l}}(\tilde{x}_{k,l}) \propto p'_{\tilde{x}_{k,l}}(\tilde{x}_{k,l})/m_{c_{k,l}\to\tilde{x}_{k,l}}(\tilde{x}_{k,l}). \tag{15.57}$$

15.3.3.4 Packet Recovery

The operations of the CSD and DL modules iterate until convergence. The final output is given by $\tilde{p}_{x_{k,l}}(x_{k,l}) = m_{c_{k,l}\to x_{k,l}}(x_{k,l})\mathcal{CN}(\hat{r}_{k,l}, v^r_{k,l})$. Based on this output and the packet structure, we first determine the non-zero positions of $\{x_{k,l}\}$ (characterized by $\{\lfloor \hat{\tau}_k \rfloor\}$) by utilizing the hard decision of the SDL submodule, then estimate the sub-symbol-level delays $\{\dot{\tau}_k\}$, and finally recover the transmitted symbols $\{s_{k,n}\}$. We note that due to the boundary effect, the effective constellations of non-zero elements in $\{x_{k,l}\}$ can be different at the beginning and at the end of a packet. To avoid this issue, we add two reference symbols at both the beginning and the end of the transmit packets.

15.3.3.5 Overall Algorithm and Complexity Analysis

Based on the procedures described in Section 15.3.3.2–15.3.3.4, we outline the whole algorithm in Algorithm 15.2. The user packets are then recovered from the output $\{\tilde{p}_{x_{k,l}}(x_{k,l})\}$. Since the constellation of $x_{k,l}$ is generally difficult to determine before a relatively reliable estimate of its activity state is obtained, we divide the whole algorithm into two stages. In the first stage, sub-SDL is ignored and the operations of CSD and SDL are iterated to refine the activity state of the effective signal $x_{k,l}$. In the second stage, all operations of CSD, SDL, and sub-SDL are iterated to refine the estimate of $x_{k,l}$.

In the first stage, since there is no feedback information on $\dot{\tau}_k$ the message $m_{c_{k,l}\to x_{k,l}}(x_{k,l})$ is given by

$$m_{c_{k,l}\to x_{k,l}}(x_{k,l}) = m_{d_{k,l}\to c_{k,l}}(0)\delta(x_{k,l}) + m_{d_{k,l}\to c_{k,l}}(1)\int_{\dot{\tau}_k}\sum_{\tilde{\mathbf{s}}_k} p(x_{k,l}|\dot{\tau}_k, \tilde{\mathbf{s}}_k)p(\dot{\tau}_k)p(\tilde{\mathbf{s}}_k), \tag{15.58}$$

where $p(\dot{\tau}_k)$ follows the uniform distribution.

The complexity of TAMP is delineated as follows. The algorithm consists of three modules: CSD, SDL, and sub-SDL. The CSD operation is based on BiGAMP; hence, its complexity per iteration is $\mathcal{O}(MK_aL)$ (Parker et al., 2014a). The complexity of SDL is $\mathcal{O}\left(K_aN(L+N)\right)$. The complexity of sub-SDL is $\mathcal{O}\left(K_aL\right)$. Consequently, $\mathcal{O}\left(MK_aL + K_aN(L+N)\right)$ is the total complexity per iteration; i.e., it is linear in the number of receiver antennas, the number of active users, and the width of observation window.

15.3.4 Numerical Results

We provide numerical results to validate the effectiveness of the proposed scheme. In simulations, we set the symbol interval $T_s = 1\,\mu s$, the length of observation window $LT_s = 256\,\mu s$ and the length of the transmit signals $NT_s = 64\,\mu s$. The data packets of each user are generated according to a Poisson point process

Algorithm 15.2 TAMP Algorithm

Input: Received signal Y, $\{p_{h_{m,k}}\}$, λ, effective shaping filter $g(t)$, constellation $\mathcal{S}$
Initialization: Initialize $\{m_{c_{k,l} \to x_{k,l}}(x_{k,l})\}$ according to (15.53) and (15.58).
// First Stage: CSD-SDL Iteration
For $i_1 = 1, \ldots, ItNum_1$
  For $i = 1, \ldots, ItNum$
    $\forall\, k, l$: Compute $\hat{r}_{k,l}(i)$ and $v^r_{k,l}(i)$ with $m_{c_{k,l} \to x_{k,l}}(x_{k,l})$ by using BiGAMP;
    $\forall\, k, l$: Compute the *a posteriori* estimate of $x_{k,l}$ based on $\hat{r}_{k,l}(i)$ and $v^r_{k,l}(i)$;
  End
  $\forall\, k, l$: Compute $m_{c_{k,l} \to x_{k,l}}(x_{k,l})$ according to (15.40), (15.46)–(15.52), and (15.44);
End
// Second Stage: CSD-SDL-subSDL Iteration
For $i_2 = 1, \ldots, ItNum_2$
  $\forall\, k, l$: Compute $\hat{r}_{k,l}$ and $v^r_{k,l}$ by using BiGAMP;
  $\forall\, k, l$: Perform SDL operations according to (15.40) and (15.46)–(15.52);
  $\forall\, k, l$: Perform sub-SDL operations according to (15.42), (15.54)–(15.57);
  $\forall\, k, l$: Compute $m_{c_{k,l} \to x_{k,l}}(x_{k,l})$ according to (15.44);
End
Output:
$\tilde{p}_{x_{k,l}}(x_{k,l}) \propto m_{c_{k,l} \to x_{k,l}}(x_{k,l}) \mathcal{CN}(\hat{r}_{k,l}, v^r_{k,l})$.

with rate $\lambda = 1/2000$. The data of all active users are QPSK modulated. We set the bandwidth to 1 MHz, and the power spectrum density of AWGN at the AP to -169 dBm/Hz. We mainly focus on the case that perfect power control is adopted to compensate for path-loss and shadowing components, such that $\beta_1 = \beta_2 = \cdots = \beta_K \equiv -128$ dB.

As a performance benchmark, Turbo-BiGAMP in Ding et al. (2019) is adopted, where the ISI effect of sub-symbol delay is integrated into the transmit symbols to form an effective transmit symbol and approximating the related messages by Bernoulli–Gaussian distributions. Moreover, to show the impact of sub-SDL, the proposed TAMP without sub-SDL iteration, marked as Reduced TAMP, is evaluated. We also include BiGAMP in Parker et al. (2014a), the single-user bound with known user delay (but unknown CSI), and LMMSE receiver with known user activity, delay and CSI for performance comparison.

Figure 15.7 shows the detection performance of the proposed design. The packet error rate (PER) defined below is adopted to measure the detection performance of the whole packet:

$$\text{PER} = 1 - \frac{\text{No. of transmitted packets correctly recovered}}{\text{No. of total transmitted packets}}.$$

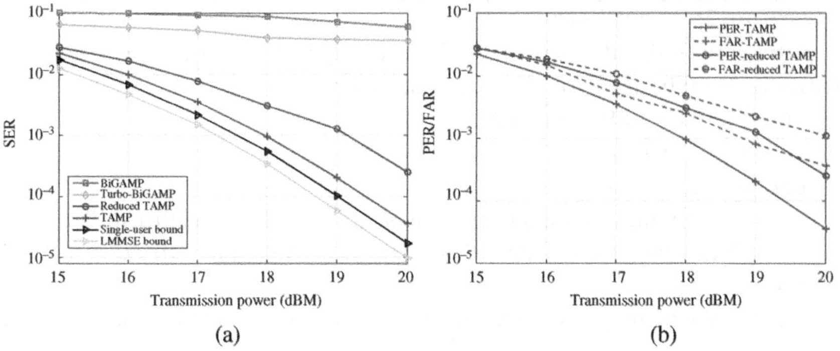

Figure 15.7 Detection performance of the proposed design with $K = 200$, $M = 40$. (a) PER performance. (b) PER and FAR performance.

Recall that the proposed receiver directly recovers the received packets without explicit user activity detection. In practice, user activity detection can be done by verifying the user identification information inserted in a data packet after its recovery. For this reason, the miss detection rate (MDR) of active users is equal to the PER in the considered scheme, since a packet recovery error corresponds to a user activity detection error.

Figure 15.7(a) shows the PER performance of various schemes. It is observed that the BiGAMP and Turbo-BiGAMP algorithms work poorly, possibly due to the insufficient characterization and utilization of the complicated messages related to $\{x_{k,l}\}$ in this case. In addition, the proposed TAMP further outperforms the Reduced TAMP by about 1.3 dB (at PER $= 10^{-2}$), and has only a 0.4 dB performance gap away from the single-user bound, verifying the effectiveness of the proposed Bayesian recursive method of sub-SDL in learning sub-symbol delay and resolving ISI.

Figure 15.7(b) further shows the false alarm rate (FAR) performance of the proposed design, with FAR defined as

$$\text{FAR} = \frac{\text{No. of incorrectly recovered packets}}{\text{No. of total transmitted packets}},$$

where the number of incorrectly recovered packets is obtained by subtracting the number of correctly recovered packets from the number of total packets recovered. It is seen that the FAR performance of the proposed design is slightly worse than the PER performance. A possible reason is that each packet can (partly) appear in several observation windows due to the fully asynchronous transmissions and sliding-window based observation/detection scheme in use, thereby leading to a larger total number of recovered packets than that of actually transmitted packets.

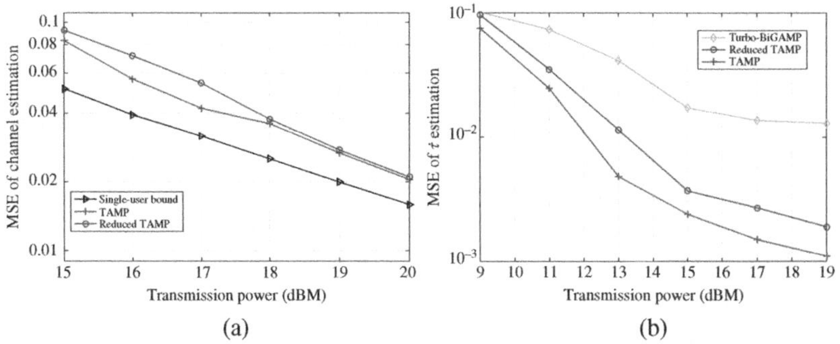

Figure 15.8 Parameter estimation of the proposed design with $K = 200$, $M = 40$. (a) Channel estimation. (b) Delay estimation.

Figure 15.8 shows the parameter estimation performance of the proposed design. Specifically, Fig. 15.8(a) shows the mean square error (MSE) performance of the channel estimation defined as $\text{E}\{\| \hat{\boldsymbol{h}}_k - \boldsymbol{h}_k\|^2\}/M$, with M being the size of $\boldsymbol{h}_k$. It is observed that again the proposed TAMP algorithm has a very good performance and performs close to the single-user bound. Figure 15.8(b) shows the MSE performance of the sub-symbol-level delay estimation defined as $\text{E}\{\| \hat{\dot{\tau}}_k - \dot{\tau}_k\|^2\}$. We see that the proposed TAMP outperforms all other schemes in all simulated SNR range.

15.3.5 Summary

We proposed an asynchronous random access scheme for massive connectivity. A low-complexity yet efficient algorithm named TAMP was developed based on judicious iterative receiver design and sophisticated applications of message passing principles. Numerical results corroborated that TAMP achieves a very good performance. As the proposed asynchronous multiple access scheme requires neither granting nor timing, it is especially suitable for massive connectivity with low-cost devices.

15.4 Conclusion

In this chapter, we proposed both synchronous and asynchronous random access scheme for massive connectivity. Based on judicious iterative receiver design and message passing principles, a novel JUICESD and TAMP algorithm were developed in the synchronous and the asynchronous system, respectively. Numerical results demonstrate that the proposed algorithms achieve a significant performance improvement over the existing alternatives.

References

L Bahl, J Cocke, F Jelinek, and J Raviv. Optimal decoding of linear codes for minimizing symbol error rate (Corresp.). *IEEE Transactions on Information Theory*, 20(2):284–287, Mar. 1974.

Xiaoming Chen, Derrick Wing Kwan Ng, Wei Yu, Erik G Larsson, Naofal Al-Dhahir, and Robert Schober. Massive access for 5G and beyond. *IEEE Journal on Selected Areas in Communications*, 39(3):615–637, Mar. 2021.

Linglong Dai, Bichai Wang, Yifei Yuan, Shuangfeng Han, I Chih-Lin, and Zhaocheng Wang. Non-orthogonal multiple access for 5G: Solutions, challenges, opportunities, and future research trends. *IEEE Communications Magazine*, 53(9):74–81, Sept. 2015.

Roberto B Di Renna and Rodrigo C de Lamare. Dynamic message scheduling based on activity-aware residual belief propagation for asynchronous mMTC. *IEEE Wireless Communications Letters*, 10(6):1290–1294, Mar. 2021.

Z Ding, Y Liu, J Choi, Q Sun, M Elkashlan, I Chih-Lin, and H V Poor. Application of non-orthogonal multiple access in LTE and 5G networks. *IEEE Communications Magazine*, 55(2):185–191, Feb. 2017.

Tian Ding, Xiaojun Yuan, and Soung Chang Liew. Sparsity learning-based multiuser detection in grant-free massive-device multiple access. *IEEE Transactions on Wireless Communications*, 18(7):3569–3582, Jul. 2019.

Yang Du, Binhong Dong, Wuyong Zhu, Pengyu Gao, Zhi Chen, Xiaodong Wang, and Jun Fang. Joint channel estimation and multiuser detection for uplink grant-free NOMA. *IEEE Wireless Communications Letters*, 7(4):682–685, Aug. 2018.

Fayezeh Ghavimi and Hsiao-Hwa Chen. M2M communications in 3GPP LTE/LTE-A networks: Architectures, service requirements, challenges, and applications. *IEEE Communications Surveys & Tutorials*, 17(2):525–549, Oct. 2015.

Gabor Hannak, Martin Mayer, Alexander Jung, Gerald Matz, and Norbert Goertz. Joint channel estimation and activity detection for multiuser communication systems. In *Proceedings of the IEEE ICC Workshops*, pages 2086–2091, London, UK, Jun. 2015.

M Hasan, E Hossain, and D Niyato. Random access for machine-to-machine communication in LTE-advanced networks: Issues and approaches. *IEEE Communications Magazine*, 51(6):86–93, Jun. 2013.

S Jiang, C Xu, X Yuan, Z Han, and W Wang. Bayesian receiver design for asynchronous massive connectivity. In *Proceedings of the IEEE ICC*, pages 1–6, Seoul, Korea, May 2022.

L Liu and Y Liu. An efficient algorithm for device detection and channel estimation in asynchronous IOT systems. In *Proceedings of the IEEE ICASSP2021*, pages 4815–4819, Toronto, Ontario, Canada, Jun. 2021.

Liang Liu and Wei Yu. Massive connectivity with massive MIMO –Part I: Device activity detection and channel estimation. *IEEE Transactions on Signal Processing*, 66(11):2933–2946, Jun. 2018.

Yuanwei Liu, Zhijin Qin, Maged Elkashlan, Zhiguo Ding, Arumugam Nallanathan, and Lajos Hanzo. Nonorthogonal multiple access for 5G and beyond. *Proceedings of the IEEE*, 105(12):2347–2381, Dec. 2017.

Liang Liu, Erik G Larsson, Wei Yu, Petar Popovski, Cedomir Stefanovic, and Elisabeth De Carvalho. Sparse signal processing for grant-free massive connectivity: A future paradigm for random access protocols in the Internet of Things. *IEEE Signal Processing Magazine*, 35(5):88–99, Sep. 2018.

Jason T Parker, Philip Schniter, and Volkan Cevher. Bilinear generalized approximate message passing –Part I: Derivation. *IEEE Transactions on Signal Processing*, 62(22):5839–5853, Nov. 2014a.

Jason T Parker, Philip Schniter, and Volkan Cevher. Bilinear generalized approximate message passing –Part II: Applications. *IEEE Transactions on Signal Processing*, 62(22):5854–5867, Nov. 2014b.

Sundeep Rangan. Generalized approximate message passing for estimation with random linear mixing. *arXiv:1010.5141*, Aug. 2012.

Sundeep Rangan, Philip Schniter, Erwin Riegler, Alyson Fletcher, and Volkan Cevher. Fixed points of generalized approximate message passing with arbitrary matrices. *IEEE Transactions on Information Theory*, 62(12):7464–7474, Dec. 2016.

Peng Sun, Zhongyong Wang, and Philip Schniter. Joint channel-estimation and equalization of single-carrier systems via bilinear AMP. *IEEE Transactions on Signal Processing*, 66(10):2772–2785, May 2018.

Bichai Wang, Linglong Dai, Talha Mir, and Zhaocheng Wang. Joint user activity and data detection based on structured compressive sensing for NOMA. *IEEE Communications Letters*, 20(7):1473–1476, Jul. 2016.

Chao Wei, Huaping Liu, Zaichen Zhang, Jian Dang, and Liang Wu. Approximate message passing-based joint user activity and data detection for NOMA. *IEEE Communications Letters*, 21(3):640–643, Mar. 2017.

Fan Wei, Wen Chen, Yongpeng Wu, Jun Ma, and Theodoros A Tsiftsis. Message-passing receiver design for joint channel estimation and data decoding in uplink grant-free SCMA systems. *IEEE Transactions on Wireless Communications*, 18(1):167–181, Jan. 2019.

Jianwen Zhang, Xiaojun Yuan, and Ying Jun Angela Zhang. Blind signal detection in massive MIMO: Exploiting the channel sparsity. *IEEE Transactions on Communications*, 66(2):700–712, Feb. 2018.

W Zhu, M Tao, X Yuan, and Y Guan. Deep-learned approximate message passing for asynchronous massive connectivity. *IEEE Transactions on Wireless Communications*, 20(8):5434–5448, Aug. 2021.

Justin Ziniel and Philip Schniter. Efficient high-dimensional inference in the multiple measurement vector problem. *IEEE Transactions on Signal Processing*, 61(2):340–354, Jan. 2013.

16

Grant-Free Random Access via Covariance-Based Approach

Ya-Feng Liu[1], Wei Yu[2], Ziyue Wang[1,3], Zhilin Chen[2], and Foad Sohrabi[4]

[1] *Institute of Computational Mathematics and Scientific/Engineering Computing, Academy of Mathematics and Systems Science, Chinese Academy of Sciences, Beijing, China*
[2] *The Edward S. Rogers Sr. Department of Electrical and Computer Engineering, University of Toronto, Toronto, Ontario, Canada*
[3] *School of Mathematical Sciences, University of Chinese Academy of Sciences, Beijing, China*
[4] *Nokia Bell Labs, Murray Hill, NJ, USA*

16.1 Introduction

Massive machine-type communication (mMTC) is expected to play a crucial role in the fifth-generation (5G) cellular systems and beyond (Bockelmann et al., 2016). One of the main challenges in mMTC is massive random access, in which a massive number of devices with sporadic data traffic wish to connect to the network in the uplink (Chen et al., 2021a). Conventional cellular systems provide random access for human-type communications by employing a set of orthogonal sequences, from which every active device randomly and independently selects one sequence to transmit as a pilot for requesting access (Dahlman et al., 2013). However, when the number of active devices is comparable to the number of available orthogonal sequences, this uncoordinated random access approach inevitably leads to collisions (with high probability), which will result in a (severe) delay of the data transmission stage because multiple rounds of retransmission signaling are required to resolve the collisions. As such, the above random access scheme is generally not suitable for mMTC.

To reduce the communication latency, grant-free random access schemes are proposed (Liu et al., 2018), where the active devices directly transmit the data signals after transmitting their preassigned non-orthogonal signature sequences without first obtaining permissions from the base stations (BSs). The BSs first identify the active devices based on the signatures, then decode the data. In this paradigm, no handshake is needed. However, the non-orthogonality of the

Next Generation Multiple Access, First Edition.
Edited by Yuanwei Liu, Liang Liu, Zhiguo Ding, and Xuemin Shen.
© 2024 The Institute of Electrical and Electronics Engineers, Inc. Published 2024 by John Wiley & Sons, Inc.

signature sequences would cause both intra-cell and inter-cell interference, which pose unique challenges in the task of device activity detection. This chapter studies the theory and algorithms for device activity detection in the grant-free random access protocol.

There are generally two mathematical optimization formulations of the device activity detection problem. In the first formulation, the device activity detection problem is formulated as a compressed sensing (CS) problem, in which the instantaneous channel state information (CSI) and the device activity are jointly recovered by exploiting the sparsity in the device activity pattern (Senel and Larsson, 2018, Liu and Yu, 2018, Chen et al., 2018). When the CSI is not needed (e.g., when the data are embedded in the pilot sequence (Senel and Larsson, 2018, Chen et al., 2019) and the BSs are equipped with a large number of antennas, it is also possible to jointly estimate the device activities (and the channel large-scale fading components) by exploiting the channel statistical information via maximum likelihood estimation (MLE). This approach is proposed in Haghighatshoar et al. (2018) and termed as the covariance-based approach because the detection relies on the sample covariance matrix of the received signal. As compared to the CS approach, this covariance-based approach has the advantage of being able to detect many more active devices due to its quadratic scaling law (Fengler et al., 2021, Haghighatshoar et al., 2018, Chen et al., 2022). It is worthwhile remarking that even in the situation where the CSI is needed, the covariance-based approach can still play an important role, e.g., in a three-phase protocol (Kang and Yu, 2022), where in the first phase, the BSs apply the covariance-based approach to detect device activities; in the second phase, the BSs transmit a common feedback message to all the active devices to schedule them in orthogonal transmission slots; and finally, in the third phase, the BSs estimate channels and detect data from the active devices. Since the users are scheduled in orthogonal channels in the three-phase protocol (Kang and Yu, 2022), the channel estimation performance is expected to be better than that of the grant-free protocol (Liu et al., 2018) based on non-orthogonal pilots.

This chapter focuses on the covariance-based approach for the device activity detection problem. We mainly study two questions. The first question is how many active devices can be successfully identified out of a large number of potential devices by the covariance-based approach given a pilot sequence length and assuming a fixed set of non-orthogonal pilot sequences. The answer to the above question leads to a theoretical characterization of the detection performance of the covariance-based approach. The second question is how to efficiently and correctly identify the active devices. To answer the above question, we present several computationally efficient algorithms for solving the device activity detection problem.

The rest of this chapter is organized as follows. Sections 16.2 and 16.3 study the theory and algorithms for the covariance-based approach for the device activity detection problem in the single-cell and multi-cell massive multiple-input multiple-output (MIMO) systems, respectively. Section 16.4 discusses some practical issues and presents two interesting extensions. Finally, Section 16.5 concludes this chapter and lists some possible directions for future research.

16.2 Device Activity Detection in Single-Cell Massive MIMO

16.2.1 System Model and Problem Formulation

In this section, we consider an uplink single-cell massive random access scenario with N single-antenna devices communicating with a BS equipped with M antennas. We assume a block fading channel model, i.e., the channel coefficients remain constant for a coherence interval. We also assume that the user traffic is sporadic, i.e., only $K \ll N$ devices are active during each coherence interval. For the purpose of device identification, each device n is preassigned a unique signature sequence $\mathbf{s}_n = [s_{1n}, s_{2n}, \dots, s_{Ln}]^T \in \mathbb{C}^L$, where L is the sequence length. In the pilot phase, we assume that all the active devices transmit their signature sequences synchronously at the same time. (We consider a more practical asynchronous scenario in Section 16.4.2.) The objective is to detect which subset of devices are active based on the received signal at the BS.

Let $a_n \in \{0, 1\}$ denote the activity of device n in a given coherence interval, i.e., $a_n = 1$ if the device is active and $a_n = 0$ otherwise. The channel vector between the BS and device n is modeled as a random vector $\sqrt{g_n}\mathbf{h}_n$, where $g_n \geq 0$ is the large-scale fading component due to path-loss and shadowing, and $\mathbf{h}_n \in \mathbb{C}^M$ is the Rayleigh fading component following $\mathcal{CN}(\mathbf{0}, \mathbf{I})$. The received signal $\mathbf{Y} \in \mathbb{C}^{L\times M}$ at the BS in the pilot phase can be expressed as

$$\mathbf{Y} = \sum_{n=1}^{N} a_n \mathbf{s}_n \sqrt{g_n}\mathbf{h}_n^T + \mathbf{W} = \mathbf{S}\boldsymbol{\Gamma}^{\frac{1}{2}}\mathbf{H} + \mathbf{W}, \tag{16.1}$$

where $\boldsymbol{\Gamma} = \text{diag}(\gamma_1, \gamma_2, \dots, \gamma_N) \in \mathbb{R}^{N\times N}$ with $\gamma_n = a_n g_n$ is a diagonal matrix indicating both the device activity a_n and the large-scale fading component g_n, $\mathbf{S} = [\mathbf{s}_1, \mathbf{s}_2, \dots, \mathbf{s}_N] \in \mathbb{C}^{L\times N}$ is the signature sequence matrix (which is assumed to be known at the BS), $\mathbf{H} = [\mathbf{h}_1, \mathbf{h}_2, \dots, \mathbf{h}_N]^T \in \mathbb{C}^{N\times M}$ is the channel matrix, and $\mathbf{W} \in \mathbb{C}^{L\times M}$ is the normalized effective independent and identically distributed (i.i.d.) Gaussian noise with variance σ_w^2. We let $\boldsymbol{\gamma} = [\gamma_1, \gamma_2, \cdots, \gamma_N]^T \in \mathbb{R}^N$ denote the diagonal entries of $\boldsymbol{\Gamma}$ and use $\boldsymbol{\gamma}$ and $\boldsymbol{\Gamma}$ interchangeably throughout this section.

Following the approach proposed in Fengler et al. (2021) and Haghighatshoar et al. (2018), we first use MLE to estimate γ from $\mathbf{Y}$, then thereafter obtain the device activity indicator a_n from γ. The idea is to treat γ as a set of deterministic but unknown parameters, and to model $\mathbf{Y}$ as an observation that follows the conditional distribution $p(\mathbf{Y} \mid \gamma)$ based on the statistics of $\mathbf{h}_n$ and $\mathbf{W}$. To compute the likelihood $p(\mathbf{Y} \mid \gamma)$, we first observe from (16.1) that given γ, the columns of $\mathbf{Y}$, denoted by $\mathbf{y}_m \in \mathbb{C}^L$, $1 \leq m \leq M$, are independent due to the i.i.d. channel coefficients over different antennas. In particular, each column $\mathbf{y}_m$ follows a complex Gaussian distribution as $\mathbf{y}_m \sim \mathcal{CN}(\mathbf{0}, \boldsymbol{\Sigma})$, where the covariance matrix is given by

$$\boldsymbol{\Sigma} = \mathbb{E}\left[\mathbf{y}_m \mathbf{y}_m^H\right] = \mathbf{S}\boldsymbol{\Gamma}\mathbf{S}^H + \sigma_w^2 \mathbf{I} = \sum_{n=1}^{N} \gamma_n \mathbf{s}_n \mathbf{s}_n^H + \sigma_w^2 \mathbf{I}. \tag{16.2}$$

Due to the independence of the columns of $\mathbf{Y}$, the likelihood function $p(\mathbf{Y} \mid \gamma)$ can be computed as

$$\begin{aligned} p(\mathbf{Y} \mid \gamma) &= \prod_{m=1}^{M} \frac{1}{|\pi \boldsymbol{\Sigma}|} \exp\left(-\mathbf{y}_m^H \boldsymbol{\Sigma}^{-1} \mathbf{y}_m\right) \\ &= \frac{1}{|\pi \boldsymbol{\Sigma}|^M} \exp\left(-\operatorname{tr}\left(\boldsymbol{\Sigma}^{-1} \mathbf{Y}\mathbf{Y}^H\right)\right), \end{aligned} \tag{16.3}$$

where $|\cdot|$ denotes the determinant of a matrix. The maximization of $\log p(\mathbf{Y} \mid \gamma)$ is equivalent to the minimization of $-\frac{1}{M} \log p(\mathbf{Y} \mid \gamma)$, so the MLE problem can be formulated as

$$\min_{\gamma} \quad \log |\boldsymbol{\Sigma}| + \operatorname{tr}\left(\boldsymbol{\Sigma}^{-1} \widehat{\boldsymbol{\Sigma}}\right) \tag{16.4a}$$

$$\text{s.t.} \quad \gamma \geq 0, \tag{16.4b}$$

where

$$\widehat{\boldsymbol{\Sigma}} = \frac{1}{M} \mathbf{Y}\mathbf{Y}^H = \frac{1}{M} \sum_{m=1}^{M} \mathbf{y}_m \mathbf{y}_m^H \tag{16.5}$$

is the sample covariance matrix of the received signal averaged over different antennas, and the constraint $\gamma \geq 0$ is due to the fact that $\gamma_n = a_n g_n$. Throughout this chapter, we focus on the massive MIMO regime where M is large, which ensures that the sample covariance matrix $\widehat{\boldsymbol{\Sigma}}$ in (16.5) is a good approximation of the true covariance matrix in (16.2).

We observe from (16.4) that the MLE problem depends on $\mathbf{Y}$ through the sample covariance matrix $\widehat{\boldsymbol{\Sigma}}$. For this reason, the approach based on solving the formulation in (16.4) is termed as the covariance-based approach in the literature. As M increases, $\widehat{\boldsymbol{\Sigma}}$ tends to the true covariance matrix of $\mathbf{Y}$, but the size of the optimization problem does not change. As such, the complexity of solving (16.4) does not scale with M. This is a desirable property especially for the massive MIMO systems.

It is noteworthy to mention an alternative way to model the device activity detection problem is through the following non-negative least squares (NNLS) formulation (Fengler et al., 2021, Haghighatshoar et al., 2018):

$$\min_{\gamma} \left\| \mathbf{\Sigma} - \widehat{\mathbf{\Sigma}} \right\|_F^2 \tag{16.6a}$$

$$\text{s.t. } \gamma \geq 0. \tag{16.6b}$$

The above NNLS formulation tries to match the true covariance matrix $\mathbf{\Sigma}$ and the sample covariance matrix $\widehat{\mathbf{\Sigma}}$ as much as possible under the Frobenius norm metric. The optimization problem (16.6) is convex. However, it has been shown in Fengler et al. (2021) and Chen et al. (2022) that the detection performance of the NNLS formulation (16.6) is much worse than that of the MLE formulation (16.4). Therefore, we focus on the MLE formulation (16.4) in this chapter.

16.2.2 Phase Transition Analysis

In this section, we assume that the MLE problem (16.4) is solved to global optimality and analyze the asymptotic properties of the true MLE solution $\hat{\gamma}^{(M)}$ in the massive MIMO regime where $M \to \infty$. Although the global minimizer of (16.4) may not be easily found in practice due to its nonconvex nature, simulation results show that the analysis still provides useful insights into the performance of practical algorithms for solving the problem (16.4). In Section 16.2.3, we present efficient algorithms for solving the MLE problem (16.4).

For notational clarity, let γ^0 denote the true parameter to be estimated. We aim to answer the following theoretical question in this section: what are the conditions on the system parameters N, K, and L such that the MLE solution $\hat{\gamma}^{(M)}$ can approach the true parameter γ^0 as $M \to \infty$? The answer to this question helps identify the desired operating regime in the space of N, K, and L for getting an accurate estimate $\hat{\gamma}^{(M)}$ via MLE with massive MIMO. The phase transition analysis result in this section is mainly from Chen et al. (2022).

Since the Fisher information matrix $\mathbf{J}(\gamma)$ plays a key role in the analysis of MLE, we first provide an explicit expression for $\mathbf{J}(\gamma)$.

Theorem 16.1 Consider the likelihood function in (16.3), where γ is the parameter to be estimated, and define $\mathbf{P} = \mathbf{S}^H \left(\mathbf{S}\mathbf{\Gamma}\mathbf{S}^H + \sigma_w^2 \mathbf{I}\right)^{-1} \mathbf{S}$. The associated $N \times N$ Fisher information matrix of γ is given by

$$\mathbf{J}(\gamma) = M\left(\mathbf{P} \odot \mathbf{P}^*\right), \tag{16.7}$$

where $\odot$ is the element-wise product, and $(\cdot)^*$ is the conjugate operation.

Next, we present a necessary and sufficient condition such that $\hat{\gamma}^{(M)}$ can approach γ^0 in the large M limit.

Theorem 16.2 Consider the MLE problem (16.4) for device activity detection with given signature sequence matrix $\mathbf{S} \in \mathbb{C}^{L\times N}$ and noise variance σ_w^2, and let $\hat{\gamma}^{(M)}$ be a sequence of solutions of (16.4) as M increases. Let γ^0 be the true parameter whose $N - K$ zero entries are indexed by $\mathcal{I}$, i.e.,

$$\mathcal{I} = \{i \mid \gamma_i^0 = 0\}. \tag{16.8}$$

Define

$$\mathcal{N} = \left\{\mathbf{x} \in \mathbb{R}^N \mid \mathbf{x}^T \mathbf{J}(\gamma^0)\mathbf{x} = 0\right\}, \tag{16.9}$$

$$\mathcal{C} = \left\{\mathbf{x} \in \mathbb{R}^N \mid x_i \geq 0, i \in \mathcal{I}\right\}, \tag{16.10}$$

where x_i is the i-th entry of $\mathbf{x}$. Then a necessary and sufficient condition for the consistency of $\hat{\gamma}^{(M)}$, i.e., $\hat{\gamma}^{(M)} \to \gamma^0$ as $M \to \infty$, is that the intersection of $\mathcal{N}$ and $\mathcal{C}$ is the zero vector, i.e., $\mathcal{N} \cap \mathcal{C} = \{\mathbf{0}\}$.

The sets $\mathcal{N}$ and $\mathcal{C}$ in Theorem 16.2 can be interpreted as follows: $\mathcal{N}$ is the null space of $\mathbf{J}(\gamma^0)$, which contains all directions $\mathbf{x}$ from γ^0 along which the likelihood function stays unchanged, i.e., $p(\mathbf{Y} \mid \gamma^0) = p(\mathbf{Y} \mid \gamma^0 + t\mathbf{x})$ holds for any sufficiently small positive t and any $\mathbf{x} \in \mathcal{N}$; $\mathcal{C}$ is a cone, which contains vectors whose coordinates indexed by $\mathcal{I}$ are always nonnegative – in other words, directions $\mathbf{x}$ from γ^0 along which $\gamma^0 + t\mathbf{x} \in [0, +\infty)^N$ holds for any sufficiently small positive t. The condition $\mathcal{N} \cap \mathcal{C} = \{\mathbf{0}\}$ guarantees that the likelihood function $p(\mathbf{Y} \mid \gamma)$ in the feasible neighborhood of γ^0 is not identical to $p(\mathbf{Y} \mid \gamma^0)$, so that the true parameter γ^0 is uniquely identifiable in the feasible region via the likelihood function maximization. See Chen et al. [2022, Fig. 1] (and the related discussion) for an illustration of the condition $\mathcal{N} \cap \mathcal{C} = \{\mathbf{0}\}$.

Since there is generally no closed-form characterization of $\mathcal{N} \cap \mathcal{C}$, we cannot analytically verify the condition $\mathcal{N} \cap \mathcal{C} = \{\mathbf{0}\}$ for a given γ^0 and $\mathbf{J}(\gamma^0)$. However, by noting that $\mathcal{N}$ and $\mathcal{C}$ are both convex sets, we can numerically test whether the condition $\mathcal{N} \cap \mathcal{C} = \{\mathbf{0}\}$ holds. By further exploiting the positive semidefiniteness of $\mathbf{J}(\gamma^0)$, the following theorem turns the verification of $\mathcal{N} \cap \mathcal{C} = \{\mathbf{0}\}$ into a linear program (LP).

Theorem 16.3 Given $\mathbf{S}$, σ_w^2, and γ^0, let $\mathbf{J}(\gamma^0)$ be the Fisher information matrix in (16.7); let $\mathbf{A} \in \mathbb{R}^{(N-K)\times(N-K)}$ be a submatrix of $\mathbf{J}(\gamma^0)$ indexed by $\mathcal{I}$; let $\mathbf{C} \in \mathbb{R}^{K\times K}$ be a submatrix of $\mathbf{J}(\gamma^0)$ indexed by $\mathcal{I}^c$, where $\mathcal{I}^c$ is the complement of $\mathcal{I}$ with respect to $\{1, 2, \ldots, N\}$; and let $\mathbf{B} \in \mathbb{R}^{(N-K)\times K}$ be a submatrix of $\mathbf{J}(\gamma^0)$ with rows and columns indexed by $\mathcal{I}$ and $\mathcal{I}^c$, respectively. Then the condition $\mathcal{N} \cap \mathcal{C} = \{\mathbf{0}\}$

(in Theorem 16.2) is equivalent to: (i) $\mathbf{C}$ is invertible; and (ii) the following problem is feasible

$$\text{find } \mathbf{x} \tag{16.11a}$$

$$\text{s.t. } (\mathbf{A} - \mathbf{B}\mathbf{C}^{-1}\mathbf{B}^T)\mathbf{x} > \mathbf{0}, \tag{16.11b}$$

where vector $\mathbf{x} \in \mathbb{R}^{N-K}$.

Theorem 16.3 offers a way of identifying the phase transition of the MLE problem numerically. More specifically, suppose that $\mathbf{S}$ and $\mathcal{I}$ are generated randomly according to some distribution for any fixed N, L, and K (e.g., $\mathbf{S}$ is Gaussian and the elements in $\mathcal{I}$ are uniformly selected from $\{1, 2, \ldots, N\}$). We can then use (16.3) to test the consistency of the MLE solution for each realization of $\mathbf{S}$ and $\mathcal{I}$. This enables us to numerically characterize the region in the space of N, L, and K such that $\hat{\gamma}^{(M)}$ can approach γ^0 in the large M limit. We present some simulation results on the phase transition of the MLE problem in Section 16.2.4.

The last theorem in this section shows the quadratic scaling law of the covariance-based approach, i.e., the maximum number of active devices K that can be correctly detected by the covariance-based approach increases quadratically with the length of the signature sequence L. Intuitively, the covariance-based approach tries to match the sample covariance matrix and the true covariance matrix (by taking the derivative of the objective in (1.4) with respect to $\mathbf{\Sigma}$), hence the number of effective observations in the covariance-based approach is in the order of L^2. The technical reason behind this quadratic scaling law is that the set $\mathcal{N}$ defined in (16.9) is equivalent to the null space of the matrix $\widehat{\mathbf{S}} = [\mathbf{s}_1^* \otimes \mathbf{s}_1, \ldots, \mathbf{s}_N^* \otimes \mathbf{s}_N] \in \mathbb{C}^{L^2 \times N}$ (where $\otimes$ is the Kronecker product), and the matrix $\widehat{\mathbf{S}}$ enjoys the null space property under such a scaling law (Chen et al., 2022).

Theorem 16.4 Let $\mathbf{S} \in \mathbb{C}^{L \times N}$ be the signature sequence matrix whose columns are uniformly drawn from the sphere of radius $\sqrt{L}$ in an i.i.d. fashion. There exist some constants c_1 and c_2 whose values do not depend on K, L, and N such that if

$$K \leq c_1 L^2 / \log^2(eN/L^2), \tag{16.12}$$

then $\mathcal{N} \cap \mathcal{C} = \{\mathbf{0}\}$ (in Theorem 16.2) holds true with probability at least $1 - \exp(-c_2 L)$.

16.2.3 Coordinate Descent Algorithms

The optimization problem (16.4) is not convex due to the fact that $\text{tr}(\mathbf{\Sigma}^{-1}\widehat{\mathbf{\Sigma}})$ is convex whereas $\log|\mathbf{\Sigma}|$ is concave in γ. However, various practical algorithms are designed and shown to have excellent performance in terms of computational

Algorithm 16.1 Coordinate Descent Algorithm for Solving Problem (16.4)

1: Initialize $\boldsymbol{\gamma} = \mathbf{0}$, $\boldsymbol{\Sigma}^{-1} = \sigma_w^{-2}\mathbf{I}$;
2: **repeat** [*one iteration*]
3: Randomly select a permutation $i_1, i_2, \dots, i_N$ of the coordinate indices $\{1, 2, \dots, N\}$;
4: **for** $n = 1$ to N **do**
5: $d = \max\left\{\frac{\mathbf{s}_{i_n}^H\boldsymbol{\Sigma}^{-1}\hat{\boldsymbol{\Sigma}}\boldsymbol{\Sigma}^{-1}\mathbf{s}_{i_n} - \mathbf{s}_{i_n}^H\boldsymbol{\Sigma}^{-1}\mathbf{s}_{i_n}}{(\mathbf{s}_{i_n}^H\boldsymbol{\Sigma}^{-1}\mathbf{s}_{i_n})^2}, -\gamma_{i_n}\right\}$;
6: $\gamma_{i_n} \leftarrow \gamma_{i_n} + d$;
7: $\boldsymbol{\Sigma}^{-1} \leftarrow \boldsymbol{\Sigma}^{-1} - \frac{d\boldsymbol{\Sigma}^{-1}\mathbf{s}_{i_n}\mathbf{s}_{i_n}^H\boldsymbol{\Sigma}^{-1}}{1 + d\mathbf{s}_{i_n}^H\boldsymbol{\Sigma}^{-1}\mathbf{s}_{i_n}}$;
8: **end for**
9: **until** $\|\text{Proj}(\boldsymbol{\gamma} - \nabla f(\boldsymbol{\gamma})) - \boldsymbol{\gamma}\|_2 < \epsilon$;
10: Output $\boldsymbol{\gamma}$.

efficiency and detection error probability for solving problem (16.4). Examples of these algorithms include coordinate descent (CD) (Haghighatshoar et al., 2018, Chen et al., 2019) and accelerated variants (Wang et al., 2021b, Dong et al., 2022), expectation-maximization/minimization (EM) (i.e., sparse Bayesian learning) (Wipf and Rao, 2007), gradient descent (Wang et al., 2021a), sparse iterative covariance-based estimation (SPICE) (Stoica et al., 2011), etc. Among the above algorithms, the CD algorithm that iteratively updates the variable associated with each device is popular for solving the MLE problem in the covariance-based approach. The reason for its popularity is that each of its subproblems (i.e., the optimization of the original objective with respect to only one of the variables) admits a closed-form solution (Haghighatshoar et al., 2018), which makes it easily implementable. In this section, we introduce the CD algorithm and its accelerated variant for solving problem (16.4).

16.2.3.1 Coordinate Descent Algorithm

The basic idea of the CD algorithm for solving problem (16.4) is to update each coordinate of the unknown variable $\boldsymbol{\gamma}$ iteratively (while keeping the others fixed) until convergence. In particular, by fixing all the other variables except γ_{i_n}, the problem reduces to a univariate optimization problem, which admits a closed-form solution due to its special structure (Haghighatshoar et al., 2018), shown in lines 5 and 6 of Algorithm 16.1.

In addition to the coordinate update strategy (i.e., how to update the selected variable), the coordinate selection strategy (i.e., selecting which coordinate to update) also plays a vital role in the CD algorithm. Two commonly used strategies are random permutation (which randomly permutes all coordinates and then updates the coordinate one by one according to the order in the permutation) and random selection (which randomly picks one coordinate from all coordinates at a time). Note that in Algorithm 16.1, we adopt the random permutation strategy; see line 3 of Algorithm 16.1. Based on our experience of solving problem (16.4), CD equipped with the random permutation strategy is more efficient than that equipped with the random selection strategy. Due to the randomness in the coordinate selection strategy, the CD algorithm sometimes is called random CD in the literature.

The dominant complexity of (random) CD Algorithm 16.1 is the matrix-vector multiplications in lines 5–7, whose complexity is $\mathcal{O}(L^2)$. Note that in line 7, a rank-one update of $\boldsymbol{\Sigma}^{-1}$ is used to reduce the computational cost and improve the computational efficiency. The overall complexity of Algorithm 16.1 is $\mathcal{O}(INL^2)$, where I is the total number of iterations. Because the complexity of the CD algorithm is linear in N and quadratic in L, it is particularly suitable for low-latency mMTC scenarios where N is often large and L is often small.

Once Algorithm 16.1 returns $\boldsymbol{\gamma}$, in order to do the detection, we still need to employ the element-wise thresholding to determine a_n from γ_n, the n-th entry of $\boldsymbol{\gamma}$, using a threshold l_{th}, i.e., $a_n = 1$ if $\gamma_n \geq l_{th}$ and $a_n = 0$ otherwise. The probabilities of missed detection and false alarm can be trade-off by setting different values for l_{th}.

16.2.3.2 Active Set Coordinate Descent Algorithm

In this section, we present a computationally more efficient active set CD algorithm for solving the device activity detection problem in (16.4). Note that most of the devices are inactive (i.e., $K \ll N$). The basic idea of the active set CD algorithm is to exploit the sparsity in the device activity pattern to avoid unnecessary computations on the inactive devices and improve the computational efficiency of the CD algorithm (i.e., Algorithm 16.1). In particular, at each iteration, the active set CD algorithm first selects a small subset of all devices, termed as the active set, which contains a number of devices that contribute the most to the deviation from the first-order optimality condition of the optimization problem (16.4), then applies the CD algorithm to update the selected variables in the active set.

We first present the first-order optimality condition of the optimization problem (16.4). Let $f(\boldsymbol{\gamma})$ denote the objective function of problem (16.4). Then, for any $n = 1, 2, \ldots, N$, the gradient of $f(\boldsymbol{\gamma})$ with respect to γ_n is

$$\left[\nabla f(\boldsymbol{\gamma})\right]_n = \mathbf{s}_n^H \boldsymbol{\Sigma}^{-1} \mathbf{s}_n - \mathbf{s}_n^H \boldsymbol{\Sigma}^{-1} \widehat{\boldsymbol{\Sigma}} \boldsymbol{\Sigma}^{-1} \mathbf{s}_n.$$

The first-order (necessary) optimality condition of problem (16.4) is

$$\left[\nabla f(\boldsymbol{\gamma})\right]_n \begin{cases} = 0, & \text{if } \gamma_n > 0; \\ \geq 0, & \text{if } \gamma_n = 0, \end{cases} \quad \forall\, n, \tag{16.13}$$

Algorithm 16.2 Active Set CD Algorithm for Solving Problem (16.4)

1: Initialize $\gamma^0 = \mathbf{0}$, $k = 0$, $\boldsymbol{\delta}^0 > \mathbf{0}$, and $\varepsilon > 0$;
2: **repeat** [*one iteration*]
3: Update $\boldsymbol{\delta}^k$;
4: Select the active set $\mathcal{A}^k$ according to (16.4);
5: Apply lines 5–7 of Algorithm 16.1 to update all coordinates in $\mathcal{A}^k$ *only once* in the order of a random permutation;
6: Set $k \leftarrow k + 1$;
7: **until** $\|\text{Proj}(\gamma^k - \nabla f^k) - \gamma^k\|_2 < \varepsilon$;
8: Output γ^k.

which is equivalent to $\text{Proj}(\gamma - \nabla f(\gamma)) - \gamma = \mathbf{0}$, where $\text{Proj}(\cdot)$ denotes the projection operator (onto the feasible region of the corresponding problem). It can be shown that the complexity of computing $\nabla f(\gamma)$ is $\mathcal{O}(NL^2)$.

The selection strategy of the active set $\mathcal{A}^k$ at a given feasible point γ^k in Wang et al. (2021a) is mainly based on the degree of the violation of the first-order optimality condition (16.13), which is given by

$$\mathcal{A}^k = \left\{ n \;\middle|\; \gamma_n^k > 0 \text{ and } \left|\nabla f_n^k\right| > \delta_n^k \right\} \cup \left\{ n \;\middle|\; \gamma_n^k = 0 \text{ and } \nabla f_n^k < -\delta_n^k \right\}, \tag{16.14}$$

where ∇f_n^k denotes $\left[\nabla f(\gamma^k)\right]_n$ and $\boldsymbol{\delta}^k \in \mathbb{R}_+^N$ is a threshold vector that changes with iteration. The choice of the threshold vector $\boldsymbol{\delta}^k$ in (16.14) is important in balancing the competing goals of improving the objective function and reducing the computational cost at the k-th iteration. A method that works well in practice is to choose a relatively large $\boldsymbol{\delta}^0$ at first and update $\boldsymbol{\delta}^k$ by a multiplicative factor of less than one at each iteration.

The active set CD algorithm for solving problem (16.4) is summarized as Algorithm 16.2. Note that in line 5 of Algorithm 16.2, lines 5–7 of Algorithm 16.1 are used to update all coordinates in the selected active set $\mathcal{A}^k$. It is worth remarking here that it is also possible to choose other algorithms. In other words, the active set strategy can be used to accelerate any algorithm (that does not properly exploit the sparsity in the device activity pattern) for solving problem (16.4), such as those mentioned at the beginning of Section 16.2.3.

16.2.4 Performance Evaluation

In this section, we present some simulation results to validate the accuracy of the phase transition analysis, and compare the existing algorithms for solving the activity detection problem in (16.4). We use the same system parameters as in Chen et al. (2022). More specifically, we consider a single cell of radius 1000 m and the channel path-loss is modeled as $128.1 + 37.6\log_{10}(d)$. We consider the worst-case scenario that all devices are located in the cell edge such that the large-scale fading components g_n's are the same for all devices. The power spectrum density of the background noise is -169 dBm/Hz over 10 MHz and the transmit power of each device is set to 25 dBm. All signature sequences of length L are uniformly drawn from the sphere of radius $\sqrt{L}$ in an i.i.d. fashion as required in Theorem 16.4.

We solve the LP in (16.3) to numerically test the condition $\mathcal{N} \cap \mathcal{C} = \{\mathbf{0}\}$ in Theorem 16.2 under a variety of choices of L and K, given $N = 900$ or $N = 3600$. Figure 16.1 plots the region of $(L^2/N, K/N)$ in which the condition is satisfied or not. The result is obtained based on 100 random realizations of $\mathbf{S}$ and γ^0 for each given K and L. The error bars indicate the range beyond which either all realizations or no realization satisfy the condition. To validate the prediction by Theorem 16.2, we also run the CD algorithm to solve the MLE problem (16.4) by replacing the sample covariance matrix with the true covariance matrix (implying that $M \to \infty$). We then identify the region of $(L^2/N, K/N)$ in which the active devices can be perfectly detected, thus obtaining the phase transition curve empirically. We observe that the curves obtained by Theorem 16.2 and by the CD algorithm match pretty well. We also observe from Fig. 16.1 that K is approximately proportional to L^2, which verifies the scaling law in Theorem 16.4.

Next, we compare the efficiency of existing algorithms (including CD, active set CD, EM (Wipf and Rao, 2007), and SPICE (Stoica et al., 2011)) for solving

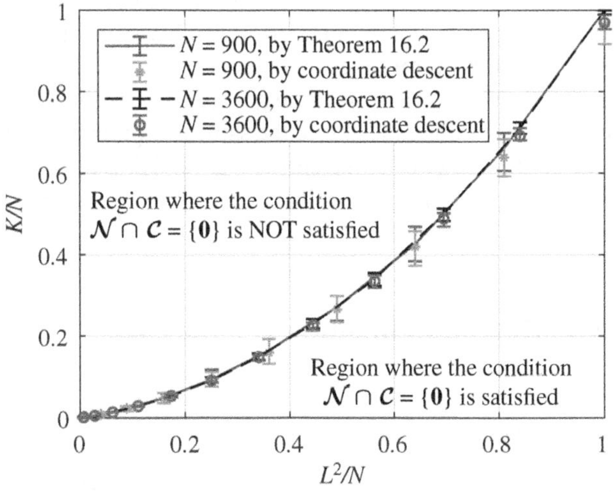

Figure 16.1 Phase transition of the covariance-based approach for device activity detection in the single-cell scenario.

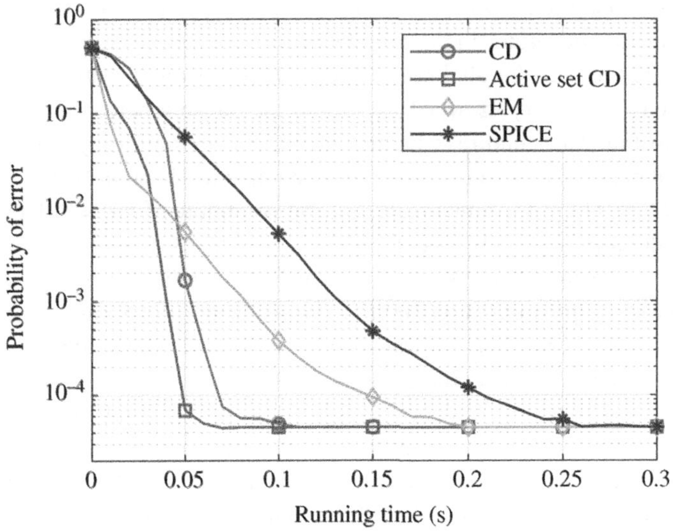

Figure 16.2 Comparison of the probability of error of existing algorithms versus running time in the single-cell scenario.

problem (16.4). Figure 16.2 plots the decrease of the probability of error as the algorithms run with $M = 128$, $L = 60$, $N = 3000$, and $K = 100$. The result is obtained by averaging over 1000 random realizations of $\mathbf{S}$, $\boldsymbol{\gamma}^0$, and the channel matrix $\mathbf{H}$. For each realization, we record the variable $\boldsymbol{\gamma}$ and calculate the corresponding probability of error when the algorithms run to a fixed moment. We observe from Fig. 16.2 that CD can take less time to achieve a better probability of error than EM and SPICE. It can also be observed from Fig. 16.2 that active set CD is more efficient than CD (due to the active set selection strategy). We also observe from Fig. 16.2 that all of these four algorithms can achieve the same low probability of error if their running time is allowed to be sufficiently long, implying that they converge to the same solution to problem (16.4) (albeit the problem is generally nonconvex).

16.3 Device Activity Detection in Multi-Cell Massive MIMO

16.3.1 System Model and Problem Formulation

In this section, we consider the multi-cell case, i.e., an uplink massive MIMO system consisting of B cells, where each cell contains one BS equipped with M antennas and N single-antenna devices. We assume that a cloud radio access network

(C-RAN) architecture is used for inter-cell interference mitigation, in which all B BSs are connected to a central unit (CU) via fronthaul links such that the signals received at the BSs can be jointly processed at the CU. Similar to the single-cell scenario, we also assume that only $K \ll N$ devices are active in each cell during any coherence interval. For device identification, each device n in cell j is preassigned a unique signature sequence $\mathbf{s}_{jn} \in \mathbb{C}^L$ with L being the sequence length. Let a_{jn} be a binary variable with $a_{jn} = 1$ for active devices and $a_{jn} = 0$ for inactive devices. The channel between device n in cell j and BS b is denoted as $\sqrt{g_{bjn}}\mathbf{h}_{bjn}$, where $g_{bjn} \geq 0$ is the large-scale fading coefficient, and $\mathbf{h}_{bjn} \in \mathbb{C}^M$ is the i.i.d. Rayleigh fading component that follows $\mathcal{CN}(\mathbf{0}, \mathbf{I})$.

Assume that all active devices synchronously transmit their preassigned signature sequences to the BSs in the uplink pilot stage. Then, the received signal at BS b can be expressed as

$$\begin{aligned}\mathbf{Y}_b &= \sum_{n=1}^{N} a_{bn}\mathbf{s}_{bn}\sqrt{g_{bbn}}\mathbf{h}_{bbn}^T + \sum_{j\neq b}\sum_{n=1}^{N} a_{jn}\mathbf{s}_{jn}\sqrt{g_{bjn}}\mathbf{h}_{bjn}^T + \mathbf{W}_b \\ &= \mathbf{S}_b\mathbf{A}_b\sqrt{\mathbf{G}_{bb}}\mathbf{H}_{bb} + \sum_{j\neq b}\mathbf{S}_j\mathbf{A}_j\sqrt{\mathbf{G}_{bj}}\mathbf{H}_{bj} + \mathbf{W}_b, \end{aligned} \tag{16.15}$$

where $\mathbf{S}_j = [\mathbf{s}_{j1}, \mathbf{s}_{j2}, \dots, \mathbf{s}_{jN}] \in \mathbb{C}^{L\times N}$ is the signature sequence matrix of the devices in cell j, $\mathbf{A}_j = \text{diag}(a_{j1}, a_{j2}, \dots, a_{jN})$ is a diagonal matrix that indicates the activity of all devices in cell j, $\mathbf{G}_{bj} = \text{diag}(g_{bj1}, g_{bj2}, \dots, g_{bjN})$ contains the large-scale fading components between the devices in cell j and BS b, $\mathbf{H}_{bj} = [\mathbf{h}_{bj1}, \mathbf{h}_{bj2}, \dots, \mathbf{h}_{bjN}]^T \in \mathbb{C}^{N\times M}$ is the Rayleigh fading channel between the devices in cell j and BS b, and $\mathbf{W}_b$ is the additive Gaussian noise that follows $\mathcal{CN}(\mathbf{0}, \sigma_w^2\mathbf{I})$ with σ_w^2 being the variance of the background noise normalized by the transmit power.

For notational simplicity, let $\mathbf{S} = [\mathbf{S}_1, \mathbf{S}_2, \dots, \mathbf{S}_B] \in \mathbb{C}^{L\times BN}$ denote the signature matrix of all devices, and let $\mathbf{G}_b = \text{diag}(\mathbf{G}_{b1}, \mathbf{G}_{b2}, \dots, \mathbf{G}_{bB}) \in \mathbb{R}^{BN\times BN}$ denote the matrix containing large-scale fading components between all devices and BS b. Let $\mathbf{A} = \text{diag}(\mathbf{A}_1, \mathbf{A}_2, \dots, \mathbf{A}_B) \in \mathbb{R}^{BN\times BN}$ be a diagonal matrix that indicates the activity of all devices, and let $\mathbf{a} \in \mathbb{R}^{BN}$ denote its diagonal entries. We use $\mathbf{A}$ and $\mathbf{a}$ interchangeably throughout this section.

The device activity detection problem in the multi-cell massive MIMO scenario is to detect the active devices from the received signals $\mathbf{Y}_b$, $b = 1, 2, \dots, B$. In this section, we assume that the large-scale fading coefficients are known, i.e., the matrices $\mathbf{G}_b$ for all b are known at the BSs. This assumption holds true if all devices are stationary so that their large-scale fadings are fixed and can be obtained before the detection. In this case, the device activity detection problem is equivalent to estimating the activity indicator vector $\mathbf{a}$.

Note that for each BS b, the Rayleigh fading components and noises are both i.i.d. Gaussian over the antennas. Therefore, for a given $\mathbf{a}$, the columns of the received signal $\mathbf{Y}_b$ in (16.15) denoted by $\mathbf{y}_{bm}$, $m = 1, 2, \ldots, M$, are i.i.d. Gaussian vectors, that is $\mathbf{y}_{bm} \sim \mathcal{CN}(\mathbf{0}, \mathbf{\Sigma}_b)$, where the covariance matrix $\mathbf{\Sigma}_b$ is given by

$$\mathbf{\Sigma}_b = \frac{1}{M}\mathbb{E}\left[\mathbf{Y}_b\mathbf{Y}_b^H\right] = \mathbf{S}\mathbf{G}_b\mathbf{A}\mathbf{S}^H + \sigma_w^2\mathbf{I}. \tag{16.16}$$

Since the received signals $\mathbf{Y}_b$, $b = 1, 2, \ldots, B$, are independent due to the i.i.d. Rayleigh fading channels, the likelihood function can be computed as

$$p(\mathbf{Y}_1, \mathbf{Y}_2, \ldots, \mathbf{Y}_B \mid \mathbf{a}) = \prod_{b=1}^{B} p(\mathbf{Y}_b \mid \mathbf{a}).$$

Hence, the MLE problem is equivalent to the minimization of $-\frac{1}{M}\sum_{b=1}^{B} \log p(\mathbf{Y}_b \mid \mathbf{a})$. Thus, the overall problem formulation (Chen et al., 2021b) is

$$\min_{\mathbf{a}} \quad \sum_{b=1}^{B}\left(\log|\mathbf{\Sigma}_b| + \operatorname{tr}\left(\mathbf{\Sigma}_b^{-1}\widehat{\mathbf{\Sigma}}_b\right)\right) \tag{16.17a}$$

$$\text{s.t.} \quad a_{bn} \in [0, 1],\ \forall\, b, n, \tag{16.17b}$$

where $\widehat{\mathbf{\Sigma}}_b = \mathbf{Y}_b\mathbf{Y}_b^H/M$ is the sample covariance matrix of the received signals at BS b.

16.3.2 Phase Transition Analysis

In this section, we aim to characterize the asymptotic detection performance of the solution to problem (16.17) as the number of antennas M tends to infinity and in particular, reveal how the number of cells B (and the inter-cell interference) affects the detection performance. To be specific, we want to answer the following question: given the system parameters L, B, and N, how many active devices can be correctly detected by solving the MLE problem (16.17) as $M \to \infty$?

We first present a necessary and sufficient condition for the consistency of the MLE estimator via solving problem (16.17) (Chen et al., 2021b), which can be seen as an extension of Theorems 16.1 and 16.2 in the single-cell scenario to the multi-cell scenario.

Theorem 16.5 Consider the MLE problem (16.17) with a given signature sequence matrix $\mathbf{S}$, large-scale fading component matrices $\{\mathbf{G}_b\}$, and noise variance σ_w^2. Let $\hat{\mathbf{a}}^{(M)}$ be the solution to (16.17) when the number of antennas M is given, and let $\mathbf{a}^0$ be the true activity indicator vector whose $B(N-K)$ zero entries are indexed by $\mathcal{I}$, i.e., $\mathcal{I} = \{i \mid a_i^0 = 0\}$.

(i) The Fisher information matrix of $\mathbf{a}$ is given by

$$\mathbf{J}(\mathbf{a}) = M\sum_{b=1}^{B}\left(\mathbf{P}_b \odot \mathbf{P}_b^*\right), \tag{16.18}$$

where $\mathbf{P}_b = \sqrt{\mathbf{G}_b}\mathbf{S}^H\left(\mathbf{S}\mathbf{G}_b\mathbf{A}\mathbf{S}^H + \sigma_w^2\mathbf{I}\right)^{-1}\sqrt{\mathbf{S}\mathbf{G}_b}$.

(ii) Define

$$\mathcal{N} = \left\{\mathbf{x} \in \mathbb{R}^{\mathrm{BN}} \mid \mathbf{x}^T\mathbf{J}(\mathbf{a}^0)\mathbf{x} = 0\right\}, \tag{16.19}$$

$$C = \left\{\mathbf{x} \in \mathbb{R}^{\mathrm{BN}} \mid x_i \geq 0 \text{ if } i \in \mathcal{I},\, x_i \leq 0 \text{ if } i \notin \mathcal{I}\right\}. \tag{16.20}$$

Then, a necessary and sufficient condition for $\hat{\mathbf{a}}^{(M)} \to \mathbf{a}^0$ as $M \to \infty$ is that the intersection of $\mathcal{N}$ and C is the zero vector, i.e., $\mathcal{N} \cap C = \{\mathbf{0}\}$.

Notice that in the multi-cell scenario, the large-scale fading coefficients are involved in the definition of the Fisher information matrix in (16.18), which is different from the single-cell scenario. Therefore, the large-scale fading coefficients play a central role in the phase transition analysis in the multi-cell scenario, i.e., the feasible set of the system parameters under which the condition $\mathcal{N} \cap C = \{\mathbf{0}\}$ (in Theorem 16.5) holds true. To establish the scaling law result in the multi-cell scenario, we first specify the assumption on the large-scale fading coefficients.

Assumption 16.1 *The multi-cell system consists of B hexagonal cells with radius R. The BSs are in the center of the corresponding cells. In this system, the large-scale fading components are inversely proportional to the distance raised to the power α. (Rappaport, 2002), i.e.,*

$$g_{bjn} = P_0\left(\frac{d_0}{d_{bjn}}\right)^{\alpha}, \tag{16.21}$$

where P_0 is the received power at the point with distance d_0 from the transmitting antenna, d_{bjn} is the BS-device distance between device n in cell j and BS b, and α is the path-loss exponent.

Now, we present an analytic scaling law result by establishing a sufficient condition for $\mathcal{N} \cap C = \{\mathbf{0}\}$ (in Theorem 16.5) (Wang et al., 2023).

Theorem 16.6 Let $\mathbf{S} \in \mathbb{C}^{L\times BN}$ be the signature sequence matrix whose columns are uniformly drawn from the sphere of radius $\sqrt{L}$ in an i.i.d. fashion. Under Assumption 16.1 with $\alpha > 2$, there exist positive constants c_1 and c_2 independent of system parameters K, L, N, and B, such that if

$$K \leq c_1 L^2/\log^2(eBN/L^2), \tag{16.22}$$

then the condition $\mathcal{N} \cap C = \{\mathbf{0}\}$ (in Theorem 16.5) holds with probability at least $1 - \exp(-c_2 L)$.

Theorem 16.6 shows that, with a sufficiently large M, the maximum number of active devices that can be correctly detected in each cell by solving the MLE problem (16.17) scales as $\mathcal{O}(L^2)$ as shown in (16.22). The scaling law in (16.22) in the multi-cell scenario is approximately the same as Theorem 16.4 in the single-cell scenario (Fengler et al., 2021, Chen et al., 2022), which provides important insights that solving the MLE problem (16.17) can detect almost as many active devices in each cell in the multi-cell scenario as solving problem (16.4) in the single-cell scenario. Notice that $\alpha > 2$ in Assumption 16.1 holds true for most channel models and application scenarios, see Rappaport [2002, Chap. 4]. Therefore, the inter-cell interference is not a limiting factor of the phase transition because B affects K only through $\log(B)$ in (16.22).

16.3.3 Coordinate Descent Algorithms

We now apply the CD algorithm to the multi-cell case. CD is one of the most efficient algorithms for solving problem (16.17) (Chen et al., 2021b). At each iteration, the algorithm randomly permutes the indices of all variables and then updates the variables one by one according to their order in the permutation. For any particularly given coordinate (b, n), the CD algorithm needs to solve the following one-dimensional subproblem

$$\min_{d} \sum_{j=1}^{B} \left(\log\left(1 + d\, g_{jbn} \mathbf{s}_{bn}^H \boldsymbol{\Sigma}_j^{-1} \mathbf{s}_{bn}\right) - \frac{d\, g_{jbn} \mathbf{s}_{bn}^H \boldsymbol{\Sigma}_j^{-1} \widehat{\boldsymbol{\Sigma}}_j \boldsymbol{\Sigma}_j^{-1} \mathbf{s}_{bn}}{1 + d\, g_{jbn} \mathbf{s}_{bn}^H \boldsymbol{\Sigma}_j^{-1} \mathbf{s}_{bn}} \right) \tag{16.23a}$$

$$\text{s.t.} \;\; d \in [-a_{bn}, 1 - a_{bn}] \tag{16.23b}$$

in order to possibly update the variable a_{bn}. The closed-form solution for the above problem generally does not exist, which is different from the single-cell case. Fortunately, problem (16.23) can be transformed into a polynomial root-finding problem of degree $2B - 1$, which can further be solved by computing the eigenvalues of the companion matrix formed using the coefficients of the corresponding polynomial function (McNamee, 2007). The computational complexity of this approach to solve problem (16.23) is $\mathcal{O}(B^3)$. The CD algorithm is summarized in Algorithm 16.3. The overall complexity of Algorithm 16.3 is $\mathcal{O}\left(IBN\left(BL^2 + B^3\right)\right)$, where I is the total number of iterations.

Below we briefly mention two ways of further accelerating Algorithm 16.3. Notice that, at each iteration, Algorithm 16.3 treats all coordinates equally and tries to update all of them. However, due to the sparsity of the solution to problem (16.17), there are a lot of coordinates (b, n) for which problem (16.23) has to be solved but a_{bn} does not change, i.e., the corresponding solution to problem (16.23) will be zero. Such computations are unnecessary and slow down Algorithm 16.3. The active set selection strategy can be used to reduce this kind of unnecessary computations and further improve the computational efficiency of Algorithm 16.3. This accelerated version of Algorithm 16.3 is called active set

Algorithm 16.3 Coordinate Descent Algorithm for Solving Problem (16.17)

Initialize $\mathbf{a} = \mathbf{0}$, $\boldsymbol{\Sigma}_b^{-1} = \sigma_w^{-2}\mathbf{I}$, $b = 1, 2, \ldots, B$, and $\varepsilon > 0$;
repeat [*one iteration*]
  Randomly select a permutation $\{i_1, i_2, \ldots, i_{BN}\}$ of the coordinate indices $\{1, 2, \ldots, BN\}$;
  for $n = 1$ to BN **do**
    Solve problem (16.23) to obtain d;
    $a_{i_n} \leftarrow a_{i_n} + d$;
    $\boldsymbol{\Sigma}_b^{-1} \leftarrow \boldsymbol{\Sigma}_b^{-1} - \frac{d g_{bi_n} \boldsymbol{\Sigma}_b^{-1} \mathbf{s}_{i_n} \mathbf{s}_{i_n}^H \boldsymbol{\Sigma}_b^{-1}}{1 + d g_{bi_n} \mathbf{s}_{i_n}^H \boldsymbol{\Sigma}_b^{-1} \mathbf{s}_{i_n}}$, $b = 1, 2, \ldots, B$;
  end for
until $\|\text{Proj}(\mathbf{a} - \nabla f(\mathbf{a})) - \mathbf{a}\|_2 < \varepsilon$;
Output $\mathbf{a}$.

CD in Section 16.3.4. More details along this direction can be found in Wang et al. (2021b).

Another way of accelerating Algorithm 16.3 is to inexactly or approximately solve the subproblem in (16.23). More specifically, for a given cell b, instead of considering all B cells as in (16.23), it is reasonable (and desirable) to only consider a cluster of cells that are close to cell b (and neglect those that are far away from cell b). In this case, the degree of the polynomial function associated with the derivative of the objective function of (16.23) will be much smaller, which improves the efficiency of solving the corresponding subproblem. This accelerated version of Algorithm 16.3 is called clustering-based CD in Section 16.3.4. More details along this direction can be found in Ganesan et al. (2021).

16.3.4 Performance Evaluation

In this section, we present some simulation results to validate the accuracy of the phase transition analysis, and compare the existing CD types of algorithms for solving the multi-cell device activity detection problem in (16.17). We use the same system parameters as in Chen et al. (2021b). More specifically, we consider a multi-cell system consisting of hexagonal cells and all potential devices within each cell are uniformly distributed. In the simulations, the radius of each cell is 500 m; the channel path-loss is modeled as $128.1 + 37.6\log_{10}(d)$ as in Assumption 16.1, where d is the corresponding BS-device distance in km; the transmit power of each device is set to 23 dBm, and the background noise power is −169 dBm/Hz over 10 MHz. All signature sequences of length L are uniformly drawn from the sphere of radius $\sqrt{L}$ in an i.i.d. fashion.

We numerically test the condition $\mathcal{N} \cap C = \{\mathbf{0}\}$ in Theorem 16.5 under a variety of choices for L and K, given $N = 240$ and $B = 1, 2, 7$. Figure 16.3 plots

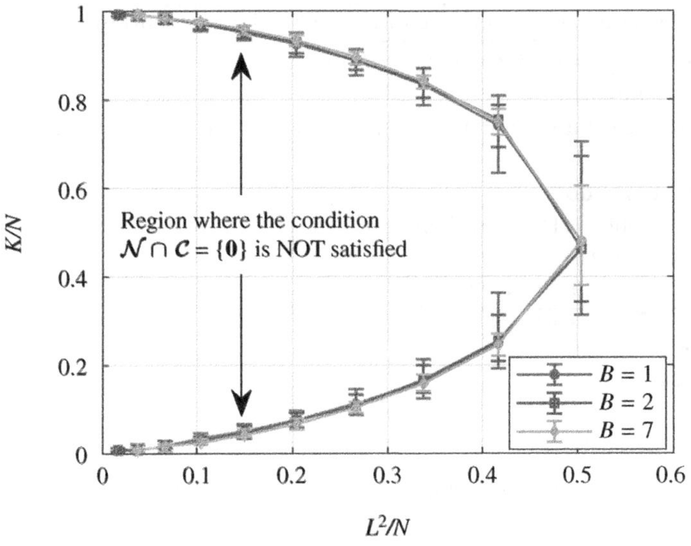

Figure 16.3 Phase transition of the covariance-based approach for device activity detection in the multi-cell scenario with B cells.

the region of $(L^2/N, K/N)$ in which the condition is satisfied or not. The result is obtained based on 100 random realizations of $\mathbf{S}$ and $\mathbf{a}^0$ for each given K and L. We observe from Fig. 16.3 that: (i) the curves with different B's overlap with each other, implying that the phase transition for $\mathcal{N} \cap \mathcal{C} = \{\mathbf{0}\}$ is almost independent of B (under Assumption 16.1 with the path-loss exponent $\alpha > 2$); the maximum number of identifiable active devices K is approximately proportional to L^2. These observations are consistent with the phase transition analysis in Theorem 16.6.

Next, we compare the efficiency of CD and its accelerated versions (i.e., active set CD (Wang et al., 2021b) and clustering-based CD (Ganesan et al., 2021)). Figure 16.4 plots the decrease in the probability of error as the algorithms run with $B = 7$, $M = 128$, $L = 50$, $N = 1000$, and $K = 50$. The result is obtained by averaging over 1000 Monte Carlo runs, and the number of clusters of clustering-based CD is chosen to be 2. It can be observed from Fig. 16.4 that active set CD and clustering-based CD are significantly more efficient than CD. It can also be observed from Fig. 16.4 that CD and active set CD can achieve the same low probability of error if their running time is allowed to be sufficiently long, but the probability of error of clustering-based CD (due to the coarse approximation) is slightly worse than those of CD and active set CD. The simulation results show that active set CD has the best efficiency and detection performance among the compared algorithms.

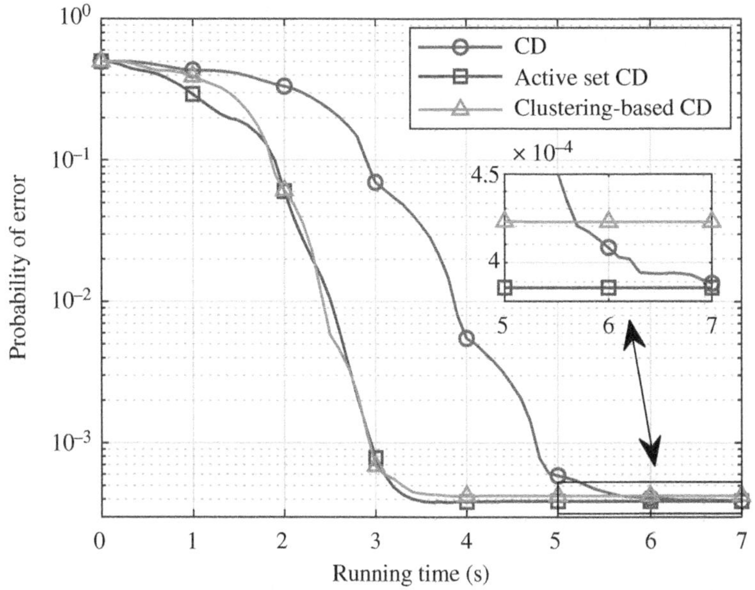

Figure 16.4 Comparison of the probability of error of existing algorithms versus running time in the multi-cell scenario.

16.4 Practical Issues and Extensions

In this section, we discuss two practical issues in the previous system models and problem formulations and present two interesting extensions.

16.4.1 Joint Device Data and Activity Detection

This section considers a grant-free massive random access scenario for mMTC with very small data payloads as first investigated in Senel and Larsson (2018), in which each device maintains a unique set of preassigned 2^J signature sequences. When a device is active, it sends J bits of data by transmitting one sequence from the set. By detecting which sequences are received, the BS acquires both the identity of the active devices as well as the J-bit messages from each of the active devices.

The joint device data and activity detection problem can be formulated as the MLE problem in both single-cell and multi-cell scenarios. We illustrate the problem formulation in the single-cell scenario. The considered scenario is the same as that in Section 16.2, except that each device n now has a unique signature sequence set $\mathcal{S}_n = \left\{\mathbf{s}_{n,1}, \mathbf{s}_{n,2}, \dots, \mathbf{s}_{n,Q}\right\}$, where $\mathbf{s}_{n,q} \in \mathbb{C}^L$, $1 \le q \le Q \triangleq 2^J$, and L is the signature sequence length. When device n is active and needs to send J bits of data, it selects one sequence from $\mathcal{S}_n$ to transmit.

Using the same technique as in Section 16.2.1, the joint device data and activity detection problem can be formulated as (Chen et al., 2019)

$$\min_{\gamma} \quad \log\left|\mathbf{S}\boldsymbol{\Gamma}\mathbf{S}^H + \sigma_w^2\mathbf{I}\right| + \mathrm{tr}\left(\left(\mathbf{S}\boldsymbol{\Gamma}\mathbf{S}^H + \sigma_w^2\mathbf{I}\right)^{-1}\hat{\boldsymbol{\Sigma}}\right) \tag{16.24a}$$

$$\text{s.t.} \quad \boldsymbol{\gamma} \geq \mathbf{0}, \tag{16.24b}$$

where $\mathbf{S} = [\mathbf{S}_1, \mathbf{S}_2, \dots, \mathbf{S}_N] \in \mathbb{C}^{L\times NQ}$ with $\mathbf{S}_n = [\mathbf{s}_{n,1}, \mathbf{s}_{n,2}, \dots, \mathbf{s}_{n,Q}] \in \mathbb{C}^{L\times Q}$ and $\boldsymbol{\gamma} = [\boldsymbol{\gamma}_1^T, \boldsymbol{\gamma}_2^T, \dots, \boldsymbol{\gamma}_N^T]^T \in \mathbb{C}^{NQ}$ with $\boldsymbol{\gamma}_n = [\gamma_{n,1}, \gamma_{n,2}, \dots, \gamma_{n,Q}]^T \in \mathbb{C}^Q$. Problem (16.24) takes the same form as problem (16.4) and hence it can be efficiently solved by the CD algorithm and its accelerated active set variant, i.e., Algorithms 16.1 and 16.2.

16.4.2 Device Activity Detection in Asynchronous Systems

This section considers a more practical grant-free massive random access scenario where all active devices asynchronously transmit their preassigned signature sequences to the BS (Liu and Liu, 2021). We adopt the same notations as in Section 16.2. We introduce a new notation $\tau_n \in \{0, 1, \dots, \tau_{\max}\}$ to denote the delay of the transmitted packet of each active device $n \in \mathcal{K}$, which means that each active device $n \in \mathcal{K}$ starts to transmit its signature sequence $\mathbf{s}_n$ at the beginning of the $(\tau_n + 1)$-th time slot. In the above, $\tau_{\max}$ is the maximum allowed delay for all the devices and is assumed to be known at the BS. However, the delay of each active device is unknown and needs to be estimated.

Given the delay τ_n, define the effective signature sequence of device n as

$$\bar{\mathbf{s}}_{n,\tau_n} = [\underbrace{0, \dots, 0}_{\tau_n}, \mathbf{s}_n, \underbrace{0, \dots, 0}_{\tau_{\max}-\tau_n}]^T, \quad n = 1, 2, \dots, N. \tag{16.25}$$

In this case, the received signal $\mathbf{Y} \in \mathbb{C}^{(L+\tau_{\max})\times M}$ from time slot 1 to time slot $L + \tau_{\max}$ is expressed as

$$\mathbf{Y} = \sum_{n=1}^{N} a_n \bar{\mathbf{s}}_{n,\tau_n} \sqrt{g_n}\mathbf{h}_n^T + \mathbf{W}. \tag{16.26}$$

Then, the joint device and delay detection problem of estimating $\mathcal{K}$ and associated $\{\tau_k\}_{k\in\mathcal{K}}$ boils down to the sequence detection problem where the sequences are given in (16.25).

Using the same technique as in Section 16.2.1, the joint activity and delay detection problem can be formulated as (Wang et al., 2022)

$$\min_{\gamma} \quad \log|\boldsymbol{\Sigma}| + \mathrm{tr}\left(\boldsymbol{\Sigma}^{-1}\hat{\boldsymbol{\Sigma}}\right) \tag{16.27a}$$

$$\text{s.t.} \quad \gamma \geq 0, \tag{16.27b}$$

$$\|\gamma_n\|_0 \leq 1, \; n = 1, 2, \ldots, N, \tag{16.27c}$$

where $\hat{\boldsymbol{\Sigma}} = \frac{1}{M}\mathbf{Y}\mathbf{Y}^H$ is the sample covariance of the received signal in (16.26), $\boldsymbol{\Sigma} = \mathbf{S}\boldsymbol{\Gamma}\mathbf{S}^H + \sigma_w^2\mathbf{I}$, and $\boldsymbol{\Gamma} = \mathrm{diag}(\gamma)$, where $\gamma = [\gamma_1^T, \gamma_2^T, \ldots, \gamma_N^T]^T$, and

$$\gamma_n = [\gamma_{n,0}, \gamma_{n,1}, \ldots, \gamma_{n,\tau_{\max}}]^T.$$

In (16.27c), $\|\gamma_n\|_0$ denotes the number of nonzero elements in the vector γ_n and the constraint is because there is at most one possible delay for each device. Although $\{\bar{\mathbf{s}}_{n,\tau_n}\}$ are no longer i.i.d., so a phase transition analysis would be challenging, problem (16.27) can still be efficiently solved by the CD algorithm with constraint (16.27c) explicitly enforced during the CD iterations (Wang et al., 2022) and by the penalty-based algorithm (Li et al., 2022) which penalizes constraint (16.27c) and solves an equivalent penalty formulation.

16.5 Conclusions

This chapter studies the device activity detection problem for grant-free massive random access with massive MIMO. The covariance-based approach is employed to formulate the device activity detection problem as an MLE problem in both single-cell and multi-cell scenarios. In this chapter, we analyze the asymptotic detection performance of the covariance-based approach as the number of antennas at the BS(s) goes to infinity, including a phase transition analysis. We also present efficient CD types of algorithms for solving the nonconvex detection problem. Finally, we discuss some practical issues in the device activity detection problem and present two extensions of practical interest.

We conclude this chapter with a brief discussion of how the analyses carried out in this chapter may be extended to account of realistic system assumptions. For example, most of the existing phase transition analysis (e.g., Theorems 16.4 and 16.6) crucially relies on the assumption that the signature/pilot sequences of devices are uniformly and randomly drawn from a sphere in an i.i.d. fashion. It would be interesting to extend the current phase transition analysis to more practical ways of generating the signature sequences, for instance, each entry of the device's signature sequence is uniformly drawn from the discrete set $\{\pm 1 \pm j\}$, where j is the imaginary unit. Moreover, the chapter focuses on the massive MIMO system. However, low-resolution ADCs are often employed in the massive MIMO system to reduce hardware cost and power consumption. In this case, the BSs can only observe a coarsely quantized version of the received signals in (16.1) or (16.15). The extension of the covariance-based approach to the massive MIMO system with low-resolution ADCs and the study on how the quantization errors affect the detection performance would be of great interest.

References

C Bockelmann, N Pratas, H Nikopour, K Au, T Svensson, Č Stefanović, P Popovski, and A Dekorsy. Massive machine-type communications in 5G: Physical and MAC-layer solutions. *IEEE Communications Magazine*, 54(9):59–65, Sept. 2016. doi: 10.1109/MCOM.2016.7565189.

Z Chen, F Sohrabi, and W Yu. Sparse activity detection for massive connectivity. *IEEE Transactions on Signal Processing*, 66(7):1890–1904, Apr. 2018. doi: 10.1109/TSP.2018. 2795540.

Z Chen, F Sohrabi, Y-F Liu, and W Yu. Covariance based joint activity and data detection for massive random access with massive MIMO. In *Proceedings of the IEEE International Conference on Communications (ICC)*, pages 1–6, Shanghai, China, May 2019. doi: 10.1109/ICC.2019.8761672.

X Chen, D W K Ng, W Yu, E G Larsson, N Al-Dhahir, and R Schober. Massive access for 5G and beyond. *IEEE Journal on Selected Areas in Communications*, 39(3):615–637, Mar. 2021a. doi: 10.1109/JSAC.2020.3019724.

Z Chen, F Sohrabi, and W Yu. Sparse activity detection in multi-cell massive MIMO exploiting channel large-scale fading. *IEEE Transactions on Signal Processing*, 69:3768–3781, Jun. 2021b. doi: 10.1109/TSP.2021.3090679.

Z Chen, F Sohrabi, Y-F Liu, and W Yu. Phase transition analysis for covariance based massive random access with massive MIMO. *IEEE Transactions on Information Theory*, 68(3):1696–1715, Mar. 2022. doi: 10.1109/TIT.2021.3132397.

E Dahlman, S Parkvall, and J Skold. *4G: LTE/LTE-Advanced for Mobile Broadband*. 2nd ed. Academic Press, 2013.

J Dong, J Zhang, Y Shi, and J H Wang. Faster activity and data detection in massive random access: A multiarmed bandit approach. *IEEE Internet of Things Journal*, 9(15):13664–13678, Aug. 2022. doi: 10.1109/JIOT.2022.3142185.

A Fengler, S Haghighatshoar, P Jung, and G Caire. Non-Bayesian activity detection, large-scale fading coefficient estimation, and unsourced random access with a massive MIMO receiver. *IEEE Transactions on Information Theory*, 67(5): 2925–2951, May 2021. doi: 10.1109/TIT.2021.3065291.

U K Ganesan, E Björnson, and E G Larsson. Clustering-based activity detection algorithms for grant-free random access in cell-free massive MIMO. *IEEE Transactions on Communications*, 69(11):7520–7530, Nov. 2021. doi: 10.1109/TCOMM.2021.3102635.

S Haghighatshoar, P Jung, and G Caire. Improved scaling law for activity detection in massive MIMO systems. In *Proceedings of the IEEE International Symposium on Information Theory (ISIT)*, pages 381–385, Vail, CO, USA, Jun. 2018. doi: 10.1109/ISIT.2018.8437359.

J Kang and W Yu. Scheduling versus contention for massive random access in massive MIMO systems. *IEEE Transactions on Communications*, 70(9):5811–5824, Sept. 2022. doi: 10.1109/TCOMM.2022.3190904.

Y Li, Q Lin, Y-F Liu, B Ai, and Y-C Wu. Asynchronous activity detection for cell-free massive MIMO: From centralized to distributed algorithms. *IEEE Transactions on Wireless Communications*, 2022. doi: 10.1109/TWC.2022.3211967.

L Liu and Y-F Liu. An efficient algorithm for device detection and channel estimation in asynchronous IoT systems. In *Proceedings of the IEEE International Conference on Acoustics, Speech, and Signal Processing (ICASSP)*, pages 4815–4819, Toronto, ON, Canada, Jun. 2021. doi: 10.1109/ICASSP39728.2021.9413870.

L Liu and W Yu. Massive connectivity with massive MIMO —Part I: Device activity detection and channel estimation. *IEEE Transactions on Signal Processing*, 66(11):2933–2946, Jun. 2018. doi: 10.1109/TSP.2018.2818082.

L Liu, E G Larsson, W Yu, P Popovski, Č Stefanović, and E De Carvalho. Sparse signal processing for grant-free massive connectivity: A future paradigm for random access protocols in the Internet of Things. *IEEE Signal Processing Magazine*, 35(5):88–99, Sept. 2018. doi: 10.1109/MSP.2018.2844952.

J M McNamee. *Numerical Methods for Roots of Polynomials-Part I*. Amsterdam, The Netherlands: Elsevier, 2007.

T S Rappaport. *Wireless Communications: Principles and Practice.* 2nd ed. Upper Saddle River, NJ: Prentice-Hall, 2002.

K Senel and E G Larsson. Grant-free massive MTC-enabled massive MIMO: A compressive sensing approach. *IEEE Transactions on Communications*, 66(12): 6164–6175, Dec. 2018. doi: 10.1109/TCOMM.2018.2866559.

P Stoica, P Babu, and J Li. SPICE: A sparse covariance-based estimation method for array processing. *IEEE Transactions on Signal Processing*, 59(2):629–638, 2011. doi: 10.1109/TSP.2010.2090525.

Z Wang, Z Chen, Y-F Liu, F Sohrabi, and W Yu. An efficient active set algorithm for covariance based joint data and activity detection for massive random access with massive MIMO. In *Proceedings of the IEEE International Conference on Acoustiscs, Speech, and Signal Processing (ICASSP)*, pages 4840–4844, Toronto, ON, Canada, Jun. 2021a. doi: 10.1109/ICASSP39728.2021.9413525.

Z Wang, Y-F Liu, Z Chen, and W Yu. Accelerating coordinate descent via active set selection for device activity detection for multi-cell massive random access. In *Proceedings of the IEEE Workshop on Signal Processing Advances in Wireless Communications (SPAWC)*, pages 366–370, Lucca, Italy, Sept. 2021b. doi: 10.1109/SPAWC51858.2021.9593150.

Z Wang, Y-F Liu, and L Liu. Covariance-based joint device activity and delay detection in asynchronous mMTC. *IEEE Signal Processing Letters*, 29:538–542, Jan. 2022. doi: 10.1109/LSP.2022.3144853.

Z Wang, Y-F Liu, Z Wang, and W Yu. Scaling law analysis for covariance based activity detection in cooperative multi-cell massive MIMO. In *Proceedings of the IEEE International Conference on Acoustics, Speech, and Signal Processing (ICASSP)*, pages 1–5, Rhodes, Greece, Jun. 2023. doi: 10.1109/ICASSP49357.2023.10096461.

D P Wipf and B D Rao. An empirical Bayesian strategy for solving the simultaneous sparse approximation problem. *IEEE Transactions on Signal Processing*, 55(7):3704–3716, Jul. 2007. doi: 10.1109/TSP.2007.894265.

17

Deep Learning-Enabled Massive Access

Ying Cui[1], Bowen Tan[1], Wang Liu[1], and Wuyang Jiang[2]

[1] *IoT Thrust Information Hub, Hong Kong University of Science and Technology (Guangzhou), Guangzhou, China*
[2] *Department of Communications and Signals, School of Urban Railway Transportation, Shanghai University of Engineering Science, Shanghai, China*

17.1 Introduction

Massive machine-type communication (mMTC) has become one of the main use cases of the fifth-generation (5G) and beyond 5G (B5G) wireless networks for supporting the Internet of Things (IoT) in various fields. Grant-free non-orthogonal multiple access (NOMA), also termed grant-free access, has been identified by the third-generation partnership project (3GPP) as a promising solution to serve enormous energy-limited and sporadically active IoT devices with small data payloads.

The basic idea of grant-free access is as follows (Liu et al., 2018). Devices are preassigned specific non-orthogonal pilots that represent only device identification (ID) for data non-embedding schemes (that do not embed data into pilots) or device ID and payload data for data embedding schemes (that embed data into pilots) (Chen et al., 2022). Active devices directly send their pilots in the pilot transmission phase, and the base station (BS) detects the sent pilots from the received pilot signal. Pilot detection is specifically referred to as device activity detection for data non-embedding schemes or joint device activity and data detection for data embedding schemes. They work in a similar way (Jiang and Cui, 2020a) and are called activity detection wherever further specifications are unnecessary. If data are not embedded into pilots, active devices immediately send their data in a subsequent phase, named the data transmission phase, without waiting for a BS grant, and the BS estimates channel states for active devices from the received pilot signal and detects data sent by the active devices from the received data signal based on the estimated channel states of the active devices. By enabling devices to transmit data in an arrive-and-go manner without waiting for BS scheduling or

Next Generation Multiple Access, First Edition.
Edited by Yuanwei Liu, Liang Liu, Zhiguo Ding, and Xuemin Shen.
© 2024 The Institute of Electrical and Electronics Engineers, Inc. Published 2024 by John Wiley & Sons, Inc.

grant, grant-free access enjoys desirable benefits for IoT devices such as overhead, latency, and energy reduction. However, the resulting activity detection and channel estimation at the BS, involving colliding devices with non-orthogonal pilots, are incredibly challenging. Fundamental problems, including activity detection, channel estimation, and pilot design, must be resolved to unleash the potential of grant-free access fully.

Existing activity detection and channel estimation methods mainly rely on compressed sensing (CS), statistical estimation, and deep learning techniques. CS-based approaches include approximate message passing (AMP) (Donoho et al., 2009, Liu and Yu, 2018), group least absolute shrinkage and selection operator (LASSO) (Qin et al., 2013, Cui et al., 2021), etc. Statistical estimation-based approaches include maximum likelihood (ML) estimation (Fengler et al., 2021a, Chen et al., 2022) and maximum a posterior probability (MAP) estimation (Jiang and Cui, 2020a, 2022), etc. CS-based and statistical estimation-based approaches usually have stronger theoretical guarantees but cannot flexibly exploit activity and channel features (especially analytically intractable features) to improve accuracy and may suffer from high computational complexity. Deep learning techniques, which can extract complex features from large data samples and achieve parallelizable inference, offer opportunities for remedying these limitations.

Non-orthogonal pilots significantly influence activity detection and channel estimation accuracies. Moreover, different non-orthogonal pilots may fit different activity detection and channel estimation methods or activity and channel features. However, analytical models of the relationship among pilot structures, activity and channel features, and activity detection and channel estimation accuracies are still unknown. What's more, pilot design for massive IoT devices involves a prohibitively large number of pilot symbols and is incredibly challenging. Most existing theoretical studies do not consider pilot designs or simply adopt Gaussian pilots (Liu and Yu, 2018, Chen et al., 2022, Fengler et al., 2021a, Jiang and Cui, 2020a, 2022). In some work, pilots are designed using standard discrete Fourier transform (Fengler et al., 2021b) and Reed–Muller sequences (Wang et al., 2019) which are not tailored for particular activity and channel features. Deep learning techniques, which can discover intricate structures in large data samples and handle large-scale optimizations using backpropagation, provide potential solutions for pilot design.

17.1.1 Existing Work

Deep learning-based activity detection and channel estimation methods can be generally classified into two categories, i.e., data-driven methods and model-driven methods. Specifically, data-driven methods primarily utilize basic neural network architectures, including fully connected neural networks (Li et al., 2019, Zhang et al., 2020, Kim et al., 2020, Sun et al., 2022), convolutional neural networks (CNNs) (Wu et al., 2021), and recurrent neural networks (RNNs) (Ahn et al., 2020, Miao et al., 2020, Emir et al., 2021). In Ahn et al. (2020), Wu et al. (2021), Yu et al. (2022), and Kim et al. (2020), data-driven methods are designed based on residual network architecture (ResNet). In Liang et al. (2022), a generative adversarial network (GAN) is used for designing data-driven methods. These data-driven methods are applicable to a relatively small number of devices (around 100 devices) and are usually ineffective for a more significant number of devices (above 1000 devices), which is typical for mMTC.

Model-driven models aim to support massive devices and achieve high accuracy by leveraging state-of-the-art activity detection and channel estimation methods. In particular, Group LASSO-based (Li et al., 2020, Sabulal and Bhashyam, 2020, Shi et al., 2020, 2022, Cui et al., 2021, Bai et al., 2022, Johnston and Wang, 2022, Shiv et al., 2022), AMP-based (Qiang et al., 2020, Cui et al., 2021, Shao et al., 2021, Sun et al., 2022, Zhu et al., 2021, Johnston and Wang, 2022), and MAP-based (Cui et al., 2021) methods are developed by unfolding the underlying Group LASSO, AMP, and MAP algorithms and learning their algorithm parameters from data samples. More specifically, parallel coordinate descent (PCD) method (Li et al., 2020), alternating direction method of multipliers (ADMM) (Sabulal and Bhashyam, 2020, Cui et al., 2021, Johnston and Wang, 2022), and iterative shrinkage-thresholding algorithm (ISTA) (Shi et al., 2020, 2022, Bai et al., 2022, Johnston and Wang, 2022, Shiv et al., 2022), proposed to solve the Group LASSO problem, are adopted to design Group LASSO-based methods. The learned algorithm parameters adapt well to activity and channel features and are much more effective than manually tuned ones. Besides, in Cui et al. (2021), additional network modules are incorporated to further improve the performances of the unfolded algorithms. Therefore, these model-driven methods achieve higher accuracy than the underlying methods used alone.

Pilot design is originally investigated in Li et al. (2019) and Cui et al. (2021) using a deep auto-encoder, which consists of an encoder and a decoder. The encoder mimics the generation of the received pilot signal, and the decoder mimics the process of activity detection or channel estimation. Specifically, Li et al. (2019) considers data-drive decoders, whereas (Cui et al., 2021) studies model-driven decoders. The proposed auto-encoder-based methods in Li et al. (2019) and Cui et al. (2021) can be used to design pilots for specific activity detection or channel estimation methods or to design pilots and activity detection or channel estimation methods jointly.

17.1.2 Main Contribution

This chapter investigates deep learning-based approaches for activity detection, channel estimation, and pilot design. A common goal is to effectively utilize state-of-the-art methods with theoretical guarantees and activity and channel features from data samples. Specifically, a model-driven activity detection approach, a model-driven channel estimation approach, and an auto-encoder-based pilot design approach are presented, primarily based on the results (Li et al., 2019, Cui et al., 2021). The main contributions are summarized as follows.

- The model-driven activity detection approach consists of three parts: an approximation part, a correction part, and a thresholding module. The model-driven channel estimation approach consists of two parts: an approximation part and a correction part. Each approximation part implements a particular method or unfolds a specific iterative algorithm. Each correction part reduces the difference between the underlying method and the approximation part or improves the performance. The model-driven activity detection (channel estimation) approach can significantly benefit from the underlying state-of-the-art methods with theoretical performance guarantee via the basic structures of the approximation part and wisely utilize activity and channel features from data samples via the adjustable parameters of the approximation part and the freely adjustable correction part. Moreover, the model-driven activity detection (channel estimation) approach is flexible enough to include data-driven ones as special cases.
- Several representative instances of the model-driven approaches are presented. Specifically, the instances of the model-driven activity detection approach include the covariance-based, MAP-based, and AMP-based activity detection methods, and the instances of the model-driven channel estimation approach include the Group LASSO-based and AMP-based channel estimation methods. These instances elegantly utilize state-of-the-art activity detection and channel estimation methods and neural network architectures and demonstrate promising gains over the underlying methods in accuracy and computation time.
- The auto-encoder-based approach consists of an encoder that mimics the generation of the received pilot signal and a model-driven decoder that mimics the process of activity detection or channel estimation. More specifically, the model-driven decoder can be a neural network implementation corresponding to the model-driven activity detection or channel estimation approach. The auto-encoder-based pilot design approach can be used to design pilots for specific activity detection or channel estimation methods or to jointly design pilots and activity detection or channel estimation methods. The resulting pilots obtained by extracting the weights of the trained encoder adapt well to activity and channel features learned from data samples and function effectively with the corresponding activity detection or channel estimation method implemented by the decoder.

17.1.3 Notation

Boldface small letters (e.g., $\mathbf{x}$), boldface capital letters (e.g., $\mathbf{X}$), non-boldface letters (e.g., x or X), and calligraphic letters (e.g., $\mathcal{X}$) represent vectors, matrices, scalar constants, and sets, respectively. The notation $X(i,j)$ denotes the (i,j)th element of matrix $\mathbf{X}$, $\mathbf{X}_{i,:}$ represents the ith row of matrix $\mathbf{X}$, $\mathbf{X}_{:,i}$ represents the ith column of matrix $\mathbf{X}$, and $x(i)$ represents the ith element of vector $\mathbf{x}$. Superscript ^{H}, superscript ^{T}, and superscript * denote transpose conjugate, transpose, and conjugation, respectively. The notation $|||\cdot|||_F$ represents the Frobenius norm of a matrix, vec$(\cdot)$ represents the column vectorization of a matrix, tr$(\cdot)$ represents the trace of a matrix, Cov$(\cdot)$ represents the covariance matrix of a random vector, diag$(\cdot)$ represents the diagonal matrix formed by a vector, $\odot$ represents the Khatri–Rao product between two matrices, $\mathbb{I}[\cdot]$ denotes the indicator function, and $\Re(\cdot)$ and $\Im(\cdot)$ represent the real part and imaginary part, respectively. $\mathbf{0}_{m\times n}$ and $\mathbf{I}_{n\times n}$ represent the $m \times n$ zero matrix and the $n \times n$ identity matrix, respectively. The complex and real fields are denoted by $\mathbb{C}$ and $\mathbb{R}$, respectively.

17.2 System Model

Consider a single cell with one M-antenna base station (BS) and N single-antenna IoT devices. Let $\mathcal{M} \triangleq \{1, 2, \dots, M\}$ and $\mathcal{N} \triangleq \{1, 2, \dots, N\}$ denote the sets of antenna indices and device indices, respectively. Consider a narrow-band system with flat fading[1] and study one coherence block. For all $m \in \mathcal{M}$ and $n \in \mathcal{N}$, let $H(n, m) \in \mathbb{C}$ denote the state of the wireless channel between the mth antenna at the BS and device n. Let $\mathbf{H} \triangleq (H(n, m))_{n\in\mathcal{N},m\in\mathcal{M}} \in \mathbb{C}^{N\times M}$ denote the channel matrix. Consider the massive access scenario arising from mMTC, where very few devices among a large number of potential devices are active and access the BS in each coherence block. For all $n \in \mathcal{N}$, let $\alpha(n) \in \{0, 1\}$ represent the active state of device n, where $\alpha(n) = 1$ means that device n accesses the channel (i.e., is active), and $\alpha(n) = 0$ otherwise. In the considered massive access scenario, $\sum_{n\in\mathcal{N}} \alpha(n) \ll N$, i.e., $\boldsymbol{\alpha} \triangleq (\alpha_n)_{n\in\mathcal{N}} \in \{0, 1\}^N$ is sparse.

A grant-free access scheme without embedding data into pilots[2] is adopted. Specifically, each device n is assigned a specific pilot sequence $\mathbf{a}_n \in \mathbb{C}^L$ of length L, where $\mathcal{L} \triangleq \{1, 2, \dots, L\}$. Let $\mathbf{A} \in \mathbb{C}^{L\times N}$ with $\mathbf{A}_{:,n} = \mathbf{a}_n$, $n \in \mathcal{N}$ denote the pilot matrix, which is known to the BS. As $L \ll N$, it is not possible to assign mutually orthogonal pilot sequences to devices. Assume that time and frequency are

1 The proposed approaches can be extended to a wide-band system with frequency-selective fading (Jiang et al., 2022).

2 The proposed approaches can be extended to a data-embedding grant-free access scheme (Chen et al., 2022, Jiang and Cui, 2020a).

perfectly synchronized.[3] In the pilot transmission phase of each coherence block, all active devices synchronously send their pilots. The received pilot signal over the L signal dimensions and M antennas at the BS, denoted as $\mathbf{Y} \in \mathbb{C}^{L\times M}$, is given by:

$$\mathbf{Y} = \mathbf{A}\boldsymbol{\Gamma}\mathbf{H} + \mathbf{Z} = \mathbf{A}\mathbf{X} + \mathbf{Z}, \tag{17.1}$$

where $\boldsymbol{\Gamma} \triangleq \mathrm{diag}(\boldsymbol{\alpha}) \in \mathbb{C}^{N\times N}$, $\mathbf{X} \triangleq \boldsymbol{\Gamma}\mathbf{H} \in \mathbb{C}^{N\times M}$, and $\mathbf{Z} \in \mathbb{C}^{L\times M}$ is the additive white Gaussian noise (AWGN) at the BS with all elements independently and identically distributed (i.i.d.) according to $\mathcal{CN}(0, \sigma^2)$.

Grant-free access encounters three challenging problems, namely, device activity detection, channel estimation, and pilot design, due to the presence of colliding devices with non-orthogonal pilots.

- *Device activity detection*: The BS tries to estimate the device activities $\alpha(n)$, $n \in \mathcal{N}$ (i.e., $\boldsymbol{\Gamma}$) from $\mathbf{Y}$, given knowledge of $\mathbf{A}$. Since $\alpha(n) = \mathbb{I}[\|\mathbf{X}_{n,:}\|_2 > 0]$, $n \in \mathcal{N}$, device activities can be directly estimated or indirectly obtained from the estimation of $\mathbf{X}$.
- *Channel estimation*: The BS aims to estimate the channel states of all active devices $\mathbf{H}_{n,:}$, $n \in \mathcal{N}$ with $\alpha(n) = 1$ from $\mathbf{Y}$, given knowledge of $\mathbf{A}$. Since $\mathbf{H}_{n,:} = \mathbf{X}_{n,:}$ for all $n \in \mathcal{N}$ with $\alpha(n) = 1$, channel estimation can be achieved by estimating $\mathbf{X}$ from $\mathbf{Y}$.
- *Pilot design*: Design pilot matrix $\mathbf{A}$ to maximally retain the information on device activities and channel states in the received pilot signal $\mathbf{Y}$.

In the following, for given pilots, channel estimation is first investigated, followed by activity detection, since channel estimation methods can also be applied to activity detection. Then, pilot design is studied.

17.3 Model-Driven Channel Estimation

This section presents a model-driven channel estimation approach based on estimating $\mathbf{X}$ from $\mathbf{Y}$, assuming that $\mathbf{A}$ is given. The neural network consists of two parts, the approximation and correction parts, as illustrated in Fig. 17.1.

The approximation part is used to approximate a particular channel estimation method. It unfolds U iterations of the corresponding algorithm using U building blocks. The algorithm parameters are usually made tunable. Moreover, per-iteration approximations with tunable approximation parameters can be adopted to simplify operations for reducing computational complexity. The input

3 The proposed approaches can be extended to time and frequency asynchronous cases (Liu et al., 2023).

to the approximation part is $\mathbf{Y}$ and $\mathbf{A}$, and the output is $\mathbf{X}^{(U)}$. Note that $\mathbf{X}^{(U)}$ can be treated as an approximation of the estimation obtained by the underlying method, and the approximation error generally decreases with U.

The correction part is used to reduce the difference between the obtained approximate estimation $\mathbf{X}^{(U)}$ and the actual $\mathbf{X}$. It consists of V fully connected layers with trainable parameters. In particular, the first correction layer has $2M$ neurons with $\left(\Re(\mathbf{X}^{(U)}_{:,m}), \Im(\mathbf{X}^{(U)}_{:,m})\right)$, $m \in \mathcal{M}$ as the input. The last correction layer has M neurons with $\left(\Re(\hat{\mathbf{X}}_{:,m}), \Im(\hat{\mathbf{X}}_{:,m})\right)$, $m \in \mathcal{M}$ as the output. In each of the first $V-1$ correction layers, rectified linear unit (ReLU) is chosen as the activation function. In the last correction layer, there is no activation function. V influences the training error and generalization error. Notice that besides fully connected neural networks, other neural network architectures, such as ResNet, may also be used for designing the correction part.

It is worth noting that U and V should be jointly chosen so that the model-driven channel estimation approach can achieve a higher accuracy (i.e., a smaller gap between $\hat{\mathbf{X}}$ and $\mathbf{X}$) and a shorter computation time than the underlying method. When U is sufficiently large and $V = 0$, the proposed model-driven channel estimation approach reduces to a particular method. When $U = 0$ and $V > 0$, the proposed model-driven channel estimation approach degrades to a purely data-driven one.

The training procedure for the model-driven channel estimation approach is as follows. Consider I training samples $(\mathbf{Y}^{[i]}, \mathbf{X}^{[i]})$, $i = 1, \dots, I$. Let $\hat{\mathbf{X}}^{[i]}$ denote the output corresponding to input $\mathbf{Y}^{[i]}$. The mean-square error (MSE) loss function is adopted to measure the difference between $\hat{\mathbf{X}}^{[i]}$ and $\mathbf{X}^{[i]}$:

$$\text{Loss}\left((\mathbf{X}^{[i]})_{i=1,\dots,I}, (\hat{\mathbf{X}}^{[i]})_{i=1,\dots,I}\right) = \frac{1}{\text{MNI}} \sum_{i=1}^{I} |||\mathbf{X}^{[i]} - \hat{\mathbf{X}}^{[i]}|||_F^2$$

The adaptive moment estimation (ADAM) algorithm is used for training.

Two instances of the model-driven channel estimation approach, i.e., the GROUP LASSO-based channel estimation method and the AMP-based channel estimation method, are presented below. Their correction parts follow the general design guideline. Therefore, the following elaborations focus on their approximation parts.

17.3.1 GROUP LASSO-Based Channel Estimation

This subsection presents the GROUP LASSO-based channel estimation method, which is illustrated in Fig. 17.1(a). First, the GROUP LASSO formulation for estimating $\mathbf{X}$ is given below (Qin et al., 2013):

$$\min_{\mathbf{X}} \frac{1}{2} |||\mathbf{A}\mathbf{X} - \mathbf{Y}|||_F^2 + \lambda \sum_{n \in \mathcal{N}} ||\mathbf{X}_{n,:}||_2 \tag{17.2}$$

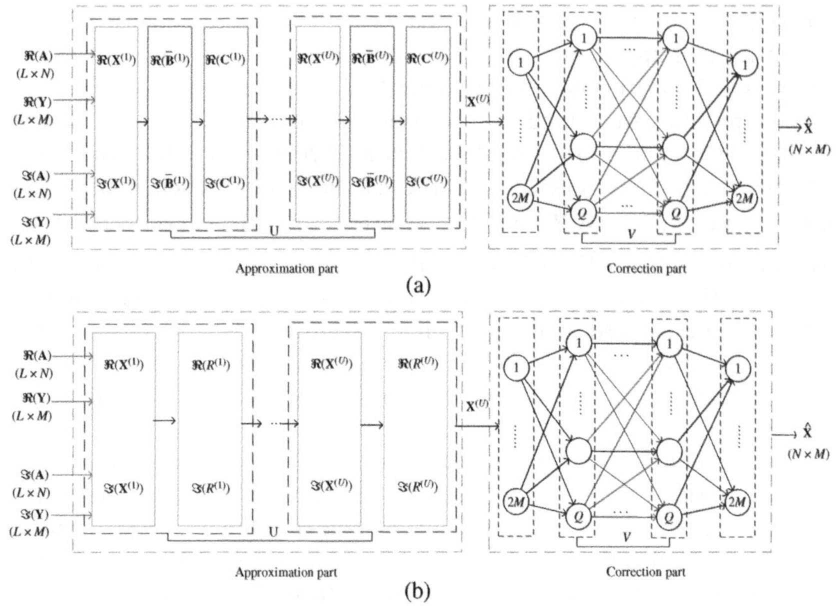

Figure 17.1 Proposed model-driven channel estimation approach. (a) GROUP LASSO-based channel estimation method. The approximation part unfolds U iterations of Algorithm 17.1 and (b) AMP-based channel estimation method. The approximation part unfolds U iterations of Algorithm 17.2.

where $\lambda \geq 0$ is a regularization parameter, and $\sum_{n \in \mathcal{N}} \|\mathbf{X}_{n,:}\|_2$ is the regularization term that prompts sparsity. Note that GROUP LASSO does not rely on any information of sparse signals (besides sparsity) or noise, and the difference between an optimal solution of the problem in (17.2) and the actual $\mathbf{X}$ varies with λ. The problem in (17.2) is convex and has a large dimension. Applying standard convex optimization methods for solving it may not be computationally efficient (Cui et al., 2021). By carefully exploiting structural properties of the problem in (17.2) and using ADMM (Boyd et al., 2011), a fast algorithm, i.e., Algorithm 17.1, that allows parallel computations and can utilize the parallelizable neural network architecture, is developed (Cui et al., 2021). It can be shown that $\mathbf{X}^{(k)}$ converges to an optimal solution of the problem in (17.2), as $k \to \infty$ (Cui et al., 2021).

Algorithm 17.1 has two parameters, i.e., $\lambda > 0$ and $\rho > 0$, which influence the channel estimation accuracy and convergence speed, respectively. The computational complexity is $\mathcal{O}$(LNM). As $\mathbf{X}_{n,:}^{(k)} \in \mathbb{C}^{1\times M}, n \in \mathcal{N}$ can be computed in parallel, and the sizes of $\overline{\mathbf{B}}^{(k)}, \mathbf{C}^{(k)} \in \mathbb{C}^{L\times M}$ are usually not large (due to $L \ll N$), the computation time of Algorithm 17.1 can be greatly reduced, compared to the block coordinate descent (BCD) algorithm in Qin et al. (2013).

Algorithm 17.1 ADMM for GROUP LASSO [Cui et al., 2021]

Set $\mathbf{X}^{(0)} = \mathbf{0}_{N\times M}$, $\overline{\mathbf{AX}}^{(k)} = \mathbf{0}_{L\times M}$, $\overline{\mathbf{B}}^{(0)} = \mathbf{0}_{L\times M}$, $\mathbf{C}^{(0)} = \mathbf{0}_{L\times M}$, and $k = 0$.
repeat
  for $n \in \mathcal{N}$ **do**
    Compute $\mathbf{t}_n = \mathbf{A}_{:,n}^H \left(\mathbf{A}_{:,n}\mathbf{X}_{n,:}^{(k)} + \overline{\mathbf{B}}^{(k)} - \overline{\mathbf{AX}}^{(k)} - \mathbf{C}^{(k)} \right)$.
    Update $\mathbf{X}_{n,:}^{(k+1)} = \dfrac{\max\left\{ 1 - \dfrac{\lambda}{\rho \|\mathbf{t}_n\|_2}, 0 \right\}}{\mathbf{A}_{:,n}^H \mathbf{A}_{:,n}} \mathbf{t}_n$.
  end for
  Compute $\overline{\mathbf{AX}}^{(k+1)} = \frac{1}{N} \sum_{n\in\mathcal{N}} \mathbf{A}_{:,n}\mathbf{X}_{n,:}^{(k+1)}$.
  Update $\overline{\mathbf{B}}^{(k+1)} = \frac{1}{N+\rho} \left(\mathbf{Y} + \rho \overline{\mathbf{AX}}^{(k+1)} + \rho \mathbf{C}^{(k)} \right)$.
  Update $\mathbf{C}^{(k+1)} = \mathbf{C}^{(k)} + \overline{\mathbf{AX}}^{(k+1)} - \overline{\mathbf{B}}^{(k+1)}$.
  Set $k = k + 1$.
until some stopping criterion is met.

The approximation part of the GROUP LASSO-based channel estimation method unfolds U iterations of Algorithm 17.1. Note that the operations for complex numbers in Algorithm 17.1 are readily implemented with operations for real numbers using a standard neural network. As illustrated in Fig. 17.1(a), each building block of the approximation part of the GROUP LASSO-based channel estimation method realizes one iteration of Algorithm 17.1. The output of the approximation part $\mathbf{X}^{(U)}$ is the estimate of $\mathbf{X}$ at the Uth iteration of Algorithm 17.1. $\lambda > 0$ and $\rho > 0$ are set tunable parameters.

17.3.2 AMP-Based Channel Estimation

This subsection presents the AMP-based channel estimation method, which is based on a slight extension of the state-of-the-art AMP algorithm with the minimum mean squared error (MMSE) denoiser (Liu and Yu, 2018).

The AMP algorithm in (Liu and Yu, 2018) is designed for the scenario where $\alpha(n)$, $n \in \mathcal{N}$ are assumed to be i.i.d. Bernoulli random variables with probability $\epsilon \in (0, 1)$ being 1, nonzero elements of $\mathbf{X}$ are i.i.d. complex Gaussian random variables with zero mean and unit variance,[4] and ϵ is assumed to be known. Here, a slightly generalized version of it, i.e., Algorithm 17.2, is presented. In particular, $\alpha(n)$, $n \in \mathcal{N}$ are assumed to be independent Bernoulli random variables, and

4 Unit variance is considered here for simplicity. The proposed AMP-based channel estimation method can be extended to the case where non-zero elements of $\mathbf{X}_{n,:}$ are i.i.d. complex Gaussian random variables with zero mean and (known) non-unit variance, and $\mathbf{X}_{n,:}$, $n \in \mathcal{N}$ are independent and have possibly different variances.

Algorithm 17.2 Generalization of AMP Liu and Yu [2018]

Set $\mathbf{X}^{(0)} = \mathbf{0}_{M\times N}$, $\mathbf{R}^{(0)} = \mathbf{Y}$ and $k = 0$.
repeat
  Compute $\tau = \frac{1}{\mathrm{ML}}|||\mathbf{R}^{(k)}|||_F^2$.
  for $n \in \mathcal{N}$ **do**
    Compute $\mathbf{v}_n = (\mathbf{R}^{(k)})^H \mathbf{A}_{:,n} + (\mathbf{X}_{n,:}^{(k)})^H$ and $t_n = \frac{1-\epsilon(n)}{\epsilon(n)}\left(\frac{\tau+1}{\tau}\right)^M \exp\left(\frac{-\|\mathbf{v}_n\|_2^2}{\tau(\tau+1)}\right)$.
    Update $\mathbf{X}_{n,:}^{(k+1)} = \frac{1}{(\tau+1)(t_n+1)}\mathbf{v}_n^H$.
  end for
  Update

$$\mathbf{R}^{(k+1)} = \mathbf{Y} - \mathbf{A}\mathbf{X}^{(k+1)} + \frac{N}{L(\tau+1)}\mathbf{R}^{(k)}\sum_{n\in\mathcal{N}}\frac{1}{(t_n+1)}\left(\mathbf{I}_{M\times M} + \frac{t_n \mathbf{v}_n \mathbf{v}_n^H}{\tau(\tau+1)(t_n+1)}\right).$$

  Set $k = k + 1$.
until some stopping criterion is met.

the probability of $\alpha(n)$ being 1 is $\epsilon(n) \in (0, 1)$. Thus, ϵ is replaced with $\epsilon(n)$ when being used for device n. In Algorithm 17.2, $\mathbf{X}^{(k)}$ represents the estimation of $\mathbf{X}$ at the kth iteration, and $\mathbf{R}^{(k)}$ represents the corresponding residual. Algorithm 17.2 has N parameters, i.e., $\epsilon(n)$, $n \in \mathcal{N}$, which influence the channel estimation accuracy. The computational complexity of Algorithm 17.2 is $\mathcal{O}$(LNM). Besides, most computations with respect to $n \in \mathcal{N}$ can be conducted in parallel.

The approximation part of the AMP-based channel estimation method unfolds Algorithm 17.2. Similarly, the operations for complex numbers in Algorithm 17.2 are readily implemented with operations for real numbers using a standard neural network. As illustrated in Fig. 17.1(b), each building block of the approximation part of the AMP-based channel estimation method realizes one iteration of Algorithm 17.2. The output of the approximation part $\mathbf{X}^{(U)}$ is the estimate of $\mathbf{X}$ at the Uth iteration of Algorithm 17.2. Note that $\epsilon(n)$, $n \in \mathcal{N}$ are set tunable parameters.

17.4 Model-Driven Activity Detection

This section presents a model-driven activity detection approach, which is mainly based on estimating $\boldsymbol{\alpha}$ from $\mathbf{Y}$ and can also base on estimating $\mathbf{X}$ from $\mathbf{Y}$, assuming that $\mathbf{A}$ is given. The neural network consists of three parts, an approximation, a correction part, and a thresholding module, as illustrated in Fig. 17.2. The rest of the section focuses on directly estimating $\boldsymbol{\alpha}$ from $\mathbf{Y}$. Nevertheless, it is worth noting that the model-driven channel estimation approach, which is based on estimating

$\mathbf{X}$ from $\mathbf{Y}$, together with a thresholding module for producing binary $\boldsymbol{\alpha}$ from the estimation of $\mathbf{X}$, can serve as a model-driven activity detection approach, as illustrated in Fig. 17.2(c).

The approximation part is used to approximate a particular activity detection method. It has U building blocks, used for (approximately) unfolding U iterations of the corresponding algorithm. The algorithm parameters (and parameters in the approximations) are generally set as tunable. The input to the approximation part is $\mathbf{Y}$ and $\mathbf{A}$, and the output is $\boldsymbol{\alpha}^{(U)}$. Generally, the gap between $\boldsymbol{\alpha}^{(U)}$ and the actual $\boldsymbol{\alpha}$ reduces with U.

The correction part is used to reduce the difference between the obtained approximate estimation $\boldsymbol{\alpha}^{(U)}$ and the actual $\boldsymbol{\alpha}$. It consists of V fully connected layers with trainable parameters. Specifically, in each of the first $V-1$ correction layers, ReLU is chosen as the activation function. The last correction layer has N neurons and uses the sigmoid function as the activation function for producing the output $\tilde{\boldsymbol{\alpha}} \in [0,1]^N$. Similarly, V influences the training error and generalization error, and neural network architectures other than fully connected layers can also be used for designing the correction part.

As for the model-driven channel estimation approach, U and V are jointly chosen so that the proposed model-driven activity detection approach can achieve a higher detection accuracy (i.e., a smaller gap between $\tilde{\boldsymbol{\alpha}}$ and $\boldsymbol{\alpha}$) and a shorter computation time than the underlying method. The same rules for selecting U and V also apply here.

The training procedure for the model-driven activity detection approach is presented below. Consider I training samples $(\mathbf{Y}^{[i]}, \boldsymbol{\alpha}^{[i]}), i = 1, \dots, I$. Let $\tilde{\boldsymbol{\alpha}}^{[i]}$ denote the output corresponding to input $\mathbf{Y}^{[i]}$. The binary cross-entropy loss function is adopted to measure the difference between $\boldsymbol{\alpha}^{[i]}$ and $\tilde{\boldsymbol{\alpha}}^{[i]}$:

$$\begin{aligned}&\text{Loss}\left((\boldsymbol{\alpha}^{[i]})_{i=1,\dots,I}, (\tilde{\boldsymbol{\alpha}}^{[i]})_{i=1,\dots,I}\right)\\&= -\frac{1}{NI}\sum_{i=1}^{I}\sum_{n\in\mathcal{N}}\left(\alpha(n)^{[i]}\log(\tilde{\alpha}(n)^{[i]} + \left(1-\alpha(n)^{[i]}\right)\log\left(1-\tilde{\alpha}(n)^{[i]}\right)\right) \qquad (17.3)\end{aligned}$$

The ADAM algorithm is used for training.

The hard thresholding module parameterized by the threshold γ converts $\tilde{\boldsymbol{\alpha}} \in [0,1]^N$ to the final output of the model-driven activity detection approach $\hat{\boldsymbol{\alpha}} \triangleq (\hat{\alpha}(n))_{n\in\mathcal{N}} \in \{0,1\}^N$, where $\hat{\alpha}(n) = \mathbb{I}[\tilde{\alpha}(n) \geq \gamma], n \in \mathcal{N}$. Let $P_E(\gamma) \triangleq \frac{1}{NI}\sum_{i=1}^{I} \|\boldsymbol{\alpha}^{[i]} - \hat{\boldsymbol{\alpha}}^{[i]}\|_1$ denote the error rate for the given threshold γ. The optimal threshold $\gamma^* = \arg\min_\gamma P_E(\gamma)$ is chosen for the hard thresholding module.

Two instances of the model-driven activity detection approach, i.e., the covariance-based activity detection method and the MAP-based activity detection method, are presented below. Note that the neural networks corresponding to the GROUP LASSO-based and AMP-based channel estimation methods in

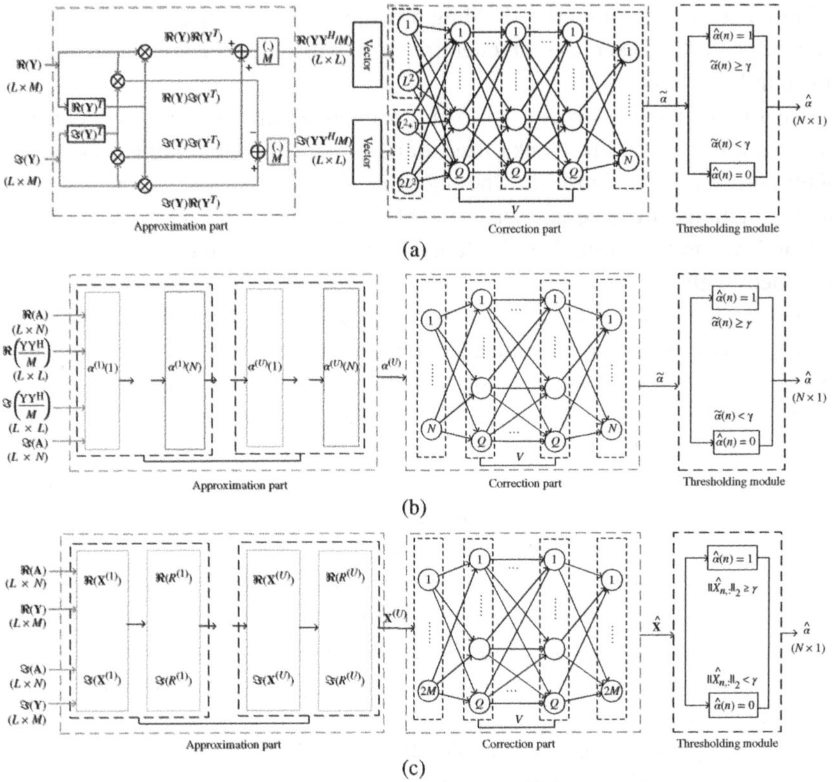

Figure 17.2 Proposed model-driven activity detection approach. (a) Covariance-based activity detection method ($U = 0$), (b) MAP-based activity detection method. The operations for generating $\Re(\mathbf{YY}^H/M)$ and $\Im(\mathbf{YY}^H/M)$ are the same as those in Fig. 17.2(b) and are omitted here. The approximation part unfolds U iterations of Algorithm 17.3, and (c) AMP-based activity detection method. The approximation part unfolds U iterations of Algorithm 17.2.

Section 17.3 followed by a thresholding module can also be viewed as instances for the model-driven activity detection approach. Note that the training process remains the same as in Section 17.3. Figure 17.2(c) illustrates the AMP-based activity detection method, based on the AMP-based channel estimation method in Section 17.3.2, as an example.

17.4.1 Covariance-Based Activity Detection

This subsection presents the covariance-based activity detection method, which is motivated by the covariance-based estimation method for jointly sparse support recovery in Pal and Vaidyanathan (2015) but successfully remedies its defects.

First, some results in Pal and Vaidyanathan (2015) are briefly reviewed below. By (17.1), $\mathbf{YY}^H/M = (\mathbf{AXX}^H\mathbf{A}^H + \mathbf{AXZ}^H + \mathbf{ZX}^H\mathbf{A}^H + \mathbf{ZZ}^H)/M$, which can be equivalently expressed as:

$$\text{vec}(\mathbf{YY}^H/M) = \mathbf{A}^* \odot \mathbf{Ar} + \text{vec}(\mathbf{E}_1) + \text{vec}(\mathbf{E}_2) \tag{17.4}$$

Here, $\text{vec}(\mathbf{YY}^H/M) \in \mathbb{R}^{L^2}$, $\mathbf{r} \triangleq (r(n))_{n\in\mathcal{N}} \in \mathbb{R}^N$ with $r(n) = \|\mathbf{X}_{n,:}\|_2^2/M$, $\mathbf{E}_1 \triangleq (E_1(k,l))_{k,l\in\mathcal{L}} \in \mathbb{C}^{L\times L}$ with

$$E_1(k,l) = \sum_{i,j\in\mathcal{N},i\neq j} A(k,i)A^*(l,j)\sum_{m\in\mathcal{M}} X(i,m)X^*(j,m)$$

and $\mathbf{E}_2 \triangleq (\mathbf{AXZ}^H + \mathbf{ZX}^H\mathbf{A}^H + \mathbf{ZZ}^H)/M$. For any given $\mathbf{A}$, if the nonzero elements of $\mathbf{X}$ are i.i.d. random variables with zero mean, then $\mathbf{Y}_{:,m}$, $m \in \mathcal{M}$ are i.i.d. random vectors, and $\mathbf{YY}^H/M$ can be viewed as the empirical covariance of $\mathbf{Y}_{:,m}, m \in \mathcal{M}$. Moreover, $\mathbf{YY}^H/M \to \text{Cov}(\mathbf{Y}_{:,m})$, $\mathbf{E}_1 \to \mathbf{0}_{L\times L}$, and $\mathbf{E}_2 \to \sigma^2\mathbf{I}_{L\times L}$, as $M \to \infty$. Thus, when the nonzero elements of $\mathbf{X}$ are i.i.d. random variables with zero mean and $M \to \infty$, (17.4) provides noiseless linear measurements of $\mathbf{r}$ which relates to $\boldsymbol{\alpha}$ by $\alpha(n) = \mathbb{I}[r(n) > 0]$. In Pal and Vaidyanathan (2015), LASSO is applied to estimate $\mathbf{r}$ in the case of very large M. It will be seen in Section 17.6.2 that the covariance-based method via LASSO in Pal and Vaidyanathan (2015) does not work well for small M (as $\mathbf{E}_1$ is non-negligible, and $\mathbf{E}_2$ is non-diagonal at small M), whereas the covariance-based activity detection method can perfectly resolve this issue.

Next, the covariance-based activity detection method is presented. Note that

$$\Re(\mathbf{YY}^H)/M = (\Re(\mathbf{Y})\Re(\mathbf{Y}^T) + \Im(\mathbf{Y})\Im(\mathbf{Y}^T))/M \tag{17.5}$$

$$\Im(\mathbf{YY}^H)/M = (\Im(\mathbf{Y})\Re(\mathbf{Y}^T) - \Re(\mathbf{Y})\Im(\mathbf{Y}^T))/M \tag{17.6}$$

The computational complexity for calculating $\Re(\mathbf{YY}^H)/M$ and $\Im(\mathbf{YY}^H)/M$, i.e., $\mathbf{YY}^H/M$, is $\mathcal{O}(\text{ML}^2)$. Note that $\Re(\mathbf{YY}^H)/M$ and $\Im(\mathbf{YY}^H)/M$ can be obtained from the input $\mathbf{Y}$, and the approximation part is no longer needed, i.e., $U = 0$. The correction part performs the activity detection based on the empirical covariance matrix $\mathbf{YY}^H/M$. The first correction layer has $2L^2$ neurons with $\text{vec}(\Re(\mathbf{YY}^H)/M)$ as the input of the first L^2 neurons and $\text{vec}(\Im(\mathbf{YY}^H)/M)$ as the input of the last L^2 neurons.

17.4.1.1 MAP-Based Activity Detection

This subsection presents the MAP-based activity detection method, which is based on a modification of the coordinate descent (CD) algorithm proposed for the MAP estimation in Jiang and Cui (2020b).

In Jiang and Cui (2020b), $\alpha(n)$, $n \in \mathcal{N}$ are independent Bernoulli random variables, and the probability of $\alpha(n)$ being 1 is $\epsilon(n) \in (0,1)$; $\epsilon(n)$, $n \in \mathcal{N}$ are known; and nonzero elements of $\mathbf{X}$ are i.i.d. complex Gaussian random variables with

Algorithm 17.3 MAP [Cui et al., 2021]

Set $(\mathbf{\Sigma}^{-1})^{(0)} = \frac{1}{\sigma^2}\mathbf{I}_{L\times L}$, $\boldsymbol{\alpha}^{(0)} = \mathbf{0}_{N\times 1}$, and $k = 0$.
repeat
  for $n \in \mathcal{N}$ **do**
    Compute $\mathbf{c}_n = (\mathbf{\Sigma}^{-1})^{(k)}\mathbf{A}_{:,n}$, $\eta_n = \mathbf{A}_{:,n}^H\mathbf{c}_n$, $\nu_n = \frac{1}{M}\mathbf{c}_n^H\mathbf{Y}\mathbf{Y}^H\mathbf{c}_n$,
    $\phi_n = \frac{\log\left(\frac{\epsilon(n)}{1-\epsilon(n)}\right)}{M}$, and $\Delta_n = \left(\eta_n^2 + 2\eta_n\phi_n\right)^2 + 4\eta_n^2\phi_n\left(\eta_n - \nu_n - \phi_n\right)$.
    Compute $(d(n))^{(k)} = \max\left\{\frac{\eta_n^2 - 2\phi_n\eta_n - \sqrt{\Delta_n}}{2\phi_n\eta_n^2}, -(\alpha(n))^{(k)}\right\}$.
    Update $(\alpha(n))^k = (\alpha(n))^{(k)} + (d(n))^{(k)}$.
    Update $(\mathbf{\Sigma}^{-1})^{(k+1)} = (\mathbf{\Sigma}^{-1})^{(k)} - \frac{(d(n))^{(k)}}{1+(d(n))^{(k)}\eta_n}\mathbf{c}_n\mathbf{c}_n^H$.
  end for
  Update $(\mathbf{\Sigma}^{-1})^{(k+1)} = (\mathbf{\Sigma}^{-1})^{(k)}$.
  Set $k = k + 1$.
until some stopping criterion is met.

zero mean and unit variance.[5] Then, the prior distribution of $\boldsymbol{\alpha}$ is given by $p(\boldsymbol{\alpha}) \triangleq \prod_{n=1}^{N} \epsilon(n)^{\alpha(n)}(1-\epsilon(n))^{1-\alpha(n)}$. The MAP estimation problem for activity detection is formulated as follows (Jiang and Cui, 2020b):[6]

$$\min_{\boldsymbol{\alpha}\geq 0} \log|\mathbf{A}\mathbf{\Gamma}\mathbf{A} + \sigma^2\mathbf{I}_{L\times L}| + \frac{1}{M}\,\mathrm{tr}((\mathbf{A}\mathbf{\Gamma}\mathbf{A} + \sigma^2\mathbf{I}_{L\times L})^{-1}\mathbf{Y}\mathbf{Y}^H)$$
$$-\frac{1}{M}\sum_{n\in\mathcal{N}}(\alpha(n)\log\epsilon(n) + (1-\alpha(n))\log(1-\epsilon(n))) \tag{17.7}$$

Note that in the objective function, the first two terms correspond to the log-likelihood of $\mathbf{Y}$, and the last term corresponds to the log-prior distribution of $\boldsymbol{\alpha}$ (Jiang and Cui, 2020b). This problem is nonconvex. A modification of the CD algorithm for the MAP estimation in Jiang and Cui (2020b) is presented in Algorithm 17.3. In Algorithm 17.3, $\boldsymbol{\alpha}^{(k)}$ represents the estimate at the kth iteration. It can be shown that show that $\boldsymbol{\alpha}^{(k)}$ converges to a stationary point (which can be a locally optimal point for some initial point) of the problem in (17.7), as $k \to \infty$. The computational complexity of Algorithm 17.3 is $\mathcal{O}(\mathrm{NL}^2)$. Algorithm 17.3 has parameters $\epsilon(n)$, $n \in \mathcal{N}$, which influence the detection accuracy.

The MAP-based activity detection method is presented below. Similarly, $\Re(\mathbf{Y}\mathbf{Y}^H/M)$, $\Im(\mathbf{Y}\mathbf{Y}^H/M)$, $\Re(\mathbf{A})$, and $\Im(\mathbf{A})$ can be obtained from the input $\mathbf{Y}$.

5 Unit variance is considered here for simplicity. The proposed MAP-based activity detection method can be extended to the general case elaborated in (17.1).
6 Continuous relaxation from $\boldsymbol{\alpha} \in \{0,1\}^N$ to $\boldsymbol{\alpha} \succcurlyeq 0$ is adopted for tractability and generalization. The output of the approximation part of the MAP-based activity detection method satisfies $\tilde{\boldsymbol{\alpha}} \in [0,1]^N$.

The approximation part of the MAP-based activity detection method unfolds U iterations of Algorithm 17.3. Note that the operations for complex numbers in Algorithm 17.3 are readily implemented with operations for real numbers using a neural network. As illustrated in Fig. 17.2 (b), each building block of the approximation part of the MAP-based activity detection method realizes one iteration of Algorithm 17.3. The output of the approximation part $\boldsymbol{\alpha}^{(U)}$ is the estimation of $\boldsymbol{\alpha}$ obtained at the Uth iteration of Algorithm 17.3. Note that $\epsilon(n)$, $n \in \mathcal{N}$ are set tunable parameters. The first correction layer has N neurons with $\boldsymbol{\alpha}^{(U)}$ as the input.

17.5 Auto-Encoder-Based Pilot Design

This section presents an auto-encoder-based pilot design approach, which can be used to design pilots for specific activity detection or channel estimation methods or to jointly design pilots and activity detection or channel estimation methods. As shown in Fig. 17.3, the auto-encoder-based pilot design approach relies on the deep auto-encoder structure. Specifically, it consists of an encoder that mimics the generation of the received pilot signal in (17.1) and a model-driven decoder that mimics the process of activity detection or channel estimation.

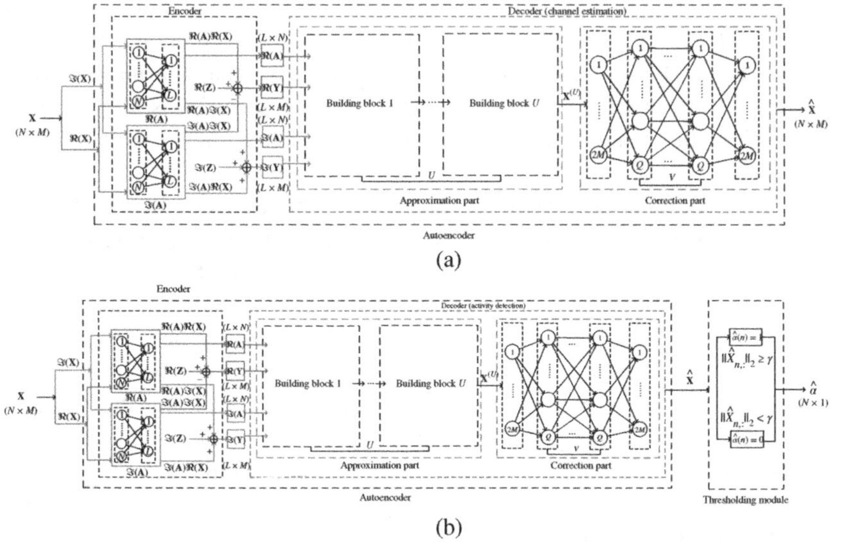

Figure 17.3 Proposed auto-encoder-based pilot design. (a) Auto-encoder-based pilot design for (or together with) channel estimation and auto-encoder-based pilot design for (or together with) activity detection.

First, the encoder is elaborated. The encoder has input $\mathbf{X} \in \mathbb{C}^{N\times M}$, i.e., $\Re(\mathbf{X}) \in \mathbb{R}^{N\times M}$ and $\Im(\mathbf{X}) \in \mathbb{R}^{N\times M}$, and output $\mathbf{Y} \in \mathbb{C}^{L\times M}$, i.e, $\Re(\mathbf{Y}) \in \mathbb{R}^{L\times M}$ and $\Im(\mathbf{Y}) \in \mathbb{R}^{L\times M}$. To mimic the generation of the received pilot signal in (17.1) using a standard auto-encoder for real numbers in deep learning, (17.1) is first equivalently expressed as (Li et al., 2019, Cui et al., 2021):

$$\Re(\mathbf{Y}) = \Re(\mathbf{A})\Re(\mathbf{X}) - \Im(\mathbf{A})\Im(\mathbf{X}) + \Re(\mathbf{Z}) \tag{17.8}$$

$$\Im(\mathbf{Y}) = \Im(\mathbf{A})\Re(\mathbf{X}) + \Re(\mathbf{A})\Im(\mathbf{X}) + \Im(\mathbf{Z}) \tag{17.9}$$

Two fully connected neural networks, each with two layers, are built to implement the multiplications with matrices $\Re(\mathbf{A}) \in \mathbb{R}^{L\times N}$ and $\Im(\mathbf{A}) \in \mathbb{R}^{L\times N}$ in (17.8) and (17.9), respectively. For each neural network, there are N neurons and L neurons in the input layer and the output layer, respectively; the weight of the connection from the nth neuron in the input layer to the lth neuron in the output layer corresponds to the (l, n)th element of the corresponding matrix; and no activation functions are used in the output layer. The elements of $\Re(\mathbf{Z}) \in \mathbb{R}^{L\times M}$ and $\Im(\mathbf{Z}) \in \mathbb{R}^{L\times M}$ are generated independently according to $\mathcal{N}(0, \frac{\sigma^2}{2})$. As shown in Fig. 17.1, when $\Re(\mathbf{X}) \in \mathbb{R}^{N\times M}$ and $\Im(\mathbf{X}) \in \mathbb{R}^{N\times M}$ are input to the encoder, $\Re(\mathbf{Y}) \in \mathbb{R}^{L\times M}$ in (17.8) and $\Im(\mathbf{Y}) \in \mathbb{R}^{L\times M}$ in (17.9) can be readily obtained.

Next, the decoder is introduced. The decoder can be a neural network implement corresponding to the model-driven channel estimation approach in Section 17.3 as illustrated in Fig. 17.3(a) or the model-driven activity detection in Section 17.4 as illustrated in Fig. 17.3(b). Note that the model-driven approaches incorporate data-driven approaches as special cases by setting $U = 0$. Thus, the model-driven approaches are considered here without loss of generality.

Finally, the training process is presented. The auto-encoder is trained using the same loss function and training algorithm as the corresponding model-driven approach in Sections 17.3 or 17.4 but different training samples, i.e., $\mathbf{X}^{[i]}, i = 1, \dots, I$ for channel estimation or $(\mathbf{X}^{[i]}, \boldsymbol{\alpha}^{[i]}), i = 1, \dots, I$ for activity detection. To design pilots for a specific channel estimation or activity detection method, only the encoder is adjusted, and the decoder corresponding to the channel estimation or activity detection method is fixed during the training process. To jointly design pilots and a channel estimation or activity detection method, the encoder and the respective decoder are simultaneously adjusted during the training process. After training, the pilot design can be obtained by extracting the encoder weights, and the decoder can be used for channel estimation or activity detection. Note that the obtained pilots and the decoder should be jointly utilized, as the pilots designed for one particular method may not be effective for another.

17.6 Numerical Results

This section presents numerical experiments on device activity detection and channel estimation for massive grant-free access. $H(n, m), n \in \mathcal{N}, m \in \mathcal{M}$, and $Z(l, m), l \in \mathcal{L}, m \in \mathcal{M}$ are generated according to i.i.d. $\mathcal{CN}(0, 1)$ and $\mathcal{CN}(0, \sigma^2)$, respectively, where $\sigma^2 = 0.1$. To demonstrate the ability of the proposed approaches in effectively exploiting activity features, two types of correlated cases of $\boldsymbol{\alpha}$, i.e., the correlated case with a single active group and the correlated case with i.i.d group activity, are studied to test the accuracies of channel estimation and device activity detection, respectively. In the correlated case with a single active group, N devices are divided into G groups of the same size N/G; the active states of the devices within each group are the same; only one of the G groups is selected to be active uniformly at random; and $1/G$ can be viewed as the access probability p. In the correlated case with i.i.d. group activity, N devices are divided into G groups of the same size N/G; the active states of the devices within each group are the same; and all groups access the channel in an i.i.d. manner with group access probability p. Two choices for N, i.e., $N = 100$ and $N = 1000$, are considered.

The proposed deep learning-based approaches are compared with state-of-the-art methods. Specifically, all the baseline schemes adopt the same set of pilot sequences for the N devices whose entries are independently generated from $\mathcal{CN}(0, 1)$ and then normalized such that $\|\mathbf{a}_n\|_2 = \sqrt{L}, n \in \mathcal{N}$. For a fair comparison, $\|\mathbf{a}_n\|_2 = \sqrt{L}, n \in \mathcal{N}$ are set in the training process. The proposed auto-encoder pilot design approach is adopted together with the proposed model-driven channel estimation activity detection approaches. For each proposed one, the pilots are obtained from the trained encoder, and the respective jointly trained decoder is used for channel estimation or activity detection. The training setup is shown in Table 17.1. When the value of a loss function on the validation set does not change for five epochs, the training process is stopped, and the corresponding parameters of the auto-encoder are saved.

17.6.1 Channel Estimation

This subsection presents numerical results on channel estimation. The proposed auto-encoder-based pilot design approach with the GROUP LASSO-based decoder and with the AMP-based decoder, namely, GROUP LASSO-NN and AMP-NN, and three baseline schemes, namely, GROUP LASSO (Qin et al., 2013), AMP (Liu and Yu, 2018), and ML-MMSE Fengler et al. (2021a), are evaluated. GROUP LASSO conducts channel estimation using the BCD algorithm (Qin et al., 2013); AMP performs channel estimation using the AMP algorithm with MMSE denoiser based on the known access probability p (Liu and Yu, 2018); ML-MMSE uses the CD

Table 17.1 Training setup.

Training parameter	Value
Size of training samples	9×10^3
Size of validation samples	1×10^3
Size of testing samples	1×10^3
Maximization epochs	1×10^5
Learning rate	1×10^{-4}
Batch size	32

Table 17.2 Parameters and computational complexities of channel estimation methods.

	Method	Parameter	Complexity
Baselines	GROUP LASSO (BCD)	200 iterations	$\mathcal{O}(\text{LNM})$
	AMP	50 iterations	$\mathcal{O}(\text{LNM})$
	ML-MMSE (CD)	55 iterations	$\mathcal{O}(\text{NL}^2)$
Proposed methods	GROUP LASSO-NN (ADMM)	$U = 200, V = 3$	$\mathcal{O}(\text{LNM})$
	AMP-NN	$U = 50, V = 3$	$\mathcal{O}(\text{LNM})$

algorithm for the ML estimation in (Fengler et al., 2021a) to detect device activities and then uses the standard MMSE method to estimate the channel of the devices that have been detected to be active. The choices of parameters and computational complexities are shown in Table 17.2.[7] Considering computation times, GROUP LASSO-NN is evaluated only at $N = 100$, but AMP-NN is evaluated at $N = 100$ and $N = 1000$.[8] The MSEs $\frac{1}{NT}\sum_{t=1}^{T} \|\|\mathbf{X}^{[t]} - \hat{\mathbf{X}}^{[t]}\|\|_F^2$ and computation times (on the same server) are evaluated using the same set of $T = 10^3$ testing samples.

Figures 17.4(a)–(c) and 17.5 (a)–(c) illustrate the MSE versus the pilot length L, access probability $p\,(= 1/G)$, and number of antennas M in the correlated case with a single active group at $N = 100$ and $N = 1000$, respectively. The following observations can be made from these figures. First, the MSE of each scheme

7 Note that when increasing the numbers of iterations for GROUP LASSO and AMP, the corresponding computation times will increase, but the MSEs are still larger than those of GROUP LASSO-NN and AMP-NN, respectively. GROUP LASSO-NN and AMP-NN can achieve smaller MSEs with shorter computation times than GROUP LASSO and AMP, respectively.

8 When N is large, it takes a large number of iterations for Algorithm 17.1 to converge to a reasonable estimate. Thus, we do not adopt GROUP LASSO-NN at $N = 1000$.

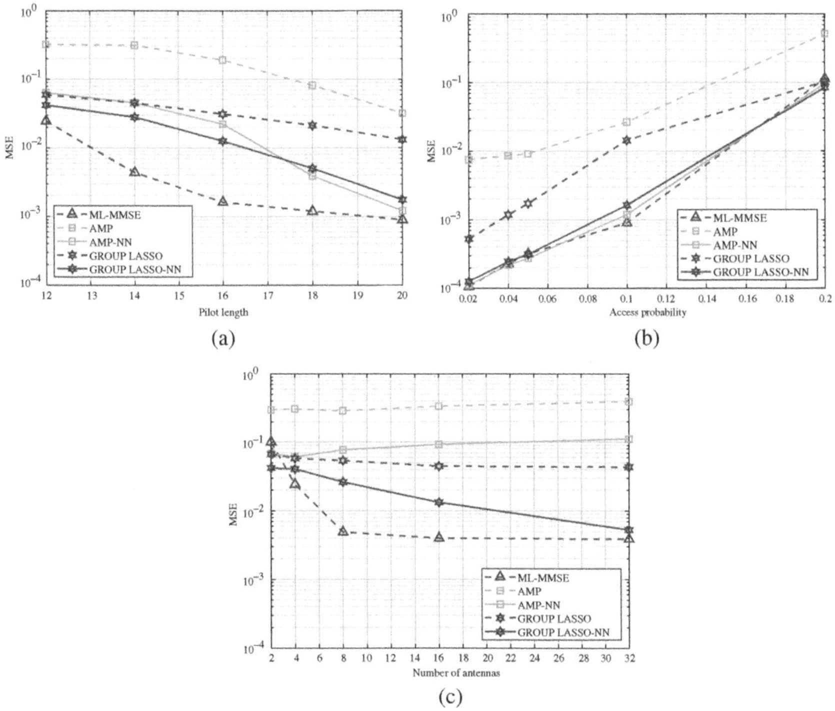

Figure 17.4 Channel estimation in the correlated case with a single active group at $N = 100$. (a) MSE versus L at $p = 0.1$, $M = 4$, (b) MSE versus p at $L = 20$, $M = 4$, and (c) MSE versus M at $L = 12$, $p = 0.1$.

continuously decreases with L and increases with p. Secondly, the MSEs of all schemes except for AMP and AMP-NN reduce with M at $N = 100$, and the MSE of each scheme decreases with M at $N = 1000$. The trend for AMP at $N = 100$ is abnormal, as AMP is unsuitable for small N. The proposed GROUP LASSO-NN and AMP-NN consistently outperform the underlying GROUP LASSO and AMP, respectively, demonstrating the advantage of the proposed model-driven approach in designing effective pilots and adjusting the tunable parameters of the approximation part and the correction part for improving channel estimation accuracy. GROUP LASSO-NN outperforms AMP-NN when N and L are small, while AMP-NN outperforms GROUP LASSO-NN when N and L are large. When $N = 100$, ML-MMSE has the smallest MSE, at the cost of computation time increase (due to ML) (Fengler et al., 2021a). When $N = 1000$, AMP-NN achieves the smallest MSE since the underlying AMP functions well for large N, and the encoder and the decoder with tunable parameters effectively utilize the activity features for further improving channel estimation accuracy.

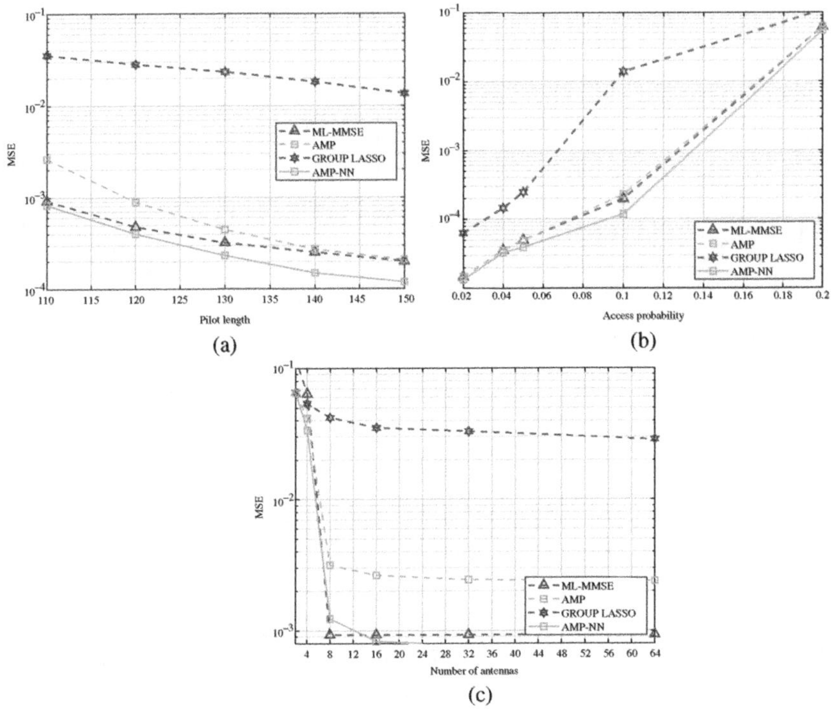

Figure 17.5 Channel estimation in the correlated case with a single active group at $N = 1000$. (a) MSE versus L at $p = 0.1$, $M = 16$, (b) MSE versus p at $L = 150$, $M = 16$, and (c) MSE versus M at $L = 110$, $p = 0.1$.

Figure 17.6 illustrates the computation time versus the pilot length L in the correlated case with a single active group at $N = 100$ and $N = 1000$. The following observations can be made from Fig. 17.6. The computation time of each scheme increases with L. GROUP LASSO-NN has a shorter computation time than GROUP LASSO. The gain derives from the fact that Algorithm 17.1 allows parallel computation. GROUP LASSO and ML-MMSE have much longer computation times than the other schemes, especially at large N. Thus, they may not apply to channel estimation in practical mMTC (with large N). AMP-NN and AMP have similar computation times, as they have the same number of iterations, and $V = 3$ is quite small.

17.6.2 Device Activity Detection

This subsection presents numerical results on device activity detection. The proposed auto-encoder-based pilot design approach with the covariance-based,

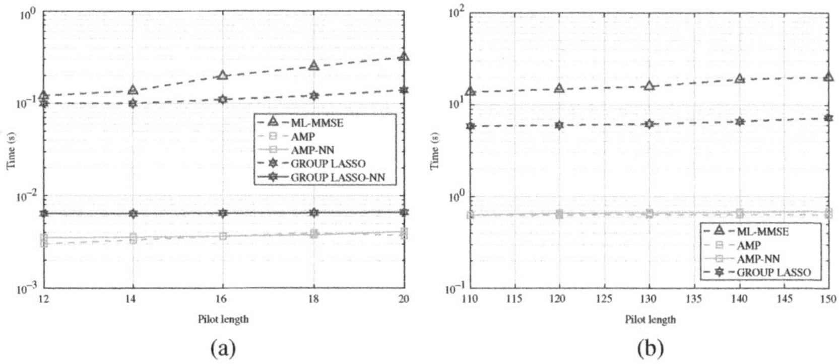

Figure 17.6 Computation time for channel estimation in the correlated case with a single active group. (a) Computation time versus L at $M = 4$ and $N = 100$ and (b) computation time versus L at $M = 16$ and $N = 1000$.

MAP-based, and AMP-based decoders, namely, NN, MAP-NN, and AMP-NN, and four baseline schemes, namely, AMP (Liu and Yu, 2018), ML (Fengler et al., 2021a), GROUP LASSO (which adopts the BCD algorithm) (Qin et al., 2013), and covariance-based LASSO (which adopts the CD algorithm) (Pal and Vaidyanathan, 2015), are evaluated. In particular, NN and MAP-NN are assessed only at $N = 100$.[9] The other schemes are evaluated at $N = 100$ and $N = 1000$. The choices of parameters and computational complexities are shown in Table 17.3.

Table 17.3 Parameters and computational complexities of activity detection methods.

	Method	**Parameter**	**Complexity**
Baselines	GROUP LASSO (BCD)	200 iterations	$\mathcal{O}(LNM)$
	AMP	50 iterations	$\mathcal{O}(LNM)$
	ML (CD)	55 iterations	$\mathcal{O}(NL^2)$
	Covariance-based LASSO (CD)	200 iterations	$\mathcal{O}(LNM)$
Proposed methods	NN	$U = 0, V = 3$	$\mathcal{O}(ML^2)$
	MAP-NN	$U = 55, V = 3$	$\mathcal{O}(NL^2)$
	AMP-NN	$U = 50, V = 3$	$\mathcal{O}(LNM)$

9 When N is large, training the proposed NN needs massive samples, and it takes a large number of iterations for Algorithm 17.3 to converge to a reasonable estimate. Numerical results show that i.i.d. Gaussian pilots are pretty suitable for Algorithm 17.3, and the proposed approach cannot find better pilots for the MAP-based decoder.

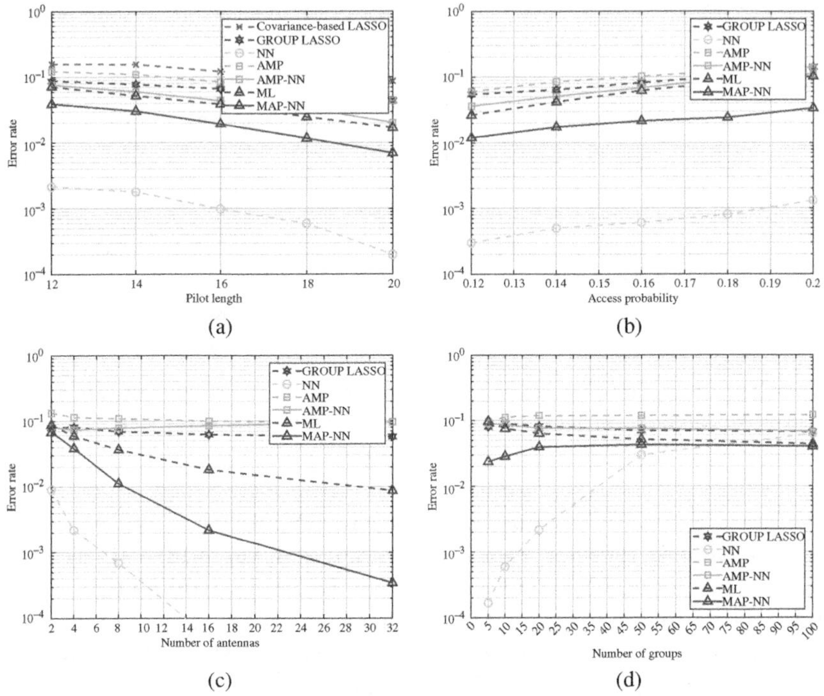

Figure 17.7 Device activity detection in the correlated case with i.i.d. group activity at $N = 100$. (a) Error rate versus L at $p = 0.1$, $M = 4$, $G = 20$, (b) error rate versus p at $L = 20$, $M = 4$, $G = 20$, (c) error rate versus M at $L = 12$, $p = 0.1$, $G = 20$, and (d) error rate versus G at $L = 12$, $M = 4$, $p = 0.1$.

The average error rates of device activity detection $\frac{1}{NT}\sum_{t=1}^{T}\|\boldsymbol{\alpha}^{[t]} - \hat{\boldsymbol{\alpha}}^{[t]}\|_1$ and computation times (on the same server) are evaluated over $T = 10^3$ testing samples.

Figures 17.7(a)–(d) and 17.8(a)–(d) illustrate the error rate versus the pilot length L, access probability p, number of antennas M, and number of groups G in the correlated case with i.i.d. group activity at $N = 100$ and $N = 1000$, respectively. Figure 17.7(a) shows that covariance-based LASSO performs much worse than Group LASSO and AMP at small M, as explained in Section 17.4.1. Given its unsatisfactory accuracy, covariance-based LASSO is no longer shown in the remaining figures. The following observations can be made from all these figures. For each scheme, the trends with respect to L, p, and M are similar to those in Section 17.6.1. The focus here is on the trends with respect to G. Note that correlation decreases with G. When $N = 100$ and $N = 1000$, the error rate of GROUP LASSO almost does not change with G, and the error rate of ML continuously decreases with G. When $N = 100$, the error rates of NN and

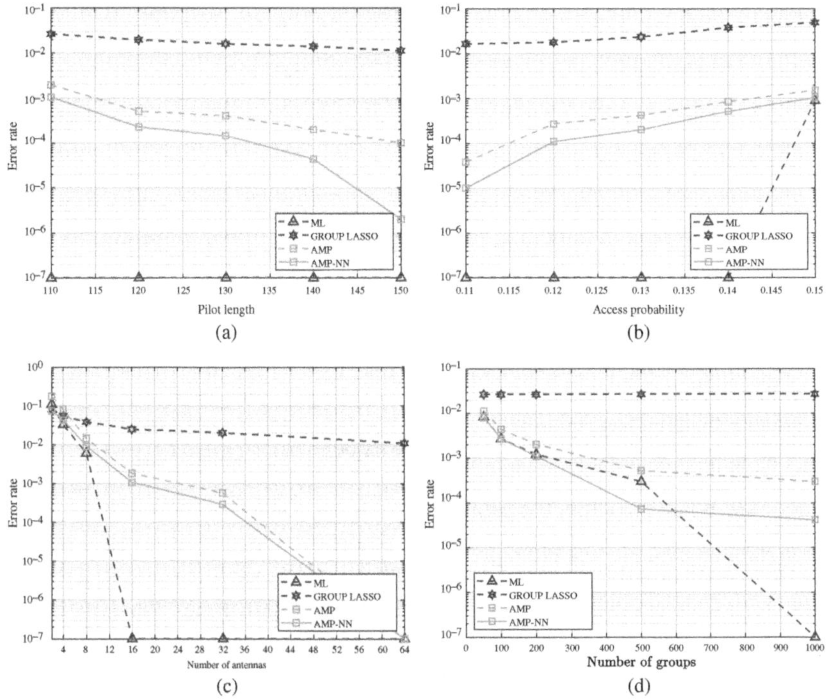

Figure 17.8 Device activity detection in the correlated case with i.i.d. group activity at $N = 1000$. (a) Error rate versus L at $p = 0.1$, $M = 16$, $G = 200$, (b) error rate versus p at $L = 150$, $M = 16$, $G = 200$, (c) error rate versus M at $L = 110$, $p = 0.1$, $G = 200$, and (d) error rate versus G at $L = 110$, $M = 16$, $p = 0.1$.

MAP-NN increase with G, as the proposed model-driven approach cannot utilize much correlation to improve activity detection accuracy at large G. The proposed MAP-NN and AMP-NN outperform ML and AMP, respectively, under all considered parameters, demonstrating the advantage of the proposed model-driven approach in designing effective pilots and model-driven decoders with adjustable parameters for improving activity detection accuracy. When $N = 100$, NN outperforms MAP-NN, as it can better utilize the correlation information in device activities. When $N = 1000$, AMP-NN achieves the lowest error rate, as it can use correlation information to improve activity detection accuracy.

Figure 17.9 illustrates the computation time versus the pilot length L in the correlated case with i.i.d. group activity at $N = 100$ and $N = 1000$. The following observations can be made from Fig. 17.9. The computation times of the proposed AMP-NN and MAP-NN are similar to those of AMP and ML, respectively. Because AMP-NN and AMP have the same number of iterations, MAP-NN and ML have the same number of iterations, and $V = 3$ in AMP-NN and MAP-NN is

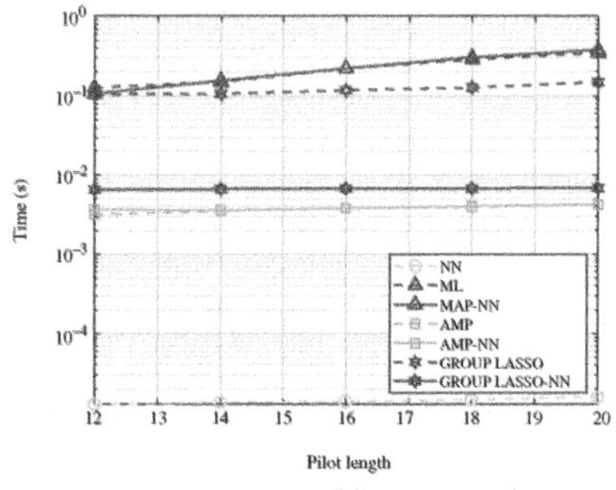

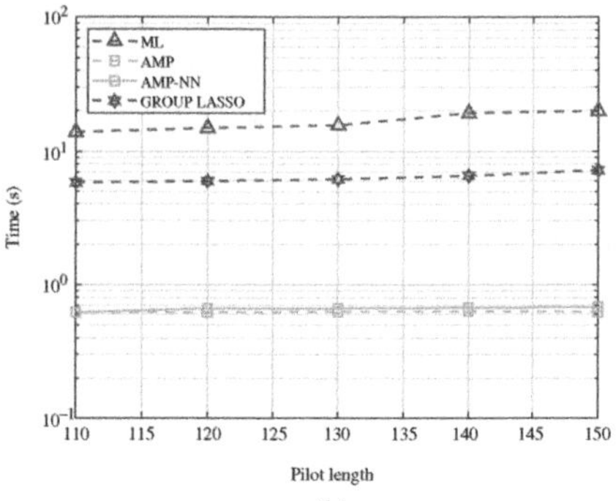

Figure 17.9 Computation time for device activity detection in the correlated case with i.i.d. group activity. (a) Computation time versus L at $M = 4$ and $N = 100$ and (b) computation time versus L at $M = 16$ and $N = 1000$.

quite small. When $N = 100$, the proposed NN has the shortest computation time, as the covariance-based decoder does not require an approximation part. When $N = 1000$, AMP and the proposed AMP-NN have the shortest computation times. GROUP LASSO and ML have much longer computation times than the other schemes, especially at large N. Thus, they may not be suitable for device activity detection in practical mMTC (with large N).

17.7 Conclusion

This chapter presents deep learning-based approaches for activity detection, channel estimation, and pilot design. The model-driven approaches for activity detection and channel estimation can significantly benefit from the underlying state-of-the-art methods with theoretical guarantees via the basic structures of the approximation parts and wisely learn activity and channel features from data samples via the adjustable parameters of approximation parts and the freely adjustable correction parts to further improve accuracies. Moreover, the model-driven approaches are flexible enough to include data-driven ones as special cases. The covariance-based, MAP-based, and AMP-based activity detection methods and the Group LASSO-based and AMP-based channel estimation methods are presented as representative instances of the model-driven activity detection approach and the model-driven channel estimation approach, respectively. Furthermore, the auto-encoder-based pilot design approach can be used to design pilots for specific activity detection or channel estimation methods or to jointly design pilots and activity detection or channel estimation methods. The numerical results show that the deep learning-based approaches can achieve higher accuracies with shorter computation times than the state-of-the-art methods, including Group LASSO, ML, and AMP, indicating the practical value of deep learning-enabled massive access.

References

Yongjun Ahn, Wonjun Kim, and Byonghyo Shim. Deep neural network-based joint active user detection and channel estimation for mMTC. In *Proceedings of the IEEE ICC*, pages 1–6, Jun. 2020.

Yanna Bai, Wei Chen, Bo Ai, Zhangdui Zhong, and Ian J Wassell. Prior information aided deep learning method for grant-free NOMA in mMTC. *IEEE Journal on Selected Areas in Communications*, 40(1):112–126, Jan. 2022.

S Boyd, N Parikh, E Chu, B Peleato, and J Eckstein. Distributed optimization and statistical learning via the alternating direction method of multipliers. *Foundations and Trends in Machine Learning*, 3(1):1–122, Jan. 2011.

Zhilin Chen, Foad Sohrabi, Ya-Feng Liu, and Wei Yu. Phase transition analysis for covariance-based massive random access with massive MIMO. *IEEE Transactions on Information Theory*, 68(3):1696–1715, Mar. 2022.

Ying Cui, Shuaichao Li, and Wanqing Zhang. Jointly sparse signal recovery and support recovery via deep learning with applications in MIMO-based grant-free random access. *IEEE Journal on Selected Areas in Communications*, 39(3):788–803, Mar. 2021.

D L Donoho, A Maleki, and A Montanari. Message-passing algorithms for compressed sensing. *Proceedings of the National Academy of Sciences of the United States of America*, 06(45):18914–18919, Nov. 2009.

Ahmet Emir, Ferdi Kara, Hakan Kaya, and Halim Yanikomeroglu. DeepMuD: Multi-user detection for uplink grant-free NOMA IoT networks via deep learning. *IEEE Wireless Communications Letters*, 10(5):1133–1137, May 2021.

Alexander Fengler, Saeid Haghighatshoar, Peter Jung, and Giuseppe Caire. Non-Bayesian activity detection, large-scale fading coefficient estimation, and unsourced random access with a massive MIMO receiver. *IEEE Transactions on Information Theory*, 67(5):2925–2951, May 2021a.

Alexander Fengler, Peter Jung, and Giuseppe Caire. Pilot-based unsourced random access with a massive MIMO receiver in the quasi-static fading regime. In *Proceedings of the IEEE SPAWC*, pages 356–360, Sept. 2021b.

Dongdong Jiang and Ying Cui. MAP-Based pilot state detection in grant-free random access for mMTC. In *Proceedings of the IEEE SPAWC*, pages 1–5, May 2020a.

Dongdong Jiang and Ying Cui. ML estimation and MAP estimation for device activities in grant-free random access with interference. In *Proceedings of the IEEE WCNC*, pages 1–6, May 2020b.

Dongdong Jiang and Ying Cui. ML and MAP device activity detections for grant-free massive access in multi-cell networks. *IEEE Transactions on Wireless Communications*, 21(6):3893–3908, Jun. 2022.

Wuyang Jiang, Yuhang Jia, and Ying Cui. Statistical device activity detection for OFDM-based massive grant-free access. *IEEE Transactions on Wireless Communications*, 22(6):3805–3820, Nov. 2022.

Jeremy Johnston and Xiaodong Wang. Model-based deep learning for joint activity detection and channel estimation in massive and sporadic connectivity. *IEEE Transactions on Wireless Communications*, 21(11):9806–9817, Nov. 2022.

Wonjun Kim, Yongjun Ahn, and Byonghyo Shim. Deep neural network-based active user detection for grant-free NOMA systems. *IEEE Transactions on Communications*, 68(4):2143–2155, Apr. 2020.

Shuaichao Li, Wanqing Zhang, Ying Cui, Hei Victor Cheng, and Wei Yu. Joint design of measurement matrix and sparse support recovery method via deep auto-encoder. *IEEE Signal Processing Letters*, 26(12):1778–1782, Dec. 2019.

Shuaichao Li, Wanqing Zhang, and Ying Cui. Jointly sparse signal recovery via deep auto-encoder and parallel coordinate descent unrolling. In *Proceedings of the IEEE WCNC*, pages 1–6, May 2020.

Shuang Liang, Yinan Zou, and Yong Zhou. GAN-based joint activity detection and channel estimation for grant-free random access. In *Proceedings of the IEEE ICASSP*, pages 4413–4417, May 2022.

Liang Liu and Wei Yu. Massive connectivity with massive MIMO-Part I: Device activity detection and channel estimation. *IEEE Transactions on Signal Processing*, 66(11):2933–2946, Jun. 2018.

Liang Liu, Erik G Larsson, Wei Yu, Petar Popovski, Cedomir Stefanovic, and Elisabeth De Carvalho. Sparse signal processing for grant-free massive connectivity: A future paradigm for random access protocols in the Internet of Things. *IEEE Signal Processing Magazine*, 35(5):88–99, Sept. 2018.

Wang Liu, Ying Cui, Feng Yang, Lianghui Ding, and Sun Jun. MLE-based device activity detection under Rician fading for massive grant-free access with perfect and imperfect synchronization. *IEEE Transactions on Wireless Communications*, Major revision, 2023.

Xiaqing Miao, Dongning Guo, and Xiangming Li. Grant-free NOMA with device activity learning using long short-term memory. *IEEE Wireless Communications Letters*, 9(7):981–984, Jul. 2020.

Piya Pal and P P Vaidyanathan. Pushing the limits of sparse support recovery using correlation information. *IEEE Transactions on Signal Processing*, 63(3):711–726, Dec. 2015.

Yiyang Qiang, Xiaodan Shao, and Xiaoming Chen. A model-driven deep learning algorithm for joint activity detection and channel estimation. *IEEE Communications Letters*, 24(11):2508–2512, Nov. 2020.

Zhiwei Qin, Katya Scheinberg, and Donald Goldfarb. Efficient block-coordinate descent algorithms for the group LASSO. *Mathematical Programming Computation*, 5(2):143–169, Mar. 2013.

Anand P Sabulal and Srikrishna Bhashyam. Joint sparse recovery using deep unfolding with application to massive random access. In *Proceedings of the IEEE ICASSP*, pages 5050–5054, May 2020.

Xiaodan Shao, Xiaoming Chen, Yiyang Qiang, Caijun Zhong, and Zhaoyang Zhang. Feature-aided adaptive-tuning deep learning for massive device detection. *IEEE Journal on Selected Areas in Communications*, 39(7):1899–1914, Jul. 2021.

Yandong Shi, Shuhao Xia, Yong Zhou, and Yuanming Shi. Sparse signal processing for massive device connectivity via deep learning. In *Proceedings of the IEEE ICC Wkshps*, pages 1–6, Jun. 2020.

Yandong Shi, Hayoung Choi, Yuanming Shi, and Yong Zhou. Algorithm unrolling for massive access via deep neural networks with theoretical guarantee. *IEEE Transactions on Wireless Communications*, 21(2):945–959, Feb. 2022.

U K Sreeshma Shiv, Srikrishna Bhashyam, Chirag Ramesh Srivatsa, and Chandra R Murthy. Learning-based sparse recovery for joint activity detection and channel estimation in massive random access systems. *IEEE Wireless Communications Letters*, 11(11):2295–2299, Nov. 2022.

Zhuo Sun, Nan Yang, Chunhui Li, Jinhong Yuan, and Tony Q S Quek. Deep learning-based transmit power control for device activity detection and channel estimation in massive access. *IEEE Wireless Communications Letters*, 11(1):183–187, Jan. 2022.

Jue Wang, Zhaoyang Zhang, and Lajos Hanzo. Joint active user detection and channel estimation in massive access systems exploiting Reed–Muller sequences. *IEEE Journal on Selected Topics in Signal Processing*, 13(3):739–752, Mar. 2019.

Xiaofu Wu, Suofei Zhang, and Jun Yan. A CNN architecture for learning device activity from MMV. *IEEE Communications Letters*, 25(9):2933–2937, Sept. 2021.

Hanxiao Yu, Zesong Fei, Zhong Zheng, Neng Ye, and Zhu Han. Deep learning based user activity detection and channel estimation in grant-free NOMA. *IEEE Transactions on Wireless Communications*, 22(4):2202–2214, 2022.

Wanqing Zhang, Shuaichao Li, and Ying Cui. Jointly sparse support recovery via deep auto-encoder with applications in MIMO-based grant-free random access for mMTC. In *Proceedings of the IEEE SPAWC*, pages 1–5, May 2020.

Weifeng Zhu, Meixia Tao, Xiaojun Yuan, and Yunfeng Guan. Deep-learned approximate message passing for asynchronous massive connectivity. *IEEE Transactions on Wireless Communications*, 20(8):5434–5448, Aug. 2021.

18

Massive Unsourced Random Access

Volodymyr Shyianov, Faouzi Bellili, Amine Mezghani, and Ekram Hossain

Department of Electrical and Computer Engineering, University of Manitoba, Winnipeg, Manitoba, Canada

18.1 Introduction

Massive random access in which a base station (BS) is serving a large number of contending users has recently attracted considerable attention. This surge of interest is fuelled by the need to satisfy the soaring demand for wireless connectivity in many envisioned IoT applications such as massive machine-type communication (mMTC). MTC has two distinct features (Dutkiewicz et al., 2017) that make them drastically different from human-type communications (HTC) around which previous cellular systems have mainly evolved: (i) machine-type devices (MTDs) require sporadic access to the network without any prior scheduling at the BS, i.e., *grantless* or *grant-free* access (Yuan et al., 2016), and (ii) MTDs usually transmit small data payloads using short-packet signaling. The sporadic access renders the overall mMTC traffic generated by an unknown and random subset of active MTDs (at each given transmission). This calls for the development of scalable random access protocols that are able to accommodate a massive number of MTDs. Short-packet transmissions, however, make the traditional grant-based access (with the associated scheduling overhead) fall short in terms of spectrum efficiency and latency. When latency requirements are critical, spending additional time/frequency resources on user identification and channel estimation would also be prohibitive, hence the necessity to develop completely incoherent transmission protocols. In applications that involve a large number of identical MTDs (e.g., sensors) the BS might be interested in the transmitted messages only and not the IDs of the active users, thereby leading to the so-called unsourced random access (URA) problem. In presence of identical MTDs the codebook can also be hard-wired into all the devices, thereby simplifying production and

Next Generation Multiple Access, First Edition.
Edited by Yuanwei Liu, Liang Liu, Zhiguo Ding, and Xuemin Shen.
© 2024 The Institute of Electrical and Electronics Engineers, Inc. Published 2024 by John Wiley & Sons, Inc.

standardization. Grant-free transmissions, incoherent communications, and the same codebook for all the users are common features that make the design of efficient URA protocols particularly compelling.

18.2 URA with Single-Antenna Base Station

URA with a single-antenna BS has been carefully investigated from both information-theoretic and algorithmic perspectives. The information-theoretic work in Polyanskiy (2017) introduced a random coding existence bound for URA using a random Gaussian codebook with maximum-likelihood decoding at the BS. Moreover, popular multiple access schemes, e.g., ALOHA, coded slotted ALOHA, and treating interference as noise (TIN) were compared against the established fundamental limit, showing that none of them achieves the optimal predicted performance. Other works have focused more on the algorithmic aspect of the problem and most of them rely on compressed sensing-based (CS-based) encoding/decoding paradigms in conjunction with a slotted transmission framework. The coded/coupled compressive sensing (CCS) scheme (Amalladinne et al., 2021), introduced the idea of inserting some redundancy bits to couple the information sequences being transmitted by each user across the slots (i.e., concatenated coding). At the BS, an inner tree-decoder is used to stitch the slot-wise decoded sequences of each user, which can be obtained by means of any CS algorithm, such as approximate message passing (AMP)(Donoho et al., 2009) or chirp-based CS (Calderbank and Thompson, 2020). In this context, a lower complexity CCS-based URA scheme has been proposed in Fengler et al. (2021a) wherein the authors exploit the concept of sparse regression codes (SPARCs) (Joseph and Barron, 2012) to reduce the size of the required codebook matrix. Another scheme, dubbed sparse Kronecker product (SKP), that is based on a Kronecker product of two codewords one of which is sparse was also introduced (Han et al., 2021).

18.2.1 System Model and Problem Formulation

Consider a single-cell network consisting of K single-antenna devices which are being served by a single-antenna base station located at the center of a cell of radius R. We assume sporadic device activity thereby resulting in a small number, $K_a \ll K$, of devices being active at each transmission time. Assume the devices communicate to the base station in an uncoordinated way. Every active device/user wishes to transmit a message of B information bits over the channel in a single communication round. The codewords transmitted by active devices are drawn uniformly at random from a common codebook $C = \left\{\mathbf{c}_1, \mathbf{c}_2, \dots, \mathbf{c}_{2^B}\right\} \subset \mathbb{C}^n$.

The device activity and codeword selection are modeled by a set of $2^B K$ Bernoulli random variables $\delta_{b,k}$ for $k = 1, \ldots, K$ and $b = 1, \ldots, 2^B$:

$$\delta_{b,k} = \begin{cases} 1 & \text{if user } k \text{ is active and transmits codeword } \mathbf{c}_b, \\ 0 & \text{otherwise,} \end{cases} \tag{18.1}$$

Over a block-fading Gaussian multiple access channel (MAC), the uplink signal received by the single-antenna BS can be expressed as follows:

$$\mathbf{y} = \sum_{k=1}^{K} \sum_{b=1}^{2^B} \sqrt{g_k} \tilde{h}_k \delta_{b,k} \mathbf{c}_b + \mathbf{w}. \tag{18.2}$$

The random noise vector, $\mathbf{w}$, is modeled by a complex circular Gaussian random vector with independent and identically distributed (i.i.d.) components, i.e., $\mathbf{w} \sim \mathcal{CN}(\mathbf{0}, \sigma_w^2 \mathbf{I})$. In addition, $\tilde{h}_k$ stands for the small-scale fading coefficient between the kth user and the BS antenna. Under the block fading assumption, the small-scale fading channel coefficients remain constant over the entire observation window which is smaller than the coherence time. Besides, g_k is the large-scale fading coefficient of user k given by (in dB scale):

$$g_k \,[\text{dB}] = -\alpha - 10\beta \log_{10}\left(r_k\right), \tag{18.3}$$

where α is the fading coefficient measured at distance $d = 1$ m and β is the pathloss exponent. For convenience, we also define the effective channel coefficient by lumping the large- and small-scale fading coefficients together in one quantity, denoted as $h_k \triangleq \sqrt{g_k}\tilde{h}_k$, thereby yielding the following equivalent model:

$$\mathbf{y} = \sum_{k=1}^{K} \sum_{b=1}^{2^B} h_k \delta_{b,k} \mathbf{c}_b + \mathbf{w}. \tag{18.4}$$

The channel model in (18.4) can be rewritten in a more succinct matrix–vector form as follows:

$$\mathbf{y} = \mathbf{C}\boldsymbol{\Delta}\mathbf{h} + \mathbf{w}, \tag{18.5}$$

where $\mathbf{C} = \left[\mathbf{c}_1, \mathbf{c}_2, \ldots, \mathbf{c}_{2^B}\right] \in \mathbb{C}^{n \times 2^B}$ is the known codebook matrix, which is common to all the users and $\mathbf{h} = [h_1, h_2, \ldots, h_K]^\mathsf{T} \in \mathbb{C}^K$ is the multiuser channel vector which incorporates the small- and large-scale fading coefficients. The matrix $\boldsymbol{\Delta} \in \{0, 1\}^{2^B \times K}$ contains only K_a nonzero columns each of which having a single nonzero entry. Observe here that both $\boldsymbol{\Delta}$ and $\mathbf{h}$ are unknown to the receiver. Hence, by defining $\mathbf{x} \triangleq \boldsymbol{\Delta}\mathbf{h}$, it follows from (18.5) that:

$$\mathbf{y} = \mathbf{C}\mathbf{x} + \mathbf{w}. \tag{18.6}$$

Notice that the vector $\mathbf{x}$ in (18.6) is sparse (i.e., has many zeroes), which brings up a close connection between URA and compressed sensing (CS). As will be seen in

Section 18.2.2, CS plays a vital role in most of the algorithmic solutions for URA. We now define the URA code over the channel in (18.6).

Definition 18.1 ***(URA code)*** Let $W_k \in [2^B] \triangleq \{1, 2, \dots, 2^B\}$ denote the message of user k. An URA code for the channel in (18.6) is a pair of maps, an encoding map $f : [2^B] \to \mathbb{C}^n$, such that $f(W_k) = \mathbf{c}_{b_k}$. By recalling that K_a stands for the number of active users, the decoding[1] map $g : \mathbb{C}^n \to \binom{[2^B]}{K_a}$ outputs a list of K_a decoded messages with the probability of error being defined as:

$$\epsilon = \frac{1}{K_a} \sum_{k=1}^{K_a} \Pr(E_k), \tag{18.7}$$

and E_k is the event defined as

$$E_k \triangleq \left\{ W_k \notin g(\mathbf{y}) \right\} \cup \left\{ W_k = W_{k'} \text{ for some } k' \neq k \right\}. \tag{18.8}$$

Observe here that ϵ depends solely on the number of active users, K_a, instead of the total number of users K. To gain better insights, it is instructive to investigate **Definition** 18.1 through a special case of the channel model in (18.4) when $h_k = 1\ \forall k$ – known as the Gaussian multiple access channel (GMAC) – with the codebook being real-valued, i.e., $\mathbf{C} \in \mathbb{R}^{n \times 2^B}$. The following theorem guarantees the existence of the URA code (i.e., maps f and g) as long as the parameters $(2^B, n, \epsilon)$ satisfy a well-defined constraint, known as the random coding existence bound.

Theorem 18.1 ***(Existence)*** (Polyanskiy, 2017) Let P be the power constraint and fix $P' < P$. There exists an (M, n, ϵ) URA code for the K_a-user GMAC satisfying the power constraint P and

$$\epsilon \leq \sum_{t=1}^{K_a} \frac{t}{K_a} \min(p_t, q_t) + p_0,$$

where

$$\begin{aligned}
p_0 &= \frac{1}{M}\binom{K_a}{2} + K_a \mathbb{P}\left[\frac{1}{n}\sum_{j=1}^{n} w_j^2 > \frac{P}{P'}\right] \\
p_t &= e^{-nE(t)}, \\
E(t) &= \max_{0 \leq \rho, \rho_1 \leq 1} -\rho\rho_1 t R_1 - \rho_1 R_2 + E_0(\rho, \rho_1) \\
E_0 &= \rho_1 a + \frac{1}{2}\log(1 - 2b\rho_1) \\
a &= \frac{\rho}{2}\log(1 + 2P' t\lambda) + \frac{1}{2}\log(1 + 2P' t\mu)
\end{aligned}$$

1 The notation $\binom{[2^B]}{K_a}$ stands for choosing K_a different elements from the set $[2^B]$.

$$
\begin{aligned}
b &= \rho\lambda - \frac{\mu}{1+2P't\mu}, \mu = \frac{\rho\lambda}{1+2P't\lambda} \\
\lambda &= \frac{P't - 1 + \sqrt{D}}{4\left(1+\rho_1\rho\right)P't}, \\
D &= \left(P't-1\right)^2 + 4P't\frac{1+\rho\rho_1}{1+\rho} \\
R_1 &= \frac{1}{n}\log M - \frac{1}{nt}\log(t!), \\
R_2 &= \frac{1}{n}\binom{K_a}{t} \\
q_t &= \inf_{\gamma} \mathbb{P}\left[I_t < \gamma\right] + \exp\{n(tR_1 + R_2) - \gamma\} \\
I_t &= \min_{S_0} i_t\left(c\left(S_0\right); \mathbf{y} \mid c\left(S_0^c\right)\right)
\end{aligned}
$$

$$
i_t(\mathbf{a};\mathbf{y} \mid \mathbf{b}) = \frac{n}{2}\log(1+P't) + \frac{\log e}{2}\left(\frac{\|\mathbf{y}-\mathbf{b}\|_2^2}{1+P't} - \|\mathbf{y}-\mathbf{a}-\mathbf{b}\|_2^2\right)
$$

In the above, $c(S) = \sum_{j\in S}\mathbf{c}_j$, $S_0 \subset [K_a]$, and $S_0^c \subset [M]\backslash[K_a]$ are generic subsets of size t. It is also important to note that the derivation of the bound assumes that any collision automatically results in an error. This probability is, however, rather small even for moderate values of B:

$$
\mathbb{P}\left[\cup_{i\neq j}\left\{W_j = W_i\right\}\right] \leq \frac{1}{M}\binom{K_a}{2} \tag{18.9}
$$

Theorem 18.1, is used to evaluate the performance of existing algorithmic solutions. The most common metric is the minimum energy-per-bit required to achieve certain a target error probability at a fixed blocklength and data payload. Algorithmically speaking the URA problem becomes increasingly hard when the number of active users grows large and the best-known schemes, such as SKP coding from Han et al. (2021), are still several dBs away from the achievability bound predicted by **Theorem 3.1**.

18.2.2 Algorithmic Solutions

18.2.3 Slotted Transmission Framework

By revisiting (18.2), we see that the number of codewords grows exponentially with the blocklength n. Indeed, for a fixed rate $R = \frac{B}{n}$, the number of codewords, being $2^B = 2^{nR}$, becomes extremely large even for moderate values of n. This makes any attempt to directly use standard CS algorithms computationally prohibitive. Practical approaches have been introduced to alleviate this computational burden, including the slotted transmissions framework which we describe in this section. From now on, unless explicitly stated otherwise, by "slotted transmission

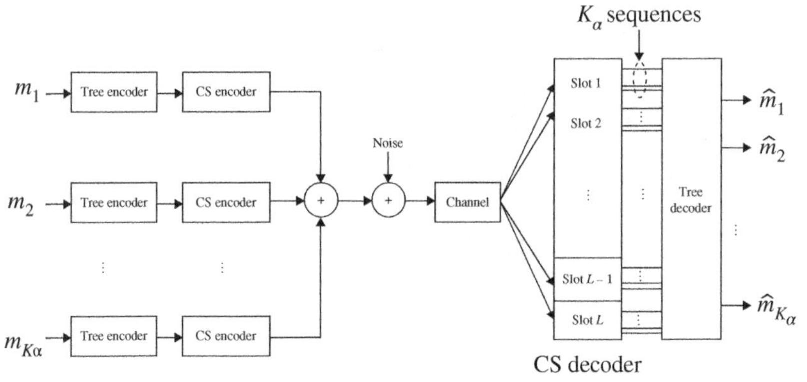

Figure 18.1 The architecture of a slotted transmission framework.

framework" we refer to the scheme depicted in Fig. 18.1. The use of slotted transmissions is driven by the need to alleviate the inherent prohibitive computational burden of the underlying index coding problem. More specifically, in the slotted CSS scheme (Amalladinne et al., 2021), the binary message of each user is partitioned into multiple binary information sequences (or chunks). Then concatenated coding is used to couple the different sequences before using a random Gaussian codebook for actual transmissions over multiple slots. On the receiver side, inner CS-based decoding is first performed to recover the slot-wise transmitted sequences up to an unknown permutation. An outer tree-based decoder is then used to stitch the decoded binary sequences across different slots (i.e., removing the effect of the underlying permutation).

(1) *Tree encoding (Concatenated coding):* The B-bit binary message $\mathbf{m}$ of each user, is split into L chunks, where the lth chunk $\mathbf{m}(l)$ consists of b_l information bits such that $\sum_{l=1}^{L} b_l = B$. The total message can then be written as a concatenation of the L chunks, i.e., $\mathbf{m} = \mathbf{m}(1)\mathbf{m}(2)\ldots\mathbf{m}(L)$. The tree-encoder appends l_j parity bits to the lth chunk, which brings the total length of each tree-encoded chunk to $b_l + l_j = B/L$ for every l (in the sequel, we define $J \triangleq \frac{B}{L}$). The first chunk is then chosen to consist of information bits only (no parity bits) i.e., $l_1 = 0$ and that $b_1 = \frac{B}{L} = J$. For the subsequent chunks, the parity bits are constructed to satisfy a random linear parity constraint associated with the information bits in the preceding chunks. In fact, the vector of parity bits in chunk l, denoted as $\mathbf{p}(l)$, is constructed as follows:

$$\mathbf{p}(l) = \sum_{j=1}^{l} \mathbf{m}(j)\mathbf{G}_{j,l}, \tag{18.10}$$

where $\mathbf{G}_{j,l}$ is a random binary matrix of size $b_j \times l_j$. Each entry of $\mathbf{G}_{j,l}$ is an independent Bernoulli trial with parameter $1/2$ and all the arithmetic is performed modulo-2. With this, the tree-encoded message can be written as follows:

$$\tilde{\mathbf{m}} = \underbrace{\mathbf{m}(1)}_{\tilde{\mathbf{m}}(1)}\ \underbrace{\mathbf{m}(2)\mathbf{p}(2)}_{\tilde{\mathbf{m}}(2)} \cdots \underbrace{\mathbf{m}(L)\mathbf{p}(L)}_{\tilde{\mathbf{m}}(L)}. \tag{18.11}$$

(2) *CS encoding:* Let the matrix $\mathbf{A} \in \mathbb{C}^{\frac{n}{L} \times 2^J}$ denote the common codebook for all the users (over all slots). That is, the columns of $\mathbf{A} = [\mathbf{a}_1, \mathbf{a}_2, \ldots, \mathbf{a}_{2^J}]$ form a set of codewords that each $\{k\text{th}\}_{k=1}^{K_a}$ active user chooses from in order to encode its $\{l\text{th}\}_{l=1}^{L}$ sequence before transmitting it over the $\{l\text{th}\}_{l=1}^{L}$ slot. Notice here that, in such a slotted transmission framework, the size of the codebook is 2^J. This is much smaller than the original codebook of size 2^B which was actually used to prove the random coding achievability bound in Theorem 18.1. Slotting is, however, a necessary step toward alleviating the prohibitive computational burden as mentioned previously.

After tree-encoding, we obtain L $J-$bit tree-coded chunks, which are in turn channel encoded separately using the codebook, $\mathbf{A} \in \mathbb{C}^{\frac{n}{L} \times 2^J}$, that will serve as the sensing matrix for sparse recovery. Conceptually, we operate on a per-slot basis by associating to every possible $J-$bit tree-encoded sequence a different column in the codebook matrix $\mathbf{A}$. Thus, one can view this matrix as a set of potentially transmitted codewords over the duration of a given slot. The multiuser CS encoder can be visualized as an abstract multiplication of $\mathbf{A}$ by an index vector $\mathbf{v}$. The positions of nonzero coefficients in $\mathbf{v}$ are nothing but the decimal representations of the tree-encoded binary sequences/chunks being transmitted by the active users over a given slot. Thus, the slotted transmission of the B-bit packets of all the active users gives rise to L small-size compressed sensing instances (one per each slot). Each kth user transmits the codeword associated to its J-bit tree-encoded sequence over the channel which in turn gets a multiplicative complex coefficient h_k and an additive white Gaussian noise. Hence, the overall baseband model over each slot reduces to the MAC model discussed in Section 18.2.1. Indeed, by recalling (18.6), the received signal over the lth slot is given by:

$$\mathbf{y}_l = \mathbf{A}\mathbf{x}_l + \mathbf{w}_l, \quad l = 1, \ldots, L. \tag{18.12}$$

18.2.3.1 Decoding

As depicted in Fig. 18.1, decoding is performed in two stages. In the first stage, the inner CS-based decoder operates over each slot separately and returns all the CS-encoded messages by further mapping them into tree-encoded chunks. Those tree-encoded chunks $\tilde{\mathbf{m}}_k(l)$ for all l are then stitched together by enforcing the linear parity constraints introduced by the tree-encoder so as to obtain the full messages $\tilde{\mathbf{m}}_k$. The redundancy bits are then removed to recover the original messages $\mathbf{m}_k$ for all k.

18.2.3.2 CS Decoding

Casting the URA problem as a sparse regression/CS recovery problem has been a major stepping stone toward developing efficient algorithmic solutions. As a separate theory, CS is not new and has been an active area of research for more than a decade now. The theory of sparse reconstruction from noisy observations, however, has been a hot research topic in statistics since the 90s and we refer the theoretically inclined reader to Chapters 7–9 in Wainwright (2019), for elaborate discussions on the theoretical guarantees for sparse reconstruction. Here we focus on a particular CS algorithm that has been most useful in the context of URA.

(1) *Approximate message passing:* Among a plethora of CS recovery techniques, the AMP algorithm (Donoho et al., 2009) has attracted considerable attention within the framework of massive random access mainly due to the existence of simple scalar equations that track its dynamics, as rigorously analyzed in Bayati and Montanari (2011). The derivation of the AMP algorithm follows from Gaussian and quadratic approximations of the well-known sum-product algorithm. Recalling the linear input–output relationship of the channel in (18.12), and focusing on a single slot while dropping the slot-index l, i.e.,

$$\mathbf{y} = \mathbf{A}\mathbf{x} + \mathbf{w}, \tag{18.13}$$

and starting from the initial condition $\hat{\mathbf{x}}^0 = \mathbf{0}, \hat{\tau}^0 = 1$, the AMP algorithm proceeds to construct a sequence of estimates $\hat{\mathbf{x}}$ as follows:

$$\hat{\mathbf{x}}^{t+1} = \eta(\hat{\mathbf{x}}^t + \mathbf{A}^{\mathrm{T}}\mathbf{r}^t; \hat{\tau}^t), \tag{18.14}$$

$$\mathbf{r}^t = \mathbf{y} - \mathbf{A}\hat{\mathbf{x}}^t + \mathbf{b}^t\mathbf{r}^{t-1}. \tag{18.15}$$

The choice of the denoising function $\eta(.;.)$, in (18.14) depends on the prior distribution of the vector $\mathbf{x}$. The prior distribution depends, however, on the fading model and can thus be tailored to the particular application at hand. One common choice is to utilize a soft-thresholding function, which is a min–max prior among a class of sparse distributions with sparsity parameter ϵ. Another popular choice is the Bernoulli–Gaussian prior, which arises naturally in Rayleigh fading scenarios. Here, we provide a simple illustrative example, under no fading, i.e., $h_k = 1$ for all k, wherein we derive an MMSE denoiser under a Bernoulli prior with sparsity parameter ϵ. In this case, the underlying denoising function $\eta(.;.)$ is the conditional mean estimator and is applied to the following scalar model (due to the decoupling property of the AMP in the large system limit) in which $n_j \sim \mathcal{N}(0, \hat{\tau}^t)$:

$$z_j^t = x_j + n_j^t \tag{18.16}$$

Now, taking x_j as Bernoulli-distributed with parameter $\varepsilon = K_a/2^J$. The optimal threshold detector for x_j would be to decide $x_j = 1$ if:

$$p(1|z_j^t) > p(0|z_j^t), \tag{18.17}$$

which is equivalent to the following decision rule. Decide $x_j = 1$, if:

$$\mathbb{E}\{x_j|z_j^t\} > \frac{1}{2}. \tag{18.18}$$

This yields an obvious choice for the denoiser:

$$\eta(z^t; \hat{\tau}^t) = \mathbb{E}\{x_j|z_j^t\}, \tag{18.19}$$

which for the case of a Bernoulli prior can be expressed as follows:

$$\eta(z^t; \hat{\tau}^t) = \frac{\varepsilon e^{-\frac{(z^t-1)^2}{2\hat{\tau}^t}}}{\varepsilon e^{-\frac{(z^t-1)^2}{2\hat{\tau}^t}} + (1-\varepsilon)e^{-\frac{z^{t2}}{2\hat{\tau}^t}}}, \tag{18.20}$$

The prescription for the reaction terms is as follows:

$$b^t = \frac{1}{n}\sum_{j=1}^{2^J} \eta_j'(\hat{\mathbf{x}}^{t-1} + \mathbf{A}^{\mathrm{T}}\mathbf{r}^{t-1}; \hat{\tau}^{t-1}), \tag{18.21}$$

$$\hat{\tau}^t = \sigma_w^2 + \frac{\hat{\tau}^{t-1} b^t}{\delta}, \tag{18.22}$$

where $\delta = \frac{n}{2^J}$. At convergence, $x_j > 1/2$ is set to 1 and $x_j < 1/2$ is set to 0.

(2) *State evolution:* The distribution of the tth AMP iterate in (18.14) admits a precise high-dimensional limit characterized by the following recursion:

$$\tau^{t+1} = \sigma_w^2 + \frac{1}{\delta}\mathbb{E}\{[\eta(x + \sqrt{\tau^t}Z) - x]^2\}. \tag{18.23}$$

With the binary distribution for x, the recursion is given by:

$$\tau^{t+1} = \sigma_w^2 + \frac{\mathrm{MSE}^t}{\delta}, \tag{18.24}$$

where

$$\mathrm{MSE}^t = \varepsilon - \frac{\varepsilon^2}{\sqrt{2\pi\tau^t}} \int_{-\infty}^{\infty} \frac{e^{-(r-1)^2/\tau^t}}{(1-\varepsilon)e^{-(r-1)^2/2\tau^t} + \varepsilon e^{-r^2/2\tau^t}} dr. \tag{18.25}$$

The asymptotic symbol-error probability is then derived in the same way as for the BPSK MAP detector:

$$P_e = (1-\varepsilon)Q\left(\frac{r^*}{\sqrt{\tau^t}}\right) + \varepsilon\left(1 - Q\left(\frac{r^*-1}{\sqrt{\tau^t}}\right)\right), \tag{18.26}$$

where

$$r^* = \frac{1 - \ln\left(\frac{\varepsilon}{1-\varepsilon}\right) 2\tau^t}{2}. \tag{18.27}$$

The above symbol-error probability could be written in terms of false alarm and miss-detection probabilities:

$$P_e = (1 - \varepsilon)P_{\text{FA}} + \varepsilon P_M. \tag{18.28}$$

18.2.3.3 Tree Decoding

After performing CS decoding, one obtains a list of all the decoded messages in each slot. We fix a particular chunk from the list in slot 1 and view it as a root of a tree. Once a given chunk was selected, there are K_a choices for the second chunk in slot 2. At the last stage, there are K_a^L possible candidate paths. Most of the candidate paths would not satisfy the parity constraint, and so would be trimmed at the early stages of the decoding process. For correct stitching, at the end of the process one must necessarily obtain K_a paths, one for each user, where each path corresponds to a single decoded message $\mathbf{m}$.

18.2.4 Sparse Kronecker Product (SKP) Coding

In SKP coding (Han et al., 2021), the original B-bit message $\mathbf{m}_k$ of the kth user is encoded as a Kronecker product of two codewords, where one of the codewords is sparse and the other is forward-error-correction (FEC) coded. The decoding is performed iteratively where matrix factorization is performed for the decomposition of the Kronecker product and soft-input soft-output decoding for the sparse code and the forward FEC code. For every user k, the B-bit message $\mathbf{m}_k$ is split into two parts. The first part, consisting of B_a-bits, $\mathbf{m}_k^{(a)}$ is encoded into a length n_a vector, $\mathbf{a}_k(m_k^{(a)})$. The second part, consisting of $B - B_a$ bits, $\mathbf{m}_k^{(x)}$ is encoded into a length $n_x = \lfloor \frac{n}{n_a} \rfloor$ vector $\mathbf{x}_k(m_k^{(x)})$. The sequence, to be transmitted over the channel, is obtained as a Kronecker product of two vectors. The received signal, at the BS antenna, can be expressed as follows:

$$\mathbf{y} = \sum_{k=1}^{K_a} \mathbf{a}_k\left(\mathbf{m}_k^{(a)}\right) \otimes \mathbf{x}_k\left(\mathbf{m}_k^{(x)}\right) + \mathbf{w}. \tag{18.29}$$

By further defining the matrices $\mathbf{A} = [\mathbf{a}_1, \dots, \mathbf{a}_{K_a}]$ and $\mathbf{X} = [\mathbf{x}_1, \dots, \mathbf{x}_{K_a}]$, the vector $\mathbf{y} \in \mathbb{C}^{n_x n_a}$ can be reshaped into the matrix $\mathbf{Y} \in \mathbb{C}^{n_a \times n_x}$ to obtain the following system model:

$$\mathbf{Y} = \sum_{k=1}^{K_a} \mathbf{a}_k \mathbf{x}_k^T + \mathbf{W} = \mathbf{A}\mathbf{X} + \mathbf{W}. \tag{18.30}$$

The structure of the matrices $\mathbf{A}$ and $\mathbf{X}$ is determined from the two-component codes. At the receiver, matrix factorization is needed to recover the two-component codes and could be efficiently performed with a message passing algorithm such as Bi-AMP (Parker et al., 2014, Akrout et al., 2020). To aid matrix factorization,

A can be chosen sparse, while **X** is simply FEC-coded with a code having a good performance at short blocklength, such as tail-baiting convolutional code. Other choices for **A** and **X** are possible and can lead to large performance gains if tailored to the particular decoding algorithm at hand.

18.2.5 Numerical Discussion

In this section, we discuss the performance of the most important single-antenna URA schemes in terms of minimum required energy-per-bit to achieve a target error probability. Let $\frac{E_b}{N_0} = \frac{P}{BN_0}$ where $P = \mathbb{E}[\| \mathbf{c}_b \|^2]$. The total blocklength is fixed to $n = 30,000$ and the minimum $\frac{E_b}{N_0}$ is determined such that the per user error probability (PUEP), denoted as ϵ, satisfies $\epsilon \leq 0.05$. The two schemes discussed above, namely SKP coding and AMP-based CS decoding with BP-based tree-decoding, are compared to the bound in Theorem 18.1, see Amalladinne et al. (2021) and Che et al. (2022) for more details. As can be seen from Fig. 18.2, both algorithmic approaches deviate from the random coding bound by at least a few

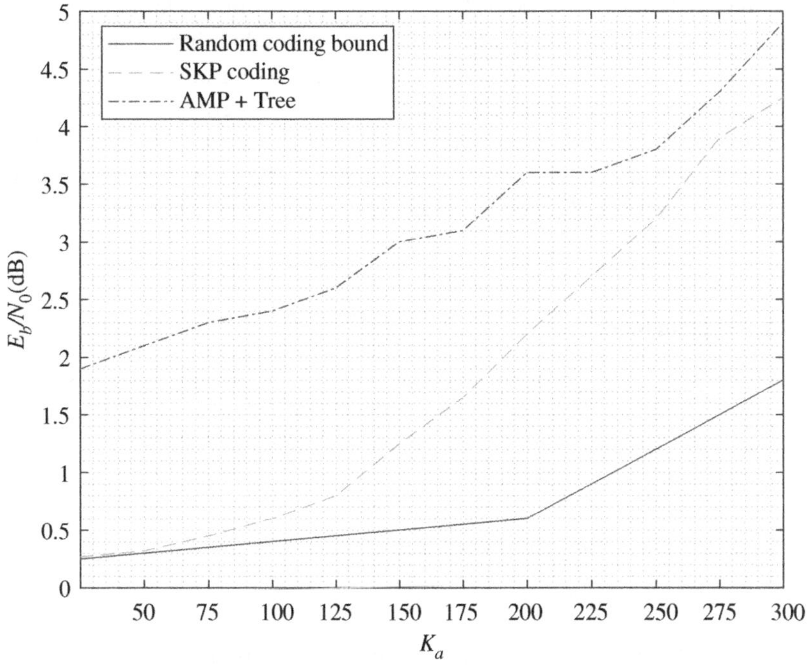

Figure 18.2 Minimum required E_b/N_0 versus the number of active users for the two main URA algorithmic solutions using a single-antenna base station from Amalladinne et al. (2021) and Che et al. (2022) together with the random coding bound.

dBs when the number of users becomes large. For a small number of active users, however, SKP-based URA gets as close as 0.1 dB to the optimum performance bound with $K_a = 75$ for instance. This indicates that a fruitful research direction for SKP coding would be to explore other single-code structures for the two code components, and optimally exploit them in the matrix factorization-based decoder within the AMP framework. Tree-based encoding/decoding appears to sacrifice too much spectral efficiency for transmitting the concatenated coding bits needed to stitch the sequences together. This loss will become even more apparent in presence of multi-antennas at the BS where more degrees of freedom are readily available for more efficient decoding.

18.3 URA with Multi-Antenna Base Station

Once the framework has been laid out for the case of a single-antenna base station, one might think that the extension to multiple antennas is straightforward. However, the spatial selectivity introduced by the antenna array at the BS must be carefully handled. By doing so, MIMO algorithms are able to utilize additional degrees of freedom for decoding and so provide unique features that single-antenna algorithms do not have. Some of the best-known algorithms for MIMO URA are the covariance-based scheme in Fengler et al. (2021b), the clustering-based scheme in Shyianov et al. (2020), and tensor-based scheme in Decurninge et al. (2020). Other more recent methods based on beam-space decoding and bilinear reconstruction have been developed in Che et al. (2022) and Ayachi et al. (2022) and Ke et al. (2023), respectively.

18.3.1 System Model

Recall the equivalent problem formulation with a single antenna BS in (18.5). In presence of a multi-antenna BS, we have the same model at each of the $\{m^{th}\}_{m=1}^{M_r}$ antennae. Indeed, the signal received by antenna m during the lth slot, is given by:

$$\mathbf{y}_l^{(m)} = \mathbf{A}\boldsymbol{\Delta}_l\mathbf{h}^{(m)} + \mathbf{w}_l^{(m)}, \quad m = 1, \dots, M_r, \tag{18.31}$$

which can be further simplified by defining $\mathbf{x}_l^{(m)} = \boldsymbol{\Delta}_l\mathbf{h}^{(m)}$:

$$\mathbf{y}_l^{(m)} = \mathbf{A}\,\mathbf{x}_l^{(m)} + \mathbf{w}_l^{(m)}, \quad m = 1, \dots, M_r. \tag{18.32}$$

With a total of M_r antennas at the BS, each active user contributes a single non-zero coefficient in $\mathbf{x}_l^{(m)}$ thereby resulting in K_a–sparse 2^J–dimensional vector. Since K_a is much smaller than the total number of codewords 2^J in each slot, $\mathbf{x}_l^{(m)}$ has a very small sparsity ratio $\lambda \triangleq \frac{K_a}{2^J}$. Observe also that the formulation in (18.32) belongs to the MMV class in compressed sensing terminology, which can be equivalently rewritten in a more succinct matrix/matrix form as follows:

$$\mathbf{Y}_l = \mathbf{A}\mathbf{X}_l + \mathbf{W}_l, \tag{18.33}$$

in which $\mathbf{Y} = \left[\mathbf{y}_l^{(1)}, \mathbf{y}_l^{(2)}, \dots, \mathbf{y}_l^{(M_r)}\right]$ is the entire observed data matrix and $\mathbf{X}_l = \left[\mathbf{x}_l^{(1)}, \mathbf{x}_l^{(2)}, \dots, \mathbf{x}_l^{(M_r)}\right]$. With this formulation, the unknown matrix $\mathbf{X}_l$ is row-sparse (i.e., entire rows are identically zero), which is a feature to be exploited while developing the algorithmic solutions.

18.3.2 Algorithmic Solutions

18.3.3 Covariance-Based Compressed Sensing

Covariance-based compressed sensing (CB-CS) URA relies on the tree outer encoder/decoder as described in Section 18.2.2. However, the CS decoder utilizes a specific covariance structure of the data which makes it particularly suitable for massive MIMO. CB-CS aims to estimate the large-scale fading coefficient of each active user, where the nonzero large-scale fading coefficients can be automatically mapped to the codewords transmitted by the active users. As the number of antennas at the BS grows, the CB-CS method improves decoding due to the increase in the amount of data, with no extra cost in the number of parameters that need to be estimated. On the other hand, the AMP-MMV algorithm which is designed to estimate the entire small-scale channel vector, $\mathbf{X}_l$, whose dimension grows with the number of BS antennas, does not enjoy a comparable decoding improvement. We first begin by rewriting (18.33), to explicitly show the small- and large-scale coefficients of the channel which were previously lumped together:

$$\mathbf{Y}_l = \mathbf{A}\mathbf{\Delta}_l\mathbf{G}^{\frac{1}{2}}\mathbf{H} + \mathbf{W}_l. \tag{18.34}$$

where $\mathbf{G}$ is a $K \times K$ diagonal matrix containing the large-scale fading coefficients, $\{g_k\}_{k=1}^{K_a}$, of all users. By defining $\mathbf{\Gamma}_l = \mathbf{\Delta}_l\mathbf{G}$, (18.34) is further rewritten as follows:

$$\mathbf{Y}_l = \mathbf{A}\mathbf{\Gamma}_l^{\frac{1}{2}}\mathbf{H} + \mathbf{W}_l, \tag{18.35}$$

Assuming that the channel vectors are spatially white and Gaussian, again dropping the subscript l, the rows of $\mathbf{Y}$ are i.i.d. Gaussian vectors with covariance matrix given by:

$$\mathbf{\Sigma}_{\mathbf{y}} = \mathbf{A}\mathbf{\Gamma}\mathbf{A}^H + \sigma^2\mathbf{I} = \sum_{k=1}^{K} g_k\mathbf{a}_k\mathbf{a}_k^H + \sigma^2\mathbf{I}. \tag{18.36}$$

The sample covariance matrix of the data is then obtained as follows:

$$\hat{\mathbf{\Sigma}}_{\mathbf{y}} = \frac{1}{M_r}\mathbf{Y}\mathbf{Y}^H = \frac{1}{M_r}\sum_{i=1}^{M_r}\mathbf{Y}_{:,i}\mathbf{Y}_{:,i}^H, \tag{18.37}$$

where $\mathbf{Y}_{:,i}$ denotes the ith column of $\mathbf{Y}$. To detect the transmitted codewords it is only required to estimate the large-scale fading parameters. In the CB-CS method, this is done through maximum likelihood estimation. In the following, we denote by $\mathbf{g}$ the vector of diagonal elements of $\mathbf{\Gamma}$. The negative log-likelihood function of the data, $f(\mathbf{g}) = -\frac{1}{M_r} \log p(\mathbf{Y} \mid \mathbf{g})$, is given by:

$$\begin{aligned} f(\mathbf{g}) &\overset{(a)}{=} -\frac{1}{M_r} \sum_{i=1}^{M_r} \log p\left(\mathbf{Y}_{:,i} \mid \mathbf{g}\right) \\ &\propto \log \left|\mathbf{A}\mathbf{\Gamma}\mathbf{A}^H + \sigma^2 \mathbf{I}\right| + \mathrm{tr}\left(\left(\mathbf{A}\mathbf{\Gamma}\mathbf{A}^H + \sigma^2 \mathbf{I}\right)^{-1} \hat{\mathbf{\Sigma}}_{\mathbf{y}}\right), \end{aligned} \tag{18.38}$$

where (a) follows from the independence of the columns of $\mathbf{Y}$. The task is then to solve the following optimization problem:

$$\mathbf{g}_{\mathrm{ML}}^* = \arg\min_{\mathbf{g} \in \mathbb{P}_+^{K_a}} f(\mathbf{g}) \tag{18.39}$$

where $\mathbb{P}_+^{K_a} = \{\mathbf{g} \in \mathbb{R}_+^K : \|\mathbf{g}\|_0 \leq K_a\}$. Here $\|\cdot\|_0$ counts the number of nonzero coefficients in a vector. The above optimization problem is combinatorial in nature and as such cannot be solved directly without further relaxation to a suitable convex problem that could be solved by coordinate-descent optimization. The following convex relaxation:

$$\mathbf{g}_{r-\mathrm{ML}}^* = \arg\min_{\mathbf{g} \in \mathbb{R}_+^{K_{\mathrm{tot}}}} f(\mathbf{g}) \tag{18.40}$$

is then more amenable to algorithmic approaches. CB-CS is an efficient iterative procedure that runs according to the algorithmic steps in Algorithm 18.1 to solve the optimization problem in (18.40).

Algorithm 18.1 Activity Detection via Coordinate-Wise Optimization

1: Input: The sample covariance matrix $\hat{\mathbf{\Sigma}}_{\mathbf{y}} = \frac{1}{M_r}\mathbf{Y}\mathbf{Y}^{\mathsf{H}}$ of the $\frac{n}{L} \times M_r$ matrix of samples $\mathbf{Y}$.

3: Initialize: $\mathbf{\Sigma}_{\mathbf{y}} = \sigma^2 \mathbf{I}, \mathbf{g} = \mathbf{0}$.

4: for $i = 1, 2, \ldots, N_{\mathrm{iter}}$, do

5: Select an index $k \in [K]$ corresponding to the kth component of $\mathbf{g} = \left(g_1, \ldots, g_K\right)^{\mathsf{T}}$ randomly or according to a specific schedule.

6: Set $d^* = \max \left\{ \frac{\mathbf{a}_k^{\mathsf{H}} \mathbf{\Sigma}_{\mathbf{y}}^{-1} \hat{\mathbf{\Sigma}}_{\mathbf{y}} \mathbf{\Sigma}_{\mathbf{y}}^{-1} \mathbf{a}_k - \mathbf{a}_k^{\mathsf{H}} \mathbf{\Sigma}_{\mathbf{y}}^{-1} \mathbf{a}_k}{\left(\mathbf{a}_k^{\mathsf{H}} \mathbf{\Sigma}_{\mathbf{y}}^{-1} \mathbf{a}_k\right)^2}, -g_k \right\}$.

9: Update $g_k \leftarrow g_k + d^*$.

10: Update $\mathbf{\Sigma}_{\mathbf{y}}^{-1} \leftarrow \mathbf{\Sigma}_{\mathbf{y}}^{-1} - \frac{d^* \mathbf{\Sigma}_{\mathbf{y}}^{-1} \mathbf{a}_k \mathbf{a}_k^{\mathsf{H}} \mathbf{\Sigma}_{\mathbf{y}}^{-1}}{1 + d^* \mathbf{a}_k^{\mathsf{H}} \mathbf{\Sigma}_{\mathbf{y}}^{-1} \mathbf{a}_k}$

11: end for

12: Output: The resulting estimate $\mathbf{g}$.

18.3.4 Clustering-Based Method

The clustering-based URA method is also geared toward a slotted transmission framework. It utilizes correlation between the slot-wise estimated channels of different users, through expectation maximization (EM) clustering instead of concatenated coding, to stitch their decoded information chunks across the L slots. Indeed, each active user partitions its B–bit message into L equal-size information bit sequences (or chunks). As opposed to tree encoding, however, the clustering-based approach does not require concatenated coding to couple the sequences across different slots (i.e., no outer binary tree encoder). Instead, the bits are distributed uniformly between the L slots and there is no need to optimize the sizes of the L sequences/chunks. In this way, there is a total number of $J = \frac{B}{L}$ bits in each sequence (that will be conveyed during the corresponding slot). The CS encoder operates in the same manner as previously, so the receive signal follows the same system model described in (18.33). The slot-wise CS-based decoding is performed via the HyGamp algorithm (Rangan et al., 2017) which is an extension of AMP tailored to group-sparsity. In fact, group-sparsity is obtained by vectorizing (18.33) over each lth slot:

$$\text{vec}\left(\mathbf{Y}_l^T\right) = (\mathbf{A}^T \otimes \mathbf{I})\text{vec}\left(\mathbf{X}_l^{\text{T}}\right) + \text{vec}\left(\mathbf{W}_l^T\right), \tag{18.41}$$

in which $\otimes$ denotes the Kronecker product of two matrices. Then, by defining $\tilde{\mathbb{A}} \triangleq \mathbf{A}^{\text{T}} \otimes \mathbf{I} \in \mathbb{C}^{\frac{n}{L}M_r \times 2^J M_r}$, $\tilde{\mathbf{y}}_l \triangleq \text{vec}\left(\mathbf{Y}_l^{\text{T}}\right) \in \mathbb{C}^{\frac{n}{L}M_r}$, $\tilde{\mathbf{x}}_l \triangleq \text{vec}\left(\mathbf{X}_l^{\text{T}}\right) \in \mathbb{C}^{2^J M_r}$, and $\tilde{\mathbf{w}}_l \triangleq \text{vec}\left(\mathbf{W}_l^{\text{T}}\right)$, we recover the problem of estimating a group-sparse vector, $\tilde{\mathbf{x}}_l$, from its noisy linear observations:

$$\tilde{\mathbf{y}}_l = \tilde{\mathbb{A}}\,\tilde{\mathbf{x}}_l + \tilde{\mathbf{w}}_l, \quad l = 1, \ldots, L. \tag{18.42}$$

The ultimate goal at the receiver is to identify the set of B-bit messages that were transmitted by all the active users. Since the messages were partitioned into L different chunks, we obtain an instance of unsourced MAC in each slot. The inner CS-based decoder must now decode, in each slot, the J-bit sequences of all the K_a active users. The outer clustering-based decoder will put together the slot-wise decoded sequences of each user, so as to recover all the original transmitted B-bit messages.

For each lth slot, the task is then to first reconstruct $\tilde{\mathbf{x}}_l$ from $\tilde{\mathbf{y}}_l = \tilde{\mathbb{A}}\,\tilde{\mathbf{x}}_l + \tilde{\mathbf{w}}_l$ given $\tilde{\mathbf{y}}_l$ and $\tilde{\mathbb{A}}$. Solving the joint activity detection and channel estimation problem can be efficiently done by means of the HyGAMP CS algorithm (Rangan et al., 2017) which can also utilize the EM concept (Dempster et al., 1977) to learn the unknown hyperparameters of the model. In particular, incorporating the EM algorithm inside HyGAMP allows one to learn the variances of the additive noise and the postulated prior, which are both required to execute HyGAMP itself. HyGAMP makes use of large-system Gaussian and quadratic approximations for the messages of loopy belief propagation on the underlying factor graph. As opposed to GAMP (Rangan, 2011), HyGAMP is able to accommodate the group sparsity structure in $\tilde{\mathbf{x}}_l$ by using a dedicated latent Bernoulli random variable, ϵ_j, for each $\{j\text{th}\}_{j=1}^{2^J}$ group, $\tilde{\mathbf{x}}_{j,l}$, in $\tilde{\mathbf{x}}_l$. HyGAMP finds the MMSE and MAP estimates, $\{\hat{\mathbf{x}}_{j,l}\}_{j=1}^{2^J}$ and $\{\hat{\epsilon}_j\}_{j=1}^{2^J}$ of $\{\tilde{\mathbf{x}}_{j,l}\}_{j=1}^{2^J}$ and $\{\epsilon_j\}_{j=1}^{2^J}$. The latter will be in turn used to decode the transmitted sequences in each slot (up to some unknown permutations). Further, by clustering the MMSE estimates of the active users' reconstructed channels, it is possible to recover those unknown permutations and correctly stitch the slot-wise decoded sequences pertaining to each user. For this reason, we denote the K_a reconstructed channels over each lth slot (i.e., the nonzero blocks in the entire reconstructed vector $\hat{\mathbf{x}}_l = \left[\hat{\mathbf{x}}_{1,l}, \hat{\mathbf{x}}_{2,l}, \dots, \hat{\mathbf{x}}_{2^J,l}\right]^{\mathsf{T}}$ as $\{\hat{\mathbf{h}}_{k,l}\}_{k=1}^{K_a}$. By denoting, the residual estimation noise as $\hat{\mathbf{w}}_{k,l}$, it follows that:

$$\hat{\mathbf{h}}_{k,l} = \bar{\mathbf{h}}_k + \hat{\mathbf{w}}_{k,l}, \quad k = 1, \dots, K_a, \quad l = 1, \dots, L, \tag{18.43}$$

in which $\bar{\mathbf{h}}_k \triangleq \left[\Re\{\mathbf{h}_k\}, \Im\{\mathbf{h}_k\}\right]^T$ with $\mathbf{h}_k \triangleq \left[h_{k,1}, h_{k,2}, \dots, h_{k,M_r}\right]^T$ is the true complex-valued channel vector for user k. The outer clustering-based decoder takes the LK_a reconstructed channels in (18.43) – which are slot-wise permuted – and returns one cluster per active user, that contains its L noisy channel estimates.

One approach to cluster the reconstructed channels into K_a different groups is to resort to the Gaussian mixture expectation-maximization procedure which consists in fitting a Gaussian mixture distribution to the data points in (18.43) under the assumption of Gaussian residual noises. Note here the fact that the reconstruction noise is Gaussian is common to all AMP algorithms including HyGAMP. Moreover, it is necessary to carefully design the EM-based clustering procedure so as to enforce the following two constraints that are very specific to the URA problem: (i) each cluster must have exactly L data points, and (ii) channels reconstructed over the same slot must not be assigned to the same cluster. These two constraints can be enforced by the Hungarian algorithm, see Shyianov et al. (2020) for elaborate discussions.

18.3.5 Tensor-Based Modulation

The tensor-based modulation URA scheme (Decurninge et al., 2020) leverages a structured design of the codebook $C = \left\{\mathbf{c}_1, \mathbf{c}_2, \dots, \mathbf{c}_{2^B}\right\} \subset \mathbb{C}^n$ based on the tensor product of smaller size codebook components. The B-bit message is also split into chunks and encoded using smaller codebooks. The two-step decoding process is used to identify the transmitted symbols from each of the smaller codebooks and map them back to the corresponding chunks of the B-bit message. To that end, the codebook C is constructed as follows:

$$C = \left\{ \mathbf{c}_1 \otimes \cdots \otimes \mathbf{c}_L \; : \; \mathbf{c}_1 \in C_1, \ldots, \mathbf{c}_L \in C_L \right\}, \tag{18.44}$$

where L is the number of codebook components. It is the rank of the corresponding tensor codeword and $C_l \subset \mathbb{C}^{n_l}$ is the codebook used during the lth slot, such that $\prod_{l=1}^{L} n_l = n$. Hence, the received signal in (18.33) can be written, after vectorization, entirely in terms of the Kronecker products of the codewords from the codebook components and the channel vector:

$$\mathbf{y} = \sum_{k=1}^{K_a} \mathbf{c}_{1,k} \otimes \cdots \otimes \mathbf{c}_{L,k} \otimes \mathbf{h}_k + \mathbf{w} \in \mathbb{C}^{nM_r}, \tag{18.45}$$

It is important to note that the codewords from C are only unique up to multiplications of component codewords by scalars $\alpha_1, \ldots, \alpha_L$, such that $\prod_{l=1}^{L} \alpha_l = 1$. To account for this scalar indeterminacy, each component codebook should either (*i*) embed at least a single pilot symbol or (*ii*) rely on a Grassmannian codebook design (Zheng and Tse, 2002). The decoding is done in two steps, the first step being multiuser separation followed by single-user demapping.

Multiuser separation: First, least-squares estimation is performed to obtain the approximate version $\hat{\mathbf{z}}_{l,k}$ of the transmitted codewords $\mathbf{c}_{l,k}$ and the estimate of the channel vector $\hat{\mathbf{h}}_k$, as follows:

$$\left\{ \hat{\mathbf{z}}_{l,k}, \hat{\mathbf{h}}_k \right\} = \underset{\substack{\{\mathbf{z}_{l,k} \in \mathbb{C}^{n_l}\}_{l=1}^{L} \\ \mathbf{h}_k \in \mathbb{C}^{M_r}}}{\arg\min} \left\| \mathbf{y} - \sum_{k=1}^{K_a} \mathbf{z}_{1,k} \otimes \cdots \otimes \mathbf{z}_{L,k} \otimes \mathbf{h}_k \right\|_2^2 . \tag{18.46}$$

Single-user demapping: The second step consists in demapping each user separately, i.e., solving the following K_a optimization problems:

$$\underset{\substack{\{\mathbf{c}_{l,k} \in C_l\}_{l=1}^{L} \\ \mathbf{h}_k \in \mathbb{C}^{M_r}}}{\arg\min} \left\| \hat{\mathbf{z}}_{1,k} \otimes \cdots \otimes \hat{\mathbf{z}}_{L,k} \otimes \hat{\mathbf{h}}_k - \mathbf{c}_{1,k} \otimes \cdots \otimes \mathbf{c}_{L,k} \otimes \mathbf{h}_k \right\|_2^2 . \tag{18.47}$$

These can actually be easily solved for each slot and each user separately as follows:

$$\hat{\mathbf{c}}_{l,k} = \underset{\mathbf{c}_{l,k} \in C_l}{\arg\max} \frac{\left| \mathbf{c}_{l,k}^H \hat{\mathbf{z}}_{l,k} \right|}{\left\| \hat{\mathbf{z}}_{l,k} \right\| \left\| \mathbf{c}_{l,k} \right\|}, \quad k = 1, 2 \ldots, K_a. \tag{18.48}$$

18.3.6 Bilinear Methods

Another useful framework for MIMO URA relies on solving an overall sparse Bilinear regression problem at the CS decoding stage, instead of solving L slot-wise sparse linear regression problems independently. With the assumption of block fading, the users' channels stay constant over the duration of the entire transmission period. Therefore, we rewrite (18.33) to explicitly show the unknowns $\mathbf{\Delta}_l$ and $\mathbf{H}$

$$\mathbf{Y}_l = \mathbf{A}\,\mathbf{\Delta}_l\,\mathbf{H} + \mathbf{W}_l \quad \text{for} \quad l = 1, \ldots, L, \tag{18.49}$$

wherein $\mathbf{H}$, as opposed to $\mathbf{\Delta}_l$ is common to all slots. At this point, we have a total of L observation matrices, each of which involves two unknown matrices that we wish to reconstruct (namely, $\mathbf{H}$ and $\mathbf{\Delta}_l$). To further exploit the fact that $\mathbf{H}$ is common to all $\mathbf{Y}_l$'s in (18.49), the slot-wise observation/assignment matrices are stacked as follows:

$$\mathbf{Y} = \begin{bmatrix} \mathbf{Y}_1 \\ \cdots\cdots\cdots \\ \vdots \\ \cdots\cdots\cdots \\ \mathbf{Y}_L \end{bmatrix} \quad \text{and} \quad \mathbf{\Delta} = \begin{bmatrix} \mathbf{\Delta}_1 \\ \cdots\cdots\cdots \\ \vdots \\ \cdots\cdots\cdots \\ \mathbf{\Delta}_L \end{bmatrix} \tag{18.50}$$

Then, by letting $\mathcal{A} \triangleq (\mathbf{A} \otimes \mathbf{I})$, it follows from (18.49) that:

$$\mathbf{Y} = \mathcal{A}\mathbf{\Delta}\mathbf{H} + \mathbf{W}, \tag{18.51}$$

in which $\mathbf{W}$ is constructed in the same way as $\mathbf{Y}$. One way to reconstruct $\mathbf{\Delta}$, which would provide us with the decoded messages, is to use a dedicated bilinear AMP algorithm.

18.3.6.1 Bilinear Vector Approximate Message Passing

The Bi-VAMP algorithm (Ayachi et al., 2022, Akrout et al., 2020) uses a probabilistic observation model to jointly recover the two unknown matrices $\mathbf{V} \triangleq \mathcal{A}\mathbf{\Delta}$ and $\mathbf{H}$ from their noise-corrupted product, i.e.,

$$\mathbf{Y} = \mathbf{V}\mathbf{H} + \mathbf{W}. \tag{18.52}$$

Essentially the Bi-VAMP algorithm (Akrout et al., 2020) works by passing second-order statistics (i.e., means and covariance matrices) from the $\mathbf{V}$ block with its associated prior, arising from the priors on $\mathbf{\Delta}_l$, to the $\mathbf{H}$ block and vice versa. The Bi-linear constraint, given the estimate of either $\mathbf{V}$ or $\mathbf{H}$, is then enforced in the Bi-LMMSE block. In principle, Bi-VAMP enables the use of different priors on the columns of $\mathbf{H}$ as well as on the rows of $\mathbf{V}$. However, the specific structure of $\mathbf{\Delta}$ enforces a prior on the columns of each $\mathbf{\Delta}_l$. As such, Bi-VAMP was appropriately modified (Akrout et al., 2020) to accommodate the mURA-induced prior on the columns of $\{\mathbf{\Delta}_l\}_{l=1}^{L}$ by appropriate linear transformations of the second-order statistics.

18.3.7 Numerical Discussion

In this section, we discuss the performance of several multi-antenna URA schemes also in terms of the minimum required energy-per-bit $\frac{E_b}{N_0}$ as defined in Section 18.2.5. Here, we fix the total blocklength is fixed to $n = 3200$, the number of antennas to $M_r = 50$, and the minimum $\frac{E_b}{N_0}$ is determined such that the target PUEP, ϵ, satisfies $\epsilon \leq 0.1$. As can be seen in Fig. 18.3, the tensor-based scheme has significant performance gains over the clustering- and CB-CS-based URA schemes. This is hardly surprising since the tensor-based scheme has a lot of common features with the SKP-based URA with a single-antenna base station. The main takeaway here is that both SKP in single-antenna and tensor-based in multi-antenna URA settings exhibit large performance gains as compared to other approaches. The investigation of the interconnect between the two approaches is an important research direction in URA and can lead to the design of a nearly optimal algorithmic solution for URA. Generalization of the optimum error bound, given in Theorem 18.1, to the multi-antenna fading channels is also an interesting research

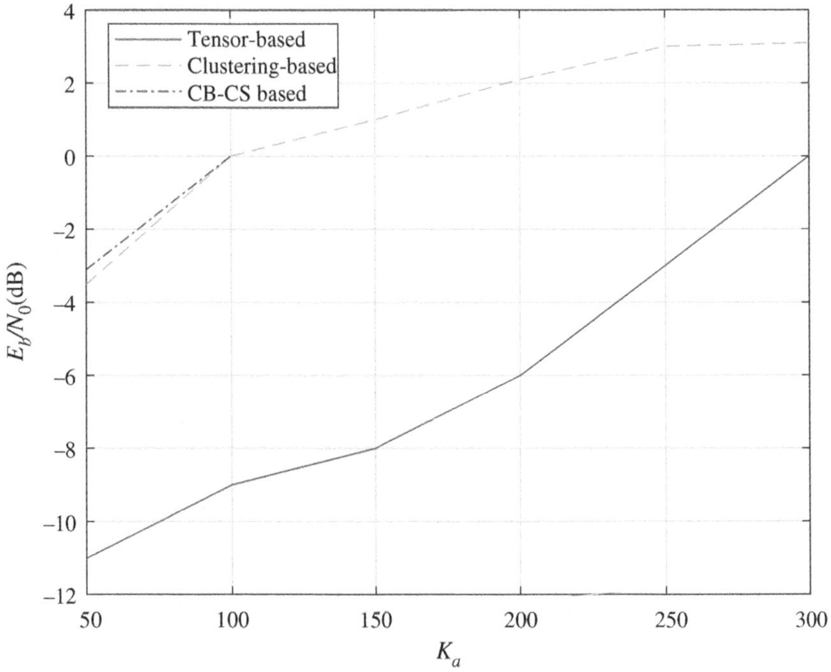

Figure 18.3 Minimum required E_b/N_0 versus number of users for the three main algorithmic URA solutions (Shyianov et al., 2020, Fengler et al., 2021b, Decurninge et al., 2020) using a multi-antenna base station with $M_r = 50$.

direction that will reveal the gap to the best possible achievable performance of all developed schemes.

References

Mohamed Akrout, Anis Housseini, Faouzi Bellili, and Amine Mezghani. Bilinear generalized vector approximate message passing. *arXiv preprint arXiv:2009.06854*, 2020.

Vamsi K Amalladinne, Asit Kumar Pradhan, Cynthia Rush, Jean-Francois Chamberland, and Krishna R Narayanan. Unsourced random access with coded compressed sensing: Integrating amp and belief propagation. *IEEE Transactions on Information Theory*, 68(4):2384–2409, 2021.

Ramzi Ayachi, Mohamed Akrout, Volodymyr Shyianov, Faouzi Bellili, and Amine Mezghani. Massive unsourced random access based on bilinear vector approximate message passing. In *ICASSP 2022-2022 IEEE International Conference on Acoustics, Speech and Signal Processing (ICASSP)*, pages 5283–5287. IEEE, 2022.

Mohsen Bayati and Andrea Montanari. The dynamics of message passing on dense graphs, with applications to compressed sensing. *IEEE Transactions on Information Theory*, 57(2):764–785, 2011.

Robert Calderbank and Andrew Thompson. CHIRRUP: A practical algorithm for unsourced multiple access. *Information and Inference: A Journal of the IMA*, 9(4):875–897, 2020.

Jingze Che, Zhaoyang Zhang, Zhaohui Yang, Xiaoming Chen, Caijun Zhong, and Derrick Wing Kwan Ng. Unsourced random massive access with beam-space tree decoding. *IEEE Journal on Selected Areas in Communications*, 40(4):1146–1161, 2022.

Alexis Decurninge, Ingmar Land, and Maxime Guillaud. Tensor-based modulation for unsourced massive random access. *IEEE Wireless Communications Letters*, 10(3):552–556, 2020.

Arthur P Dempster, Nan M Laird, and Donald B Rubin. Maximum likelihood from incomplete data via the EM algorithm. *Journal of the Royal Statistical Society: Series B: Methodological*, 39(1):1–22, 1977.

David L Donoho, Arian Maleki, and Andrea Montanari. Message-passing algorithms for compressed sensing. *Proceedings of the National Academy of Sciences of the United States of America*, 106(45):18914–18919, 2009.

Eryk Dutkiewicz, Xavier Costa-Perez, Istvan Z Kovacs, and Markus Mueck. Massive machine-type communications. *IEEE Network*, 31(6):6–7, 2017.

Alexander Fengler, Peter Jung, and Giuseppe Caire. Sparcs for unsourced random access. *IEEE Transactions on Information Theory*, 67(10):6894–6915, 2021a.

Alexander Fengler, Saeid Haghighatshoar, Peter Jung, and Giuseppe Caire. Non-Bayesian activity detection, large-scale fading coefficient estimation, and unsourced random access with a massive MIMO receiver. *IEEE Transactions on Information Theory*, 67(5):2925–2951, 2021b.

Zeyu Han, Xiaojun Yuan, Chongbin Xu, Shuchao Jiang, and Xin Wang. Sparse Kronecker-product coding for unsourced multiple access. *IEEE Wireless Communications Letters*, 10(10):2274–2278, 2021.

Antony Joseph and Andrew R Barron. Least squares superposition codes of moderate dictionary size are reliable at rates up to capacity. *IEEE Transactions on Information Theory*, 58(5):2541–2557, 2012.

Malong Ke, Zhen Gao, Mingyu Zhou, Dezhi Zheng, Derrick Wing Kwan Ng, and H Vincent Poor. Next-generation massive URLLC with massive MIMO: A unified semi-blind detection framework for sourced and unsourced random access. *arXiv preprint arXiv:2303.04414*, 2023.

Jason T Parker, Philip Schniter, and Volkan Cevher. Bilinear generalized approximate message passing—Part I: Derivation. *IEEE Transactions on Signal Processing*, 62(22):5839–5853, 2014.

Yury Polyanskiy. A perspective on massive random-access. In *2017 IEEE International Symposium on Information Theory (ISIT)*, pages 2523–2527. IEEE, 2017.

Sundeep Rangan. Generalized approximate message passing for estimation with random linear mixing. In *2011 IEEE International Symposium on Information Theory Proceedings*, pages 2168–2172. IEEE, 2011.

Sundeep Rangan, Alyson K Fletcher, Vivek K Goyal, Evan Byrne, and Philip Schniter. Hybrid approximate message passing. *IEEE Transactions on Signal Processing*, 65(17):4577–4592, 2017.

Volodymyr Shyianov, Faouzi Bellili, Amine Mezghani, and Ekram Hossain. Massive unsourced random access based on uncoupled compressive sensing: Another blessing of massive MIMO. *IEEE Journal on Selected Areas in Communications*, 39(3):820–834, 2020.

Martin J Wainwright. *High-Dimensional Statistics: A Non-Asymptotic Viewpoint*, volume 48. Cambridge University Press, 2019.

Yifei Yuan, Zhifeng Yuan, Guanghui Yu, Chien-hwa Hwang, Pei-kai Liao, Anxin Li, and Kazuaki Takeda. Non-orthogonal transmission technology in LTE evolution. *IEEE Communications Magazine*, 54(7):68–74, 2016.

Lizhong Zheng and David N C Tse. Communication on the Grassmann manifold: A geometric approach to the noncoherent multiple-antenna channel. *IEEE Transactions on Information Theory*, 48(2):359–383, 2002.

Part III

Other Advanced Emerging MA Techniques for NGMA

Part III

19

Holographic-Pattern Division Multiple Access

Ruoqi Deng, Boya Di, and Lingyang Song

School of Electronics, Peking University, Beijing, China

The space-division multiple access (SDMA) scheme is envisioned as one of the promising candidates to handle the exponentially increasing data transmission demands by exploiting the spatial diversity (Larsson et al., 2014). It was first proposed in multi-base station (BS) systems by exploiting spatial diversity among different BSs. Based on different spatial signatures of the BSs, different BSs can transmit signals in the same time slots without interfering with each other (Xu and Li, 1994). Evolving from multi-BS systems, multiple-input multiple-output (MIMO) systems based on the SDMA scheme have been proposed (Gesbert et al., 2003), where a BS simultaneously transmits multiple beams in different directions (Mietzner et al., 2009). By scaling up MIMO by orders of magnitude, massive MIMO systems with large-scale phased arrays have drawn great attention due to their capabilities of highly directional beamforming and significant capacity enhancement (Wang et al., 2016).

Nowadays, due to the enormous growth in the number of mobile devices (Wang et al., 2019), the next-generation wireless communications look forward to enhancing capacity and massive connectivity significantly over high-frequency bands through ultra-massive MIMO systems (Rusek et al., 2012). However, the inherent defects of phased arrays limit the future development of ultra-massive MIMO systems. Specifically, as the physical dimensions of phased arrays relying on costly components (such as phase shifters) scale up, the implementation of ultra-massive MIMO systems in practical engineering becomes prohibitive from both cost and power consumption perspectives (Sohrabi and Yu, 2016). Hence, there is an urgent need to develop novel multiple-access technologies to meet the exponentially increasing data demands while covering the shortages of traditional SDMA scheme.

Next Generation Multiple Access, First Edition.
Edited by Yuanwei Liu, Liang Liu, Zhiguo Ding, and Xuemin Shen.
© 2024 The Institute of Electrical and Electronics Engineers, Inc. Published 2024 by John Wiley & Sons, Inc.

Owing to the recent breakthrough of the reconfigurable metamaterial-based antennas, it is now possible to regulate electromagnetic waves via software instead of costly hardware components (Zheludev, 2010). As one of the representative metamaterial antennas, reconfigurable holographic surfaces (RHSs) inlaid with numerous metamaterial radiation elements serving antennas integrated with transceivers have been proposed (Hwang, 2020). Specifically, the feeds of the RHS are embedded in the bottom layer of the RHS to generate the incident electromagnetic waves, enabling an ultrathin structure. The RHS utilizes the metamaterial radiation elements to construct a *holographic pattern* based on the holographic interference principle (Sleasman et al., 2015). Each element can thus control the *radiation amplitude* of the incident electromagnetic waves electrically to generate desired directional beams (Deng et al., 2021).

It is worth mentioning that RHSs are different from another hardware technology for wireless communication enhancement called reconfigurable intelligent surfaces (RISs) (Liu et al., 2021), which can reflect the incident signals and generate directional beams toward receivers directly. Unlike the feeds of RHSs embedded in meta-surfaces, the feeds of RISs are on the outside of the meta-surface due to their reflection characteristic (Di et al., 2020). Due to their different physical structures, RISs are widely used as a relay such that two channels (i.e., the channel from the transmitter to the RIS and the channel from the RIS to the receiver) should be considered in RIS-aided communications (Nemati et al., 2021). Differently, RHSs can serve as transmit and receive antennas directly, and only one channel (i.e., the channel between the RHS and the receiver/transmitter) needs to be considered in RHS-aided communications. Benefitting from the above characteristics, RHSs provide a powerful yet lightweight solution to fulfill the challenging visions in next-generation wireless communications and have attracted wide attention from industry (Smith et al., 2017). Two commercial prototypes of RHSs have already been proposed by Pivotal Commwave and Kymeta. Pivotal Commwave has provided customized RHS systems from 1 to 70 GHz to extend the coverage of cellular communications (Pivotal Staff, 2019). Kymeta together with Microsoft has developed the commercial RHS for satellite communications to respond to natural disasters (Stevenson et al., 2020).

In this chapter, we propose a new type of SDMA, called holographic-pattern division multiple access (HDMA), by utilizing the superposition of different holographic patterns to serve multiple users. Specifically, in the HDMA, the transmitter superposes all holographic patterns corresponding to the receivers in different directions to generate a superposed holographic pattern, and thus, the transmitted data are mapped onto the superposed holographic pattern. When the electromagnetic waves generated by the feeds excite this holographic pattern, the signals intended for the receivers can be differentiated, i.e., the RHS can generate multiple desired directional beams toward different receivers.

The rest of this chapter is organized as follows. In Section 19.1, we introduce the basics of an RHS and the principle of HDMA. In Section 19.2, we introduce an HDMA wireless communication system and give the channel model. In Section 19.3, we present the multiuser holographic beamforming scheme and the asymptotic capacity of the HDMA system. In Section 19.4, we derive a closed-form optimal holographic pattern to achieve the asymptotic capacity. In Section 19.5, based on the derived optimal holographic pattern, we evaluate the performance of the HDMA system. Finally, the conclusions are drawn in Section 19.6.

19.1 Overview of HDMA

In this section, we introduce the structure and properties of an RHS, based on which the principle of HDMA is illustrated.

19.1.1 RHS Basics

An RHS is a leaky-wave antenna mainly consisting of three parts, i.e., K feeds, a waveguide, and N sub-wavelength metamaterial radiation elements. Specifically, as shown in Fig. 19.1, the feeds are embedded in the bottom layer of the RHS and generate the incident electromagnetic waves, which are also called reference waves, carrying intended signals for users (Di, 2021). The waveguide playing the role of guiding structure then guides the reference wave to propagate on it. During the propagation process, the reference wave radiates its energy through the radiation elements into free space and the radiated wave is also known as the leaky wave (Smith et al., 2017), where the waveform of the leaky wave is the same as that of the transmit signal.

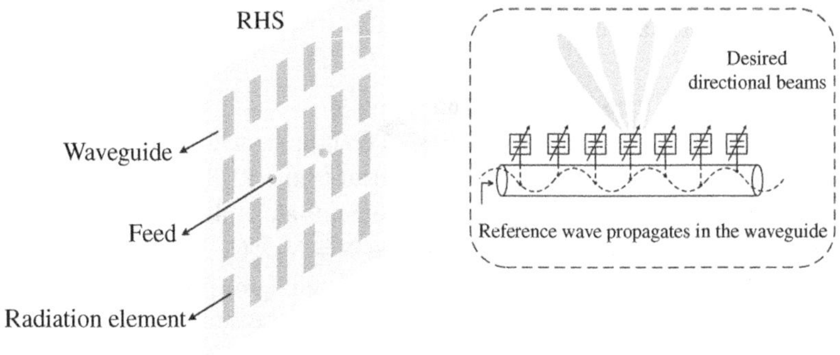

Figure 19.1 Physical structure of RHS.

Since the electromagnetic response of the radiation elements can be intelligently controlled via a simple diode-based amplitude controller, the RHS utilizes the radiation elements to construct a holographic pattern, which records the interference between the reference wave and the desired object wave (Sleasman et al., 2015). The leaked wave from each element can then be shaped independently according to the holographic pattern and the superposition of the leaked waves from different elements finally generates the desired directional beams.

Therefore, the basic working principles and specific implementation of an RHS differ from those of a phased array. As for basic working principles, the traditional phased array realizes beamforming by controlling the phase of the electromagnetic wave propagating along each antenna of the array. Differently, the RHS can achieve amplitude-controlled beamforming by controlling the radiation amplitude of the reference wave propagating along the waveguide. As for specific implementations, unlike the phased array with complex phase-shifting circuits with numerous phase shifters, the metamaterial radiation elements of the RHS can be fabricated by loading RF switches or tunable materials rather than costly hardware components. In addition, the phased array utilizes the method of parallel feeding where each radiation element has an equal long feeding line, thereby providing in-phase feed. The RHS utilizes the method of series feeding where radiation elements are located progressively farther and farther away from the feed point. The reference wave generated by the feeds excites the radiation elements one by one.

19.1.2 Principle of HDMA

The main characteristic of the HDMA is to map the transmitted data onto a holographic pattern according to the holographic interference principle. Utilizing the holographic pattern, the RHS can control the radiation amplitude of the leaky wave at each element effectively to obtain the desired signal transmission directions. The specific principles of constructing the holographic pattern and realizing HDMA are elaborated as below.

19.1.2.1 Holographic Pattern Construction

As shown in Fig. 19.2, we adopt the Cartesian coordinate where the $y-z$ plane coincides with the RHS, and the x-axis is vertical to the RHS. The RHS is centered at the origin. The number of radiation elements along the y-axis and z-axis is denoted as N_y and N_z, respectively, satisfying $N = N_y \times N_z$. The inter-element spacing along the y-axis and z-axis is denoted as d_y and d_z, respectively. For notational convenience, it is assumed that both N_y and N_z are odd numbers. Therefore, the position vector of the (n_y, n_z)th element is $\mathbf{r}_{n_y,n_z} = \left[0, n_y d_y, n_z d_z\right]^T$, where $n_y = \{0, \pm 1, \pm 2, \dots, \pm \frac{N_y-1}{2}\}$ and $n_z = \{0, \pm 1, \pm 2, \dots, \pm \frac{N_z-1}{2}\}$.

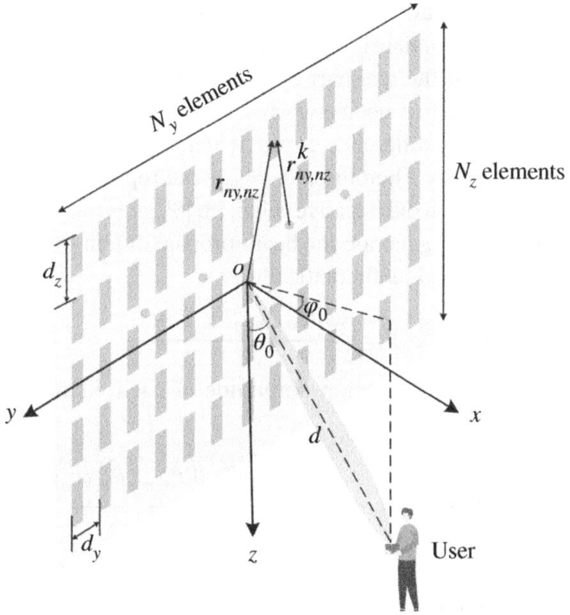

Figure 19.2 Geometrical relation between RHS and object beam.

At the (n_y, n_z)th radiation element, the reference wave generated by feed k and the wave propagating in free space with a direction of (θ_0, φ_0) can be given by (Johnson et al., 2016)

$$\Psi_{\text{ref}}\left(\mathbf{r}^k_{n_y,n_z}\right) = \exp\left(-j\mathbf{k}_s \cdot \mathbf{r}^k_{n_y,n_z}\right), \tag{19.1}$$

$$\Psi_{\text{obj}}\left(\mathbf{r}_{n_y,n_z}, \theta_0, \varphi_0\right) = \exp\left(-j\mathbf{k}_f \cdot \mathbf{r}_{n_y,n_z}\right), \tag{19.2}$$

where $\mathbf{k}_s$ is the propagation vector of the reference wave, $\mathbf{r}^k_{n_y,n_z}$ is the distance vector from the feed k to the (n_y, n_z)th radiation element, and $\mathbf{k}_f$ is the desired directional propagation vector in free space. The interference between the reference wave and the object wave is defined as

$$\Psi_{\text{intf}}\left(\mathbf{r}^k_{n_y,n_z}, \theta_0, \varphi_0\right) = \Psi_{\text{obj}}\left(\mathbf{r}_{n_y,n_z}, \theta_0, \varphi_0\right)\Psi^*_{\text{ref}}\left(\mathbf{r}^k_{n_y,n_z}\right). \tag{19.3}$$

The information contained in Ψ_{intf}, which is also called holographic pattern, is supposed to be recorded by the radiation elements. When the holographic pattern is excited by the reference wave, we have $\Psi_{\text{intf}}\Psi_{\text{ref}} \propto \Psi_{\text{obj}}|\Psi_{\text{ref}}|^2$, and thus, the wave propagating in the direction (θ_0, φ_0) is generated.

To construct the holographic pattern given in (19.3), each radiation element controls the radiation amplitude of the reference wave instead of conventional phase shifting. Specifically, the phase of the reference wave at each element is determined by the position of the element as given in (19.1). The radiation elements whose reference waves are in phase with the object wave are tuned to radiate strongly (large amplitude), while the radiation elements that are out of phase are detuned so as not to radiate (small amplitude)(Johnson et al., 2016).

Note that the real part of the interference (i.e., $\mathrm{Re}\left[\Psi_{\mathrm{intf}}\right]$), i.e., cosine value of the phase difference between the reference wave and the object wave, decreases as the phase difference grows, which exactly meets the above amplitude control requirements. Therefore, $\mathrm{Re}\left[\Psi_{\mathrm{intf}}\right]$ can represent the radiation amplitude of each element. To avoid negative value, $\mathrm{Re}\left[\Psi_{\mathrm{intf}}\right]$ is normalized to [0,1]. The holographic pattern $\boldsymbol{m}$ to generate the beam propagating in the direction $\left(\theta_0, \varphi_0\right)$ can then be parameterized mathematically by

$$\boldsymbol{m}\left(\theta_0, \varphi_0\right) = \frac{\mathrm{Re}\left[\Psi_{\mathrm{intf}}\left(\theta_0, \varphi_0\right)\right] + 1}{2}, \tag{19.4}$$

and thus, the radiation amplitude of each element in the holographic pattern $\boldsymbol{m}\left(\theta_0, \varphi_0\right)$ is

$$m\left(\mathbf{r}^k_{n_y,n_z}, \theta_0, \varphi_0\right) = \frac{\mathrm{Re}\left[\Psi_{\mathrm{intf}}\left(\mathbf{r}^k_{n_y,n_z}, \theta_0, \varphi_0\right)\right] + 1}{2}. \tag{19.5}$$

19.1.2.2 HDMA Transmission Model

The HDMA utilizes the superposition of the holographic pattern to map all transmitted data to a single holographic pattern on the RHS, and thus, the signals intended for the users can be differentiated, i.e., the RHS can generate multiple desired directional beams pointing to the users. Specifically, the holographic pattern of the RHS can be calculated as a weighted summation of the radiation amplitude distribution corresponding to each object beam according to (19.5).

Denote the users served by the RHS as $\mathcal{Q} = \{1, 2, \dots, Q\}$, and their zenith and azimuth angles in terms of the origin as θ_q and φ_q, respectively. The holographic pattern $\boldsymbol{m}$, i.e., the normalized radiation amplitude of each radiation element can then be given by

$$m_{n_y,n_z} = \sum_{q=1}^{Q}\sum_{k=1}^{K} a_{q,k} m\left(\mathbf{r}^k_{n_y,n_z}, \theta_q, \varphi_q\right), \tag{19.6}$$

where $a_{q,k}$ is the amplitude ratio for the beam pointing to user q from feed k satisfying $\sum_{q=1}^{Q}\sum_{k=1}^{K} a_{q,k} = 1$. This constraint is set to guarantee that the radiation amplitude of each RHS element m_{n_y,n_z} lies in [0,1].

An Illustrative Example for HDMA Figure 19.3 shows an example of data mapping according to the HDMA. For simplicity, we assume that the BS transmits signals to two users via an RHS.[1] The feeds of the RHS are assumed to be located close

1 The specific transmission model for general HDMA wireless communication systems will be presented in Section 19.2.

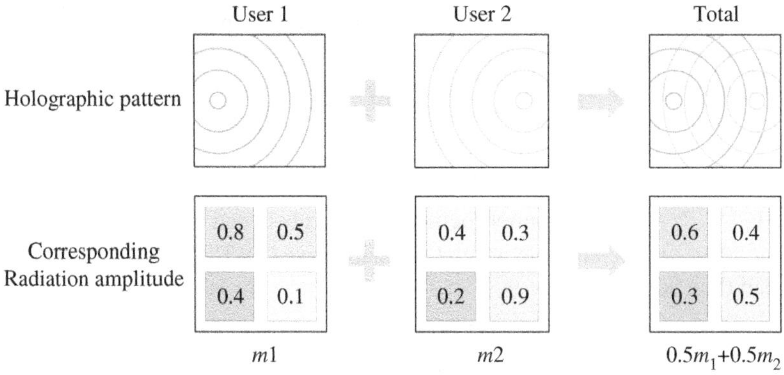

Figure 19.3 Illustration of HDMA.

to the origin, where $\mathbf{r}^k_{n_y,n_z} \approx \mathbf{r}_{n_y,n_z}, \forall k$. Denote the zenith and azimuth angles of the user q $(q = 1{,}2)$ as (θ_q, φ_q). Based on (19.5), we have $m\left(\mathbf{r}^1_{n_y,n_z}, \theta_q, \varphi_q\right) = m\left(\mathbf{r}^2_{n_y,n_z}, \theta_q, \varphi_q\right) = \cdots = m\left(\mathbf{r}^K_{n_y,n_z}, \theta_q, \varphi_q\right) = m\left(\mathbf{r}_{n_y,n_z}, \theta_q, \varphi_q\right)$, and thus, the index k can be omitted. The normalized radiation amplitude of each radiation element in the holographic pattern $\boldsymbol{m}_q$ corresponding to user q can then be denoted as $\{m\left(\mathbf{r}_{n_y,n_z}, \theta_q, \varphi_q\right)\}$. In the HDMA, users 1 and 2 are multiplexed on a holographic pattern, i.e., the RHS maps the intended signals for these two users onto a single holographic pattern superposed by holographic patterns $\boldsymbol{m}_1$ and $\boldsymbol{m}_2$, which can be given by

$$
\begin{gathered}
\boldsymbol{m} = a_1 \boldsymbol{m}_1 + a_2 \boldsymbol{m}_2 \Leftrightarrow \\
m_{n_y,n_z} = a_1 m\left(\mathbf{r}_{n_y,n_z}, \theta_1, \varphi_1\right) + a_2 m\left(\mathbf{r}_{n_y,n_z}, \theta_2, \varphi_2\right).
\end{gathered}
\tag{19.7}
$$

Denote the signals transmitted to users as $\mathbf{s} = \left[s_1, s_2\right]^T$, and the precoding vector at the BS as $\mathbf{V} \in \mathbb{C}^{K\times 2}$. Since the RHS is a leaky-wave antenna, the leaky-wave effect makes the amplitude of the reference wave carrying the signals gradually decrease during its propagation process (Gomez-Tornero et al., 2005). Therefore, in HDMA, s_1 and s_2 are superposed as $\mathbf{x} = \mathbf{MVs}$, where $\mathbf{M} \in \mathbb{C}^{N_yN_z\times K}$ is composed of the following elements:

$$
M^k_{n_y,n_z} = \sqrt{\eta} \cdot m_{n_y,n_z} \cdot e^{-\alpha|\mathbf{r}^k_{n_y,n_z}|} \cdot e^{-j\mathbf{k}_s \cdot \mathbf{r}^k_{n_y,n_z}},
\tag{19.8}
$$

where η as the efficiency of each RHS element, i.e., the ratio of the power accepted by each RHS element P_a to the total power P_t of the reference waves from all feeds,[2]

2 The efficiency of an RHS element η is determined by its size and hardware structure. For an RHS element with a given size and hardware structure, η is a constant (Sleasman et al., 2015).

i.e., $\eta = P_a/P_t$. α is the attenuation constant on the radiation property, and $e^{-j\mathbf{k}_s \cdot \mathbf{r}^k_{n_y,n_z}}$ denotes the phase of the reference wave when it propagates to the (n_y, n_z)th radiation element from the feed k. The received signal at user q can then be given by

$$y_q = \mathbf{H}_q\mathbf{x} + n_q = \mathbf{H}_q\mathbf{MVs} + n_q, \tag{19.9}$$

where $\mathbf{H}_q \in \mathbb{C}^{1\times N_yN_z}$ is the channel coefficient matrix between user q and the RHS, $n_q \sim \mathcal{CN}(\mathbf{0}, \sigma^2)$ is the additive white Gaussian noise (AWGN). The data rate of user q can then be given by

$$R_q = \log_2\left(1 + \frac{|\mathbf{H}_q\mathbf{MV}_q|^2}{\sigma^2 + \sum_{q'\neq q}|\mathbf{H}_q\mathbf{MV}_{q'}|^2}\right), \tag{19.10}$$

where $\mathbf{V}_q$ is the qth column of $\mathbf{V}$.

In the HDMA, the RHS can generate two beams pointing to these two users through the superposed holographic pattern $\boldsymbol{m}$. For each user q, the interference from the other user can be alleviated by the precoder $\mathbf{V}$ at the BS such as the zero-forcing (ZF) precoder and the minimum mean squared error (MMSE) precoder (Hoydis et al., 2013), and thus, the quality-of-service of each user can be guaranteed. After user q receives the signal y_q, it down-converts the received signal to the baseband and then recovers the final signal.

19.2 System Model

In this section, we introduce an HDMA wireless communication system where a BS with an extremely large-scale RHS serves multiple users via holographic beamforming, based on which the channel model is constructed.

19.2.1 Scenario Description

As shown in Fig. 19.4, we consider a downlink HDMA wireless communication system, where a BS equipped with an extremely large-scale RHS transmits to multiple single-antenna users denoted as $\mathcal{Q} = \{1,2, \dots, Q\}$. By leveraging the principle of HDMA, the RHS maps all transmitted data to a single holographic pattern. When the reference wave carrying the transmitted signals excites the holographic pattern, the RHS can achieve beamforming without relying on complex phase-shifting circuits and bulky mechanics, and such a beamforming technique is also known as holographic beamforming.

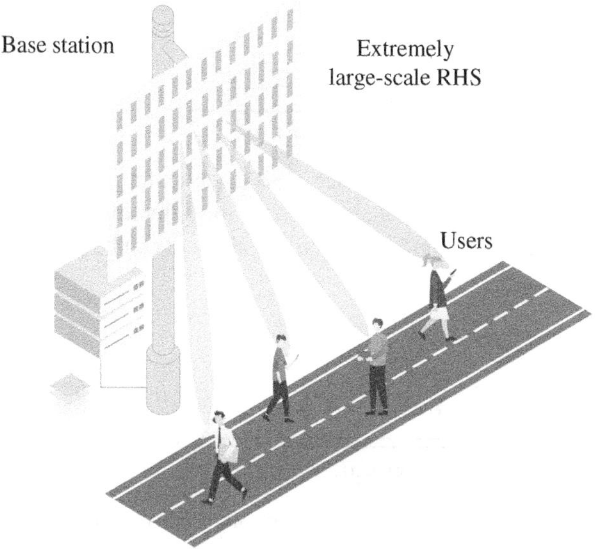

Figure 19.4 HDMA wireless communication system aided by an extremely large-scale RHS.

19.2.2 Channel Model

Note that for an extremely large-scale RHS satisfying $N_y \to \infty$ and $N_z \to \infty$, the distance between the RHS and the user will be shorter than the Rayleigh distance, i.e., the users are likely to be located in the near-field region of the RHS (Lu and Zeng, 2021). The signal propagation distance and its phase are different with respect to each element, and thus, the conventional uniform plane wave (UPW) model in the far-field region does not hold. The general spherical wave model is utilized to describe the variation of the signal propagation distances and its phase for different elements (Friedlander, 2019).

Denote the distance between each user q and the origin as d_q, the zenith and azimuth angles as θ_q and φ_q, and thus, the position vector of user q is $\mathbf{d}_q = \left[d_q \sin\theta_q \cos\varphi_q, d_q \sin\theta_q \sin\varphi_q, d_q \cos\theta_q\right]^T$. The distance vector from the (n_y, n_z)th element to user q can then be given by $\mathbf{d}^q_{n_y,n_z} = \mathbf{d}_q - \mathbf{r}_{n_y,n_z}$. As a theoretical analysis of the fundamental performance limits and asymptotic behaviors, we adopt the free-space line-of-sight (LoS) propagation. Denote the size of each radiation element as $A = \sqrt{A} \times \sqrt{A}$, and thus, the surface domain of the (n_y, n_z)th element is $A_{n_y,n_z} = \left[n_y d_y - \frac{\sqrt{A}}{2}, n_y d_y + \frac{\sqrt{A}}{2}\right] \times \left[n_z d_z - \frac{\sqrt{A}}{2}, n_z d_z + \frac{\sqrt{A}}{2}\right]$.

The channel power gain between user q and the (n_y, n_z)th element can be expressed as[3]

$$|h^q_{n_y,n_z}|^2 = \iint_{A_{n_y,n_z}} G_q \left(\frac{\lambda}{4\pi |\mathbf{d}^q_{n_y,n_z}|} \right)^2 dydz = \frac{A\lambda^2 G_q}{(4\pi)^2 |\mathbf{d}^q_{n_y,n_z}|^2}, \tag{19.11}$$

where λ is the wavelength of the signal in free space, G_q is the antenna gain of each user. Therefore, considering the variation of signal phase for different elements, the channel coefficient between user q and the (n_y, n_z)th element can be given by

$$\begin{aligned} h^q_{n_y,n_z} &= \frac{\lambda\sqrt{G_q}\sqrt{A} \cdot e^{-j\mathbf{k}_f \cdot \mathbf{d}^q_{n_y,n_z}}}{4\pi \|\mathbf{d}^q_{n_y,n_z}\|} \\ &= \frac{\lambda\sqrt{G_q}\sqrt{A} \cdot e^{-j\mathbf{k}_f \cdot \mathbf{d}^q_{n_y,n_z}}}{4\pi\sqrt{d_q^2 - 2d_q\Phi_q n_y d_y - 2d_q\Theta_q n_z d_z + n_y^2 d_y^2 + n_z^2 d_z^2}}, \end{aligned} \tag{19.12}$$

where $\mathbf{k}_f$ is the propagation vector in free space, $\Phi_q = \sin\theta_q \sin\varphi_q$, and $\Theta_q = \cos\theta_q$.

19.3 Multiuser Holographic Beamforming

In this section, we present the multiuser holographic beamforming scheme for HDMA. Following that, the capacity of the HDMA wireless communication system is given.

To serve the users simultaneously, as shown in Fig. 19.5, the BS first encodes the data streams for different users via a digital beamformer $\mathbf{V} \in \mathbb{C}^{K\times Q}$ at baseband, since the RHS does not have any digital processing capability. The processed

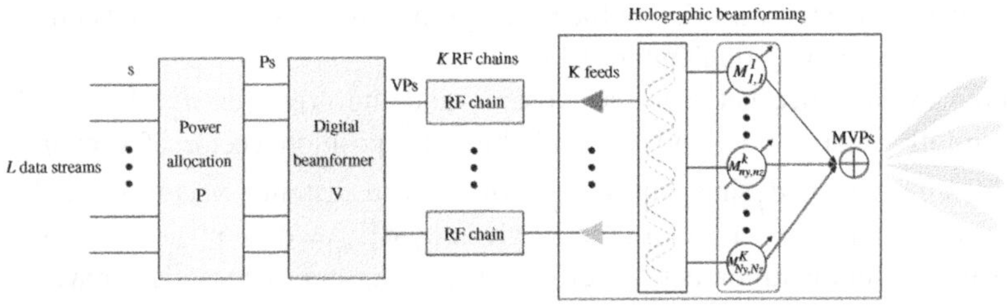

Figure 19.5 HDMA transmission block diagram.

3 Since the size of each radiation element is far smaller than the signal propagation distance between an element and the user, the variation of the signal propagation distance across different points in a radiation element is negligible.

signals are then up-converted to the carrier frequency by passing through K radio frequency (RF) chains. Specifically, each RF chain is connected with a feed of the RHS. After up-converting the transmitted signals to the carrier frequency, each RF chain sends the up-converted signals to its connected feed. The feed then transforms the high-frequency current into the electromagnetic wave, which is also called reference wave, propagating on the RHS. The reference waves will be transformed into leaky waves through radiation elements on the RHS and leak out energy into free space for radiation, where the radiation amplitude of the reference wave at each radiation element is controlled via a holographic beamformer $\mathbf{M} \in \mathbb{C}^{N_y N_z \times K}$ to generate desired directional beams pointing to the users. Based on the principle of HDMA, the holographic beamforming matrix $\mathbf{M} \in \mathbb{C}^{N_y N_z \times K}$ is then formed by the following elements

$$M^k_{n_y,n_z} = \sqrt{\eta} \cdot m_{n_y,n_z} \cdot e^{-\alpha |\mathbf{r}^k_{n_y,n_z}|} \cdot e^{-j\mathbf{k}_s \cdot \mathbf{r}^k_{n_y,n_z}}, \tag{19.13}$$

where m_{n_y,n_z} is the holographic pattern given in (19.6), and $e^{-j\mathbf{k}_s \cdot \mathbf{r}^k_{n_y,n_z}}$ denotes the phase of the reference wave when it propagates to the (n_y, n_z)th radiation element from feed k.

Without loss of generality, it is assumed that the signals are transmitted under a power allocation matrix[4] $\mathbf{P}$ satisfying $\mathrm{Tr}\{\mathbf{P}\mathbf{P}^H\} = P$, where P is the total transmit power at the BS. Denote the intended signal vector for Q users as $\mathbf{s} \in \mathbb{C}^{Q\times 1}$ satisfying $\mathbb{E}\left[\mathbf{s}\mathbf{s}^H\right] = \mathbf{I}_Q$. Therefore, the transmitted signals at the BS are $\mathbf{VPs}$. After the holographic beamforming by the RHS integrated with the BS, the received signal at the users can then be given by

$$\mathbf{y} = \mathbf{HMVPs} + \mathbf{n}, \tag{19.14}$$

where $\mathbf{H} \in \mathbb{C}^{Q\times N_y N_z}$ is the channel matrix composed of element $h^q_{n_y,n_z}$ as given in (19.12), $\mathbf{M}$ is the holographic beamforming matrix as given in (19.13), and $\mathbf{n} \sim \mathcal{CN}\left(\mathbf{0}, \sigma^2\right)$ is the AWGN.

The capacity of the HDMA wireless communication system can be achieved by dirty paper coding (DPC) (Tran et al., 2013), where the interference among the users can be eliminated by utilizing the DPC scheme at the BS. Specifically, the capacity can be given by (Zhang et al., 2021)

$$C = \sum_{q=1}^{Q} \log_2 \left(1 + \frac{P_q}{\sigma^2} \left[\mathbf{G}\mathbf{G}^H\right]_{q,q}\right), \tag{19.15}$$

4 Since we mainly focus on the holographic pattern design for asymptotic capacity analysis, the power allocation matrix $\mathbf{P}$ is assumed to be a constant matrix in the following.

where $P_q = \left[\mathbf{P}\mathbf{P}^H\right]_{q,q}$, the equivalent channel matrix $\mathbf{G} = \mathbf{H}\mathbf{M}$ consists of the following elements:

$$[\mathbf{G}]_{q,k} = \sum_{n_y=-\frac{N_y-1}{2}}^{\frac{N_y-1}{2}} \sum_{n_z=-\frac{N_z-1}{2}}^{\frac{N_z-1}{2}} h^q_{n_y,n_z} M^k_{n_y,n_z}. \tag{19.16}$$

Based on the channel model given in (19.12) and the holographic beamforming matrix given in (19.13), the following proposition about the specific expression of $[\mathbf{G}]_{q,k}$ can be derived.

Proposition 19.1 The equivalent channel $[\mathbf{G}]_{q,k}$ can be expressed as

$$[\mathbf{G}]_{q,k} = \frac{\lambda\sqrt{\eta}\sqrt{G_q}\sqrt{A}e^{-j\mathbf{k}_f\cdot\mathbf{d}_q}}{4\pi d_y d_z} \iint_{\mathcal{A}} \frac{e^{-\alpha\sqrt{(y-y_k)^2+(z-z_k)^2}} - j\omega_{q,k}}{\sqrt{d_q^2 - 2d_q\Phi_q y - 2d_q\Theta_q z + y^2 + z^2}}$$

$$\cdot\left[\sum_{q'=1}^{Q}\sum_{k'=1}^{K} \frac{a_{q',k'}\left(1+\cos\omega^{q',k'}_{n_y,n_z}\right)}{2}\right] dydz, \tag{19.17}$$

where $\mathcal{A} = \left\{(y,z)\,||y| \le \frac{N_y d_y}{2}, |z| \le \frac{N_z d_z}{2}\right\}$ is the integral domain, $\left(y_k, z_k\right)$ is the position of feed k, and

$$\omega^{q,k}_{n_y,n_z} = \mathbf{k}_f\cdot\mathbf{r}_{n_y,n_z} - \mathbf{k}_s\cdot\mathbf{r}^k_{n_y,n_z} = |\mathbf{k}_f|\cdot\left(y\Phi_q + z\Theta_q\right) - |\mathbf{k}_s|$$
$$\cdot\sqrt{(y-y_k)^2 + (z-z_k)^2}. \tag{19.18}$$

Proof: Based on (19.12) and (19.13), the equivalent channel $[\mathbf{G}]_{q,k}$ can be given by

$$[G]_{q,k} = \sum_{n_y=-\frac{N_y-1}{2}}^{\frac{N_y-1}{2}} \sum_{n_z=-\frac{N_z-1}{2}}^{\frac{N_z-1}{2}} \frac{\lambda\sqrt{\eta}\sqrt{G_q}\sqrt{A}\cdot e^{-j\mathbf{k}_f\cdot\mathbf{d}^q_{n_y,n_z}}\cdot m_{n_y,n_z}\cdot e^{-\alpha|\mathbf{r}^k_{n_y,n_z}|}\cdot e^{-j\mathbf{k}_s\cdot\mathbf{r}^k_{n_y,n_z}}}{4\pi\sqrt{d_q^2 - 2d_q\Phi_q n_y d_y - 2d_q\Theta_q n_z d_z + n_y^2 d_y^2 + n_z^2 d_z^2}}$$

$$= \frac{\lambda\sqrt{\eta}\sqrt{G_q}\sqrt{A}e^{-j\mathbf{k}_f\cdot\mathbf{d}_q}}{4\pi} \sum_{n_y=-\frac{N_y-1}{2}}^{\frac{N_y-1}{2}} \sum_{n_z=-\frac{N_z-1}{2}}^{\frac{N_z-1}{2}} \frac{e^{-j\left(\mathbf{k}_f\cdot\mathbf{r}_{n_y,n_z}-\mathbf{k}_s\cdot\mathbf{r}^k_{n_y,n_z}\right)}\cdot m_{n_y,n_z}\cdot e^{-\alpha\|\mathbf{r}^k_{n_y,n_z}\|}}{\sqrt{d_q^2 - 2d_q\Phi_q n_y d_y - 2d_q\Theta_q n_z d_z + n_y^2 d_y^2 + n_z^2 d_z^2}}$$

$$= \frac{\lambda\sqrt{\eta}\sqrt{G_q}\sqrt{A}e^{-j\mathbf{k}_f\cdot\mathbf{d}_q}}{4\pi} \sum_{n_y,n_z} \frac{e^{-\alpha\sqrt{(n_y d_y-y_k)^2+(n_z d_z-z_k)^2}-j\omega^{q,k}_{n_y,n_z}}\left[\sum_{q',k'}\frac{a_{q',k'}\left(1+\cos\omega^{q',k'}_{n_y,n_z}\right)}{2}\right]}{\sqrt{d_q^2 - 2d_q\Phi_q n_y d_y - 2d_q\Theta_q n_z d_z + n_y^2 d_y^2 + n_z^2 d_z^2}}. \tag{19.19}$$

Define the function $g(y,z)$ as

$$g(y,z) = \frac{e^{-\alpha d_q \sqrt{(y-y_k)^2+(z-z_k)^2}}}{\sqrt{d_q^2 - 2d_q\Phi_q y - 2d_q\Theta_q z + y^2 + z^2}} \cdot e^{-j\omega_{n_y,n_z}^{k,q}}$$

$$\cdot \left[\sum_{q'=1}^{Q} \sum_{k'=1}^{K} \frac{a_{q',k'}\left(1+\cos\omega_{n_y,n_z}^{q',k'}\right)}{2} \right]. \tag{19.20}$$

Since $d_y \ll 1, d_z \ll 1$, based on the definition of double integral, we have $\sum_{n_y,n_z} g\left(n_y d_y, n_z d_z\right) \approx \frac{1}{d_y d_z} \iint_{|y|\le \frac{N_y d_y}{2}, |z| \le \frac{N_z d_z}{2}} g(y,z)\, dydz$, where $y = n_y d_y, z = n_z d_z$. $[\mathbf{G}]_{q,k}$ can then be expressed as

$$[\mathbf{G}]_{q,k} = \frac{\lambda\sqrt{\eta}\sqrt{G_q}\sqrt{A}e^{-j\mathbf{k}_f\cdot\mathbf{d}_q}}{4\pi d_y d_z} \iint_{\mathcal{A}} \frac{e^{-\alpha\sqrt{(y-y_k)^2+(z-z_k)^2}-j\omega_{q,k}}}{\sqrt{d_q^2 - 2d_q\Phi_q y - 2d_q\Theta_q z + y^2 + z^2}}$$

$$\cdot \left[\sum_{q'=1}^{Q} \sum_{k'=1}^{K} \frac{a_{q',k'}\left(1+\cos\omega_{n_y,n_z}^{q',k'}\right)}{2} \right] dydz, \tag{19.21}$$

where $\mathcal{A} = \left\{ (y,z) \,\middle|\, |y| \le \frac{N_y d_y}{2}, |z| \le \frac{N_z d_z}{2} \right\}$. This completes the proof. □

19.4 Holographic Pattern Design

In this section, we develop a holographic pattern optimization scheme to achieve the asymptotic capacity of the HDMA wireless communication system aided by an extremely large-scale RHS.

Based on the expression of the capacity given in (19.15), the holographic pattern optimization problem can be formulated as

$$\max_{\{a_{q,k}\}} \sum_{q=1}^{Q} \log_2\left(1 + \frac{P_q}{\sigma^2}\left[\mathbf{G}\mathbf{G}^H\right]_{q,q}\right), \tag{19.22a}$$

$$\text{s.t.} \sum_{q=1}^{Q}\sum_{k=1}^{K} a_{q,k} = 1. \tag{19.22b}$$

Remark 19.1 According to the above holographic pattern optimization problem, it can be seen that the number of variables to be optimized (i.e., $\{a_{q,k}\}$) is $Q \times K$. Differently, in a traditional SDMA system with K RF chains and N

transmit antennas, the size of the phase-controlled analog beamformer is $N \times K$, indicating that $N \times K$ phase shifters need to be optimized. Since the number of users is less than the number of antenna elements, i.e., $Q < N$ in general, the precoding complexity of the HDMA system is less than that of the traditional SDMA system. Besides, different from the traditional SDMA system relying on complex phase-shifting circuits, HDMA with RHS is realized by metamaterial radiation elements with controllable radiation amplitude. Such metamaterial radiation elements can be fabricated by loading low-power and low-complexity RF switches such as PIN diodes, and their radiation amplitude can be changed by controlling the biased voltage applied to these RF switches (Pivotal Staff, 2019). Thus, the HDMA system is envisioned to be of lower complexity than the traditional SDMA system.

To solve this problem, we first present the following two lemmas about the properties of the equivalent channel matrix $\mathbf{G}$ when the RHS is extremely large.

Lemma 19.1 For an extremely large-scale RHS satisfying $N_y \to \infty$ and $N_z \to \infty$, $[\mathbf{G}]_{q,k}$ can be approximated by

$$[\mathbf{G}]_{q,k} = \frac{a_{q,k}\lambda\sqrt{\eta}\sqrt{G_q}\sqrt{A}\cdot e^{-j\mathbf{k}_f\cdot\mathbf{d}_q}}{16\pi d_y d_z}\iint_A \frac{e^{-\alpha\sqrt{(y-y_k)^2+(z-z_k)^2}}}{\sqrt{d_q^2-2d_q\Phi_q y-2d_q\Theta_q z+y^2+z^2}}dydz. \tag{19.23}$$

Proof: Note that $e^{-j\omega_{n_y,n_z}^{q,k}}\cdot\left[\sum_{q'=1}^{Q}\sum_{k'=1}^{K}\frac{a_{q',k'}\left(1+\cos\omega_{n_y,n_z}^{q',k'}\right)}{2}\right]$ can be rewritten as

$$\begin{aligned}
&\left(\cos\omega_{n_y,n_z}^{q,k}+j\sin\omega_{n_y,n_z}^{q,k}\right)\cdot\left[\sum_{q'=1}^{Q}\sum_{k'=1}^{K}\frac{a_{q',k'}\left(1+\cos\omega_{n_y,n_z}^{q',k'}\right)}{2}\right]\\
&=\frac{a_{q,k}}{4}+a_{q,k}\left(\frac{\cos\omega_{n_y,n_z}^{q,k}}{2}+\frac{\cos 2\omega_{n_y,n_z}^{q,k}}{4}\right)\\
&+\sum_{q'\neq q}\sum_{k'\neq k}a_{q',k'}\left[\frac{\cos\omega_{n_y,n_z}^{q,k}}{2}+\frac{\cos\left(\omega_{n_y,n_z}^{q,k}+\omega_{n_y,n_z}^{q',k'}\right)}{4}+\frac{\cos\left(\omega_{n_y,n_z}^{q,k}-\omega_{n_y,n_z}^{q',k'}\right)}{4}\right]\\
&+j\cdot\sum_{q',k'}a_{q',k'}\left[\frac{\sin\omega_{n_y,n_z}^{q,k}}{2}+\frac{\sin\left(\omega_{n_y,n_z}^{q,k}+\omega_{n_y,n_z}^{q',k'}\right)}{4}+\frac{\sin\left(\omega_{n_y,n_z}^{q,k}-\omega_{n_y,n_z}^{q',k'}\right)}{4}\right].
\end{aligned} \tag{19.24}$$

We need to prove that for an extremely large-scale RHS satisfying $N_y \to \infty$ and $N_z \to \infty$, the integral of the items containing sine and cosine functions can be neglected, such that only the first item $\frac{a_{q,k}}{4}$ remains. For brevity, we only give proof for the item $\frac{\cos \omega_{n_y,n_z}^{q,k}}{2}$. The proof for the other items can be obtained similarly.

For an extremely large-scale RHS, utilizing the polar coordinate transformation $y - y_k = \rho \cos \vartheta$, $z - z_k = \rho \sin \vartheta$, the integral in (19.17) corresponding to the item $\frac{\cos \omega_{n_y,n_z}^{q,k}}{2}$ can be expressed as

$$\begin{aligned} &\frac{1}{2}\iint_{\mathbb{R}^2} \frac{e^{-\alpha\sqrt{(y-y_k)^2+(z-z_k)^2}}}{\sqrt{d_q^2 - 2d_q\Phi_q y - 2d_q\Theta_q z + y^2 + z^2}} \cdot \cos \omega_{n_y,n_z}^{q,k}\, dydz \\ &= \frac{1}{2}\int_0^{2\pi}\int_0^{+\infty} h_\vartheta(\rho)\cos\left(B_q\rho + \eta_{q,k}\right) d\rho d\vartheta, \end{aligned} \tag{19.25}$$

where $h_\vartheta(\rho) = \frac{\rho e^{-\alpha\rho}}{\sqrt{\left(\rho\cos\vartheta + y_k - d_q\Phi_q\right)^2 + \left(\rho\sin\vartheta + z_k - d_q\Theta_q\right)^2 + d_q^2\Psi_q^2}}$, $\Psi_q = \sin\theta_q \cos\varphi_q$, $B_q = |\mathbf{k}_s| - |\mathbf{k}_f| \cdot \left(\Phi_q \cos\vartheta + \Theta_q \sin\vartheta\right)$, $\eta_{q,k} = |\mathbf{k}_f|\left(y_k\Phi_q + z_k\Theta_q\right)$. For a fixed ϑ, by utilizing integration by parts we have

$$\begin{aligned} &\int_0^{+\infty} h_\vartheta(\rho)\cos\left(B_q\rho + \eta_{q,k}\right) d\rho \overset{(a)}{=} -\frac{1}{B_q}\int_0^{+\infty} h_\vartheta(\rho)\sin\left(B_q\rho + \eta_{q,k}\right) d\rho \\ &= \frac{1}{B_q^2} h'_\vartheta(0)\cos\eta_{q,k} - \frac{1}{B_q^2}\int_0^{+\infty} h''_\vartheta(\rho)\cos\left(B_q\rho + \eta_{q,k}\right) d\rho \\ &= \frac{1}{B_q^2 d_q^2} \cdot \frac{\cos\eta_{q,k}}{\left[\left(y_k/d_q - \Phi_q\right)^2 + \left(z_k/d_q - \Theta_q\right)^2 + \Psi_q^2\right]^{\frac{3}{2}}} \\ &- \frac{1}{B_q^3}\int_0^{+\infty} h''_\vartheta(\rho)\, d\sin\left(B_q\rho + \eta_{q,k}\right), \end{aligned} \tag{19.26}$$

where in *(a)* we utilize $h_\vartheta(0) = h_\vartheta(+\infty) = 0$. Since $B_q d_q \geq d_q\left(|\mathbf{k}_s| - |\mathbf{k}_f|\right)$ has a magnitude of 10^3, the first item in (19.26) has a magnitude of 10^{-6} and the second item has a magnitude of 10^{-9}. The integral corresponding to $\frac{\cos \omega_{n_y,n_z}^{q,k}}{2}$ can then be neglected. This completes the proof. □

The following lemma can then be derived from Lemma 19.1.

Lemma 19.2 For an extremely large-scale RHS satisfying $N_y \to \infty$ and $N_z \to \infty$, $\left[\mathbf{G}\mathbf{G}^H\right]_{q,q}$ can be given by

$$\left[\mathbf{G}\mathbf{G}^H\right]_{q,q} = \frac{\lambda^2\eta G_q A}{256\pi^2 d_y^2 d_z^2}\sum_{k=1}^{K} a_{q,k}^2 I_{q,k}^2, \tag{19.27}$$

where $I_{q,k}$ is defined as

$$I_{q,k} = \iint_A \frac{e^{-\alpha\sqrt{(y-y_k)^2+(z-z_k)^2}}}{\sqrt{d_q^2 - 2d_q\Phi_q y - 2d_q\Theta_q z + y^2 + z^2}} dydz. \tag{19.28}$$

The holographic pattern optimization problem given in (19.22) can then be rewritten as

$$\max_{\{a_{q,k}\}} \sum_{q=1}^{Q} \log_2 \left(1 + \frac{P_q\lambda^2\eta G_q A}{256\sigma^2\pi^2 d_y^2 d_z^2} \sum_{k=1}^{K} a_{q,k}^2 I_{q,k}^2\right), \tag{19.29a}$$

$$\text{s.t.} \sum_{q=1}^{Q}\sum_{k=1}^{K} a_{q,k} = 1. \tag{19.29b}$$

Proposition 19.2 The capacity can be upper bounded by

$$C \leq \sum_{q=1}^{Q} \log_2 \left(1 + \frac{P_q\lambda^2\eta G_q A}{256\sigma^2\pi^2 d_y^2 d_z^2} \left(\sum_{k=1}^{K} a_{q,k}\right)^2 \left(\max_{1\leq k\leq K} I_{q,k}\right)^2\right), \tag{19.30}$$

where the equality holds when $a_{q,k}$ satisfies[5]

$$\begin{cases} a_{q,k} \neq 0 \ , \ k = k_q^*, \\ a_{q,k} = 0 \ , \ k \neq k_q^*, \end{cases} \tag{19.31}$$

where for each user q, $k_q^* = \arg\max_{1\leq k\leq K} I_{q,k}$.

Proof: Without loss of generality, we assume that $I_{q,1} \geq I_{q,2} \geq \cdots \geq I_{q,K}$. Therefore, we have

$$\begin{aligned}
\left(\sum_{k=1}^{K} a_{q,k}\right)^2 I_{q,1}^2 - \sum_{k=1}^{K} a_{q,k}^2 I_{q,k}^2 &= I_{q,1}^2 \left(a_{q,1} + \sum_{k=1}^{K} a_{q,k}\right)\left(-a_{q,1} + \sum_{k=1}^{K} a_{q,k}\right) - \sum_{k=2}^{K} a_{q,k}^2 I_{q,k}^2 \\
&= I_{q,1}^2 \left(a_{q,1} + \sum_{k=1}^{K} a_{q,k}\right) \sum_{k=2}^{K} a_{q,k} - \sum_{k=2}^{K} a_{q,k}^2 I_{q,k}^2 \\
&\geq I_{q,1}^2 \sum_{k=2}^{K} a_{q,k} \cdot \left(\sum_{k=1}^{K} a_{q,k}\right) - \sum_{k=2}^{K} a_{q,k}^2 I_{q,k}^2 \\
&\geq I_{q,1}^2 \sum_{k=2}^{K} a_{q,k}^2 - \sum_{k=2}^{K} a_{q,k}^2 I_{q,k}^2 \geq 0.
\end{aligned} \tag{19.32}$$

5 For simplicity, we assume that for each fixed q, there exists a unique k maximizing $\{I_{q,k}\}_{k=1}^{K}$.

where the equality holds when $a_{q,k}$ satisfies $\begin{cases} a_{q,k} \neq 0 \ , \ k = 1, \\ a_{q,k} = 0 \ , \ k \neq 1 \end{cases}$. This completes the proof. □

Remark 19.2 Proposition 19.2 reveals that in an HDMA wireless communication system, the asymptotic capacity can be achieved if and only if the holographic pattern $\boldsymbol{m}$ is formed by the following elements

$$m_{n_y,n_z} = \sum_{q=1}^{L} a_q^* m\left(\mathbf{r}_{n_y,n_z}^{k_q^*}, \theta_q, \varphi_q\right), \tag{19.33}$$

where a_q^* will be given in (19.36), $m\left(\mathbf{r}_{n_y,n_z}^{k_q^*}, \theta_q, \varphi_q\right)$ is given in (19.5).

For notational convenience, we define $I_q = \frac{P_q \lambda^2 G_q A}{256\sigma^2 \pi^2 d_y^2 d_z^2}\left(I_{q,k_q^*}\right)^2$ and $a_q = a_{l,k_q^*}$. Therefore, based on (19.29), the holographic pattern optimization problem can be simplified as

$$\max_{\{a_q\}} \sum_{q=1}^{Q} \log_2\left(1 + I_q a_q^2\right), \tag{19.34a}$$

$$\text{s.t.} \sum_{q=1}^{L} a_q = 1. \tag{19.34b}$$

To derive the optimal solution of $\{a_q^*\}$, we adopt the Lagrangian dual form to relax the constraint (19.34b) with multiplier. Denote β as the Lagrangian multiplier associated with the constraint (19.34b). The Lagrangian associated with holographic pattern design can be given by

$$\mathcal{L}\left(a_q, \beta\right) = \sum_{q=1}^{Q} \log_2\left(1 + I_q a_q^2\right) - \beta\left(\sum_{q=1}^{Q} a_q - 1\right). \tag{19.35}$$

By setting $\partial\mathcal{L}/\partial a_q = 0$, the optimal $\{a_q^*\}$ and β^* can be obtained by solving the following system of equations:

$$\begin{cases} a_q^* = \frac{1}{\beta \ln 2} + \sqrt{\frac{1}{(\beta \ln 2)^2} - \frac{1}{I_q}}, \\ \sum_{q=1}^{Q} a_q^* = 1. \end{cases} \tag{19.36}$$

Therefore, the holographic pattern $\boldsymbol{m}$ can be obtained by substituting $\{a_q^*\}$ into (19.6), i.e., the holographic pattern $\boldsymbol{m}$ is formed by the following elements

$$m_{n_y,n_z} = \sum_{q=1}^{Q} a_q^* m\left(\mathbf{r}_{n_y,n_z}^{k_q^*}, \theta_q, \varphi_q\right), \tag{19.37}$$

where $m\left(\mathbf{r}_{n_y,n_z}^{k_q^*}, \theta_q, \varphi_q\right)$ is given in (19.5).

Proposition 19.3 When I_q satisfies $\min_q I_q \geq 12(Q-1)^2$, the optimal solution given by (19.36) is the global optimal solution to problem (19.22).[6]

Proof: We first prove that the solution given by (19.36) is the only interior maximum point of problem (19.22). Since the Hessian matrix of $\sum_{q=1}^{Q} \log_2\left(1+I_q a_q^2\right)$ with respect to a_q is $\operatorname{diag}\left\{\frac{1-I_q a_q^2}{\left(1+I_q a_q^2\right)^2} \cdot \frac{2I_q}{\ln 2}\right\}$, the maximum point of problem (19.22) must satisfy $1 - I_q a_q^2 \leq 0, \forall q$. For each user q, the equation $\partial \mathcal{L}/\partial a_q = 0$ has two solutions $a_q = \frac{1}{\beta \ln 2} \pm \sqrt{\frac{1}{(\beta \ln 2)^2} - \frac{1}{I_q}}$, and $1 - I_q a_q^2$ can then be given by

$$1 - I_q a_q^2 = 2\sqrt{\frac{I_q}{(\beta \ln 2)^2} - 1}\left(\mp\sqrt{\frac{I_q}{(\beta \ln 2)^2}} - \sqrt{\frac{I_q}{(\beta \ln 2)^2} - 1}\right). \tag{19.38}$$

Therefore, the maxima is achieved only when $a_q = \frac{1}{\beta \ln 2} + \sqrt{\frac{1}{(\beta \ln 2)^2} - \frac{1}{I_q}}$.

We now prove that when $\min_q I_q \geq 12(Q-1)^2$, the solution given by (19.36) is the global maximum point of problem (19.22), i.e., the maxima can not be achieved on the boundary of the feasible region $\{\sum_{q=1} a_q = 1, a_q \geq 0, \forall q\}$. We prove this by contradiction. Suppose there exists an optimal solution a_q such that $a_q = 0$ for at least one user q. Without loss of generality, we assume that $I_1 \leq I_2 \leq \cdots \leq I_Q$. Note that if $I_{q_1} \geq I_{q_2}$, then $a_{q_1} \geq a_{q_2}$, otherwise exchanging the values of a_{q_1} and a_{q_2} will lead to an increase of the capacity, which contradicts the optimality of a_q. Therefore, $a_1 \leq a_2 \leq \cdots \leq a_Q$, and thus, $a_1 = 0$ and $a_Q \geq \frac{1}{Q-1}$. Define another feasible solution $b_q = \begin{cases} \frac{a_Q}{2} & q = 1, Q, \\ a_q, & 2 \leq q \leq Q-1 \end{cases}$. When $I_1 \geq 12(Q-1)^2$, we have

$$\begin{aligned}
&\sum_{q=1}^{Q} \log_2\left(1+I_q b_q^2\right) - \sum_{q=1}^{Q} \log_2\left(1+I_q a_q^2\right) \\
&= \log_2\left(1+I_1 b_1\right)^2 + \log_2\left(1+I_Q b_Q^2\right) - \log_2\left(1+I_1 a_1^2\right) - \log_2\left(1+I_Q a_Q^2\right) \\
&= \log_2\left[\left(1+\frac{1}{4}I_1 a_Q^2\right)\left(1+\frac{1}{4}I_Q a_Q^2\right)\right] - \log_2\left(1+I_Q a_Q^2\right) \\
&> \log_2\left(1+\frac{1}{4}I_Q a_Q^2 + \frac{1}{16}I_Q a_Q^2 \cdot I_1 a_Q^2\right) - \log_2\left(1+I_Q a_Q^2\right) \\
&\geq \log_2\left(1+\frac{1}{4}I_Q a_Q^2 + \frac{1}{16}I_Q a_Q^2 \cdot 12\right) - \log_2\left(1+I_Q a_Q^2\right) = 0,
\end{aligned} \tag{19.39}$$

6 When $N_y = N_z = 100$, under the parameter settings given in Section 19.5, the magnitude of $\min_q I_q$ is about 10^5. Therefore, for an extremely large-scale RHS satisfying $N_y \to \infty$ and $N_z \to \infty$, the condition $\min_q I_q \geq 12(Q-1)^2$ can be easily satisfied.

indicating that the capacity corresponding to $\{b_q\}$ is larger than that of $\{a_q\}$, which contradicts the optimality of a_q. Therefore, the maxima can not be achieved on the boundary of the feasible region and the solution given by (19.36) is the global optimal solution to problem (19.22). □

Based on (19.36) and Proposition 19.2, we have the following proposition.

Proposition 19.4 For an extremely large-scale RHS satisfying $N_y \to \infty$ and $N_z \to \infty$, the asymptotic capacity of the HDMA wireless communication system can be given by

$$\tilde{C} = \sum_{q=1}^{Q} \log_2 \left(1 + I_q (a_q^*)^2\right) = \sum_{q=1}^{Q} \log_2 \left(1 + \frac{P_q \lambda^2 G_q A}{256\sigma^2 \pi^2 d_y^2 d_z^2} (a_q^*)^2 \left(I_{q,k_q^*}\right)^2\right), \quad (19.40)$$

where a_q^* is given in (19.36) and I_{q,k_q^*} can be obtained by (19.28).

19.5 Performance Analysis and Evaluation

In this section, we first analyze the relation between the sum rate with ZF precoding and the capacity of the HDMA wireless communication system. It is demonstrated that for an extremely large-scale RHS, the optimal holographic pattern proposed in Section 19.4 can make the sum rate with ZF precoding achieve the asymptotic capacity. Following that, we evaluate the performance of the HDMA wireless communication system to validate the theoretical analysis.

19.5.1 Relation Between Sum Rate and Capacity

Since the ZF precoder can obtain a near-optimal solution with low complexity, we consider ZF precoding together with power allocation at the BS to alleviate the inter-user interference (Rusek et al., 2012). The qth column of the ZF precoder $\mathbf{V} \in \mathbb{C}^{K\times Q}$ can be given by

$$\mathbf{V}_q = \frac{\mathbf{W}_q}{|\mathbf{W}_q|}, \quad (19.41)$$

where $\mathbf{W}_q$ is the qth column of matrix $\mathbf{G}^H\left(\mathbf{G}\mathbf{G}^H\right)^{-1}$ and $\mathbf{G} = \mathbf{HM}$. Based on the expression of $\mathbf{V}$, the following proposition can be derived.

Proposition 19.5 The data rate of user q can be given by

$$R_q = \log_2 \left(1 + \frac{P_q}{\sigma^2 \left[\left(\mathbf{G}\mathbf{G}^H\right)^{-1}\right]_{q,q}}\right). \quad (19.42)$$

Proof: According to (19.14) and (19.41), the data rate of user q is $R_q = \log_2\left(1+\frac{P_q}{\sigma^2}\left(\mathbf{G}_q\mathbf{V}_q\right)\left(\mathbf{G}_q\mathbf{V}_q\right)^H\right)$, where $\mathbf{G}_q$ is the qth row of $\mathbf{G}$. Based on (19.41), $\left(\mathbf{G}_q\mathbf{V}_q\right)\left(\mathbf{G}_q\mathbf{V}_q\right)^H$ can be rewritten as

$$\left(\mathbf{G}_q\mathbf{V}_q\right)\left(\mathbf{G}_q\mathbf{V}_q\right)^H = \mathbf{G}_q\mathbf{V}_q\mathbf{V}_q^H\mathbf{G}_q^H = \mathbf{G}_q\frac{\mathbf{W}_q}{|\mathbf{W}_q|}\frac{\mathbf{W}_q^H}{|\mathbf{W}_q|}\mathbf{G}_q^H. \tag{19.43}$$

Since $\mathbf{W} = \mathbf{G}^H\left(\mathbf{G}\mathbf{G}^H\right)^{-1}$, we have $\mathbf{G}_q\mathbf{W}_q\mathbf{W}_q^H\mathbf{G}_q^H = \left[\mathbf{G}\mathbf{G}^H\left(\mathbf{G}\mathbf{G}^H\right)^{-1}\right]_{q,q}\left[\mathbf{G}\mathbf{G}^H\left(\mathbf{G}\mathbf{G}^H\right)^{-1}\right]_{q,q}^H = 1$. Therefore, $\left(\mathbf{G}_q\mathbf{V}_q\right)\left(\mathbf{G}_q\mathbf{V}_q\right)^H$ can be further expressed as

$$\begin{aligned}\left(\mathbf{G}_q\mathbf{V}_q\right)\left(\mathbf{G}_q\mathbf{V}_q\right)^H &= \frac{1}{|\mathbf{W}_q|^2} = \frac{1}{\left[\mathbf{W}^H\mathbf{W}\right]_{q,q}}\\ &= \frac{1}{\left[\left(\mathbf{G}\mathbf{G}^H\right)^{-1}\mathbf{G}\mathbf{G}^H\left(\mathbf{G}\mathbf{G}^H\right)^{-1}\right]_{q,q}} = \frac{1}{\left[\left(\mathbf{G}\mathbf{G}^H\right)^{-1}\right]_{q,q}}.\end{aligned} \tag{19.44}$$

This completes the proof. □

Therefore, the sum rate with ZF precoding of the HDMA communication system can be given by

$$R = \sum_{q=1}^{Q} R_q = \sum_{q=1}^{Q}\log_2\left(1+\frac{P_q}{\sigma^2\left[\left(\mathbf{G}\mathbf{G}^H\right)^{-1}\right]_{q,q}}\right). \tag{19.45}$$

Based on Proposition 19.2, the asymptotic capacity is determined by $\{a_{q,k}\}$, which is related to k_q^* and $I_{q,k}$. The properties about k_q^* and $I_{q,k}$ are given in the following lemma.

Lemma 19.3 Define the distance between user q and feed k as d_k^q. For an extremely large-scale RHS satisfying $N_y \to \infty$ and $N_z \to \infty$, $I_{q,k}$ given in (19.28) decreases as d_k^q grows, and thus, k_q^* can be given by

$$k_q^* = \arg\min_{1\le k\le K} d_k^q = \arg\min_{1\le k\le K}\sqrt{\left(y_k - d_q\Phi_q\right)^2 + \left(z_k - d_q\Theta_q\right)^2 + \left(d_q\sin\theta_q\cos\phi_q\right)^2}. \tag{19.46}$$

Proof: Based on the expression of $I_{q,k}$ given in (19.28), for an extremely large-scale RHS satisfying $N_y \to \infty$ and $N_z \to \infty$, $I_{q,k}$ can be rewritten as $I_{q,k} = \iint_{\mathbb{R}^2}\frac{e^{-\alpha\sqrt{u^2+v^2}}dudv}{\sqrt{\left(u+y_k-d_q\Phi_q\right)^2+\left(v+z_k-d_q\Theta_q\right)^2+\left(d_q\Psi_q\right)^2}}$, where $\Psi_q = \sin\theta_q\cos\varphi_q$. Utilizing the polar coordinates, we define $y_k - d_q\Phi_q = t_{q,k}\cos\delta_{q,k}$ and $z_k - d_q\Theta_q = t_{q,k}\sin\delta_{q,k}$. We then have $I_{q,k} = \int_0^{2\pi}\int_0^{+\infty}\frac{\rho e^{-\alpha\rho}}{\sqrt{\rho^2+t_{q,k}^2+\left(d_q\Psi_q\right)^2+2\rho t_{q,k}\cos\left(\delta_{q,k}-\vartheta\right)}}d\vartheta d\rho$.

Due to the periodicity of the cosine function, the value of $\delta_{q,k}$ does not influence $I_{q,k}$, and thus, we assume that $\delta_{q,k}=0$. $I_{q,k}$ can then be simplified as $I_{q,k}=\int_{-\infty}^{+\infty}\left(\int_{-\infty}^{+\infty}\frac{e^{-\alpha\sqrt{u^2+v^2}}}{\sqrt{\left(u+t_{q,k}\right)^2+v^2+\left(d_q\Psi_q\right)^2}}du\right)dv$. The derivative of $I_v\left(t_{q,k}\right)=\int_{-\infty}^{+\infty}\frac{e^{-\alpha\sqrt{u^2+v^2}}}{\sqrt{\left(u+t_{q,k}\right)^2+v^2+\left(d_q\Psi_q\right)^2}}du$ can then be given by

$$
\begin{aligned}
I_v'\left(t_{q,k}\right) &= \int_{-\infty}^{+\infty}\frac{d}{dt_{q,k}}\left[\frac{e^{-\alpha\sqrt{u^2+v^2}}}{\sqrt{\left(u+t_{q,k}\right)^2+v^2+\left(d_q\Psi_q\right)^2}}\right]du \\
&= \int_{-\infty}^{+\infty}\frac{-\left(u+t_{q,k}\right)e^{-\alpha\sqrt{u^2+v^2}}}{\left[\sqrt{\left(u+t_{q,k}\right)^2+v^2+\left(d_q\Psi_q\right)^2}\right]^{\frac{3}{2}}}du \\
&= \int_{-\infty}^{+\infty}\frac{-ue^{-\alpha\sqrt{(u-t_{q,k})^2+v^2}}}{\left[\sqrt{u^2+v^2+\left(d_q\Psi_q\right)^2}\right]^{\frac{3}{2}}}du \\
&= \int_{0}^{+\infty}\frac{u\left(e^{-\alpha\sqrt{\left(u+t_{q,k}\right)^2+v^2}}-e^{-\alpha\sqrt{\left(u-t_{q,k}\right)^2+v^2}}\right)}{\left[\sqrt{u^2+v^2+\left(d_q\Psi_q\right)^2}\right]^{\frac{3}{2}}}du.
\end{aligned} \tag{19.47}
$$

Since $\left(u+t_{q,k}\right)^2\geq\left(u-t_{q,k}\right)^2$ when $u\geq 0$ and $t_{q,k}\geq 0$, $I_v'\left(t_{q,k}\right)\leq 0$, and thus, $I_{q,k}$ decreases as $t_{q,k}$ grows. Therefore, $I_{q,k}$ decreases as $d_{q,k}=\sqrt{t_{q,k}^2+\left(d_q\Psi_q\right)^2}$ grows. This completes the proof. □

Combining Proposition 19.2 and Lemma 19.3 also gives the following proposition about the relation between the sum rate with ZF precoding and the asymptotic capacity of the HDMA wireless communication system.

Proposition 19.6 In an HDMA wireless communication system aided by an extremely large-scale RHS, the sum rate with ZF precoding can achieve the asymptotic capacity when the following conditions are satisfied: (i) The feed closest to each user is different, i.e., $k_{q_1}^*\neq k_{q_2}^*, \forall q_1\neq q_2$; (ii) The holographic pattern $\boldsymbol{m}$ is formed by the following elements

$$
m_{n_y,n_z}=\sum_{q=1}^{Q}a_q^* m\left(\mathbf{r}_{n_y,n_z}^{k_q^*},\theta_q,\varphi_q\right), \tag{19.48}
$$

where a_q^* is given in (19.36), $m\left(\mathbf{r}_{n_y,n_z}^{k_q^*},\theta_q,\varphi_q\right)$ is given in (19.5).

Proof: When the feed closest to each user is different, without loss generality, it can be assumed that $k_q^* = q, \forall q$. Bases on Proposition 19.3, we have

$$\begin{cases} a_{q,k} \neq 0\ ,\ k = q, \\ a_{q,k} = 0\ ,\ k \neq q. \end{cases} \Rightarrow \begin{cases} [\mathbf{G}]_{q,k} \neq 0\ ,\ k = q, \\ [\mathbf{G}]_{q,k} = 0\ ,\ k \neq q. \end{cases} \tag{19.49}$$

Therefore, $\mathbf{G}$ is a diagonal matrix, and thus, $\mathbf{G}\mathbf{G}^H$ is also a diagonal matrix satisfying $\left[\left(\mathbf{G}\mathbf{G}^H\right)^{-1}\right]_{q,q} = \left[\left(\mathbf{G}\mathbf{G}^H\right)\right]_{q,q}^{-1}$. According to (19.45), the sum rate with ZF precoding of the HDMA wireless communication system can be expressed as

$$R = \sum_{q=1}^{Q} R_q = \sum_{q=1}^{Q} \log_2 \left(1 + \frac{P_q}{\sigma^2 \left[\left(\mathbf{G}\mathbf{G}^H\right)^{-1}\right]_{q,q}} \right) = \sum_{q=1}^{Q} \log_2 \left(1 + \frac{P_q}{\sigma^2} \left[\left(\mathbf{G}\mathbf{G}^H\right)\right]_{q,q} \right) = C. \tag{19.50}$$

Since $\boldsymbol{m}$ given in (19.37) is the optimal holographic pattern maximizing the capacity, the sum rate with ZF precoding can achieve the asymptotic capacity. □

19.5.2 Performance Evaluation

In this section, we evaluate the performance of the HDMA wireless communication system to validate the theoretical analysis. Major simulation parameters are set up based on the existing works (Zhang et al., 2021), (Xu et al., 2009) and 3GPP specifications (3GPP, 2017) as given in Table 19.1.

Figure 19.6 depicts the capacity of the HDMA wireless communication system C versus the physical dimension of the RHS, where the optimal holographic pattern given in (19.37) is adopted. We observe that not only for an extremely large-scale RHS, but also for a normal-sized RHS, the optimal holographic pattern can make

Table 19.1 Simulation parameters.

Parameters	Values
Transmit power of the UT P (W)	10
Carrier frequency f (GHz)	30
Element spacing of the RHS d_y and d_z (cm)	0.25
Propagation vector in the free space k_f	200π
Propagation vector on the RHS k_s	$200\sqrt{3}\pi$
Antenna gain of each user G_q	1
Noise figure over Ka band (dB)	1.2

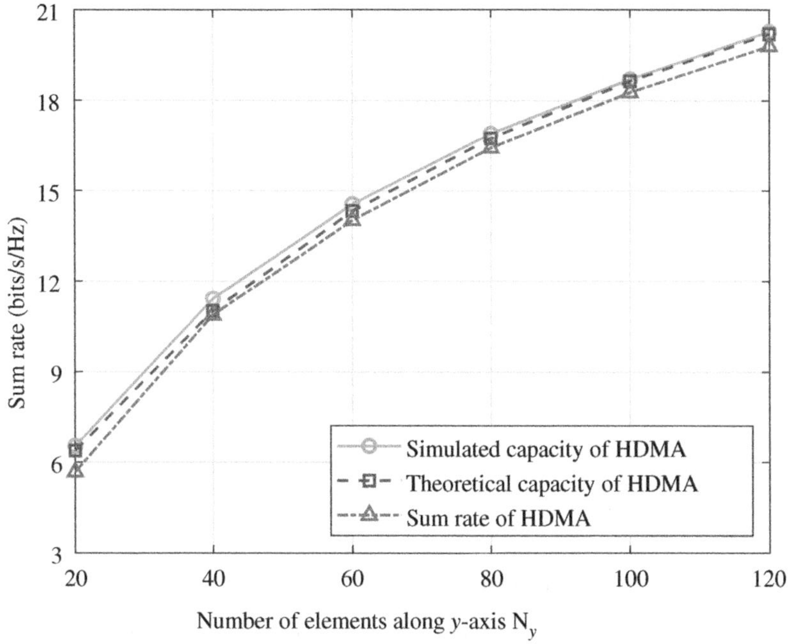

Figure 19.6 Sum rate versus number of RHS elements.

the sum rate with ZF precoding of the HDMA system achieve the capacity C. The accuracy of the approximation of the equivalent channel $[\mathbf{G}]_{q,k}$ given in Lemma 19.1 is also verified.

Figure 19.7 evaluates the cost-efficiency of the HDMA system and the traditional SDMA system achieved by a phased array, where the cost-efficiency metric is defined as the ratio of the sum rate to the hardware cost. In general, since a phased array requires high-priced electronic components such as phase shifters at each antenna element, the hardware cost ratio β of a phased array to the RHS of the same number of elements is about 2∼10 (Pivotal Staff, 2019). Specifically, in Fig. 19.7, we set the element spacing and that of the phased array as the same (i.e., $d_y = d_y^p, d_z = d_z^p$), such that the number of elements in the RHS and that of the phased array is also the same. We observe that with the same element spacing, the cost-efficiency of the HDMA system is greater than that of the traditional SDMA system due to the low hardware cost of the RHS element. Moreover, the advantages of the RHS in cost savings become overwhelming when the number of elements and the cost ratio β are large.

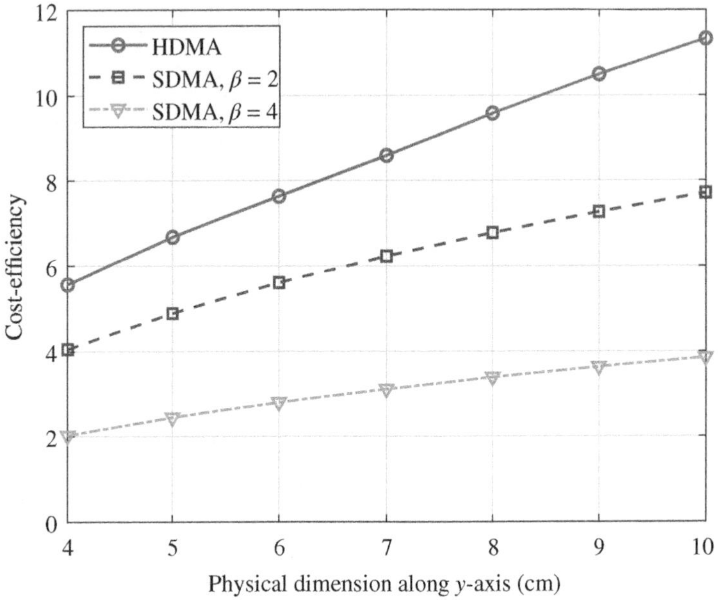

Figure 19.7 Cost-efficiency versus the physical dimension $N_y d_y$ with different cost ratio β.

19.6 Summary

In this chapter, the concept of HDMA for future wireless communications has been presented, where the holographic pattern construction of the RHS and the HDMA principles have been elaborated. Specifically, utilizing the superposition of the holographic patterns corresponding to the receivers, the intended signals can be mapped to a superposed holographic pattern. Based on the multiuser holographic beamforming scheme for HDMA, the asymptotic capacity and sum rate of an HDMA wireless communication system aided by an extremely large-scale RHS have been analyzed. The closed-form optimal holographic pattern to achieve asymptotic capacity has been derived. Simulation results show that the HDMA provides a more cost-effective solution for pursuing high data rate compared with the traditional SDMA with phased array owing to the low hardware cost of the RHS.

References

3GPP. Study on new radio access technology: Radio Frequency (RF) and co-existence aspects. Technical Specification (TS) 38.803, 3rd Generation Partnership Project (3GPP), Version 14.2.0, 09 2017. URL https://portal.3gpp.org/desktopmodules/Specifications/SpecificationDetails.aspx?specificationId=3069.

Ruoqi Deng, Boya Di, Hongliang Zhang, Yunhua Tan, and Lingyang Song. Reconfigurable holographic surface: Holographic beamforming for metasurface-aided wireless communications. *IEEE Transactions on Vehicular Technology*, 70(6):6255–6259, 2021.

Boya Di. Reconfigurable holographic metasurface aided wideband OFDM communications against beam squint. *IEEE Transactions on Vehicular Technology*, 70(5):5099–5103, 2021.

Boya Di, Hongliang Zhang, Lingyang Song, Yonghui Li, Zhu Han, and H Vincent Poor. Hybrid beamforming for reconfigurable intelligent surface based multi-user communications: Achievable rates with limited discrete phase shifts. *IEEE Journal on Selected Areas in Communications*, 38(8):1809–1822, 2020.

Benjamin Friedlander. Localization of signals in the near-field of an antenna array. *IEEE Transactions on Signal Processing*, 67(15):3885–3893, 2019.

David Gesbert, Mansoor Shafi, Da-shan Shiu, Peter J Smith, and Ayman Naguib. From theory to practice: An overview of MIMO space-time coded wireless systems. *IEEE Journal on Selected Areas in Communications*, 21(3):281–302, 2003.

J L Gomez-Tornero, F D Quesada-Pereira, and A Alvarez-Melcon. Analysis and design of periodic leaky-wave antennas for the millimeter waveband in hybrid waveguide-planar technology. *IEEE Transactions on Antennas and Propagation*, 53(9):2834–2842, 2005.

Jakob Hoydis, Stephan Ten Brink, and Merouane Debbah. Massive MIMO in the UL/DL of cellular networks: How many antennas do we need? *IEEE Journal on Selected Areas in Communications*, 31(2):160–171, 2013.

Ruey-Bing Raybeam Hwang. Binary meta-hologram for a reconfigurable holographic metamaterial antenna. *Scientific Reports*, 10(1):1–10, 2020.

Mikala C Johnson, Steven L Brunton, Nathan B Kundtz, and Nathan J Kutz. Extremum-seeking control of the beam pattern of a reconfigurable holographic metamaterial antenna. *JOSA A*, 33(1):59–68, 2016.

Erik G Larsson, Ove Edfors, Fredrik Tufvesson, and Thomas L Marzetta. Massive MIMO for next generation wireless systems. *IEEE Communications Magazine*, 52(2):186–195, 2014.

Yuanwei Liu, Xiao Liu, Xidong Mu, Tianwei Hou, Jiaqi Xu, Marco Di Renzo, and Naofal Al-Dhahir. Reconfigurable intelligent surfaces: Principles and opportunities. *IEEE Communications Surveys & Tutorials*, 23(3):1546–1577, 2021.

Haiquan Lu and Yong Zeng. Communicating with extremely large-scale array/surface: Unified modeling and performance analysis. *IEEE Transactions on Wireless Communications*, 21(6):4039–4053, 2021.

Jan Mietzner, Robert Schober, Lutz Lampe, Wolfgang H Gerstacker, and Peter A Hoeher. Multiple-antenna techniques for wireless communications-a comprehensive literature survey. *IEEE Communications Surveys & Tutorials*, 11(2):87–105, 2009.

Mahyar Nemati, Behrouz Maham, Shiva Raj Pokhrel, and Jinho Choi. Modeling RIS empowered outdoor-to-indoor communication in mmWave cellular networks. *IEEE Transactions on Communications*, 69(11):7837–7850, 2021.

Pivotal Staff. *Holographic Beam Forming and Phased Arrays*, pages 68–73. Kirkland, WA, USA: Pivotal Commware, Inc., 2019.

Fredrik Rusek, Daniel Persson, Buon Kiong Lau, Erik G Larsson, Thomas L Marzetta, Ove Edfors, and Fredrik Tufvesson. Scaling up MIMO: Opportunities and challenges with very large arrays. *IEEE Signal Processing Magazine*, 30(1):40–60, 2012.

Timothy Sleasman, Mohammadreza F Imani, Wangren Xu, John Hunt, Tom Driscoll, Matthew S Reynolds, and David R Smith. Waveguide-fed tunable metamaterial element for dynamic apertures. *IEEE Antennas and Wireless Propagation Letters*, 15:606–609, 2015.

David R Smith, Okan Yurduseven, Laura Pulido Mancera, Patrick Bowen, and Nathan B Kundtz. Analysis of a waveguide-fed metasurface antenna. *Physical Review Applied*, 8(5):054048, 2017.

Foad Sohrabi and Wei Yu. Hybrid digital and analog beamforming design for large-scale antenna arrays. *IEEE Journal of Selected Topics in Signal Processing*, 10(3):501–513, 2016.

Ryan A Stevenson, Mohsen Sazegar, and Phillip Sullivan. Enabling a hyper-connected world: Advanced antenna design using liquid crystals and LCD manufacturing. In *Proceedings of the International Display Workshops 72*, 2020.

Le-Nam Tran, Markku Juntti, Mats Bengtsson, and Bjorn Ottersten. Weighted sum rate maximization for MIMO broadcast channels using dirty paper coding and zero-forcing methods. *IEEE Transactions on Communications*, 61(6):2362–2373, 2013.

Dongming Wang, Yu Zhang, Hao Wei, Xiaohu You, Xiqi Gao, and Jiangzhou Wang. An overview of transmission theory and techniques of large-scale antenna systems for 5G wireless communications. *Science China Information Sciences*, 59:1–18, 2016.

Tianyu Wang, Shaowei Wang, and Zhi-Hua Zhou. Machine learning for 5G and beyond: From model-based to data-driven mobile wireless networks. *China Communications*, 16(1):165–175, 2019.

Guanghan Xu and San-Qi Li. Throughput multiplication of wireless LANs for multimedia services: SDMA protocol design. In *1994 IEEE GLOBECOM. Communications: The Global Bridge*, volume 3, pages 1326–1332. IEEE, 1994.

Feng Xu, Ke Wu, and Xiupu Zhang. Periodic leaky-wave antenna for millimeter wave applications based on substrate integrated waveguide. *IEEE Transactions on Antennas and Propagation*, 58(2):340–347, 2009.

Hongliang Zhang, Boya Di, Zhu Han, H Vincent Poor, and Lingyang Song. Reconfigurable intelligent surface assisted multi-user communications: How many reflective elements do we need? *IEEE Wireless Communications Letters*, 10(5):1098–1102, 2021.

Nikolay I Zheludev. The road ahead for metamaterials. *Science*, 328(5978):582–583, 2010.

20

Over-the-Air Computation

Yilong Chen[1], Xiaowen Cao[1], Jie Xu[1], Guangxu Zhu[2], Kaibin Huang[3], and Shuguang Cui[1,2,4]

[1] *School of Science and Engineering (SSE) and Future Network of Intelligence Institute (FNii), The Chinese University of Hong Kong (Shenzhen), Shenzhen, China*
[2] *Data-Driven Information System Lab, Shenzhen Research Institute of Big Data, Shenzhen, China*
[3] *Department of Electrical and Electronic Engineering, The University of Hong Kong, Hong Kong SAR, China*
[4] *Peng Cheng Laboratory, Shenzhen, China*

20.1 Introduction

The advancements in artificial intelligence of things (AIoT) are expected to revolutionize the way we live and work through providing ubiquitous connectivity to everything. Toward this end, wireless communications have witnessed a paradigm shift from human-type communications toward machine-type communications. Particularly, it is predicted by the global system for mobile communications association (GSMA) that the number of Internet of things (IoT) wireless devices (WDs) will reach 75 billion by 2025. Providing wireless connectivity to such a gigantic number of WDs poses a grand challenge to wireless system designs. This has prompted an increasing number of researchers to depart from the traditional task-agnostic design principle that isolates communication from the subsequent tasks (or applications), and to explore the designs of task-oriented communication technologies crossing disciplines such as machine learning, computing, and wireless communications (Letaief et al., 2019).

Aligned with this trend, we are particularly interested in a specific class of IoT applications, which requires an access point (AP) to aggregate data distributed at WDs with wireless connectivity, termed wireless data aggregation (WDA). Such applications may include vehicle platooning, drone swarm control, and distributed sensing and learning, which share a common task of computing a function of distributed data generated by WDs. Such data may include artificial intelligence (AI)

Next Generation Multiple Access, First Edition.
Edited by Yuanwei Liu, Liang Liu, Zhiguo Ding, and Xuemin Shen.
© 2024 The Institute of Electrical and Electronics Engineers, Inc. Published 2024 by John Wiley & Sons, Inc.

model updates in distributed learning, accelerations, and velocities in vehicle platooning or drone swarms, and temperature/humidity/chemical levels in sensing. The applications are either data-intensive (e.g., distributed learning) or latency critical (e.g., vehicle platooning).

Conventionally, orthogonal multiple access (OMA) (e.g., time division multiple access (TDMA) and orthogonal frequency division multiple access (OFDMA)) and non-orthogonal multiple access (NOMA) (Ding et al., 2017) can be implemented for WDA, in which the AP first decodes each WD's messages individually by treating those from other WDs as harmful interference, and then aggregates them in a separate, subsequent step. However, due to the decoupling of communication and computation, these conventional multiple-access schemes may lead to significant transmission delays, thus limiting the efficiency of WDA. To overcome this issue, over-the-air computation (AirComp) (Zhu et al., 2021b, Cao et al., 2020, Chen et al., 2022, Zhu and Huang, 2019) has emerged as a new multiple access technique. In contrast to the conventional schemes that aim to decode multiple WDs' individual messages by mitigating the harmful inter-user interference (Ding et al., 2017), AirComp exploits the waveform superposition property of wireless channels to compute functions based on multiple WDs' individual data (Abari et al., 2016), in which the inter-user interference is harnessed as a beneficial factor for computation, thus leading to highly efficient WDA, as shown in Fig. 20.1.

It is noteworthy that the underlying principle of AirComp can be traced back to the well-known physical-layer networking coding (PNC) techniques that also exploit the waveform superposition property of multiple-access channels

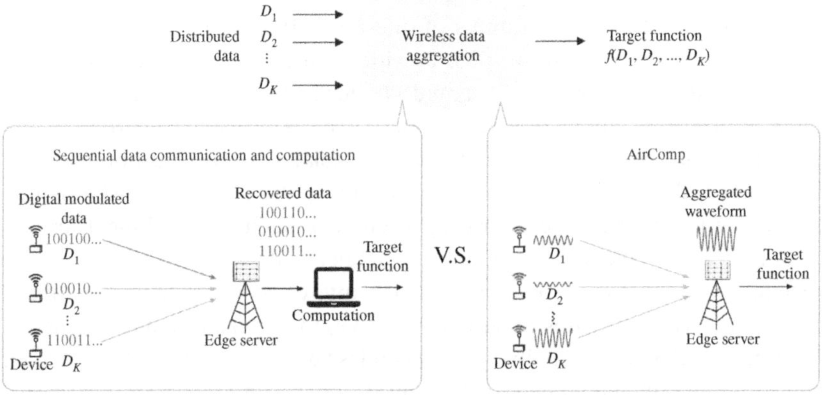

Figure 20.1 Two paradigms for WDA: sequential data communication and computation versus AirComp with integrated communication and computation. Source: Zhu et al., 2021b/IEEE.

(MACs) but for a different purpose, namely to relay messages for increasing the network throughput (Zhang et al., 2006). Despite the similarity, AirComp differs from PNC in terms of design objectives, techniques, and applications. In particular, AirComp designs are aimed at supporting high-accuracy and low-latency WDA for next-generation IoT applications in 5G and beyond. To this end, many new techniques have been designed ranging from power control to beamforming to over-the-air federated learning (Cao et al., 2022a, 2022b). In view of growing interests in AirComp, this chapter provides an overview of the new technology covering its basic principles, advanced techniques in power control and beamforming, and future research directions.

20.1.1 Notations

Boldface letters are used for vectors (lower-case) and matrices (upper-case). For a vector $\boldsymbol{a}$, $\boldsymbol{a}^{\dagger}$, $\boldsymbol{a}^{H}$, and $||\boldsymbol{a}||$ denote its conjugate, conjugate transpose, and Euclidean norm, respectively. For a square matrix $\boldsymbol{A}$, $\text{tr}(\boldsymbol{A})$ denotes its trace. For an arbitrary-size matrix $\boldsymbol{A}$, $\boldsymbol{A}^{H}$ and $||\boldsymbol{A}||_F$ denote its conjugate transpose and Frobenius norm, respectively. $\boldsymbol{I}$ denotes the identity matrix whose dimension will be clear from the context. $\mathbb{C}^{m\times n}$ denotes the $m \times n$ dimensional complex space. $\mathbb{E}[\cdot]$ denotes the statistic expectation.

20.2 AirComp Fundamentals

In general, AirComp can be classified into two categories, namely coded and uncoded AirComp. Due to its ease of implementation and its optimality in Gaussian channels with independent Gaussian sources (Gastpar, 2008), uncoded AirComp is particularly appealing.

We particularly focus on uncoded AirComp in this chapter. For ease of exposition, the fundamentals of AirComp are described under a single-antenna setting in this section. As mentioned, the basic idea of AirComp is to exploit the analog-wave superposition property of a MAC. As a result, the signals simultaneously transmitted by synchronized WDs are added over the air and arrive at the AP as a weighted sum, called the aggregated signal, with weights being the channel coefficients.

As shown in Fig. 20.2, the two essential operations for AirComp are analog amplitude modulation and channel pre-compensation at each WD. The former modulates the data values into the magnitudes of the carrier signals, and the latter inverts the heterogeneous channel fading of different links. As a result, each received signal component is the transmitted data scaled by a pre-determined factor. Setting the factor uniform for all signals, called magnitude alignment, renders

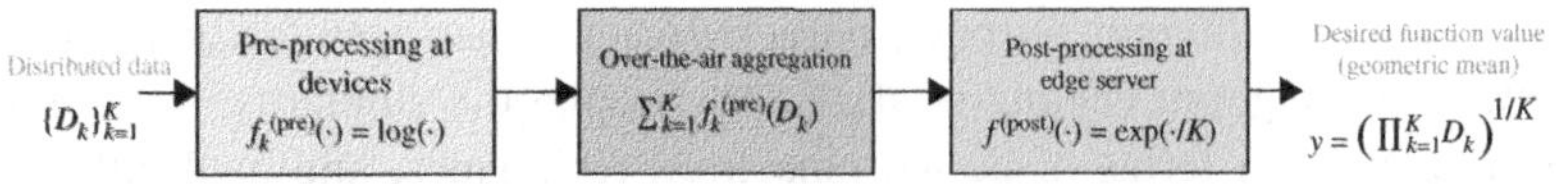

Figure 20.2 Illustration of the basic principle for AirComp (with an example of geometric mean computation). Source: Adapted from Zhu et al., 2021b.

the aggregated signal the desired average of transmitted distributed data, realizing the AirComp of an average function. Essentially, AirComp can be understood as a joint source and channel design in contrast to the classic separation-based design featuring sequential communication and computation.

With appropriate data pre-/post-processing, the capability of AirComp can go beyond averaging to compute a class of so-called nomographic functions, which can generally be expressed as a post-processed summation of multiple pre-processed data values (Zhu and Huang, 2019). Typical functions in this class include arithmetic mean, weighted sum, geometric mean, polynomial, and Euclidean norm. For example, to compute the geometric mean, the pre-processing is a logarithm function and the post-processing is an exponential function, as illustrated in Fig. 20.2. Note that the preprocessing needed for AirComp at IoT devices is usually computing the most elementary functions (e.g., multiplication, logarithm function, and square function) (see (Zhu and Huang, 2019) for more details), which thus can be easily deployed at the low-end edge devices with low computation power and energy supply.

Interestingly, it is proved in Buck (1976) that any continuous function of n variables can be decomposed into a summation form of at most $(2n + 1)$ nomographic component functions. The result implies that any function of n variables is AirComputable via at most $(2n + 1)$ channel uses. However, when the number of component functions exceeds n, it is more efficient to perform WDA using a conventional multiple-access scheme (e.g., TDMA) for sequential communication and computation, which requires only n channel uses for TDMA. Otherwise, AirComp is more efficient.

Strict time synchronization in WDs' transmissions poses a key challenge for AirComp implementation, but can be overcome using the rich set of existing synchronization techniques. For instance, uplink synchronization in 4G long-term evolution (LTE) systems relies on a so-called "timing advance" mechanism, which can be used to facilitate AirComp in practice. Specifically, each WD estimates the propagation delay and then transmits ahead of time (with a negative time offset equal to the delay) so that the signal always arrives at the AP within the allocated time slot regardless of the WD's location. The propagation delay can be estimated by measuring the difference between the transmit and receive timestamps of a broadcast pilot signal, subject to an error due to the timing (synchronization) offset between WD and AP. The timing offset is inversely proportional to the bandwidth of the synchronization channel. For example, a typical bandwidth of 1 MHz reins in the timing offset/error to be within 0.1μs (Ghosh et al., 2010). Consider the implementation of AirComp in a popular orthogonal frequency-division multiplexing (OFDM) system; the timing offset simply introduces a phase shift to the received symbol if the offset is shorter than the cyclic prefix (CP), which can thus be compensated by sub-channel equalization. The typical CP length in LTE systems is 5 μs, which is far longer than the typical timing offset, i.e., 0.1 μs (Ghosh et al., 2010). Thus, the time synchronization for broadband AirComp is feasible.

Besides time synchronization, frequency synchronization is another practical challenge faced by AirComp due to the carrier frequency offset (CFO) issue among an AP and different WDs. The conventional approach to distributed frequency synchronization relies on deploying high-precision local oscillators at distributed nodes, which is expensive and thus may not be practical for low-end WDs (e.g., sensors). Alternatively, low-cost solutions such as AirShare (Abari et al., 2015) can be used to deal with this issue. The key idea is to broadcast a shared clock over the air and feed it to the WDs as a reference clock. This tackles the CFO problem in a distributed network without deploying high-precision local oscillators at WDs.

Due to the employed analog modulation, AirComp is exposed to signal distortion caused by, e.g., channel fading, noise, and channel estimation error. The distortion of the received functional values can be suitably measured using the mean squared error (MSE) with respect to the noiseless ground truth, which is a commonly used performance metric for AirComp and termed the computation error. Another performance metric related to AirComp is computation rate, referring to the number of received functional values per channel use. By definition, single-antenna AirComp has a unit computation rate, while the rate of multiple-input multiple-output (MIMO) AirComp is multiplied by the spatial multiplexing gain. With a given computation rate, how to minimize the computation MSE is a critical technical problem. In the following, we introduce power control and beamforming techniques to deal with this issue.

20.3 Power Control for AirComp

This section presents the power control design for AirComp in a single-input single-output (SISO) MAC, in different scenarios including static channels, fading channels, and the presence of imperfect channel state information (CSI), respectively.

20.3.1 Static Channels

To start with, we consider an uncoded AirComp system with static channels, in which an AP aims to compute the function value of distributed data from a set $\mathcal{K} \triangleq$

$\{1, \ldots, K\}$ of $K \geq 1$ WDs. It is assumed that both the AP and the WDs are equipped with a single antenna. Let s_k denote the transmitted message by WD $k \in \mathcal{K}$, where s_k's are independent complex random variables with zero mean and unit variance.[1] The AP is interested in computing the averaging function of s_k's, i.e.,

$$f = \frac{1}{K} \sum_{k=1}^{K} s_k. \tag{20.1}$$

Let h_k denote the complex channel coefficient from WD $k \in \mathcal{K}$ to the AP, b_k denote the complex transmit coefficient at WD k. The received signal at the AP is given by

$$y = \sum_{k=1}^{K} h_k b_k s_k + z, \tag{20.2}$$

where z denotes the additive white Gaussian noise (AWGN) at the AP that is a circularly symmetric complex Gaussian (CSCG) random variable with zero mean and variance σ_z^2. The AP adopts a complex denoising factor w for data aggregation. Accordingly, the AP recovers the average function from the received signal as

$$\hat{f} = \frac{wy}{K} = \frac{1}{K} \left(\sum_{k=1}^{K} w h_k b_k s_k + wz \right). \tag{20.3}$$

We use the MSE as the performance metric of AirComp, which is expressed as follows to characterize the distortion of $\hat{f}$ with respect to the ground truth average f.

$$\text{MSE} = \mathbb{E}_{\{s_k\},z}[|\hat{f} - f|^2] = \frac{1}{K^2} \left(\underbrace{\sum_{k=1}^{K} |w h_k b_k - 1|^2}_{\text{Signal misalignment error}} + \underbrace{|w|^2 \sigma_z^2}_{\text{Noise-induced error}} \right). \tag{20.4}$$

The MSE above consists of two terms, including the signal misalignment error and the noise-induced error. The trade-off between them needs to be properly balanced. Let P_k denote the maximum transmit power at WD $k \in \mathcal{K}$. Accordingly, we have $\mathbb{E}_{\{s_k\}}[|b_k s_k|^2] = |b_k|^2 \leq P_k, \forall k \in \mathcal{K}$. Our objective is to minimize MSE in (20.4) by jointly optimizing the transmit coefficients $\{b_k\}$ at the WDs and the denoising factor w at the AP, subject to the individual power budgets at the WDs.

The MSE minimization problem in the SISO case is formulated as problem (P1) in the following. Here, the constant coefficient $\frac{1}{K^2}$ in (20.4) is dropped

1 In this chapter, we consider the independent sources for ease of analysis, while it can be extended into the case with correlated sources (see, e.g., (Zhu and Huang, 2019)).

for notational convenience, and with phase alignment, the minimum MSE is achieved by setting $w \geq 0$ and $b_k = \tilde{b}_k \frac{h_k^\dagger}{|h_k|}$, where $\tilde{b}_k \geq 0$ denotes the transmit amplitude of WD $k \in \mathcal{K}$.

$$(\text{P1}): \min_{\{\tilde{b}_k\}, w \geq 0} \sum_{k=1}^{K} (w|h_k|\tilde{b}_k - 1)^2 + w^2\sigma_z^2$$
$$\text{s.t.}\ 0 \leq \tilde{b}_k \leq \sqrt{P_k}, \forall k \in \mathcal{K}.$$

Due to the coupling between $\{b_k\}$ and w in the objective function, (P1) is nonconvex, which is thus difficult to be optimally solved.

To solve problem (P1), we define $\rho_k \triangleq \sqrt{P_k}|h_k|$ as the channel quality indicator for WD $k \in \mathcal{K}$. Without loss of generality, we assume that $\rho_1 \leq \cdots \leq \rho_k \leq \cdots \leq \rho_K$, $\rho_0 \to 0$, and $\rho_{K+1} \to \infty$. We also define $F_k(w) \triangleq \sum_{i=1}^{k} (w\sqrt{P_i}|h_i| - 1)^2 + w^2\sigma_z^2$.

Theorem 20.1 ***(Optimal Solution to Problem (P1))*** The optimal solution of w to (P1) is given by

$$w^* = \tilde{w}_{k^*} = \frac{\sum_{i=1}^{k^*} \sqrt{P_i}|h_i|}{\sum_{i=1}^{k^*} P_i|h_i|^2 + \sigma_z^2}, k^* = \arg\max_{k \in \mathcal{K}} (\tilde{w}_k), \tag{20.5}$$

where $\tilde{w}_k = \frac{\sum_{i=1}^{k} \sqrt{P_i}|h_i|}{\sum_{i=1}^{k} P_i|h_i|^2 + \sigma_z^2}$. Furthermore, it holds that $\rho_k \leq \frac{1}{w^*}$ for WDs $k \in \{1, \ldots, k^*\}$ and $\rho_k \geq \frac{1}{w^*}$ for WDs $k \in \{k^* + 1, \ldots, K\}$.

The optimal solution of $\{b_k\}$ to (P1) is given by

$$b_k^* = \begin{cases} \sqrt{P_k}\frac{h_k^\dagger}{|h_k|}, k \in \{1, \ldots, k^*\}, \\ \frac{h_k^\dagger}{w^*|h_k|^2}, k \in \{k^* + 1, \ldots, K\}. \end{cases} \tag{20.6}$$

Proof: See Cao et al. (2020). □

Remark 20.1 ***(Threshold-Based Optimal Power Control)*** Based on Theorem 20.1, we have the following insights on the optimal power control for AirComp over WDs in static channels. The optimal power-control policy over WDs has a threshold-based structure. The threshold is specified by the denoising factor w^* and applied on the derived quality indicator ρ_k accounting for both the channel power gain and power budget of WD $k \in \mathcal{K}$. It is shown that for each WD $k \in \{k^* + 1, \ldots, K\}$ with its quality indicator exceeding the threshold, i.e., $\rho_k \geq \frac{1}{w^*}$, the channel-inversion power control is applied with $\tilde{b}_k^* = \frac{1}{w^*|h_k|}$; while for each WD $k \in \{1, \dots, k^*\}$ with $\rho_k \leq \frac{1}{w^*}$, the full power transmission is deployed with $\tilde{b}_k^* = \sqrt{P_k}$. As shown in (20.5), the optimal threshold w^* is a monotonically decreasing function with respect to the noise variance σ_z^2. This suggests that a larger σ_z^2 leads to more WDs transmitting with full power and vice versa. The result is intuitive by noting that w^* also plays another role as the denoising factor: a large σ_z^2 requires a small w^* for suppressing the dominant noise-induced error.

Remark 20.2 ***(Asymptotic Analysis)*** Next, we analyze the derived optimal power-control policy to problem (P1) in two extreme regimes, namely the high and low signal-to-noise ratio (SNR) regimes, respectively. In the high SNR regime, i.e., $\sigma_z^2 \to 0$, the MSE is dominated by the signal misalignment error. Thus, we can expect that the channel-inversion power control is optimal as it minimizes the signal misalignment error. Aligned with the intuition, it follows that $k^* = 1$ and $w^* = \frac{1}{\rho_1}$ should be the global minimizer. Therefore, the optimal power control is given by $\tilde{b}^* = \sqrt{P_1}\frac{|h_1|}{|h_k|}, \forall k \in \mathcal{K}$. In the low SNR regime with $\sigma_z^2 \to \infty$, the noise-induced error becomes dominant, and thereby it is desirable to minimize w to suppress the noise-induced error, i.e., $w^* \leq \frac{1}{\rho_K}$ must hold. Then, we have $k^* = K$ and $w^* = \tilde{w}_K$. In this case, all WDs should transmit with full power to achieve the optimal w^*, i.e., the optimal power control in this case is $\tilde{b}^* = \sqrt{P_k}, \forall k \in \mathcal{K}$.

Figure 20.3 shows the MSE of AirComp versus the receive SNR, where the receive SNRs are equal for $k = 20$ WDs. For comparison, we consider two benchmark schemes, namely the full power transmission, and the traditional channel inversion modified by including a truncation operation that cuts off the channel if its gain is below a prespecified threshold. It is observed that the achieved MSE by all schemes decreases with the average receive SNR increasing, and the proposed power control outperforms the other two benchmark schemes throughout the whole receive SNR regime. With low receive SNR, the full power transmission has the same performance as the proposed power control, and both of them outperform the traditional channel inversion scheme. This is because the full power transmission significantly suppresses the noise-induced error that is dominant for the MSE in this case. As the receive SNR increases, the performance gap between the proposed power control and the full power transmission becomes large, while the traditional channel inversion scheme performs close to optimal due to the perfect signal-magnitude alignment.

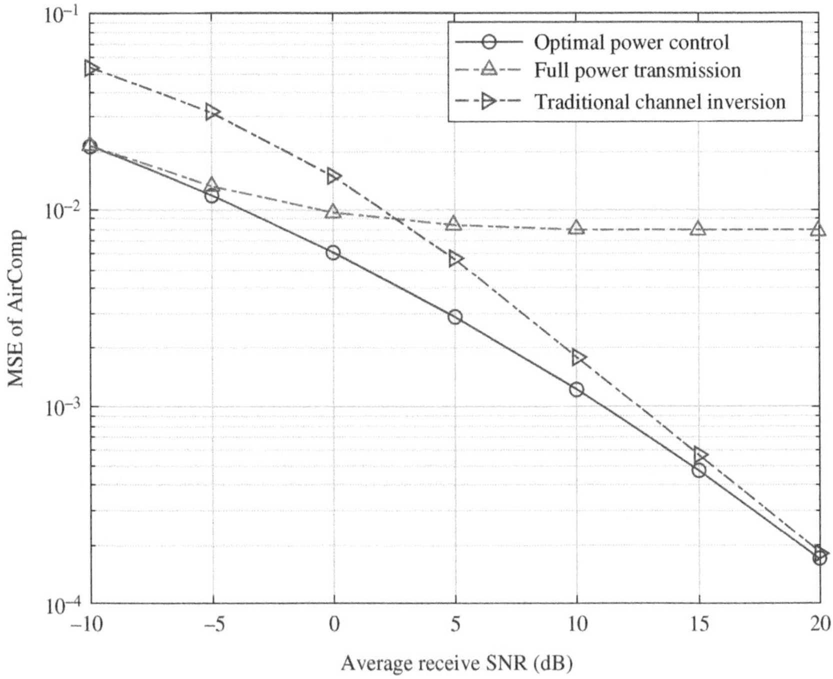

Figure 20.3 The computation MSE of AirComp versus the average receive SNR in static channels with $K = 20$. Source: Cao et al., 2020/IEEE.

20.3.2 Fading Channels

Next, we extend the power control design to the case with fading channels. We define the set of fading states as $\mathcal{M}$. In each fading state $m \in \mathcal{M}$, let $s_{k,m}$ denote the transmitted message by WD $k \in \mathcal{K}$, where $s_{k,m}$'s are independent complex random variables with zero mean and unit variance. The AP is interested in computing the averaging function of $s_{k,m}$'s, i.e.,

$$f_m = \frac{1}{K}\sum_{k=1}^{K} s_{k,m}, \forall m \in \mathcal{M}. \tag{20.7}$$

In fading state $m \in \mathcal{M}$, let $h_{k,m}$ denote the complex channel coefficient from WD $k \in \mathcal{K}$ to the AP, $b_{k,m}$ denote the complex transmit coefficient at WD k. The AP adopts the denoising factor w_m for data aggregation. Accordingly, the AP recovers the average function from the received signal as

$$\hat{f}_m = \frac{w_m y_m}{K} = \frac{1}{K}\left(\sum_{k=1}^{K} w_m h_{k,m} b_{k,m} s_{k,m} + w_m z\right), \forall m \in \mathcal{M}. \tag{20.8}$$

By using the stationary and ergodic nature of the fading channels, the ensemble-average MSE between $\{\hat{f}_m\}$ and $\{f_m\}$ over fading channels is given by

$$\begin{aligned} \text{MSE} &= \mathbb{E}_m\left[\mathbb{E}_{\{s_{k,m}\},z}[|\hat{f}_m - f_m|^2]\right] \\ &= \mathbb{E}_m\left[\frac{1}{K^2}\left(\sum_{k=1}^{K}|w_m h_{k,m} b_{k,m} - 1|^2 + |w_m|^2\sigma_z^2\right)\right]. \end{aligned} \tag{20.9}$$

Our objective is to minimize the average MSE by jointly optimizing $\{b_{k,m}\}$ and $\{w_m\}$, subject to the individual power budget $\mathbb{E}_m\left[\mathbb{E}_{\{s_{k,m}\}}[|b_{k,m}s_{k,m}|^2]\right] = \mathbb{E}_m[|b_{k,m}|^2] \leq P_k$ at each WD k.

The average MSE minimization problem with fading channels is formulated as problem (P2) in the following, in which we set $w_m \geq 0$ and $b_k = \tilde{b}_{k,m}\frac{h_{k,m}^\dagger}{|h_{k,m}|}$, $\forall k \in \mathcal{K}, m \in \mathcal{M}$, which are optimal for MSE minimization.

$$\begin{aligned} \text{(P2)}: \min_{\{\tilde{b}_{k,m}\geq 0\},\{w_m\geq 0\}} \ & \mathbb{E}_m\left[\sum_{k=1}^{K}(w_m|h_{k,m}|\tilde{b}_{k,m} - 1)^2 + w_m^2\sigma_z^2\right] \\ \text{s.t.}\ & \mathbb{E}_m[\tilde{b}_{k,m}^2] \leq P_k, \forall k \in \mathcal{K}. \end{aligned}$$

Since problem (P2) satisfies the Slater's condition, strong duality holds between (P2) and its dual problem (Boyd and Vandenberghe, 2004). Therefore, we can solve (P2) by equivalently solving its dual problem.

Theorem 20.2 ***(Optimal Solution to Problem (P2))*** The optimal solution of $\{w_m\}$ to problem (P2), denoted by $\{w_m^{\text{opt}}\}$, is obtained via the bisection search over

$$\sum_{k=1}^{K}\frac{|h_{k,m}|^2\mu_k}{\left((w_m^{\text{opt}})^2|h_{k,m}|^2 + \mu_k^{\text{opt}}\right)^2} = \sigma_z^2, \forall m \in \mathcal{M}, \tag{20.10}$$

and the optimal solution of $\{\tilde{b}_{k,m}\}$, denoted by $\{\tilde{b}_{k,m}^{\text{opt}}\}$, is given by

$$\tilde{b}_{k,m}^{\text{opt}} = \frac{w_m^{\text{opt}}|h_{k,m}|}{(w_m^{\text{opt}})^2|h_{k,m}|^2 + \mu_k^{\text{opt}}}, \forall k \in \mathcal{K}, m \in \mathcal{M}, \tag{20.11}$$

where $\{\mu_k^{\text{opt}} \geq 0\}$ denote the optimal dual variables associated with the transmit power constraints at different WDs.

Proof: See Cao et al. (2020). □

Remark 20.3 ***(Regularized Channel Inversion Power Control):*** The optimal power control $\{\tilde{b}_{k,m}^{\text{opt}}\}$ is observed to follow an interesting regularized channel inversion structure, with denoising factors $\{w_m^{\text{opt}}\}$ and dual variables $\{\mu_k^{\text{opt}}\}$ acting as parameters for regularization. More specifically, it is observed that for any WD $k \in \mathcal{K}$, if $\mu_k^{\text{opt}} > 0$ holds, the average power constraint of WD k must be tight at the optimality due to the complementary slackness condition (i.e., $\mu_k^{\text{opt}}\left(\mathbb{E}_m[\tilde{b}_{k,m}^2] - P_k\right) = 0$, and thus this WD should use up its transmit power budget based on the regularized channel inversion power control over fading states; otherwise, with $\mu_k^{\text{opt}} = 0$, WD k should transmit with channel-inversion power control without using up its power budget.

Remark 20.4 ***(Rayleigh Fading)*** With Rayleigh fading, where the channel coefficient h_k's are independent CSCG random variables with zero mean and variance σ_h^2, it must hold at the optimal solution to problem (P2) that $\mathbb{E}_m[\tilde{b}_{k,m}^2] = P_k$. In other words, the average power constraints must be tight for all the K WDs. Intuitively, this is due to the fact that in this case, deep channel fading may occur over time, and thus sufficiently large transmit power is required for implementing the regularized channel inversion power control.

Figure 20.4 shows the MSE of AirComp versus the receive SNR, where the receive SNRs are equal for $k = 20$ WDs. For comparison, we consider two benchmark schemes, namely the uniform power control and traditional channel inversion. In the uniform power control scheme, each WD transmits with fixed power equal to the expected power budget regardless of the channel state. The traditional channel inversion implements its counterpart for static case separately for each fading state with a uniform instantaneous power constraint set to be its expected value. It is observed that, for all schemes, the MSE decreases as the receive SNR increases. The proposed power control performs consistently well throughout the whole receive SNR regime. It is further observed that the performance gap between the proposed and uniform power control is small in the low-receive-SNR regime while the gap becomes large in the high-receive-SNR regime. This shows the effectiveness of power control optimization, especially in the high-receive-SNR regime. Particularly, when the receive SNR is 20 dB, the proposed power control achieves up to two orders of magnitude AirComp error reduction, as compared to the uniform power control. It is interesting to note that such a phenomenon is in sharp contrast to the power control for rate maximization in conventional wireless communication systems over fading channels (e.g., single-user point-to-point channels), where the adaptive power control (e.g., water-filling) is more crucial in the low to moderate SNR regimes.

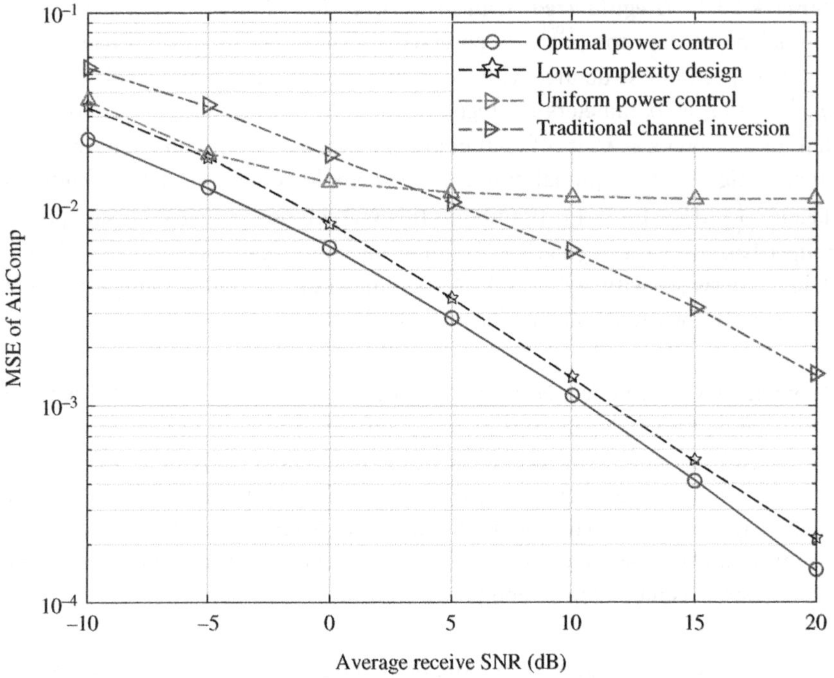

Figure 20.4 The average MSE of AirComp versus the average receive SNR in fading channels with $K = 20$. Source: Cao et al., 2020/IEEE.

20.3.3 Effect of Imperfect CSI

Furthermore, we investigate the effect of imperfect CSI on the performance of AirComp by considering static channels, where the AP is interested in computing the averaging function in (20.1). We assume that the AP only accesses imperfect CSI to coordinate the transceiver design, due to the channel estimation errors. Let $\hat{h}_k$ denote the estimated complex channel coefficient for WD k. Then, we have (Yoo and Goldsmith, 2006)

$$\hat{h}_k = h_k + e_k, \forall k \in \mathcal{K}, \tag{20.12}$$

where e_k denotes the complex channel estimation error that is a CSCG random variable with zero mean and variance $\sigma^2_{e,k}$. Based on the estimated CSI $\{\hat{h}_k\}$, the AP adopts the denoising factor w for data aggregation. Accordingly, the AP recovers the average function from the received signal as

$$\hat{f} = \frac{wy}{K} = \frac{1}{K}\left(\sum_{k=1}^{K} w(\hat{h}_k - e_k)b_k s_k + wz\right). \tag{20.13}$$

The computation MSE between $\hat{f}$ and f is given by

$$\begin{aligned}\text{MSE} &= \mathbb{E}_{\{s_k\},\{e_k\},z}[|\hat{f} - f|^2] \\ &= \frac{1}{K^2}\left(\underbrace{\sum_{k=1}^{K}|w\hat{h}_k b_k - 1|^2}_{\text{Signal misalignment error}} + \underbrace{\sum_{k=1}^{K}|w|^2\sigma_{e,k}^2|b_k|^2}_{\text{CSI-related error}} + \underbrace{|w|^2\sigma_z^2}_{\text{Noise-induced error}}\right),\end{aligned} \tag{20.14}$$

which consists of three terms, including the signal misalignment error, the noise-induced error, and the CSI-related error (due to channel estimation errors). This is different from (20.4) with perfect CSI, where only the first two terms are available.

The computation MSE minimization problem with channel estimation errors is formulated as problem (P3) in the following, where similar to problem (P1), we consider $w \geq 0$ and $b_k = \tilde{b}_k \frac{h_k^\dagger}{|h_k|}, \forall k \in \mathcal{K}$.

$$\begin{aligned}(\text{P3}):\ &\min_{\{\tilde{b}_k\},w\geq 0} \sum_{k=1}^{K}\left((w\hat{h}_k\tilde{b}_k - 1)^2 + w^2\sigma_{e,k}^2\tilde{b}_k^2\right) + w^2\sigma_z^2 \\ &\text{s.t. } 0 \leq \tilde{b}_k \leq \sqrt{P_k}, \forall k \in \mathcal{K}.\end{aligned}$$

To deal with problem (P3) with imperfect CSI, we define $\tilde{\rho}_k \triangleq \sqrt{P_k}\frac{|\hat{h}_k|^2+\sigma_{e,k}^2}{|\hat{h}_k|}$ as the channel quality indicator for WD $k \in \mathcal{K}$. Without loss of generality, we assume that $\tilde{\rho}_1 \leq \cdots \leq \tilde{\rho}_k \leq \cdots \leq \tilde{\rho}_K$, $\tilde{\rho}_0 \to 0$, and $\tilde{\rho}_{K+1} \to \infty$. We also define $F_k(w) \triangleq \sum_{i=1}^{k}((w\sqrt{P_i}|\hat{h}_i| - 1)^2 + w^2P_i\sigma_{e,i}^2) + \sum_{j=k+1}^{K}\frac{\sigma_{e,j}^2}{|\hat{h}_j|^2+\sigma_{e,j}^2} + w^2\sigma_z^2$.

Theorem 20.3 ***(Optimal Solution to Problem (P3))*** The optimal solution of w to problem (P3) is given by

$$w^* = w_{k^*}^*, k^* = \arg\min_{k\in\{0\}\cup\mathcal{K}} F_k(w_k^*), \tag{20.15}$$

where $w_k^* = \max\left(\frac{1}{\tilde{\rho}_{k+1}}, \min\left(\frac{\sum_{i=1}^{k}\sqrt{P_i}|\hat{h}_i|}{\sum_{i=1}^{k}P_i(|\hat{h}_i|^2+\sigma_{e,i}^2)+\sigma_z^2}, \frac{1}{\tilde{\rho}_k}\right)\right)$.

The optimal solution of $\{b_k\}$ to (P3) is obtained as

$$b_k^* = \begin{cases}\sqrt{P_k}\dfrac{\hat{h}_k^\dagger}{|\hat{h}_k|}, k \in \{1, \ldots, k^*\}, \\[2ex] \dfrac{\hat{h}_k^\dagger}{w^*(|\hat{h}_k|^2 + \sigma_{e,k}^2)}, k \in \{k^*+1, \ldots, K\}.\end{cases} \tag{20.16}$$

Proof: See Chen et al. (2022). □

The optimal power control policy has a similar threshold-based channel inversion structure to the case without CSI error in (20.6). The only difference is that the channel inversion here is regularized, where regularization depends on the channel estimation error $\sigma_{e,k}^2$.

It is also interesting to analyze the computation MSE when each WD has asymptotically high transmit power (i.e., $P_k \to \infty, \forall k \in \mathcal{K}$).

Proposition 20.1 When $P_k \to \infty, \forall k \in \mathcal{K}$, it follows that

$$\text{MSE} \to \frac{1}{K^2} \sum_{k=1}^{K} \frac{\sigma_{e,k}^2}{|\hat{h}_k|^2 + \sigma_{e,k}^2}. \tag{20.17}$$

Proof: See Chen et al. (2022). □

Proposition 20.1 shows that due to the existence of channel estimation errors $\{\sigma_{e,k}^2\}$, a nonzero computation MSE becomes inevitable even when the transmit powers at WDs become infinity. This is different from the case with perfect CSI (i.e., $\{\sigma_{e,k}^2 = 0\}$), in which MSE $\to 0$ when $P_k \to \infty, \forall k \in \mathcal{K}$.

Figure 20.5 shows the MSE of AirComp versus the transmit power P, where the transmit power P and the CSI error $\sigma_e^2 = 0.1$ are equal for $k = 20$ WDs. For comparison, we consider three benchmark schemes. Besides the benchmarks with full power transmission and channel inversion power control in Section 1.3.1, we also consider the benchmark ignoring CSI errors, where the AP and WDs optimize the transceiver design via solving problem (P1) (or problem (P3) by ignoring the channel estimation errors). It is observed that the proposed design outperforms the other benchmarks in the whole transmit power regime. In the low power regime (e.g., $P \leq 0$ dB), the full power transmission is observed to perform close to the proposed design, as it can efficiently suppress the noise-induced error that is dominant in MSE in this case. In the high power regime (e.g., $P \geq 15$ dB), the channel inversion power control is observed to perform close to the proposed design, due to the efficient signal magnitude alignment. When P becomes large, the computation MSE achieved by the proposed design is observed to approach the lower bound, as indicated in Proposition 20.1. By contrast, the benchmark ignoring CSI errors is observed to perform the worst and even lead to an increased MSE when P becomes large. This is because the CSI errors are amplified by the high transmit power, hence degrading the AirComp performance.

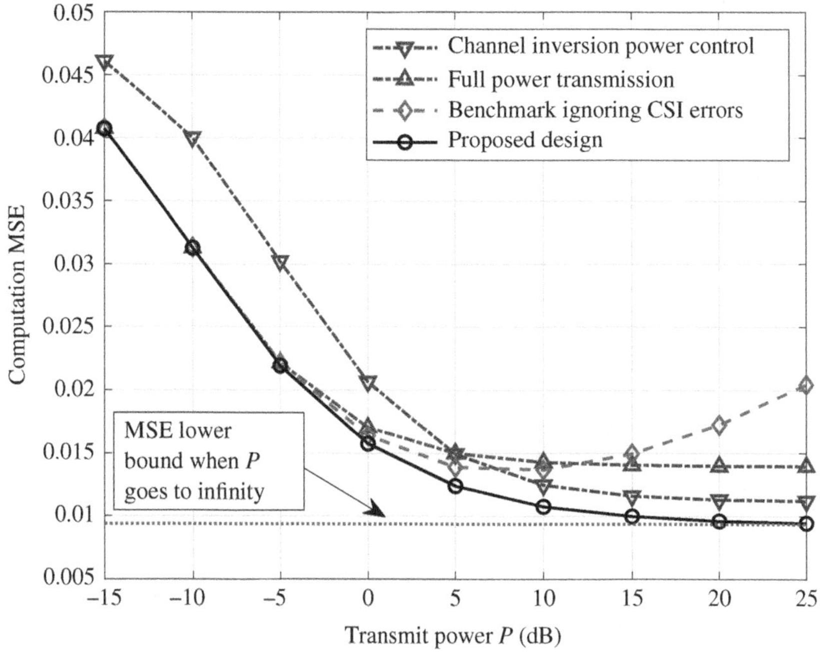

Figure 20.5 The computation MSE versus the transmit power P when $N_r = 1$, $K = 20$, and $\sigma_e^2 = 0.1$. Source: Chen et al., 2022/IEEE.

20.4 Beamforming for AirComp

Multi-antenna techniques are efficient in enhancing the AirComp performance by providing multiplexing diversity and array gains. This section focuses on AirComp in single-input multiple-output (SIMO) and MIMO channels, in which beamforming is employed to further reduce the computation MSE.

20.4.1 SIMO AirComp

We consider an uncoded AirComp system in static SIMO channels under imperfect CSI. It is assumed that the AP is equipped with $N_r \geq 1$ receive antennas and each WD is equipped with one single transmit antenna. The AP is interested in computing the averaging function in (20.1).

We assume that the AP only accesses imperfect CSI to coordinate the transceiver design, due to the channel estimation errors. Let $\boldsymbol{h}_k \in \mathbb{C}^{N_r \times 1}$ and $\hat{\boldsymbol{h}}_k \in \mathbb{C}^{N_r \times 1}$ denote the channel vector and the estimated channel vector from WD $k \in \mathcal{K}$ to the AP. Then, we have

$$\hat{\boldsymbol{h}}_k = \boldsymbol{h}_k + \boldsymbol{e}_k, \forall k \in \mathcal{K}, \tag{20.18}$$

where $\boldsymbol{e}_k \in \mathbb{C}^{N_r \times 1}$ denotes the channel estimation error that is a CSCG random vector with zero mean and covariance $\sigma_{e,k}^2 \boldsymbol{I}$.

Based on the estimated CSI $\{\hat{\boldsymbol{h}}_k\}$, the AP adopts the receive beamforming vector $\boldsymbol{w} \in \mathbb{C}^{N_r \times 1}$ for data aggregation. Accordingly, the AP recovers the average function from the received signal as

$$\hat{f} = \frac{\boldsymbol{w}^H \boldsymbol{y}}{K} = \frac{1}{K}\left(\sum_{k=1}^{K} \boldsymbol{w}^H(\hat{\boldsymbol{h}}_k - \boldsymbol{e}_k) b_k s_k + \boldsymbol{w}^H \boldsymbol{z}\right), \tag{20.19}$$

where $\boldsymbol{z} \in \mathbb{C}^{N_r \times 1}$ denotes the AWGN at the AP that is a CSCG random vector with zero mean and covariance $\sigma_z^2 \boldsymbol{I}$.

The computation MSE between $\hat{f}$ and f is given by

$$\begin{aligned} \text{MSE} &= \mathbb{E}_{\{s_k\},\{\boldsymbol{e}_k\},\boldsymbol{z}}[|\hat{f} - f|^2] \\ &= \frac{1}{K^2}\left(\sum_{k=1}^{K}(|\boldsymbol{w}^H \hat{\boldsymbol{h}}_k b_k - 1|^2 + \|\boldsymbol{w}\|^2 \sigma_{e,k}^2 |b_k|^2) + \|\boldsymbol{w}\|^2 \sigma_z^2\right). \end{aligned} \tag{20.20}$$

Our objective is to minimize MSE in (20.20) by jointly optimizing the transmit coefficients $\{b_k\}$ at the WDs and the receive beamforming vector $\boldsymbol{w}$ at the AP, subject to the individual power budgets at the WDs (i.e., $\mathbb{E}_{\{s_k\}}[|b_k s_k|^2] = |b_k|^2 \le P_k, \forall k \in \mathcal{K}$). The optimization problem is formulated as

$$\begin{aligned} \text{(P4)}: \min_{\{b_k\},\boldsymbol{w}} & \sum_{k=1}^{K}\left(|\boldsymbol{w}^H \hat{\boldsymbol{h}}_k b_k - 1|^2 + \|\boldsymbol{w}\|^2 \sigma_{e,k}^2 |b_k|^2\right) + \|\boldsymbol{w}\|^2 \sigma_z^2 \\ \text{s.t.} & \ |b_k|^2 \le P_k, \forall k \in \mathcal{K}. \end{aligned}$$

To deal with the coupling of transmit coefficients $\{b_k\}$ and receive combining vector $\boldsymbol{w}$ in this case, we propose an efficient solution based on alternating optimization, where $\{b_k\}$ and $\boldsymbol{w}$ are updated alternately with the other given.

Theorem 20.4 For problem (P4) with given $\boldsymbol{w}$, the optimal solution of $\{b_k\}$ is given by

$$b_k^\star = \tilde{b}_k^\star \frac{\hat{\boldsymbol{h}}_k^H \boldsymbol{w}}{|\boldsymbol{w}^H \hat{\boldsymbol{h}}_k|}, \tilde{b}_k^\star = \min\left(\sqrt{P_k}, \frac{|\boldsymbol{w}^H \hat{\boldsymbol{h}}_k|}{|\boldsymbol{w}^H \hat{\boldsymbol{h}}_k|^2 + \|\boldsymbol{w}\|^2 \sigma_{e,k}^2}\right), \forall k \in \mathcal{K}. \tag{20.21}$$

Proof: See Chen et al. (2022). □

The optimized transmit coefficient or equivalently power control solution in (20.21) is observed to follow a similar threshold-based regularized channel inversion power control as in (20.16). For the WDs with sufficiently large transmit power and/or good channel quality (by viewing $|\boldsymbol{w}^H \hat{\boldsymbol{h}}_k|^2$ as the equivalent channel power gain), their transmit powers follow a regularized channel inversion structure; for the other WDs with limited transmit power and/or poor channel quality, the full power transmission is adopted.

Theorem 20.5 For problem (P4) with given $\{b_k\}$, the optimal solution of $\boldsymbol{w}$ is given by

$$\boldsymbol{w}^\star = \left(\sum_{k=1}^{K} |b_k|^2 (\hat{\boldsymbol{h}}_k \hat{\boldsymbol{h}}_k^H + \sigma_{e,k}^2 \boldsymbol{I}) + \sigma_z^2 \boldsymbol{I} \right)^{-1} \sum_{k=1}^{K} \hat{\boldsymbol{h}}_k b_k. \tag{20.22}$$

Proof: See Chen et al. (2022). □

The optimized receive beamforming solution in (20.22) is observed to have a sum minimum mean squared error (MMSE) structure. This is in order to better aggregate the signals from all the WDs to facilitate the functional computation.

The alternating-optimization-based algorithm for solving (P4) is implemented in an iterative manner. In each iteration, we first obtain the transmit coefficients as $\{b_k^\star\}$ in (20.21) under given $\boldsymbol{w}$, and then update the receive beamforming vector as $\boldsymbol{w}^\star$ based on (20.22) under given $\{b_k\}$. Notice that in each iteration, the optimality is achieved, and as a result, the updated computation MSE is ensured to be monotonically nonincreasing. As the computation MSE in (P4) is lower bounded, the convergence of our proposed alternating-optimization-based algorithm can be guaranteed.

It is interesting to discuss the computation MSE in the cases with sufficiently high transmit powers or a massive number of receive antennas, for which we have the following two propositions, whose proofs can be referred to Chen et al. (2022).

Proposition 20.2 Under any given receive beamforming vector $\boldsymbol{w}$, if $P_k \to \infty$, $\forall k \in \mathcal{K}$, then we have

$$\text{MSE} \to \frac{1}{K^2} \sum_{k=1}^{K} \frac{\|\boldsymbol{w}\|^2 \sigma_{e,k}^2}{|\boldsymbol{w}^H \hat{\boldsymbol{h}}_k|^2 + \|\boldsymbol{w}\|^2 \sigma_{e,k}^2} \geq \frac{1}{K^2} \sum_{k=1}^{K} \frac{\sigma_{e,k}^2}{\|\hat{\boldsymbol{h}}_k\|^2 + \sigma_{e,k}^2}. \tag{20.23}$$

It follows from Proposition 20.2 that with a finite number of receive antennas, a nonzero MSE becomes inevitable even when the WDs employ extremely high transmit powers.

Proposition 20.3 If $N_r \to \infty$ and $\boldsymbol{h}_i$'s (and equivalently $\hat{\boldsymbol{h}}_i$'s) are independent and identically distributed (IID) random vectors, then we have MSE $\to 0$.

Proposition 20.3 shows that increasing the number of receive antennas is efficient to combat against the imperfect CSI, thus showing the benefit of massive antennas in AirComp.

Figure 20.6 shows the MSE of AirComp versus the transmit power P for the SIMO case with $N_r = 10$, where the transmit power P and the CSI error $\sigma_e^2 = 0.1$ are equal for $k = 20$ WDs. It is observed that the proposed design outperforms the other benchmarks in the whole transmit power regime. In the low power regime, the full power transmission is observed to perform close to the proposed design, while in the high power regime, the channel inversion power control is observed to perform close to the proposed design. When P becomes large, the computation MSE achieved by the proposed design is observed to approach the lower bound, as indicated in Proposition 20.2. By contrast, the benchmark ignoring CSI errors is observed to perform the worst and even leads to an increased MSE when P becomes large.

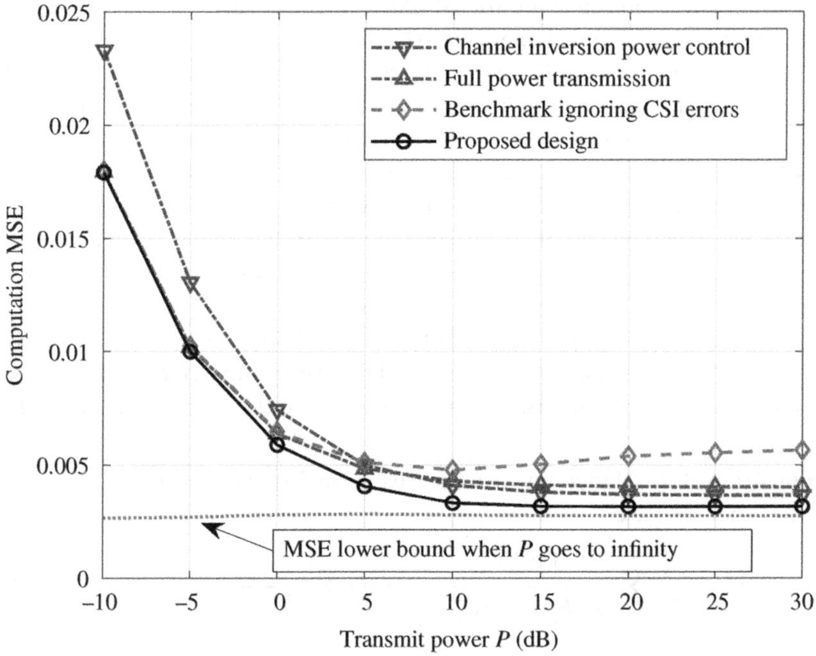

Figure 20.6 The computation MSE versus the transmit power P when $N_r = 10$, $K = 20$, and $\sigma_e^2 = 0.1$. Source: Chen et al., 2022/IEEE.

20.4.2 MIMO AriComp

Furthermore, we consider the AirComp system in static MIMO channels, by assuming perfect CSI for each decryption. It is assumed that the AP is equipped with $N_r \geq 1$ receive antennas and each WD is equipped with $N_t \geq 1$ transmit antennas. Let $\boldsymbol{s}_k \in \mathbb{C}^{L\times 1}$ denote the transmitted message by WD $k \in \mathcal{K}$ consisting of L heterogeneous time-varying parameters of the environment, e.g., temperature, pollution, and humidity, where $\boldsymbol{s}_k$'s are independent complex random variables with zero mean and covariance $\boldsymbol{I}$. The AP aims at computing the L averaging functions of L corresponding types of data of $\boldsymbol{s}_k$'s, i.e.,

$$\boldsymbol{f} = \frac{1}{K}\sum_{k=1}^{K} \boldsymbol{s}_k. \tag{20.24}$$

Let $\boldsymbol{H}_k \in \mathbb{C}^{N_r\times N_t}$ denote the channel matrix from WD $k \in \mathcal{K}$ to the AP, $\boldsymbol{B}_k \in \mathbb{C}^{N_t\times L}$ denote the transmit beamforming matrix at WD k, and $\boldsymbol{W} \in \mathbb{C}^{N_r\times L}$ denote the receive beamforming matrix at the AP. Accordingly, the AP recovers the average function from the received signal as

$$\hat{\boldsymbol{f}} = \frac{\boldsymbol{W}^H\boldsymbol{y}}{K} = \frac{1}{K}\left(\sum_{k=1}^{K} \boldsymbol{W}^H\boldsymbol{H}_k\boldsymbol{B}_k\boldsymbol{s}_k + \boldsymbol{W}^H\boldsymbol{z}\right), \tag{20.25}$$

where $\boldsymbol{z} \in \mathbb{C}^{N_r\times 1}$ denotes the AWGN at the AP that is a CSCG random vector with zero mean and covariance $\sigma_z^2\boldsymbol{I}$.

The computation MSE between $\hat{\boldsymbol{f}}$ and $\boldsymbol{f}$ is given by

$$\begin{aligned}\text{MSE} &= \mathbb{E}_{\{\boldsymbol{s}_k\},\boldsymbol{z}}[\|\hat{\boldsymbol{f}}_m - \boldsymbol{f}_m\|^2] \\ &= \frac{1}{K^2}\left(\sum_{k=1}^{K} \text{tr}(\boldsymbol{W}^H\boldsymbol{H}_k\boldsymbol{B}_k - \boldsymbol{I})(\boldsymbol{W}^H\boldsymbol{H}_k\boldsymbol{B}_k - \boldsymbol{I})^H + \sigma_z^2\text{tr}\boldsymbol{W}^H\boldsymbol{W}\right).\end{aligned} \tag{20.26}$$

Our objective is to minimize MSE in (20.26) by jointly optimizing the transmit beamforming matrices $\{\boldsymbol{B}_k\}$ at the WDs and the receive beamforming matrix $\boldsymbol{W}$ at the AP, subject to the individual power budgets at the WDs (i.e., $\mathbb{E}_{\{\boldsymbol{s}_k\}}[\|\boldsymbol{B}_k\boldsymbol{s}_k\|^2] = \|\boldsymbol{B}_k\|_F^2 \leq P_k, \forall k \in \mathcal{K}$). Furthermore, we can write $\boldsymbol{W} = w\boldsymbol{V}$, where w is a nonnegative denoising factor, and $\boldsymbol{V}$ is a tall unitary matrix.

The computation MSE minimization problem is formulated as problem (P5) in the following.

$$\begin{aligned}\text{(P5)}:\ \min_{\{\boldsymbol{B}_k\},\boldsymbol{W}}\ &\sum_{k=1}^{K} \text{tr}(\boldsymbol{W}^H\boldsymbol{H}_k\boldsymbol{B}_k - \boldsymbol{I})(\boldsymbol{W}^H\boldsymbol{H}_k\boldsymbol{B}_k - \boldsymbol{I})^H + \sigma_z^2\text{tr}\boldsymbol{W}^H\boldsymbol{W} \\ \text{s.t.}\ &\|\boldsymbol{B}_k\|_F^2 \leq P_k, \forall k \in \mathcal{K}, \\ &\boldsymbol{W}^H\boldsymbol{W} = w^2\boldsymbol{I},\end{aligned}$$

which is nonconvex due to the coupling between transmit and receive beamformers in the objective function, and the orthogonality constraint on the receive beamformer.

While problem (P5) is more difficult to solve, the authors in Zhu and Huang (2019) developed an efficient solution based on zero-forcing transmit beamforming and an efficient receive beamforming based on differential geometry. The solution reveals that the optimal receive beamformer can be approximated by the weighted centroids of a cluster of points on a Grassmann manifold, each corresponding to the subspace of an individual MIMO channel. In this case, both multiplexing and array gains of MIMO are exploited to enhance the computation rate (with functions/symbols) and reduce the computation error.

20.5 Extension

20.5.1 Multicell AirComp Networks

In next-generation IoT, the relevance of AirComp to different types of applications and its being a promising low-latency solution suggest the need to consider its large-scale deployment in a multicell network. Multicell AirComp can be implemented in two modes, namely hierarchical AirComp and coordinated AirComp.

- *Hierarchical AirComp via relaying*: In hierarchical AirComp, a global server further aggregates AirComp results output by local servers through backhaul links to scale up the aggregation gain, e.g., training a larger model or exploiting data from a large-scale network in the context of distributed edge learning. As an attempt, a new hierarchical AirComp network over a large area has been investigated in Wang et al. (2022), in which a set of intermediate amplify-and-forward (AF) relays are exploited to facilitate the massive data aggregation from a large number of WDs.
- *Coordinated AirComp*: On the other hand, coordinated AirComp aims to support coexisting WDA tasks in different cells, each of which is characterized by its application, data type, and target function. The coexisting tasks in coordinated AirComp are exposed to inter-cell interference. This calls for interference management by multicell coordination to balance the errors in the coexisting tasks. While multicell AirComp is an open area, an initial attempt has been made in Cao et al. (2021) to understand the performance limit of coordinated AirComp by quantifying the Pareto boundary of the multicell MSE region.

20.5.2 Intelligent Reflecting Surface (IRS)-Assisted AirComp

IRS has emerged as another key technology toward 6G, which involves the use of a metasurface for manipulating the wireless propagation environment to enhance the system performance. IRS consists of a number of reflecting elements whose phases and amplitudes can be adjusted to form reflective beamformers to enhance the desired signal strength and mitigate the undesired interference. In particular, IRS can be exploited to increase the signal coverage of AirComp networks, and also enhance the signal strength of worst-case WDs to minimize signal misalignment errors for reducing the computation errors. The joint optimization of power control at WDs, reflective beamforming at IRSs, and the receive processing at the AP is a new design issue to be tackled for the IRS-enabled AirComp networks.

20.5.3 Unmanned Aerial Vehicle (UAV)-Enabled AirComp

The AirComp performance is limited by the worst-case WD in the poorest channel condition, thus making the large-scale implementation challenging. Employing unmanned aerial vehicle (UAV)s as new aerial APs is becoming a new solution to enable large-scale AirComp. In the UAV-enabled AirComp, the UAV trajectory design is becoming a new design degree of freedom (DoF) for optimizing the AirComp performance, together with the conventional joint transmit and receive design. In particular, UAVs can first fly around the distributed WDs on the ground to collect their data, and thus enhance the worst-case channels of these WDs. Furthermore, the WDs can be separated into different groups, such that AirComp can be implemented over each group, and then aggregated together. There may exist a trade-off between computation efficiency and latency.

20.5.4 Over-the-Air FEEL (Air-FEEL)

Recent years have witnessed the spreading of AI algorithms from the cloud to the network edge, resulting in an active area called distributed edge learning. Among others, federated edge learning (FEEL) is particularly appealing due to its privacy-preserving feature, and thus has been envisioned as one of the candidate 6G techniques to enable ubiquitous brain-inspired intelligence and abundant applications such as autonomous driving, intelligent health, and industrial IoT.

In particular, FEEL corresponds to implementing distributed stochastic gradient descent (SGD) over wireless networks, which allows distributed edge devices to collaboratively train a shared AI model by using their local data. Due to the high dimensionality of each model/gradient update (usually comprising millions to billions of parameters), the frequent communication (for model/gradient uploading) and computation (for aggregation) from the edge devices to the edge server become the performance bottleneck for FEEL, especially when the number of edge devices sharing the same wireless medium becomes large.

Due to the benefit introduced by the communication-computation integration, AirComp has found successful applications in FEEL to motivate the so-called over-the-air FEEL (Zhu et al., 2021a), in which multiple edge devices are allowed to transmit their local model/gradient updates concurrently, such that the edge server can exploit the AirComp for "one-shot" aggregation. As compared to conventional FEEL with digital OMA or NOMA, the Air-FEEL is expected to significantly reduce the communication and computation latencies for model uploading and aggregation, thus enhancing the training efficiency. Power control can also be exploited to enhance the efficiency of Air-FEEL to combat against the AirComp error (Cao et al., 2022a,2022b).

20.6 Conclusion

This chapter introduced AirComp as a potentially scalable solution for massive IoT. Breaking from the classic design principle of isolating communication from its applications, AirComp explores a new task-oriented design approach to boost spectrum efficiency and reduce multiple-access latency for massive IoT via seamless integration of sensing, computation, control, and AI. In particular, this chapter introduced the power control and beamforming techniques for AirComp systems with single-antenna and multi-antenna setups, respectively, and discussed advanced AirComp techniques (e.g., hierarchical AirComp and coordinated AirComp) and applications related to other emerging techniques (e.g., IRS, UAV, and FEEL) for future works.

References

Omid Abari, Hariharan Rahul, Dina Katabi, and Mondira Pant. AirShare: Distributed coherent transmission made seamless. In *2015 IEEE Conference on Computer Communications (INFOCOM)*, pages 1742–1750, Apr. 2015. doi: 10.1109/INFOCOM.2015.7218555.

Omid Abari, Hariharan Rahul, and Dina Katabi. Over-the-air function computation in sensor networks. *arXiv:1612.02307*, Dec. 2016.

Stephen Boyd and Lieven Vandenberghe. *Convex Optimization*. Cambridge University Press, 2004.

R Creighton Buck. Approximate complexity and functional representation. Technical report. Wisconsin University Madison Mathematics Research Center, 1976.

Xiaowen Cao, Guangxu Zhu, Jie Xu, and Kaibin Huang. Optimized power control for over-the-air computation in fading channels. *IEEE Transactions on Wireless Communications*, 19(11):7498–7513, Nov. 2020. ISSN 1558-2248. doi: 10.1109/TWC.2020.3012287.

Xiaowen Cao, Guangxu Zhu, Jie Xu, and Kaibin Huang. Cooperative interference management for over-the-air computation networks. *IEEE Transactions on Wireless Communications*, 20(4):2634–2651, Apr. 2021. ISSN 1558-2248. doi: 10.1109/TWC.2020.3043787.

Xiaowen Cao, Guangxu Zhu, Jie Xu, and Shuguang Cui. Transmission power control for over-the-air federated averaging at network edge. *IEEE Journal on Selected Areas in Communications*, 40(5):1571–1586, May 2022a. ISSN 1558-0008. doi: 10.1109/JSAC.2022.3143217.

Xiaowen Cao, Guangxu Zhu, Jie Xu, Zhiqin Wang, and Shuguang Cui. Optimized power control design for over-the-air federated edge learning. *IEEE Journal on Selected Areas in Communications*, 40(1):342–358, Jan. 2022b. ISSN 1558-0008. doi: 10.1109/JSAC.2021.3126060.

Yilong Chen, Guangxu Zhu, and Jie Xu. Over-the-air computation with imperfect channel state information. In *2022 IEEE 23rd International Workshop on Signal Processing Advances in Wireless Communication (SPAWC)*, pages 1–5, Jul. 2022. doi: 10.1109/SPAWC51304.2022.9834026.

Zhiguo Ding, Yuanwei Liu, Jinho Choi, Qi Sun, Maged Elkashlan, I Chih-Lin, and H Vincent Poor. Application of non-orthogonal multiple access in LTE and 5G networks. *IEEE Communications Magazine*, 55(2):185–191, Feb. 2017. ISSN 1558-1896. doi: 10.1109/MCOM.2017.1500657CM.

Michael Gastpar. Uncoded transmission is exactly optimal for a simple Gaussian "sensor" network. *IEEE Transactions on Information Theory*, 54(11):5247–5251, Nov. 2008. ISSN 1557-9654. doi: 10.1109/TIT.2008.929967.

Arunabha Ghosh, Jun Zhang, Jeffrey G Andrews, and Rias Muhamed. *Fundamentals of LTE*. Pearson Education, 2010.

Khaled B Letaief, Wei Chen, Yuanming Shi, Jun Zhang, and Ying-Jun Angela Zhang. The roadmap to 6G: AI empowered wireless networks. *IEEE Communications Magazine*, 57(8):84–90, Aug. 2019. ISSN 1558-1896. doi: 10.1109/MCOM.2019.1900271.

Feng Wang, Jie Xu, Vincent K N Lau, and Shuguang Cui. Amplify-and-forward relaying for hierarchical over-the-air computation. *IEEE Transactions on Wireless Communications*, 21(12):10529–10543, Dec. 2022. ISSN 1558-2248. doi: 10.1109/TWC.2022.3185074.

Taesang Yoo and A Goldsmith. Capacity and power allocation for fading MIMO channels with channel estimation error. *IEEE Transactions on Information Theory*, 52(5):2203–2214, May 2006. ISSN 1557-9654. doi: 10.1109/TIT.2006.872984.

Shengli Zhang, Soung Chang Liew, and Patrick P Lam. Hot topic: Physical-layer network coding. In *Proceedings of the 12th Annual International Conference on Mobile Computing and Networking*, MobiCom '06, page 358–365. New York, NY, USA: Association for Computing Machinery, 2006. ISBN 1595932860. doi: 10.1145/1161089.1161129.

Guangxu Zhu and Kaibin Huang. MIMO over-the-air computation for high-mobility multimodal sensing. *IEEE Internet of Things Journal*, 6(4):6089–6103, Aug. 2019. ISSN 2327-4662. doi: 10.1109/JIOT.2018.2871070.

Guangxu Zhu, Yuqing Du, Deniz Gündüz, and Kaibin Huang. One-bit over-the-air aggregation for communication-efficient federated edge learning: Design and convergence analysis. *IEEE Transactions on Wireless Communications*, 20(3):2120–2135, Mar. 2021a. ISSN 1558-2248. doi: 10.1109/TWC.2020.3039309.

Guangxu Zhu, Jie Xu, Kaibin Huang, and Shuguang Cui. Over-the-air computing for wireless data aggregation in massive IoT. *IEEE Wireless Communications*, 28(4):57–65, Aug. 2021b. ISSN 1558-0687. doi: 10.1109/MWC.011.2000467.

21

Multi-Dimensional Multiple Access for 6G: Efficient Radio Resource Utilization and Value-Oriented Service Provisioning

Wudan Han, Jie Mei, and Xianbin Wang

Department of Electrical and Computer Engineering, Western University, London, Ontario, Canada

21.1 Introduction

Multiple access (MA) techniques in wireless communications provide the ability to share a common communication channel or radio resource pool for supporting network access of coexisting users and devices simultaneously. In their critical roles, MA techniques have fundamental impacts to the radio resource utilization, quality-of-service (QoS) provisioning, and overall network operation performance.

In the era of 6G, the rapid proliferation of wireless devices, the ever-growing wireless applications, and their diverse QoS requirements have brought many new design challenges for next-generation multiple access (NGMA) techniques. These challenges are primarily focused on achieving individualized tailored QoS provisioning with limited but extremely complex radio resources, while considering overall network operation performance.

Specifically, the use of large-scale antenna arrays, mmWave bands, and non-orthogonal radio resource utilization techniques has increased both the dimensionality and heterogeneity of radio resource in future wireless networks. MA techniques for 6G hinge on highly accurate and efficient multi-dimensional resource sharing among coexisting devices and users as well as effective network orchestration to fully utilize scarce resources including time, frequency, power, space, and code domains.

In this chapter, a novel multiple access technique for future generation of wireless networks, termed multi-dimensional multiple access (MDMA), is presented based on a series of publications from the authors of Liu et al. (2021a), Liu et al. (2021b), Han et al. (2021), Mei et al. (2022), and Wang et al. (2023). The overall goal of MDMA is to create a more flexible and intelligent multiple access technique to fully utilize the heterogeneous radio resources in different domains and

Next Generation Multiple Access, First Edition.
Edited by Yuanwei Liu, Liang Liu, Zhiguo Ding, and Xuemin Shen.
© 2024 The Institute of Electrical and Electronics Engineers, Inc. Published 2024 by John Wiley & Sons, Inc.

maximize the network operational objectives, which include the individualized QoS requirements of each user, the overall network operational performance, and eventually the value-realization of the network.

21.1.1 Difficulties of Existing Multiple Access Techniques

In the rapidly evolving landscape of wireless communications, existing MA techniques struggle to keep pace with the demands of an increasingly connected world. The changing operating frequencies, bandwidth, signal processing capabilities and QoS requirements have fundamentally influenced MA design.

In retrospect, orthogonal MA schemes, such as frequency division multiple access (FDMA), time division multiple access (TDMA), and code division multiple access (CDMA), have empowered 1G to 3G in the practical implementation and commercialization by exploiting single-domain degree of freedom (DoF). However, due to the rigid single-domain resource division, these MA techniques have inherent shortages in supporting dynamic data traffic with diverse QoS requirements. The introduction of orthogonal frequency division multiple access (OFDMA) in 4G long-term evolution (LTE) and IEEE 802.11ax (Wi-Fi 6) expanded MA design into both frequency and time domains, leading to improved resource utilization flexibility and capability of supporting more diverse date traffic (Khorov et al., 2020). Moreover, the multi-antenna aided multiple input multiple output (MIMO) technique has further enhanced the spatial-domain multiplexing and network capacity. With the advancement of signal processing and hardware capabilities, non-orthogonal multiple access (NOMA) has been under intensive research and development to further boost the spectral efficiency and massive connectivity. Despite the historical successes of these MA schemes, there is an urgent need for designing more flexible, intelligent, and objective-driven MA schemes for 6G to comprehensively address following critical issues.

21.1.1.1 Lack of Diverse and Individualized Service Provisioning Capabilities

As future networks are envisioned as a multipurpose platform for diverse applications and services, individualized service provisioning becomes indispensable to NGMA schemes in order to meet user-specific performance requirements. These requirements not only encompass data rate, latency, reliability, and efficiency, but also vary significantly depending on user preferences, device capabilities, and resource constraints. Specifically, wireless devices possess different hardware capabilities (i.e., signal processing and computing capacities, storage limitations, and power/battery supplies), incurring diverse radio resource utilization costs while employing different MA schemes. By incorporating individualized service provisioning as design goals, more efficient resource utilization and better user experience can be achieved with the new MDMA.

21.1.1.2 Lack of Flexibility and Adaptability in Coping with Heterogeneous Network Scenarios

Current 5G networks devise scenario-specific paradigms, including enhanced mobile broadband (eMBB), ultra-reliable, and low-latency communications (uRLLC), and massive machine-type communications (mMTC) to guide the network operation and QoS provisioning (Sharma and Wang, 2020). Each scenario has its resource allocation principles, priorities, and optimization goals in terms of MA design. For example, eMBB targets high throughput and spectral efficiency, whereas mMTC emphasizes energy efficiency and massive connectivity. Though this classification greatly simplifies the 5G network operation, ultimately, each service and application has unique set of requirements and many emerging applications won't fit exactly into these categories. Moreover, in circumstances of coexisting service classes with competing needs, more sophisticated MA schemes are required, along with advanced interference management and situation-aware decision-makings. Therefore, future networks call for more flexible and situation-aware MA schemes to address these evolving requirements.

21.1.1.3 Lack of Opportunistic Resource Orchestration and Utilization Capabilities

Opportunistic resource orchestration and utilization are essential in achieving optimal network performance and QoS provisioning for each user with constrained radio resource. Resource orchestration in MA design involves resource partitioning and multiplexing. The former allows wireless systems to gain finer resource granularity to effectively improve multitasking performance and network capacity, while the latter can be leveraged to exploit multiuser diversity to increase resource utilization efficiency. However, conventional MA schemes usually adopt static and predetermined patterns of resource orchestration, which failed to accommodate the dynamic nature of wireless networks such as user mobility and fluctuating traffic demands. For example, for a user in a TDMA system with no data to transmit during its assigned time slot, the time resource remains underutilized, leading to performance degradation at system level. Furthermore, resources can be classified as replenishable and non-replenishable categories, which helps the network adopt different strategies to opportunistically exploit them. For resources that cannot be restored or regenerated once used up, MA schemes must make intelligent and situation-aware scheduling decisions to ensure efficient resource utilization.

To overcome above challenges, there have been research efforts that focus on unifying multi-dimensional resource utilization in a non-orthogonal paradigm. Liu et al. (2020) shows the hybrid MA method by engaging both NOMA and orthogonal multiple access (OMA) can effectively balance the massive access and energy efficiency requirements of MTC devices. On the other hand, multi-antenna aided NOMA scheme (i.e., MIMO–NOMA) provides a generalized platform to jointly utilizes spatial division multiple access (SDMA) and power-domain NOMA (Liu et al., 2018), (Chen et al., 2016), (Choi, 2016), (Ali et al., 2017), where the optimal beamforming, modulation, and successive interference cancelation (SIC) decoding order designs still remain as open issues. On a parallel pathway, rate-splitting (RSMA) treats SDMA and NOMA as two special cases by utilizing linearly precoded rate-splitting at the transmitter and SIC at the receivers (Clerckx et al., 2021) and (Mao et al., 2022). Additionally, multi-carrier NOMA is another promising direction (Liu et al., 2021a) and (Han et al., 2021), which combines the principles of NOMA with multi-carrier transmission techniques (e.g., OFDM). These works reveal the dominant trend of improving MA techniques by multi-dimensional resource orchestration. However, most designs do not adequately address the aforementioned limitations, which hinders their ability to cater to ever-evolving requirements of both network operators and end-users for 6G networks.

21.1.2 Embracing 6G with Multi-Dimensional Multiple Access

MDMA technique provides a promising solution by harnessing the capability of intelligent multi-dimensional resource allocation to revolutionize the path to hyperconnected 6G networks. By employing a comprehensive approach of resource orchestration across multiple domains, including time, frequency, space, and power, MDMA aims to overcome the challenges posed by heterogeneous service requirements, dynamic network situations, and increasingly constrained radio resources. More precisely, MDMA simultaneously leverages advanced signal processing, beamforming, and adaptive resource allocation strategies with the following objectives and principles:

Firstly, MDMA consolidates different existing MA schemes in a generalized non-orthogonal paradigm, including OMA as a special case, as well as power-domain and spatial-domain NOMA, to empower more suitable MA schemes based on users' specific requirements and network conditions. Due to hardware impairment and communication channel impact, conventional orthogonal MA schemes inevitably exhibit certain degrees of non-orthogonality in different resource domains. The joint consideration of multi-domain resource utilization and channel access leads to the opportunity of optimizing QoS provisioning and network operation. To meet the diverse and individualized QoS demands stemming from various application domains, MDMA provides a generalized platform accommodating different application scenarios for future 6G communications. Moreover, by taking into account of user-specific situation-dependent resource utilization cost, MDMA delivers cost-gain aware QoS satisfactions, ensuring that each user's unique demands in terms of latency, data rate, reliability, and efficiency are effectively met.

Secondly, MDMA does not rely on fixed and predetermined frameworks as adopted by RSMA and NOMA, but adapts itself flexibly to heterogeneous network scenarios by dynamically adjusting resource allocation and optimization strategies. MDMA further empowers the service provider to exploit multiuser diversity and intelligently multiplex users across frequency, time, power, and space domains in any combinations. This innovative approach allows MDMA to concurrently utilize as many resource domains as possible, ensuring robust performance even in highly overloaded situations. As such, MDMA presents a flexible and scenario-adaptive solution capable of accommodating stringent demands of emerging service types and applications for 6G networks.

Lastly, MDMA not only considers the varying availability and characteristics of different resources, but also assesses the implications resulted from the resource utilization process in order to incorporate with more intelligent designs, such as the advanced interference mitigation schemes, power control strategies, and transmission scheduling/prioritization mechanisms. Therefore, MDMA can opportunistically and efficiently allocate resources with situation awareness by performing timely evaluations of current resource conditions, user QoS requirements, and value-realization goals at network level.

The rest of this chapter is organized as follows: Section 21.2 introduces the design principle and methodology for proposed MDMA scheme. Section 21.3 elaborates on value-oriented implementation of MDMA with a dual-perspective approach, including both individual and system level performance optimization strategies. Section 21.4 presents the proposed multi-dimensional resource allocation solutions. Section 21.5 provides simulation results to evaluate and validate proposed MDMA schemes and finally, Section 21.6 concludes this chapter.

21.2 Principle of MDMA

The principle of MDMA design revolves around establishing a unified and flexible framework that integrates different multiple access schemes in a generalized paradigm, including OMA, power-domain, and spatial-domain NOMA. This comprehensive approach enables MDMA to intelligently multiplexes users in frequency, time, power, and space domains with any resource combinations. By fully utilizing the available multi-dimensional resources, MDMA can accommodate diverse QoS requirements from different application scenarios for 6G communications.

In this section, we introduce a generalized system model to explain the key concepts and methodologies in MDMA design. As shown in Fig. 21.1, we consider a downlink scenario in a single-cell cellular network, where a base

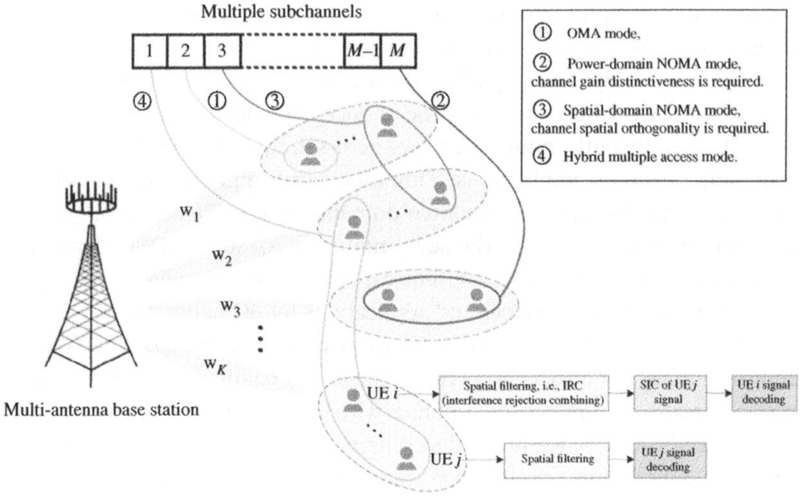

Figure 21.1 Illustration of MDMA scheme via multi-dimensional multiplexing for heterogeneous coexisting users: (1) basic OMA mode; (2) and (3) are power-domain and spatial-domain NOMA modes, respectively; (4) hybrid multiple access mode.

station (BS) equipped with N_t transmit antennas is deployed at the cell center, and serves a set of $\mathcal{K}$ of K single receiving-antenna user equipments (UEs). The total available bandwidth W is divided into M orthogonal subchannels (SCs), allowing for independent modulation and parallel data transmission of each SC. In principle, MDMA allows BS to adaptively choose suitable multiple access modes for each UE according to their resource conditions, constraints, and diverse QoS requirements. By leveraging multiuser diversity and maximizing the separation distance among the shared resource blocks allocated to heterogeneous users, MDMA can reduce or even avoid non-orthogonal interference among coexisting users and support them by multi-dimensional resource allocation strategies.

21.2.1 Core Concepts and Mechanisms of Achieving MDMA

First of all, the proposed MDMA scheme empowers BS to flexibly serve each UE by suitable and beneficial multiple access mode based on their service requirements. To determine the MA mode in the multi-dimensional multiplexing of MDMA, it is essential to evaluate UEs' disparate multi-dimensional resource constraints and the mutual resource-sharing conflicts. As an example, there could be four possible MA modes for each SC, namely, *OMA* mode, *power-domain NOMA* mode, *spatial-domain NOMA* mode, and *hybrid MA* mode, with each MA mode multiplexing UEs in different resource domains. If there is only one UE monopolizing the SC, then we consider *OMA* mode is used. If more than one UEs are multiplexed by the SC and distinguished by their effective channel gain difference only, then *power-domain NOMA* mode is used; Otherwise, the coexisting UEs on the same SC are differentiated by their spatial orthogonality only, we regard it as *spatial-domain NOMA*. Lastly, if UEs on the SC utilize both spatial and power domain NOMA, we view it as *hybrid MA* mode.

To maximize the separation distance among resource-sharing UEs, as well as to reduce the overall non-orthogonality, there are several designing guidelines for MDMA. Firstly, to leverage the multi-antenna techniques of BS, the spatial domain is coarsely divided into B beamspaces[1] based on the information of UEs' spatial direction and their unique channel characteristics. More precisely, UEs with (nearly) aligned spatial direction but distinctive propagation distances are clustered into same beamspace. UEs with large spatial orthogonality (or small channel response correlation) are classified as different beamspaces. The spatial resolution can be flexibly adjusted by changing the number of beamspaces and the number of BS's antennas. Secondly, to balance the system performance and computational complexity, we consider *power-domain NOMA* mode can be employed only by UEs within the same beamspace, whereas *spatial-domain NOMA* mode can be used only by UEs from different beamspaces with sufficient spatial orthogonality. Without loss of generality, we denote the set of UEs associated with the bth beamspace as $\mathcal{B}_b$, where $\bigcup_{b=1}^{B} \mathcal{B}_b = \mathcal{K}$ and $\mathcal{B}_b \cap \mathcal{B}_{b'} = \emptyset$, $\forall b \neq b' \in \mathcal{B}$. we denote $\boldsymbol{h}_{k,m} \in \mathbb{C}^{N_t \times 1}$ as the channel vector between the BS and UE k in bth beamspace on mth SC, which captures path loss, shadowing, and multipath fading effects.

To evaluate and compare UEs' performance gains of each possible MA mode, we include a signal transmission model that comprehensively analyzes the interference and UEs' achievable data rate under different conditions. Let $\mathcal{U}_m$ denote the set of UEs[2] allocated to mth SC, then depending on the total number of UEs allocated to the SC and their resource-sharing situations, we can mathematically formulate these four MA modes. We adopt two binary variables, α_m and β_m, to indicate if spatial-domain or power-domain NOMA is employed for mth SC respectively,

$$\alpha_m = \begin{cases} 1, & \text{if } \sum_{b=1}^{B} \mathbf{1}\left\{\left|\mathcal{B}_b \cap \mathcal{U}_m\right| \geq 1\right\} > 1, \\ 0, & \text{otherwise.} \end{cases} \tag{21.1}$$

$$\beta_m = \begin{cases} 1, & \text{if } \exists b \in \mathcal{B} : \left|\mathcal{B}_b \cap \mathcal{U}_m\right| \geq 2, \\ 0, & \text{otherwise.} \end{cases} \tag{21.2}$$

1 In this chapter, the term "cluster" and "beamspace" are used interchangeably.
2 In (Mei et al. (2022)), such UE set is termed *UE coalition* and will be introduced with details in the case study of Section 21.4.

where $\mathbf{1}\{\cdot\}$ is the indicator function. Hence, four candidate MA modes for mth SC can be determined jointly by α_m and β_m

$$(\alpha_m, \beta_m) = \begin{cases} (0,0), & \text{if OMA is set,} \\ (0,1), & \text{if power domain NOMA is set,} \\ (1,0), & \text{if spatial domain NOMA is set,} \\ (1,1), & \text{if hybrid MA is set.} \end{cases} \tag{21.3}$$

Let $s_{k,m}$ denote the SC allocation indicator, where $s_{k,m} = 1$ indicates UE k is assigned to mth SC; otherwise, $s_{k,m} = 0$. Furthermore, each UE is served by a dedicated beamforming vector $\mathbf{w}_k \in \mathbb{C}^{N_t \times 1}$ and zero-forcing beamforming (ZFBF) is assumed to serve UEs associated with different beamspaces, while UEs clustered in the same beamspace share the same beamforming vector. Consider p_k as the downlink transmission power for UE k, then, the signal-to-interference-plus-noise-ratio (SINR) of UE k on mth SC is expressed as

$$\gamma_{k,m} = \frac{p_k \left| \boldsymbol{h}_{k,m}^{\mathrm{H}} \mathbf{w}_k \right|^2}{I_{k,m} + N_0}, \text{ for } s_{k,m} = 1, \tag{21.4}$$

where N_0 denotes the additive white Gaussian noise (AWGN), and the term $I_{k,m}$ in the denominator represents the overall non-orthogonal interference, given by four different situations for each MA mode

$$I_{k,m} = \begin{cases} 0, & \text{for OMA mode,} \\ I_{k,m}^{\mathrm{PD}}, & \text{for power domain NOMA mode,} \\ I_{k,m}^{\mathrm{SD}}, & \text{for spatial domain NOMA mode,} \\ I_{k,m}^{\mathrm{PD}} + I_{k,m}^{\mathrm{SD}}, & \text{for hybrid MA mode.} \end{cases} \tag{21.5}$$

There is no additional interference for OMA mode. However, in case of power-domain NOMA mode, UEs with smaller channel gains need to decode their own signals in presence of interference from UEs with larger channel gains. Such interference in power domain (PD), $I_{k,m}^{\mathrm{PD}}$, is given by

$$I_{k,m}^{\mathrm{PD}} = \sum_{b=1}^{B} \mathbf{1}\left\{k \in \mathcal{B}_b\right\} \cdot \sum_{i \in \mathcal{S}_k} p_i \left| \boldsymbol{h}_{k,m}^{H} \mathbf{w}_i \right|^2, \tag{21.6}$$

where $\mathcal{S}_k$ is the set of UEs who have better channel gain than UE k, and given by $\mathcal{S}_k = \left\{ i | i \in \mathcal{U}_m \cap \mathcal{B}_b, \| \boldsymbol{h}_{i,m} \|^2 > \| \boldsymbol{h}_{k,m} \|^2 \right\}$.

For interference in case of the spatial domain NOMA, $I_{k,m}^{\mathrm{SD}}$ is calculated by

$$I_{k,m}^{\mathrm{SD}} = \sum_{b=1}^{B} \mathbf{1}\left\{k \notin \mathcal{B}_b\right\} \cdot \sum_{i \in \mathcal{U}_m \cap \mathcal{B}_b} p_i \left| \boldsymbol{h}_{k,m}^{H} \mathbf{w}_i \right|^2. \tag{21.7}$$

For the case of hybrid MA mode, the incurred interference includes both spatial and power domains, namely, $I_{k,m} = I^{\mathrm{PD}}_{k,m} + I^{\mathrm{SD}}_{k,m}$. Overall, the data rate of UE k can be derived by

$$r_k = \sum_{m=1}^{M} s_{k,m} \cdot \frac{W}{M} \log_2 \left(1 + \gamma_{k,m}\right), \quad k \in \mathcal{U}_m. \tag{21.8}$$

21.2.2 Enabling Blocks of Individualized Service Provisioning in MDMA

One of the critical aspects in achieving individualized service provisioning is to understand "*resource utilization cost*," which evaluates and quantify the amount of efforts required for the receivers to obtain desired communication services. For instance, key factors that contribute to resource utilization cost may include signal processing, power consumption, and computational complexity, etc. By carefully reviewing these factors that influence UE performance trade-offs and resource usage, we create a comprehensive and flexible metric that facilitate a fair comparison for diverse UE types and different MA modes of MDMA.

To start with, utilizing power-domain NOMA mode requires SIC process at the "near-UEs" (i.e., with larger channel gains), which inevitably introduces extra computational complexity and power consumption. Moreover, each layer of SIC implementation adds decoding latency and error propagation effects, thereby lowering the QoS and QoE performance especially for delay-sensitive and high-reliability UEs. Therefore, the *resource utilization cost* of power-domain NOMA can be formulated

$$\psi^{\mathrm{PD}}_{k,m} = \sum_{b=1}^{B} \mathbf{1}\left\{k \in \mathcal{B}_b\right\} \cdot \sum_{i \in \mathcal{X}_k} \left[\rho_0 - \rho_1 \lg\left(\gamma^{\mathrm{SIC}}_{k,m}\right)\right], \tag{21.9}$$

where $\psi^{\mathrm{PD}}_{k,m}$ contains two parts: the first term ρ_0 denotes the energy consumption (i.e., circuit power) of SIC processing; Moreover, higher SINR values for the SIC signal detection and decoding operations will yield lower complexity and error probability that can be reflected by achievable bit error rate (BER). Therefore, the second part "$-\rho_1 \lg\left(\gamma^{\mathrm{SIC}}_{k,m}\right)$" is designed as an increasing function of the inverse SINR of SIC, where ρ_1 is a positive scalar and $\gamma^{\mathrm{SIC}}_{k,m}$ denotes the experienced SINR of UE k's SIC operations when it detects the signal of UE i, i.e., "far-UE," in presence of interference from its desired signal

$$\gamma^{\mathrm{SIC}}_{k,m} = \frac{p_i |\boldsymbol{h}^H_{k,m} \mathbf{w}_i|^2}{p_k |\boldsymbol{h}^H_{k,m} \mathbf{w}_k|^2 + I^{\mathrm{SD}}_{k,m} + N_0}. \tag{21.10}$$

Besides, $\mathcal{X}_k = \{i | i \in \mathcal{U}_m \cap \mathcal{B}_b, \|\boldsymbol{h}_{k,m}\|^2 > \|\boldsymbol{h}_{i,m}\|^2\}$ denotes the set of UEs with worse channel gain than UE k, and as shown, there would be no additional utilization cost in the power domain for "far-UEs" that do not perform SIC.

Likewise, we can formulate resource utilization cost for spatial domain NOMA mode by analyzing the impact of the spatial channel correlation between desired signal and interference signal. High channel correlations imply that the signals from different users have similar spatial characteristics, which lead to increased difficulties in distinguishing and separating these signals. For UE $k \in \mathcal{U}_m$ sharing mth SC with UEs in different beamspaces, its spatial domain resource utilization cost can be expressed as

$$\psi_{k,m}^{\mathrm{SD}} = \sum_{b=1}^{B} \mathbf{1}\{k \notin \mathcal{B}_b\} \cdot \sum_{i \in \mathcal{U}_m \cap \mathcal{B}_b} \rho_2 \cdot \frac{\left|\boldsymbol{h}_{k,m}^H \boldsymbol{h}_{i,m}\right|^2}{\|\boldsymbol{h}_{k,m}\|_2 \cdot \|\boldsymbol{h}_{i,m}\|_2}, \tag{21.11}$$

where the weight ρ_2 is a positive scalar related to the incurred costs experienced by the receiver UE for using spatial domain NOMA.

Overall, we can quantify the multi-dimensional resource utilization cost experienced by UE k as

$$g_k = \sum_{m=1}^{M} s_{k,m} \cdot g_{k,m} = \sum_{m=1}^{M} s_{k,m} \cdot \left(\alpha_m \cdot \psi_{k,m}^{\mathrm{SD}} + \beta_m \cdot \psi_{k,m}^{\mathrm{PD}}\right). \tag{21.12}$$

It is worth mentioning that UEs' hardware constraints are also taken into account for designing a more inclusive MDMA. For example, legacy and low-cost devices may not incorporate advanced signal processing techniques like SIC. For IoT users or sensor nodes who perform low complexity operations and prefer to preserve battery life, their MA modes need to be carefully designed.

21.3 Value-Oriented Operation of MDMA

21.3.1 Value-Oriented Operation Paradigm

This section concentrates on deriving the optimization objectives for implementing value-oriented MDMA. In this context, the objective of "value realization" of 6G MDMA design can be delineated from the perspectives of end-users at the individual level and network operators at the system level.

21.3.1.1 Individual Level

Due to the increasingly diversified service requirements and distinct perceived resource values, it is essential to accurately define "value realization" in a more user-centric and situation-dependent manner.

From the perspective of end-users, value realization entails meeting individualized QoS requirements with improve user experience. While conventional evaluation metrics such as latency, data rate, reliability, and efficiency provide effective measurement of users' QoS, they do not accurately account for the impact of resource utilization costs. Accordingly, to bridge the gap in value-oriented service provisioning, a cost-gain aware utility formulation for users' perceived value of allocated radio resources is proposed. Such design can be leveraged to optimize value realization for individual users while reducing the imposed resource utilization costs. Additionally, we intend to capture the increasing discrepancies among devices' hardware capabilities in 6G, ranging from low-budget to high-end user equipment. As illustrated in Fig. 21.2, incorporating these factors into resource allocation decision-makings enables MDMA to deliver tailored services of each user.

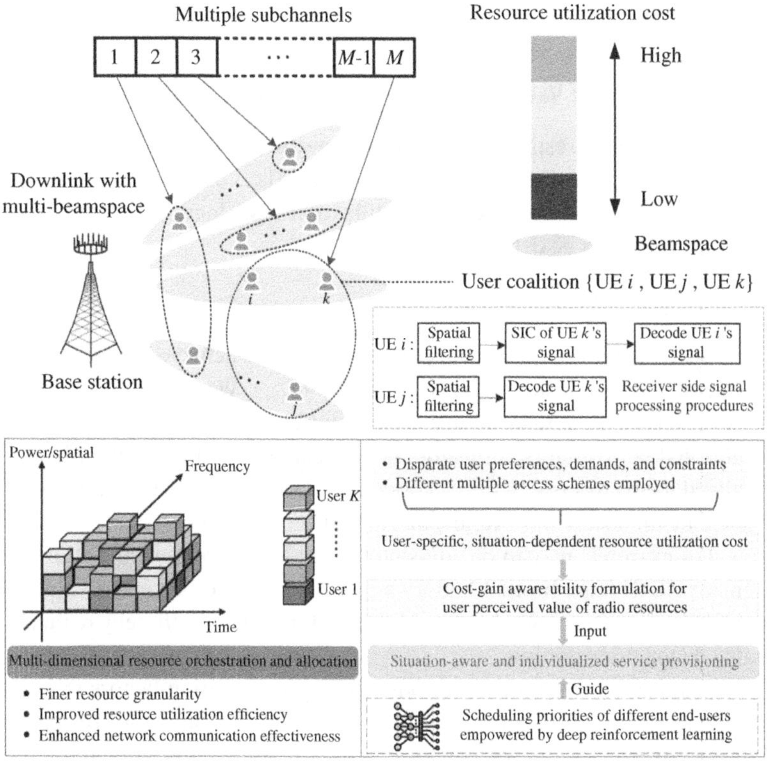

Figure 21.2 Illustration of the MDMA design incorporated with resource utilization cost awareness for individualized service provisioning in 6G communication.

21.3.1.2 System Level

From the perspective of 6G network operators, value realization involves improving operational efficiency and enhancing the communication effectiveness of the entire network.

To evaluate the operational efficiency, one of the key aspects is to optimize the energy consumption in terms of overall resource usage. For instance, energy efficiency (EE) is commonly adopted as the network optimization objective and defined as the ratio of network sum throughput to the total transmit power consumption, i.e., bits/Joule (Fang et al., 2017); (Zhu et al., 2017). However, such design emphasizing on single performance indicator (e.g., data rate) becomes less effective in 6G owing to the heterogeneity of concurrent multi-domain communication requirements. Therefore, we evaluate the communication effectiveness by the aggregated individual-level achieved values of all served end-users. The operational efficiency goal of 6G MDMA can be designed by maximizing overall prioritized values and QoS satisfactions of different users while reducing the operational cost.

21.3.2 Individual Level Value Realization: User-Centric Perspective

At the individual level, we establish the cost-gain aware utility model by two metrics: (i) performance gain of communication service (i.e., UE's QoS satisfaction level) and (ii) multi-dimensional resource utilization costs incurred. For UE $k \in \mathcal{K}$, its utility function can be defined as

$$u_k = r_k/R_{\max} - w_k \cdot g_k, \ k \in \mathcal{U}_m, \tag{21.13}$$

where the linear model $r_k/R_{\max}$ is adopted to reflect the achieved performance gain in contrast to UE's desired QoS (denoted by $R_{\max}$). We consider $R_{\max}$ can be obtained when it experiences no inter-user interference (IUI) with given SNR 30 dB. Additional performance indicators of UEs can also be included with similar normalized forms in Eq. (21.13). Secondly, the radio resource utilization cost g_k is weighted by the factor w_k to reflect disparate user preferences, demands, and constraints. For example, w_k can be adjusted to penalize power consumption for energy-sensitive UEs (e.g., limited battery supply). Thus, the perceived value of individual UE can be formulated by above weighted subtraction, thereby realizing individualized QoS provisioning.

This approach can be integrated with other resource allocation factors to facilitate MDMA design, including fairness and user prioritization. More precisely, with "fairness," we aims to improve the overall performance gains while reducing the inequality and discrepancy among UEs' utilities, such as max–min fairness approach (Han et al., 2021)

$$
\begin{aligned}
\mathcal{P}_1 : \ & \max_{\boldsymbol{\Omega}} \min \ \left\{u_k | k \in \mathcal{K}\right\} \\
\text{s.t.} \quad & \text{C1:} \sum_{m=1}^{M} s_{k,m} = 1, s_{k,m} = \{0,1\}, k \in \mathcal{K}, \\
& \text{C2:} \ 1 \le \left|\mathcal{U}_m\right| \le L_{\max}, \bigcup_{m=1}^{M} \mathcal{U}_m = \mathcal{K}, \bigcap_{m=1}^{M} \mathcal{U}_m = \emptyset, \\
& \text{C3:} \ R_{\text{th}} \le r_k \le R_{\max}, \forall k \in \mathcal{K}, \\
& \text{C4:} \sum_{k \in \mathcal{K}} p_k \le P_{\max}, \\
& \text{C5:} \sum_{k \in \mathcal{U}_m} \sum_{b=1}^{B} \sum_{i \in \mathcal{U}_m \cap \mathcal{B}_b} \mathbf{1}\left\{k \in \mathcal{K}_{\text{no-sic}}\right\} \cdot \\
& \qquad \qquad \mathbf{1}\left\{\| \boldsymbol{h}_{k,m} \|_2 > \| \boldsymbol{h}_{i,m} \|_2\right\} = 0, \forall m \in \mathcal{M}.
\end{aligned}
\tag{21.14}
$$

where $\boldsymbol{\Omega} = \left\{(\alpha_m, \beta_m), s_{k,m}, p_k | \forall k \in \mathcal{K}, m \in \mathcal{M}\right\}$ defines the multi-dimensional resource allocation decision including MA mode determination, SC assignment and transmit power allocation for each UE. C5 indicates the inherent constraints of UEs' hardware capability required for SIC operation, where $\mathcal{K}_{\text{no-sic}}$ denotes the set of UEs that cannot perform SIC.

Alternatively, we can optimize the aggregated user utilities as the MDMA designing objective shown in (Mei et al. (2022)). This design can be also extended and used for "user prioritization" by considering a situation-aware scheduling priority factor χ_k carried by each individual UE while optimizing the summed utilities

$$
\begin{aligned}
\mathcal{P}_2 : \ & \max_{\boldsymbol{\Omega}} \left\{ \sum_{m=1}^{M} \sum_{k=1}^{|\mathcal{U}_m|} \chi_k u_k \right\} \\
\text{s.t.} \quad & \text{C1–C5.}
\end{aligned}
\tag{21.15}
$$

where χ_k can be learned by timely discovery of user channel condition, traffic pattern, and QoS requirements (Wang et al., 2023). As a result, MDMA can intelligently balance the differentiated service provisioning of end-users and the overall network operational goals in practical implementation.

21.3.3 System Level Value Realization: Network Operator Perspective

At the system level, the value realization objective can be designed by jointly considering the overall operational efficiency and multi-dimensional non-orthogonality incurred in MDMA. As shown in Liu et al. (2021a), a novel metric called "I-QoSE" is proposed to measure the resource allocation decision values. Specifically, "I-QoSE" is defined as

$$
f_{\text{I-QoSE}} = \alpha_{\text{EE}} \frac{\eta_{\text{EE}}}{\eta_{\text{EE}}^{\max}} + \alpha_{\text{TO}} \left(1 - \frac{\varsigma_{\text{total}}}{\varsigma_{\text{total}}^{\max}}\right), \tag{21.16}
$$

where the first part of above weighted sum indicates the normalized EE level and η_{EE} is defined as the ratio of sum-rate over the total power consumption,

$$\eta_{\text{EE}} = \frac{\sum_{k=1}^{K} r_k}{P_{\text{total}}} = \frac{\sum_{k=1}^{K} r_k}{\sum_{k=1}^{K} p_k + P_{\text{h}}}, \tag{21.17}$$

where P_{total} is comprised of the total transmit power allocated to all UEs, and the circuit power consumption related to hardware design of BS, i.e., $P_h = \xi_0 N_t$ and ξ_0 denotes the circuit power consumption coefficient associated with each antenna at BS. The second part "total non-orthogonality" ς_{total} measures overall non-orthogonal interference suffered by all UEs particularly in spatial and power domains

$$\varsigma_{\text{total}}^{\text{spatial}} = \sum_{m=1}^{M} \sum_{k \neq i \in \mathcal{U}_m} \left(\alpha_m \frac{\left| \boldsymbol{h}_{k,m}^H \boldsymbol{h}_{i,m} \right|}{\left\| \boldsymbol{h}_{k,m} \right\| \cdot \left\| \boldsymbol{h}_{i,m} \right\|} \right), \text{for } k \in \mathcal{B}_b, i \in \mathcal{B}_{b'}, \text{ and } b \neq b', \tag{21.18}$$

$$\varsigma_{\text{total}}^{\text{power}} = \sum_{m=1}^{M} \sum_{k \neq i \in \mathcal{U}_m} \left(\beta_m \frac{\min \left\{ \boldsymbol{h}_{k,m}, \boldsymbol{h}_{i,m} \right\}}{\max \left\{ \boldsymbol{h}_{k,m}, \boldsymbol{h}_{i,m} \right\}} \right), \text{for } k, i \in \mathcal{B}_b. \tag{21.19}$$

The smaller $\varsigma_{\text{total}}^{\text{spatial}}$ value, the less channel correlations in spatial domain NOMA multiplexing, the less difficulties for receiver UEs to distinguish overlapped signals. Similarly, the smaller $\varsigma_{\text{total}}^{\text{power}}$ value, the more distinctive channel gains among superimposed signals using power-domain NOMA, the better performance gains it may yield in comparison to OMA. Therefore, we can achieve system level value realization by maximizing $f_{\text{I-QoSE}}$ in Eq. (21.16)

$$\begin{aligned} \mathcal{P}_3 : & \max_{\boldsymbol{\Omega}} \ \{ f_{\text{I-QoSE}} \} \\ \text{s.t.} & \quad \text{C1–C5.} \end{aligned} \tag{21.20}$$

Another way of formulating system level value realization is proposed in (Wang et al. (2023) and Han et al. (2023)), where the value of each resource allocation for MDMA (denoted by u_{EDI}) is formulated to reflect overall communication effectiveness by the tradeoff between the mean return and the variance of all UEs' utilities

$$u_{\text{EDI}} = \text{mean} \left[u_k | \forall k \in \mathcal{K} \right] - \omega \cdot \text{Var} \left[u_k | \forall k \in \mathcal{K} \right], \tag{21.21}$$

where mean $[\cdot]$ takes the expectation of obtained utilities of all UEs, and function Var $[\cdot]$ calculates the distribution variance to measure the risk of discrepancy in UEs' QoS satisfaction levels. ω is the risk-aversion factor to adjust the trade-off. Instead of relying on the instantaneous measurement, we aim to maximize the system-level long-term communication effectiveness in a time-slotted system while minimizing the network operational costs

$$
\begin{aligned}
\mathcal{P}_4:\ & \max_{\Omega(t),\forall t} \bar{u}_{\text{EDI}} \triangleq \lim_{T\to\infty}\frac{1}{T}\sum_{t=1}^{T}\mathbb{E}\left[u_{\text{EDI}}(t)\right] \\
\text{s.t.:}\ & \text{C1}: \sum_{k\in\mathcal{K}} s_{k,m}(t) \le L_{\max}, s_{k,m}(t) = \{0,1\}, \forall k, m, \\
& \text{C2}: \sum_{m=1}^{M} s_{k,m}(t) = 1, \forall k, \\
& \text{C3}: R_k^{\text{th}} \le R_k(t) \le R_k^{\max}, \forall k, \\
& \text{C4}: P_{\text{tot}}(t) = \sum_{k=1}^{K} p_k(t) \le P_{\text{ins}}^{\max}, \\
& \text{C5}: \overline{P}_{\text{tot}} \triangleq \lim_{T\to\infty}\frac{1}{T}\sum_{t=1}^{T}\mathbb{E}\left[P_{\text{tot}}(t)\right] \le P_{\text{avg}}^{\max},
\end{aligned}
\tag{21.22}
$$

where C4 and C5 are the network operational cost requirements, confined by the instantaneous peak values and time averaged power expenditure of BS's total transmit power consumption respectively.

21.4 Multi-Dimensional Resource Utilization in Value-Oriented MDMA

In this section, we provide a case study of MDMA that focus on the maximization of the summation of cost-gain aware utilities of all UEs (Mei et al., 2022), i.e., maximizing $\mathcal{P}_2$ with $\chi_k = 1, \forall k \in \mathcal{K}$. Considering the target problem has nonconvex mixed-integer nonlinear structure, $\mathcal{P}_2$ is decomposed into two subproblems: UE coalition formation and real-time multi-dimensional resource allocation.

21.4.1 User Coalition Formation Approach

In MDMA, UEs are multiplexed by multi-dimensional resource in different MA modes. To efficiently determine resource utilization methods for each UE, we introduce the concept of *UE coalition*, namely, several UEs with distinct multi-dimensional resource constraints can coordinate as resource-sharing coalitions to utilize the same SC in a cost-effective and less conflicting manner. Grouping UEs as coalitions can be computationally expensive and induces prohibitive signaling overheads, as different UE sets can result in different MA modes with distinct resource utilization costs. Moreover, UEs' real-time channel condition, location, and concurrent QoS requirements are critical contributing factors in order to reduce the overall multi-dimensional resource utilization cost (Eq. (21.12)). Therefore, we propose to approximate g_k by a new metric termed "intensity of radio resource-sharing conflicts," which calculation mostly relies on UEs' long-term channel features and hardware characteristics, such that coalition formation can be executed with a coarse time granularity rather than each time slot. Let $\mathcal{U}_n$ denotes the nth UE coalition and $\tilde{g}_k^n$ as the intensity of radio resource-sharing conflicts experienced by UE k when it joins coalition $n \in \mathcal{N}$, then we have

$$\tilde{g}_k^n = \rho_0 - \rho_1 \cdot \sum_{b=1}^{B} \mathbf{1}\left\{k \in \mathcal{B}_b\right\} \cdot \sum_{i \in \mathcal{U}_n \cap \mathcal{B}_b} \mathbf{1}\left\{\mathrm{PL}(d_k) > \mathrm{PL}(d_i)\right\} \cdot$$
$$\lg\left[\frac{\mathrm{PL}(d_k)}{\mathrm{PL}(d_i) + N_0/P_{\max}}\right] + \rho_2 \cdot \sum_{b=1}^{B} \mathbf{1}\left\{k \notin \mathcal{B}_b\right\} \cdot \sum_{i \in \mathcal{U}_n \cap \mathcal{B}_b} \left|\boldsymbol{a}(\theta_k)^H \boldsymbol{a}(\theta_i)\right|, \tag{21.23}$$

As a result, we can define the objective of the coalition formation as

$$\mathcal{P}_5: \quad \min_{\{\mathcal{U}_n, \alpha_n, \beta_n\}} \left\{\sum_{n \in \mathcal{N}} \sum_{k \in \mathcal{U}_n} w_k \cdot \tilde{g}_k^n\right\}$$
$$\text{s.t.: C2,}$$
$$\tilde{\text{C}}5: \sum_{k \in \mathcal{U}_n} \sum_{b=1}^{B} \sum_{i \in \mathcal{U}_n \cap \mathcal{B}_b} \mathbf{1}\left\{k \in \mathcal{K}_{\text{no-sic}}\right\} \cdot \mathbf{1}\left\{\mathrm{PL}(d_k) > \mathrm{PL}(d_i)\right\} = 0. \tag{21.24}$$

To reduce implementation complexity while guarantee the system performance, we propose to use matching theory to solve the coalition formation problem $\mathcal{P}_5$ to jointly satisfy UEs' diversity preference and matching exchange stability.

Definition 21.1 ***(Coalition Formation)*** The coalition formation can be formulated by a two-sided many-to-one matching Π, where Π is a function mapping the UE set $\mathcal{K}$ to N disjoint un-ordered sets, i.e., the UE coalitions $\mathcal{N} = \{\mathcal{U}_1, \mathcal{U}_2, \ldots, \mathcal{U}_n\}$, such that:

1) $\Pi(k) \in \mathcal{N}, \forall k \in \mathcal{K}$ and $|\Pi(k)| = 1$.
2) $\mathcal{U}_n = \Pi^{-1}(n) \subseteq \mathcal{K}, \forall n \in \mathcal{N}$ and the cardinality of coalition $\mathcal{U}_n$ is denoted by $l \triangleq |\mathcal{U}_n|$, satisfying $1 \leq l \leq L_{\max}, l \in \mathbb{N}^+$.
3) $n = \Pi(k)$ if and only if $k \in \Pi^{-1}(n)$.

21.4.1.1 Preferences Over Matchings

To achieve matching stability and optimality, we make use of each user's preferences over alternative matchings, which describes the dynamic interactions and competition behaviors between UEs and coalitions. Specifically, each UE $k \in \mathcal{K}$ uses $\succ_k$ to establish the *preference relation* over two different coalitions

$$n \succ_k n' \Leftrightarrow \tilde{g}_k^n < \tilde{g}_k^{n'} \text{ and } x_n = 0, \; \forall k \in \mathcal{K}, \forall n \in \mathcal{N}. \tag{21.25}$$

Namely, UEs prefer to choose their peers with the least amount of resultant resource-sharing conflicts. The constraint of $\tilde{\text{C}}5$ states that UEs admitted to the same coalition shall guarantee either sufficient spatial orthogonality for spatial domain NOMA, or distinctive channel gain difference as well as hardware processing capability for power domain NOMA. Otherwise, such UE group is defined as *forbidden coalition* denoted by the binary variable x_n.

The matching algorithm is designed with two phases. Firstly, a feasible and valid matching is established by allowing UEs to selfishly and rationally form the coalitions. This phase terminates if all UEs and coalitions are matched with no forbidden coalitions. Then, the obtained matching will be stabilized and optimized by creating *rotation sequence*, which utilizes the inter-dependency of UEs' selections to reduce the total weighted resource sharing conflicts among all UEs, i.e., $\sum_{n\in\mathcal{N}}\sum_{k\in\mathcal{U}_n} w_k \cdot \tilde{g}_k^n$ in $\mathcal{P}_5$. The process ends if we have traversed all rotation sequences with no further improvement. Please refer to Mei et al. (2022) for detailed matching algorithm design.

21.4.2 Real-Time Multi-Dimensional Resource Allocation

After deriving the UE coalition results and its corresponding multiple access mode $\left(\mathcal{U}_n^*, \alpha_n^*, \beta_n^*\right)$, the target problem $\mathcal{P}_2$ is reduced to a joint subchannel and power allocation problem, expressed by

$$\begin{aligned} \mathcal{P}_6 : \quad & \max_{\{s_{n,m}, p_k\}} \left\{ \sum_{n\in\mathcal{N}} \sum_{k\in\mathcal{U}_n} u_k \right\} \\ \text{s.t.:} \quad & \text{C1, C3, C4.} \end{aligned} \tag{21.26}$$

However, $\mathcal{P}_6$ is non-concave and difficult to solve efficiently for the global optimal solution. A low-complexity solution is developed using the successive convex approximation (SCA) and the Lagrange dual decomposition methods.

Lemma 21.1 ***(Transformation of Problem $\mathcal{P}_6$)*** $\mathcal{P}_6$ can be approximated by the following concave maximization problem[3], where the transmit power of each UE p_k is replaced by an equivalent form, $p_k = \sum_{m=1}^{M} s_{n,m} \cdot e^{z_{k,m}}, k \in \mathcal{U}_n$ and the term $\bar{r}_k$ is the lower bound of the data rate of UE k in Eq. (21.8), given by

$$\bar{r}_k = \sum_{m=1}^{M} s_{n,m} \cdot \frac{W}{M} \cdot \left[a_k \log_2\left(\gamma_k\right) + b_k\right] \le r_k, k \in \mathcal{U}_n. \tag{21.27}$$

3 The detailed proof of Lemma 1.1 is available in (Mei et al. (2022)).

where the constants $\{a_k, b_k\}$ are chosen as specified below,

$$a_k = \frac{\gamma_k^{\text{th}}}{1+\gamma_k^{\text{th}}}, \tag{21.27a}$$

$$b_k = \log_2\left(1+\gamma_k^{\text{th}}\right) - \frac{\gamma_k^{\text{th}}}{1+\gamma_k^{\text{th}}}\log_2\left(\gamma_k^{\text{th}}\right), \tag{21.27b}$$

γ_k^{th} is the lower bound of SINR for UE k, given by $\gamma_k^{\text{th}} = 2^{MR_{\text{th}}/W} - 1$.

Therefore, applying $\bar{r}_k$ to subproblem $\mathcal{P}_6$, we can obtain an standard concave maximization problem $\tilde{\mathcal{P}}_6$ in variables $\{s_{n,m}, z_{k,m} \,|\forall n \in \mathcal{N}, k \in \mathcal{U}_n, m \in \mathcal{M}\}$

$$\begin{aligned}
\tilde{\mathcal{P}}_6 : \quad & \max_{\{s_{n,m}, z_{k,m}\}} \left\{ \sum_{n\in\mathcal{N}} \sum_{k\in\mathcal{U}_n} \left(\bar{r}_k/R_{\max} - w_k \cdot \sum_{m=1}^{M} s_{n,m} \cdot g_{k,m} \right) \right\} \\
\text{s.t.:} \quad & p_k = \sum_{m=1}^{M} s_{n,m} \cdot e^{z_{k,m}}, \forall k \in \mathcal{U}_n, n \in \mathcal{N}, \\
& \bar{r}_k = \sum_{m=1}^{M} s_{n,m} \cdot \frac{W}{M} \cdot \left[a_{k,m}\log_2\left(\gamma_{k,m}\right) + b_{k,m}\right], \forall k \in \mathcal{U}_n, n \in \mathcal{N}, \\
& \text{C1, C3} : R_{\text{th}} \le \bar{r}_k \le R_{\max}, \forall k \in \mathcal{K}, \text{and C4.}
\end{aligned} \tag{21.28}$$

21.4.2.1 Lagrange Dual Decomposition Method

Given that $\tilde{\mathcal{P}}_6$ is mixed-integer optimization problem, we first form the partial Lagrangian of $\tilde{\mathcal{P}}_6$ without considering the constraint of C1,

$$\begin{aligned}
L(\boldsymbol{\eta}, \boldsymbol{\mu}, \nu, \{s_{n,m}, z_{k,m}\}) = & \sum_{n\in\mathcal{N}} \sum_{k\in\mathcal{U}_n} \left[\bar{r}_k/R_{\max} - w_k \cdot \sum_{m=1}^{M} s_{n,m} \cdot g_{k,m} \right] \\
& + \sum_{n\in\mathcal{N}} \sum_{k\in\mathcal{U}_n} \left[\eta_k \cdot \left(\bar{r}_k - R_{\text{th}}\right) - \mu_k \cdot \left(\bar{r}_k - R_{\max}\right)\right] \\
& - \nu \cdot \left(\sum_{n\in\mathcal{N}} \sum_{k\in\mathcal{U}_n} \sum_{m=1}^{M} s_{n,m} \cdot e^{z_{k,m}} - P_{\max} \right),
\end{aligned} \tag{21.29}$$

where $\boldsymbol{\eta} = \{\eta_k, \forall k\}$ and $\boldsymbol{\mu} = \{\mu_k, \forall k\}$ are two Lagrange multiplier vectors associated with the data rate constraints of each UE in inequality C3, and ν is the Lagrange multiplier related to the maximum transmission power of BS. Then, the Lagrange dual function $J(\boldsymbol{\eta}, \boldsymbol{\mu}, \nu)$ is derived by solving the following problem

$$\begin{aligned}
J(\boldsymbol{\eta}, \boldsymbol{\mu}, \nu) = & \max_{\{s_{n,m}, z_{k,m}\}} L\left(\boldsymbol{\eta}, \boldsymbol{\mu}, \nu, \{s_{n,m}, z_{k,m}\}\right), \\
& \text{s.t. : C1.}
\end{aligned} \tag{21.30a}$$

To do so, firstly, by setting $s_{n,m} = 1, \forall n, m,$ and ignoring the constraint in problem (21.30a), we can obtain M^2 individual decomposed problems defined as

$$J_{n,m}(\boldsymbol{\eta}, \boldsymbol{\mu}, v) = \max_{\{z_{k,m}|k \in \mathcal{U}_n\}} \sum_{k \in U_n} \Big[\left(1/R_{\max} + \eta_k - \mu_k\right) \cdot \tfrac{W}{M} \cdot a_k \log_2\left(\gamma_k\right) - v \cdot e^{z_{k,m}} - w_k \cdot g_{k,m} \Big], \forall n \in \mathcal{N}, 1 \le m \le M. \tag{21.30b}$$

For nth UE coalition assigned to mth SC, we can derive the optimal power allocation of UEs by solving problem (21.30a); Then, with consideration of C1, the optimal solution of problem (21.30a) and dual function value $J(\boldsymbol{\eta}, \boldsymbol{\mu}, v)$ can be derived.

Lemma 21.2 ***(Optimal Solution to Problem (21.30a))*** The optimal subchannel allocation for UE coalitions with fixed $(\boldsymbol{\eta}, \boldsymbol{\mu}, v)$ can be decided based on the following criterion,

$$s^*_{n,m} = \arg\ \max_{\{n' \in \mathcal{N}, 1 \le m' \le M\}} \left\{ \sum_{n', m'} J_{n',m'}(\boldsymbol{\eta}, \boldsymbol{\mu}, v) \right\}, \quad \text{subject to C1,} \tag{21.31}$$

which fits a maximum weight matching for bipartite graphs and can be solved by the Hungarian algorithm. Then, the corresponding power allocation of UEs in each coalition can be calculated by

$$p^*_k = \sum_{m=1}^{M} s^*_{n,m} \cdot \exp\left(z^*_{k,m}\right). \tag{21.32}$$

Now, the corresponding dual problem to $\tilde{\mathcal{P}}_6$ is

$$\min_{\{\boldsymbol{\eta}, \boldsymbol{\mu}, v\}} \{J(\boldsymbol{\eta}, \boldsymbol{\mu}, v)\}, \text{ subject to } \boldsymbol{\eta} \ge \mathbf{0}, \boldsymbol{\mu} \ge \mathbf{0}, v \ge 0. \tag{21.33}$$

Solving problem (21.33) we can obtain the optimal value for $\tilde{\mathcal{P}}_6$ due to the strong duality. Given that the dual function $J(\boldsymbol{\eta}, \boldsymbol{\mu}, v)$ is the point-wise infimum of a set of affine functions for Lagrange multipliers, it's a typical convex function that can be solved by standard subgradient methods (Boyd and Vandenberghe, 2004). Please refer to (Mei et al. (2022)) for the complete algorithm design. Finally, we obtain the real-time multi-dimensional resource allocation results including optimal UE coalition formation, MA mode selection, subchannel, and power allocation scheme, $\boldsymbol{\Omega}^* = \left\{(\mathcal{U}^*_n, \alpha^*_n, \beta^*_n), s^*_{n,m}, p^*_k\right\}$.

21.5 Numerical Results and Analysis

In this section, we use a system-level simulation platform to evaluate the proposed value-oriented MDMA scheme. For simulation parameters and deployments, we consider a single cell and set the maximum transmit power of BS as 33 dBm, the bandwidth of subchannel as 10 MHz, the number of subchannels as 12, noise variance as −140 dBm/Hz and the maximum number of UEs of for each coalition is 3. Moreover, the default number of UEs is 15. Value-oriented MDMA framework entails enhancing user's QoS experience and improving the network operational efficiency. We first validate the necessity of implementing value-oriented paradigm in MDMA for QoS/resource-aware diverse service provisioning. Then, comparing with exiting state-of-art MA schemes, we illustrate the effectiveness of the proposed value-oriented MDMA.

21.5.1 Performance Evaluation of Value-Oriented Paradigm in MDMA

To further evaluate the benefits of value-oriented paradigm in MDMA framework, we choose the conventional resource-utilization-efficiency-oriented MDMA

Table 21.1 MA mode selection for value-oriented MDMA and MD-IMA.

Number of UEs	OMA mode	Power-domain NOMA mode	Spatial-domain NOMA mode	Hybrid NOMA mode
		Value-oriented MDMA scheme		
15	61.13%[a)]	2.47%	36.4%	0%
20	28.41%	8.97%	62.62%	0%
25	12.88%	15.67%	71.45%	0%
30	10.21%	17.29%	72.15%	0.35%
35	7.09%	18.58%	73.33%	5.49%
		MD-IMA (Resource-utilization-efficiency-oriented MDMA)		
15	61.40%	6.93%	31.67%	0%
20	28.20%	11.90%	59.90%	0%
25	12.48%	17.00%	70.52%	0%
30	3.83%	18.03%	78.13%	0%
35	0%	20.57%	78.23%	1.2%

a) It means the proportion of UEs that selecting a particular multiple access (MA) mode (expressed as a percentage of UEs).

scheme, referred to as multi-dimensional intelligent multiple access (MD-IMA) in Liu et al. (2021a), as the baseline MDMA scheme.

As validated by Table 21.1, in comparison to MD-IMA, the proposed value-oriented MDMA exhibits contextual agility through its flexible MA mode selections, aiming to satisfy diverse QoS requirements and maintain fairness under dynamic changes of traffic loads (in terms of the number of UEs). For example, when traffic load is high, the value-oriented approach still attempts to serve specific UEs with OMA mode to ensure their strict QoS requirements.

Figure 21.3 illustrates that the simulated user-level value-realization achieved by the value-oriented MDMA and MD-IMA scheme. As shown, the position of each dot on the vertical and horizontal axes indicates values for the QoS satisfaction and resource utilization cost at user-side, respectively. Furthermore, a darker color indicates a higher probability density than areas marked with lighter colors. Value-oriented MDMA outperforms MD-IMA, which can be attributed to two main factors. First, the MD-IMA scheme is solely focused on enhancing the overall QoS performance for UEs, which leads to poorer service quality for users with worse channel conditions. Conversely, the value-oriented MDMA scheme possesses a situational understanding of individual UE's resource constraints and utilization costs, thereby offering better service quality with reduced resource utilization costs at the user end.

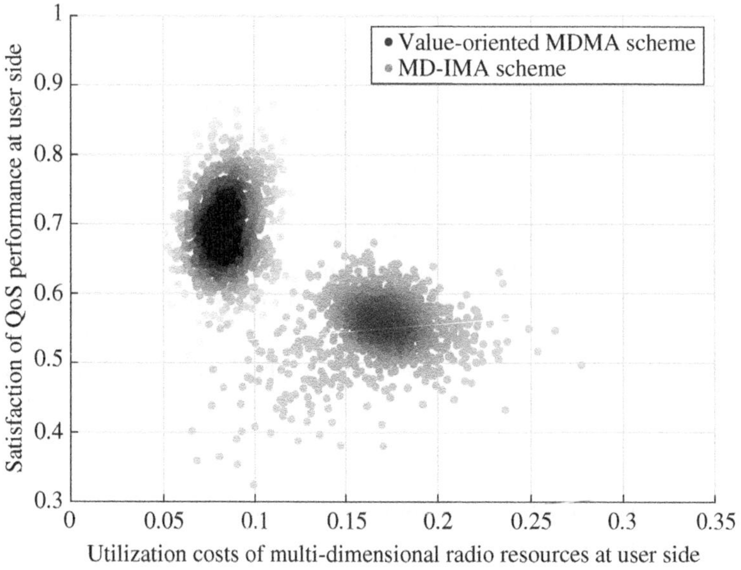

Figure 21.3 The performance of individual-level value realization for each end-user. Individual-level value realization for each end-user is expressed by joint probability density of user's satisfaction of QoS performance and resource utilization cost.

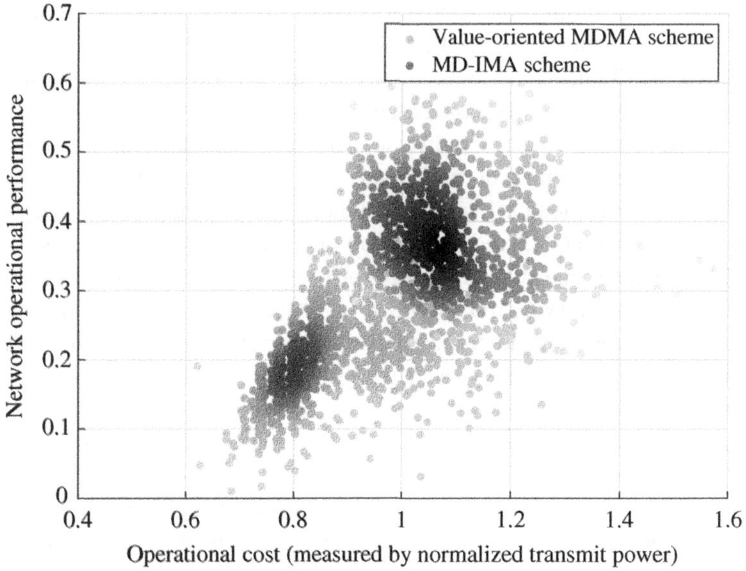

Figure 21.4 The performance of system-level value realization for network operator. System-level value realization is expressed by joint probability density of network performance and network operational cost.

Figure 21.4 shows the performance of system level value realization for network operator. In detail, the performance of system level value realization is measured by two conflicting metrics: (i) network operational performance reflecting the overall level of user QoS satisfaction in the serviced area, which is calculated by Eq. (21.21); (ii) the system operational cost that is defined as the normalized total transmit power of the base station. As we can see, the value-oriented MDMA scheme significantly outperforms the baseline MD-IMA scheme. The MD-IMA scheme slightly outperforms the proposed scheme in term of operational cost. The reason is that the proposed scheme is prone to ensure overall satisfaction level of specific QoS requirements from different end-users at cost of using more transmit power than usual at some network conditions. However, the baseline scheme only focuses on optimizing the performance of radio resource utilization in whole network.

21.5.2 Performance Comparison of Value-Oriented MDMA and State-of-the-Art MA Schemes

Furthermore, to illustrate the effectiveness of value-oriented MDMA, this section compares the performance of our scheme with the following two stat-of-art MA schemes:

- *MIMO–NOMA*: This MA scheme is derived from the work presented in (Wang et al. (2020)), where all UEs are organized into multiple clusters. Within each cluster, UEs share a single beamforming vector and are served by power-domain NOMA.
- *RSMA*: In this scheme, multiple UEs share the same subchannel. BS divides a portion of each UE's message (i.e., the private message) and constructs a common message to broadcast to these UEs. The remaining private messages and the common message are transmitted via SDMA. Using SIC, each UE decodes the common message, followed by the decoding of the intended private message, which is then combined with part of the decoded common message (Mao et al., 2022) (Yang et al., 2021). To align RSMA with our framework, we assign UEs to each subchannel according to the paradigm described in (Han et al. (2021)).

Figure 21.5 shows the cumulative distribution function (CDF), which offers an overview of the performance of user's value-realization utility function achieved by three different MA schemes under the deployment of 25 UEs. In detail, the curves of average utility function at individual level for RSMA and MIMO–NOMA range from a minimum value of 0 to a maximum of 0.76, and 0.17 to 0.54, respectively. In contrast, the proposed MDMA scheme significantly

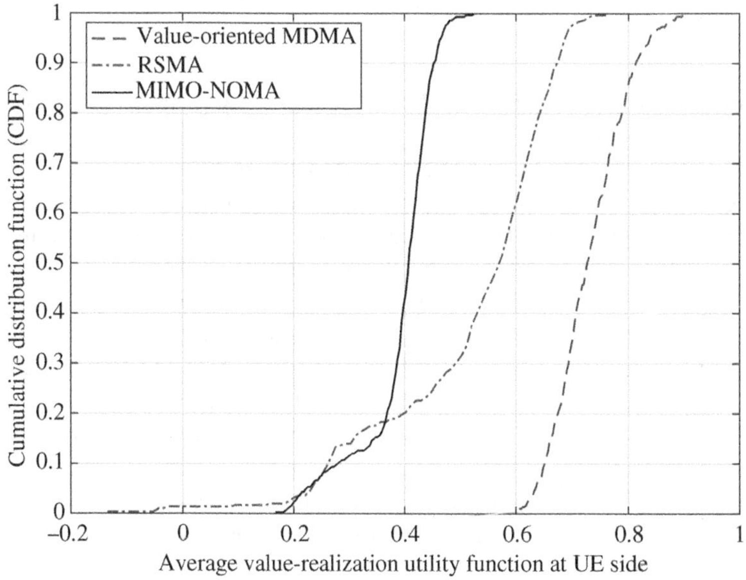

Figure 21.5 CDF of UE's value-realization utility function defined in formula (21.13). This utility function reflects the cost-efficiency of service provisioning. The higher value means that can achieve the satisfied QoS performance with less utilization cost of UE.

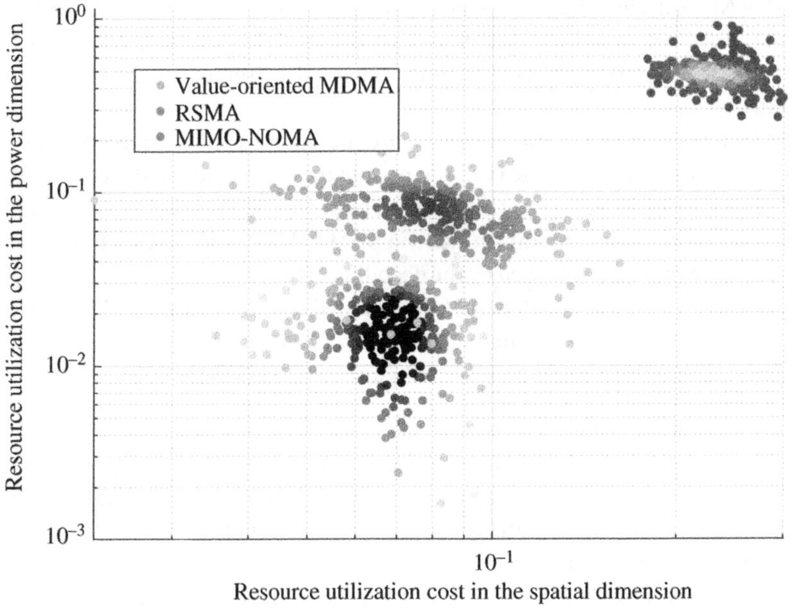

Figure 21.6 Scatter plot of resource utilization cost in different dimensions. The performance of UE's resource utilization cost defined in (21.9) and (21.11), which is the processing cost of using power/spatial-domain radio resources at the UE side.

outperforms the former approaches with notable numerical gains, as evidenced by the increased upper bound.

Figure 21.6 shows the RSMA and MIMO–NOMA schemes are less effective in non-orthogonal interference mitigation process. With the MIMO–NOMA, UEs in the same beamspace experience (nearly) aligned spatial directions with a large disparity in channel strengths, while UEs in different beamspaces experience orthogonal channels. As a result, UEs located in the edge of served area may have severe co-channel interference. RSMA shows a higher utilization cost than the other two MA schemes for three main reasons: (i) UEs need to perform SIC to decode their private messages with increased cost in the power-domain, particularly for UEs with poor channel conditions; (ii) RSMA's performance also depends on the precoder design for broadcasting the common message, which is a challenging problem; (iii) The UE-subcarrier (SC) matching for multi-carrier RSMA remains as an open issue, as different UE combinations sharing one SC may lead to significant performance variations.

Figure 21.7 can be viewed as a "cost-gain" comparison of several multiple access (MA) schemes under various user deployment scenarios, ranging from underloaded to overloaded network situations. In comparison to RSMA and

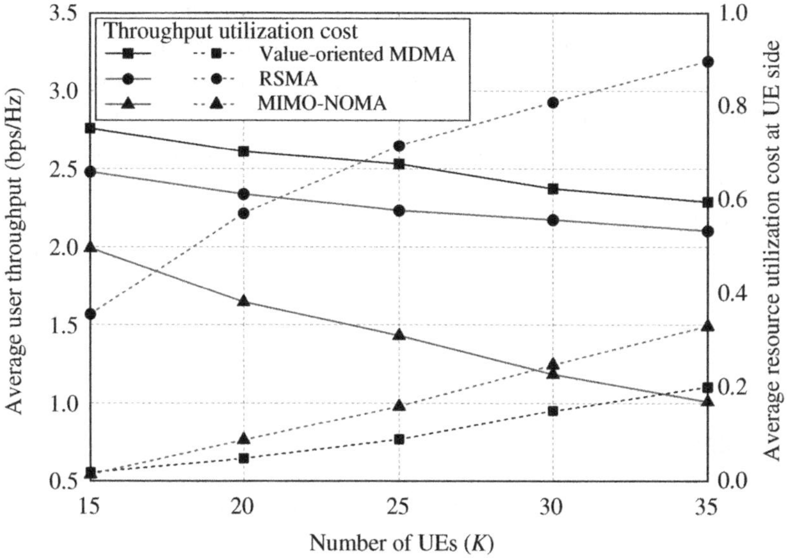

Figure 21.7 Average downlink throughput and resource utilization cost of UE under different MA schemes with respect to the number of UEs.

MIMO–NOMA, the value-oriented MDMA scheme exhibits the lowest utilization cost and highest throughput, with its "cost-gain" performance remaining robust across varying network loads and situation-dependent resource conditions. This is because RSMA and MIMO–NOMA employ static and predetermined MA modes for UEs under all scenarios, leading to sub-optimal utilization of available multi-dimensional radio resources. The results also highlight the importance of intelligently adapting MA modes for UEs based on each end-user device's specific demands and radio resource conditions. Overall, the outcome aligns with our expectations for the value-oriented MDMA scheme.

21.6 Conclusion

This chapter has provided a comprehensive overview of the innovative MDMA technique, an intelligent and flexible approach that addresses the emerging challenges faced by NGMA schemes in the 6G era. By unifying different multiple access schemes in a cohesive paradigm and dynamically adapting to evolving network scenarios, MDMA offers a promising solution to enhance massive connectivity, satisfy diverse service requirements, and fulfill heterogeneous application scenarios and vertical industries for 6G networks.

Throughout the chapter, we have elaborated the fundamental principles of MDMA design, its significant advantages over traditional multiple access schemes, and its potential to achieve individualized service provisioning as well as value-oriented communications. Furthermore, the optimization objectives for value-realization of MDMA are presented from the perspectives of end-user at the individual level and network operators at the system level. Afterwards, the performance of multi-dimensional resource allocation strategies in MDMA has been compared with those in state-of-the-art multiple access schemes. Finally, we demonstrated that MDMA greatly enhances the overall resource utilization efficiency and empowers the intelligent value-driven network operations for 6G communications.

References

Shipon Ali, Ekram Hossain, and Dong In Kim. Non-orthogonal multiple access (NOMA) for downlink multiuser MIMO systems: User clustering, beamforming, and power allocation. *IEEE Access*, 5:565–577, 2017. doi: 10.1109/ACCESS.2016.2646183.

S Boyd and L Vandenberghe. *Convex Optimization*. Cambridge, UK: Cambridge University Press, 2004.

Zhiyong Chen, Zhiguo Ding, and Xuchu Dai. Beamforming for combating inter-cluster and intra-cluster interference in hybrid NOMA systems. *IEEE Access*, 4:4452–4463, 2016. doi: 10.1109/ACCESS.2016.2598380.

Jinho Choi. On the power allocation for MIMO-NOMA systems with layered transmissions. *IEEE Transactions on Wireless Communications*, 15(5):3226–3237, 2016. doi: 10.1109/TWC.2016.2518182.

Bruno Clerckx, Yijie Mao, Robert Schober, Eduard A Jorswieck, David J Love, Jinhong Yuan, Lajos Hanzo, Geoffrey Ye Li, Erik G Larsson, and Giuseppe Caire. Is NOMA efficient in multi-antenna networks? A critical look at next generation multiple access techniques. *IEEE Open Journal of the Communications Society*, 2:1310–1343, 2021. doi: 10.1109/OJCOMS.2021.3084799.

Fang Fang, Haijun Zhang, Julian Cheng, Sébastien Roy, and Victor C M Leung. Joint user scheduling and power allocation optimization for energy-efficient NOMA systems with imperfect CSI. *IEEE Journal on Selected Areas in Communications*, 35(12):2874–2885, 2017. doi: 10.1109/JSAC.2017.2777672.

Wudan Han, Jie Mei, and Xianbin Wang. User-centric multi-dimensional multiple access in 6G communications. In *2021 IEEE International Conference on Communications Workshops (ICC Workshops)*, pages 1–6, 2021. doi: 10.1109/ICCWorkshops50388.2021.9473721.

Wudan Han, Xianbin Wang, and Jie Mei. Edi-driven multi-dimensional resource allocation for inclusive 6G communications. *IEEE Networking Letters*, 5(2):110–114, 2023. doi: 10.1109/LNET.2023.3262704.

Evgeny Khorov, Ilya Levitsky, and Ian F Akyildiz. Current status and directions of IEEE 802.11be, the future Wi-Fi 7. *IEEE Access*, 8:88664–88688, 2020. doi: 10.1109/ACCESS.2020.2993448.

Yuanwei Liu, Hong Xing, Cunhua Pan, Arumugam Nallanathan, Maged Elkashlan, and Lajos Hanzo. Multiple-antenna-assisted non-orthogonal multiple access. *IEEE Wireless Communications*, 25(2):17–23, 2018. doi: 10.1109/MWC.2018.1700080.

Yanan Liu, Xianbin Wang, and Jie Mei. Hybrid multiple access and service-oriented resource allocation for heterogeneous QoS provisioning in machine type communications. *Journal of Communications and Information Networks*, 5(2):225–236, 2020. doi: 10.23919/JCIN.2020.9130438.

Yanan Liu, Xianbin Wang, Gary Boudreau, Akram Bin Sediq, and Hatem Abou-Zeid. A multi-dimensional intelligent multiple access technique for 5G beyond and 6G wireless networks. *IEEE Transactions on Wireless Communications*, 20(2):1308–1320, 2021a. doi: 10.1109/TWC.2020.3032631.

Yanan Liu, Xianbin Wang, Jie Mei, Gary Boudreau, Hatem Abou-Zeid, and Akram Bin Sediq. Situation-aware resource allocation for multi-dimensional intelligent multiple access: A proactive deep learning framework. *IEEE Journal on Selected Areas in Communications*, 39(1):116–130, 2021b. doi: 10.1109/JSAC.2020.3036969.

Yijie Mao, Onur Dizdar, Bruno Clerckx, Robert Schober, Petar Popovski, and H Vincent Poor. Rate-splitting multiple access: Fundamentals, survey, and future research trends. *IEEE Communications Surveys & Tutorials*, 24(4):2073–2126, 2022. doi: 10.1109/COMST.2022.3191937.

Jie Mei, Wudan Han, Xianbin Wang, and H Vincent Poor. Multi-dimensional multiple access with resource utilization cost awareness for individualized service provisioning in 6G. *IEEE Journal on Selected Areas in Communications*, 40(4):1237–1252, 2022. doi: 10.1109/JSAC.2022.3145909.

Shree Krishna Sharma and Xianbin Wang. Toward massive machine type communications in ultra-dense cellular IoT networks: Current issues and machine learning-assisted solutions. *IEEE Communications Surveys & Tutorials*, 22(1):426–471, 2020. doi: 10.1109/COMST.2019.2916177.

Jue Wang, Ye Li, Chen Ji, Qiang Sun, Shi Jin, and Tony Q. S. Quek. Location-based MIMO-NOMA: Multiple access regions and low-complexity user pairing. *IEEE Transactions on Communications*, 68(4):2293–2307, Apr. 2020. doi: 10.1109/TCOMM.2020.2968896.

Xianbin Wang, Jie Mei, Shuguang Cui, Cheng-Xiang Wang, and Xuemin (Sherman) Shen. Realizing 6G: The operational goals, enabling technologies of future networks, and value-oriented intelligent multi-dimensional multiple access. *IEEE Network*, 37(1):10–17, 2023. doi: 10.1109/MNET.001.2200429.

Zhaohui Yang, Mingzhe Chen, Walid Saad, and Mohammad Shikh-Bahaei. Optimization of rate allocation and power control for rate splitting multiple access (RSMA). *IEEE Transactions on Communications*, 69(9):5988–6002, Sept. 2021. doi: 10.1109/TCOMM.2021.3091133.

Jianyue Zhu, Jiaheng Wang, Yongming Huang, Shiwen He, Xiaohu You, and Luxi Yang. On optimal power allocation for downlink non-orthogonal multiple access systems. *IEEE Journal on Selected Areas in Communications*, 35(12):2744–2757, 2017. doi: 10.1109/JSAC.2017.2725618.

22

Efficient Federated Meta-Learning Over Multi-Access Wireless Networks

Sheng Yue and Ju Ren

Department of Computer Science and Technology, Tsinghua University, Beijing, China

22.1 Introduction

The integration of artificial intelligence (AI) and Internet-of-Things (IoT) has led to a proliferation of research efforts on edge intelligence, aiming at pushing AI frontiers to the wireless network edge proximal to IoT devices and data sources (Park et al., 2019). It is expected that edge intelligence will reduce time-to-action latency down to milliseconds for IoT applications while minimizing network bandwidth and offering security guarantees (Plastiras et al., 2018). However, a general consensus is that a single IoT device can hardly realize edge intelligence due to its limited computational and storage capabilities. Accordingly, it is natural to rely on collaboration in edge learning, whereby IoT devices work together to accomplish computation-intensive tasks (Zhang et al., 2019).

Building on a synergy of federated learning (McMahan et al., 2017) and meta-learning (Finn et al., 2017), federated meta-learning (FML) has been proposed under a common theme of fostering edge–edge collaboration (Chen et al., 2019, Jiang et al., 2019, Lin et al., 2020). In FML, IoT devices join forces to learn an *initial shared model* under the orchestration of a central server such that current or new devices can quickly adapt the learned model to their local datasets via one or a few gradient descent steps. Notably, FML can keep all the benefits of the federated learning paradigm (such as simplicity, data security, and flexibility), while giving a more personalized model for each device to capture the differences among tasks (Fallah et al., 2020b). Therefore, FML has emerged as a promising approach to tackle the heterogeneity challenges in federated learning and to facilitate efficient edge learning (Kairouz et al., 2021).

Despite its promising benefits, FML comes with new challenges. On one hand, the number of participated devices can be enormous. Selecting devices

Next Generation Multiple Access, First Edition.
Edited by Yuanwei Liu, Liang Liu, Zhiguo Ding, and Xuemin Shen.
© 2024 The Institute of Electrical and Electronics Engineers, Inc. Published 2024 by John Wiley & Sons, Inc.

uniformly and randomly, as is often done in the existing methods, can result in slow convergence speed (Yue et al., 2021a, Fallah et al., 2020b). Although recent studies (Nishio and Yonetani, 2019, Nguyen et al., 2020, Chen et al., 2020b) have characterized the convergence of federated learning and proposed nonuniform device selection mechanisms, they cannot be directly applied to FML problems due to the bias and high-order information in the stochastic gradients. On the other hand, the performance of FML in a wireless environment is highly related to its wall-clock time, including the computation time (determined by local data sizes and devices' CPU types) and communication time (depending on channel gains, interference, and transmission power) (Dinh et al., 2020, Liu et al., 2023). If not properly controlled, a large wall-clock time can cause unexpected training delay and communication inefficiency. In addition, the deep neural network (DNN) model training, which involves a large number of data samples and epochs, usually induces high computational cost, especially for sophisticated model structures consisting of millions of parameters (Dinh et al., 2020). Due to the limited power capacity of IoT devices, energy consumption should also be properly managed to ensure system sustainability and stability (Chen et al., 2020b). In a nutshell, for the purpose of efficiently deploying FML in today's wireless systems, the strategy of device selection and resource allocation must be carefully crafted not only to accelerate the learning process, but also to control the wall-clock time of training and energy cost in edge devices. Unfortunately, despite their importance, there are limited studies on these aspects in the current literature.

In this chapter, we tackle the above-mentioned challenges in two steps: (i) we develop an algorithm (called NUFM) with a nonuniform device selection scheme to improve the convergence rate of the FML algorithm; (ii) based on NUFM, we propose a resource allocation strategy (called URAL) that jointly optimizes the convergence speed, wall-clock time, and energy consumption in the context of multi-access wireless systems. More specifically, we first rigorously quantify the contribution of each device to the convergence of FML via deriving a tight lower bound on the reduction of one-round global loss. Based on the quantitative results, we present a nonuniform device selection scheme that maximizes the loss reduction per round, followed by the NUFM algorithm. Then, we formulate a resource allocation problem for NUFM over wireless networks, capturing the trade-offs among convergence, wall-clock time, and energy cost. To solve this problem, we exploit its special structure and decompose it into two sub-problems. The first one is to minimize the computation time via controlling devices' CPU-cycle frequencies, which is solved optimally based on the analysis of the effect of device heterogeneity on the objective. The second sub-problem aims at optimizing the resource block allocation and transmission power management. It is a nonconvex mixed-integer non-linear programming (MINLP) problem, and deriving a closed-form solution is a challenging task. By deconstructing the problem step by step, we devise an iterative method to solve the formulated MINLP and provide a convergence guarantee.

In summary, our main contributions are listed as follows:

- We provide a theoretical characterization of contribution of an individual device to the convergence of FML in each round, via establishing a tight lower bound on the one-round reduction of expected global loss. Using this quantitative result, we develop NUFM, a fast-convergent FML algorithm with nonuniform device selection;
- To embed NUFM in the context of multi-access wireless systems, we formulate a resource allocation problem, capturing the trade-offs among the convergence, wall-clock time, and energy consumption. By decomposing the original problem into two sub-problems and deconstructing sub-problems step by step, we propose a joint device selection and resource allocation algorithm (namely URAL) to solve the problem effectively with theoretical performance guarantees;
- To reduce the computational complexity, we further integrate our proposed algorithms with two first-order approximation techniques in Fallah et al. (2020b), by which the complexity of a one-step update in NUFM can be reduced from $O(d^2)$ to $O(d)$. We also show that our theoretical results hold in these cases;
- We provide extensive simulation results on challenging real-world benchmarks (i.e., Fashion-MNIST, CIFAR-10, CIFAR-100, and ImageNet) to demonstrate the efficacy of our methods.

22.2 Related Work

Federated learning (FL) (McMahan et al., 2017) has been proposed as a promising technique to facilitate edge-edge collaborative learning (Fadlullah and Kato, 2020). However, due to the heterogeneity in devices, models, and data distributions, a shared global model often fails to capture the individual information of each device, leading to performance degradation in inference or classification (Wu et al., 2020, Kairouz et al., 2021, Yang et al., 2019).

Very recently, based on the advances in meta-learning (Finn et al., 2017), FML has garnered much attention, which aims to learn a personalized model for each device to cope with the heterogeneity challenges (Chen et al., 2019, Jiang et al., 2019, Lin et al., 2020, Fallah et al., 2020b, Yue et al., 2021a). Chen et al. (2019) first introduce an FML method called FedMeta, integrating the model-agnostic meta-learning (MAML) algorithm (Finn et al., 2017) into the federated learning framework. They show that FML can significantly improve the performance of FedAvg (McMahan et al., 2017). Jiang et al. (2019) analyze the connection between

FedAvg and MAML, and empirically demonstrate that FML enables better and more stable personalized performance. From a theoretical perspective, Lin et al. (2020) analyze the convergence properties and computational complexity of FML with strongly convex loss functions and exact gradients. Fallah et al. (2020b) further provide the convergence guarantees in non-convex cases with stochastic gradients. Different from the above gradient descent–based approaches, another recent work (Yue et al., 2021a) develops an ADMM-based FML method and gives its convergence guarantee under non-convex cases. However, due to the reason that devices are uniformly and randomly selected, the existing FML algorithms often suffer from slow convergence and low communication efficiency (Fallah et al., 2020b). Further, deploying FML in practical wireless systems calls for effective resource allocation strategies (Dinh et al., 2020, Yue et al., 2022), which is still an open problem in the existing FML literature.

There exists a significant body of works placing interests in convergence improvement and resource allocation for FL (Ren et al., 2020a,b, Zhu et al., 2019, Zeng et al., 2020, Shi et al., 2020b, Nguyen et al., 2020, Chen et al., 2020a,b, Yu et al., 2019, Dinh et al., 2020, Xu and Wang, 2020, Wang et al., 2019, Karimireddy et al., 2020, Luo et al., 2021, Luping et al., 2019, Li et al., 2019, Sim et al., 2019, Shi et al., 2020a, Yang et al., 2020). Regarding the convergence improvement, (Nguyen et al., 2020) propose a fast-convergent FL algorithm, called FOLB, which achieves a near-optimal lower bound for the overall loss decrease in each round. Note that, while the idea of NUFM is similar to Nguyen et al. (2020), the lower bound for FML is derived from a completely different technical path due to the inherent complexity in the local update. To minimize the convergence time, a probabilistic device selection scheme for FL is designed in Chen et al. (2020a), which assigns high probabilities to the devices with large effects on the global model. Ren et al. (2020b) investigate a batchsize selection strategy for accelerating the FL training process. Karimireddy et al. (2020) employ the variance reduction technique to develop a new FL algorithm. Based on momentum methods, (Yu et al., 2019) give an FL algorithm with linear speedup property. Regarding the resource allocation in FL, (Dinh et al., 2020) embed FL in wireless networks, considering the trade-offs between training time and energy consumption, under the assumption that all devices participate in the whole training process. From a long-term perspective, (Xu and Wang, 2020) empirically investigate the device selection scheme jointly with bandwidth allocation for FL, using the Lyapunov optimization method. Chen et al. (2020b) investigate a device selection problem with “hard” resource constraints to enable the implementation of FL over wireless networks. Wang et al. (2019) propose a control algorithm to determine the best trade-off between the local update and global aggregation under a resource budget.

Although extensive research has been carried out on FL, researchers have not treated FML in much detail. In particular, the existing FL acceleration techniques cannot be directly applied to FML due to the high-order information and biased stochastic gradients in the local update phases.[1] At the same time, device selection and resource allocation require crafting jointly (Chen et al., 2019), rather than simply plugging the existing strategies in.

22.3 Preliminaries and Assumptions

In this section, we introduce FML, including the learning problem, standard algorithm, and assumptions for theoretical analysis.

22.3.1 Federated Meta-Learning Problem

We consider a set $\mathcal{N}$ of user devices that are all connected to a server. Each device $i \in \mathcal{N}$ has a labeled dataset $\mathcal{D}_i = \{x_i^j, y_i^j\}_{j=1}^{D_i}$ that can be accessed only by itself. Here, the tuple $(x_i^j, y_i^j) \in \mathcal{X} \times \mathcal{Y}$ is a data sample with input x_i^j and label y_i^j, and follows an unknown underlying distribution P_i. Define θ as the model parameter, such as the weights of a DNN model. For device i, the loss function of a model parameter $\theta \in \mathbb{R}^d$ is defined as $\ell_i(\theta; x, y)$, which measures the error of model θ in predicting the true label y given input x.

FML looks for a good model initialization (also called meta-model) such that the well-performed models of different devices can be quickly obtained via one or a few gradient descent steps. More specially, FML aims to solve the following problem

$$\min_{\theta \in \mathbb{R}^d} F(\theta) := \frac{1}{n} \sum_{i \in \mathcal{N}} f_i(\theta - \alpha \nabla f_i(\theta)), \tag{22.1}$$

where f_i represents the expected loss function over the data distribution of device i, i.e., $f_i(\theta) := \mathbb{E}_{(x,y)\sim P_i}\left[\ell_i(\theta; x, y)\right]$, $n = |\mathcal{N}|$ is the number of devices, and α is the stepsize. The advantages of this formulation are twofold: (i) It gives a personalized solution that can capture any heterogeneity between the devices; (ii) the meta-model can quickly adapt to new devices via slightly updating it with respect to their own data. Clearly, FML well fits edge learning cases, where edge devices have insufficient computing power and limited data samples.

Next, we review the standard FML algorithm in the literature.

1 Different from FML, FL is a first-order method with unbiased stochastic gradients.

Table 22.1 Key notations.

Notation	Definition	Notation	Definition
θ	Model parameter	$F_i(\cdot)$	Meta–loss function
α, β	Stepsize and meta–learning rate	L_i, ρ_i	Lipschitz continuous parameters
i, j	Indexes of user devices and data samples	σ_G, σ_H	Upper bounds of variances
k	Index of training rounds	γ_G, γ_H	Similarity parameters
t	Index of local update steps	u_i	Contribution to global loss reduction of device i
τ	Number of local update steps	v_i	CPU-cycle frequency of device i
P_i	Underlying data distribution of device i	p_i	Transmission power of device i
$\mathcal{D}_i, D_i$	Dataset and its size of device i	$z_{i,m}$	Binary variable indicating if device i accesses RB m
$\mathcal{N}$	Set of devices	η_1, η_2	Weight parameters
n	Number of devices	c_i	CPU cycles for computing one sample by device i
$\mathcal{N}_k$	Set of participating devices in round k	h_i, I_m	Channel gain of device i and inference in RB m
n_k	Number of participating devices in round k	B, N_0	Bandwidth of each RB and noise power spectral density
x, y	Input and corresponding label	$\mathcal{M}, M$	Set and number of RBs
$l_i(\theta; \mathrm{x}, \mathrm{y})$	Loss of model θ on sample (x, y)	E	Total energy consumption
$f_i(\cdot)$	Expected loss function	T	Total latency

22.3.2 Standard Algorithm

Similar to federated learning, standard FML algorithm solves (22.1) in two repeating steps: *local update* and *global aggregation* (Fallah et al., 2020b), as detailed below.

- *Local update*: At the beginning of each round k, the server first sends the current global model θ^k to a fraction of devices $\mathcal{N}_k$ chosen uniformly at random with pre-set size n_k. Then, each device $i \in \mathcal{N}_k$ updates the received model based on its meta-function $F_i(\theta) := f_i(\theta - \alpha \nabla f_i(\theta))$ by running τ ($\geq$1) steps of stochastic gradient descent locally (also called mini-batch gradient descent), i.e.,

$$\theta_i^{k,t+1} = \theta_i^{k,t} - \beta\tilde{\nabla}F_i(\theta_i^{k,t}), \text{ for } 0 \le t \le \tau - 1, \tag{22.2}$$

where θ_t^k denotes the local model of device i in the t-th step of the local update in round k with $\theta_i^{k,0} = \theta^k$, and $\beta > 0$ is the meta-learning rate. In (22.2), the stochastic gradient $\tilde{\nabla}F_i(\theta)$ is given by

$$\tilde{\nabla}F_i(\theta) := (I - \alpha\tilde{\nabla}^2 f_i(\theta, \mathcal{D}_i''))\tilde{\nabla}f_i\left(\theta - \alpha\tilde{\nabla}f_i(\theta, \mathcal{D}_i), \mathcal{D}_i'\right), \tag{22.3}$$

where $\mathcal{D}_i$, $\mathcal{D}_i'$, and $\mathcal{D}_i''$ are independent batches,[2] and for any batch $\mathcal{D}$, $\tilde{\nabla}f_i(\theta, \mathcal{D})$ and $\tilde{\nabla}^2 f_i(\theta, \mathcal{D})$ are the unbiased estimates of $\nabla f_i(\theta)$ and $\nabla^2 f_i(\theta)$ respectively, i.e.,

$$\tilde{\nabla}f_i(\theta, \mathcal{D}) := \frac{1}{|\mathcal{D}|}\sum_{(x,y)\in\mathcal{D}} \nabla\ell_i(\theta; x, y), \tag{22.4}$$

$$\tilde{\nabla}^2 f_i(\theta, \mathcal{D}) := \frac{1}{|\mathcal{D}|}\sum_{(x,y)\in\mathcal{D}} \nabla^2\ell_i(\theta; x, y). \tag{22.5}$$

- *Global aggregation*: After updating the local model parameter, each selected device sends its local model $\theta_i^k = \theta_i^{k,\tau-1}$ to the server. The server updates the global model by averaging over the received models, i.e.,

$$\theta^{k+1} = \frac{1}{n_k}\sum_{i\in\mathcal{N}_k} \theta_i^k. \tag{22.6}$$

It is easy to see that the main difference between FL and FML lies in the local update phase: In FL, local update is done using the unbiased gradient estimates while FML uses the biased one consisting of high-order information. Besides, federated learning can be considered as a special case of FML, i.e., FML under $\alpha = 0$.

22.3.3 Assumptions

In this subsection, we list the standard assumptions for the analysis of FML algorithms (Fallah et al., 2020b, Lin et al., 2020, Yue et al., 2021a).

Assumption 22.1 ***(Smoothness)*** The expected loss function f_i corresponding to device $i \in \mathcal{N}$ is twice continuously differentiable and L_i-smooth, i.e.,

$$\|\nabla f_i(\theta_1) - \nabla f_i(\theta_2)\| \le L_i \|\theta_1 - \theta_2\|, \ \forall\theta_1, \theta_2 \in \mathbb{R}^d. \tag{22.7}$$

Besides, its gradient is bounded by a positive constant ζ_i, i.e., $\|\nabla f_i(\theta)\| \le \zeta_i$.

Assumption 22.2 ***(Lipschitz Hessian)*** The Hessian of function f_i is ρ_i-Lipschitz continuous for each $i \in \mathcal{N}$, i.e.,

$$\|\nabla^2 f_i(\theta_1) - \nabla^2 f_i(\theta_2)\| \le \rho_i \|\theta_1 - \theta_2\|, \ \forall\theta_1, \theta_2 \in \mathbb{R}^d. \tag{22.8}$$

2 We slightly abuse the notation $\mathcal{D}_i$ as a batch of the local dataset of the ith device.

Assumption 22.3 ***(Bounded Variance)*** Given any $\theta \in \mathbb{R}^d$, the following facts hold for stochastic gradient $\nabla \ell_i(\theta; x, y)$ and Hessians $\nabla^2 \ell_i(\theta; x, y)$ with $(x, y) \in \mathcal{X} \times \mathcal{Y}$

$$\mathbb{E}_{(x,y)\sim P_i}\left[\|\nabla \ell_i(\theta; x, y) - \nabla f_i(\theta)\|^2\right] \le \sigma_G^2 \tag{22.9}$$

$$\mathbb{E}_{(x,y)\sim P_i}\left[\left\|\nabla^2 \ell_i(\theta; x, y) - \nabla^2 f_i(\theta)\right\|^2\right] \le \sigma_H^2. \tag{22.10}$$

Assumption 22.4 ***(Similarity)*** For any $\theta \in \mathbb{R}^d$ and $i, j \in \mathcal{N}$, there exist non-negative constants $\gamma_G \ge 0$ and $\gamma_H \ge 0$ such that the gradients and Hessians of the expected loss functions $f_i(\theta)$ and $f_j(\theta)$ satisfy the following conditions

$$\|\nabla f_i(\theta) - \nabla f_j(\theta)\| \le \gamma_G \tag{22.11}$$

$$\|\nabla^2 f_i(\theta) - \nabla^2 f_j(\theta)\| \le \gamma_H. \tag{22.12}$$

Assumption 22.2 implies the high-order smoothness of $f_i(\theta)$ dealing with the second-order information in the local update step (22.2). Assumption 22.4 indicates that the variations of gradients between different devices are bounded by some constants, which captures the similarities between devices' tasks corresponding to non-IID data. It holds for many practical loss functions (Zhang et al., 2020), such as logistic regression and hyperbolic tangent functions. In particular, ψ_i^g and ψ_i^h can be roughly seen as a distance between data distributions P_i and P_j (Fallah et al., 2020a).

22.4 Nonuniform Federated Meta-Learning

Due to the uniform selection of devices in each round, the convergence rate of the standard FML algorithm is naturally slow. In this section, we present a non-uniform device selection scheme to tackle this challenge.

22.4.1 Device Contribution Quantification

We begin with quantifying the contribution of each device to the reduction of one-round global loss using its dataset size and gradient norms. For convenience, define $\zeta := \max_i \zeta_i$, $L := \max_i L_i$, and $\rho := \max_i \rho_i$. We first provide necessary lemmas before giving the main result.

Lemma 22.1 If Assumptions 22.1 and 22.2 hold, local meta-function F_i is smooth with parameter $L_F := (1 + \alpha L)^2 L + \alpha \rho \zeta$.

Proof: [of Proof]. We expand the expression of $\| \nabla F_i(\theta_1) - \nabla F_i(\theta_2) \|$ into two parts by triangle inequality, followed by bounding each part via Assumptions 1 and 2. The detailed proof is presented in Appendix A of the technical report (Yue et al., 2021b). □

Lemma 22.1 gives the smoothness of the local meta-function F_i and global loss function F.

Lemma 22.2 Suppose that Assumptions 22.1–22.3 are satisfied, and $\mathcal{D}_i$, $\mathcal{D}'_i$, and $\mathcal{D}''_i$ are independent batches with sizes D_i, D'_i, and D''_i respectively. For any $\theta \in \mathbb{R}^d$, the following holds

$$\left\| \mathbb{E}\left[\tilde{\nabla} F_i(\theta) - \nabla F_i(\theta)\right] \right\| \leq \frac{\alpha \sigma_G L(1 + \alpha L)}{\sqrt{D_i}} \tag{22.13}$$

$$\mathbb{E}\left[\|\tilde{\nabla} F_i(\theta) - \nabla F_i(\theta)\|^2\right] \leq \sigma_{F_i}^2, \tag{22.14}$$

where $\sigma_{F_i}^2$ is denoted as

$$\sigma_{F_i}^2 := 6\sigma_G^2(1 + \alpha L)^2 \left(\frac{1}{D'_i} + \frac{(\alpha L)^2}{D_i} \right) + \frac{6(\alpha \sigma_G \sigma_H)^2}{D''_i} \left(\frac{1}{D'_i} + \frac{(\alpha L)^2}{D_i} \right) + \frac{3(\alpha \zeta \sigma_H)^2}{D''_i}. \tag{22.15}$$

Proof: We first obtain

$$\begin{aligned} \| \mathbb{E}[\tilde{\nabla} F_i(\theta) - \nabla F_i(\theta)] \| \leq \| &(I - \alpha \nabla^2 f_i(\theta)) \mathbb{E}[\delta_2^*] + \mathbb{E}[\delta_1^* \delta_2^*] \\ &+ \mathbb{E}[\delta_1^*] \nabla f_i(\theta - \alpha \nabla f_i(\theta)) \|, \\ \mathbb{E}[\| \tilde{\nabla} F_i(\theta) - \nabla F_i(\theta) \|^2] \leq 3 &\left(\mathbb{E}[\| \delta_1^* \|^2] \mathbb{E}[\| \delta_2^* \|^2] \right. \\ &\left. + \zeta^2 \mathbb{E}[\| \delta_1^* \|^2] + (1 + \alpha L)^2 \mathbb{E}[\| \delta_2^* \|^2] \right), \end{aligned}$$

where δ_1^* and δ_2^* are given by

$$\begin{aligned} \delta_1^* &= \alpha \left(\nabla^2 f_i(\theta) - \tilde{\nabla}^2 f_i(\theta, \mathcal{D}''_i) \right) \\ \delta_2^* &= \tilde{\nabla} f_i \left(\theta - \alpha \tilde{\nabla} f_i(\theta, \mathcal{D}_i), \mathcal{D}'_i \right) - \nabla f_i \left(\theta - \alpha \nabla f_i(\theta) \right). \end{aligned}$$

Then, we derive the results via bounding the first and second moments of δ_1^* and δ_2^*. The detailed proof is presented in Appendix B of the technical report (Yue et al., 2021b). □

Lemma 22.2 shows that the stochastic gradient $\tilde{\nabla} F_i(\theta)$ is a biased estimate of $\nabla F_i(\theta)$, revealing the challenges in analyzing FML algorithms.

Lemma 22.3 If Assumptions 22.1, 22.2, and 22.4 are satisfied, then for any $\theta \in \mathbb{R}^d$ and $i, j \in \mathcal{N}$, we have

$$\|\nabla F_i(\theta) - \nabla F_j(\theta)\| \leq (1 + \alpha L)^2 \gamma_G + \alpha \zeta \gamma_H. \tag{22.16}$$

Proof: We divide the bound of $\|\nabla F_i(\theta) - \nabla F_j(\theta)\|$ into two independent terms, followed by bounding the two terms separately. The detailed proof is presented in Appendix C of the technical report (Yue et al., 2021b). □

Lemma 22.3 characterizes the similarities between the local meta-functions, which is critical for analyzing the one-step global loss reduction because it relates the local meta-function and global objective. Based on Lemmas 22.1–22.3, we are now ready to give our main result.

Theorem 22.1 Suppose that Assumptions 22.1–22.4 are satisfied, and $\mathcal{D}_i$, $\mathcal{D}'_i$, and $\mathcal{D}''_i$ are independent batches. If the local update and global aggregation follow (22.2) and (22.6) respectively, then the following fact holds true for $\tau = 1$

$$\begin{aligned}\mathbb{E}[F(\theta^k) - F(\theta^{k+1})] \geq \beta \mathbb{E}\Bigg[\frac{1}{n_k} \sum_{i \in \mathcal{N}_k} \Bigg(\left(1 - \frac{L_F \beta}{2}\right) \|\tilde{\nabla} F_i(\theta^k)\|^2 \\ - \left(\sqrt{(1 + \alpha L)^2 \gamma_G + \alpha \zeta \gamma_H} + \sigma_{F_i} \right) \\ \times \sqrt{\mathbb{E}\left[\|\tilde{\nabla} F_i(\theta^k)\|^2 \mid \mathcal{N}_k\right]} \Bigg) \Bigg],\end{aligned} \tag{22.17}$$

where the outer expectation of RHS is taken with respect to the selected user set $\mathcal{N}_k$ and data sample sizes, and the inner expectation is only regarding data sample sizes.

Proof: Using the smoothness condition of F_i, we express the lower bound of loss reduction by

$$\mathbb{E}[F(\theta^k) - F(\theta^{k+1})] \geq \mathbb{E}[G^k],$$

where G^k is defined as

$$G^k := \beta \nabla F(\theta^k)^\top \left(\frac{1}{n_k} \sum_{i \in \mathcal{N}_k} \tilde{\nabla} F_i(\theta^k) \right) - \frac{L_F \beta^2}{2} \left\| \frac{1}{n_k} \sum_{i \in \mathcal{N}_k} \tilde{\nabla} F_i(\theta^k) \right\|^2.$$

The key step to derive the desired result is providing a tight lower bound for the product of $\nabla F(\theta^k)$ and $\tilde{\nabla} F(\theta^k)$. The detailed proof is presented in Appendix G of the technical report (Yue et al., 2021b). □

Theorem 22.1 provides a lower bound on the one-round reduction of the global objective function F based on the device selection. It implies that different user selection has varying impacts on the objective improvement and quantifies the contribution of each device to the objective improvement, depending on the variance of local meta-function, task similarities, smoothness, and learning rates. It therefore provides a criterion for selecting users to accelerate the convergence.

From Theorem 22.1, we have the following corollary, which simplifies the above result and extends it to multi-step cases.

Corollary 22.1 Suppose that Assumptions 22.1–22.4 are satisfied, and $\mathcal{D}_i$, $\mathcal{D}'_i$, and $\mathcal{D}''_i$ are independent batches with $D_i = D'_i = D''_i$. If the local update and global aggregation follow (22.2) and (22.6) respectively, then the following fact holds true with $\beta \in [0,1/L_F)$

$$\mathbb{E}\left[F(\theta^k) - F(\theta^{k+1})\right] \geq \frac{\beta}{2}\mathbb{E}\left[\frac{1}{n_k}\sum_{i\in\mathcal{N}_k}\sum_{t=0}^{\tau-1}\left(\|\tilde{\nabla}F_i(\theta_i^{k,t})\|^2\right.\right.$$
$$\left.\left.-2\left(\lambda_1 + \frac{\lambda_2}{\sqrt{D_i}}\right)\sqrt{\mathbb{E}\left[\|\tilde{\nabla}F_i(\theta_i^{k,t})\|^2 \mid \mathcal{N}_k\right]}\right)\right], \tag{22.18}$$

where positive constants λ_1 and λ_2 satisfy that

$$\lambda_1 \geq \sqrt{(1+\alpha L)^2\gamma_G + \alpha\zeta\gamma_H} + \beta\tau\sqrt{35(\gamma_G^2 + 2\sigma_F^2)} \tag{22.19}$$

$$\lambda_2^2 \geq 6\sigma_G^2\left(1+(\alpha L)^2\right)\left((\alpha\sigma_H)^2 + (1+\alpha L)^2\right)$$
$$+ 3(\alpha\zeta\sigma_H)^2. \tag{22.20}$$

Proof: The proof is similar to Theorem 22.1, with additional tricks in bounding the product of $\nabla F(\theta^k)$ and $\tilde{\nabla}F(\theta^k)$. The detailed proof is presented in Appendix I of the technical report (Yue et al., 2021b). □

Corollary 22.1 implies that the device with a large gradient naturally accelerates the global loss decrease, but a small dataset size degrades the process due to corresponding high variance. Besides, as the device dissimilarities become large, the lower bound (22.18) weakens.

Motivated by Corollary 22.1, we study the device selection in the following Section 22.4.2.

22.4.2 Device Selection

To improve the convergence speed, we aim to maximize the lower bound (22.18) on the one-round objective reduction. Based on Corollary 22.1, we define the contribution u_i^k device i to the convergence in round k as

$$u_i^k := \sum_{t=0}^{\tau-1} \| \tilde{\nabla} F_i(\theta_i^{k,t}) \|^2 - 2\left(\lambda_1 + \frac{\lambda_2}{\sqrt{D_i}}\right) \| \tilde{\nabla} F_i(\theta_i^{k,t}) \|, \tag{22.21}$$

where we replace the second moment of $\tilde{\nabla} F_i(\theta_i^{k,t})$ by its sample value $\| \tilde{\nabla} F_i(\theta_i^{k,t}) \|^2$. Then, the device selection problem in round k can be formulated as

$$\begin{aligned} \max_{\{z_i\}} \quad & \sum_{i \in \mathcal{N}} z_i u_i^k \\ \text{s.t.} \quad & \sum_{i \in \mathcal{N}} z_i = n_k \\ & z_i \in \{0,1\},\ \forall i \in \mathcal{N}. \end{aligned} \tag{22.22}$$

In (22.22), z_i is a binary variable: $z_i = 1$ for selecting device i in this round; $z_i = 0$, otherwise. The solution of (22.22) can be found by the *Select* algorithm introduced in Cormen et al. (2009, Chapter 9) with the worst-case complexity $O(n)$ (the problem (22.22) is indeed a *selection problem*). Accordingly, our device selection scheme is presented as follows.

Device selection: Following the local update phase, instead of selecting devices uniformly at random, each device i first computes its contribution scalar u_i^k locally and sends it to the server. After receiving $\{u_i^k\}_{i \in \mathcal{N}}$ from all devices, the server runs the *Select* algorithm and finds the optimal device set denoted by $\mathcal{N}_k^*$. Notably, although constants λ_1 and λ_2 in (22.22) consist of unknown parameters such as L, γ_G, and γ_H, they can be either estimated during training as in Wang et al. (2019) or directly tuned as in our simulations.

Based on the device selection scheme, we propose the *Nonuniform Federated Meta-Learning (NUFM)* algorithm (depicted in Algorithm 22.1). In particular, although NUFM requires an additional communication phase to upload u_i^k to the server, the communication overhead can be negligible because u_i^k is just a scalar.

22.5 Federated Meta-Learning Over Wireless Networks

In this section, we extend NUFM to the context of multi-access wireless systems, where the bandwidth for uplink transmission and the power of IoT devices are limited. First, we present the system model followed by the problem formulation. Then, we decompose the original problem into two sub-problems and devise solutions for each of them with theoretical performance guarantees.

22.5.1 System Model

As illustrated in Fig. 22.1, we consider a wireless multiuser system, where a set $\mathcal{N}$ of n end devices joint forces to carry out FML aided by an edge server. Each round consists of two stages: the computation phase and the communication phase. In the computation phase, each device $i \in \mathcal{N}$ downloads the current global model and computes the local model based on its local dataset; in the communication phase, the selected devices transmit the local models to the edge server via a limited number of wireless channels. After that, the edge server runs the global aggregation and starts the next round. Here, we do not consider the downlink communication due to the asymmetric uplink-downlink settings in wireless networks. That is, the transmission power at the server (e.g., a base station) and the downlink communication bandwidth are generally sufficient for global meta-model transmission. Thus, the downlink time is usually neglected, compared to the uplink data transmission time (Dinh et al., 2020). Since we focus

---

Algorithm 22.1 Nonuniform Federated Meta-Learning (NUFM)

Input: α, β, λ_1, λ_2

Server initializes model θ^0 and sends it to all devices;
for round $k = 0$ to $K - 1$ **do**
  for each device $i \in \mathcal{N}$ **do**
    //Local update
    Initialize $\theta_i^{k,0} \leftarrow \theta^k$ and $u_i^k \leftarrow 0$;
    for local step $t = 0$ to $\tau - 1$ **do**
      Compute stochastic gradient $\tilde{\nabla} F_i(\theta_i^{k,t})$ by (22.3) using batches $\mathcal{D}_i$, $\mathcal{D}_i'$, and $\mathcal{D}_i''$
      Update local model $\theta_i^{k,t+1}$ by (22.2);
      Update contribution scalar u_i^k by $u_i^k \leftarrow u_i^k + \|\tilde{\nabla} F_i(\theta_i^{k,t})\|^2 - 2\left(\lambda_1 + \frac{\lambda_2}{\sqrt{D_i}}\right)\|\tilde{\nabla} F_i(\theta_i^{k,t})\|$;
    end for
    Set $\theta_i^k = \theta_i^{k,\tau}$ and send u_i^k to server;
  end for
  //Device selection
  Once receiving $\{u_i^k\}_{i\in\mathcal{N}}$, server computes optimal device selection $\mathcal{N}_k^*$ by solving (22.22);
  //Global aggregation
  After receiving local models $\{\theta_i^k\}_{i\in\mathcal{N}_k^*}$, server computes the global model by (22.6);
end for
return θ^K

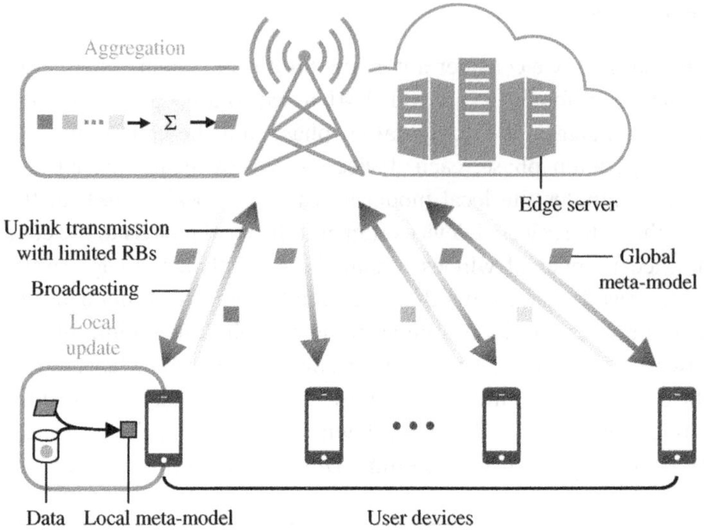

Figure 22.1 The architecture of federated meta-learning over a wireless network with multiple user devices and an edge server. Due to limited communication resources, only part of user devices can upload their local models in each training round.

on the device selection and resource allocation problem in each round, we omit subscript k for brevity throughout this section.

22.5.1.1 Computation Model

We denote c_i as the CPU cycles for device i to update the model with one sample, which can be measured offline as a priori knowledge (Dinh et al., 2020). Assume that the batch size of device i used in local update phase (22.2) is D_i. Then, the number of CPU cycles required for device i to run a one-step local update is $c_i D_i$. We denote the CPU-cycle frequency of device i as ν_i. Thus, the CPU energy consumption of device i in the computation during the local update phase can be expressed by

$$E_i^{cp}(\nu_i) := \frac{\iota_i}{2}\tau_i c_i D_i \nu_i^2, \tag{22.23}$$

where $\iota_i/2$ is the effective capacitance coefficient of the computing chipset of device i (Burd and Brodersen, 1996). The computational time of device i in a round can be denoted as

$$T_i^{cp}(\nu_i) := \frac{\tau_i c_i D_i}{\nu_i}. \tag{22.24}$$

For simplicity, we set $\tau = 1$ in the following.

22.5.1.2 Communication Model

We consider a multi-access protocol for devices, i.e., the orthogonal frequency division multiple access (OFDMA) technique whereby each device can occupy one uplink resource block (RB) in a communication round to upload its local model. There are M RBs in the system, denoted by $\mathcal{M} = \{1,2,\dots,M\}$. The achievable transmission rate of device i is Chen et al. (2020a)

$$r_i(\boldsymbol{z}_i, p_i) := \sum_{m\in\mathcal{M}} z_{i,m} B \log_2 \left(1 + \frac{h_i p_i}{I_m + BN_0}\right), \tag{22.25}$$

with B being the bandwidth of each RB, h_i the channel gain, N_0 the noise power spectral density, p_i the transmission power of device i, and I_m the interference caused by the devices that are located in other service areas and use the same RB. In (22.25), $z_{i,m} \in \{0,1\}$ is a binary variable associated with the mth RB allocation for device i: $z_{i,m} = 1$ indicates that RB m is allocated to device i, and $z_{i,m} = 0$ otherwise. Each device can only occupy one RB at maximum while each RB can be accessed by at most one device, thereby we have

$$\sum_{m\in\mathcal{M}} z_{i,m} \le 1 \quad \forall i \in \mathcal{N} \tag{22.26}$$

$$\sum_{i\in\mathcal{N}} z_{i,m} \le 1 \quad \forall m \in \mathcal{M}. \tag{22.27}$$

Due to the fixed dimension of model parameters, we assume that the model sizes of devices are constant throughout the learning process, denoted by S. If device i is selected, the time duration of transmitting the model is given by

$$T_i^{\text{co}}(\boldsymbol{z}_i, p_i) := \frac{S}{r_i(\boldsymbol{z}_i, p_i)}, \tag{22.28}$$

where $\boldsymbol{z}_i := \{z_{i,m} \mid m \in \mathcal{M}\}$. Besides, the energy consumption of the transmission is

$$E_i^{\text{co}}(\boldsymbol{z}_i, p_i) := \sum_{m\in\mathcal{M}} z_{i,m} T_i^{\text{co}}(\boldsymbol{z}_i, p_i) p_i. \tag{22.29}$$

If no RB is allocated to device i in current round, its transmission power and energy consumption is zero.

22.5.2 Problem Formulation

For ease of exposition, we define $\boldsymbol{v} := \{v_i \mid i \in \mathcal{N}\}$, $\boldsymbol{p} := \{p_i \mid i \in \mathcal{N}\}$, and $\boldsymbol{z} := \{z_{i,m} \mid i \in \mathcal{N}, m \in \mathcal{M}\}$. Recall the procedure of NUFM. The total energy consumption $E(\boldsymbol{z}, \boldsymbol{p}, \boldsymbol{v})$ and wall-clock time $T(\boldsymbol{z}, \boldsymbol{p}, \boldsymbol{v})$ in a round can be expressed by

$$E(\boldsymbol{z}, \boldsymbol{p}, \boldsymbol{v}) := \sum_{i\in\mathcal{I}} \left(E_i^{\text{cp}}(v_i) + E_i^{\text{co}}(\boldsymbol{z}_i, p_i)\right) \tag{22.30}$$

$$T(\mathbf{z}, \mathbf{p}, \mathbf{v}) := \max_{i \in \mathcal{N}} T_i^{\text{cp}}(v_i) + \max_{i \in \mathcal{N}} \sum_{m \in \mathcal{M}} z_{i,m} T_i^{\text{co}}(\mathbf{z}_i, p_i), \tag{22.31}$$

where we neglect the communication time for transmitting the scalar u_i. The total contribution to the convergence is

$$U(\mathbf{z}) = \sum_{i \in \mathcal{N}} \sum_{m \in \mathcal{M}} z_{i,m} u_i, \tag{22.32}$$

where u_i is given in (22.21).[3]

We consider the following non-convex mixed-integer non-linear programming (MINLP) problem

$$\text{(P)} \max_{\mathbf{z}, \mathbf{p}, \mathbf{v}} \quad U(\mathbf{z}) - \eta_1 E(\mathbf{z}, \mathbf{p}, \mathbf{v}) - \eta_2 T(\mathbf{z}, \mathbf{p}, \mathbf{v})$$

$$\text{s.t.} \quad 0 \le p_i \le p_i^{\max}, \ \forall i \in \mathcal{N} \tag{22.33}$$

$$0 \le v_i \le v_i^{\max}, \ \forall i \in \mathcal{N} \tag{22.34}$$

$$z_{i,m} \in \{0,1\}, \ \forall i \in \mathcal{N}, m \in \mathcal{M} \tag{22.35}$$

$$\sum_{i \in \mathcal{N}} z_{i,m} \le 1, \ \forall m \in \mathcal{M} \tag{22.26}$$

$$\sum_{m \in \mathcal{M}} z_{i,m} \le 1, \ \forall i \in \mathcal{N}, \tag{22.27}$$

where $\eta_1 \ge 0$ and $\eta_2 \ge 0$ are weight parameters to capture the Pareto-optimal trade-offs among convergence, latency, and energy consumption, the values of which depend on specific scenarios. Constraints (22.33) and (22.34) give the feasible regions of devices' transmission power levels and CPU-cycle frequencies, respectively. Constraints (22.26) and (22.27) restrict that each device can only access one uplink RB while each RB can be allocated to one device at most.

In this formulation, we aim to maximize the convergence speed of FML, while minimizing the energy consumption and wall-clock time in each round. Notably, our solution can adapt to the problem with hard constraints on energy consumption and wall-clock time as in Chen et al. (2020b) via setting "virtual devices" (see Lemmas 22.4 and 22.6).

Next, we provide a joint device selection and resource allocation algorithm to solve this problem.

22.5.3 A Joint Device Selection and Resource Allocation Algorithm

Substituting (22.30), (22.31), and (22.32) into problem (P), we can easily decompose the original problem into the following two sub-problems (SP1) and (SP2), as

3 One can regularize u_i via adding a large enough constant in (22.21) to keep it positive.

follows:

$$
\begin{aligned}
\text{(SP1)}\ \min_{\boldsymbol{v}}\ & g_1(\boldsymbol{v}) = \eta_1 \sum_{i\in\mathcal{N}} \frac{\iota_i}{2} c_i D_i v_i^2 + \eta_2 \max_{i\in\mathcal{N}} \frac{c_i D_i}{v_i} \\
\text{s.t.}\ & 0 \le v_i \le v_i^{\max},\ \forall i \in \mathcal{N}. \\
\text{(SP2)}\ \max_{\boldsymbol{z},\boldsymbol{p}}\ & g_2(\boldsymbol{z},\boldsymbol{p}) = \sum_{i\in\mathcal{N}}\sum_{m\in\mathcal{M}} z_{i,m} u_i \\
& - \sum_{i\in\mathcal{N}}\sum_{m\in\mathcal{M}} \eta_1 \frac{S p_i}{B \log_2\left(1 + \frac{h_i p_i}{I_m + BN_0}\right)} \\
& - \eta_2 \max_{i\in\mathcal{N}} \sum_{m\in\mathcal{M}} z_{i,m} \frac{S}{B \log_2\left(1 + \frac{h_i p_i}{I_m + BN_0}\right)} \\
\text{s.t.}\ & 0 \le p_i \le p_i^{\max},\ \forall i \in \mathcal{N} \\
& \sum_{i\in\mathcal{N}} z_{i,m} \le 1,\ \forall m \in \mathcal{M} \\
& \sum_{m\in\mathcal{M}} z_{i,m} \le 1,\ \forall i \in \mathcal{N} \\
& z_{i,m} \in \{0,1\},\ \forall i \in \mathcal{N}, m \in \mathcal{M}.
\end{aligned}
$$

(SP1) aims at controlling the CPU-cycle frequencies for devices to minimize the energy consumption and latency in the computational phase. (SP2) controls the transmission power and RB allocation to maximize the convergence speed while minimizing the transmission cost and communication delay. We provide the solutions to these two sub-problems separately.

22.5.3.1 Solution to (SP1)

Denote the optimal CPU-cycle frequencies of (SP1) as $\boldsymbol{v}^* = \{v_i^*\}_{i\in\mathcal{N}}$. We first give the following lemma to offer insights into this sub-problem.

Lemma 22.4 If device j is the straggler of all devices with optimal CPU frequencies $\boldsymbol{v}^*$, i.e., $j = \arg\max_{i\in\mathcal{N}} c_i D_i / v_i^*$, the following holds true for any $\eta_1, \eta_2 > 0$

$$
v_i^* = \begin{cases} \min\left\{ \sqrt[3]{\dfrac{a_2}{2a_1}}, \min\limits_{i\in\mathcal{N}} \dfrac{c_j D_j v_j^{\max}}{c_i D_i} \right\} & \text{, if } i = j \\ \dfrac{c_i D_i v_j^*}{c_j D_j} & \text{, otherwise.} \end{cases} \tag{22.36}
$$

The positive constants a_1 and a_2 in (22.36) are defined as

$$
a_1 := \eta_1 \sum_{i\in\mathcal{N}} \frac{\iota_i (c_i D_i)^3}{2(c_j D_j)^2} \tag{22.37}
$$

$$
a_2 := \eta_2 c_j D_j. \tag{22.38}
$$

Proof: The derivation of the results involves two steps, i.e., expressing v_i^* by v_j^* and deriving v_j^* by solving the corresponding optimization problem. The detailed proof is presented in Appendix D of the technical report (Yue et al., 2021b). □

Lemma 22.4 implies that if the straggler (a device with the lowest computational time) can be determined, then the optimal CPU-cycle frequencies of all devices can be derived as closed-form solutions. Intuitively, due to the contradiction in minimizing the energy consumption and computational time, if the straggler is fixed, then the other devices can use the smallest CPU-cycle frequencies as long as the computational time is shorter than that of the straggler. It leads to the following Theorem.

Theorem 22.2 Denote $\boldsymbol{v}^*_{\text{straggler}:j}$ as the optimal solution (i.e., (22.36)) under the assumption that j is the straggler. Then, the global optimal solution of (SP1) can be obtained by

$$\boldsymbol{v}^* = \arg\min_{\boldsymbol{v}\in\mathcal{V}} g_1(\boldsymbol{v}), \tag{22.39}$$

where $\mathcal{V} := \{\boldsymbol{v}^*_{\text{straggler}:j}\}_{j\in\mathcal{N}}$, and g_1 is the objective function in (SP1).

Proof: The result can be directly obtained from Lemma 22.4. We omit it for brevity. □

Theorem 22.2 shows that the optimal solution of (SP1) is the fixed-straggler solution in Lemma 22.4 corresponding to the minimum objective g_1. Thus, (SP1) can be solved with computational complexity $O(n)$ by comparing the achievable objective values corresponding to different stragglers.

22.5.3.2 Solution to (SP2)

Similar to Section 22.5.3.1, we denote the optimal solutions of (SP2) as $\boldsymbol{z}^* = \{z^*_{i,m} \mid i \in \mathcal{N}, m \in \mathcal{M}\}$ and $\boldsymbol{p}^* = \{p_i^* \mid i \in \mathcal{N}\}$ respectively, $\mathcal{N}^* := \{i \in \mathcal{N} \mid \sum_{m\in\mathcal{N}} z^*_{i,m} = 1\}$ as the optimal set of selected devices and, for each $i \in \mathcal{N}^*$, RB block allocated to i as m_i^*, i.e., $z^*_{i,m_i^*} = 1$.

It is challenging to derive a closed-form solution for (SP2) because it is a non-convex MINLP problem with non-differentiable "max" operator in the objective. Thus, in the following, we develop an iterative algorithm to solve this problem and show that the algorithm will converge to a local minimum.

We begin with analyzing the properties of the optimal solution in the next lemma.

Lemma 22.5 Denote the transmission delay regarding $\boldsymbol{z}^*$ and $\boldsymbol{p}^*$ as δ^*, i.e.,

$$\delta^* := \max_{i\in\mathcal{N}} \sum_{m\in\mathcal{M}} z^*_{i,m} \frac{S}{B\log_2\left(1+\frac{h_i p^*_i}{I_m+BN_0}\right)}. \tag{22.40}$$

The following relation holds

$$\delta^* \geq \frac{S}{B\log_2\left(1+\frac{h_i p^{\max}_i}{I_m+BN_0}\right)}, \tag{22.41}$$

and p^*_i can be expressed by

$$p^*_i = \begin{cases} \frac{\left(I_{m^*_i}+BN_0\right)\left(2^{\frac{S}{B\delta^*}}-1\right)}{h_i} &, \text{if } i\in\mathcal{N}^* \\ 0 &, \text{otherwise.} \end{cases} \tag{22.42}$$

Proof: We prove (22.41) by contradiction and (22.42) by solving the corresponding transformed problem. The detailed proof is presented in Appendix E of the technical report (Yue et al., 2021b). □

Lemma 22.5 indicates that the optimal transmission power can be derived as a closed-form via (22.42), given the RB allocation and transmission delay. Lemma 22.5 also implies that for any RB allocation strategy $\tilde{\boldsymbol{z}}$ and transmission delay $\tilde{\delta}$ (not necessarily optimal), Eq. (22.42) provides the "optimal" transmission power under $\tilde{\boldsymbol{z}}$ and $\tilde{\delta}$ as long as (22.41) is satisfied. Based on that, we have the following result.

Theorem 22.3 Denote $\mu_{i,m} := (I_m + BN_0)(2^{\frac{S}{B\delta^*}} - 1)/h_i$. Given transmission delay δ^*, the optimal RB allocation strategy can be obtained by

$$\boldsymbol{z}^* = \arg\max_{\boldsymbol{z}} \left\{ \sum_{i,m} z_{i,m}\left(u_i - e_{i,m}\right),\ \text{s.t. } (22.26) - (22.35) \right\}, \tag{22.43}$$

where

$$e_{i,m} := \begin{cases} \eta_1 \delta^* \mu_{i,m} &, \text{if } \mu_{i,m} \leq p^{\max}_i \\ u_i + 1 &, \text{otherwise.} \end{cases} \tag{22.44}$$

Proof: From Lemma 22.5, Eq. (22.43) holds if $\mu_{i,m} \leq p^{\max}_i$. On the other hand, when $\mu_{i,m} > p^{\max}_i$, if device i is selected, the transmission delay will be larger than δ^* (see the proof of Lemma 22.5), which is contradictory to the given condition. Thus, when $\mu_{i,m} > p^{\max}_i$, we set $e_{i,m} = u_i + 1$, ensuring device i not to be selected. □

Theorem 22.3 shows the optimal RB allocation strategy can be obtained by solving (22.43), given transmission delay δ^*. Naturally, problem (22.43) can be equivalently transformed to a bipartite matching problem. Consider a *Bipartite Graph* $\mathcal{G}$ with source set $\mathcal{N}$ and destination set $\mathcal{M}$. For each $i \in \mathcal{N}$ and $m \in \mathcal{M}$, denote the weight of the edge from node i to node j as $w_{i\to j}$: If $u_i - e_{i,m} > 0$, $w_{i\to j} = e_{i,m} - u_i$; otherwise, $w_{i\to j} = \infty$. Therefore, maximizing (22.43) is equivalent to finding a matching in $\mathcal{G}$ with the minimum sum of weights. It means that we can obtain the optimal RB allocation strategy under fixed transmission delay via *Kuhn–Munkres* algorithm with the worst complexity of $O(Mn^2)$ (Weisstein, 2011).

We proceed to show how to iteratively approximate the optimal δ^*, $\boldsymbol{p}^*$, and $\boldsymbol{z}^*$.

Lemma 22.6 Let j denote the communication straggler among all selected devices with respect to RB allocation $\boldsymbol{z}^*$ and transmission power $\boldsymbol{p}^*$, i.e., for any $i \in \mathcal{N}^*$,

$$\underbrace{\sum_{m\in\mathcal{M}} z^*_{i,m} \frac{S}{B\log_2\left(1+\frac{h_i p^*_i}{I_m+BN_0}\right)}}_{T_i^{\rm co}(\boldsymbol{z}^*_i, p^*_i)} \le \underbrace{\sum_{m\in\mathcal{M}} z^*_{j,m} \frac{S}{B\log_2\left(1+\frac{h_j p^*_j}{I_m+BN_0}\right)}}_{T_j^{\rm co}(\boldsymbol{z}^*_j, p^*_j)}. \tag{22.45}$$

Then, the following holds true:

1. Define function $f_4(p) := b_1\left((1+p)\log_2(1+p)\ln 2 - p\right) - \eta_2$. Then, $f_4(p)$ is monotonically increasing with respect to $p \ge 0$, and has unique zero point $\tilde{p}^0_j \in (0, b_2]$, where b_1 and b_2 are denoted as follows

$$b_1 := \eta_1 \sum_{i\in\mathcal{N}^*} \frac{I_{m^*_i} + BN_0}{h_i} \tag{22.46}$$

$$b_2 := 2^{\left(1+\sqrt{\max\{\frac{\eta_2}{b_1},1\}-1}\right)/\ln 2}; \tag{22.47}$$

2. Denote $(SNR)_j := (I_{m^*_j} + BN_0)/h_j$. For $i \in \mathcal{N}^*$, we have

$$p^*_i = \begin{cases} \min\left\{\tilde{p}^0_j, \min\limits_{i\in\mathcal{N}^*} \dfrac{h_i p^{\max}_i}{I_{m^*_i} + BN_0}\right\} & \text{, if } i = j \\ \dfrac{(SNR)_i}{(SNR)_j} p^*_j & \text{, otherwise.} \end{cases} \tag{22.48}$$

Proof: We obtain the first result by analyzing the property of f_4 and derive (22.48) via solving the corresponding optimization problem. The detailed proof is presented in Appendix F of the technical report (Yue et al., 2021b). □

Lemma 22.6 indicates that, given optimal RB allocation strategy $\boldsymbol{z}^*$ and straggler, the optimal transmission power can be derived by (22.48), different from Lemma 22.5 that requires the corresponding transmission delay δ^*. Notably, in (22.48), we can obtain zero point $\tilde{p}_j^0$ of f_4 with any required tolerance ϵ by *Bisection method* in $\log_2(\frac{b_2}{\epsilon})$ iterations at most.

Similar to Theorem 22.2, we can find the optimal transmission power by the following theorem, given the RB allocation.

Theorem 22.4 Denote $\boldsymbol{p}^*_{\text{straggler}:j}$ as the optimal solution under the assumption that j is the communication straggler, given fixed RB allocation $\boldsymbol{z}^*$. The corresponding optimal transmission power is given by

$$\boldsymbol{p}^* = \arg\max_{\boldsymbol{p}\in\mathcal{P}} g_2(\boldsymbol{z}^*, \boldsymbol{p}), \tag{22.49}$$

where $\mathcal{P} := \{\boldsymbol{p}^*_{\text{straggler}:j}\}_{j\in\mathcal{N}}$ and g_2 is the objective function defined in (SP2).

Proof: We can easily obtain the result from Lemma 22.6, thereby omitting it for brevity. □

Define the communication time corresponding to $\boldsymbol{z}$ and $\boldsymbol{p}$ as $T^{\text{co}}(\boldsymbol{z}, \boldsymbol{p}) := \max_{i\in\mathcal{N}} \sum_{m\in\mathcal{M}} z_{i,m} T_i^{\text{co}}(\boldsymbol{z}_i, p_i)$. Based on Theorems 22.3 and 22.4, we have the following *Iterative Solution (IVES)* algorithm to solve (SP2).

IVES: We initialize transmission delay δ^0 (based on (22.42)) as follows

$$\delta^0 = \max_{i\in\mathcal{N}} \frac{S}{B\log_2\left(1 + \frac{h_i p_i^{\max}}{I_m + BN_0}\right)}. \tag{22.50}$$

In each iteration t, we first compute an RB allocation strategy $\boldsymbol{z}^t$ via solving (22.43) by the *Kuhn–Munkres* algorithm. Then, based on $\boldsymbol{z}^t$, we find the corresponding transmission power $\boldsymbol{p}^t$ by (22.49) and update the transmission delay by $\delta^{t+1} = T^{\text{co}}(\boldsymbol{z}^t, \boldsymbol{p}^t)$ before the next iteration. The details of IVES are depicted in Algorithm 22.2.

Using IVES, we can solve (SP2) in an iterative manner. In the following theorem, we provide the convergence guarantee for IVES.

Theorem 22.5 If we solve (SP2) by IVES, then $\{g_2(\boldsymbol{z}^t, \boldsymbol{p}^t)\}$ monotonically increases and converges to a unique point.

Algorithm 22.2 Iterative Solution (IVES)

Input: $\eta_1, \eta_2, S, B, \{h_i\}, \{I_m\}$

1: Initialize $t = 0$ and δ_0 by (22.50);

2: **while** not done **do**

3: Compute RB allocation strategе $\boldsymbol{z}^t$ under δ^t using Kuhn-Munkres algorithm based on (22.43);

4: Compute transmission power $\boldsymbol{p}^t$ by (22.49);

5: Update $\boldsymbol{\delta}^{t+1} = T^{co}(\boldsymbol{z}^t, \boldsymbol{p}^t)$;

6: **end while**

7: **return** $\boldsymbol{z}^*, \boldsymbol{p}^*$

Proof: The result is derived via proving $g(\boldsymbol{z}^t, \boldsymbol{p}^t) \leq g(\boldsymbol{z}^{t+1}, \hat{\boldsymbol{p}}^{t+1})$ and $g(\boldsymbol{z}^{t+1}, \hat{\boldsymbol{p}}^{t+1}) \leq g(\boldsymbol{z}^{t+1}, \boldsymbol{p}^{t+1})$. The detailed proof is presented in Appendix F of the technical report (Yue et al., 2021b). □

Although IVES solves (SP2) iteratively, we observe that it can converge extremely fast (often with only two iterations) in the simulation, achieving a low computation complexity.

Combining the solutions of (SP1) and (SP2), we provide the *User selection and Resource Allocation (URAL)* in Algorithm 22.3 to solve the original problem (P). The URAL can simultaneously optimize the convergence speed, training time, and energy consumption via jointly selecting devices and allocating resources. Further, URAL can be directly integrated into the NUFM paradigm in the device selection phase to facilitate the deployment of FML in wireless networks.

Algorithm 22.3 User Selection and Resource Allocation (URAL) Algorithm

Input: $\eta_1, \eta_2, S, B, \{h_i\}, \{I_m\}$ $\{\iota_i\}, \{c_i\}, \{D_i\}$

1: Compute v^* by (22.39) *//Solve (SP1)*

2: Compute $\boldsymbol{z}^*$ and $\boldsymbol{p}^*$ by IVES *//Solve (SP2)*

3: **return** $v^*, \boldsymbol{z}^*, \boldsymbol{p}^*$

22.6 Extension to First-Order Approximations

Due to the computation of Hessian in local update (22.2), it may cause high computational cost for resource-limited IoT devices. In this section, we address this challenge.

There are two common methods used in the literature to reduce the complexity in computing Hessian (Fallah et al., 2020b):

1. replacing the stochastic gradient, $\tilde{\nabla} F_i(\theta)$, by

$$\tilde{G}_i(\theta) := \tilde{\nabla} f_i\left(\theta - \alpha\tilde{\nabla} f_i(\theta, \mathcal{D}_i), \mathcal{D}_i'\right); \tag{22.51}$$

2. replacing the Hessian-gradient product, $\tilde{\nabla}^2 f_i(\theta, \mathcal{D}_i'')\tilde{\nabla} f_i\left(\theta - \alpha\tilde{\nabla} f_i(\theta, \mathcal{D}_i), \mathcal{D}_i'\right)$, by

$$\tilde{H}_i(\theta) := \frac{\tilde{\nabla} f_i\left(\theta + \epsilon\tilde{g}_i, \mathcal{D}_i''\right) - \tilde{\nabla} f_i\left(\theta - \epsilon\tilde{g}_i, \mathcal{D}_i''\right)}{2\epsilon}, \tag{22.52}$$

where $\tilde{g}_i = \tilde{\nabla} f_i\left(\theta - \alpha\tilde{\nabla} f_i(\theta, \mathcal{D}_i), \mathcal{D}_i'\right)$.

By doing so, the computational complexity of a one-step local update can be reduced from $O(d^2)$ to $O(d)$, while not sacrificing too much learning performance. Next, we show that our results in Theorem 22.1 hold in the above two cases.

Corollary 22.2 Suppose that Assumptions 22.1–22.4 are satisfied, and $\mathcal{D}_i$, $\mathcal{D}_i'$, and $\mathcal{D}_i''$ are independent batches. If the local update and global aggregation follow (22.2) and (22.6) respectively, we have for $\tau = 1$

$$\begin{aligned}\mathbb{E}[F(\theta^k) - F(\theta^{k+1})] \geq \beta\mathbb{E}\Bigg[\frac{1}{n_k}\sum_{i\in\mathcal{N}_k}\Bigg(\left(1 - \frac{L_F\beta}{2}\right)\|\tilde{\nabla} F_i(\theta^k)\|^2 \\ - \left(\sqrt{(1+\alpha L)^2\gamma_G + \alpha\zeta\gamma_H} + \tilde{\sigma}_{F_i}\right) \\ \times\sqrt{\mathbb{E}\left[\|\tilde{\nabla} F_i(\theta^k)\|^2 \mid \mathcal{N}_k\right]}\Bigg)\Bigg],\end{aligned} \tag{22.53}$$

where $\tilde{\sigma}_{F_i}$ is defined as follows

$$\tilde{\sigma}_{F_i}^2 = \begin{cases} 4\sigma_G^2\left(\frac{1}{D_i'} + \frac{(\alpha L)^2}{D_i}\right) + 2(\alpha L\zeta)^2 & \text{, if using (22.51),} \\ 6\sigma_G^2\left(\frac{\alpha^2}{\epsilon^2 D_i''} + \left(1 + 2(\alpha L)^2\right)\right. & \\ \left.\cdot\left(\frac{1}{D_i'} + \frac{(\alpha L)^2}{D_i}\right)\right) + 2(\alpha\rho\epsilon)^2\zeta^4 & \text{, if using (22.52).}\end{cases} \tag{22.54}$$

Proof: The detailed proof is presented in Appendix I of the technical report (Yue et al., 2021b). □

Corollary 22.2 indicates that NUFM can be directly combined with the first-order approximation techniques to reduce computational cost. Further, similar to Corollary 22.1, Corollary 22.2 can be extended to the multi-step cases.

22.7 Simulation

This section evaluates the performance of our proposed algorithms by comparing with existing baselines in real-world datasets. We first present the experimental setup, including the datasets, models, parameters, baselines, and environment. Then we provide our results from various aspects.

22.7.1 Datasets and Models

We evaluate our algorithms on four widely-used datasets, namely Fashion-MNIST (Xiao et al., 2017), CIFAR-10 (Krizhevsky, 2009), CIFAR-100 (Krizhevsky, 2009), and ImageNet (Deng et al., 2009). Specifically, the data are distributed among $n = 100$ devices as follows: (i) Each device has samples from two random classes; (ii) the number of samples per class follows a truncated Gaussian distribution $\mathcal{N}(\mu, \sigma^2)$ with $\mu = 5$ and $\sigma = 5$. We select 50% devices at random for training with the rest for testing. For each device, we divide the local dataset into a support set and a query set. We consider 1-shot 2-class classification tasks, i.e., the support set contains only 1 labeled example for each class. We set the stepsizes as $\alpha = \beta = 0.001$. We use a convolutional neural network (CNN) with max-pooling operations and the Leaky Rectified Linear Unit (Leaky ReLU) activation function, containing three convolutional layers with sizes 32, 64, and 128, respectively, followed by a fully connected layer and a softmax layer. The strides are set as 1 for convolution operation and 2 for the pooling operation.

22.7.2 Baselines

To compare the performance of NUFM, we first consider two existing algorithms, i.e., FedAvg (McMahan et al., 2017) and Per-FedAvg (Fallah et al., 2020b). Further, to validate the effectiveness of URAL in multi-access wireless networks, we use two baselines, called *Greedy* and *Random*. In each round, the *Greedy* strategy determines the CPU-cycle frequency v_i^g and transmission power p_i^g for device $i \in \mathcal{N}$ by greedily minimizing its individual objective, i.e.,

$$v_i^g = \arg\min_{v_i} \left\{ \eta_1 \frac{l_i c_i D_i v_i^2}{2} + \eta_2 \frac{c_i D_i}{v_i}, \text{ s.t. (22.34)} \right\} \tag{22.55}$$

$$p_i^g = \arg\min_{p_i} \left\{ \sum_{m \in \mathcal{M}} \frac{z_{i,m}^g (\eta_1 p_i + \eta_2) S}{B \log_2 \left(\frac{1 + h_i p_i}{I_m + BN_0} \right)}, \text{ s.t. (22.33)} \right\}, \tag{22.56}$$

Table 22.2 Parameters in simulation.

Parameter	Value
Step size (α) and meta-learning rate (β)	0.001
# edge devices (n)	100
# participating devices (n_k) in Experiments 1 and 2	An integer varying among $\{20,30,40\}$
# local updates (τ)	An integer varying among $\{1,2,\ldots,10\}$
Hyper-parameters (λ_1 and λ_2)	1
Weight parameters (η_1 and η_2)	Real numbers varying among $\{0.5, 1, 1.5, 2.0, 2.5\}$
# RBs	An integer varying among $\{1, 5, 10, 20, 30, 40, 50\}$
CPU cycles of devices (c_i)	Real numbers following $U(0, 0.25)$
Maximum CPU-cycle frequencies of devices ($v_i^{\max}$)	Real numbers following $U(0, 2)$
Maximum transmission powers of devices ($p_i^{\max}$)	Real numbers following $U(0, 1)$
Inference in RBs (I_m)	Real numbers following $U(0, 0.8)$
Model size (S), Bandwidth (B),	1
Noise power spectral density (N_0)	1
effective capacitance coefficients of devices (ι_i)	Real numbers following $U(0, 1)$
Channel gains of devices	Real numbers following $U(0.1, h^{\max})$
	where $h^{\max}$ varies among $\{0.25, 0.5, 0.75, \ldots, 2.0\}$

where $\{z_{i,m}^g\}_{m\in\mathcal{M}}$ is selected at random (i.e., randomly allocating RBs to selected devices). The *Random* strategy decides the CPU-cycle frequencies, transmission powers, and RB allocation for the selected devices uniformly at random from the feasible regions.

22.7.3 Implementation

We implement the code in TensorFlow Version 1.14 on a server with two Intel® Xeon® Golden 5120 CPUs and one Nvidia® Tesla-V100 32G GPU. The parameters used in the simulation can be found in Table 22.2.

22.7.4 Experimental Results

22.7.4.1 Convergence Speed

To demonstrate the improvement of NUFM on the convergence speed, we compare the algorithms on different benchmarks with the same initial model and learning rate. We vary the number of participated devices n_k from 20 to 40, and set the numbers of local update steps and total communication rounds as $\tau = 1$ and $K = 50$, respectively. We let $\lambda_1 = \lambda_2 = 1$. As illustrated in Fig. 22.2 and Table 22.3, NUFM significantly improve the convergence speed and corresponding test accuracy of the existing FML approach on all datasets.[4] Clearly, it validates the effectiveness of our proposed device selection scheme that maximizes the lower bound of one-round global loss reduction. Interestingly, Fig. 22.2 also indicates that NUFM converges more quickly with relatively fewer participated devices. For example, in round 19, the loss achieved by NUFM with $n_k = 20$ decreases by more than 9% and 20% over those with $n_k = 30$ and $n_k = 40$ on Fashion-MNIST, respectively. The underlying rationale is that relatively fewer "good" devices

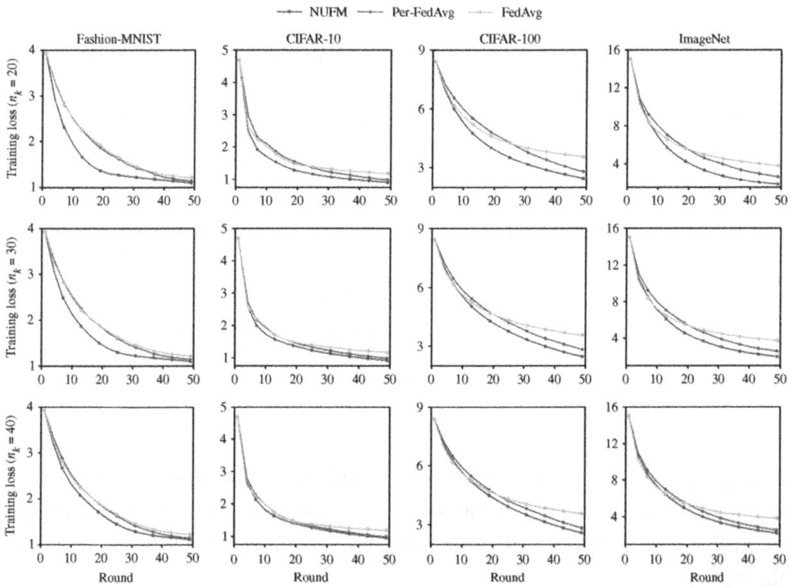

Figure 22.2 Comparison of convergence rates with different numbers of participated devices. NUFM significantly accelerates the convergence of the existing FML approach, especially with fewer participated devices. In addition, with more participating devices, the advantages of NUFM weaken, leading to smaller gaps between NUFM and the existing algorithms.

4 To make the graphs more legible, we draw symbols every two points in Fig. 22.2.

Table 22.3 Test accuracy after 50 rounds of training.

Algorithm	Fashion-MNIST (%)	CIFAR-10 (%)	CIFAR-100 (%)	ImageNet (%)
NUFM	**68.04**	**58.80**	**23.95**	**34.04**
Per-FedAvg	62.75	58.22	21.49	30.98
FedAvg	61.04	54.31	10.13	12.14

The bold score in each benchmark is the highest among algorithms.

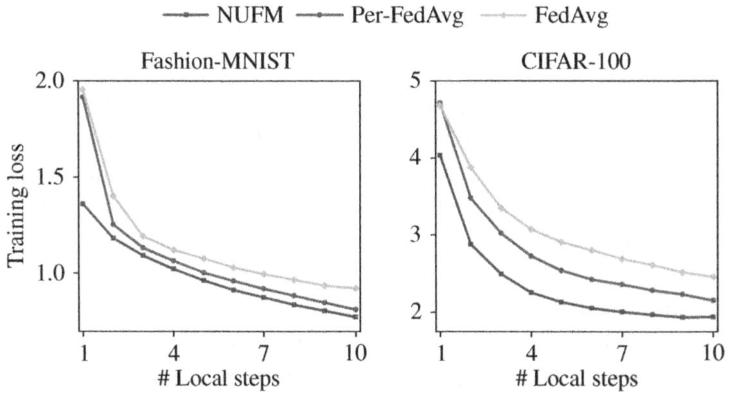

Figure 22.3 Effect of local update steps on convergence rates. Fewer local steps lead to larger gaps between NUFM and the existing methods.

can provide a larger lower bound on the one-round global loss decrease (note that in (22.17) the lower bound takes the average of the selected devices). More selected devices in each round generally require more communication resources. Thus, the results reveal the potential of NUFM in applications to resource-limited wireless systems.

22.7.4.2 Effect of Local Update Steps

To show the effect of local update steps on the convergence rate of NUFM, we present results with varying numbers of local update steps $\tau = 1, 2, \dots, 10$ in each round. For clarity of illustration, we compare the loss under different numbers of local steps in the round 19 on Fashion-MNIST and CIFAR-100. Figure 22.3 shows that fewer local update steps lead to a larger gap between the baselines and NUFM, which verifies the theoretical result that a small number of local steps can slow the convergence of FedAvg and Per-FedAvg Fallah et al. [2020b, Theorem 4.5]. It also implies that NUFM can improve the computational efficiency of local devices.

22.7.4.3 Performance of URAL in Wireless Networks

We evaluate the performance of URAL by comparing with four baselines, namely NUFM-*Greedy*, NUFM-*Random*, RU-*Greedy*, and RU-*Random*, as detailed below.

1. *NUFM-Greedy*: select devices by NUFM, decide CPU-cycle frequencies, RB allocation, and transmission power by *Greedy* strategy;
2. *NUFM-Random*: select devices by NUFM, decide CPU-cycle frequencies, RB allocation, and transmission power by *Random* strategy;
3. *RU-Greedy*: select devices uniformly at random, decide CPU-cycle frequencies, RB allocation, and transmission power by *Greedy* strategy;
4. *RU-Random*: select devices uniformly at random, decide CPU-cycle frequencies, RB allocation, and transmission power by *Random* strategy.

We simulate a wireless system consisting of $M = 20$ RBs and let the channel gain h_i of device i follow a uniform distribution $U(h^{\min}, h^{\max})$ with $h^{\min} = 0.1$ and $h^{\max} = 1$. We set $S = 1, B = 1$, and $n_0 = 1$. The interference of RB m is drawn from $I_m \sim U(0, 0.8)$. We set $\iota_i \sim U(0, 1), c_i \sim U(0, 0.25), p_i^{\max} \sim U(0, 1)$, and $v_i^{\max} \sim U(0, 2)$ for each $i \in \mathcal{N}$ to simulate the device heterogeneity. In the following experiments, we run the algorithms on FMNIST with local update steps $\tau = 1$ and $\eta_1 = \eta_2 = 1$.

As shown in Fig. 22.4, URAL can significantly reduce energy consumption and wall-clock time, as compared with the baselines. However, it is counter-intuitive that the *Greedy* strategy is not always better than *Random*. There are two reasons. On one hand, the energy cost and wall-clock time depend on the selection of

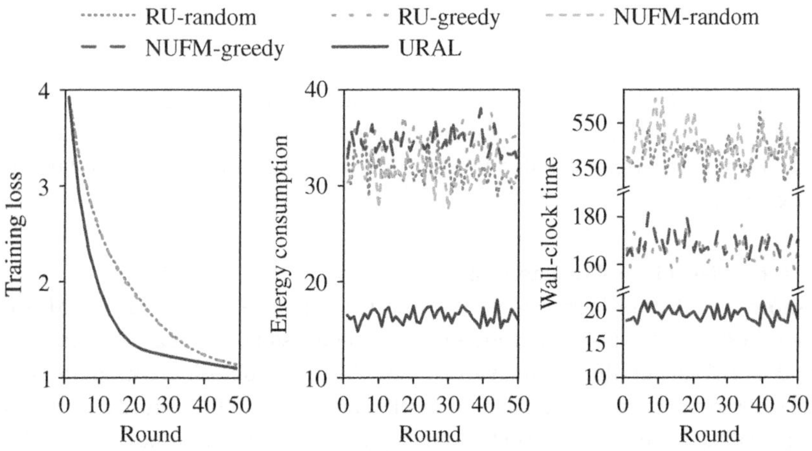

Figure 22.4 Comparison of convergence, energy cost, and wall-clock training time. URAL can achieve a great convergence speed and short wall-clock time with low energy consumption.

weight parameters η_1 and η_2. The results in Fig. 22.4 imply that, when $\eta_1 = \eta_2 = 1$, *Greedy* pays more attention to the wall-clock time than to energy consumption. Accordingly, *Greedy* achieves much lower average delay than that of *Random*, but sacrificing parts of the energy. On the other hand, the wall-clock time and energy cost require joint control with RB allocation. Although *Greedy* minimizes the individual objectives (22.55)–(22.56), improper RB allocation can cause arbitrary performance degradation. Different from *Greedy* and *Random*–based baselines, since URAL aims to maximize the joint objective (P) via co-optimizing the CPU-cycle frequencies, transmission power, and RB allocation strategy, it can alleviate the above-mentioned issues, and achieve a better delay and energy control. At the same time, Fig. 22.4 indicates that URAL converges as fast as NUFM-*Greedy* and NUFM-*Random* (the corresponding lines almost overlap), which select devices greedily to accelerate convergence (as in NUFM). Thus, URAL can have an excellent convergence rate.

22.7.4.4 Effect of Resource Blocks

In Fig. 22.5, we test the performance of URAL under different numbers of RBs. We vary the number of RBs M from 1 to 50. More RBs enable more devices to be selected in each round, leading to larger energy consumption. As shown in Fig. 22.5, URAL keeps stable wall-clock time with the increase of RBs. Meanwhile, URAL can control the power in devices to avoid serious waste of energy. It is counter-intuitive that the convergence speed does not always decrease with the increase of RBs, especially for URAL. The reason is indeed the same as that in

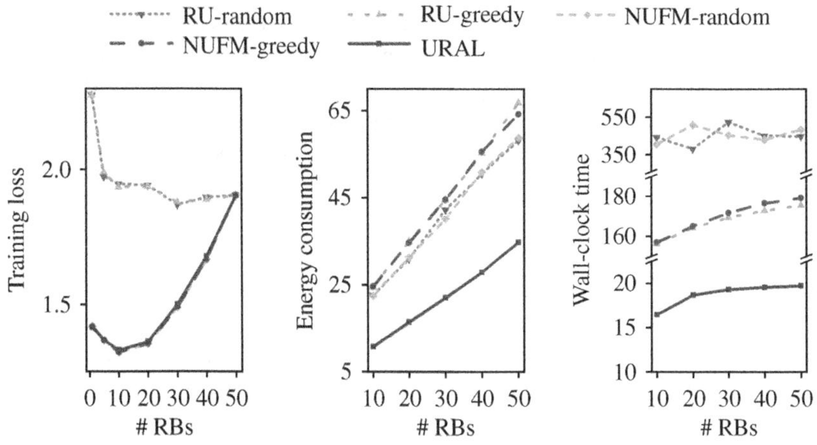

Figure 22.5 Comparison of convergence, energy cost, and wall-clock time under different numbers of RBs. URAL can well control the energy cost and wall-clock time with more available RBs. Meanwhile, it can achieve fast convergence with only a small number of RBs.

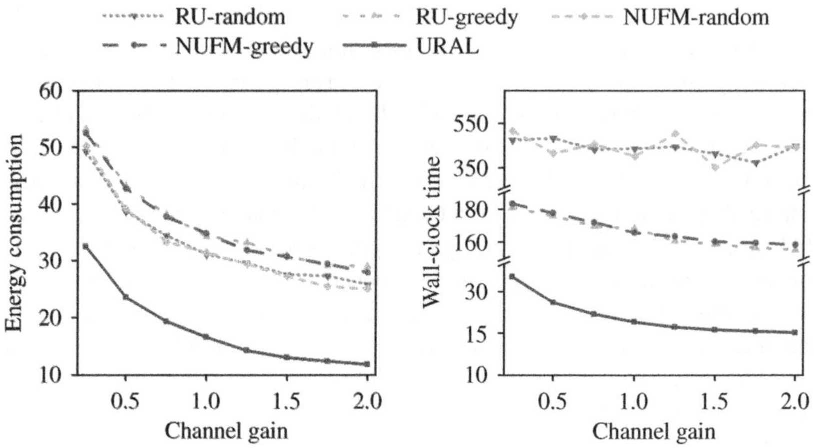

Figure 22.6 Effect of channel gains on performance. Worse channel conditions would induce larger transmission power and longer wall-clock time.

Section 22.7.4.1. That is, too few selected devices can slow the convergence due to insufficient information provided in each round while a large number of participated devices may weaken the global loss reduction as shown in (22.1). Therefore, URAL can adapt to the practical systems with constrained wireless resources via achieving fast convergence with only a small set of devices.

22.7.4.5 Effect of Channel Quality

To investigate the effect of channel conditions on performance, we set the number of RBs $M = 20$ and vary the maximum channel gain $h^{\max}$ from 0.25 to 2, and show the corresponding energy consumption and wall-clock training time in Fig. 22.6. The results indicate that the energy consumption and latency decrease as channel quality improves, because devices can use less power to achieve a relatively large transmission rate.

22.7.4.6 Effect of Weight Parameters

We study how weight parameters η_1 and η_2 affect the average energy consumption and wall-clock time of URAL in Fig. 22.7. We first fix $\eta_2 = 1$ and vary η_1 from 0.5 to 2.5. As expected, the total energy consumption decreases with the increase of η_1, with the opposite trend for wall-clock time. Then we vary η_2 with $\eta_1 = 1$. Similarly, a larger η_2 leads to less latency and more energy cost. It implies that we can control the levels of wall-clock training time and energy consumption by tuning the weight parameters. In particular, even with a large η_1 or η_2, the wall-clock time and energy cost can be controlled at low levels. Meanwhile, the convergence rate achieved by URAL is robust to η_1 and η_2. Thus, URAL can make full use of the resources

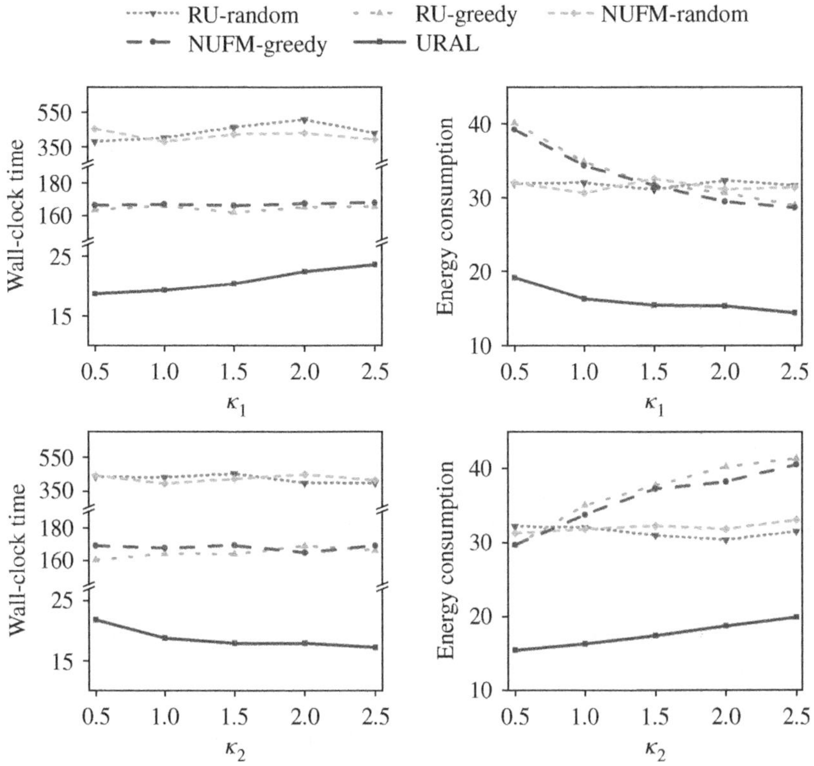

Figure 22.7 Effect of weight parameters η_1 and η_2. A large value of η_1 leads to lower energy consumption but longer wall-clock time (the average wall-clock time is 18.63 when $\eta_1 = 0.5$; it is 21.53 when $\eta_1 = 2.5$). The opposite is true for η_2.

(including datasets, bandwidth, and power) and achieve great trade-offs among the convergence rate, latency, and energy consumption.

22.8 Conclusion

In this chapter, we have proposed an FML algorithm, called NUFM, that maximizes the theoretical lower bound of global loss reduction in each round to accelerate the convergence. Aiming at effectively deploying NUFM in wireless networks, we have presented a device selection and resource allocation strategy (URAL), which jointly controls the CPU-cycle frequencies and RB allocation to optimize the trade-off between energy consumption and wall-clock training time. Moreover, we have integrated the proposed algorithms with two first-order approximation techniques to further reduce the computational complexity in IoT devices. Extensive simulation have been presented to demonstrate that the proposed methods outperform the baseline algorithms.

Future work will investigate the trade-off between the local update and global aggregation in FML to minimize the convergence time and energy cost from a long-term perspective. In addition, how to characterize the convergence properties and communication complexity of the NUFM algorithm requires further research.

References

Thomas D Burd and Robert W Brodersen. Processor design for portable systems. *Journal of VLSI Signal Processing Systems for Signal, Image, and Video Technology*, 13(2):203–221, 1996.

Fei Chen, Mi Luo, Zhenhua Dong, Zhenguo Li, and Xiuqiang He. Federated meta-learning with fast convergence and efficient communication. *ArXiv reprints arXiv: 1802.07876*, 2019.

Mingzhe Chen, H Vincent Poor, Walid Saad, and Shuguang Cui. Convergence time optimization for federated learning over wireless networks. *IEEE Transactions on Wireless Communications*, 20(4):2457–2471, 2020a.

Mingzhe Chen, Zhaohui Yang, Walid Saad, Changchuan Yin, H Vincent Poor, and Shuguang Cui. A joint learning and communications framework for federated learning over wireless networks. *IEEE Transactions on Wireless Communications*, 20(1):269–283, 2020b.

Thomas H Cormen, Charles E Leiserson, Ronald L Rivest, and Clifford Stein. *Introduction to Algorithms*. MIT Press, 2009.

Jia Deng, Wei Dong, Richard Socher, Li-Jia Li, Kai Li, and Li Fei-Fei. ImageNet: A large-scale hierarchical image database. In *Proceedings of the IEEE CVPR*, pages 248–255. Miami, Florida, USA, 2009.

Canh T Dinh, Nguyen H Tran, Minh N H Nguyen, Choong Seon Hong, Wei Bao, Albert Y Zomaya, and Vincent Gramoli. Federated learning over wireless networks: Convergence analysis and resource allocation. *IEEE/ACM Transactions on Networking*, 29(1):398–409, 2020.

Zubair Md Fadlullah and Nei Kato. HCP: Heterogeneous computing platform for federated learning based collaborative content caching towards 6G networks. *IEEE Transactions on Emerging Topics in Computing*, 10(1):112–123, 2020.

Alireza Fallah, Aryan Mokhtari, and Asuman Ozdaglar. On the convergence theory of gradient-based model-agnostic meta-learning algorithms. In *Proceedings of the AISTATS*, pages 1082–1092. Virtual Conference, 2020a.

Alireza Fallah, Aryan Mokhtari, and Asuman Ozdaglar. Personalized federated learning with theoretical guarantees: A model-agnostic meta-learning approach. In *Proceedings of the NeurIPS*, pages 1–12. Virtual Conference, 2020b.

Chelsea Finn, Pieter Abbeel, and Sergey Levine. Model-agnostic meta-learning for fast adaptation of deep networks. In *Proceedings of the ICML*, pages 1126–1135. Sydney, Australia, 2017.

Yihan Jiang, Jakub Konečný, Keith Rush, and Sreeram Kannan. Improving federated learning personalization via model agnostic meta learning. *ArXiv reprints arXiv: 1909.12488*, 2019.

Peter Kairouz, H Brendan McMahan, Brendan Avent, Aurélien Bellet, Mehdi Bennis, Arjun Nitin Bhagoji, Kallista Bonawitz, Zachary Charles, Graham Cormode, Rachel Cummings, et al. Advances and open problems in federated learning. *ArXiv reprints arXiv: 1912.04977*, 2021.

Sai Praneeth Karimireddy, Satyen Kale, Mehryar Mohri, Sashank Reddi, Sebastian Stich, and Ananda Theertha Suresh. SCAFFOLD: Stochastic controlled averaging for federated learning. In *Proceedings of the ICML*, pages 5132–5143. Virtual Conference, 2020.

Alex Krizhevsky. Learning multiple layers of features from tiny images. Technical Report TR-2009. University of Toronto, 2009.

Jeffrey Li, Mikhail Khodak, Sebastian Caldas, and Ameet Talwalkar. Differentially private meta-learning. In *Proceedings of the ICLR*. New Orleans, Louisiana, USA, 2019.

Sen Lin, Guang Yang, and Junshan Zhang. A collaborative learning framework via federated meta-learning. In *Proceedings of the IEEE ICDCS*, pages 289–299. Singapore, Singapore, 2020.

Jiagang Liu, Ju Ren, Yongmin Zhang, Xuhong Peng, Yaoxue Zhang, and Yuanyuan Yang. Efficient dependent task offloading for multiple applications in MEC-cloud system. *IEEE Transactions on Mobile Computing*, 22(4):2147–2162, 2023.

Bing Luo, Xiang Li, Shiqiang Wang, Jianwei Huang, and Leandros Tassiulas. Cost-effective federated learning design. In *Proceedings of the IEEE INFOCOM*. Vancouver, Canada, 2021.

Wang Luping, Wang Wei, and Li Bo. CMFL: Mitigating communication overhead for federated learning. In *Proceedings of the IEEE ICDCS*, pages 954–964. Dallas, Texas, USA, 2019.

Brendan McMahan, Eider Moore, Daniel Ramage, Seth Hampson, and Blaise Aguera y Arcas. Communication-efficient learning of deep networks from decentralized data. In *Proceedings of the AISTATS*, pages 1273–1282. Fort Lauderdale, Florida, USA, 2017.

Hung T Nguyen, Vikash Sehwag, Seyyedali Hosseinalipour, Christopher G Brinton, Mung Chiang, and H Vincent Poor. Fast-convergent federated learning. *IEEE Journal on Selected Areas in Communications*, 39(1):201–218, 2020.

Takayuki Nishio and Ryo Yonetani. Client selection for federated learning with heterogeneous resources in mobile edge. In *Proceedings of the IEEE ICC*, pages 1–7, 2019.

Jihong Park, Sumudu Samarakoon, Mehdi Bennis, and Mérouane Debbah. Wireless network intelligence at the edge. *Proceedings of the IEEE*, 107(11):2204–2239, 2019.

George Plastiras, Maria Terzi, Christos Kyrkou, and Theocharis Theocharidcs. Edge intelligence: Challenges and opportunities of near-sensor machine learning applications. In *Proceedings of the IEEE ASAP*, pages 1–7. Milan, Italy, 2018.

Jinke Ren, Yinghui He, Dingzhu Wen, Guanding Yu, Kaibin Huang, and Dongning Guo. Scheduling for cellular federated edge learning with importance and channel awareness. *IEEE Transactions on Wireless Communications*, 19(11):7690–7703, 2020a.

Jinke Ren, Guanding Yu, and Guangyao Ding. Accelerating DNN training in wireless federated edge learning systems. *IEEE Journal on Selected Areas in Communications*, 39(1):219–232, 2020b.

Wenqi Shi, Sheng Zhou, and Zhisheng Niu. Device scheduling with fast convergence for wireless federated learning. In *Proceedings of the IEEE ICC*, pages 1–6. Virtual Conference, 2020a.

Wenqi Shi, Sheng Zhou, Zhisheng Niu, Miao Jiang, and Lu Geng. Joint device scheduling and resource allocation for latency constrained wireless federated learning. *IEEE Transactions on Wireless Communications*, 20(1):453–467, 2020b.

Khe Chai Sim, Françoise Beaufays, Arnaud Benard, Dhruv Guliani, Andreas Kabel, Nikhil Khare, Tamar Lucassen, Petr Zadrazil, Harry Zhang, Leif Johnson, et al. Personalization of end-to-end speech recognition on mobile devices for named entities. In *Proceedings of the IEEE ASRU*, pages 23–30. Sentosa, Singapore, 2019.

Shiqiang Wang, Tiffany Tuor, Theodoros Salonidis, Kin K Leung, Christian Makaya, Ting He, and Kevin Chan. Adaptive federated learning in resource constrained edge computing systems. *IEEE Journal on Selected Areas in Communications*, 37(6):1205–1221, 2019.

Eric W Weisstein. Hungarian maximum matching algorithm. https://mathworld.wolfram.com/, 2011.

Qiong Wu, Kaiwen He, and Xu Chen. Personalized federated learning for intelligent IoT applications: A cloud-edge based framework. *IEEE Open Journal of the Computer Society*, 1:35–44, 2020.

Han Xiao, Kashif Rasul, and Roland Vollgraf. Fashion-MNIST: A novel image dataset for benchmarking machine learning algorithms. *ArXiv reprints arXiv: 1708.07747*, 2017.

Jie Xu and Heqiang Wang. Client selection and bandwidth allocation in wireless federated learning networks: A long-term perspective. *IEEE Transactions on Wireless Communications*, 20(2):1188–1200, 2020.

Qiang Yang, Yang Liu, Tianjian Chen, and Yongxin Tong. Federated machine learning: Concept and applications. *ACM Transactions on Intelligent Systems and Technology*, 10(2):1–19, 2019.

Zhaohui Yang, Mingzhe Chen, Walid Saad, Choong Seon Hong, and Mohammad Shikh-Bahaei. Energy efficient federated learning over wireless communication networks. *IEEE Transactions on Wireless Communications*, 20(3):1935–1949, 2020.

Hao Yu, Rong Jin, and Sen Yang. On the linear speedup analysis of communication efficient momentum SGD for distributed non-convex optimization. In *Proceedings of the ICML*, pages 7184–7193. Long Beach, California, USA, 2019.

Sheng Yue, Ju Ren, Jiang Xin, Sen Lin, and Junshan Zhang. Inexact-ADMM based federated meta-learning for fast and continual edge learning. In *Proceedings of the ACM MobiHoc*, page 91–100. Shanghai, China, 2021a.

Sheng Yue, Ju Ren, Jiang Xin, Deyu Zhang, Yaoxue Zhang, and Weihua Zhuang. Efficient federated meta-learning over multi-access wireless networks. *arXiv preprint arXiv:2108.06453*, 2021b.

Sheng Yue, Ju Ren, Nan Qiao, Yongmin Zhang, Hongbo Jiang, Yaoxue Zhang, and Yuanyuan Yang. TODG: Distributed task offloading with delay guarantees for edge computing. *IEEE Transactions on Parallel and Distributed Systems*, 33(7):1650–1665, 2022.

Qunsong Zeng, Yuqing Du, Kaibin Huang, and Kin K Leung. Energy-efficient radio resource allocation for federated edge learning. In *Proceedings of the IEEE ICC Workshops*, pages 1–6, 2020.

Xingzhou Zhang, Yifan Wang, Sidi Lu, Liangkai Liu, Lanyu xu, and Weisong Shi. OpenEI: An open framework for edge intelligence. In *Proceedings of the IEEE ICDCS*, pages 1840–1851. Dallas, Texas, USA, 2019.

Xinwei Zhang, Mingyi Hong, Sairaj Dhople, Wotao Yin, and Yang Liu. FedPD: A federated learning framework with optimal rates and adaptivity to non-iid data. *ArXiv reprints arXiv: 2005.11418*, 2020.

Guangxu Zhu, Yong Wang, and Kaibin Huang. Broadband analog aggregation for low-latency federated edge learning. *IEEE Transactions on Wireless Communications*, 19(1):491–506, 2019.

Index

Next Generation Multiple Access, First Edition.
Edited by Yuanwei Liu, Liang Liu, Zhiguo Ding, and Xuemin Shen.
© 2024 The Institute of Electrical and Electronics Engineers, Inc. Published 2024 by John Wiley & Sons, Inc.

Printed and bound by CPI Group (UK) Ltd, Croydon, CR0 4YY
16/04/2025
14658761-0002